Advances in Neurobiology

Volume 24

Series Editor
Arne Schousboe, Department of Drug Design & Pharmacology,
University of Copenhagen, Copenhagen, Denmark

More information about this series at http://www.springer.com/series/8787

M. Mohamed Essa • M. Walid Qoronfleh
Editors

Personalized Food Intervention and Therapy for Autism Spectrum Disorder Management

Springer

Editors
M. Mohamed Essa
Food Science and Nutrition, CAMS
Sultan Qaboos University
Muscat, Oman

M. Walid Qoronfleh
World Innovation Summit for Health WISH
Qatar Foundation
Doha, Qatar

ISSN 2190-5215 ISSN 2190-5223 (electronic)
Advances in Neurobiology
ISBN 978-3-030-30404-1 ISBN 978-3-030-30402-7 (eBook)
https://doi.org/10.1007/978-3-030-30402-7

This Springer imprint is published by the registered company Springer Nature Switzerland AG
The registered company address is: Gewerbestrasse 11, 6330 Cham, Switzerland

Foreword by Brian J. Wilkinson

It is a great pleasure to write a foreword for the book *Personalized Food Intervention and Therapy for Autism Spectrum Disorder Management* edited by Drs. Mohamed Essa and M. Walid Qoronfleh, who have gained extensive experience in their respective scientific fields, including natural compounds and their potential beneficial effects on brain diseases and human health. It is highly commendable for these two scientists to author this book at this time when people are looking for something that would give them a good perspective on autism and the impact of food natural products on this disease spectrum.

Nowadays, there is great interest in investigations of food ingredients for their potential health benefits among the people suffering from autism spectrum disorder (ASD). We frequently fail to recognize the real benefit of traditional medicines and food natural ingredients despite the fact that they have been used for many centuries and are still being used widely in all cultures across the globe. This book provides a compelling perspective on the potential benefits of many edible items, including natural compounds, herbs, vegetables, fruits, grains, seeds, and nuts, for which substantial evidence now exist in the literature of their effectiveness against ASD. The individual chapters that are covered and written by experienced authors in their fields provide an additional value to the reader. The chapters are organized under three parts, covering the following: (1) Autism Spectrum Disorder from Background to Interplay of Genetic, Epigenetic, Environmental Risk Factors and Nutraceuticals, (2) Specific Foods and Nutrient Qualities in Autism, and (3) Food and Dietary Intervention and Therapy in Autism. This organization provides the reader a better basic understanding of pharmacological constituents, including their potential benefits against ASD.

Why this book is timely? Firstly, a greater proportion of the population in most developing and developed countries have become open-minded for the use of food ingredients and traditional medicines in spite of the use of allopathic medicines for treatment of diseases. Secondly, there is a huge increase in the prevalence of ASD globally, and until now, there are no proper medications to deal with symptoms or effectively manage comorbidities, behavior, performance, and food intolerances. Thirdly, misunderstanding and confusion exist among the parents of an ASD individual and their caregivers, and this book provides up-to-date clarification to the public.

Another notable factor in this book is that several chapters provide a better understanding of the mechanisms of the effects of various phytochemical constituents, such as sulforaphane, curcumin polyphenols (resveratrol), and flavonoids (luteolin). For instance, there is now an extensive literature on omega-3 fatty acids from nuts, showing they function as precursors for anti-inflammatory agents such as neuroprotectin D1 and resolvins, which might exert significant benefits on many neurological disorders including ASD.

In conclusion, I appreciate the efforts of the editors to publish this book, which is a great resource of composite information on food and autism. It should also be a good resource for parents with ASD children, dieticians, nutritionists, nutrition researchers, and other readers with interests in human health.

Brian J. Wilkinson
School of Biological Sciences
Illinois State University
Normal, IL, USA

Foreword by Magid Abou-Gharbia

I was delighted when I received a request from Dr. M. Walid Qoronfleh to write a brief foreword for his book. I have known Dr. Qoronfleh for many years, and I have been impressed by his accomplishments in the biomedical field. During the last decade, Dr. Qoronfleh and I were among the team of Arab Expatiate Scientists (AES) who worked with Qatar Foundation to help in promoting Qatar research culture. In the process, we embarked on several new research initiatives in several important research areas. In the biomedical area, we also worked together on putting together and writing the planning document that was used to establish Qatar Biomedical Research Institute (QBRI), where significant biomedical research is currently being conducted today.

As a drug discovery researcher for over three decades in the pharmaceutical industry and academia, my research teams have always explored innovative approaches for design and synthesis of novel therapeutics for treatment of many diseases including central nervous system (CNS) disorders. We successfully discovered ten marketed drugs. Examples of CNS disorder drugs include the antidepressants Effexor® and Pristiq®, the sedative Sonata®, and the analgesic Dezocine®. While pursuing our drug discovery research, we were conscious of the importance of natural products therapeutic potential as well as the role of nutrition in preventing and ameliorating disease symptoms. Our efforts led to the discovery of natural products drugs like Mylotarg® and Toresil®.

I am pleased to provide a foreword to *Personalized Food Intervention and Therapy for Autism Spectrum Disorder Management*. Having such book is a good step in the direction toward increasing public awareness of the benefit of nutrition in preventing neurological diseases such as autism. I believe this book will serve as a good introduction to the benefit of personalized food in disease prevention and disease therapy.

Magid Abou-Gharbia
Moulder Center for Drug Discovery Research
School of Pharmacy, Temple University
Philadelphia, PA, USA

Foreword by Theodore DeFrank

Having known Dr. Qoronfleh for many years, it is an honor to submit a foreword for the book *Personalized Food Intervention and Therapy for Autism Spectrum Disorder Management*. I worked with Dr. Qoronfleh during our years together at Pierce Biotechnology, where he was the Director of R&D, developing new technologies and reagents for proteomics research. He was instrumental in positioning Pierce to be a leader in this emerging field in the advent of creating personalized medicine as an approach to therapeutic treatment. After I moved to Active Motif, a life science research company specializing in epigenetics, we continued our dialog on how environmental factors affect gene expression and hence one's health. The book edited by Drs. Mohamed Essa and M. Walid Qoronfleh further advances this topic to the field of autism spectrum disorders.

Theodore DeFrank
Active Motif Inc.
Carlsbad, CA, USA

Preface

Synopsis of the Book

Nature has bestowed on mankind an overabundance of edible vegetables, fruits, nuts, and seeds. The availability of a variety of nutrients and bioactive compounds in these natural products play a crucial role in the health and disease status of human beings. Hippocrates said "*Let food be thy medicine and medicine be thy food.*" Awareness and specific interest in natural medicine have substantially increased and continue to rise. The practice has taken hold in our society, and this wide acceptance emphasizes an integrative health approach that addresses body, mind, and spirit. Food and its active phytochemicals are used in prevention, cure, and/or management of various neurological disorders, neurodevelopmental disorders, neurodegenerative diseases, and other neuronal dysfunctions. In the recent decade, numerous studies were conducted that proved naturally occurring phyto-compounds (polyphenolics, flavonoids, and antioxidants found in fruits, vegetables, herbs, nuts, etc.) could potentially offer benefit and impact the severity of neurodevelopmental disorders. Certainly, individualized nutritional approaches with dietary management during various disease states including autism have seen a steep increase as of late. Autism spectrum disorder (ASD) is a debilitating neurodevelopment disorder characterized by stereotyped interests and behaviors, and abnormalities in verbal and non-verbal communication. It is a multifactorial disorder resulting from interactions between genetic, environmental, and immunological factors. Excito-toxicity and oxidative stress are potential contributing mechanisms, which are likely to serve as a converging point to these risk factors. The potential use of antioxidants against free radical toxicity has been an emerging field in the management of many neurodevelopmental conditions including autism. The supportive role of antioxidants and anti-inflammatory agents to reduce the severity of autism via the promotion of functional neurogenesis and neuro-protection in the pathological child brain has great promise. The present idea for this book comprehends the recent studies describing the therapeutic roles of antioxidants and other active pharmacological compounds in ASD and other neurologic disorders, while highlighting the scope of using antioxidants to promote neurogenesis and improve other symptoms in ASD. The molecular mechanisms behind the curative effects may rely mainly on the action of phytonutrients on distinct signaling pathways associated with protein folding and neuro-inflammation. This book is aimed to advocate a new line of research approach to define the mechanisms by which antioxidant-rich food offers possible therapeutic strategies to ASD. The book focuses on implications of traditional and processed foods for ASD intervention and management. Many publications exist on the benefit of diet in relation to ASD. However, a comprehensive collection of various aspects of food active pharmacological ingredients for intervention, dietary approaches, nutritional management, and neurotherapeutics, with respect to ASD in the form of a book is lacking. This book provides a comprehensive collection of research studies that will benefit students at various levels, researchers in several disciplines (such as alternative medicine, nutrition, neuroscience, agriculture, food science, and medicine), and many others who are interested in this discipline. The book also can be used as a required or recommended text for related courses taught at universities globally.

Muscat, Oman
Doha, Qatar

Musthafa Mohamed Essa
M. Walid Qoronfleh

Acknowledgements

We are grateful to the book chapters' authors for sharing their expertise to provide evidence-based information on the role or benefits of foods and natural products in autism.

We are indebted to our families for their understanding and unconditional support during this endeavor allowing us to spend extra time on completing the book. Many thanks are due to our teachers who have inspired us to investigate the potential health and medicinal benefits of food/natural products and to those colleagues for many stimulating discussions.

Special thanks to The Editing Refinery, USA, for providing the wonderful proofreading and editing services.

Thank you to the staff of Springer Nature Publishers for their patience and assistance at various stages of the book publication.

We as editors gratefully want to acknowledge our respective institutions for their continued support that allowed us to finish this book in a successful manner.

We declare there exists no potential conflict of interest.

Contents

Part I Autism Spectrum Disorder from Background to Interplay of Genetic, Epigenetic, Environmental Risk Factors and Nutraceuticals

Overview and Introduction to Autism Spectrum Disorder (ASD) 3
Nader Al-Dewik, Rana Al-Jurf, Meghan Styles, Sona Tahtamouni, Dalal Alsharshani, Mohammed Alsharshani, Amal I. Ahmad, Azhar Khattab, Hilal Al Rifai, and M. Walid Qoronfleh

New Horizons for Molecular Genetics Diagnostic and Research in Autism Spectrum Disorder 43
Nader Al-Dewik and Mohammed Alsharshani

Genomics of Autism 83
Khalid A. Fakhro

Neuropsychopathology of Autism Spectrum Disorder: Complex Interplay of Genetic, Epigenetic, and Environmental Factors 97
Ranjana Bhandari, Jyoti K. Paliwal, and Anurag Kuhad

Maternal Prenatal Exposures in Pregnancy and Autism Spectrum Disorder: An Insight into the Epigenetics of Drugs and Diet as Key Environmental Influences 143
Kholoud N. Bastaki, Sura Alwan, and Farah R. Zahir

Psychological Comorbidities in Autism Spectrum Disorder 163
Eman Shaltout, Nader Al-Dewik, Muthanna Samara, Hisham Morsi, and Azhar Khattab

Role of Oxidative Stress and Antioxidants in Autism 193
Thamilarasan Manivasagam, Selvaraj Arunadevi, Mustafa Mohamed Essa, Chidambaram SaravanaBabu, Anupom Borah, Arokiasamy Justin Thenmozhi, and M. Walid Qoronfleh

The Regulation of Reactive Neuroblastosis, Neuroplasticity, and Nutraceuticals for Effective Management of Autism Spectrum Disorder 207
G. P. Poornimai Abirami, Risna Kanjirassery Radhakrishnan, Esther Johnson, Syed Aasish Roshan, Ajisha Yesudhas, Suhadha Parveen, Abir Biswas, Vijaya Roobini Ravichandran, Anusuyadevi Muthuswamy, and Mahesh Kandasamy

Part II Specific Foods and Nutrient Qualities in Autism

Vegetables. 225
Sawsan G. Mohammed and M. Walid Qoronfleh

Fruits 279
Sawsan G. Mohammed and M. Walid Qoronfleh

Grains. 377
Sawsan G. Mohammed, Saoud Bossa, and M. Walid Qoronfleh

Nuts. 395
Sawsan G. Mohammed and M. Walid Qoronfleh

Seeds. 421
Sawsan G. Mohammed and M. Walid Qoronfleh

The Role of Gluten in Autism. 469
T. Sumathi, T. Manivasagam, and A. Justin Thenmozhi

Food Color and Autism: A Meta-Analysis 481
Prabasheela Bakthavachalu, S. Meenakshi Kannan, and M. Walid Qoronfleh

Food Selection and Preferences of Omani Autistic Children 505
Najma M. Al-Kindi, Yahya M. Al-Farsi, Buthaina Al-Bulushi, Amanat Ali, Syed Gauhar Alam Rizvi, and Musthafa Mohamed Essa

Part III Food, Dietary Intervention, and Therapy in Autism

Overview of Nutritional Therapy for Autism Spectrum Disorder 527
Carla Vartanian

Importance of Nutrition Intervention in Autistic Patients 535
Tahra ElObeid, Joyce Moawad, and Zumin Shi

Dietary Approaches to the Management of Autism Spectrum Disorders 547
Richard E. Hartman and Dhira Patel

Protein Nutrition in Autism 573
Saravana Babu Chidambaram, Abid Bhat, Arehally Marappa Mahalakshmi, Bipul Ray, Sunanda Tuladhar, B. S. Sushmitha, B. Saravanan, Manivasagam Thamilarasan, Arokiasamy Justin Thenmozhi, Musthafa Mohamed Essa, Gilles J. Guillemin, and M. Walid Qoronfleh

Autism and Gut–Brain Axis: Role of Probiotics 587
Saravana Babu Chidambaram, Sunanda Tuladhar, Abid Bhat, Arehally Marappa Mahalakshmi, Bipul Ray, Musthafa Mohamed Essa, Muhammed Bishir, Srinivasa Rao Bolla, Nandakumar Dalavaikodihalli Nanjaiah, Gilles J. Guillemin, and M. Walid Qoronfleh

Natural Products and Their Therapeutic Effect on Autism Spectrum Disorder 601
Satarupa Deb, Banashree Chetia Phukan, Ankumoni Dutta, Rajib Paul, Pallab Bhattacharya, Thamilarasan Manivasagam, Arokiasamy Justin Thenmozhi, Chidambaram Saravana Babu, Musthafa Mohamed Essa, and Anupom Borah

Dietary Phytochemicals as Neurotherapeutics for Autism Spectrum Disorder: Plausible Mechanism and Evidence 615
Ranjana Bhandari, Jyoti K. Paliwal, and Anurag Kuhad

Regulation of Dietary Amino Acids and Voltage-Gated Calcium Channels in Autism Spectrum Disorder 647
Shubham Singh, Supraj Raja Sangam, and Rajagopal Senthilkumar

Bioactive Metabolites from Marine Ascidians: Future Treatment for Autism Spectrum Disorder 661
Manigandan Venkatesan, Velusamy Arumugam, Rathinam Ayyasamy, Selvakumar Murugesan, Nishakavya Saravanan, Umamaheswari Sundaresan, Saravanan Ramachandran, Thamilarasan Manivasagam, Arokiasamy Justin Thenmozhi, and M. Walid Qoronfleh

Reality-Based Technologies for Children with Autism Spectrum Disorder: A Recommendation for Food Intake Intervention 679
Bilikis Banire, Kamran Khowaja, Bilal Mansoor, Marwa Qaraqe, and Dena Al Thani

Editors' Biography

Musthafa Mohamed Essa, PhD is an Associate Professor of Nutrition at Sultan Qaboos University, Oman, holding visiting A/Prof position in Neuropharmacology group, ASAM, Macquarie University, Sydney, Australia. He is an Editor-in-Chief of *International Journal of Nutrition, Pharmacology, Neurological Diseases* published by Wolters and Kluwer, USA. He is also acting as an Associate Editor for BMC Complementary and Alternative Medicine and Editor for Frontiers in Biosciences. Dr. Essa is involved in the editorial/reviewer board of various well-known journals such as Frontiers in Neurology, Biochemie, and PLOS one. Dr. Essa is conducting research on the effect of dietary supplementation of natural products on neurodegenerative diseases. He is an expert in the field of Nutritional Neuroscience/Neuropharmacology and published more than 120 papers, 46 book chapters, and 11 books. He has strong international collaborations with institutes in the USA, Australia, Qatar, and India. In 2015, Dr. Essa founded a new foundation named "Food and Brain Research Foundation" to support research in nutritional neuroscience.

Dr. Essa is recipient of so many awards from local and international bodies. In July 2017, his book titled *Food and Parkinson's Disease* was awarded as the Best Book in the World by GOURMAND Cookbook Awards. In 2015, one of his books titled *Food and Brain Health* was awarded as the Best Book in the World by GOURMAND Cookbook Awards, and the same book was awarded the Best Book in the World of past 20 years by GOURMAND Cookbook Awards. In 2015 and 2016, Dr. Essa received the National Research Award under health sector from The Research Council, Oman. In 2013 and 2017, he was awarded as Distinguished Researcher in Sultan Qaboos University, Oman. One of his scientific images won the Best of Show, Best Overall in all categories and individual as well in the "Science of Art-Photo Contest" in ISN-ASN International Meeting in Mexico, 2013. Further, in 2010, Dr. Essa received the Professional from Developing Country Award from IMFAR, USA. In 2009, he was awarded ASN Young Investigator Educational Enhancement Award (ASN-YIEE-2009).

He has over 15 years of research and teaching experience, and till now he has received many research grants from local and international agencies in the area of

Neuropharmacology and Nutritional Neurosciences (approximately US$1.31 million). He has guided 3 PhD students and 9 master's students. Currently, he is guiding 2 PhD students in the area of Nutritional Neuroscience.

M. Walid Qoronfleh, PhD, MBA is currently the Director of Healthcare Research and Policy at Qatar Foundation–World Innovation Summit for Health (WISH). Dr. Qoronfleh leads the WISH Doha-based research unit which is engaged in research areas that are of immediate global and local healthcare relevance to Qatar. The research is undertaken with a view to positively impact health policy. These areas include autism, mental health, dementia, patient safety, non-communicable diseases (cancer, diabetes, and obesity), precision medicine, and bioethics.

Immediately prior to WISH, Walid was the Biotechnology Development Director at Hamad Bin Khalifa University–Qatar Biomedical Research Institute (HBKU-QBRI). Dr. Qoronfleh has over 20 years of scientific, technology, business, and commercial experience. He has held several senior management and executive positions with increasing responsibilities at GlaxoSmithKline (GSK), Sanofi, US National Institutes of Health–National Cancer Institute (NIH-NCI), ThermoFisher, NextGen Sciences (VP, Personalized Medicine), and SDIX (Executive Director, Diagnostics business, Nasdaq: SDIX) and Executive Director and PI at the University of Michigan—Ann Arbor. Dr. Qoronfleh has been awarded two pioneering grants. As a PI at the University of Michigan—Ann Arbor, he was awarded $3.2 million to build a state-wide medical research infrastructure by the Michigan Economic Development Corporation (MEDC). Previously, while at ThermoFisher-Senior Director R&D, as a co-applicant (co-PI/Investigator) he was awarded a 7-year NIH grant with the Medical College of Wisconsin (MCW) to establish a proteomic center ($16.0 million).

Dr. Qoronfleh is also involved on different levels in Government-University-Industry partnerships through research, technology, and commercialization as well as higher education, research capacity building, and grant program management. Dr. Qoronfleh brings expertise in economic development and business improvement in knowledge-based economy (Leadership, Entrepreneurship, and Innovation), where he also lectures locally and at HEC Paris in Qatar. Dr. Qoronfleh was an active member of the Arab Expatriate Scientists group in Qatar. He is also one of the main co-authors for the planning document for the establishment of Qatar Biomedical Research Institute (QBRI), which was developed in collaboration with leading Arab expatriate scientists and scientists in Qatar (2007–2008).

Dr. Qoronfleh is the founder of three biotechnology companies in the USA (HT Expression, Q-Pharma, and Minerva Bioscience), and he is the co-founder and the managing director of the US-based boutique consulting company Q3CG. Dr. Qoronfleh has published over 100 combined peer-reviewed abstracts and papers. He is also an editor and an *ad hoc* reviewer for various scientific journals and a frequent speaker at national and international conferences. He holds postgraduate degrees from the University of Louisville School of Medicine, Penn State University, and the University of Wisconsin–Madison.

Part I
Autism Spectrum Disorder from Background to Interplay of Genetic, Epigenetic, Environmental Risk Factors and Nutraceuticals

Psychiatrist Leo Kanner published a paper describing "*children whose condition differed so markedly and uniquely from anything reported so far.*" L. Kanner. Autistic disturbances of affective contact. *Nervous Child*, 2, 217–250 (1943).

1.1 Overview and Reflection

Autism spectrum disorder (ASD) is one of the most common and challenging neurodevelopmental disorders in children. The prevalence rate of ASD has reached over 1% worldwide prompting governments, health providers, and schools to develop programs and policies to address this challenging disorder. In 2018, recent CDC estimates in the USA cited by Autism Speaks recized prevalence among the nation's children to 1 in 59 children. In this section, we present chapters that discuss the causes, as well as environmental factors, genetic mutations, epigenetics role, and neural pathways that are implicated in ASD and criteria that are commonly used for its diagnosis and future genomics/genetic testing that can aid in the diagnosis of this disorder. Finally, we provide practical steps that can be used to reduce the incidence and severity of ASD.

ASD is characterized by deficits in communication and social interaction, as well as the presence of repetitive and restrictive behavior. ASD often manifests with a wide range of comorbidities including morphological (macrocephaly), physiological (gastrointestinal and/or sleep problems), and psychiatric (anxiety) conditions. The most common proposed causes of ASD are physiological and metabolic disorders, involving immunity, oxidative stress, and mitochondrial dysfunction. There is no pharmacological cure to ASD.

The number of individuals diagnosed with ASD has risen dramatically over the past 40 years. Regardless of the reasons, this increase in diagnosis has led to large-scale research initiatives, awareness campaigns, and the need for government support. Currently, ASD is diagnosed clinically based on the severity of a heterogeneous list of social, communicative, and behavioral deficits; however, there is no effective medical diagnosis for ASD, however, early diagnosis (12–18 months) using universal standardized assessment screening tools and intervention coupled with remedial services is highly beneficial to patients.

Separate from High-Income Countries (HICs), public health knowledge or campaigns concerning ASD impact remain poorly undeveloped and implemented, respectively. To date, 86.5% of all cases of ASD have been reported in HICs representing only 20% of the world population. Insufficient population-based studies have been conducted in Low- to Medium-Income Countries (LMICs), which may well underrepresent the impact of ASD. There are also difficulties in measuring the prevalence of ASD, stemming from population awareness, selection of studies and diagnostic capabilities, as well as cross-cultural appropriateness and comparability of ASD screening, measurement, and epidemiological data. The uneven rates of diagnosis have also led to variations in ASD prevalence by race and ethnicity including the MENA and Gulf regions. Therefore, the knowledge gap between evidence and action in the care of ASD in LMICs has remained considerably wide.

The ASD economic impact is substantial, which includes direct medical/non-medical billings and indirect productivity costs. The approximate lifetime cost of caring for an individual with ASD inflicted with comorbid conditions is ~$2.2 million in the USA, and £1.5 million in the UK; though the cost drops to $1.4 million in the USA and £0.9.2 million in the UK for ASD otherwise. In addition, if unrecognized or untreated, ASD can contribute to poor educational attainment and difficulty with employment, leading to negative economic implications. A 2015 US study estimates the total economic impact of ASD, based on direct medical, non-medical, and productivity costs combined, totaled $268 billion—ranging from 0.9% to 2% of gross domestic product (GDP). This amount is expected to rise to $461 billion (ranging from 0.9.9% to 3.6% of GDP) by 2025. These costs are on par with recent estimates for "silent epidemics" such as diabetes. Indeed, the burden of ASD appears to exceed the cost of the traditional enemies of health. However, what sets ASD apart from other non-communicable diseases—such as heart disease, cancer, stroke, and hypertension—are the significantly higher non-medical costs when compared to direct medical costs. There is also suggestion that comorbidities of ASD tend to amplify burden to the society and afflicted individuals alike.

Overview and Introduction to Autism Spectrum Disorder (ASD)

Nader Al-Dewik, Rana Al-Jurf, Meghan Styles, Sona Tahtamouni, Dalal Alsharshani, Mohammed Alsharshani, Amal I. Ahmad, Azhar Khattab, Hilal Al Rifai, and M. Walid Qoronfleh

Abstract Autism spectrum disorder (ASD) is a complex neurodevelopmental disorder generally manifesting in the first few years of life and tending to persist into adolescence and adulthood. It is characterized by deficits in communication and social interaction and restricted, repetitive patterns of behavior, interests, and activities. It is a disorder with multifactorial etiology. In this chapter, we will focus on the most important and common epidemiological studies, pathogenesis, screening, and diagnostic tools along with an explication of genetic testing in ASD.

Keywords Autism spectrum disorder · Epidemiology · Pathogenesis · Screening tools · Diagnostic tools · Genetic testing

---

N. Al-Dewik (✉)
Clinical and Metabolic Genetics Section, Pediatrics Department, Hamad General Hospital (HGH), Women's Wellness and Research Center (WWRC) and Interim Translational Research Institute (iTRI), Hamad Medical Corporation (HMC), Doha, Qatar

College of Health and Life Sciences, Hamad Bin Khalifa University (HBKU), Doha, Qatar

Faculty of Health and Social Care Sciences, Kingston University, St. George's University of London, London, UK
e-mail: naldewik@hamad.qa; Nader.Al-Dewik@kingston.ac.uk

R. Al-Jurf
Department of Biomedical Science, College of Health Science, Qatar University, Doha, Qatar

M. Styles
Health Profession Awareness Program, Health Facilities Development, Hamad Medical Corporation (HMC), Doha, Qatar

S. Tahtamouni
Child Development Center, Hamad Medical Corporation, Doha, Qatar

D. Alsharshani
College of Health and Life Sciences, Hamad Bin Khalifa University (HBKU), Doha, Qatar

M. Alsharshani
Diagnostic Genetics Division (DGD), Department of Laboratory Medicine and Pathology (DLMP), Hamad Medical Corporation (HMC), Doha, Qatar

M. M. Essa, M. W. Qoronfleh (eds.), *Personalized Food Intervention and Therapy for Autism Spectrum Disorder Management*, Advances in Neurobiology 24,
https://doi.org/10.1007/978-3-030-30402-7_1

1 Introduction

The first description of ASD was in 1943 by the child psychiatrist Leo Kanner who identified 11 children with extreme withdrawal and incapability to form normal relationships with others. He labeled them as having early infantile autism [1].

Leo Kanner had characterized individuals with this condition as desiring extreme loneliness and adhering to a strict routine. They could distract themselves for hours with simple, repetitive tasks and were easily distressed by the slightest deviation from what they were accustomed doing. These children displayed variation in their social abilities, with some unable to speak or even communicate altogether [1].

A year after Kanner's report, pediatrician Hans Asperger independently documented four of his patients who displayed similar characteristics but differed markedly in their intellectual abilities by demonstrating advanced aptitudes in science and math. Despite this difference, Asperger described his patients as also having autism [2].

The core symptoms of ASD rarely occur in isolation and are typically associated with other comorbidities such as multiple psychiatric disorders like anxiety and depression, attention-deficit/hyperactivity disorder (ADHD), epilepsy, gastrointestinal problems, sleep disorders, feeding disorder, learning disability (LD), intellectual disability (ID), and obsessive-compulsive disorder (OCD) which will all be discussed thoroughly in the chapter on psychological comorbidities.

The term "spectrum" represents the variability in severity of symptoms ranging from mildly autistic, high-functioning individuals to severely impaired cases requiring long-term specialist support, often seen in affected children. ASD was found to affect about 1–2% of the general population [3, 4].

2 Epidemiology

The global prevalence of ASD was previously reported in 2010 as 7.6 per 1000 (1 in 132) [5]. The overall prevalence of ASD in Europe, Asia, and the USA ranges from 2 to 25 per 1000, or approximately 1 in 40 to 1 in 500 [6–19].

A. I. Ahmad · A. Khattab
Qatar Rehabilitation Institute (QRI), Hamad Medical Corporation (HMC), Doha, Qatar

H. Al Rifai
Department of Pediatrics and Neonatology, Newborn Screening Unit, Hamad Medical Corporation, Doha, Qatar

M. Walid Qoronfleh
Research and Policy Department, World Innovation Summit for Health (WISH), Qatar Foundation, Doha, Qatar

In 2010 and 2012, the estimated prevalence of ASD among 8-year-old children was 14.7 per 1000, approximately 1 in 68 overall (1 in 42 boys and 1 in 190 girls) [10, 11, 20, 21]. The prevalence of ASD estimates varies widely: by geographical location and racial/ethnic groups. In 2014, the prevalence slightly increased to 16.8 per 1000, 1 in 59 overall, with an estimation of 1 in 38 boys and 1 in 151 girls. This was thought to be largely due to improved screening and diagnosis among black and Hispanic children [6]. Using a new surveillance case definition consistent with the Diagnostic and Statistical Manual of Mental Disorders, Fifth Edition (DSM-5), diagnostic criteria (published in 2013) had little effect on the prevalence estimate. In 2016, the estimated NHIS prevalence of ASD was 25 per 1000 (95% CI 22.3 to 28.1 per 1000), approximately 1 in 40 children overall, 1 in 26 boys and 1 in 93 girls [13].

The prevalence of ASD has significantly increased over time, mainly since the late 1990s [22, 23]. Several studies have proposed that modifications of case definition and increased awareness explain the obvious rise [5, 24–27]. This along with other factors may play an important role like earlier detection, availability of more specialized developmental services, diagnostic substitution (increase in ASD prevalence along with declines in the prevalence of others such as learning disorders, developmental language disorder, and/or ID), as well as a true increase in prevalence [19, 24, 25, 28–33]. It is concluded that ASD affects about 1% of the general population.

Male-to-female ratio: ASD is four times more common in boys than girls [6, 34]. In a systematic review and meta-analyses of 54 studies including > 13,700,000 patients, the overall male-to-female ratio was 4.2 (95% CI 3.8–4.6) [34]. However, in higher quality studies that screened the general population for cases of ASD, the male-to-female ratio was closer to 3, suggesting that ASD may be underdiagnosed in girls.

Frequency in siblings: The estimated prevalence of ASD in siblings of children with ASD who do not have other medical conditions or syndromes ranges from 3 to 10% [35–44]. However, other studies have suggested that the prevalence of ASD in siblings of children with ASD may be as high as 20% [41, 45, 46].

Interestingly, younger male siblings of a child with ASD are more often affected than younger female siblings. But the risk of recurrence appears to increase when the indexed patient is a girl (younger brothers of girls with ASD, 17%). The recurrence of ASD among children in more than 1.5 million families with two children aged from 4 to 18 years between 2008 and 2016 showed that the overall prevalence of ASD was at 1.25%. Among the families in which the older child had ASD, the risk of recurrence varied according to the sex of the siblings as follows: younger brothers of boys with ASD, 13%; younger sisters of girls with ASD, 8%; and younger sisters of boys with ASD, 4% [46].

The risk of ASD in siblings of children with ASD without an identifiable etiology is 7% if the affected child is female, 4% if the affected child is male, and ≥ 30% if there are two or more affected children in the family according to the 2013 American College of Medical Genetics and Genomics practice guideline [41, 47, 48].

It was also discovered that siblings of children with ASD may have symptoms of ASD even if they do not meet criteria for diagnosis of ASD (occasionally named the "broad ASD phenotype"). Symptoms of ASD or associated neurodevelopmental abnormalities were found to be more common among siblings of children with ASD than siblings of children without ASD in observational studies [49–53].

3 Pathogenesis

The pathogenesis of ASD has not been understood completely. Many factors seem to be involved in the onset of this disorder. These include epigenetic interactions between genetic and environmental factors.

The general consensus is that ASD is caused by genetic factors that alter brain development, more specifically, neural connectivity, thereby disturbing the social communication development pathways, leading to restricted interests and repetitive behaviors [54–56]. This consensus is supported by the "epigenetic theory" in which an aberrant gene is switched "on" during early fetal development and alters the other genes' expression without changing their primary DNA sequence [57, 58].

3.1 Genetic Factors

The complexity of ASD and the diversity of clinical presentations are possibly due to interactions among numerous genes or gene combinations along with epigenetic influences, i.e., exposure to environmental modifiers that cause fluctuating gene expression [40, 59–63]. ASD has been also associated with polygenic variants, single-nucleotide variants, copy-number variants, and rare inherited variants [56, 61, 64, 65]. A solid genetic influence on the development of ASD is supported by the unequal sex distribution, increased prevalence in siblings, high concordance in monozygotic twins, and increased risk of ASD with increased relatedness [40, 60, 61, 66–70].

Sandin et al. 2014 showed, through a large population-based study, that the cumulative risk of ASD by the age of 20 years was found to be around 59% for monozygotic twins, 13% for full siblings and dizygotic twins, 9% for maternal half-siblings, 7% for paternal half-siblings, and 3% for cousins [68]. Though male predominance proposes to be X-linked, several studies indicated that the male-to-male transmission in a number of families eliminates X-linkage as the sole mode of inheritance [40, 71]. It was also found that the prevalence among siblings of ASD patients is higher than the prevalence in the general population but much lower than would be anticipated for single-gene disorders [36, 38, 40, 42].

The correlation between clinical phenotypes and specific genetic profiles continues to be examined. Linkage studies and whole-exome sequencing (WES) have identified many genetic variations predisposing one to ASD [47, 64, 72]. The genomics of ASD is discussed in a separate chapter. On the other hand, neurobiological elements such as neuroimaging, electrophysiology, and autopsy studies in

ASD patients have proposed that brain anomalies, particularly atypical neural connectivity, play an vital role in the development of ASD [22, 73]. In addition, ASD children may have enhanced head development during infancy and enlarged overall brain size [74, 75].

In comparison to individuals without ASD, ASD individuals have different total and regional gray and white matter volumes, brain chemical concentrations, neural network anatomy, sulcal and gyral anatomy, brain lateralization, and cortical structure and organization [76–83]. Furthermore, cortical changes appear to result from abnormal neuronal differentiation during prenatal development [76]. Individuals with ASD utilize atypical and distinct forms of connectivity, cognitive approaches, and brain regions to manage information during tasks necessitating social interactions (e.g., faces, eye gaze, speech) and social and nonsocial rewards compared to individuals without ASD [84–95].

A chapter 2 in this book has been dedicated to genetics and diagnostics of autism. Genetic changes including chromosome abnormalities, genetic variations, transcriptional epigenetics, and noncoding RNA topics are also highlighted.

3.2 Environmental Factors

Several environmental elements or factors are thought to contribute to autism, including (1) advanced parental age, (2) prenatal exposures, (3) perinatal risk factors, (4) maternal medication, (5) smoking and alcohol use, (6) vaccination, (7) nutrition, (8) toxic exposures, (9) socioeconomic status (SES), and (10) gut microbiota (GM) disequilibrium which may also explain some ASD cases [96]. Undoubtedly, both genetic and environmental factors influence autism.

- **Advanced Parental Age** Both paternal and maternal ages were found to be connected to increased risk of having a child with ASD [96–103]. This association could be attributed to the amplified risk of having de novo spontaneous mutations (more often paternal), DNA methylation changes in the sperm, and/or alterations in genetic imprinting in the advanced parental age [104, 105].
- **Prenatal Exposure** Prenatal exposure such as sex hormone alterations, maternal obesity, diabetes, hypertension, infections (rubella, cytomegalovirus, and influenza), in vitro fertilization (IVF) pregnancy, immune activities, and ultrasound may all be considered a liability and contribute to ASD risk [96, 106–119].

It is well known that some viral congenital infections, such as rubella and cytomegalovirus, being correlated with brain calcification, microcephaly, and ASD, can interfere with brain development. Due to the increasing number of babies born with microcephaly, structural brain abnormalities, and neurological alterations in regions affected by Zika virus (ZIKV), investigations have been carried out in order to understand this process better. The maternal immune system directly influences the fetal central nervous system (CNS), and complications during pregnancy have been associated with neurodevelopmental disorders. ZIKV could be considered a possible risk factor for ASD due to neuroimmunological aspects [120].

- **Perinatal Risk Factors** Perinatal risk factors for instance prematurity, cesarean delivery, low birth weight, low Apgar score, and hypoxia may have a role in enhanced autism risk [96, 121–123].
- **Maternal Medication** The potential role or use of drugs during gestational period in the development of ASD has been investigated. A large comparative systematic review and meta-analysis has found valproic acid (VPA) to be associated with the increased risk of ASD [124]. On the other hand, several meta-analyses of observational and systematic review studies suggested that antenatal selective serotonin reuptake inhibitor (SSRI) exposure is not correlated with an increased risk of ASD [125, 126]. Maternal prenatal drug exposure during pregnancy is detailed in another chapter.
- **Smoking and Alcohol Usage** Both smoking and alcohol abuse are a form or a subcategory of drugs that have been found to be reliably associated with neurological, psychiatric, and neurodevelopmental disorders, including comorbidities of ASD, such as ADHD [127]. However, this does not occur with ASD phenotypes. On the other hand, smoking was found to be associated with an increased risk of ASD with ID [128, 129]. Two meta-analyses showed no proof of smoking as a risk factor in ASD, even after correcting for multiple confounds including SES and parental psychiatric history [130, 131].
- **Vaccines** A potential role of an association between vaccines and ASD has also been investigated. There was some speculation that certain vaccines (e.g., measles, mumps, and rubella, MMR) or vaccine constituents (e.g., thimerosal) may play a role in the development of ASD [132–138]. However, prospective studies established that these ASD findings are often present during the first year of life, hence before the first dose of MMR, suggesting lack of association with immunization. Moreover, epidemiological evidence does not support an association between immunizations and ASD [37, 139–142].
- **Maternal Nutrition Status** The depletion of essential nutrients in the mother has been correlated with adverse health outcomes for children, including increased ASD risk. By virtue of this, short inter-pregnancy interval has also been associated with an increased risk of ASD [143, 144]. An entire section in this book has been devoted to specific food and nutrient qualities.

Mazahery et al. 2016 and Magnusson et al. 2016 have showed that vitamin D deficiency during early development combines with other risks and perhaps contributes to ASD risk. Vitamin D could be used as a preventive measure to reduce ASD symptoms in diagnosed cases [145, 146]. In addition, Demarquoy et al. 2019 were recently able to establish a possible link between ASD and carnitine deficiency. They were also able to show that dietary supplementation with L-carnitine is beneficial to these patients [147].

Schmidt et al. 2014 also showed that maternal iron deficiency and low iron intake doubled the chance of having an ASD child in the presence of other ASD risk factors, where low iron intake significantly interacted with advanced maternal age and metabolic conditions; joined exposures were associated with a fivefold increase of ASD incidence [148]. Maternal deficiencies in zinc and cooper were found to possibly be contributing to overall ASD risk [149–152].

- **Pollution and Toxic Pollutant Exposure** A systematic review and meta-analysis of 23 studies showed that the risk of having ASD increases one- or twofold with PM_{10} and $PM_{2.5}$, respectively [153]. A systematic review and meta-analysis studying the link between concentrations of toxic metals (such as antimony, arsenic, cadmium, lead, manganese, mercury, nickel, silver, and thallium) in biofluids (whole blood, plasma, serum, red cells, hair, and urine) in ASD patients showed higher levels of mercury and lead in the blood and higher antimony, cadmium, lead, and mercury in the hair [154].

A review study showed there are several significant associations between all classes of pesticides and ASD risk. These effects were greatest for exposures in weeks 1–7 of pregnancy and postnatally in weeks 4–12 [155]. Another study showed that ASD risk increased by 60% with exposure to organophosphates during pregnancy. This risk was two times greater for exposure during the third trimester and three times greater with exposure to chlorpyrifos during the second trimester. Exposure to pyrethroid insecticide immediately prior to conception or during third trimester increased risk of both ASD and developmental delay around two times [156].

The role of *nonpersistent organic pollutants*, primarily phthalates and bisphenol which are used mainly in the production of plastics, has been investigated in relation to ASD. It was found that there are associations between phthalates and ASD, but the association with bisphenol was controversial [157–161]. In addition, the role of three major agents of persistent organic pollutants such as dichlorodiphenyltrichloroethane (DDT), polychlorinated biphenyls (PCBs), and polybrominated diphenyl ethers (PBDEs) has been investigated in relation to ASD [162]. It was found that the pesticide DDT had a negative impact on cognitive skills (IQ, memory) and gene expression in the hypothalamus [163, 164]. PCBs have negative effects on various intellectual, motor, and verbal outcomes of relevance to autism, and PBDEs had also negative effects on neurodevelopment [165]. However, the latter finding was not confirmed in another study [166].

- **Socioeconomic Status (SES)** Studies on the association between SES and autism have not been conclusive [167]. However, several studies in the USA found that lower SES, specifically concerning the aspects of household income and parental education, was associated with decreased risk of ASD and higher social class (higher median family income) was significantly associated with ASD [168–170]. In contrast, studies in European regions and other countries showed negative/inverse or no relationship between SES and ASD. For instance, increased risk of autism was associated with lower SES in Sweden, France, and Japan [167, 171, 172], and no association was found in Denmark and the UK [173, 174]. A recent study in China showed that children in middle-income and high-income families were less likely to have ASD than their low-income peers. Children in families with socioeconomic disadvantage, in the form of lower family income and education, had greater risk of childhood ASD [175].

- **Gut Microbiota (GM)** GM alteration in gut microflora equilibrium and its metabolites on the development of ASD symptoms is a new research frontier that is evolving at a rapid pace. Animal model studies with ASD behavioral traits revealed that there is in fact a correlation between GM dysbiosis and clinical features reported in ASD patients such as behavioral alterations, gastrointestinal tract abnormalities, and immunologic alterations [176, 177]; additionally, GM microbiota has a distinctive attribute in ASD children [178]. Thus, proliferation and/or depletion of clusters of particular bacteria controls intestinal functions and may interfere with neuro-immune communication and behavior in ASD patients [179–181].

It is worth mentioning that the influence of environmental exposure seems to depend on several factors such as the timing and duration of exposure, concentration of the toxin, mechanism of action, and diffusion in the CNS [56, 182, 183]

Gardener et al. 2011 carried out a comprehensive meta-analysis of 40 diverse observational studies of perinatal and neonatal risk factors for ASD and found only a slight hint of relationship with any single risk factor in the etiology of ASD [122]. However, some evidence suggested that a wide range of conditions, such as abnormal presentation, umbilical-cord complications, fetal distress, birth injury or trauma, multiple birth, maternal hemorrhage, summer birth, low birth weight, small size for gestational age, congenital malformation, low 5-min Apgar score, feeding difficulties, meconium aspiration, neonatal anemia, ABO or Rh incompatibility, and hyperbilirubinemia, are associated with ASD risk. A subsequent meta-analysis of observational studies supported the notion of increased risk of ASD in preterm infants [184]. Succeeding observational studies also proposed that maternal health conditions (e.g., diabetes, obesity, hypertension, preeclampsia) are linked with increased ASD risk [114, 185–189].

Further studies are still needed correlating these risk factors and ASD to inform prevention strategies.

4 Protective Factors

While the majority of research studies have investigated environmental risk factors in ASD, there is an evolving body of research studying the vital role of likely protective factors, mainly from the field of nutrition and food supplementation such as folate, fatty acids including the omega-3 group, and others such as probiotic bacteria. Current studies indicate that prenatal nutrition and food supplementation might reduce the risk of ASD in children or alleviate severity. This field is receiving substantial attention at the international level. The reader is referred to the chapters focusing on the special qualities of specific foods and nutrients.

5 Screening Tools

Screening for ASD is suggested in children (1) with delayed language/communication milestones, (2) with a regression in social or language skills, and (3) whose parents raise concerns regarding ASD. ASD-specific screening is recommended by the American Academy of Pediatrics for all children at 18 and 24 months of age because these are critical intervals for early social and language development and earlier intervention is more effective for ASD [182].

The choice of the screening test depends on the age of the child and whether he or she is being screened for the first time or has been identified through developmental surveillance or screening to be at risk of developmental problems [182]. In the former situation, a first-tier screen should be used, and in the latter, a second-tier screen. Herein, we discuss the tiered screening model. In the tiered screening model, first-tier screening is used to recognize ASD children at risk from the general population. Second-tier screening is used to discriminate ASD from other developmental disorders in children with developmental concerns. Second-tier screening tools are appropriate for use among children who have failed general developmental screening or an ASD-specific screening test, depending upon their age and the level of concern. Second-tier tools are usually more time-consuming and may necessitate more expertise to administer and interpret. Several screening tools have been developed for ASD patients; these tools are listed in Table 1.

Tool Comparison Studies Several investigations have directly compared screening tools in specific populations. A few of these studies are described below.

A study that examined multiple screening tools, including the Social Communication Questionnaire (SCQ), Infant Toddler Checklist (ITC), and key items from the Checklist for Autism in Toddlers (CHAT) in a population of 238 high-risk children (clinician concern or positive first-tier screen), found that none of the instruments adequately discriminated ASD from non-ASD [224].

In a study comparing the Modified-CHAT (M-CHAT) with the SCQ in a subsample of 39 preschool children referred for suspected ASD, the M-CHAT correctly classified 24 out of 29 children with ASD and 5 out of 10 children with non-ASD. The SCQ correctly identified 21 out of 29 children with ASD and 3 out of 10 children with non-ASD. Both instruments were more accurate in evaluating children with lower intellectual and adaptive functioning [225].

In a study comparing Parent's Observations of Social Interactions (POSI) with the M-CHAT in 232 children (16 to 36 months) from primary care and specialty clinic populations, the POSI had higher sensitivity (83 vs. 50%) but lower specificity (75 vs. 84%). In another study comparing the M-CHAT with the POSI in 217 children (18 to 48 months) from a developmental clinic, the sensitivity of the POSI was higher (89 vs. 71%), but specificity was not significantly different (54 vs. 62%) [204]. Noteworthy is the fact that the M-CHAT has been validated only in children up to 30 months of age, so it had not been used as designated. In addition, a revised version of the M-CHAT i.e. M-CHAT, Revised with Follow-Up (M-CHAT-R/F) was released after this study.

Table 1 Screening tools for ASD

Tool	Age	Description	Sensitivity/specificity	Validation	Comment	References
M-CHAT--R/F	16–30 months	20 parent-report items Takes approximately 5 min to administer and 2 min to score	Sensitivity: 85% Specificity: 99%	16,071 children in primary care practices	Validated as first-tier screening method[a] Children are stratified into low, medium, or high risk of ASD Children at medium risk require structured follow-up questions for additional information before referral for diagnostic evaluation Follow-up interview takes approximately 5–10 min	[190–194]
ESAT	14–15 months	14-item questionnaire: First stage: 4 items Second stage: 14-item questionnaire	Sensitivity: 90% highly sensitive but has low specificity	478 children (14- to 15-month-old) 153 older children with ASD 76 older children with ADHD	The parents of the older children were instructed to answer based on their child's presentation at age 14 months	[195–197]
			Positive predictive value (PPV): 25%	31,724 children in a random population	False-positive screens included children with language disorder and ID, but not typically developing children Children with negative screens were not systematically followed up, precluding calculation of sensitivity and specificity	

STAT	24–36 months	It is an interactive measure that can be used for screening in children ages 24–36 months It was designed to discriminate between autism and other developmental disorders. Although it is primarily a second-stage screen, the STAT is used by primary care providers for enhanced screening in some programs 12 observed activities during 20-min play session Requires training for administration and scoring	Sensitivity: 92–95% Specificity: 73–85%	52 children with ASD and other developmental disorders and 71 high-risk children	Not validated as a first-tier screen[a] Language comprehension is not required	[198–201]
ITC	6–24 months	24-item questionnaire, a component of the Communication and Symbolic Behavior Scales-Developmental Profile (CSBS-DP)	Sensitivity and specificity of 88.9% PPV of 71–79% and negative predictive value of 88–99% for 9–24-month-olds PPV: 75% for 32–36 months of age	10,479 infants screened at 1 year during health supervision visit	The CSBS-DP is a broadband screen for communication delays but has not been validated as a screen for ASD	[202, 203]
POSI	18–35 months	7-item parent-report items, a component of the Survey of Well-Being of Young Children (SWYC) Takes ≤ 5 min to complete	Sensitivity: 83% Specificity: 74%	232 children (16–36 months) from primary care and specialty clinics	Additional studies in community samples are necessary before the POSI can be recommended as a first-tier screen[a]	[204, 205]
			Sensitivity: 89% Specificity: 54%	217 children (18–48 months) from specialty clinic		

(continued)

Table 1 (continued)

Tool	Age	Description	Sensitivity/specificity	Validation	Comment	References
SCQ	4–40 years	It was developed from the Autism Diagnostic Interview-Revised (ADI-R) 40 parent-report items (yes/no) Takes < 10 min to administer and < 5 min to score	A cutoff score of 15 on the SCQ had a sensitivity and specificity of 85% and 75%, respectively	200 high-risk patients aged 4–40 years	Different cutoff scores may be needed for verbal and nonverbal individuals since several items related to verbal language are not included in the final score for nonverbal individuals Additional studies are necessary before the SCQ can be used as a first-tier screen[a] Nonverbal children may require different cutoff scores The SCQ is used primarily as a second-stage screen in research	[206–211]
			A cutoff score of 15 on the SCQ had a sensitivity and specificity of 71% and 79%, respectively	151 children (mean age 5 years)		
			A cutoff score of 15 on the SCQ had a sensitivity and specificity of 64% and 75%, respectively	808 children aged 18–48 months		
	5–≥11 years	Validated the Arabic version of the SCQ	A cutoff score of 15 had a sensitivity and specificity of 79.6% and 96.65, respectively In a cutoff score of 12, the values for sensitivity and specificity were 0.893 and 0.893, respectively In cutoffs ranging from 11 to 15, the sensitivity varied between 90.3% and 79.6%, whereas the specificity varied between 85.4% and 96.6%	206 children with ASD and 206 typically developing children		
CAST	4–11 years	37 parent-report items	Accuracy varied with case definition	1925 children	Additional studies are necessary before the CAST can be recommended as a first-tier screen[a]	[26, 212–214]

ASSQ	7–16 years	27-item checklist to be completed by parents or teachers Takes approximately 10 min to complete	Varies depending upon the cutoff score: A cutoff score of 19 for parent ratings is associated with a sensitivity and specificity of 62% and 90%, respectively A cutoff score of 22 for teacher ratings is associated with a sensitivity and specificity of 70% and 91%, respectively A cutoff score of $\geq$ 17 (combined parent and teacher ratings) provided a sensitivity and specificity of 91% and 86%, respectively	9430 children (7–9 years)	Validated as first-tier screen[a] Designed as a first-tier screen for children with high-functioning ASD Clinicians can individualize cutoff scores to meet their diagnostic needs	[215, 216]
AQ-Child	4–11 years	Parent-report measure It consists of 50 questions assessing social skills, attention, communication, and imagination	Sensitivity: 95% Specificity: 95%	540 children with ASD, 1225 children from the general population	Additional studies are necessary before the AQ-Child can be recommended as a first-tier screening tool[a]	[217–219]
DBC-ASA	4–18 years	29 parent-report items from the DBC-P	In a cutoff score of 17, the DBC-ASA had a sensitivity and specificity of 86% and 69%, respectively	180 children with ASD and 180 controls (matched for age, sex, and IQ range)	For use in children with ID	[220, 221]

(continued)

Table 1 (continued)

Tool	Age	Description	Sensitivity/specificity	Validation	Comment	References
DBC-ES	18–48 months	17 parent-report items from the DBC-P	In a cutoff score of ≥ 10.5, the DBC-ES had a sensitivity and specificity of 88% and 69%, respectively	60 children with developmental delay and ASD and 60 controls (developmental delay without ASD)	For use in children with developmental delay	[222, 223]
			In a cutoff score of ≥ 11, the DBC-ES had a sensitivity and specificity of 83% and 48%, respectively PPV of 70–84% and negative predictive value of 42–70%	207 children aged 20–51 months from the community		

M-CHAT-R/F Modified Checklist for Autism in Toddlers Revised with Follow-Up, *ESAT* Early Screening of Autistic Traits, *STAT* Screening Tool for Autism in Toddlers and Young Children, *ITC* Infant Toddler Checklist, *POSI* Parent's Observations of Social Interactions, *SCQ* Social Communication Questionnaire, *CAST* Childhood Autism Spectrum Test, *ASSQ* Autism Spectrum Screening Questionnaire, *AQ-Child* Autism Spectrum Quotient-Children's Version, *DBC-ASA* Developmental Behavior Checklist-Autism Screening Algorithm, *IQ* intelligence quotient, *DBC-P* Developmental Behavior Checklist for Pediatrics, *DBC-ES* Developmental Behavior Checklist-Early Screen

[a]First-tier screening tools are used to identify children at risk for ASD from a general population; second-tier screening tools are used to discriminate ASD from other developmental disorders in children with developmental concerns

In a study investigating the validity of identifying ASD in a sample of 49 children with ID, the DBC-ASA showed similar sensitivity (94 vs. 92%) and lower specificity (46 vs. 62%) compared to the SCQ. Six of the seven children with false-positive scores on the Developmental Behavior Checklist-Autism Screening Algorithm (DBC-ASA) had elevated problem-behavior scores [221].

6 Diagnostic Tools

Several diagnostic tools have been employed in combination with clinical judgment to reach a diagnosis in ASD [182, 226, 227]. Diagnostic tools for ASD are commonly administered by a trained specialist.

These diagnostic tools are offered as self-completion questionnaires and a formal diagnostic interview. However, their diagnostic accuracy has not been well evaluated [228–230]. The tools that are suggested as part of international guidelines are included in Table 2 [182, 226, 249–251].

In clinical practice, the Asperger Syndrome Diagnostic Scale is also used for higher functioning children along with the Autism Diagnostic Interview-Revised (ADI-R), Autism Diagnostic Observation Schedule-Second Edition (ADOS-2), Childhood Autism Rating Scale-Second Edition (CARS-2), and Gilliam Autism Rating Scale (GARS). In a systematic review of observational studies evaluating diagnostic accuracy in children < 6 years of age, who underwent multidisciplinary evaluation for ASD (the reference standard), none of the studies using GARS, Developmental Dimensional and Diagnostic Interview (3di), or Diagnostic Interview for Social and Communication Disorder (DISCO) met the inclusion criteria [229]. Among studies of ADI-R, CARS, and ADOS, there was substantial variation in sensitivity and specificity, likely related to differences in study populations and methodology. The remaining studies that were available were too few in number to permit meaningful direct comparison. However, when the summary statistics were compared, ADOS was most sensitive (94% compared with 80% for CARS and 52% for ADI-R). The three tools had similar specificity (ranging from 80 to 88%).

Diagnostic tools for ASD must be used in conjunction with clinical judgment for a number of reasons. The administration protocols that are used in research studies may not be achievable in clinical practice. In the studies included in the systematic review, ASD was diagnosed according to criteria from DSM-4 or earlier classifications that do not directly correlate with DSM-5 criteria [182]. In addition, the versions of the tools evaluated in published studies may have been updated after publication.

Ancillary Testing Ancillary testing is necessary to assess functional impairment, define the child's strengths and weaknesses for education planning, identify associated conditions (e.g., intellectual impairment, language impairment), and

Table 2 Diagnostic tools for ASD

Assessments	Tools	Age	Description	References
Autism-specific diagnostic interviews	ADI-R	≥ 2 years	A semi-structured interview with parent/caregiver 96-item interviews Takes around 2–3 h	[231]
	DISCO[a]	All ages	A clinical semi-structured interview with parent/primary caregiver > 300, including 93 for diagnosis Takes around 2–3 h	[232, 233]
	CARS/ CARS-2	≥ 2 years	Combination of interview and direct observation 15-item behavioral rating scale A hybrid, collecting information from a variety of people and situations, including reports from parents and teachers alongside school and clinic observations Takes around 5–15 min	[234, 235]
	3di[a]	Early childhood to adulthood	A structured interview with parent/ caregiver Covers other mental states along with demography, family background, developmental history, and motor skills Takes around 1.5–2 h	[236]
Autism diagnostic observational assessment	ADOS-2	≥ 12 months	A semi-structured behavioral observation Four modules It uses a combination of standardized play, activities, and verbal interview Takes around 30–45 min to administer and a further 20 min to determine the scores	[237–241]
Tools to identify an increased likelihood of ASD	GARS/ GARS-2	3–22 years	A 42-item checklist Takes around 5–10 min to complete and score	[242–246]
	PIA-CV	≤ 3 years	A 118-item structured interview for parents Consists of 11 domains: social relating, affective responses, imitation, peer interactions, object play, imaginative play, language understanding, nonverbal communication, motoric behaviors, sensory responses, and need for sameness Takes around 30–45 min	[247, 248]

ADI-R, Autism Diagnostic Interview-Revised; *DISCO*, Diagnostic Interview for Social and Communication Disorder; *CARS-2*, Childhood Autism Rating Scale-Second Edition; *3di*, Developmental Dimensional and Diagnostic Interview; *ADOS-2*, Autism Diagnostic Observation Schedule-Second Edition; *GARS*, Gilliam Autism Rating Scale; *PIA-CV*, Parent Interview for Autism-Clinical Version

[a]Used predominantly outside the USA

evaluate conditions characterized by symptoms that mimic ASD. Generally, it includes [226, 251–253] the following:

1. Speech, language, and communication assessment—the speech and language assessment provides a profile of language and communication skills and may differentiate ASD from developmental language disorder, language-based learning disorder, and social (pragmatic) communication disorder. The speech and language evaluation includes assessment of:
   (a) Formal language functions (e.g., vocabulary, grammar, syntax)
   (b) Prosodic features (e.g., rate, rhythm, volume, emotional expressiveness)
   (c) Pragmatic language functions (e.g., nonverbal communication [facial expressions, gestures, body language, prosody], nonliteral language [e.g., metaphor, humor], content of conversations [appropriateness of topic for the social situation], ability to stay on topic [topic maintenance])

      Pragmatic language tests are subject to observer interpretation. Individuals with ASD may perform successfully in the 1:1 testing situation but not in real-time situations (e.g., classroom discussion, peer interaction).
2. Developmental/intelligence testing with separate estimates for verbal and nonverbal skills.
3. Assessment of adaptive skills to document associated ID and to help establish priorities for when intervention is planned; functional impairment is one of the diagnostic criteria for both ASD and ID. In addition, in the USA, overall levels of function determine eligibility for services in many states.
4. Sensorimotor and/or occupational therapy evaluation for treatment planning.
5. Vision and hearing assessment (if not already performed).
6. Lead testing (if not already performed).
7. Other tests—ancillary testing may also include specific tests for certain conditions associated with ASD as indicated by the initial clinical evaluation.

7 Diagnostic Criteria

The diagnostic criteria for ASD differ geographically. The most common diagnostic criteria are mentioned below:

1. The DSM which was updated in 2013 (DSM-5) [254].
2. The World Health Organization International Classification of Diseases, 10th revision (ICD-10) [255]. A new version was recently released in 2018, i.e., the 11th revision (ICD-11) [256].

Clinical diagnosis of ASD is made in children who meet the established diagnostic criteria for ASD based on available history and observation of behavior. There are two major sets of diagnostic criteria, both of which center on atypical social communication and interaction and restricted, repetitive patterns of behavior, activities, and interests: the DSM and the ICD.

DSM-Fifth Edition (DSM-5) Criteria According to the DSM-5 criteria, a diagnosis of ASD requires all of the following [254]:

- Persistent deficits in social communication and social interaction in multiple settings, demonstrated by deficits in all three of the following (either currently or by history):
  - Social-emotional reciprocity (e.g., failure to produce mutually enjoyable and agreeable conversations or interactions because of a lack of mutual sharing of interests and a lack of awareness or understanding of the thoughts or feelings of others)
  - Nonverbal communicative behaviors used for social interaction (e.g., difficulty coordinating verbal communication with its nonverbal aspects [eye contact, facial expressions, gestures, body language, and/or prosody/tone of voice])
  - Developing, maintaining, and understanding relationships (e.g., difficulty in adjusting behavior to social setting, lack of ability to display socially acceptable behaviors, lack of interest in socializing, difficulty making friends even when interested in having friendships)
- Restricted, repetitive patterns of behavior, interests, or activities; demonstrated by ≥ 2 of the following (either currently or by history):
  - Stereotyped or repetitive movements, use of objects, or speech (e.g., stereotypies such as rocking, flapping, or spinning); echolalia (repeating parts of speech, repeating scripts from movies or prior conversations)
  - Insistence on sameness, unwavering adherence to routines, or ritualized patterns of verbal or nonverbal behavior (e.g., ordering toys into a line)
  - Highly restricted, fixated interests that are abnormal in strength or focus (e.g., preoccupation with certain objects [trains, vacuum cleaners, or parts of trains or vacuum cleaners]); perseverative interests (e.g., excessive focus on a topic such as dinosaurs or natural disasters)
  - Increased or decreased response to sensory input or unusual interest in sensory aspects of the environment (e.g., adverse response to particular sounds, apparent indifference to temperature, excessive touching/smelling of objects)
- The symptoms must impair function (e.g., social, academic, completing daily routines).
- The symptoms must be present in the early developmental period. However, they may become apparent only after social demands exceed limited capacity. In later life, symptoms may be masked by learned strategies.
- The symptoms are not better explained by ID or global developmental delay.

ASD may occur with or without medical, genetic, neurodevelopmental, mental, or behavioral disorders (e.g., intellectual impairment, language impairment, epilepsy, fetal valproate or alcohol exposure). The presence or absence of these conditions is specified as part of the DSM-5 diagnosis of ASD (e.g., ASD with accompanying intellectual impairment, ASD without accompanying language impairment). Some accompanying conditions are identified during the comprehensive evaluation; others may require additional testing.

ICD-10th Revision (ICD-10) Criteria The ICD-10 criteria for the diagnosis of pervasive developmental disorders are provided in the ICD-10 [255]. Although the 11th edition of ICD was released in 2018, the clinical descriptions and diagnostic guidelines in the ICD-10 should be used until January 2022, when the transition to ICD-11 is scheduled to occur.

Associated conditions and syndromes: A number of neurodevelopmental conditions and genetic syndromes are associated with ASD. Approximately 45% of patients with ASD have ID, as many as 50% have ADHD, and as many as 30% have epilepsy [22, 257]. The risk of epilepsy increases in patients with more severe ID [182].

Up to 25% of cases of ASD are associated with a clinically or molecularly defined syndrome (e.g., tuberous sclerosis complex [TSC], valproate embryopathy, 15q chromosome duplication) [64, 182, 258–263]. Associated syndromes are more common in patients with global developmental delay or ID [182].

Clinical Syndromes Clinically defined syndromes commonly associated with ASD [22, 64, 182, 226, 264] include the following:

- *TSC* is an inherited neurocutaneous disorder that is characterized by the development of a variety of benign tumors in multiple organs. Associated clinical features include hypopigmented macules, angiofibromas, shagreen patches, seizures, and cognitive deficits. Approximately 40% of patients with TSC also have ASD; however, only 0.4–4% of patients with ASD have TSC [265–268]. Patients with comorbid TSC and ASD often have epilepsy [265, 267, 269, 270].
- *Fragile X syndrome (FXS)* is an X-linked disorder that is often associated with ID. Characteristic features of the classic phenotype include a long, narrow face, prominent forehead and chin, large ears, testicular enlargement in adolescence, macrocephaly, arched palate, and hyperextensible joints. In a systematic review, 30% of males with FXS had features of ASD [266]. However, FXS is rarely found in patients with ASD [271–273].
- *Chromosome 15 q11-q13 duplication syndrome* is described by hypotonia, joint laxity, global (especially motor) developmental delays, seizures, speech delay, social deficits, stereotypies, and a variable pattern of mild facial dysmorphisms [274, 275]. 15q11-q13 duplication has been reported in approximately 1–2% of children with ASD, usually those with moderate to profound ID [226, 276–279].
- *Angelman syndrome (AS)* is a neurodevelopmental disorder characterized by severe ID, postnatal microcephaly, and movement or balance problems. It is caused by the absence of the maternally inherited copy of the *UBE3A* gene, which maps onto chromosome 15q11-q13. In a systematic review, 34% of patients with AS had ASD [266].

- *Classic Rett syndrome* occurs almost exclusively in girls. It is characterized by loss of speech, replacement of purposeful hand movement with stereotypic hand movement, gait abnormalities, and an abnormal respiratory pattern. In a systematic review, approximately 60% of females with Rett syndrome had phenomenology of ASD [266].
- *CHARGE syndrome*: is characterized by coloboma of the eye, heart defects, choanal atresia, growth retardation, genitourinary anomalies, and ear abnormalities. As many as 50% of affected patients have ASD [22, 266].
- *Joubert syndrome*: is a heterogeneous syndrome characterized by hypoplasia of the cerebellar vermis, neurological symptoms (e.g., dysregulation of breathing pattern, developmental delay), retinal dystrophy, and renal anomalies. Approximately 40% of patients with Joubert syndrome also have ASD [22].
- *Smith-Lemli-Opitz syndrome* is an autosomal recessive disorder of cholesterol biosynthesis [182]. Clinical features include postnatal microcephaly, soft cleft palate/bifid uvula, micrognathia, low-set posteriorly rotated ears, poor weight gain, syndactyly of the second and third toes, abnormal genitalia, ID, hypotonia, and autistic features (e.g., deficits in social interaction and communication, repetitive and stereotyped behaviors) [280, 281]. In one case series, 10 to 12 of 14 children with Smith-Lemli-Opitz syndrome met criteria for ASD [281].
- *Timothy syndrome*: is characterized by syndactyly, congenital heart disease, multiorgan dysfunction, and cognitive abnormalities. As many as 70% of patients with Timothy syndrome also have ASD [22].
- *Macrocephaly/autism syndrome*: Clinical features of macrocephaly/autism syndrome include postnatal macrocephaly, broad forehead, frontal bossing, long philtrum, depressed nasal bridge, and ID.
- *Cowden/Bannayan-Riley-Ruvalcaba syndrome*: Clinical features include macrocephaly, birdlike facies, hypoplastic mandible and maxilla, cataract, microstomia, high-arched palate, pectus excavatum, genitourinary anomalies, skin tags, lipomas, and penile macules.

Molecularly defined syndromes may account for as many as 20% of cases of ASD [64] and are characterized by incomplete penetrance and variable expressivity, making them difficult to identify clinically. Although there is some overlap with clinically defined syndromes, examples of molecularly defined syndromes include chromosomal variations (e.g., isodicentric 15q), ASD-associated copy-number variants (e.g., 16p11.2 deletions or duplications), and pathogenic variants of ASD-risk genes (e.g., *CHD8* [chromosome helicase DNA binding protein 8]).

8 Genetic Testing

First-Tier Genetic Studies Suspected ASD cases whether or not they have dysmorphic features are evaluated by chromosomal microarray (CMA) and DNA analysis for fragile X. Karyotyping is also warranted if balanced translocation is suspected (e.g., history of $\geq$ 2 miscarriages) because CMA does not detect balanced translocations [282, 283]. However, truly balanced de novo translocations are rare [284].

Identification of a genetic diagnosis may provide information about prognosis and recurrence risk [258, 285]. It may also provide emotional relief for parents and can be crucial to the therapeutic alliance. However, few studies have evaluated the effect of genetic testing on such outcomes, and it is unclear whether or not genetic testing affects health outcomes.

However, classification and reporting of variants of unknown significance (VUS) remains a challenge. Thus, consultation with a clinical geneticist may be necessary for interpretation of CMA results, especially when novel and/or recurrent copy-number VUS are identified. Nonetheless, CMA still has the highest diagnostic rate amongst current genetic tests for ASD patients (excluding WES, which may be costly and not widely available) [286–289].

In a cohort of 933 patients who underwent genetic testing for a diagnosis of ASD, karyotype was abnormal in 2%, fragile X testing was abnormal in 0.5%, and array comparative genomic hybridization (CGH) identified abnormal deletions or duplications in 7% [287]. In another population-based sample of 258 unrelated children consecutively diagnosed with ASD, CMA yielded a molecular diagnosis in 9.3% [288]. Among the 95 children who underwent both CMA and WES, the diagnostic yield of WES was comparable to CMA (8.4%), and the combined yield of CMA and WES was 15.8%. Molecular diagnosis was more often achieved in children with more severe dysmorphology, suggesting that it may be possible to identify children with the greatest likelihood of genetic diagnosis [290].

Recently, Jang et al. 2019 also showed that employing CMA as a first-tier test in developmental delay/intellectual disability (DD/ID), autism spectrum disorders (ASD), and multiple congenital anomalies (MCA) increases diagnostic yields and the quality of clinical management in these patients [291].

The genetic testing approach to children with ASD is part of the consistent recommendations of the American College of Medical Genetics and Genomics and the International Standards for Cytogenomic Arrays Consortium [44, 282, 292–295].

Other Genetic Tests as Clinically Indicated Other types of genetic testing are performed as clinically indicated in children with dysmorphic features, microcephaly, macrocephaly, cognitive impairment/abnormailities, suspicious medical or family history, or in cases where prenatal genetic counseling is desired [40, 47]. Consultation with or referral to a clinical geneticist can be helpful in determining the appropriate studies.

Specific testing should be guided by clinical findings. As examples:

- Testing for the X-linked *MECP2* Rett mutation may be warranted for patients, particularly girls, with a history of significant developmental regression [226, 296].

- Testing for mutations in the *PTEN* gene should be completed for patients with ASD and macrocephaly (greater than 2.5 standard deviations above the mean for age and sex) to rule out hamartomatous tumor syndromes (e.g., Proteus syndrome, Cowden syndrome, including Bannayan-Riley-Ruvalcaba syndrome) [47].

9 Additional Testing as Indicated

Metabolic Testing Disorders of amino acid, carbohydrate, purine, peptide, and mitochondrial metabolism account for < 5% of cases of ASD [226, 297].

Metabolic testing in children with ASD and symptoms or signs of a metabolic disorder [226, 258, 297, 298] includes:

Lethargy, limited endurance (particularly if associated with mild illness)
Hypotonia
Recurrent vomiting and dehydration
Early seizure
Dysmorphic or coarse features
ID (or if ID cannot be excluded)
Developmental regression
Hearing impairment
Vision impairment
Unusual odors
Specific food intolerance (e.g., protein)
Inadequate or questionable adequate newborn screen

Several observational studies studying the yield of metabolic testing in the absence of signs or symptoms of metabolic disease have shown that it has a low diagnostic yield [47, 258, 299, 300].

Neuroimaging Decisions about neuroimaging in children with ASD are made on a case-by-case basis. In observational studies, the yield of magnetic resonance imaging is low in children with ASD and no other neurological findings (e.g., ID, abnormal neurologic examination, seizures, headache, focal neurologic findings) [258, 301].

Electroencephalography (EEG) EEG is done on children with ASD only if warranted by history or physical examination, specifically for clinical seizures, unusual spells, or behaviors frankly suggestive of seizures, to exclude Landau-Kleffner syndrome (acquired epileptic aphasia) in children with regression in language skills [182, 226]. Among children with ASD and staring spells, it has been found that EEG rarely yields clinically significant findings [302] and thus is not routinely recommended.

10 Conclusion and Perspective

Increasing prevalence of ASD and high rates of related comorbidities have caused serious health deterioration and placed substantial burden on supporting families, caregivers, and health care services. The economic impact associated with ASD is substantial and includes direct medical, nonmedical, and indirect productivity costs.

The sizable economic burden of ASD in the USA is predicted to be $175–268 billion, exceeding the cost of noncommunicable diseases like cancer, stroke, and heart disease together. ASD is projected to cost the USA a whopping $461 billion by 2025 [303]. Yearly direct medical and nonmedical and productivity costs together were forecasted to be $268 billion (range $162–367 billion; 0.88–2.0% of GDP) for 2015 and $461 billion (range $276–1011 billion; 0.98–3.6% of GDP) by 2025.

Studies estimate the lifetime cost of caring for ASD an individual to be $2.2 million in the USA and £1.5 million in the UK though the cost drops to $1.4 million in the USA and £0.9.2 million in the UK for ASD without comorbid conditions. In addition, if unrecognized or untreated, ASD can contribute to poor educational attainment and difficulty with employment, leading to negative economic implications. There is also a suggestion that comorbidities of ASD tend to amplify burden to the society and afflicted individuals alike [304, 305]. However, economic burden could be reduced via more investment and funding in ASD research to comprehend the causes of and develop treatments for ASD. Thus, there is a need for an increase in public, research, and government attention to ensure that all children have access to intensive early intervention and that school-based interventions support academics as well as social and language skills to explore the causes and best treatments for ASD.

In this review, we offered a summary of epidemiological studies, disease pathogenesis, screening and diagnostic tools, and genetic testing of ASD. The future for individuals suffering from less severe forms of ASD is bright. Our hope is that through early behavioral screenings, genetic testing, identification of environmental risk factors, as well as a better understanding of neural development, the number of individuals suffering from autistic phenotypes may be greatly reduced. While there is much work to be done in understanding and treating ASD, there are important steps that can be taken now. First, continued awareness programs so that children are identified and treated as early as possible. Second, prenatal and prepregnancy awareness of environmental factors, including recommendations against consanguineous marriages and information regarding optimal maternal nutrition and the importance of limiting exposure to toxins and pollutants. Finally, the expansion of genetic screening and early postnatal monitoring of infant feeding, nutrition, and eye contact that will help provide treatment as early as possible.

Progress in clinical care for those affected by ASD will continue to be driven by multidisciplinary, collaborative research efforts. Efforts are currently underway to identify functional, genetic-ontological subtypes that may provide additional utility with regard to clinical intervention. This includes investigation of broader phenotypes associated with gene disruptions that share molecular properties. Identification of measurable neurological effects of gene disruptions, such as an electrophysiological (EEG) signature, could translate to meaningful ASD biomarkers that are essential for clinical treatment trials. Development of treatment is also required to increase our knowledge about the timing of genetic expression and explore the possibility of reversing neurodevelopmental impairment. Thus, sustained comprehensive phenotyping will be vital to the success of clinical trials for genetic subtypes of ASD.

ASD is necessitating considerable joint efforts of the government and of other societal actors to perform more experimental and clinical research on children with ASD in order to move forward with advancements in clinical practice. There are great opportunities for collaborative research and innovation to contribute to a growing body of evidence and knowledge in ASD.

Acknowledgements The authors want to thank their respective institutions for their continued support. The authors declare no conflict of interest.

References

1. Kanner, L. (1943). Autistic disturbances of affective contact. *The Nervous Child, 2*, 217–250.
2. Asperger, H. (1944). Die "Autistischen Psychopathen" im Kindesalter. *Archiv f. Psychiatrie, 117*(1), 76–136.
3. Elsabbagh, M., Divan, G., Koh, Y.-J., Kim, Y. S., Kauchali, S., Marcín, C., et al. (2012). Global prevalence of autism and other pervasive developmental disorders. *Autism Research, 5*(3), 160–179.
4. Wiśniowiecka-Kowalnik, B., & Nowakowska, B. A. (2019). Genetics and epigenetics of autism spectrum disorder-current evidence in the field. *Journal of Applied Genetics, 60*(1), 37–47.
5. Baxter, A. J., Brugha, T. S., Erskine, H. E., Scheurer, R. W., Vos, T., & Scott, J. G. (2015). The epidemiology and global burden of autism spectrum disorders. *Psychological Medicine, 45*(3), 601–613.
6. Baio, J., Wiggins, L., Christensen, D. L., Maenner, M. J., Daniels, J., Warren, Z., et al. (2018). Prevalence of autism spectrum disorder among children aged 8 years - Autism and developmental disabilities monitoring network, 11 sites, United States, 2014. *MMWR Surveillance Summaries, 67*(6), 1–23.
7. Baird, G., Simonoff, E., Pickles, A., Chandler, S., Loucas, T., Meldrum, D., et al. (2006). Prevalence of disorders of the autism spectrum in a population cohort of children in South Thames: The Special Needs and Autism Project (SNAP). *Lancet, 368*(9531), 210–215.
8. Brugha, T. S., Spiers, N., Bankart, J., Cooper, S.-A., McManus, S., Scott, F. J., et al. (2016). Epidemiology of autism in adults across age groups and ability levels. *The British Journal of Psychiatry, 209*(6), 498–503.
9. Chien, I.-C., Lin, C.-H., Chou, Y.-J., & Chou, P. (2011). Prevalence and incidence of autism spectrum disorders among national health insurance enrollees in Taiwan from 1996 to 2005. *Journal of Child Neurology, 26*(7), 830–834.
10. Christensen, D. L., Baio, J., Van Naarden Braun, K., Bilder, D., Charles, J., Constantino, J. N., et al. (2016). Prevalence and characteristics of autism spectrum disorder among children aged 8 years--autism and developmental disabilities monitoring network, 11 sites, United States, 2012. *MMWR Surveillance Summaries, 65*(3), 1–23.
11. Investigators, D. D. M. N. S. Y. P. and C. f. D. C. a. P. (CDC). (2014). Prevalence of autism spectrum disorder among children aged 8 years - autism and developmental disabilities monitoring network, 11 sites, United States, 2010. *MMWR Surveillance Summaries, 63*(2), 1–21.

12. Kim, Y. S., Leventhal, B. L., Koh, Y.-J., Fombonne, E., Laska, E., Lim, E.-C., et al. (2011). Prevalence of autism spectrum disorders in a total population sample. *The American Journal of Psychiatry, 168*(9), 904–912.
13. Kogan, M. D., Vladutiu, C. J., Schieve, L. A., Ghandour, R. M., Blumberg, S. J., Zablotsky, B., et al. (2018). The prevalence of parent-reported autism spectrum disorder among US children. *Pediatrics, 142*, 6.
14. Schendel, D. E., & Thorsteinsson, E. (2018). Cumulative incidence of autism into adulthood for birth cohorts in Denmark, 1980–2012. *JAMA, 320*(17), 1811–1813.
15. Surén, P., Bakken, I. J., Aase, H., Chin, R., Gunnes, N., Lie, K. K., et al. (2012). Autism spectrum disorder, ADHD, epilepsy, and cerebral palsy in Norwegian children. *Pediatrics, 130*(1), 152–158.
16. Webb, E., Morey, J., Thompsen, W., Butler, C., Barber, M., & Fraser, W. I. (2003). Prevalence of autistic spectrum disorder in children attending mainstream schools in a Welsh education authority. *Developmental Medicine and Child Neurology, 45*(6), 377–384.
17. Williams, J. G., Allison, C., Scott, F. J., Bolton, P. F., Baron-Cohen, S., Matthews, F. E., et al. (2008a). The childhood autism spectrum test (CAST): Sex differences. *Journal of Autism and Developmental Disorders, 38*(9), 1731–1739.
18. Xu, G., Strathearn, L., Liu, B., & Bao, W. (2018). Prevalence of autism spectrum disorder among US children and adolescents, 2014–2016. *JAMA, 319*(1), 81–82.
19. Zablotsky, B., Black, L. I., Maenner, M. J., Schieve, L. A., & Blumberg, S. J. (2015). Estimated prevalence of autism and other developmental disabilities following questionnaire changes in the 2014 national health interview survey. *Natl. Health Stat. Report., 87*, 1–20.
20. Investigators, A. a. D. D. M. N. S. Y. P. and C. f. D. C. a. P. (CDC). (2009). Prevalence of autism spectrum disorders - Autism and developmental disabilities monitoring network, United States, 2006. *MMWR Surveillance Summaries, 58*(10), 1–20.
21. Investigators, D. D. M. N. S. Y. P. and C. f. D. C. a. P. (CDC). (2012). Prevalence of autism spectrum disorders--Autism and developmental disabilities monitoring network, 14 sites, United States, 2008. *MMWR Surveillance Summaries, 61*(3), 1–19.
22. Lai, M.-C., Lombardo, M. V., & Baron-Cohen, S. (2014). Autism. *Lancet, 383*(9920), 896–910.
23. Williams, J. G., Higgins, J. P. T., & Brayne, C. E. G. (2006a). Systematic review of prevalence studies of autism spectrum disorders. *Archives of Disease in Childhood, 91*(1), 8–15.
24. Fombonne, E. (2009). Epidemiology of pervasive developmental disorders. *Pediatric Research, 65*(6), 591–598.
25. Shattuck, P. T. (2006). The contribution of diagnostic substitution to the growing administrative prevalence of autism in US special education. *Pediatrics, 117*(4), 1028–1037.
26. Williams, J., Allison, C., Scott, F., Stott, C., Bolton, P., Baron-Cohen, S., et al. (2006b). The childhood Asperger syndrome test (CAST): Test-retest reliability. *Autism, 10*(4), 415–427.
27. Wing, L., & Potter, D. (2002). The epidemiology of autistic spectrum disorders: Is the prevalence rising? *Mental Retardation and Developmental Disabilities Research Reviews, 8*(3), 151–161.
28. Barbaresi, W. J., Katusic, S. K., Colligan, R. C., Weaver, A. L., & Jacobsen, S. J. (2005). The incidence of autism in Olmsted County, Minnesota, 1976–1997: Results from a population-based study. *Archives of Pediatrics & Adolescent Medicine, 159*(1), 37–44.
29. Bishop, D. V. M., Whitehouse, A. J. O., Watt, H. J., & Line, E. A. (2008). Autism and diagnostic substitution: Evidence from a study of adults with a history of developmental language disorder. *Developmental Medicine and Child Neurology, 50*(5), 341–345.
30. Croen, L. A., Grether, J. K., Hoogstrate, J., & Selvin, S. (2002). The changing prevalence of autism in California. *Journal of Autism and Developmental Disorders, 32*(3), 207–215.
31. Hertz-Picciotto, I., & Delwiche, L. (2009). The rise in autism and the role of age at diagnosis. *Epidemiology, 20*(1), 84–90.
32. Mandell, D. S., & Palmer, R. (2005). Differences among states in the identification of autistic spectrum disorders. *Archives of Pediatrics & Adolescent Medicine, 159*(3), 266–269.

33. Parner, E. T., Schendel, D. E., & Thorsen, P. (2008). Autism prevalence trends over time in Denmark: Changes in prevalence and age at diagnosis. *Archives of Pediatrics & Adolescent Medicine, 162*(12), 1150–1156.
34. Loomes, R., Hull, L., & Mandy, W. P. L. (2017). What is the male-to-female ratio in autism spectrum disorder? A systematic review and meta-analysis. *Journal of the American Academy of Child and Adolescent Psychiatry, 56*(6), 466–474.
35. Asherson, P. J., & Curran, S. (2001). Approaches to gene mapping in complex disorders and their application in child psychiatry and psychology. *The British Journal of Psychiatry, 179*, 122-8. Review.
36. Bolton, P., Macdonald, H., Pickles, A., Rios, P., Goode, S., Crowson, M., et al. (1994). A case-control family history study of autism. *Journal of Child Psychology and Psychiatry, 35*(5), 877–900.
37. Fombonne, E., Zakarian, R., Bennett, A., Meng, L., & McLean-Heywood, D. (2006). Pervasive developmental disorders in Montreal, Quebec, Canada: Prevalence and links with immunizations. *Pediatrics, 118*(1), 139–150.
38. Jorde, L. B., Hasstedt, S. J., Ritvo, E. R., Mason-Brothers, A., Freeman, B. J., Pingree, C., et al. (1991). Complex segregation analysis of autism. *American Journal of Human Genetics, 49*(5), 932–938.
39. Lauritsen, M. B., Pedersen, C. B., & Mortensen, P. B. (2005). Effects of familial risk factors and place of birth on the risk of autism: A nationwide register-based study. *Journal of Child Psychology and Psychiatry, 46*(9), 963–971.
40. Muhle, R., Trentacoste, S. V., & Rapin, I. (2004). The genetics of autism. *Pediatrics, 113*(5), 472–486.
41. Ozonoff, S., Young, G. S., Carter, A., Messinger, D., Yirmiya, N., Zwaigenbaum, L., et al. (2011). Recurrence risk for autism spectrum disorders: A baby siblings research consortium study. *Pediatrics, 128*(3), 488–495.
42. Piven, J., Gayle, J., Chase, G. A., Fink, B., Landa, R., Wzorek, M. M., et al. (1990). A family history study of neuropsychiatric disorders in the adult siblings of autistic individuals. *Journal of the American Academy of Child and Adolescent Psychiatry, 29*(2), 177–183.
43. Risch, N., Spiker, D., Lotspeich, L., Nouri, N., Hinds, D., Hallmayer, J., et al. (1999). A genomic screen of autism: Evidence for a multilocus etiology. *American Journal of Human Genetics, 65*(2), 493–507.
44. Schaefer, G. B., Mendelsohn, N. J., & Professional Practice and Guidelines Committee. (2013). Clinical genetics evaluation in identifying the etiology of autism spectrum disorders: 2013 guideline revisions. *Genetics in Medicine, 15*(5), 399–407.
45. Constantino, J. N., Zhang, Y., Frazier, T., Abbacchi, A. M., & Law, P. (2010). Sibling recurrence and the genetic epidemiology of autism. *The American Journal of Psychiatry, 167*(11), 1349–1356.
46. Palmer, N., Beam, A., Agniel, D., Eran, A., Manrai, A., Spettell, C., et al. (2017). Association of sex with recurrence of autism spectrum disorder among siblings. *JAMA Pediatrics, 171*(11), 1107–1112.
47. Schaefer, G. B., & Mendelsohn, N. J. (2013). Clinical genetics evaluation in identifying the etiology of autism spectrum disorders: 2013 guideline revisions. *Genetics in Medicine, 15*(5), 399–407.
48. Simonoff, E. (1998). Genetic counseling in autism and pervasive developmental disorders. *Journal of Autism and Developmental Disorders, 28*(5), 447–456.
49. Dalton, K. M., Nacewicz, B. M., Alexander, A. L., & Davidson, R. J. (2007). Gaze-fixation, brain activation, and amygdala volume in unaffected siblings of individuals with autism. *Biological Psychiatry, 61*(4), 512–520.
50. Gamliel, I., Yirmiya, N., Jaffe, D. H., Manor, O., & Sigman, M. (2009). Developmental trajectories in siblings of children with autism: Cognition and language from 4 months to 7 years. *Journal of Autism and Developmental Disorders, 39*(8), 1131–1144.
51. Gamliel, I., Yirmiya, N., & Sigman, M. (2007). The development of young siblings of children with autism from 4 to 54 months. *Journal of Autism and Developmental Disorders, 37*(1), 171–183.

52. Piven, J., Palmer, P., Jacobi, D., Childress, D., & Arndt, S. (1997). Broader autism phenotype: Evidence from a family history study of multiple-incidence autism families. *The American Journal of Psychiatry, 154*(2), 185–190.
53. Yirmiya, N., Gamliel, I., Shaked, M., & Sigman, M. (2007). Cognitive and verbal abilities of 24- to 36-month-old siblings of children with autism. *Journal of Autism and Developmental Disorders, 37*(2), 218–229.
54. Baron-Cohen, S. (2006). Two new theories of autism: Hyper-systemising and assortative mating. *Archives of Disease in Childhood, 91*(1), 2–5.
55. Ecker, C., Bookheimer, S. Y., & Murphy, D. G. M. (2015). Neuroimaging in autism spectrum disorder: Brain structure and function across the lifespan. *Lancet Neurology, 14*(11), 1121–1134.
56. Muhle, R. A., Reed, H. E., Stratigos, K. A., & Veenstra-VanderWeele, J. (2018). The emerging clinical neuroscience of autism spectrum disorder: A review. *JAMA Psychiatry, 75*(5), 514–523.
57. Lopez-Rangel, E., & Lewis, M. E. S. (2006). Loud and clear evidence for gene silencing by epigenetic mechanisms in autism spectrum and related neurodevelopmental disorders. *Clinical Genetics, 69*(1), 21–22.
58. Samaco, R. C., Nagarajan, R. P., Braunschweig, D., & LaSalle, J. M. (2004). Multiple pathways regulate MeCP2 expression in normal brain development and exhibit defects in autism-spectrum disorders. *Human Molecular Genetics, 13*(6), 629–639.
59. Bacchelli, E., & Maestrini, E. (2006). Autism spectrum disorders: Molecular genetic advances. *American Journal of Medical Genetics. Part C, Seminars in Medical Genetics, 142C*(1), 13–23.
60. Hallmayer, J., Cleveland, S., Torres, A., Phillips, J., Cohen, B., Torigoe, T., et al. (2011). Genetic heritability and shared environmental factors among twin pairs with autism. *Archives of General Psychiatry, 68*(11), 1095–1102.
61. Robinson, E. B., Neale, B. M., & Hyman, S. E. (2015). Genetic research in autism spectrum disorders. *Current Opinion in Pediatrics, 27*(6), 685–691.
62. Sahin, M., & Sur, M. (2015). Genes, circuits, and precision therapies for autism and related neurodevelopmental disorders. *Science, 350*, 6263.
63. Yuen, R. K. C., Thiruvahindrapuram, B., Merico, D., Walker, S., Tammimies, K., Hoang, N., et al. (2015). Whole-genome sequencing of quartet families with autism spectrum disorder. *Nature Medicine, 21*(2), 185–191.
64. Fernandez, B. A., & Scherer, S. W. (2017). Syndromic autism spectrum disorders: Moving from a clinically defined to a molecularly defined approach. *Dialogues in Clinical Neuroscience, 19*(4), 353–371.
65. Woodbury-Smith, M., & Scherer, S. W. (2018). Progress in the genetics of autism spectrum disorder. *Developmental Medicine and Child Neurology, 60*(5), 445–451.
66. Rosenberg, R. E., Law, J. K., Yenokyan, G., McGready, J., Kaufmann, W. E., & Law, P. A. (2009). Characteristics and concordance of autism spectrum disorders among 277 twin pairs. *Archives of Pediatrics & Adolescent Medicine, 163*(10), 907–914.
67. Lichtenstein, P., Carlström, E., Råstam, M., Gillberg, C., & Anckarsäter, H. (2010). The genetics of autism spectrum disorders and related neuropsychiatric disorders in childhood. *The American Journal of Psychiatry, 167*(11), 1357–1363.
68. Sandin, S., Lichtenstein, P., Kuja-Halkola, R., Larsson, H., Hultman, C. M., & Reichenberg, A. (2014). The familial risk of autism. *JAMA, 311*(17), 1770–1777.
69. Colvert, E., Tick, B., McEwen, F., Stewart, C., Curran, S. R., Woodhouse, E., et al. (2015). Heritability of autism spectrum disorder in a UK population-based twin sample. *JAMA Psychiatry, 72*(5), 415–423.
70. Sandin, S., Lichtenstein, P., Kuja-Halkola, R., Hultman, C., Larsson, H., & Reichenberg, A. (2017). The heritability of autism spectrum disorder. *JAMA, 318*(12), 1182–1184.
71. Cheng, Y., Qin, G., Dai, X., & Zhao, Y. (2007). NPY1, a BTB-NPH3-like protein, plays a critical role in auxin-regulated organogenesis in Arabidopsis. *PNAS, 104*(47), 18825–18829.
72. Yin, J., & Schaaf, C. P. (2017). Autism genetics - An overview. *Prenatal Diagnosis, 37*(1), 14–30.

73. Boddaert, N., Zilbovicius, M., Philipe, A., Robel, L., Bourgeois, M., Barthélemy, C., et al. (2009). MRI findings in 77 children with non-syndromic autistic disorder. *PLoS One, 4*(2), e4415.
74. Courchesne, E., Mouton, P. R., Calhoun, M. E., Semendeferi, K., Ahrens-Barbeau, C., Hallet, M. J., et al. (2011). Neuron number and size in prefrontal cortex of children with autism. *JAMA, 306*(18), 2001–2010.
75. Hazlett, H. C., Gu, H., Munsell, B. C., Kim, S. H., Styner, M., Wolff, J. J., et al. (2017). Early brain development in infants at high risk for autism spectrum disorder. *Nature, 542*(7641), 348–351.
76. Bauman, M. L., & Kemper, T. L. (2003). The neuropathology of the autism spectrum disorders: What have we learned? *Novartis Foundation Symposium, 251*, 112–122.
77. Chen, R., Jiao, Y., & Herskovits, E. H. (2011). Structural MRI in autism spectrum disorder. *Pediatric Research, 69*(5 Pt), 63R–68R.
78. Foster, N. E. V., Doyle-Thomas, K. A. R., Tryfon, A., Ouimet, T., Anagnostou, E., Evans, A. C., et al. (2015). Structural gray matter differences during childhood development in autism spectrum disorder: A multimetric approach. *Pediatric Neurology, 53*(4), 350–359.
79. Lainhart, J. E. (2006). Advances in autism neuroimaging research for the clinician and geneticist. *American Journal of Medical Genetics. Part C, Seminars in Medical Genetics, 142C*(1), 33–39.
80. Pelphrey, K., Adolphs, R., & Morris, J. P. (2004). Neuroanatomical substrates of social cognition dysfunction in autism. *Mental Retardation and Developmental Disabilities Research Reviews, 10*(4), 259–271.
81. Stoner, R., Chow, M. L., Boyle, M. P., Sunkin, S. M., Mouton, P. R., Roy, S., et al. (2014). Patches of disorganization in the neocortex of children with autism. *The New England Journal of Medicine, 370*(13), 1209–1219.
82. Supekar, K., Kochalka, J., Schaer, M., Wakeman, H., Qin, S., Padmanabhan, A., et al. (2018). Deficits in mesolimbic reward pathway underlie social interaction impairments in children with autism. *Brain, 141*(9), 2795–2805.
83. Tang, G., Gudsnuk, K., Kuo, S.-H., Cotrina, M. L., Rosoklija, G., Sosunov, A., et al. (2014). Loss of mTOR-dependent macroautophagy causes autistic-like synaptic pruning deficits. *Neuron, 83*(5), 1131–1143.
84. Bailey, A. J., Braeutigam, S., Jousmäki, V., & Swithenby, S. J. (2005). Abnormal activation of face processing systems at early and intermediate latency in individuals with autism spectrum disorder: A magnetoencephalographic study. *The European Journal of Neuroscience, 21*(9), 2575–2585.
85. Clements, C. C., Zoltowski, A. R., Yankowitz, L. D., Yerys, B. E., Schultz, R. T., & Herrington, J. D. (2018). Evaluation of the social motivation hypothesis of autism: A systematic review and meta-analysis. *JAMA Psychiatry, 75*(8), 797–808.
86. Dalton, K. M., Nacewicz, B. M., Johnstone, T., Schaefer, H. S., Gernsbacher, M. A., Goldsmith, H. H., et al. (2005). Gaze fixation and the neural circuitry of face processing in autism. *Nature Neuroscience, 8*(4), 519–526.
87. Emerson, R. W., Adams, C., Nishino, T., Hazlett, H. C., Wolff, J. J., Zwaigenbaum, L., et al. (2017). Functional neuroimaging of high-risk 6-month-old infants predicts a diagnosis of autism at 24 months of age. *Science Translational Medicine, 9*, 393.
88. Gervais, H., Belin, P., Boddaert, N., Leboyer, M., Coez, A., Sfaello, I., et al. (2004). Abnormal cortical voice processing in autism. *Nature Neuroscience, 7*(8), 801–802.
89. Kasai, K., Hashimoto, O., Kawakubo, Y., Yumoto, M., Kamio, S., Itoh, K., et al. (2005). Delayed automatic detection of change in speech sounds in adults with autism: A magnetoencephalographic study. *Clinical Neurophysiology, 116*(7), 1655–1664.
90. McPartland, J., Dawson, G., Webb, S. J., Panagiotides, H., & Carver, L. J. (2004). Event-related brain potentials reveal anomalies in temporal processing of faces in autism spectrum disorder. *Journal of Child Psychology and Psychiatry, 45*(7), 1235–1245.
91. Pierce, K., Haist, F., Sedaghat, F., & Courchesne, E. (2004). The brain response to personally familiar faces in autism: Findings of fusiform activity and beyond. *Brain, 127*(Pt), 12.

92. Piggot, J., Kwon, H., Mobbs, D., Blasey, C., Lotspeich, L., Menon, V., et al. (2004). Emotional attribution in high-functioning individuals with autistic spectrum disorder: A functional imaging study. *Journal of the American Academy of Child and Adolescent Psychiatry, 43*(4), 473–480.
93. Teder-Sälejärvi, W. A., Pierce, K. L., Courchesne, E., & Hillyard, S. A. (2005). Auditory spatial localization and attention deficits in autistic adults. *Brain Research. Cognitive Brain Research, 23*(2-3), 221–234.
94. Wang, A. T., Dapretto, M., Hariri, A. R., Sigman, M., & Bookheimer, S. Y. (2004). Neural correlates of facial affect processing in children and adolescents with autism spectrum disorder. *Journal of the American Academy of Child and Adolescent Psychiatry, 43*(4), 481–490.
95. Williams, J. H. G., Waiter, G. D., Gilchrist, A., Perrett, D. I., Murray, A. D., & Whiten, A. (2005). Neural mechanisms of imitation and 'mirror neuron' functioning in autistic spectrum disorder. *Neuropsychologia, 44*(4), 610–621.
96. Bölte, S., Girdler, S., & Marschik, P. B. (2018). The contribution of environmental exposure to the etiology of autism spectrum disorder. *Cellular and Molecular Life Sciences, 76*(7), 1275–1297.
97. Reichenberg, A., Gross, R., Weiser, M., Bresnahan, M., Silverman, J., Harlap, S., et al. (2006). Advancing paternal age and autism. *Archives of General Psychiatry, 63*(9), 1026–1032.
98. Croen, L. A., Najjar, D. V., Fireman, B., & Grether, J. K. (2007). Maternal and paternal age and risk of autism spectrum disorders. *Archives of Pediatrics & Adolescent Medicine, 161*(4), 334–340.
99. Durkin, M. S., Maenner, M. J., Newschaffer, C. J., Lee, L.-C., Cunniff, C. M., Daniels, J. L., et al. (2008). Advanced parental age and the risk of autism spectrum disorder. *American Journal of Epidemiology, 168*(11), 1268–1276.
100. Gardener, H., Spiegelman, D., & Buka, S. L. (2009). Prenatal risk factors for autism: Comprehensive meta-analysis. *The British Journal of Psychiatry, 195*(1), 7–14.
101. Grether, J. K., Anderson, M. C., Croen, L. A., Smith, D., & Windham, G. C. (2009). Risk of autism and increasing maternal and paternal age in a large north American population. *American Journal of Epidemiology, 170*(9), 1118–1126.
102. Gabis, L., Raz, R., & Kesner-Baruch, Y. (2010). Paternal age in autism spectrum disorders and ADHD. *Pediatric Neurology, 43*(4), 300–302.
103. Sandin, S., Hultman, C. M., Kolevzon, A., Gross, R., MacCabe, J. H., & Reichenberg, A. (2012). Advancing maternal age is associated with increasing risk for autism: A review and meta-analysis. *Journal of the American Academy of Child and Adolescent Psychiatry, 51*(5), 477–4861.
104. Kong, A., Frigge, M. L., Masson, G., Besenbacher, S., Sulem, P., Magnusson, G., et al. (2012). Rate of de novo mutations and the importance of father's age to disease risk. *Nature, 488*(7412), 471–475.
105. Atsem, S., Reichenbach, J., Potabattula, R., Dittrich, M., Nava, C., Depienne, C., et al. (2016). Paternal age effects on sperm FOXK1 and KCNA7 methylation and transmission into the next generation. *Human Molecular Genetics, 25*(22), 4996–5005.
106. Atladóttir, H. O., Thorsen, P., Østergaard, L., Schendel, D. E., Lemcke, S., Abdallah, M., et al. (2010). Maternal infection requiring hospitalization during pregnancy and autism spectrum disorders. *Journal of Autism and Developmental Disorders, 40*(12), 1423–1430.
107. Atladóttir, H. Ó., Henriksen, T. B., Schendel, D. E., & Parner, E. T. (2012). Autism after infection, febrile episodes, and antibiotic use during pregnancy: An exploratory study. *Pediatrics, 130*(6), 1447–1454.
108. Abdallah, M. W., Larsen, N., Grove, J., Nørgaard-Pedersen, B., Thorsen, P., Mortensen, E. L., et al. (2013). Amniotic fluid inflammatory cytokines: Potential markers of immunologic dysfunction in autism spectrum disorders. *The World Journal of Biological Psychiatry, 14*(7), 528–538.
109. Brimberg, L., Sadiq, A., Gregersen, P. K., & Diamond, B. (2013). Brain-reactive IgG correlates with autoimmunity in mothers of a child with an autism spectrum disorder. *Molecular Psychiatry, 18*(11), 1171–1177.

110. Baron-Cohen, S., Auyeung, B., Nørgaard-Pedersen, B., Hougaard, D. M., Abdallah, M. W., Melgaard, L., et al. (2015). Elevated fetal steroidogenic activity in autism. *Molecular Psychiatry, 20*(3), 369–376.
111. Gardner, R. M., Lee, B. K., Magnusson, C., Rai, D., Frisell, T., Karlsson, H., et al. (2015). Maternal body mass index during early pregnancy, gestational weight gain, and risk of autism spectrum disorders: Results from a Swedish total population and discordant sibling study. *International Journal of Epidemiology, 44*(3), 870–883.
112. Lee, B. K., Magnusson, C., Gardner, R. M., Blomström, Å., Newschaffer, C. J., Burstyn, I., et al. (2015). Maternal hospitalization with infection during pregnancy and risk of autism spectrum disorders. *Brain, Behavior, and Immunity, 44*, 100–105.
113. Rivera, H. M., Christiansen, K. J., & Sullivan, E. L. (2015). The role of maternal obesity in the risk of neuropsychiatric disorders. *Frontiers in Neuroscience, 9*, 194.
114. Xiang, A. H., Wang, X., Martinez, M. P., Walthall, J. C., Curry, E. S., Page, K., et al. (2015). Association of maternal diabetes with autism in offspring. *JAMA, 313*(14), 1425–1434.
115. Li, M., Fallin, M. D., Riley, A., Landa, R., Walker, S. O., Silverstein, M., et al. (2016). The association of maternal obesity and diabetes with autism and other developmental disabilities. *Pediatrics, 137*(2), e20152206.
116. Meltzer, A., & Van de Water, J. (2017). The role of the immune system in autism spectrum disorder. *Neuropsychopharmacology, 42*(1), 284–298.
117. Carpita, B., Muti, D., & Dell'Osso, L. (2018). Oxidative stress, maternal diabetes, and autism spectrum disorders. *Oxidative Medicine and Cellular Longevity, 2018*, 3717215.
118. Maeyama, K., Tomioka, K., Nagase, H., Yoshioka, M., Takagi, Y., Kato, T., et al. (2018). Congenital cytomegalovirus infection in children with autism spectrum disorder: Systematic review and meta-analysis. *Journal of Autism and Developmental Disorders, 48*(5), 1483–1491.
119. Maher, G. M., O'Keeffe, G. W., Kearney, P. M., Kenny, L. C., Dinan, T. G., Mattsson, M., et al. (2018). Association of hypertensive disorders of pregnancy with risk of neurodevelopmental disorders in offspring: A systematic review and meta-analysis. *JAMA Psychiatry, 75*(8), 809–819.
120. Vianna, P., Gomes, J. d. A., Boquett, J. A., Fraga, L. R., Schuch, J. B., Vianna, F. S. L., et al. (2019). Zika Virus as a possible risk factor for autism spectrum disorder: Neuroimmunological aspects. *Neuroimmunomodulation, 25*(5–6), 320–327. https://doi.org/10.1159/000495660. Epub 2019 Jan 10.
121. Bolton, P. F., Murphy, M., Macdonald, H., Whitlock, B., Pickles, A., & Rutter, M. (1997). Obstetric complications in autism: Consequences or causes of the condition? *Journal of the American Academy of Child and Adolescent Psychiatry, 36*(2), 272–281.
122. Gardener, H., Spiegelman, D., & Buka, S. L. (2011). Perinatal and neonatal risk factors for autism: A comprehensive meta-analysis. *Pediatrics, 128*(2), 344–355.
123. Wang, C., Geng, H., Liu, W., & Zhang, G. (2017). Prenatal, perinatal, and postnatal factors associated with autism: A meta-analysis. *Medicine (Baltimore), 96*(18), e6696.
124. Veroniki, A. A., Rios, P., Cogo, E., Straus, S. E., Finkelstein, Y., Kealey, R., et al. (2017). Comparative safety of antiepileptic drugs for neurological development in children exposed during pregnancy and breast feeding: A systematic review and network meta-analysis. *BMJ Open, 7*(7), e017248.
125. Brown, H. K., Hussain-Shamsy, N., Lunsky, Y., Dennis, C.-L. E., & Vigod, S. N. (2017). The association between antenatal exposure to selective serotonin reuptake inhibitors and autism: A systematic review and meta-analysis. *The Journal of Clinical Psychiatry, 78*(1), 48–58.
126. Mezzacappa, A., Lasica, P.-A., Gianfagna, F., Cazas, O., Hardy, P., Falissard, B., et al. (2017). Risk for autism spectrum disorders according to period of prenatal antidepressant exposure: A systematic review and meta-analysis. *JAMA Pediatrics, 171*(6), 555–563.
127. Huizink, A. C., & Mulder, E. J. H. (2005). Maternal smoking, drinking or cannabis use during pregnancy and neurobehavioral and cognitive functioning in human offspring. *Neuroscience and Biobehavioral Reviews, 30*(1), 24–41.

128. Kalkbrenner, A. E., Braun, J. M., Durkin, M. S., Maenner, M. J., Cunniff, C., Lee, L.-C., et al. (2012). Maternal smoking during pregnancy and the prevalence of autism spectrum disorders, using data from the autism and developmental disabilities monitoring network. *Environmental Health Perspectives, 120*(7), 1042–1048.
129. Tran, P. L., Lehti, V., Lampi, K. M., Helenius, H., Suominen, A., Gissler, M., et al. (2013). Smoking during pregnancy and risk of autism spectrum disorder in a Finnish National Birth Cohort. *Paediatric and Perinatal Epidemiology, 27*(3), 266–274.
130. Rosen, B. N., Lee, B. K., Lee, N. L., Yang, Y., & Burstyn, I. (2015). Maternal smoking and autism spectrum disorder: A meta-analysis. *Journal of Autism and Developmental Disorders, 45*(6), 1689–1698.
131. Tang, S., Wang, Y., Gong, X., & Wang, G. (2015). A meta-analysis of maternal smoking during pregnancy and autism spectrum disorder risk in offspring. *International Journal of Environmental Research and Public Health, 12*(9), 10418–10431.
132. Wakefield, A. J., & Montgomery, S. M. (1999). Autism, viral infection and measles-mumps-rubella vaccination. *The Israel Medical Association Journal, 1*(3), 183–187.
133. Bernard, S., Enayati, A., Redwood, L., Roger, H., & Binstock, T. (2001). Autism: A novel form of mercury poisoning. *Medical Hypotheses, 56*(4), 462–471.
134. Bernard, S., Enayati, A., Roger, H., Binstock, T., & Redwood, L. (2002). The role of mercury in the pathogenesis of autism. *Molecular Psychiatry, 7*(Suppl), 2.
135. Geier, M. R., & Geier, D. A. (2003). Neurodevelopmental disorders after thimerosal-containing vaccines: A brief communication. *Experimental Biology and Medicine (Maywood, N.J.), 228*(6), 660–664.
136. McCormick, M. C. (2003). The autism "epidemic": Impressions from the perspective of immunization safety review. *Ambulatory Pediatrics, 3*(3), 119–120.
137. Geier, D. A., & Geier, M. R. (2004). Neurodevelopmental disorders following thimerosal-containing childhood immunizations: A follow-up analysis. *International Journal of Toxicology, 23*(6), 369–376.
138. Gerber, J. S., & Offit, P. A. (2009). Vaccines and autism: A tale of shifting hypotheses. *Clinical Infectious Diseases, 48*(4), 456–461.
139. Madsen, K. M., Hviid, A., Vestergaard, M., Schendel, D., Wohlfahrt, J., Thorsen, P., et al. (2002). A population-based study of measles, mumps, and rubella vaccination and autism. *The New England Journal of Medicine, 347*(19), 1477–1482.
140. Taylor, L. E., Swerdfeger, A. L., & Eslick, G. D. (2014). Vaccines are not associated with autism: An evidence-based meta-analysis of case-control and cohort studies. *Vaccine, 32*(29), 3623–3629.
141. Davidson, M. (2017). Vaccination as a cause of autism-myths and controversies. *Dialogues in Clinical Neuroscience, 19*(4), 403–407.
142. Zerbo, O., Modaressi, S., Goddard, K., Lewis, E., Fireman, B. H., Daley, M. F., et al. (2018). Vaccination patterns in children after autism spectrum disorder diagnosis and in their younger siblings. *JAMA Pediatrics, 172*(5), 469–475.
143. Kozuki, N., & Walker, N. (2013). Exploring the association between short/long preceding birth intervals and child mortality: Using reference birth interval children of the same mother as comparison. *BMC Public Health, 13*(Suppl), 3.
144. Cheslack Postava, K., & Winter, A. S. (2015). Short and long interpregnancy intervals: Correlates and variations by pregnancy timing among U.S. women. *Perspectives on Sexual and Reproductive Health, 47*(1), 19–26.
145. Magnusson, C., Lundberg, M., Lee, B. K., Rai, D., Karlsson, H., Gardner, R., et al. (2016). Maternal vitamin D deficiency and the risk of autism spectrum disorders: Population-based study. *BJPsych Open, 2*(2), 170–172.
146. Mazahery, H., Camargo, C. A., Conlon, C., Beck, K. L., Kruger, M. C., & von Hurst, P. R. (2016). Vitamin D and autism spectrum disorder: A literature review. *Nutrients, 8*(4), 236.
147. Demarquoy, C., & Demarquoy, J. (2019). Autism and carnitine: A possible link. *World Journal of Biological Chemistry, 10*(1), 7–16.

148. Schmidt, R. J., Tancredi, D. J., Krakowiak, P., Hansen, R. L., & Ozonoff, S. (2014). Maternal intake of supplemental iron and risk of autism spectrum disorder. *American Journal of Epidemiology, 180*(9), 890–900.
149. Chowanadisai, W., Graham, D. M., Keen, C. L., Rucker, R. B., & Messerli, M. A. (2013). Neurulation and neurite extension require the zinc transporter ZIP12 (slc39a12). *PNAS, 110*(24), 9903–9908.
150. Curtin, P., Austin, C., Curtin, A., Gennings, C., Arora, M., Tammimies, K., et al. (2018). Dynamical features in fetal and postnatal zinc-copper metabolic cycles predict the emergence of autism spectrum disorder. *Science Advances, 4*(5), eaat1293.
151. Velie, E. M., Block, G., Shaw, G. M., Samuels, S. J., Schaffer, D. M., & Kulldorff, M. (1999). Maternal supplemental and dietary zinc intake and the occurrence of neural tube defects in California. *American Journal of Epidemiology, 150*(6), 605–616.
152. Yasuda, H., Yoshida, K., Yasuda, Y., & Tsutsui, T. (2011). Infantile zinc deficiency: Association with autism spectrum disorders. *Scientific Reports, 1*, 129.
153. Lam, J., Sutton, P., Kalkbrenner, A., Windham, G., Halladay, A., Koustas, E., et al. (2016). A systematic review and meta-analysis of multiple airborne pollutants and autism spectrum disorder. *PLoS One, 11*(9), e0161851.
154. Saghazadeh, A., & Rezaei, N. (2017). Systematic review and meta-analysis links autism and toxic metals and highlights the impact of country development status: Higher blood and erythrocyte levels for mercury and lead, and higher hair antimony, cadmium, lead, and mercury. *Progress in Neuro-Psychopharmacology and Biological Psychiatry, 79*(Pt B), 340–368.
155. Kalkbrenner, A. E., Schmidt, R. J., & Penlesky, A. C. (2014). Environmental chemical exposures and autism spectrum disorders: A review of the epidemiological evidence. *Current Problems in Pediatric and Adolescent Health Care, 44*(10), 277–318.
156. Shelton, J. F., Geraghty, E. M., Tancredi, D. J., Delwiche, L. D., Schmidt, R. J., Ritz, B., et al. (2014). Neurodevelopmental disorders and prenatal residential proximity to agricultural pesticides: The CHARGE study. *Environmental Health Perspectives, 122*(10), 1103–1109.
157. Jeddi, M. Z., Janani, L., Memari, A. H., Akhondzadeh, S., & Yunesian, M. (2016). The role of phthalate esters in autism development: A systematic review. *Environmental Research, 151*, 493–504.
158. Kardas, F., Bayram, A. K., Demirci, E., Akin, L., Ozmen, S., Kendirci, M., et al. (2016). Increased serum phthalates (MEHP, DEHP) and bisphenol a concentrations in children with autism spectrum disorder: The role of endocrine disruptors in autism etiopathogenesis. *Journal of Child Neurology, 31*(5), 629–635.
159. Kondolot, M., Ozmert, E. N., Ascı, A., Erkekoglu, P., Oztop, D. B., Gumus, H., et al. (2016). Plasma phthalate and bisphenol a levels and oxidant-antioxidant status in autistic children. *Environmental Toxicology and Pharmacology, 43*, 149–158.
160. Rahbar, M. H., Swingle, H. M., Christian, M. A., Hessabi, M., Lee, M., Pitcher, M. R., et al. (2017). Environmental exposure to dioxins, dibenzofurans, bisphenol A, and phthalates in children with and without autism spectrum disorder living near the gulf of Mexico. *International Journal of Environmental Research and Public Health, 14*, 11.
161. Stein, T. P., Schluter, M. D., Steer, R. A., Guo, L., & Ming, X. (2015). Bisphenol A exposure in children with autism spectrum disorders. *Autism Research, 8*(3), 272–283.
162. Ye, B. S., Leung, A. O. W., & Wong, M. H. (2017). The association of environmental toxicants and autism spectrum disorders in children. *Environmental Pollution, 227*, 234–242.
163. Ribas-Fitó, N., Torrent, M., Carrizo, D., Muñoz-Ortiz, L., Júlvez, J., Grimalt, J. O., et al. (2006). In utero exposure to background concentrations of DDT and cognitive functioning among preschoolers. *American Journal of Epidemiology, 164*(10), 955–962.
164. Shutoh, Y., Takeda, M., Ohtsuka, R., Haishima, A., Yamaguchi, S., Fujie, H., et al. (2009). Low dose effects of dichlorodiphenyltrichloroethane (DDT) on gene transcription and DNA methylation in the hypothalamus of young male rats: Implication of hormesis-like effects. *The Journal of Toxicological Sciences, 34*(5), 469–482.

165. Herbstman, J. B., Sjödin, A., Kurzon, M., Lederman, S. A., Jones, R. S., Rauh, V., et al. (2010). Prenatal exposure to PBDEs and neurodevelopment. *Environmental Health Perspectives, 118*(5), 712–719.
166. Hertz-Picciotto, I., Bergman, A., Fängström, B., Rose, M., Krakowiak, P., Pessah, I., et al. (2011). Polybrominated diphenyl ethers in relation to autism and developmental delay: A case-control study. *Environmental Health, 10*(1), 1.
167. Rai, D., Lewis, G., Lundberg, M., Araya, R., Svensson, A., Dalman, C., et al. (2012). Parental socioeconomic status and risk of offspring autism spectrum disorders in a Swedish population-based study. *Journal of the American Academy of Child and Adolescent Psychiatry, 51*(5), 467–4766.
168. Bhasin, T. K., & Schendel, D. (2007). Sociodemographic risk factors for autism in a US metropolitan area. *Journal of Autism and Developmental Disorders, 37*(4), 667–677.
169. Durkin, M. S., Maenner, M. J., Meaney, F. J., Levy, S. E., DiGuiseppi, C., Nicholas, J. S., et al. (2010). Socioeconomic inequality in the prevalence of autism spectrum disorder: Evidence from a U.S. cross-sectional study. *PLoS One, 5*(7), e11551.
170. Boyle, C. A., Boulet, S., Schieve, L. A., Cohen, R. A., Blumberg, S. J., Yeargin-Allsopp, M., et al. (2011). Trends in the prevalence of developmental disabilities in US children, 1997–2008. *Pediatrics, 127*(6), 1034–1042.
171. Fujiwara, T. (2014). Socioeconomic status and the risk of suspected autism spectrum disorders among 18-month-old toddlers in Japan: A population-based study. *Journal of Autism and Developmental Disorders, 44*(6), 1323–1331.
172. Delobel-Ayoub, M., Ehlinger, V., Klapouszczak, D., Maffre, T., Raynaud, J.-P., Delpierre, C., et al. (2015). Socioeconomic disparities and prevalence of autism spectrum disorders and intellectual disability. *PLoS One, 10*(11), e0141964.
173. Larsson, H. J., Eaton, W. W., Madsen, K. M., Vestergaard, M., Olesen, A. V., Agerbo, E., et al. (2005). Risk factors for autism: Perinatal factors, parental psychiatric history, and socioeconomic status. *American Journal of Epidemiology, 161*(10), 916–925.
174. Sun, X., Allison, C., Auyeung, B., Baron-Cohen, S., & Brayne, C. (2014). Parental concerns, socioeconomic status, and the risk of autism spectrum conditions in a population-based study. *Research in Developmental Disabilities, 35*(12), 3678–3688.
175. He, P., Guo, C., Wang, Z., Chen, G., Li, N., & Zheng, X. (2018). Socioeconomic status and childhood autism: A population-based study in China. *Psychiatry Research, 259*, 27–31.
176. Hsiao, E. Y., McBride, S. W., Hsien, S., Sharon, G., Hyde, E. R., McCue, T., et al. (2013). Microbiota modulate behavioral and physiological abnormalities associated with neurodevelopmental disorders. *Cell, 155*(7), 1451–1463.
177. Coretti, L., Cristiano, C., Florio, E., Scala, G., Lama, A., Keller, S., et al. (2017). Sex-related alterations of gut microbiota composition in the BTBR mouse model of autism spectrum disorder. *Scientific Reports, 7*. https://doi.org/10.1038/srep45356
178. de Magistris, L., Familiari, V., Pascotto, A., Sapone, A., Frolli, A., Iardino, P., et al. (2010). Alterations of the intestinal barrier in patients with autism spectrum disorders and in their first-degree relatives. *Journal of Pediatric Gastroenterology and Nutrition, 51*(4), 418–424.
179. Coretti, L., Paparo, L., Riccio, M. P., Amato, F., Cuomo, M., Natale, A., et al. (2018). Gut microbiota features in young children with autism spectrum disorders. *Frontiers in Microbiology, 9*, 3146.
180. Zhang, M., Ma, W., Zhang, J., He, Y., & Wang, J. (2018). Analysis of gut microbiota profiles and microbe-disease associations in children with autism spectrum disorders in China. *Scientific Reports, 8*(1), 13981.
181. Wang, M., Zhou, J., He, F., Cai, C., Wang, H., Wang, Y., et al. (2019). Alteration of gut microbiota-associated epitopes in children with autism spectrum disorders. *Brain, Behavior, and Immunity, 75*, 192–199.
182. Johnson, C. P., & Myers, S. M. (2007). Identification and evaluation of children with autism spectrum disorders. *Pediatrics, 120*(5), 1183–1215.

183. Mazina, V., Gerdts, J., Trinh, S., Ankenman, K., Ward, T., Dennis, M. Y., et al. (2015). Epigenetics of autism-related impairment: Copy number variation and maternal infection. *Journal of Developmental and Behavioral Pediatrics, 36*(2), 61–67.
184. Agrawal, S., Rao, S. C., Bulsara, M. K., & Patole, S. K. (2018). Prevalence of autism spectrum disorder in preterm infants: A meta-analysis. *Pediatrics, 142*, 3.
185. Dachew, B. A., Mamun, A., Maravilla, J. C., & Alati, R. (2018). Pre-eclampsia and the risk of autism-spectrum disorder in offspring: Meta-analysis. *The British Journal of Psychiatry, 212*(3), 142–147.
186. Hisle-Gorman, E., Susi, A., Stokes, T., Gorman, G., Erdie-Lalena, C., & Nylund, C. M. (2018). Prenatal, perinatal, and neonatal risk factors of autism spectrum disorder. *Pediatric Research, 84*(2), 190–198.
187. Krakowiak, P., Walker, C. K., Bremer, A. A., Baker, A. S., Ozonoff, S., Hansen, R. L., et al. (2012). Maternal metabolic conditions and risk for autism and other neurodevelopmental disorders. *Pediatrics, 129*(5), 1121–1128.
188. Xiang, A. H. (2017). Association of maternal diabetes with autism in offspring. *JAMA, 317*(5), 537–538.
189. Xiang, A. H., Wang, X., Martinez, M. P., Page, K., Buchanan, T. A., & Feldman, R. K. (2018). Maternal type 1 diabetes and risk of autism in offspring. *JAMA, 320*(1), 89–91.
190. Dumont-Mathieu, T., & Fein, D. (2005). Screening for autism in young children: The modified checklist for autism in toddlers (M-CHAT) and other measures. *Mental Retardation and Developmental Disabilities Research Reviews, 11*(3), 253–262.
191. Pandey, J., Verbalis, A., Robins, D. L., Boorstein, H., Klin, A. M. I., Babitz, T., et al. (2008). Screening for autism in older and younger toddlers with the modified checklist for autism in toddlers. *Autism, 12*(5), 513–535.
192. Chlebowski, C., Robins, D. L., Barton, M. L., & Fein, D. (2013). Large-scale use of the modified checklist for autism in low-risk toddlers. *Pediatrics, 131*(4), 1121–1127.
193. Robins, D. L., Casagrande, K., Barton, M., Chen, C.-M. A., Dumont-Mathieu, T., & Fein, D. (2014). Validation of the modified checklist for Autism in toddlers, revised with follow-up (M-CHAT-R/F). *Pediatrics, 133*(1), 37–45.
194. Toh, T.-H., Tan, V. W.-Y., Lau, P. S.-T., & Kiyu, A. (2018). Accuracy of modified checklist for autism in toddlers (M-CHAT) in detecting autism and other developmental disorders in community clinics. *Journal of Autism and Developmental Disorders, 48*(1), 28–35.
195. Dietz, C., Swinkels, S., van Daalen, E., van Engeland, H., & Buitelaar, J. K. (2006). Screening for autistic spectrum disorder in children aged 14–15 months. II: Population screening with the Early Screening of Autistic Traits Questionnaire (ESAT). Design and general findings. *Journal of Autism and Developmental Disorders, 36*(6), 713–722.
196. Swinkels, S. H. N., Dietz, C., van Daalen, E., Kerkhof, I. H. G. M., van Engeland, H., & Buitelaar, J. K. (2006). Screening for autistic spectrum in children aged 14 to 15 months. I: The development of the early screening of autistic traits questionnaire (ESAT). *Journal of Autism and Developmental Disorders, 36*(6), 723–732.
197. García-Primo, P., Hellendoorn, A., Charman, T., Roeyers, H., Dereu, M., Roge, B., et al. (2014). Screening for autism spectrum disorders: State of the art in Europe. *European Child & Adolescent Psychiatry, 23*(11), 1005–1021.
198. Stone, W. L., Coonrod, E. E., & Ousley, O. Y. (2000). Brief report: Screening tool for autism in two-year-olds (STAT): Development and preliminary data. *Journal of Autism and Developmental Disorders, 30*(6), 607–612.
199. Stone, W. L., Coonrod, E. E., Turner, L. M., & Pozdol, S. L. (2004). Psychometric properties of the STAT for early autism screening. *Journal of Autism and Developmental Disorders, 34*(6), 691–701.
200. Stone, W. L., McMahon, C. R., & Henderson, L. M. (2008). Use of the screening tool for autism in two-year-olds (STAT) for children under 24 months: An exploratory study. *Autism, 12*(5), 557–573.
201. Warren, Z., Stone, W., & Humberd, Q. (2009). A training model for the diagnosis of autism in community pediatric practice. *Journal of Developmental and Behavioral Pediatrics, 30*(5), 442–446.

202. Wetherby, A. M., Brosnan-Maddox, S., Peace, V., & Newton, L. (2008). Validation of the Infant-Toddler Checklist as a broadband screener for autism spectrum disorders from 9 to 24 months of age. *Autism, 12*(5), 487–511.
203. Pierce, K., Carter, C., Weinfeld, M., Desmond, J., Hazin, R., Bjork, R., et al. (2011). Detecting, studying, and treating autism early: The one-year well-baby check-up approach. *The Journal of Pediatrics, 159*(3), 458–4651.
204. Smith, N. J., Sheldrick, R. C., & Perrin, E. C. (2013). An abbreviated screening instrument for autism spectrum disorders. *Infant Mental Health Journal, 34*(2), 149–155.
205. Floatinghospital. (2019). Parts of the SWYC - POSI at Tufts Medical Center.
206. Berument, S. K., Rutter, M., Lord, C., Pickles, A., & Bailey, A. (1999). Autism screening questionnaire: Diagnostic validity. *The British Journal of Psychiatry, 175*, 444–451. Research.
207. Eaves, L. C., Wingert, H. D., Ho, H. H., & Mickelson, E. C. R. (2006a). Screening for autism spectrum disorders with the social communication questionnaire. *Journal of Developmental and Behavioral Pediatrics, 27*(2), S95–S103.
208. Eaves, L. C., Wingert, H., & Ho, H. H. (2006b). Screening for autism: Agreement with diagnosis. *Autism, 10*(3), 229–242.
209. Charman, T., Baird, G., Simonoff, E., Chandler, S., Davison-Jenkins, A., Sharma, A., et al. (2016). Testing two screening instruments for autism spectrum disorder in UK community child health services. *Developmental Medicine and Child Neurology, 58*(4), 369–375.
210. Marvin, A. R., Marvin, D. J., Lipkin, P. H., & Law, J. K. (2017). Analysis of social communication questionnaire (SCQ) screening for children less than age 4. *Current Developmental Disorders Reports, 4*(4), 137.
211. Aldosari, M., Fombonne, E., Aldhalaan, H., Ouda, M., Elhag, S., Alshammari, H., et al. (2019). Validation of the Arabic version of the social communication questionnaire. *Autism*. 1362361318816065.
212. Scott, F. J., Baron-Cohen, S., Bolton, P., & Brayne, C. (2002). The CAST (Childhood Asperger Syndrome Test): Preliminary development of a UK screen for mainstream primary-school-age children. *Autism, 6*(1), 9–31.
213. Allison, C., Williams, J., Scott, F., Stott, C., Bolton, P., Baron-Cohen, S., et al. (2007). The childhood Asperger syndrome test (CAST): Test-retest reliability in a high scoring sample. *Autism, 11*(2), 173–185.
214. Williams, E., Thomas, K., Sidebotham, H., & Emond, A. (2008b). Prevalence and characteristics of autistic spectrum disorders in the ALSPAC cohort. *Developmental Medicine and Child Neurology, 50*(9), 672–677.
215. Ehlers, S., Gillberg, C., & Wing, L. (1999). A screening questionnaire for Asperger syndrome and other high-functioning autism spectrum disorders in school age children. *Journal of Autism and Developmental Disorders, 29*(2), 129–141.
216. Posserud, M.-B., Lundervold, A. J., & Gillberg, C. (2009). Validation of the autism spectrum screening questionnaire in a total population sample. *Journal of Autism and Developmental Disorders, 39*(1), 126–134.
217. Baron-Cohen, S., Wheelwright, S., Skinner, R., Martin, J., & Clubley, E. (2001). The autism-spectrum quotient (AQ): Evidence from Asperger syndrome/high-functioning autism, males and females, scientists and mathematicians. *Journal of Autism and Developmental Disorders, 31*(1), 5–17.
218. Voracek, M., & Dressler, S. G. (2006). Lack of correlation between digit ratio (2D:4D) and Baron-Cohen's "Reading the Mind in the Eyes" test, empathy, systemising, and autism-spectrum quotients in a general population sample. *Personality and Individual Differences, 41*(8), 1481–1491.
219. Wakabayashi, A., Baron-Cohen, S., Wheelwright, S., & Tojo, Y. (2006). The autism-spectrum quotient (AQ) in Japan: A cross-cultural comparison. *Journal of Autism and Developmental Disorders, 36*(2), 263–270.
220. Brereton, A. V., Tonge, B. J., Mackinnon, A. J., & Einfeld, S. L. (2002). Screening young people for autism with the developmental behavior checklist. *Journal of the American Academy of Child and Adolescent Psychiatry, 41*(11), 1369–1375.

221. Witwer, A. N., & Lecavalier, L. (2007). Autism screening tools: An evaluation of the social communication questionnaire and the developmental behaviour checklist-autism screening algorithm. *Journal of Intellectual & Developmental Disability, 32*(3), 179–187.
222. Gray, K. M., & Tonge, B. J. (2005). Screening for autism in infants and preschool children with developmental delay. *The Australian and New Zealand Journal of Psychiatry, 39*(5), 378–386.
223. Gray, K. M., Tonge, B. J., Sweeney, D. J., & Einfeld, S. L. (2008). Screening for autism in young children with developmental delay: An evaluation of the developmental behaviour checklist: Early screen. *Journal of Autism and Developmental Disorders, 38*(6), 1003–1010.
224. Oosterling, I. J., Swinkels, S. H., van der Gaag, R. J., Visser, J. C., Dietz, C., & Buitelaar, J. K. (2009). Comparative analysis of three screening instruments for autism spectrum disorder in toddlers at high risk. *Journal of Autism and Developmental Disorders, 39*(6), 897–909.
225. Snow, A. V., & Lecavalier, L. (2008). Sensitivity and specificity of the modified checklist for autism in toddlers and the social communication questionnaire in preschoolers suspected of having pervasive developmental disorders. *Autism, 12*(6), 627–644.
226. Filipek, P. A., Accardo, P. J., Ashwal, S., Baranek, G. T., Cook, E. H., Dawson, G., et al. (2000). Practice parameter: Screening and diagnosis of autism: Report of the Quality Standards Subcommittee of the American Academy of Neurology and the Child Neurology Society. *Neurology, 55*(4), 468–479.
227. Penner, M., Anagnostou, E., Andoni, L. Y., & Ungar, W. J. (2018). Systematic review of clinical guidance documents for autism spectrum disorder diagnostic assessment in select regions. *Autism, 22*(5), 517–527.
228. Falkmer, T., Anderson, K., Falkmer, M., & Horlin, C. (2013). Diagnostic procedures in autism spectrum disorders: A systematic literature review. *European Child & Adolescent Psychiatry, 22*(6), 329–340.
229. Randall, M., Egberts, K. J., Samtani, A., Scholten, R. J., Hooft, L., Livingstone, N., et al. (2018). Diagnostic tests for autism spectrum disorder (ASD) in preschool children. *Cochrane Database of Systematic Reviews, 7*, CD009044.
230. Vllasaliu, L., Jensen, K., Hoss, S., Landenberger, M., Menze, M., Schütz, M., et al. (2016). Diagnostic instruments for autism spectrum disorder (ASD). *Cochrane Database of Systematic Reviews, 1*, CD012036.
231. Lord, C., Rutter, M., & Le Couteur, A. (1994). Autism diagnostic interview-revised: A revised version of a diagnostic interview for caregivers of individuals with possible pervasive developmental disorders. *Journal of Autism and Developmental Disorders, 24*(5), 659–685.
232. Wing, L., & Gould, J. (1979). Severe impairments of social interaction and associated abnormalities in children: Epidemiology and classification. *Journal of Autism and Developmental Disorders, 9*(1), 11–29.
233. Wing, L., Leekam, S. R., Libby, S. J., Gould, J., & Larcombe, M. (2002). The diagnostic interview for social and communication disorders: Background, inter-rater reliability and clinical use. *Journal of Child Psychology and Psychiatry, 43*(3), 307–325.
234. Chlebowski, C., Green, J. A., Barton, M. L., & Fein, D. (2010). Using the childhood autism rating scale to diagnose autism spectrum disorders. *Journal of Autism and Developmental Disorders, 40*(7), 787–799.
235. Vaughan, C. A. (2011). Test review: E. Schopler, M. E. Van Bourgondien, G. J. Wellman, & S. R. Love Childhood Autism Rating Scale (2nd ed.). Los Angeles, CA: Western Psychological Services, 2010. *Journal of Psychoeducational Assessment, 29*(5), 489–493.
236. Skuse, D., Warrington, R., Bishop, D., Chowdhury, U., Lau, J., Mandy, W., et al. (2004). The developmental, dimensional and diagnostic interview (3di): A novel computerized assessment for autism spectrum disorders. *Journal of the American Academy of Child and Adolescent Psychiatry, 43*(5), 548–558.
237. Lord, C., Rutter, M., Goode, S., Heemsbergen, J., Jordan, H., Mawhood, L., et al. (1989). Autism diagnostic observation schedule: A standardized observation of communicative and social behavior. *Journal of Autism and Developmental Disorders, 19*(2), 185–212.

238. Lord, C., Risi, S., Lambrecht, L., Cook, E. H., Leventhal, B. L., DiLavore, P. C., et al. (2000). The autism diagnostic observation schedule-generic: A standard measure of social and communication deficits associated with the spectrum of autism. *Journal of Autism and Developmental Disorders, 30*(3), 205–223.
239. Gotham, K., Risi, S., Pickles, A., & Lord, C. (2007). The autism diagnostic observation schedule: Revised algorithms for improved diagnostic validity. *Journal of Autism and Developmental Disorders, 37*(4), 613–627.
240. Gotham, K., Risi, S., Dawson, G., Tager-Flusberg, H., Joseph, R., Carter, A., et al. (2008). A replication of the autism diagnostic observation schedule (ADOS) revised algorithms. *Journal of the American Academy of Child and Adolescent Psychiatry, 47*(6), 642–651.
241. Gotham, K., Pickles, A., & Lord, C. (2009). Standardizing ADOS scores for a measure of severity in autism spectrum disorders. *Journal of Autism and Developmental Disorders, 39*(5), 693–705.
242. Lecavalier, L. (2005). An evaluation of the gilliam autism rating scale. *Journal of Autism and Developmental Disorders, 35*(6), 795–805.
243. Mazefsky, C. A., & Oswald, D. P. (2006). The discriminative ability and diagnostic utility of the ADOS-G, ADI-R, and GARS for children in a clinical setting. *Autism, 10*(6), 533–549.
244. Montgomery, J. M., Newton, B., & Smith, C. (2008). Test review: Gilliam, J. (2006). GARS-2: Gilliam autism rating scale—Second Edition. Austin, TX: PRO-ED. *Journal of Psychoeducational Assessment, 26*(4), 395–401.
245. Pandolfi, V., Magyar, C. I., & Dill, C. A. (2010). Constructs assessed by the GARS-2: Factor analysis of data from the standardization sample. *Journal of Autism and Developmental Disorders, 40*(9), 1118–1130.
246. Sikora, D. M., Hall, T. A., Hartley, S. L., Gerrard-Morris, A. E., & Cagle, S. (2008). Does parent report of behavior differ across ADOS-G classifications: Analysis of scores from the CBCL and GARS. *Journal of Autism and Developmental Disorders, 38*(3), 440–448.
247. Stone, W. L., & Hogan, K. L. (1993). A structured parent interview for identifying young children with autism. *Journal of Autism and Developmental Disorders, 23*(4), 639–652.
248. Stone, W. L., Coonrod, E. E., Pozdol, S. L., & Turner, L. M. (2003). The parent interview for autism-clinical version (PIA-CV): A measure of behavioral change for young children with autism. *Autism, 7*(1), 9–30.
249. Ministries of Health and Education. (2016). *New Zealand autism spectrum disorder guideline*. Ministry of Health.
250. SIGN. (2016). *Assessment, diagnosis and interventions for autism spectrum disorders*. SIGN.
251. Volkmar, F., Siegel, M., Woodbury-Smith, M., King, B., McCracken, J., & State, M. (2014). Practice parameter for the assessment and treatment of children and adolescents with autism spectrum disorder. *Journal of the American Academy of Child and Adolescent Psychiatry, 53*(2), 237–257.
252. Dover, C. J., & Le Couteur, A. (2007). How to diagnose autism. *Archives of Disease in Childhood, 92*(6), 540–545.
253. Klin, A., & Volkmar, F. R. (2003). Asperger syndrome: Diagnosis and external validity. *Child and Adolescent Psychiatric Clinics of North America, 12*(1), 1–13.
254. Posar, A., Resca, F., & Visconti, P. (2015). Autism according to diagnostic and statistical manual of mental disorders 5th edition: The need for further improvements. *Journal of Pediatric Neurosciences, 10*(2), 146.
255. World Health Organization. (1992). *The ICD-10 classification of mental and behavioural disorders: Clinical descriptions and diagnostic guidelines*. Geneva: World Health Organization.
256. World Health Organization. (2018). *International classification of diseases, 11th revision (ICD-11)*. World Health Organization.
257. Leitner, Y. (2014). The co-occurrence of autism and attention deficit hyperactivity disorder in children - what do we know? *Frontiers in Human Neuroscience, 8*, 268.
258. Battaglia, A., & Carey, J. C. (2006). Etiologic yield of autistic spectrum disorders: A prospective study. *American Journal of Medical Genetics. Part C, Seminars in Medical Genetics, 142C*(1), 3–7.

259. Chakrabarti, S., & Fombonne, E. (2001). Pervasive developmental disorders in preschool children. *JAMA, 285*(24), 3093–3099.
260. Chakrabarti, S., & Fombonne, E. (2005). Pervasive developmental disorders in preschool children: Confirmation of high prevalence. *The American Journal of Psychiatry, 162*(6), 1133–1141.
261. Gillberg, C., & Coleman, M. (1996). Autism and medical disorders: A review of the literature. *Developmental Medicine and Child Neurology, 38*(3), 191–202.
262. Rutter, M., Bailey, A., Bolton, P., & Le Couteur, A. (1994). Autism and known medical conditions: Myth and substance. *Journal of Child Psychology and Psychiatry, 35*(2), 311–322.
263. Williams, G., King, J., Cunningham, M., Stephan, M., Kerr, B., & Hersh, J. H. (2001). Fetal valproate syndrome and autism: additional evidence of an association. *Developmental Medicine and Child Neurology, 43*(3), 202–206.
264. Cohen, D., Pichard, N., Tordjman, S., Baumann, C., Burglen, L., Excoffier, E., et al. (2005). Specific genetic disorders and autism: Clinical contribution towards their identification. *Journal of Autism and Developmental Disorders, 35*(1), 103–116.
265. Numis, A. L., Major, P., Montenegro, M. A., Muzykewicz, D. A., Pulsifer, M. B., & Thiele, E. A. (2011). Identification of risk factors for autism spectrum disorders in tuberous sclerosis complex. *Neurology, 76*(11), 981–987.
266. Richards, C., Jones, C., Groves, L., Moss, J., & Oliver, C. (2015). Prevalence of autism spectrum disorder phenomenology in genetic disorders: A systematic review and meta-analysis. *Lancet Psychiatry, 2*(10), 909–916.
267. Smalley, S. L. (1998). Autism and tuberous sclerosis. *Journal of Autism and Developmental Disorders, 28*(5), 407–414.
268. Wong, V. (2006). Study of the relationship between tuberous sclerosis complex and autistic disorder. *Journal of Child Neurology, 21*(3), 199–204.
269. Curatolo, P., Cusmai, R., Cortesi, F., Chiron, C., Jambaque, I., & Dulac, O. (1991). Neuropsychiatric aspects of tuberous sclerosis. *Annals of the New York Academy of Sciences, 615*, 8–16.
270. Gillberg, I. C., Gillberg, C., & Ahlsén, G. (1994). Autistic behaviour and attention deficits in tuberous sclerosis: A population-based study. *Developmental Medicine and Child Neurology, 36*(1), 50–56.
271. Hartley, T., Potter, R., Badalato, L., Smith, A. C., Jarinova, O., & Boycott, K. M. (2017). Fragile X testing as a second-tier test. *Genetics in Medicine, 19*, 12.
272. Mullegama, S. V., Klein, S. D., Nguyen, D. C., Kim, A., Signer, R., Fox, M., et al. (2017). Is it time to retire fragile X testing as a first-tier test for developmental delay, intellectual disability, and autism spectrum disorder? *Genetics in Medicine, 19*, 12.
273. Weinstein, V., Tanpaiboon, P., Chapman, K. A., Mew, N. A., & Hofherr, S. (2017). Do the data really support ordering fragile X testing as a first-tier test without clinical features? *Genetics Medicine, 19*(12), 1317–1322.
274. Battaglia, A., Parrini, B., & Tancredi, R. (2010). The behavioral phenotype of the idic(15) syndrome. *American Journal of Medical Genetics. Part C, Seminars in Medical Genetics, 154C*(4), 448–455.
275. Bolton, P. F., Dennis, N. R., Browne, C. E., Thomas, N. S., Veltman, M. W., Thompson, R. J., et al. (2001). The phenotypic manifestations of interstitial duplications of proximal 15q with special reference to the autistic spectrum disorders. *American Journal of Medical Genetics, 105*(8), 675–685.
276. Baker, P., Piven, J., Schwartz, S., & Patil, S. (1994). Brief report: Duplication of chromosome 15q11-13 in two individuals with autistic disorder. *Journal of Autism and Developmental Disorders, 24*(4), 529–535.
277. Cook, E. H., Lindgren, V., Leventhal, B. L., Courchesne, R., Lincoln, A., Shulman, C., et al. (1997). Autism or atypical autism in maternally but not paternally derived proximal 15q duplication. *American Journal of Human Genetics, 60*(4), 928–934.
278. Nurmi, E. L., Dowd, M., Tadevosyan-Leyfer, O., Haines, J. L., Folstein, S. E., & Sutcliffe, J. S. (2003). Exploratory subsetting of autism families based on savant skills improves evi-

dence of genetic linkage to 15q11-q13. *Journal of the American Academy of Child and Adolescent Psychiatry, 42*(7), 856–863.
279. Wolpert, C. M., Menold, M. M., Bass, M. P., Qumsiyeh, M. B., Donnelly, S. L., Ravan, S. A., et al. (2000). Three probands with autistic disorder and isodicentric chromosome 15. *American Journal of Medical Genetics, 96*(3), 365–372.
280. Bukelis, I., Porter, F. D., Zimmerman, A. W., & Tierney, E. (2007). Smith-Lemli-Opitz syndrome and autism spectrum disorder. *The American Journal of Psychiatry, 164*(11), 1655–1661.
281. Sikora, D. M., Pettit-Kekel, K., Penfield, J., Merkens, L. S., & Steiner, R. D. (2006). The near universal presence of autism spectrum disorders in children with Smith-Lemli-Opitz syndrome. *American Journal of Medical Genetics. Part A, 140*(14), 1511–1518.
282. Miller, D. T., Adam, M. P., Aradhya, S., Biesecker, L. G., Brothman, A. R., Carter, N. P., et al. (2010). Consensus statement: Chromosomal microarray is a first-tier clinical diagnostic test for individuals with developmental disabilities or congenital anomalies. *American Journal of Human Genetics, 86*(5), 749–764.
283. Sun, F., Oristaglio, J., Levy, S. E., Hakonarson, H., Sullivan, N., Fontanarosa, J., et al. (2015). *Genetic testing for developmental disabilities, intellectual disability, and autism spectrum disorder*. Rockville, MD: Agency for Healthcare Research and Quality (US).
284. Beaudet, A. L. (2013). The utility of chromosomal microarray analysis in developmental and behavioral pediatrics. *Child Development, 84*(1), 121–132.
285. Lord, C., Elsabbagh, M., Baird, G., & Veenstra-Vanderweele, J. (2018). Autism spectrum disorder. *Lancet, 392*(10146), 508–520.
286. Rossi, M., El-Khechen, D., Black, M. H., Farwell Hagman, K. D., Tang, S., & Powis, Z. (2017). Outcomes of diagnostic exome sequencing in patients with diagnosed or suspected autism spectrum disorders. *Pediatric Neurology, 70*, 34–43.
287. Shen, Y., Dies, K. A., Holm, I. A., Bridgemohan, C., Sobeih, M. M., Caronna, E. B., et al. (2010). Clinical genetic testing for patients with autism spectrum disorders. *Pediatrics, 125*(4), 727–735.
288. Tammimies, K., Marshall, C. R., Walker, S., Kaur, G., Thiruvahindrapuram, B., Lionel, A. C., et al. (2015). Molecular diagnostic yield of chromosomal microarray analysis and whole-exome sequencing in children with autism spectrum disorder. *JAMA, 314*(9), 895–903.
289. Yang, Y., Muzny, D. M., Reid, J. G., Bainbridge, M. N., Willis, A., Ward, P. A., et al. (2013). Clinical whole-exome sequencing for the diagnosis of Mendelian disorders. *The New England Journal of Medicine, 369*(16), 1502–1511.
290. Miles, J. H. (2015). Complex autism spectrum disorders and cutting-edge molecular diagnostic tests. *JAMA, 314*(9), 879–880.
291. Jang, W., Kim, Y., Han, E., Park, J., Chae, H., Kwon, A., et al. (2019). Chromosomal microarray analysis as a first-tier clinical diagnostic test in patients with developmental delay/intellectual disability, autism spectrum disorders, and multiple congenital anomalies: A prospective multicenter study in Korea. *Annals of Laboratory Medicine, 39*(3), 299–310.
292. Duncan, A. M., & Chodirker, B. (2011). Use of array genomic hybridization technology for constitutional genetic diagnosis in Canada. *Paediatrics & Child Health, 16*(4), 211–212.
293. Manning, M., & Hudgins, L. (2010). Array-based technology and recommendations for utilization in medical genetics practice for detection of chromosomal abnormalities. *Genetics in Medicine, 12*(11), 742–745.
294. Martin, C. L., & Ledbetter, D. H. (2017). Chromosomal microarray testing for children with unexplained neurodevelopmental disorders. *JAMA, 317*(24), 2545–2546.
295. Mefford, H. C., Batshaw, M. L., & Hoffman, E. P. (2012). Genomics, intellectual disability, and autism. *The New England Journal of Medicine, 366*(8), 733–743.
296. Sanmann, J. N., Schaefer, G. B., Buehler, B. A., & Sanger, W. G. (2012). Algorithmic approach for methyl-CpG binding protein 2 (MECP2) gene testing in patients with neurodevelopmental disabilities. *Journal of Child Neurology, 27*(3), 346–354.
297. Manzi, B., Loizzo, A. L., Giana, G., & Curatolo, P. (2008). Autism and metabolic diseases. *Journal of Child Neurology, 23*(3), 307–314.

298. Weissman, J. R., Kelley, R. I., Bauman, M. L., Cohen, B. H., Murray, K. F., Mitchell, R. L., et al. (2008). Mitochondrial disease in autism spectrum disorder patients: A cohort analysis. *PLoS One, 3*(11), e3815.
299. Campistol, J., Díez-Juan, M., Callejón, L., Fernandez-De Miguel, A., Casado, M., Garcia Cazorla, A., et al. (2016). Inborn error metabolic screening in individuals with nonsyndromic autism spectrum disorders. *Developmental Medicine and Child Neurology, 58*(8), 842–847.
300. Schiff, M., Benoist, J.-F., Aïssaoui, S., Boespflug-Tanguy, O., Boepsflug-Tanguy, O., Mouren, M.-C., et al. (2011). Should metabolic diseases be systematically screened in nonsyndromic autism spectrum disorders? *PLoS One, 6*(7), e21932.
301. Cooper, A. S., Friedlaender, E., Levy, S. E., Shekdar, K. V., Bradford, A. B., Wells, K. E., et al. (2016). The implications of brain MRI in autism spectrum disorder. *Journal of Child Neurology, 31*(14), 1611–1616.
302. Hughes, R., Poon, W.-Y., & Harvey, A. S. (2015). Limited role for routine EEG in the assessment of staring in children with autism spectrum disorder. *Archives of Disease in Childhood, 100*(1), 30–33.
303. Leigh, J. P., & Du, J. (2015). Brief report: Forecasting the economic burden of autism in 2015 and 2025 in the United States. *Journal of Autism and Developmental Disorders, 45*(12), 4135–4139.
304. Kerim, M. M., Lavelle, T. A., Helm, D. T., Thompson, D., Prestt, J., Azeem, M. W. (2016). Autism a global framework for action. https://www.wish.org.qa/wp-content/uploads/2018/01/IMPJ4495_WISH_Autism_Report_WEB.pdf
305. WISH. (2018). *Autism*. WISH.

New Horizons for Molecular Genetics Diagnostic and Research in Autism Spectrum Disorder

Nader Al-Dewik and Mohammed Alsharshani

Abstract Autism spectrum disorder (ASD) is a highly heritable, heterogeneous, and complex pervasive neurodevelopmental disorder (PND) characterized by distinctive abnormalities of human cognitive functions, social interaction, and speech development.

Nowadays, several genetic changes including chromosome abnormalities, genetic variations, transcriptional epigenetics, and noncoding RNA have been identified in ASD. However, the association between these genetic modifications and ASDs has not been confirmed yet.

The aim of this review is to summarize the key findings in ASD from genetic viewpoint that have been identified from the last few decades of genetic and molecular research.

Keywords ASD · Chromosome abnormalities · CNV · SNPs · Rare variants inherited and de novo · Epigenetic changes and noncoding RNA

N. Al-Dewik (✉)
Clinical and Metabolic Genetics Section, Pediatrics Department, Hamad General Hospital (HGH) and Women's Wellness and Research Center (WWRC), Doha, Qatar

Interim Translational Research Institute (iTRI), Hamad Medical Corporation (HMC), Doha, Qatar

College of Health and Life Sciences, Hamad Bin Khalifa University (HBKU), Doha, Qatar

Faculty of Health and Social Care Sciences, Kingston University, St George's University of London, London, UK
e-mail: naldewik@hamad.qa; Nader.Al-Dewik@kingston.ac.uk

M. Alsharshani
Diagnostic Genetics Division (DGD), Department of Laboratory Medicine and Pathology (DLMP), Hamad Medical Corporation (HMC), Doha, Qatar

M. M. Essa, M. W. Qoronfleh (eds.), *Personalized Food Intervention and Therapy for Autism Spectrum Disorder Management*, Advances in Neurobiology 24,
https://doi.org/10.1007/978-3-030-30402-7_2

1 Introduction

ASD is a clinically and genetically complex heterogeneous disorder with diverse patterns of inheritance and an underlying genetic background. Understanding of the currently well-defined genetic architecture of ASD is vital to study altered molecular pathways in ASD.

The majority of ASD cases (85%) are idiopathic and thus have no known genetic causes [1]. In these cases, it is likely that a combination of multiple genetic and nongenetic factors interacts with each other and results in ASDs [1]. That is, they are polygenic or multifactorial in nature, the result of genes plus environmental factors.

Several genes have been found to be implicated in ASD. The identified genes are known to be functionally heterogeneous, and many are found to be involved in synaptic formation, transcriptional regulation, and chromatin remodeling [2–5]. From a genetic point of view, 200–1000 genes have been found to be involved in contributing to ASD susceptibility. Several consortia such as the Simons Simplex Collection (SSC) and the Autism Sequencing Consortium (ASC) have been established to study the complexity of genetic aspects of ASD [6, 7].

The genetic contribution to ASD has been known since the 1970s, after two identical twins were found to have the same condition [8]. It has since been determined that the heritable rate is 30–99% in identical twins and the conforming rate for sibling twins is around 3–30% with estimated overall heritability at around 0.7–0.8 [9–12].

2 Inheritance Pattern

ASD has a tendency to run or aggregate in families and can also present itself in diverse patterns of inheritance and causal genetic variations. Thus, the inheritance pattern of the most ASD cases is polygenic or multifactorial (not Mendelian) [10, 11, 13]. Almost half (44%) of the subjects with ASD have co-occurring adaptive and cognitive functioning deficits. Early diagnosis (12–18 months) and intervention is highly beneficial to patients.

The most common inheritance patterns of alleles associated with ASD are a dominant variant type, while recessive inheritance patterns are rare. X-linked or de novo inheritance patterns are very rare. Approximately five times more males than females are affected by this disorder pointing towards a possible involvement of sex chromosomes and imprinting effects in the etiology of the disorder. However, no specific genes have been conclusively implicated so far.

Gender gap or sexual dimorphism has also emerged in the equation deciphering the etiology of ASD. Female protective effect is one of the explanations for them being five times less likely to have ASDs, i.e., males being four to five times more likely to have the disorder. This could be attributed to greater genetic load in females and possibly greater plasticity of the female ASD brain.

Genetic changes associated with ASD commonly lead to inheriting increased risk of acquiring the disorder rather than the condition itself. This is mainly because ASD features of another genetic syndrome can be passed on according to the mode of inheritance of that syndrome.

Research on the genetics of ASD has accelerated in recent years due to rapid advancement of DNA-decoding technologies [3]. With the identification of genetic variants related to ASD, testing at birth or in vitro may become a risk factor identification tool that leads to early intervention.

As technology advances, the list of genes linked to autism is growing. Many of these genes are vital for communication between neurons or control of the expression of other genes. Although over 400 genes have been linked strongly and 200 have been linked weakly to this disorder, the influence of each specific gene within the ASD population is very minor, and not exceed more than 2% of patients.

It is obvious that the genetic make-up of ASD is extremely diverse, with contributions from alleles (variant regions within a gene, of which an individual has two copies of, one inherited from each parent) of varying frequencies.

3 Genotype/Phenotype Correlations

One of the very key concerns that remain unresolved is the understanding of the genotype-phenotype association, taking into account the up-to-date findings that exact mutations may be correlated with great phenotype heterogeneity.

However, at any rate, genetic basis can be correlated with three phenotypic presentations with unique genetic basis: (1) ASD with syndromic phenotype described by rare, single-gene disorders, (2) severe and specific phenotype induced by de novo variations in the patient, and (3) broad autistic phenotypes induced by genetic changes in one or many genes, these changes being frequent in the general population but causing diverse clinical presentations when it passes a particular threshold by multiplex gene-gene (GxG) and gene-environment (GxE) interactions [14].

4 Genetic Changes

There are more than thousand genetic alterations/variations that have been described to be associated with ASD. However, no clear assumptions can be made to date pertaining to genetic changes involved in these heterogeneous group disorders.

Several hypotheses have been proposed to solve this dilemma regarding several susceptibility genes networking together in a sophisticated model leading into ASD.

Many Mendelian diseases were found to be linked with ASD, presenting that single genes can remarkably increase risk for ASD.

5 Monogenic Autisms

Some autistic characteristics can also be observed to be part of several well-known syndromes and disorders such as fragile X syndrome (FXS; MIM 300624), tuberous sclerosis (TS, MIM 191100), neurofibromatosis type 1 (NF1; MIM 162200), Angelman syndrome (AS; MIM 105830), Cornelia de Lange and Down syndrome (DS; MIM 190685), untreated phenylketonuria, and others (Fig. 1). Around 10% of all ASDs are classically associated with dysmorphic features and/or deformities and are named syndromic autism [15–17].

Approximately 5–10% of ASD patients have co-occurring monogenic syndromes or disorders. The overall incidence of ASD in the syndrome was documented to be significantly higher than the incidence of the syndrome in ASD cases. For instance, the highest was observed in adenylosuccinate lyase deficiency (80–100%), and the lowest was in NF1 (Fig. 1) [18].

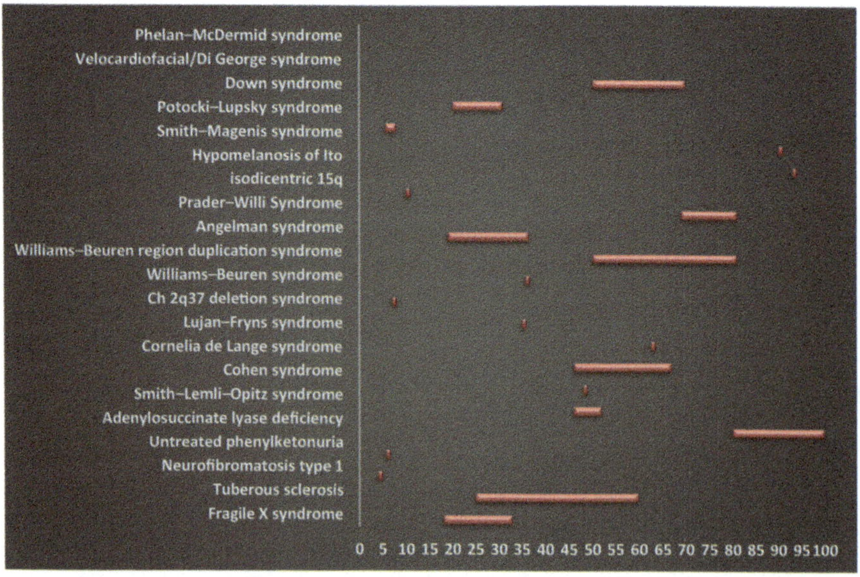

Fig. 1 Incidence of ASDs in a well-known syndrome and disorders

On the other hand, incidence of the syndrome in ASD was found to be less frequent and not exceeding 15%. For instance, FXS is the most frequent syndrome in ASD accounting for approximately 2% of ASD and resulting from the expansion of a cytosine–guanine–guanine (CGG) trinucleotide repeat at the 5′ end of the fragile X mental retardation 1 (*FMR1*) gene [18]. FXS patients are presented with severe ID and distinctive dysmorphisms, and approximately 90% of FXS males exhibit at least one ASD presentation.

In addition, Rett syndrome (RTT; MIM 312750) is originated by X-linked variations/mutations in the methyl-CpG binding protein 2 (*MECP2*) gene [19, 20]. RTT is a neurodevelopmental disorder resulting from changes in *MECP2* dosage leading to several neurobehavioral abnormalities such as *MECP2* duplication syndrome [21, 22], ASD, mild learning disabilities, X-linked ID, and infantile encephalopathy [23, 24]. Approximately 4% of females diagnosed with ASD have *MECP2* mutations, while males with *MECP2* duplications often present with ASD.

Furthermore, tuberous sclerosis complex (TSC) is caused by mutations in either the *TSC1 or TSC2* tumor suppressor (TS) genes. TSC patients presented with benign tumors in numerous parts of the body, including the brain, epilepsy, ID, behavioral abnormalities, learning difficulties, and ASD. Around 5% of ASDs have a TSC [25, 26].

Phosphatase and tensin homolog (*PTEN*)-related disorders, on the other hand, have been identified in approximately 7% of ASD patients. A group of tumor syndromes, including Cowden syndrome and Bannayan–Riley–Ruvalcaba syndrome, among others, are caused by *PTEN* germline haploinsufficiency.

Remarkably, the clinical presentations of the syndromic autism are extremely heterogeneous. This could be attributed to the differences in genetic background and epigenetic influences.

6 Chromosomal Abnormalities, Copy-Number Variation (CNV)

Chromosomal anomalies including numerical and/or structural chromosomal variations (CNV involving deletion and duplication, translocation, inversion) including 1q21, 2q37, 7q11.23, 15q11–13, 16p11.2, 17p11.2, 22q11.2, and 22q13 have been well established in genetic syndromes and identified and associated with ASD [27] (Fig. 2).

Large and submicroscopic chromosomal anomalies have been found to contribute to ASD. Several large chromosomal anomalies have been described in 5% of ASD cases and most commonly occurred at 15q11–q13, 16p11.2, 22q11.2, and 22q13.3. These regions have been involved in several neurodevelopmental disorders and contribute substantially to ASD risk [27–31] (Fig. 2).

Chromosome 15 is apparently the most prevalent area of autosomal anomalies in ASD, the duplication of 15q11–q13 being the most regularly described variation. Duplications of 15q11–q13 (Dup15q syndrome), for instance, occur in approximately

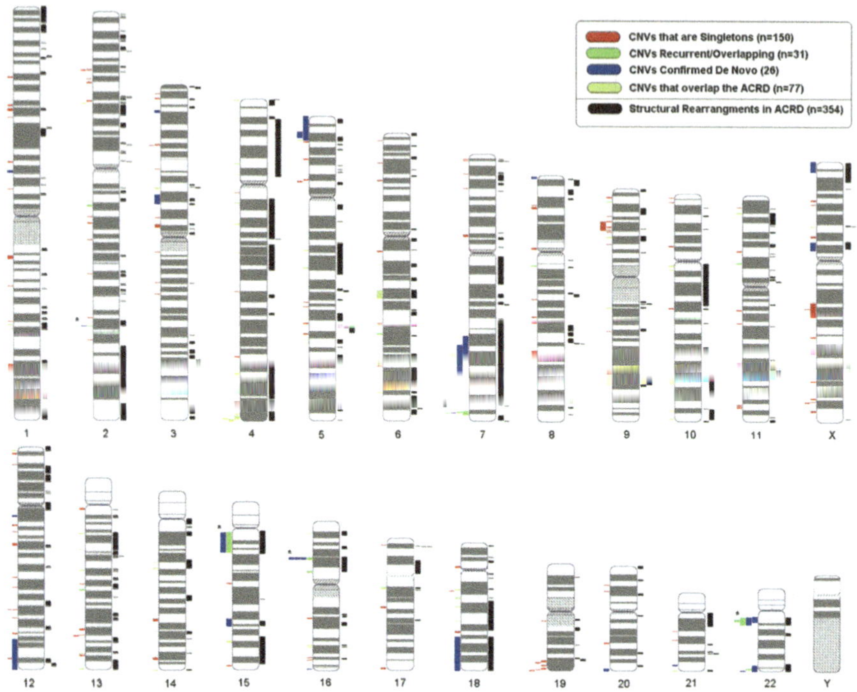

Fig. 2 Genome-wide distribution of CNVs. CNVs from the Autism Chromosome Rearrangement Database (ACRD) are plotted to the right of each chromosome (black). CNV data from the autism-specific stringent dataset of the present study are shown to the left of the chromosome and categorized as de novo (blue), overlapping/recurrent (green), CNVs overlapping with structural variation from the ACRD (yellow), and singleton CNVs (red). Note that five CNVs belong to both de novo and the recurrent categories and these are denoted by an asterisk (adopted from [60])

1–3% of all ASD cases and are known to be maternally inherited. This region encompasses at least 30 genes, of which several have been linked with ASD and other abnormalities like neurobehavioral disorders, cognitive deficits, hypotonia, language delay, and seizures. Many genes in this chromosomal region such as *GABRA5* and *GABRB3* (GABA receptors), *UBE3A* and *HERC2* (components of the proteasome complex), and *SNRPN* (ribonucleoprotein peptide N) as well as *CYFIP1* (the FMRP interacting protein) have essential functions in the brain [27, 28, 32–38] (Fig. 2).

The 15q11–q13 region is very well known for its genetic instability and has numerous low copy repeats and segmental duplications. It is recognized as a crucial area for Prader–Willi syndrome (PWS; MIM 176270)/AS (MIM 105830) and has a complex pattern of paternal and maternal imprinting. 15q11–q13 includes at least five paternally expressed genes (*MKRN3*, *MAGEL2*, *NDN*, *C15orf2*, *snoRNAs*, *and SNRPN-SNURF*) and two maternally expressed genes (*UBE3A and ATP10A*). Aberrations in 15q11–q13 dosage result in ID, developmental delay, ataxia, and seizures. Furthermore, epigenetic elements controlling 15q11–13 have been concerned with the presence of ASD and will be discussed in the epigenetic section [27, 28, 32] (Fig. 2).

CNVs in terms of microdeletions and microduplications in the 16p11.2 locus are one of the most recurrent changes in ASD and a spectrum of neurodevelopmental disorders including developmental delay, intellectual disability, epilepsy, autism, and other psychiatric disorders which are all subject to incomplete penetrance and variable expressivity (Fig. 3). The deletions were commonly found to be de novo and inherited in around 50% of ASD cases. Losses in candidate genes in the16p11.2 locus, such as *ALDOA*, *DOC2A*, *HIRIP3*, *MAPK3*, *MAZ*, *PPP4C*, *SEZ6L2*, and *TAOK2*, seem to be contributing to the ASD phenotype. On the other hand, duplications of the same region lead to more diverse phenotypes including brain malformations, schizophrenia, or attention deficit hyperactivity disorder (ADHD) [39–42]. In addition, deletion and duplication at 16p11.2 were found to give rise to two distinct phenotypes: macrocephaly with obesity and microcephaly with underweight, respectively. An animal model showed that *KCTD13* within the 16p11.2 locus is accountable for brain size phenotype [43] (Fig. 3).

The 22q11 deletion is related to DiGeorge syndrome (DGS; MIM 188400) or velocardiofacial syndrome (VCFS; MIM 192430) and other related disorders like developmental delay and neurobehavioral abnormalities. Several studies showed that the 22q11 deletion syndrome is also related to neuropsychiatric features. In addition, 22q11.2 hemizygous deletion is particularly seen in ASD, and *Tbx1* is one of the potential candidate genes responsible for 22q11.2 hemizygosity-associated ASD phenotypes, while duplications in the identical area are related with several neurodevelopmental and neurobehavioral disorders [44–51] (Fig. 3).

Deletions in the 22q13.3 region are also related to Phelan–McDermid syndrome (PHMDS; MIM 606232) characterized by severe speech delays and autistic behaviors and other phenotypes, while duplications in the identical area are related with cases of ADHD, Asperger syndrome, and hyperkinetic neuropsychiatric phenotypes [52–56] (Fig. 3).

Submicroscopic deletions and duplications have been described in few of ASD cases. These CNVs can either be inherited or de novo and were found in a variety of disorders including ASD. Rare de novo CNVs (dnCNVs) were found in around 5–10% of ASD cases, more frequently in sporadic ASD (10% in simplex families and 3% in multiplex families), and 1% controls proposing that rare dnCNVs may be important risk factors for sporadic ASD. In addition, homozygous CNVs such as 1q21.1, 7q11.23 Williams–Beuren syndrome (WBS; MIM 194050), 15q11–13, 16p11.2, and 22q11.2 DGS have been found in ASD cases. The prevalence of CNVs have very low frequency in ASD, and frequently, a particular CNV can be distinctive to a single patient [57–64].

Though the burden of dnCNVs was found to be higher in affected than healthy individuals, these CNVs can also manifest in the normal individuals which is sometimes difficult to determine if these alterations are causing the disease. It is worth mentioning that CNVs associated with ASD are not particularly central to the presentations of ASD but can be found in a broad variety of neurodevelopmental phenotypes, including ID, epilepsy, and schizophrenia. The diversity of CNV-correlated phenotypes can also present itself within the same family members [65]. In a recent study of a Chinese cohort with ASD, 17 clinically important CNVs were identified, of which 12 overlapped with recurrent autism risk loci or genes. Novel candidate genes were also identified in the rare CNV regions [66].

The most frequently recurrent CNVs identified in ASD are 1q21.1 deletion and duplication syndrome, 2q37 deletion syndrome, 3q29 deletion syndrome, 7q11.23 duplication syndrome, 15q11q13 deletion and duplication syndrome, 16p11.2 deletion and duplication syndrome, 16p12.1 deletion syndrome, 16p13.1 deletion, 17p11.2 deletion and duplication syndrome, 17q12 deletion syndrome, 17q21.31 deletion and duplication syndrome, and 22q11.2 deletion and duplication syndrome [67]. Other chromosomal abnormalities identified in ASD patients include aneuploidies: 21 (Down syndrome), X (Turner syndrome, Klinefelter syndrome, XXX syndrome), and Y (XYY syndrome) [15].

7 Single-Nucleotide Variants (SNVs): De Novo

Several studies showed that approximately 5% of ASDs are caused by de novo mutations [68–71]. These de novo variants and novel ASD susceptibility genes harboring de novo loss of function or gene-disrupting SNVs have been found to be implicated in ASD, for instance, *DYRK1A*, *POGZ*, *CHD8*, *NTNG1*, *GRIN2B*, *KATNAL2*, *TBR1*, *PTEN*, *TBL1XR1*, *GPR98*, *KIRREL3*, *and SCN2A* [68–72]. Interestingly, it was also found that some of the SNVs in *NLGN3* and *NLGN4* and other neuroligin genes can be both inherited and de novo but frequently occur de novo [73, 74] (Table 1).

8 Common Gene Variations

The largest class of genetic risks of ASD accounts for around 40–60% in simplex families and multiplex families, respectively [13, 86]. It is estimated to be derived from common variant single-nucleotide polymorphisms (SNPs) of an addition influence, nearly all of which have yet to be determined (SNPs with allele frequency more than 5% in the general population).

Many common gene variations, most of which have not been identified, are thought to affect the risk of developing ASD, but not all people with the gene variation will be affected, though most of the gene variations have only a small effect or impact. The association of these common variants is still useful. Much larger sample sizes are still needed to replicate the findings and identify many novel loci (Table 2).

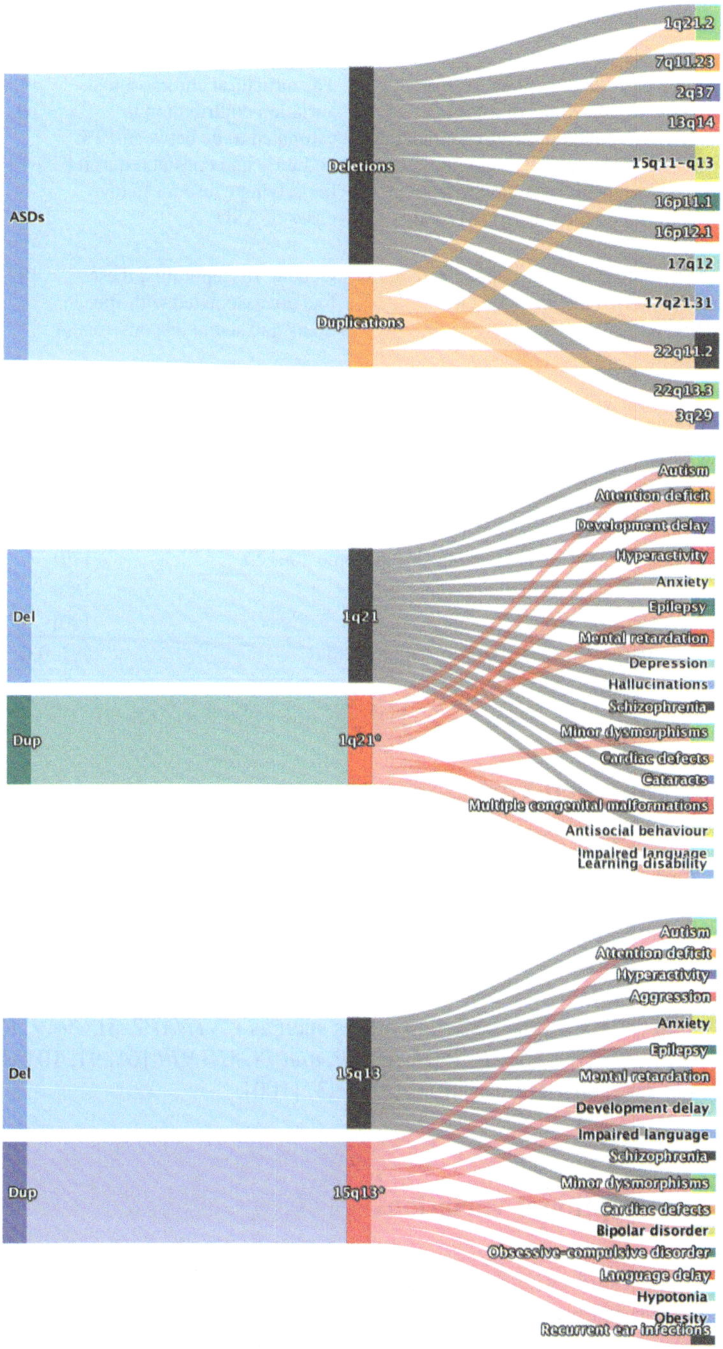

Fig. 3 ASD-associated CNVs. Deletions and duplications in the identical region can provide overlapping or specific phenotypes. These CNVs are also associated with different syndromes

Table 1 CNVs in ASD

CNVs	Abnormality	References
De novo large CNV duplications, such as duplication of 15q, deletions of 22q11.2, deletion of Xp22.3, and duplication or deletion of 16p11.2 (~600 kb, comprises around 29 genes), along with de novo submicroscopic CNVs such as duplication of 7q11.23 and microdeletion of 16p11.2.	The structural chromosomal variation contribution is estimated to be between 15% and 25% and considered as the underlying cause for many cases of ASD	[4, 42, 60, 63, 75–81]
Regions/loci on chromosomes 20p13 and 7q35 (*CNTNAP2*) and two loci mapped on chromosome 8	Genetic susceptibility, these loci are associated with speech delay and social responsiveness	[82–85]

Table 2 Common gene variations in ASD

Common variants	References
Intergenic variants between *CDH9* and *CDH10* in chromosome 5p14.1	[79, 87, 88]
Variants in 5p15.31 (between *SEMA5A* and *TAS2R1*)	[83]
Variants within the *MACROD2* gene in chromosome 20p12.1	[79]
Variants in 1p13. 2 within candidate gene *TRIM33*	[89]
DOCK4 IMMP2L and ZNF533	[90]
Variants in 7q35 (within *CNTNAP2*)	[91–93]

9 Rare Inherited Variations

In contrast, 1–5% of ASD is believed to be caused by rare genetic mutations (minor allele frequency [MAF] less than 5% of general population). These rare gene mutations occur only in a single gene of the high-risk autism-associated gene category. Synaptic genes such as neuroligin family (*NLGN3 and NLGN4*) [74], scaffolding protein family (*SHANK1, SHANK2, and SHANK3*) as well as neurexin family (*NRXN1 and NRXN3*) [94–100], and others such as *CNTNAP2*, *SLC9A9*, *BCKDK*, *AMT*, *PEX7*, *SYNE1*, *VPS13B*, *PAH*, *ADNP*, *and POMGNT1* [61, 91, 101–103] are examples of rare inherited variations (Table 3) [100].

10 Transcriptional Variations

While genomic studies have come far in comprehending the first line of ASD etiology, gene expression studies provide proof and form the foundation for discriminating the altered molecular pathways of ASDs. They could provide support in proposing novel genes' association for comprehensive genetic linkage studies via

Table 3 Rare inherited variations in ASD

Genes	Descriptions	References
Neuroligin family *NLGN3* and *NLGN4*	Rare X-linked mutations in ASD males and mental retardation in several families	[74]
CNTNAP2, SLC9A9, and BCKDK AMT, PEX7, SYNE1, VPS13B, PAH, and POMGNT1	Rare recessive mutations in consanguineous families were described in Amish families as well as Middle Eastern families with ASD and epilepsy	[61, 91, 100, 102, 103]
The scaffolding protein family (*SHANK1, SHANK2, and SHANK3*) as well as neurexin family (*NRXN1 and NRXN3*) and *ADNP*	Rare inherited variants have been documented in ASD An association of facial dysmorphism and ASD caused by mutations in *ANDP*	[94–101]

measuring genes that are dysregulated in the ASD group. Numerous biological activities have been shown to be implicated in ASD along with alterations in gene expression levels amongst various types of tissues such as postmortem brain, peripheral blood lymphocyte (PBL), gastrointestinal tissue, adult olfactory stem cells, and scalp hair follicles [104, 105].

In the first comprehensive transcriptome study on the brains of ASD patients, 444 genes were found to be differentially expressed (DE) among the cerebral cortices of ASD and control brains. Two distinct modules of co-expressed genes associated with autism were suggested. In the first model, synaptic function and neuronal projection were found to be downregulated in ASD cases, while in the second model, immune genes and glial markers were found to be overexpressed in ASD. These findings are consistent with both synaptic dysfunction and immune dysregulation in ASD as a incredibly substantial enrichment for variants genetically correlated with ASD was found in the first model, providing the genetic foundation of synaptic abnormality in ASD. On the other hand, no proof for a genetic etiology for the upregulation of genes of the second model was found [106].

Most gene expression studies have been conducted in common gene structural variants or polymorphism regions located next to high-risk areas on the chromosomes related to ASDs to investigate their effects and discover integrated gene networks for ASDs [106–108].

Alterations in gene expression patterns among ASD patients and healthy individuals have been determined in several tissues [104, 106, 109–111]. For instance, the dysregulation of *FOXP1*, *MAL*, and *C11orf30* genes has been found to be implicated in the ASD pathogenesis as several variants have been established to be linked with language delay and autism [107, 112].

The abnormal gene expression could be utilized in future as technique for ASD diagnosis and classifying (Tables 4, 5, and 6).

Table 4 Gene expression studies in ASD

Study description and number of samples	Tissue source	Study findings	References
1540 ASD unrelated patients (1020 males, 520 females; aged 7–11 years) and 1490 unrelated control children (970 males, 520 females; aged 7–11 years)	PB	mRNA levels of types I, II, and III of *NRG1* gene were significantly downregulated in ASD patient	[113]
87 ASD and 55 control male subjects; aged 1–4 years	PB	Eight Hc genes (*CHD8*, *ARID1A*, *ASH2L*, *ACTB*, *NR3C2*, *SUV420H1*, *ADPN*, and *MYO9B*) were DE in leukocytes acquired from ASD and control Four other Hc genes (*CUL3*, *SYNGAP1*, *NAA15*, and *ARID1B*) were upstream regulators of the PPI candidate genes, and *CUL3*, *SYNGAP1*, and *NAA15* were also DE in postmortem brain tissue DE in leukocytes, such as *CHD8*, *ARID1A*, *AKT1*, beta-catenin (*CTNNB1*), *SMAD3*, *CREB1*, and *NOTCH1*, and/or in postmortem brain tissue (*TCF4*, *CREB1*, *SMAD3*, *CAMK2A*, *LIMK1*, *NCOA3*, *CCNE1*, *and BRD2*)	[114]
82 ASD children (mean age 5.5 years) and 64 controls (mean age: 7.9 years) Group 1 comparison of ASD children to children of younger fathers Group 2 comparison of ASD children to children of older fathers Group 3 comparison of children of older fathers to children of younger fathers	PBL	Group 1 findings: significantly under-expressed and 641 significantly overexpressed Group 2 findings: 593 genes were underexpressed, and 145 genes were overexpressed in both, i.e., overlapped Group 3 findings: 1476 significantly underexpressed and 764 significantly overexpressed	[115]
Nine ASD and eight controls	ACG, MC, THL	28 genes showed brain region-specific decreased expression in ASD	[116]

(continued)

Table 4 (continued)

Study description and number of samples	Tissue source	Study findings	References
Part I: compared total gene expression profiling analysis between 16 ASD male (age range 4–18 years) and 16 male control subjects (age range 18–67 years) Part II: compared transcript level of one particular gene (*FOXP1*) between 83 ASD male patients with and 83 male healthy controls	LCL derived from the EBV transformation of lymphocytes of peripheral blood	202 genes were DE in the ASD group including 89 overexpressed and 113 under-expressed	[112]
16 from young postmortem males (2–14 years, 9 ASD, 7 control) and 17 adult males (15–56 years, 6 ASD, 11 control)	DLPFC	2017 genes across all autistic and control cases independent of age	[117]
Nine adults with severe autism and low to very low developmental disabilities and two adults with mild or moderate autism and no or mild cognitive abilities (Asperger syndrome or high-functioning autism) paired with 11 matched controls (age and gender)	Adult nasal olfactory stem cells	156 genes that were DE in at least one ASD patient, of which 31 were dysregulated in more than a third of the cohort	[118]
Nine ASD and nine controls	BA19 (occipital) brain tissues	876 uniquely marked genes amongst ASD and control brain tissue	[119]
60 infants and toddlers at risk for ASDs (autistic disorder and pervasive developmental disorder), 34 at risk for LD, 17 at risk for DD, and 68 TD children	PBMCs	154 probes showed significant dysregulation in ASD	[120]
27 ASD and 30 controls	DLPFC	Three genes under-expressed and one gene overexpressed in ASD samples	[121]
30 idiopathic ASD cases (24 males, 6 females) aged 3–11 years (mean age of sample 6.86 years) and 30 matched controls (age and gender)	Peripheral blood	23 DE, 10 overexpressed, 13 under-expressed	[122]
13 ASD and 13 controls	Cerebellar cortex	Seven genes	[123]
Ten ASD (3F,7M; ranged from 4 to 15 years of age), 11 controls (4F,7M; ranged between 5 and 16 years of age)	CB, BST, CG, ORC, PT, Wer	15 genes showed brain region-specific dysregulated expression in ASD samples	[124]

(continued)

Table 4 (continued)

Study description and number of samples	Tissue source	Study findings	References
20 ASD probands and 20 unaffected sibling pairs (5 proband–sibling pairs to the same gender, i.e., males, while 15 pairs of the different gender including 12 male and 3 female probands) and 18 unrelated control (11 males, 7 females) individuals	Peripheral blood	163 unique genes were significantly altered amongst probands and siblings	[125]
Group 1: 21 young adults ASD (17 males and 4 females: aged 26.7 ± 5.5 years, age range: 18–38 years) Group 2: 21 age- and gender-matched healthy controls aged 27.0 ± 5.5 years, age range: 19–39 Group 3: 21 healthy mothers having children with ASD (ASD-MO), aged 44.7 ± 6.7 years, age range: 33–58 years Group 4: ASD-MO control, aged 44.7 ± 6.7 years, age range: 31–59 years	Peripheral blood	ASD/control: 19 genes significantly dysregulated (18 overexpressed and 1 under-expressed); ASD-MO/ASD-MO control: 57 genes significantly dysregulated (17 overexpressed and 40 under-expressed) Three genes overlapped and dysregulated both in individuals with ASD and in under-expressed	[126]
18 ASD (16 males, 2 females; aged 25.61 ± 4.95) and 24 male controls (aged 32.60 ± 3.91)	Scalp hair follicles	One gene	[127]
Two separate series of sib-pairs totaling 36 children and adolescents between 4 and 18 years of age from the Italian Autism Network (ITAN) cohort	LCLs	No significant differences amongst ASD and non-affected brothers were found for RBFOX1	[128]
21 ASD adolescents and adults (20 males, 1 female) and 10 healthy controls (10 males)	Whole blood	Three genes DE: NT3, NT4 significantly under-expressed and p75(NTR) overexpressed in ASD compared to healthy controls	[129]
33 ASD boys (mean age 45.3 months; age range of 31–60 months) and 51 age-matched control boys (mean age 43.3 months; age range 28–57 months) gene expression versus Hg levels	Whole blood	11 genes	[130]

(continued)

Table 4 (continued)

Study description and number of samples	Tissue source	Study findings	References
ASD group: three females (6, 11, and 13 years old) and two males (5 and 12 years old) and five unrelated controls age- and sex-matched	LCL-derived RNAs	57 genes	[131]
51 children with ADHD, 26 children with ASD (19/26 comorbid with ADHD), and 39 controls	Whole blood cells	Two genes	[132]
37 ASD children (32 males, 5 females; average age 44.2 ± 10 months) compared to 15 (11 males, 4 females; average age 41.2 ± 6 months)	Whole blood	31 genes	[133]
19 ASD and 17 controls	STG, prefrontal cortex (BA9) and cerebellar vermis	444 genes DE in autism cortex samples, 2 genes DE in cerebellum	[134]
25 ASD gastrointestinal (ASD-GI) children (23 males and 2 females, 16 had; mean age 5.08 years) Three control groups: (1) 15 non-ASD with no chronic GI symptoms children (six males and nine females; mean age 12.263.07 years) (2) Eight non-ASD with Crohn's disease (three males and five females; mean age 12.97 years) (3) 5 non-ASD with ulcerative colitis (all females; mean age 12.06 years)	Tissue specimen from seven anatomic locations (from terminal ileum to rectum)	Ileal mucosa: ASD-GI/ controls:1409 DE Colonic mucosa: ASD-GI/ controls: 1189 DE ASD-GI (ileum and colon)/ TD Overlap between both sets ASD-GI:178 DE	[135]
15 ASD-GI children (mean onset age 13.4+/25.4 months, median age at biopsy 4.5) and 7 control-GI (median age at biopsy 4.0)	Ileum and cecum	Six genes	[136]
35 ASD (mean age 12.9 years ±12.4 SD) and 35 healthy controls (mean age 34.8 years ±9.7 SD)	LCLs	Two genes	[137]
Ten ASD and ten controls	Cerebellar cortex	Four genes	[138]

(continued)

Table 4 (continued)

Study description and number of samples	Tissue source	Study findings	References
Reanalyzed sex-specific gene expression from a recent large transcriptomic study [139]; 57 (31 males and 26 females), including 39 with both hemispheres; age, 5.7 post-conceptual weeks to 82 years	Transient prenatal structures and immature and mature forms of 16 brain regions	37, 123 genes DE in female and male, respectively	[140]

PBL peripheral blood lymphocytes, *LCL* lymphoblast cell line, *PBMCs* peripheral blood mononuclear cells, *ACG* anterior cingulate gyrus, *MC* motor cortex, *THL* thalamus, *DLPFC* dorsolateral prefrontal cortex, *CB* cerebellar, *BST* brain stem, *CG* cingulated gyrus, *ORC* orbitofrontal cortex, *PT* putamen, *Wer* Wernicke's, *STG* superior temporal gyrus

Table 5 Overlapping of gene expression changes in ASD in various tissues

Gene symbol	Expression change in ASD	References
Gene expression changes in ASD in blood/LCL and brain tissue across studies		
ABHD3	Over-expressed	[125, 141]
ANXA1	Over-expressed in Garbett et al. 2008 and Ziats et al. 2013 Under-expressed in Chien et al. 2013	[112, 134, 141, 142]
CHI3L1	Over-expressed in Garbett et al. Under-expressed in Chien et al.	[112, 141]
CMKOR1	Over-expressed	[36, 134, 141]
CTNNB1	Over-expressed in Kong et al. 2013 Under-expressed in Chow et al. 2012	[117, 125]
CX3CR1	Over-expressed	[110, 142, 143]
CXCL10	Over-expressed	[112, 117]
CXCR4	Over-expressed	[112, 117]
DNASE1L3	Under-expressed	[112, 117]
FOSL1	Over-expressed in Chow et al. 2012 Under-expressed in Ivanov et al. 2015	[117, 122]
GAD1	Under-expressed	[112, 138]
GPR56	Over-expressed	[110, 119]
GRIA3	Under-expressed	[112, 117]
HIST1H3H	Over-expressed in Chow et al. 2012 Under-expressed in Nishimura et al. 2007	[36, 117]
KIF1B	Over-expressed in Garbett et al. 2008 and Talebizadeh et al. 2014 Under-expressed in Hu et al. 2006	[131, 141, 144]
MeCP2	Over-expressed in Kuwano et al. 2011 and Zhubi et al. 2014 Under-expressed in James et al. 2014	[123, 126, 138]
NDUFB5	Over-expressed in Talebizadeh et al. 2014 Under-expressed in Anitha et al. 2012	[116, 131]
PARP9	Over-expressed in Garbett et al. 2008 Under-expressed in Glatt et al. 2012	[120, 141]
PITPNC1	Over-expressed in Nishimura et al. 2007 Garbett et al. 2008 and Voineagu et al. 2013 Under-expressed in Hu et al. 2009	[36, 104, 134, 141]

(continued)

Table 5 (continued)

Gene symbol	Expression change in ASD	References
SERPINA1	Over-expressed in Chow et al. 2012 Under-expressed in Chien et al. 2013	[112, 117]
SLC9A9	Over-expressed	[131, 134, 141]
STOM	Over-expressed in Garbett et al. 2008 Under-expressed in Glatt et al. 2012 and Ziats et al. 2013	[120, 141, 142]
SYCE1	Under-expressed	[117, 125]
TAP1	Over-expressed in Garbett et al. 2008 Under-expressed in Glatt et al. 2012	[120, 141]
TNFRSF19	Under-expressed	[112, 117]
WWTR1	Over-expressed in Garbett et al. 2008 Under-expressed in Chien et al. 2013	[112, 141]
Gene expression changes in ASD in brain tissues and intestinal biopsy across studies		
ACTG2	Under-expressed	[135, 142]
ALAD	Under-expressed	[117, 135]
LAMP2	Under-expressed	[117, 135]
SFTPA2	Over-expressed in Chow et al. 2012 Under-expressed in Walker et al. 2013	[117, 135]
Gene expression changes in ASD in blood and intestinal biopsy across studies		
ATF3	Over-expressed in Hu et al. 2006 Under-expressed in Walker et al. 2013	[135, 144]
CCL17	Over-expressed	[36, 135]
IGF2BP1	Over-expressed in Walker et al. 2013 Under-expressed in Ivanov et al. 2015	[122, 135]
IL2RA	Over-expressed	[112, 135]
MIA	Over-expressed in Nishimura et al. 2007 Under-expressed in Walker et al. 2013 (colon)	[36, 135]
UBD	Over-expressed	[112, 135]
Gene expression changes in ASD only in brain tissue across studies		
ADM	Over-expressed	[134, 141]
AHI1	Under-expressed	[134, 141]
AQP4	Over-expressed	[134, 141]
BAG3	Over-expressed	[134, 141]
C20orf7	Under-expressed	[117, 119]
C5orf16	Under-expressed	[134, 141]
CLIC1	Over-expressed	[134, 141]
CNN3	Over-expressed	[134, 141]
COL4A1	Over-expressed	[134, 141]
COX7B	Under-expressed	[116, 119]
CSDA	Over-expressed	[134, 141]
CYC1	Under-expressed	[116, 119]
DLX1	Under-expressed	[134, 141]
GADD45B	Over-expressed	[134, 141]
HIST1H1C	Over-expressed	[134, 141]
HIST1H2BD	Over-expressed	[134, 141]
HSPB1	Over-expressed	[119, 141]
IFITM2	Over-expressed	[134, 141]

(continued)

Table 5 (continued)

Gene symbol	Expression change in ASD	References
IFITM3	Over-expressed	[134, 141]
MKNK2	Over-expressed	[134, 141]
MSI2	Over-expressed	[134, 141]
MSN	Over-expressed	[134, 141]
NDUFA2	Under-expressed	[116, 119]
NDUFB3	Under-expressed	[116, 119]
NP	Over-expressed	[134, 141]
P4HA1	Over-expressed	[134, 141]
PALLD	Over-expressed	[134, 141]
PIR	Over-expressed	[134, 141]
PLEKHC1	Over-expressed	[134, 141]
RELN	Over-expressed in Khan et al. 2014 Under-expressed in Chow et al. 2012 and Zhubi et al. 2011	[117, 124, 138]
RPS21	Over-expressed in Garbett et al. 2008 Under-expressed in Ginsberg et al. 2012	[119, 141]
S100A10	Over-expressed	[134, 141]
SCARA3	Over-expressed	[134, 141]
SDC2	Over-expressed	[141, 142]
SERPINH1	Over-expressed	[117, 141, 142]
SERTAD1	Over-expressed	[134, 141]
SHANK3	Under-expressed	[121, 137]
PTTG1IP	Over-expressed	[134, 141]
TAGLN2	Over-expressed	[134, 141]
TET1	Over-expressed	[123, 138]
TIMP1	Over-expressed	[134, 141]
TMBIM1	Over-expressed	[134, 141]
TNPO1	Over-expressed	[134, 141]
YAP1	Over-expressed	[134, 141]
ZFP36L1	Over-expressed	[134, 141]
Gene expression changes in ASD in blood across studies		
ALPK1	Over-expressed	[36, 131]
ANKRD22	Over-expressed in Ivanov et al. 2015 Under-expressed in Glatt et al. 2012	[120, 122]
CD160	Over-expressed	[110, 143]
CYFIP1	Over-expressed	[36, 131]
DRD4	Over-expressed in Emanuele et al. 2010 Under-expressed in Taurines et al. 2011	[132, 145]
FAM46C	Over-expressed in Chien et al. 2013 Under-expressed in Nishimura et al. 2007	[36, 112]
GZMB	Over-expressed	[110, 112, 143]
HCK	Over-expressed in Talebizadeh et al. 2014 Under-expressed in Hu et al. 2006 and Chien et al. 2013	[112, 131, 144]
HLA-DQA1	Under-expressed	[110, 130]
IGHA1	Over-expressed	[110, 112]
IGHG1	Over-expressed in Chien et al. 2013 and Gregg et al. 2008 Under-expressed in Hu et al. 2006	[110, 112, 144]
IL2RB	Over-expressed	[110, 143]
ITGB2	Over-expressed	[110, 143]

Table 5 (continued)

Gene symbol	Expression change in ASD	References
KIR3DL2	Over-expressed	[110, 143]
KSP37	Over-expressed	[110, 143]
LRP6	Under-expressed	[112, 131]
NEURL3	Over-expressed in Chien et al. 2013 Under-expressed in Kong et al. 2013	[112, 125]
NKG7	Over-expressed	[110, 143]
P2RX5	Over-expressed	[112, 144]
PAM	Over-expressed	[110, 143]
PRF1	Over-expressed	[110, 143]
PTGDR	Over-expressed	[110, 143]
PXDN	Over-expressed in Chien et al. 2013 Under-expressed in Stamova et al. 2011	[112, 130]
SH2DIB/EAT2	Over-expressed	[110, 143]
SLC38A2	Over-expressed	[125, 144]
SPON2	Over-expressed	[110, 143]
TBX21	Over-expressed	[110, 143]
TMEM40	Under-expressed	[122, 125]

Table 6 Gene expression studies in postmortem brain tissue

Findings	Studies
Involvement of glutamate neurotransmitter system in autism	[146]
Increased expression of immune genes (mainly cytokine regulatory pathway)	[141]
Identification of distinct neuronal and immune modules	[106]
Specific enrichment of immune system and M2 microglial genes in autism brain	[107]

11 Epigenetic Variations

Epigenetic changes, including DNA methylation, histone methylation, and acetylation, are known to be altered in response to either genetic mutations or environmental exposure and regulate the expression of many genes without changing the primary DNA sequence. Genomic imprinting, for instance, is a common way of controlling of gene expression through epigenetic variations that is parent in origin.

One of the most frequent epigenetic mechanisms is DNA methylation (DNAm) which is now believed to be a vital player in ASD etiology. This imprinting mechanism regulates gene expression that can lead to apparent specific gene expression. Vital proof documented significant variations in DNAm patterns in ASD-discordant monozygotic (MZ) twins having both identical genotype and environmental interactions [147]. Furthermore, DNAm is now known to be implicated in disorders like

FXS and RS (which present along with autistic traits) [148]. Epigenetic profiling of autism showed the hypomethylated and hypermethylated regions enriched for immune system and synaptic genes, respectively [149]. Ladd-Acosta et al. 2013 and 2014 [150, 151] documented numerous crucial differentially methylated regions (DMRs) in the brains of ASDs.

In another epigenetic profiling study on ASD, brain tissue analysis demonstrated lesser temporal distinction with regard to DNA methylation and the occurrence of hypomethylated and hypermethylated regions enriched for the immune system and synaptic genes, respectively [149]. Increased paternal age has also been associated with the development of ASD. DNAm errors during spermatogenesis could describe this phenomenon. DNAm in paternal sperm has been associated with early risk of ASD in children [152].

In an animal model, sperm analysis revealed substantial reduction of methylation in older mice compared to a younger group. Comparable transcriptional abnormality and behavioral matters were noticed in their offspring as well [153].

Behnia et al. 2015 have also connected hypermethylation of the ASD candidate gene such as *OXTR* with preterm birth (PTB) that has been well known earlier as a greater ASD risk [154]. In spite of the fact that the precise impact of epigenetic alterations in the etiology of ASD is yet to be defined, that it is a crucial player in disease development is evident.

Zhubi et al. 2014 documented an increase in DNA hydroxymethylation at the *GAD1* and *RELN* promoter in cerebella from ASD patients. This modification was co-occurred by elevated binding of the methyl-CpG binding protein 2 (*MeCP2*) which contributes in gene silencing. Current evidence suggests that *GAD1* is a good candidate gene that could be epigenetically misregulated by in utero environment, predisposing one to ASD [138]. *EN2* has also been found to be involved in the regulate of pattern genesis during neurodevelopment. The *EN2* promoter had elevated levels of DNA methylation in ASD cerebella compared to asymptomatic controls [155]. DNA methylation profile in lymphoblastoid cells of autistic patients showed decreased expression of the retinoic acid-related orphan receptor alpha gene (*RORA*) and B-cell lymphoma 2 (*BCL-2*) [156].

Epigenetic dysregulations have been correlated with ASD, for instance, with the mutation of the methyl-CpG binding protein 2 (*MeCP2*) in RTT, parental imprinting of numerous chromosomal regions (transcriptional regulation of either the maternal allele or the paternal allele) such as AS, PWS, and TS and common regions with microduplications or microdeletions such as 15qllql3 [157–159]. In addition, several studies reported an association between common variant SNPs in genes directly involved in methylation and ASD [160, 161].

Several regions that are subject to genomic imprinting on chromosomes are 15q11–13, 7q21–31.31, 7q32.3–36.3, and perhaps 4q21–31, 11p11.2–13, and 13q12.3, with the loci on chromosomes 15q and 7q representing the most compelling proof for an association of genetic and epigenetic components that attribute to ASD risks. Genes in the imprinted cluster on chromosome 15q11–13 include *MKRN3*, *ZNF127AS*, *MAGE12*, *NDN*, *ATP10A*, *GABRA5*, *GABRB3*, and *GABRG3*. Genes in the imprinted cluster on chromosome 7q21.3 include *SGCE*, *PEG10*, *PPP1R9A*, *DLX5*, *CALCR*, *ASB4*, *PON1*, *PON2*, and *PON3* [162] (Table 7).

Table 7 Epigenetic biomarkers

RELN, PRRT1, ZFP57, GAD1, RORA, BCL-2, TSPAN32/C11orf21, OXTR, EN2, MTHFR, and MECP2	ASD-specific methylation biomarkers Histone acetylation changes, i.e., misregulated patterns of splicing sites and chromosome remodeling complex modifications, have also been documented in ASD	[123, 138, 149, 151, 155, 156, 163–170]
ENO2	15% of ASD patients have hypermethylation of the *ENO2*	[171]

12 Mitochondrial DNA (mtDNA) Dysfunctions

Mitochondrial DNA (mtDNA) dysfunction has been also described in ASD. Several studies have showed that mitochondrial abnormality may be one of the most frequent medical disorders correlated with ASD [172, 173].

Lombard et al. 1998 hypothesized that ASD may be a condition with aberrant mitochondrial function [174]. Clinical and biochemical studies have revealed an emerging connection amongst mtDNA abnormality and neurodevelopmental disorders, including ID [175], childhood epilepsy, and ASD [173]. Moreover, mtDNA abnormality has been correlated with some patterns of syndromic ASD [172, 175]. For instance, biochemical changes such as increased levels of creatine kinase, lactate, pyruvate, carnitine, ammonia, and alanine were described in the serum of ASD patients [175–177]. Abnormal respiratory chain enzyme activities [178] or underexpression of OXPHOS genes was described in the ASD brain [179], revealing abnormal or altered mitochondrial function.

Impairment of the *OXPHOS* pathway was documented in ASD individuals, as discussed by Napoli et al. 2014 [180] and examined by Valenti et al. 2014 [175]. Oliveira et al. 2005 revealed that 7% (7/100) of ASD children who were clinically indistinguishable from other children affected by ASD, presented with mitochondrial respiratory chain disorder [177]. Weissman et al. 2008 also suggested that abnormal mitochondrial *OXPHOS* may be a complementary key pathogenic component in a group of ASD [181].

Notwithstanding proof of changed mitochondrial function in some ASD individuals it is not known whether mitochondrial impairment is a cause or an effect of ASD. Although a mitochondrial subgroup in ASD could be found [182], Rossignol and Frye 2011 and 2012 identified that even in this subgroup the underlying genetic factor could be detected in a portion of cases (23%). In cases of non-syndromic ASD, mitochondrial impairment without mtDNA modifications has been often seen [172, 173].

Rossignol and Frye 2012 reported that Mitochondrial disease (MD) was seen in 5% of ASD children. In this ASD/MD subset, mtDNA aberrations were noticed in 23% of patients. These evidences indicate that primary MD may be present in a subset of ASD children [173].

On the other hand, mtDNA deletions in ASD individuals have been also documented [176, 183–185]. Single mtDNA deletions have a role in numerous pediatric-

and adult-onset primary MDs such as Kearns–Sayre syndrome, Pearson syndrome, and progressive external ophthalmoplegia [186]. Multiple mtDNA deletions appear often as a result of pathogenic mutations in genes accountable for intergenomic connection. However, they are frequently associated to ageing or unhealthy environmental factors as well as mtDNA has a inadequate DNA repair system [187, 188].

A recent study on 60 ASD patients and 60 healthy individuals screened for common mtDNA mutations showed that mtDNA deletions were established in 16.6% (10/60) of ASD (mtdel-ASD). In 90% of these mtdel-ASD children, rare SNVs were found in ASD-related genes (one of those was pathogenic). In the intergenomic panel of this cohort, one likely pathogenic variant was present. Mutations and variants of uncertain significance (VUS) in genes responsible for mtDNA maintenance were also found more commonly in MD patients than mtdel-ASD or other comparison groups. Only VUS were also discovered in healthy controls and in patients without an mtDNA deletion [189].

13 Noncoding RNA

Noncoding RNAs (ncRNAs) are defined as RNA transcripts that are not translated into proteins and are known for their connection with numerous disorders.

NcRNAs are commonly split into the following types: (1) small and (2) long ncRNAs. The small ncRNAs contain a diverse group of ncRNA that are (a) transfer and ribosomal RNA, (b) small nuclear RNAs (snRNA), (c) small nucleolar RNAs (snoRNA), (d) microRNAs (miRNAs), (e) piwi-interacting RNAs (piRNAs), (f) small interfering RNAs (siRNAs), and (g) small Cajal body-specific RNAs (scaRNAs).

Several studies showed that overexpression and knockdown of noncoding RNA studies have vital functions in controlling a variety of mechanisms: splicing, transcription, localization, and organization of subcellular compartments. LncRNAs are also implicated in the regulation of chromatin structure and conformation through their connection with regulatory proteins. There are several studies that have showed the implication of ncRNA in ASDs, and several databases document these findings, i.e., important types of ncRNA and their relation to ASDs. These ncRNAs have relatively high tissue specificity and regional transcriptional homogeneity in ASD when compared with controls.

MicroRNAs (miRNAs) are a group of small noncoding RNA molecules, 18–25 nucleotides in length. The ASD-related miRNAs comprise a conserved type of ncRNA that can differently control gene expression. Several miRNAs in are expressed in the brain tissues and display controlling role for a variety of biological mechanisms relevant to neurogenesis, brain maturation, and synaptic plasticity [190, 191].

Stamova et al. 2015 showed that there are distinct expression profiles of miRNA that differ significantly in different regions across the brain, specifically the superior temporal sulcus (STS) and the primary auditory cortex (PAC). There are fewer differences of miRNA and snoRNA in the two areas in the human superior temporal gyrus: the STS and the PAC in ASD brain postmortem tissue. There was an absence of the regular age-related changes, while miRNA and snoRNA differed markedly between STS and PAC in healthy individuals, and specific expression patterns were noticed [192].

An updated study showed there is sexual dimorphism and identified 20 small ncRNAs that are DE in STS of ASD females compared to control females but only eight abnormally regulated in the STS of ASD males compared to control males. Moreover, eight small ncRNAs were DE in PAC of ASD females compared to control females and three small ncRNAs abnormally regulated in PAC of ASD male compared to control males. Thus, there are commonly more impairment of regulation small ncRNA in ASD females compared to ASD males. This could be attributed to a greater genetic load in females, a female protective effect, and perhaps greater plasticity of female ASD brain. A study identified few sexually dimorphic dysregulated miRNAs [193]. These specific miRNAs are miR-219 and miR-338 that are abnormally regulated in STS of female ASD brain. They are correlated with oligodendrocyte differentiation that could relate to sexual dimorphism of white matter tracts. Also, miR-488 could correlate to more anxiety in females; finally, miR-125 and miR-181 involved in neuronal growth may be sexually dimorphic.

In another study matching the postmortem cerebellar brain tissue derived from ASD individuals with healthy age- and gender-matched individuals, 28 miRNAs were documented as being notably DE. However, there are some concerns about the statistical method in the study that may in turn indicate unreliable results [194–196].

Another study on cultured lymphoblastoid cell lines (LCL) by Talebizadeh et al. 2008 showed that 9 of the 470 miRNAs were DE in ASD samples compared with controls. In an updated study by Sarachana et al., 2010 three of the nine miRNAs were replicated with similar overexpression of miR-23a and miR-23b and under-expression of miR-132, and 43 miRNAs were DE in LCLs in ASD when compared to controls. Abu-Elneel et al. 2008 studied 466 miRNAs in 13 postmortem cerebellar cortex tissues from ASD and found that 13 and 16 miRNA were found to be under- and over-expressed, respectively. Dysregulation of miR-23a and miR-106b in the autistic cerebellar cortex was also found in LCLs. These results support the claim that abnormally regulated of miRNA in peripheral blood cells can replicate at least some miRNA alterations occurring in the brain, thus offering support to the use of LCL as a surrogate tissue to study miRNA expression in individuals with ASD. In addition, the DE miRNAs could be related to both neurological and comorbid features of ASD such as developing gastrointestinal diseases, circadian rhythm signaling, steroid hormone metabolism, and receptor signaling [194, 197, 198].

Novel and reported DE miRNAs such as miRNA-107, miRNA-106a-5p, miRNA-10a-5p, miRNA-136-5p, and miRNA-155-have been identified in ASD postmortem brains and the role of miRNA-21-3p as its transcripts revealed enrichment for ASD candidate genes and genes under-expressed in ASD cortex [199].

Mundalil Vasu et al. 2014 identified the potential of 13 miRNAs in serum as likely biomarkers of ASD. Five miRNAs (miRNA-181b-5p, miRNA-320a, miRNA-572, miRNA-130a-3p, and miRNA-19b-3p) were considered predictive.

These miRNAs have been implicated in the pathogenesis of ASD [200]. Kichukova et al. 2017 [201] have also showed that miRNA-365a-3p, miRNA-619-5p, and miRNA-664a-3p are the most overexpressed types and miRNA-3135a, miRNA-328-3p, miRNA-197-5p, miRNA-424-5p, and miRNA-500a-5p are under-expressed in the serum of ASD patients.

miRNAs can also be found in different body fluids such as plasma and saliva. Plasma has the highest number of unique miRNA species and is followed by saliva [202]. Hicks et al. 2016 showed that 14 miRNAs are DE in saliva of mild ASD with no history of neurologic disorder, preterm birth, or known chromosomal abnormality, ten miRNAs were found to be overexpressed upregulated, and four were under-expressed in ASD compared to controls. These miRNAs are widely and highly expressed in the developing human brain. Most of these miRNAs were found to be significantly correlated with Vineland neurodevelopmental scores [203].

Using a different source of tissue/cells, Nguyen et al. 2016 carried out a study on olfactory mucosal stem cells from ASD patients and controls which represented a neurologically appropriate tissue. They discovered four miRNAs markedly dysregulated in the ASD patients: one miRNA, miRNA-146a overexpressed, and three other miRNAs under-expressed, i.e., miRNA-221, miRNA-654-5p, and miRNA-656 [204].

The role of several miRNAs such as miRNA-125b, miRNA-13, miRNA-137, and miRNA-138 has been documented in regulation of dendritic spine density, structure, and morphology in ASD and other psychiatric disorders [205].

In contrast to miRNA, fairly less is known about the potential function of snoRNA and piRNAs in ASD. snoRNAs are crucial in modifications and processing of another small ncRNA [206]. Numerous snoRNAs have been found to be brain-specific, and their influence on neurological development has already been described. Remarkably, brain-specific snoRNAs are found not to be involved in the alteration of typical snoRNA targets. In animal models, snoRNAs in mice have shown the possible association of two brain-specific snoRNAs in learning and memory: MBII-48 and MBII-52. The human homolog of MBII-52 seems to be implicated in the regulation of the 5-HT2C receptor subunit mRNA, and increased blood serotonin (5-HT) levels were acknowledged as a biomarker in ASD [206–210]. On the other hand, piRNAs were identified to regulate memory-related synaptic plasticity in neurons. Thus, further to miRNA, both snoRNA and piRNA may be also implicated in the development of neurodevelopmental, psychiatric, and neurodegenerative diseases, like ASD [211].

Long ncRNAs (lncRNAs) are more than 200 nucleotides in length. LncRNAs have been recognized as being crucial to the development, maintenance, and function of the brain, more particularly, neurogenesis, synaptogenesis, and GABAergic interneuron function [212].

Ziats et al. 2013 showed that there is abnormal expression of lncRNAs in ASD patients' postmortem brain tissue. More than 200 lncRNAs were found to be DE among ASD and control. 82/222 were unique to the prefrontal cortex and 143/222 were unique to the cerebellum. In addition, the number of lncRNAs was found to be more DE in control brain rather than ASD brain tissues. This finding could explain that there are fewer specialized regions in autistic brains than in the brains of healthy subjects in imaging studies [142, 213].

These lncRNAs were enhanced for genes particularly related with neuronal migration and gene targets for miR-103/miR-107. Furthermore, intraindividual differences in the expression of the abovementioned lncRNA amongst the prefrontal cortex and cerebellum were also documented in small sample size of two ASD individuals and two controls [142].

Another study by Kerin et al. 2012 showed that lncRNAs MSNP1AS contribute to ASD risk and are highly overexpressed (12.7-fold) in the postmortem cerebral cortex of ASD individuals than in those from controls. MSNP1AS is transcribed from the antisense strand of *MSNP1* (moesin pseudogene 1) from the 5p14.1 chromosomal region. This region contains the SNP rs4307059, which was found to have a strong connection with ASDs. MSNP1AS was found to be 94% same and antisense to the X chromosome transcript MSN. MSN encodes a protein (moesin) that regulates neuronal architecture and immune response. ASD-associated rs4307059 T allele have elevated expression of MSNP1AS. MSNP1AS, which binds to MSN, was highly expressed in postmortem cerebral cortex samples from ASDs individuals. Such high levels of expression could decrease the production of MSN transcripts, the moesin level, and the number and length of neuritis [88, 214].

The influence of MSNP1AS on neuronal formation and gene expression was also further studied by DeWitt et al. 2016 using human neural progenitor cells. They showed that MSNP1AS functions precisely by modulating the translation of the MSN transcript to moesin protein but not altering the expression of the MSN transcript. This indicates that upregulation of MSNP1AS changes the expression of genes that attribute to chromatin organization jointly with genes that are implicated in translation more comprehensively [215]. The two genes *DISC1* and *ST7*, along with their antisense transcripts DISC2 and ST7OT1-2, have also been associated with ASDs [216, 217]. These findings deliver a new point of view on the profile and function of lncRNAs that are related with ASDs.

DE lncRNAs were also acknowledged in peripheral blood samples of 25 paired ASD controls. A total of 3929 lncRNAs were detected to be DE ASD peripheral lymphocytes, 2407 lncRNAs were overexpressed, and 1522 lncRNA were underexpressed. Dysregulated LncRNA affects 13 pathways, like the inflammatory, synaptic vesicle cycling, and long-term potentiation pathways, crucial in the ASD group. These DE lncRNAs were found to be transcribed from the HOX gene or HOX-related genes, signifying a crucial role of HOX gene in the development of ASD and are termed lncHOXs that could serve as a new set of biomarkers for ASD. Additionally, the DE lncRNA SHANK2-AS and BDNF-AS, transcribed from corresponding genes SHANK and BDNF, respectively, were also documented to be different in ASD patients. The impact of lncRNAs in downregulated pathways is more pathogenic than in upregulated pathways [218].

Parikshak et al. 2016 conducted postmortem genome-wide transcriptome analysis of the largest cohort of samples that showed 60 lncRNA DE among the 48 ASD and 49 controls. A set of 20 lncRNAs were found to interact with microRNA (miRNA)–protein complexes, and nine with FMRP, whose mRNA targets are enriched in ASD risk gene. Two lncRNAs *LINC00693* and *LINC00689* were found to interact with miRNA processing complexes, were naturally under-expressed during development, and were found to be overexpressed in ASD as compared to controls. Thus, dysregulation of lncRNAs is an essential module of the transcriptomic signature of ASD [219].

In 2017, Gudenas et al. 2017 identified 14 novel candidate lncRNAs associated with ASD. These lncRNAs are DE in brain tissues of ASD individuals, specifically during the development of cortex, that are vital for the synaptic signaling and transmission of signals, immune responses, and lipid transport pathways. These lncRNAs were found to be overexpressed in synaptic transmission and under-expressed in immune response and lipid transport pathways [220].

Overlying spliced lncRNAs (PTCHD1AS1, PTCHD1AS2) were identified to regulate PTCHD1 in the X-chromosome through several molecular processes, including alteration of chromatin, transcriptional regulation, and posttranscriptional alteration in ASD and intellectual or learning disability. Mutations in the X-chromosome *PTCHD1* gene were reported in seven families with ASD and in three families with intellectual disability [221]. Another candidate gene RAY1 in 7 (7q31) was also proposed in ASD. Four ncRNAs were described as potential regulatory RNAs that may be involved in ASD [216].

Velmeshev et al. 2013 identified another antisense lncRNA analogous to the SYNGAP1 locus (SYNGAP1-AS) that is significantly over-expressed in prefrontal cortex and superior temporal gyrus but not in the cerebellum of patients with ASD compared to controls [222]. A recent study investigating the role of lncRNA *MALAT1* and *AGO2* in 30 ASD patients and 41 healthy controls from peripheral whole blood samples showed a significant direct correlation between *MALAT1* and *AGO2*, indicating their interactive network. However, the altered expressions between *MALAT1* and *AGO2* were not strong enough to be significant. Thus, further studies using lager sample sizes and specific subsets of white blood cells are still needed to strengthen these findings [223] (Table 8).

Table 8 Noncoding RNAs findings in ASD

Noncoding RNA: LncRNA: MSNP1AS, Evf2, RPPH1, NEAT1, and MALAT1, C210orf121, AK128400, FTHL3, and LST1 SEC1, COAS3, SDHA PMS2L4, SYP-AS1, STXBP5-AS1, STX8, SHANK2-AS, BDNF-AS, MSNP1AS, DISC2, HAR1, and others MiRNA hsa-miR-21-3p, hsa-miR-29b, hsa-miR-219-5p, miR-146a, miR-221, miR-654-5p, and miR-656 miR-133b/miR-206 miRNA-365a-3p, miRNA-619-5p, miRNA-664a-3p miRNA-3135a, miRNA-328-3p, miRNA-197-5p, miRNA-424-5p, miRNA-500a-5p	Found to be more likely involved in the molecular function and/or regulation of the specific ASD risk genes	[103, 142, 197, 199, 214, 215, 217–220, 223–231]

14 Conclusion and Perspective

Nowadays, studies suggest that ASDs are frequently caused by genetic factors and heritability is estimated to be more than 80% instead of the previous 50% reported in studies.

The advancement of genomic technologies and the substantial multitude of efforts in the direction of sharing and reanalyzing large datasets of ASD have helped make rapid progress in identifying the risk genes of ASD and correlating genotype–phenotype based on ASD patients' genetic background. Further studies are still needed for understanding genotype–phenotype correlations along with functional validations to advance our understanding of the molecular mechanisms of ASD.

The combination of genetics with clinical phenotypes and other functional genomic studies such as transcriptomics and epigenomics will contribute to a better comprehending of the molecular processes implicated in ASD and eventually inform clinical care. In addition, there is still a need to understand more about ASD neurobiology via establishing genetic animal models of ASDs, including novel techniques such as genome editing, modulation of neuronal activity, biological network studies, and stem cell approaches.

Current genetic results have markedly advanced our understanding of the genetic underlying of ASD. However, establishment of phenotypic markers is still challenging due to phenotypic and genotypic diversity.

References

1. NHGRI. (2017). *Learning about autism* [Online]. Bethesda, MD: National Human Genome Research Institute. Retrieved from https://www.genome.gov/25522099/learning-aboutautism/
2. De Rubeis, S., & Buxbaum, J. D. (2015). Genetics and genomics of autism spectrum disorder: Embracing complexity. *Human Molecular Genetics, 24*(R1), R24–R31.
3. Geschwind, D. H., & State, M. W. (2015). Gene hunting in autism spectrum disorder: On the path to precision medicine. *Lancet Neurology, 14*(11), 1109–1120.
4. Ronemus, M., Iossifov, I., Levy, D., & Wigler, M. (2014). The role of de novo mutations in the genetics of autism spectrum disorders. *Nature Reviews Genetics, 15*(2), 133–141.
5. De Rubeis, S., He, X., Goldberg, A. P., Poultney, C. S., Samocha, K., Cicek, A. E., et al. (2014). Synaptic, transcriptional and chromatin genes disrupted in autism. *Nature, 515*(7526), 209–215.
6. SFARI. *The Simons Simplex Collection*. 2011 2011/09/16/; Available from: https://www.sfari.org/funded-project/the-simons-simplex-collection.
7. Consortium, A.S. *Advancing autism research*. 2019; Available from: https://genome.emory.edu/ASC.
8. Treffert, D. A. (1970). Epidemiology of infantile autism. *Archives of General Psychiatry, 22*(5), 431–438.
9. Bailey, A., Le Couteur, A., Gottesman, I., Bolton, P., Simonoff, E., Yuzda, E., et al. (1995). Autism as a strongly genetic disorder: Evidence from a British twin study. *Psychological Medicine, 25*(1), 63–77.
10. Colvert, E., Tick, B., McEwen, F., Stewart, C., Curran, S. R., Woodhouse, E., et al. (2015). Heritability of autism spectrum disorder in a UK population-based twin sample. *JAMA Psychiatry, 72*(5), 415–423.

11. Hallmayer, J., Cleveland, S., Torres, A., Phillips, J., Cohen, B., Torigoe, T., et al. (2011). Genetic heritability and shared environmental factors among twin pairs with autism. *Archives of General Psychiatry, 68*(11), 1095–1102.
12. Rosenberg, R. E., Law, J. K., Yenokyan, G., McGready, J., Kaufmann, W. E., & Law, P. A. (2009). Characteristics and concordance of autism spectrum disorders among 277 twin pairs. *Archives of Pediatrics & Adolescent Medicine, 163*(10), 907–914.
13. Gaugler, T., Klei, L., Sanders, S. J., Bodea, C. A., Goldberg, A. P., Lee, A. B., et al. (2014). Most genetic risk for autism resides with common variation. *Nature Genetics, 46*(8), 881–885.
14. Bonnet-Brilhault, F. (2011). Genotype/phenotype correlation in autism: Genetic models and phenotypic characterization. *Encephale, 37*(1), 68–74.
15. Devlin, B., & Scherer, S. W. (2012). Genetic architecture in autism spectrum disorder. *Current Opinion in Genetics & Development, 22*(3), 229–237.
16. Lintas, C., & Persico, A. M. (2009). Autistic phenotypes and genetic testing: State-of-the-art for the clinical geneticist. *Journal of Medical Genetics, 46*(1), 1–8.
17. Toriello, H. V. (2012). Approach to the genetic evaluation of the child with autism. *Pediatric Clinics of North America, 59*(1), 113–128. xi.
18. Persico, A. M., & Napolioni, V. (2013). Autism genetics. *Behavioural Brain Research, 251*, 95–112.
19. Pieretti, M., Zhang, F. P., Fu, Y. H., Warren, S. T., Oostra, B. A., Caskey, C. T., et al. (1991). Absence of expression of the FMR-1 gene in fragile X syndrome. *Cell, 66*(4), 817–822.
20. Amir, R. E., Van den Veyver, I. B., Wan, M., Tran, C. Q., Francke, U., & Zoghbi, H. Y. (1999). Rett syndrome is caused by mutations in X-linked MECP2, encoding methyl-CpGbinding protein 2. *Nature Genetics, 23*(2), 185–188.
21. Ramocki, M. B., Tavyev, Y. J., & Peters, S. U. (2010). The MECP2 duplication syndrome. *American Journal of Medical Genetics Part A, 152A*(5), 1079–1088.
22. Ramocki, M. B., Peters, S. U., Tavyev, Y. J., Zhang, F., Carvalho, C. M., Schaaf, C. P., et al. (2009). Autism and other neuropsychiatric symptoms are prevalent in individuals with MeCP2 duplication syndrome. *Annals of Neurology, 66*(6), 771–782.
23. Chahrour, M., & Zoghbi, H. Y. (2007). The story of Rett syndrome: from clinic to neurobiology. *Neuron, 56*(3), 422–437.
24. Lugtenberg, D., Kleefstra, T., Oudakker, A. R., Nillesen, W. M., Yntema, H. G., Tzschach, A., et al. (2009). Structural variation in Xq28: MECP2 duplications in 1% of patients with unexplained XLMR and in 2% of male patients with severe encephalopathy. *European Journal of Human Genetics, 17*(4), 444–453.
25. Wiznitzer, M. (2004). Autism and tuberous sclerosis. *Journal of Child Neurology, 19*(9), 675–679.
26. Jeste, S. S., Sahin, M., Bolton, P., Ploubidis, G. B., & Humphrey, A. (2008). Characterization of autism in young children with tuberous sclerosis complex. *Journal of Child Neurology, 23*(5), 520–525.
27. Depienne, C., Moreno-De-Luca, D., Heron, D., Bouteiller, D., Gennetier, A., Delorme, R., et al. (2009). Screening for genomic rearrangements and methylation abnormalities of the 15q11-q13 region in autism spectrum disorders. *Biological Psychiatry, 66*(4), 349–359.
28. Cook Jr., E. H., Lindgren, V., Leventhal, B. L., Courchesne, R., Lincoln, A., Shulman, C., et al. (1997). Autism or atypical autism in maternally but not paternally derived proximal 15q duplication. *American Journal of Human Genetics, 60*(4), 928.
29. Kielinen, M., Rantala, H., Timonen, E., Linna, S.-L., & Moilanen, I. (2004). Associated medical disorders and disabilities in children with autistic disorder: A population-based study. *Autism, 8*(1), 49–60.
30. Tucker, T., Giroux, S., Clément, V., Langlois, S., Friedman, J. M., & Rousseau, F. (2013). Prevalence of selected genomic deletions and duplications in a French–Canadian population-based sample of newborns. *Molecular Genetics & Genomic Medicine, 1*(2), 87.
31. Wassink, T. H., Piven, J., & Patil, S. R. (2001). Chromosomal abnormalities in a clinic sample of individuals with autistic disorder. *Psychiatric Genetics, 11*(2), 57–63.

32. Al Ageeli, E., Drunat, S., Delanoë, C., Perrin, L., Baumann, C., Capri, Y., et al. (2014). Duplication of the 15q11-q13 region: Clinical and genetic study of 30 new cases. *European Journal of Medical Genetics, 57*(1), 5–14.
33. Bucan, M., Abrahams, B. S., Wang, K., Glessner, J. T., Herman, E. I., Sonnenblick, L. I., et al. (2009). Genome-wide analyses of exonic copy number variants in a family-based study point to novel autism susceptibility genes. *PLoS Genetics, 5*(6), e1000536.
34. Hogart, A., Wu, D., LaSalle, J. M., & Schanen, N. C. (2010). The comorbidity of autism with the genomic disorders of chromosome 15q11.2-q13. *Neurobiology of Disease, 38*(2), 181–191.
35. Menold, M. M., Shao, Y., Wolpert, C. M., Donnelly, S. L., Raiford, K. L., Martin, E. R., et al. (2001). Association analysis of chromosome 15 Gabaa receptor subunit genes in autistic disorder. *Journal of Neurogenetics, 15*(3–4), 245–259.
36. Nishimura, Y., Martin, C. L., Vazquez-Lopez, A., Spence, S. J., Alvarez-Retuerto, A. I., Sigman, M., et al. (2007). Genome-wide expression profiling of lymphoblastoid cell lines distinguishes different forms of autism and reveals shared pathways. *Human Molecular Genetics, 16*(14), 1682–1698.
37. Puffenberger, E. G., Jinks, R. N., Wang, H., Xin, B., Fiorentini, C., Sherman, E. A., et al. (2012). A homozygous missense mutation in HERC2 associated with global developmental delay and autism spectrum disorder. *Human Mutation, 33*(12), 1639–1646.
38. Tan, E.-S., Yong, M.-H., Lim, E. C., Li, Z.-H., Brett, M. S., & Tan, E.-C. (2014). Chromosome 15q11-q13 copy number gain detected by array-CGH in two cases with a maternal methylation pattern. *Molecular Cytogenetics, 7*, 32.
39. Filges, I., Sparagana, S., Sargent, M., Selby, K., Schlade-Bartusiak, K., Lueder, G. T., et al. (2014). Brain MRI abnormalities and spectrum of neurological and clinical findings in three patients with proximal 16p11.2 microduplication. *American Journal of Medical Genetics. Part A, 164A*(8), 2003–2012.
40. Hanson, E., Nasir, R. H., Fong, A., Lian, A., Hundley, R., Shen, Y., et al. (2010). Cognitive and behavioral characterization of 16p11.2 deletion syndrome. *Journal of Developmental and Behavioral Pediatrics, 31*(8), 649–657.
41. Kumar, R. A., Marshall, C. R., Badner, J. A., Babatz, T. D., Mukamel, Z., Aldinger, K. A., et al. (2009). Association and mutation analyses of 16p11.2 Autism candidate genes. *PLoS One, 4*(2), e4582.
42. Weiss, L. A., Shen, Y., Korn, J. M., Arking, D. E., Miller, D. T., Fossdal, R., et al. (2008). Association between microdeletion and microduplication at 16p11.2 and autism. *The New England Journal of Medicine, 358*(7), 667–675.
43. Golzio, C., Willer, J., Talkowski, M. E., Oh, E. C., Taniguchi, Y., Jacquemont, S., et al. (2012). KCTD13 is a major driver of mirrored neuroanatomical phenotypes of the 16p11.2 copy number variant. *Nature, 485*(7398), 363–367.
44. Angkustsiri, K., Goodlin-Jones, B., Deprey, L., Brahmbhatt, K., Harris, S., & Simon, T. J. (2014). Social impairments in chromosome 22q11.2 deletion syndrome (22q11.2DS): Autism spectrum disorder or a different endophenotype? *Journal of Autism and Developmental Disorders, 44*(4), 739–746.
45. Hiramoto, T., Kang, G., Suzuki, G., Satoh, Y., Kucherlapati, R., Watanabe, Y., et al. (2011). Tbx1: Identification of a 22q11.2 gene as a risk factor for autism spectrum disorder in a mouse model. *Human Molecular Genetics, 20*(24), 4775–4785.
46. Mukaddes, N. M., & Herguner, S. (2007). Autistic disorder and 22q11.2 duplication. *The World Journal of Biological Psychiatry, 8*(2), 127–130.
47. Ou, Z., Berg, J. S., Yonath, H., Enciso, V. B., Miller, D. T., Picker, J., et al. (2008). Microduplications of 22q11.2 are frequently inherited and are associated with variable phenotypes. *Genetics in Medicine, 10*(4), 267–277.
48. Radoeva, P. D., Coman, I. L., Salazar, C. A., Gentile, K. L., Higgins, A. M., Middleton, F. A., et al. (2014). Association between autism Spectrum disorder (ASD) in individuals with Velo-cardio-facial (22q11.2 deletion) syndrome and PRODH and COMT genotypes. *Psychiatric Genetics, 24*(6), 269.

49. Squarcione, C., Torti, M. C., Di Fabio, F., & Biondi, M. (2013). 22q11 deletion syndrome: A review of the neuropsychiatric features and their neurobiological basis. *Neuropsychiatric Disease and Treatment, 9*, 1873–1884.
50. Vorstman, J. A. S., Breetvelt, E. J., Thode, K. I., Chow, E. W. C., & Bassett, A. S. (2013). Expression of autism spectrum and schizophrenia in patients with a 22q11.2 deletion. *Schizophrenia Research, 143*(1), 55–59.
51. Vorstman, J. A. S., Morcus, M. E. J., Duijff, S. N., Klaassen, P. W. J., Heineman-de Boer, J. A., Beemer, F. A., et al. (2006). The 22q11.2 deletion in children: High rate of autistic disorders and early onset of psychotic symptoms. *Journal of the American Academy of Child and Adolescent Psychiatry, 45*(9), 1104–1113.
52. Han, K., Holder, J. L., Schaaf, C. P., Lu, H., Chen, H., Kang, H., et al. (2013). SHANK3 overexpression causes manic-like behaviour with unique pharmacogenetic properties. *Nature, 503*(7474), 72–77.
53. Kolevzon, A., Angarita, B., Bush, L., Wang, A. T., Frank, Y., Yang, A., et al. (2014). Phelan-McDermid syndrome: A review of the literature and practice parameters for medical assessment and monitoring. *Journal of Neurodevelopmental Disorders, 6*(1), 39.
54. Phelan, K., & McDermid, H. E. (2012). The 22q13.3 deletion syndrome (Phelan-McDermid syndrome). *Molecular Syndromology, 2*(3–5), 186.
55. Phelan, M. C., Rogers, R. C., Saul, R. A., Stapleton, G. A., Sweet, K., McDermid, H., et al. (2001). 22q13 deletion syndrome. *American Journal of Medical Genetics, 101*(2), 91–99.
56. Prasad, C., Prasad, A. N., Chodirker, B. N., Lee, C., Dawson, A. K., Jocelyn, L. J., et al. (2000). Genetic evaluation of pervasive developmental disorders: The terminal 22q13 deletion syndrome may represent a recognizable phenotype. *Clinical Genetics, 57*(2), 103–109.
57. Sebat, J., Lakshmi, B., Malhotra, D., Troge, J., Lese-Martin, C., Walsh, T., et al. (2007). Strong association of de novo copy number mutations with autism. *Science, 316*(5823), 445–449.
58. Dong, S., Walker, M. F., Carriero, N. J., DiCola, M., Willsey, A. J., Ye, A. Y., et al. (2014). De novo insertions and deletions of predominantly paternal origin are associated with autism spectrum disorder. *Cell Reports, 9*(1), 16–23.
59. Levy, D., Ronemus, M., Yamrom, B., Lee, Y.-H., Leotta, A., Kendall, J., et al. (2011). Rare de novo and transmitted copy-number variation in autistic spectrum disorders. *Neuron, 70*(5), 886–897.
60. Marshall, C. R., Noor, A., Vincent, J. B., Lionel, A. C., Feuk, L., Skaug, J., et al. (2008). Structural variation of chromosomes in autism spectrum disorder. *American Journal of Human Genetics, 82*(2), 477–488.
61. Morrow, E. M., Yoo, S.-Y., Flavell, S. W., Kim, T.-K., Lin, Y., Hill, R. S., et al. (2008). Identifying autism loci and genes by tracing recent shared ancestry. *Science, 321*(5886), 218–223.
62. Pinto, D., Pagnamenta, A. T., Klei, L., Anney, R., Merico, D., Regan, R., et al. (2010). Functional impact of global rare copy number variation in autism spectrum disorders. *Nature, 466*(7304), 368–372.
63. Sanders, S. J., Ercan-Sencicek, A. G., Hus, V., Luo, R., Murtha, M. T., Moreno-De-Luca, D., et al. (2011). Multiple recurrent de novo CNVs, including duplications of the 7q11.23 Williams syndrome region, are strongly associated with autism. *Neuron, 70*(5), 863–885.
64. Sanders, S. J., He, X., Willsey, A. J., Ercan-Sencicek, A. G., Samocha, K. E., Cicek, A. E., et al. (2015). Insights into autism spectrum disorder genomic architecture and biology from 71 risk loci. *Neuron, 87*(6), 1215.
65. Sebat, J., Lakshmi, B., Troge, J., Alexander, J., Young, J., Lundin, P., et al. (2004). Large-scale copy number polymorphism in the human genome. *Science, 305*(5683), 525–528.
66. Fan, Y., Du, X., Liu, X., Wang, L., Li, F., & Yu, Y. (2018). Rare copy number variations in a Chinese cohort of autism spectrum disorder. *Frontiers in Genetics, 9*, 665.
67. Wiśniowiecka-Kowalnik, B., & Nowakowska, B. A. (2019). Genetics and epigenetics of autism spectrum disorder—Current evidence in the field. *Journal of Applied Genetics, 60*, 37–47.

68. Iossifov, I., Ronemus, M., Levy, D., Wang, Z., Hakker, I., Rosenbaum, J., et al. (2012). De novo gene disruptions in children on the autistic spectrum. *Neuron, 74*(2), 285–299.
69. Neale, B. M., Kou, Y., Liu, L., Ma'ayan, A., Samocha, K. E., Sabo, A., et al. (2012). Patterns and rates of exonic de novo mutations in autism spectrum disorders. *Nature, 485*(7397), 242.
70. O'Roak, B. J., Deriziotis, P., Lee, C., Vives, L., Schwartz, J. J., Girirajan, S., et al. (2011). Exome sequencing in sporadic autism spectrum disorders identifies severe de novo mutations. *Nature Genetics, 43*(6), 585.
71. Sanders, S. J., Murtha, M. T., Gupta, A. R., Murdoch, J. D., Raubeson, M. J., Willsey, A. J., et al. (2012). De novo mutations revealed by whole-exome sequencing are strongly associated with autism. *Nature, 485*(7397), 237.
72. Kong, A., Frigge, M. L., Masson, G., Besenbacher, S., Sulem, P., Magnusson, G., et al. (2012). Rate of de novo mutations and the importance of father's age to disease risk. *Nature, 488*(7412), 471–475.
73. Durand, C. M., Betancur, C., Boeckers, T. M., Bockmann, J., Chaste, P., Fauchereau, F., et al. (2007). Mutations in the gene encoding the synaptic scaffolding protein SHANK3 are associated with autism spectrum disorders. *Nature Genetics, 39*(1), 25.
74. Jamain, S., Quach, H., Betancur, C., Rastam, M., Colineaux, C., Gillberg, I. C., et al. (2003). Mutations of the X-linked genes encoding neuroligins NLGN3 and NLGN4 are associated with autism. *Nature Genetics, 34*(1), 27–29.
75. Bundey, S., Hardy, C., Vickers, S., Kilpatrick, M. W., & Corbett, J. A. (1994). Duplication of the 15q11-13 region in a patient with autism, epilepsy and ataxia. *Developmental Medicine and Child Neurology, 36*(8), 736–742.
76. Thomas, N. S., Sharp, A. J., Browne, C. E., Skuse, D., Hardie, C., & Dennis, N. R. (1999). Xp deletions associated with autism in three females. *Human Genetics, 104*(1), 43–48.
77. Fine, S. E., Weissman, A., Gerdes, M., Pinto-Martin, J., Zackai, E. H., McDonald-McGinn, D. M., et al. (2005). Autism spectrum disorders and symptoms in children with molecularly confirmed 22q11.2 deletion syndrome. *Journal of Autism and Developmental Disorders, 35*(4), 461–470.
78. Kumar, R. A., KaraMohamed, S., Sudi, J., Conrad, D. F., Brune, C., Badner, J. A., et al. (2008). Recurrent 16p11.2 microdeletions in autism. *Human Molecular Genetics, 17*(4), 628–638.
79. Anney, R., Klei, L., Pinto, D., Regan, R., Conroy, J., Magalhaes, T. R., et al. (2010). A genome-wide scan for common alleles affecting risk for autism. *Human Molecular Genetics, 19*(20), 4072–4082.
80. Cooper, G. M., Coe, B. P., Girirajan, S., Rosenfeld, J. A., Vu, T. H., Baker, C., et al. (2011). A copy number variation morbidity map of developmental delay. *Nature Genetics, 43*(9), 838–846.
81. Simons Vip, C. (2012). Simons Variation in Individuals Project (Simons VIP): A genetics-first approach to studying autism spectrum and related neurodevelopmental disorders. *Neuron, 73*(6), 1063–1067.
82. Alarcon, M., Abrahams, B. S., Stone, J. L., Duvall, J. A., Perederiy, J. V., Bomar, J. M., et al. (2008). Linkage, association, and gene-expression analyses identify CNTNAP2 as an autism-susceptibility gene. *American Journal of Human Genetics, 82*(1), 150–159.
83. Weiss, L. A., Arking, D. E., Gene Discovery Project of Johns Hopkins & the Autism Consortium, Daly, M. J., & Chakravarti, A. (2009). A genome-wide linkage and association scan reveals novel loci for autism. *Nature, 461*(7265), 802–808.
84. Geschwind, D. H. (2008). Autism: Many genes, common pathways? *Cell, 135*(3), 391–395.
85. Lowe, J. K., Werling, D. M., Constantino, J. N., Cantor, R. M., & Geschwind, D. H. (2015). Social responsiveness, an autism endophenotype: Genomewide significant linkage to two regions on chromosome 8. *The American Journal of Psychiatry, 172*(3), 266–275.
86. Klei, L., Sanders, S. J., Murtha, M. T., Hus, V., Lowe, J. K., Willsey, A. J., et al. (2012). Common genetic variants, acting additively, are a major source of risk for autism. *Molecular Autism, 3*(1), 9.

87. Ma, D., Salyakina, D., Jaworski, J. M., Konidari, I., Whitehead, P. L., Andersen, A. N., et al. (2009). A genome-wide association study of autism reveals a common novel risk locus at 5p14.1. *Annals of Human Genetics, 73*(Pt 3), 263–273.
88. Wang, K., Zhang, H., Ma, D., Bucan, M., Glessner, J. T., Abrahams, B. S., et al. (2009). Common genetic variants on 5p14.1 associate with autism spectrum disorders. *Nature, 459*(7246), 528–533.
89. Xia, K., Guo, H., Hu, Z., Xun, G., Zuo, L., Peng, Y., et al. (2014). Common genetic variants on 1p13.2 associate with risk of autism. *Molecular Psychiatry, 19*(11), 1212–1219.
90. Liang, S., Wang, X. L., Zou, M. Y., Wang, H., Zhou, X., Sun, C. H., et al. (2014). Family-based association study of ZNF533, DOCK4 and IMMP2L gene polymorphisms linked to autism in a northeastern Chinese Han population. *Journal of Zhejiang University Science B, 15*(3), 264–271.
91. Strauss, K. A., Puffenberger, E. G., Huentelman, M. J., Gottlieb, S., Dobrin, S. E., Parod, J. M., et al. (2006). Recessive symptomatic focal epilepsy and mutant contactin-associated protein-like 2. *The New England Journal of Medicine, 354*(13), 1370–1377.
92. Bakkaloglu, B., O'Roak, B. J., Louvi, A., Gupta, A. R., Abelson, J. F., Morgan, T. M., et al. (2008). Molecular cytogenetic analysis and resequencing of contactin associated protein-like 2 in autism spectrum disorders. *American Journal of Human Genetics, 82*(1), 165.
93. Peñagarikano, O., & Geschwind, D. H. (2012). What does CNTNAP2 reveal about autism spectrum disorder? *Trends in Molecular Medicine, 18*(3), 156.
94. Berkel, S., Marshall, C. R., Weiss, B., Howe, J., Roeth, R., Moog, U., et al. (2010). Mutations in the SHANK2 synaptic scaffolding gene in autism spectrum disorder and mental retardation. *Nature Genetics, 42*(6), 489–491.
95. Berkel, S., Tang, W., Trevino, M., Vogt, M., Obenhaus, H. A., Gass, P., et al. (2012). Inherited and de novo SHANK2 variants associated with autism spectrum disorder impair neuronal morphogenesis and physiology. *Human Molecular Genetics, 21*(2), 344–357.
96. Kim, H. G., Kishikawa, S., Higgins, A. W., Seong, I. S., Donovan, D. J., Shen, Y., et al. (2008). Disruption of neurexin 1 associated with autism spectrum disorder. *American Journal of Human Genetics, 82*(1), 199–207.
97. Moessner, R., Marshall, C. R., Sutcliffe, J. S., Skaug, J., Pinto, D., Vincent, J., et al. (2007). Contribution of SHANK3 mutations to autism spectrum disorder. *American Journal of Human Genetics, 81*(6), 1289–1297.
98. Sato, D., Lionel, A. C., Leblond, C. S., Prasad, A., Pinto, D., Walker, S., et al. (2012). SHANK1 deletions in males with autism spectrum disorder. *American Journal of Human Genetics, 90*(5), 879–887.
99. Vaags, A. K., Lionel, A. C., Sato, D., Goodenberger, M., Stein, Q. P., Curran, S., et al. (2012). Rare deletions at the neurexin 3 locus in autism spectrum disorder. *American Journal of Human Genetics, 90*(1), 133–141.
100. Al-Dewik, N., Mohd, H., Al-Mureikhi, M., Ali, R., Al-Mesaifri, F., Mahmoud, L., et al. (2019). Clinical exome sequencing in 509 Middle Eastern families with suspected Mendelian diseases: The Qatari experience. *American Journal of Medical Genetics Part A, 179*(6), 927–935.
101. Helsmoortel, C., Vulto-van Silfhout, A. T., Coe, B. P., Vandeweyer, G., Rooms, L., van den Ende, J., et al. (2014). A SWI/SNF-related autism syndrome caused by de novo mutations in ADNP. *Nature Genetics, 46*(4), 380–384.
102. Novarino, G., El-Fishawy, P., Kayserili, H., Meguid, N. A., Scott, E. M., Schroth, J., et al. (2012). Mutations in BCKD-kinase lead to a potentially treatable form of autism with epilepsy. *Science, 338*(6105), 394–397.
103. Yu, T. W., Chahrour, M. H., Coulter, M. E., Jiralerspong, S., Okamura-Ikeda, K., Ataman, B., et al. (2013). Using whole exome sequencing to identify inherited causes of autism. *Neuron, 77*(2), 259–273.
104. Hu, V. W., Sarachana, T., Kim, K. S., Nguyen, A., Kulkarni, S., Steinberg, M. E., et al. (2009). Gene expression profiling differentiates autism case-controls and phenotypic variants of autism spectrum disorders: Evidence for circadian rhythm dysfunction in severe autism. *Autism Research, 2*(2), 78–97.

105. Luo, R., Sanders, S. J., Tian, Y., Voineagu, I., Huang, N., Chu, S. H., et al. (2012). Genome-wide transcriptome profiling reveals the functional impact of rare de novo and recurrent CNVs in autism spectrum disorders. *American Journal of Human Genetics, 91*(1), 38–55.
106. Voineagu, I., Wang, X., Johnston, P., Lowe, J. K., Tian, Y., Horvath, S., et al. (2011). Transcriptomic analysis of autistic brain reveals convergent molecular pathology. *Nature, 474*(7351), 380–384.
107. Gupta, S., Ellis, S. E., Ashar, F. N., Moes, A., Bader, J. S., Zhan, J., et al. (2014). Transcriptome analysis reveals dysregulation of innate immune response genes and neuronal activity-dependent genes in autism. *Nature Communications, 5*, 5748.
108. Hormozdiari, F., Penn, O., Borenstein, E., & Eichler, E. E. (2015). The discovery of integrated gene networks for autism and related disorders. *Genome Research, 25*(1), 142–154.
109. Ch'ng, C., Kwok, W., Rogic, S., & Pavlidis, P. (2015). Meta-analysis of gene expression in autism spectrum disorder. *Autism Research, 8*(5), 593–608.
110. Gregg, J. P., Lit, L., Baron, C. A., Hertz-Picciotto, I., Walker, W., Davis, R. A., et al. (2008). Gene expression changes in children with autism. *Genomics, 91*(1), 22–29.
111. Ning, L. F., Yu, Y. Q., GuoJi, E. T., Kou, C. G., Wu, Y. H., Shi, J. P., et al. (2015). Meta-analysis of differentially expressed genes in autism based on gene expression data. *Genetics and Molecular Research, 14*(1), 2146–2155.
112. Chien, W. H., Gau, S. S., Chen, C. H., Tsai, W. C., Wu, Y. Y., Chen, P. H., et al. (2013). Increased gene expression of FOXP1 in patients with autism spectrum disorders. *Molecular Autism, 4*(1), 23.
113. Abbasy, S., Shahraki, F., Haghighatfard, A., Qazvini, M. G., Rafiei, S. T., Noshadirad, E., et al. (2018). Neuregulin1 types mRNA level changes in autism spectrum disorder, and is associated with deficit in executive functions. *eBioMedicine, 37*, 483–488.
114. Pramparo, T., Lombardo, M. V., Campbell, K., Barnes, C. C., Marinero, S., Solso, S., et al. (2015). Cell cycle networks link gene expression dysregulation, mutation, and brain maldevelopment in autistic toddlers. *Molecular Systems Biology, 11*(12), 841.
115. Alter, M. D., Kharkar, R., Ramsey, K. E., Craig, D. W., Melmed, R. D., Grebe, T. A., et al. (2011). Autism and increased paternal age related changes in global levels of gene expression regulation. *PLoS One, 6*(2), e16715.
116. Anitha, A., Nakamura, N., Thanseem, I., Yamada, K., Iwayama, Y., Toyota, T., et al. (2012). Brain region-specific altered expression and association of mitochondria-related genes in autism. *Molecular Autism, 3*(1), 12.
117. Chow, M. L., Pramparo, T., Winn, M. E., Barnes, C. C., Li, H.-R., Weiss, L., et al. (2012). Age-dependent brain gene expression and copy number anomalies in autism suggest distinct pathological processes at young versus mature ages. *PLoS Genetics, 8*(3), e1002592.
118. Féron, F., Gepner, B., Lacassagne, E., Stephan, D., Mesnage, B., Blanchard, M.-P., et al. (2016). Olfactory stem cells reveal MOCOS as a new player in autism spectrum disorders. *Molecular Psychiatry, 21*(9), 1215–1224.
119. Ginsberg, M. R., Rubin, R. A., Falcone, T., Ting, A. H., & Natowicz, M. R. (2012). Brain transcriptional and epigenetic associations with autism. *PLoS One, 7*(9), e44736.
120. Glatt, S. J., Tsuang, M. T., Winn, M., Chandler, S. D., Collins, M., Lopez, L., et al. (2012). Blood-based gene expression signatures of infants and toddlers with autism. *Journal of the American Academy of Child and Adolescent Psychiatry, 51*(9), 934–442.
121. Chana, G., Laskaris, L., Pantelis, C., Gillett, P., Testa, R., Zantomio, D., et al. (2015). Decreased expression of mGluR5 within the dorsolateral prefrontal cortex in autism and increased microglial number in mGluR5 knockout mice: Pathophysiological and neurobehavioral implications. *Brain, Behavior, and Immunity, 49*, 197–205.
122. Ivanov, H. Y., Stoyanova, V. K., Popov, N. T., Bosheva, M., & Vachev, T. I. (2015). Blood-based gene expression in children with autism spectrum disorder. *BioDiscovery, 17*, e8966.
123. James, S. J., Shpyleva, S., Melnyk, S., Pavliv, O., & Pogribny, I. P. (2014). Elevated 5-hydroxymethylcytosine in the Engrailed-2 (EN-2) promoter is associated with increased gene expression and decreased MeCP2 binding in autism cerebellum. *Translational Psychiatry, 4*, e460.

124. Khan, A., Harney, J. W., Zavacki, A. M., & Sajdel-Sulkowska, E. M. (2014). Disrupted brain thyroid hormone homeostasis and altered thyroid hormone-dependent brain gene expression in autism spectrum disorders. *Journal of Physiology and Pharmacology, 65*(2), 257–272.
125. Kong, S., Shimizu-Motohashi, Y., Campbell, M., Lee, I., Collins, C., Brewster, S., et al. (2013). Peripheral blood gene expression signature differentiates children with autism from unaffected siblings. *Neurogenetics, 14*(2), 143.
126. Kuwano, Y., Kamio, Y., Kawai, T., Katsuura, S., Inada, N., Takaki, A., et al. (2011). Autism-associated gene expression in peripheral leucocytes commonly observed between subjects with autism and healthy women having autistic children. *PLoS One, 6*(9), e24723.
127. Maekawa, M., Yamada, K., Toyoshima, M., Ohnishi, T., Iwayama, Y., Shimamoto, C., et al. (2015). Utility of scalp hair follicles as a novel source of biomarker genes for psychiatric illnesses. *Biological Psychiatry, 78*(2), 116–125.
128. Prandini, P., Zusi, C., Malerba, G., Itan, P., Pignatti, F., & Trabetti, E. (2014). Analysis of RBFOX1 gene expression in lymphoblastoid cell lines of Italian discordant autism spectrum disorders sib-pairs. *Molecular and Cellular Probes, 28*(5–6), 342–346.
129. Segura, M., Pedreño, C., Obiols, J., Taurines, R., Pàmias, M., Grünblatt, E., et al. (2015). Neurotrophin blood-based gene expression and social cognition analysis in patients with autism spectrum disorder. *Neurogenetics, 16*(2), 123–131.
130. Stamova, B., Green, P. G., Tian, Y., Hertz-Picciotto, I., Pessah, I. N., Hansen, R., et al. (2011). Correlations between gene expression and mercury levels in blood of boys with and without autism. *Neurotoxicity Research, 19*(1), 31–48.
131. Talebizadeh, Z., Aldenderfer, R., & Wen Chen, X. (2014). A proof-of-concept study: Exon-level expression profiling and alternative splicing in autism using lymphoblastoid cell lines. *Psychiatric Genetics, 24*(1), 1–9.
132. Taurines, R., Grünblatt, E., Schecklmann, M., Schwenck, C., Albantakis, L., Reefschläger, L., et al. (2011). Altered mRNA expression of monoaminergic candidate genes in the blood of children with attention deficit hyperactivity disorder and autism spectrum disorder. *The World Journal of Biological Psychiatry, 12*(Suppl), 1.
133. Tian, Y., Green, P. G., Stamova, B., Hertz-Picciotto, I., Pessah, I. N., Hansen, R., et al. (2011). Correlations of gene expression with blood lead levels in children with autism compared to typically developing controls. *Neurotoxicity Research, 19*(1), 1.
134. Voineagu, I., & Eapen, V. (2013). Converging pathways in autism spectrum disorders: Interplay between synaptic dysfunction and immune responses. *Frontiers in Human Neuroscience, 7*, 738.
135. Walker, S. J., Fortunato, J., Gonzalez, L. G., & Krigsman, A. (2013). Identification of unique gene expression profile in children with regressive autism spectrum disorder (ASD) and ileocolitis. *PLoS One, 8*(3), e58058.
136. Williams, B. L., Hornig, M., Buie, T., Bauman, M. L., Cho Paik, M., Wick, I., et al. (2011). Impaired carbohydrate digestion and transport and mucosal dysbiosis in the intestines of children with autism and gastrointestinal disturbances. *PLoS One, 6*(9), e24585.
137. Yasuda, Y., Hashimoto, R., Yamamori, H., Ohi, K., Fukumoto, M., Umeda-Yano, S., et al. (2011). Gene expression analysis in lymphoblasts derived from patients with autism spectrum disorder. *Molecular Autism, 2*(1), 9.
138. Zhubi, A., Chen, Y., Dong, E., Cook, E. H., Guidotti, A., & Grayson, D. R. (2014). Increased binding of MeCP2 to the GAD1 and RELN promoters may be mediated by an enrichment of 5-hmC in autism spectrum disorder (ASD) cerebellum. *Translational Psychiatry, 4*, e349.
139. Kang, H. J., Kawasawa, Y. I., Cheng, F., Zhu, Y., Xu, X., Li, M., et al. (2011). Spatiotemporal transcriptome of the human brain. *Nature, 478*(7370), 483.
140. Ziats, M. N., & Rennert, O. M. (2013). Sex-biased gene expression in the developing brain: Implications for autism spectrum disorders. *Molecular Autism, 4*, 10.
141. Garbett, K., Ebert, P. J., Mitchell, A., Lintas, C., Manzi, B., Mirnics, K., et al. (2008). Immune transcriptome alterations in the temporal cortex of subjects with autism. *Neurobiology of Disease, 30*(3), 303–311.
142. Ziats, M. N., & Rennert, O. M. (2013). Aberrant expression of long noncoding RNAs in autistic brain. *Journal of Molecular Neuroscience, 49*(3), 589–593.

143. Enstrom, A. M., Lit, L., Onore, C. E., Gregg, J. P., Hansen, R. L., Pessah, I. N., et al. (2009). Altered gene expression and function of peripheral blood natural killer cells in children with autism. *Brain, Behavior, and Immunity, 23*(1), 124–133.
144. Hu, V. W., Frank, B. C., Heine, S., Lee, N. H., & Quackenbush, J. (2006). Gene expression profiling of lymphoblastoid cell lines from monozygotic twins discordant in severity of autism reveals differential regulation of neurologically relevant genes. *BMC Genomics, 7*, 118.
145. Emanuele, E., Boso, M., Cassola, F., Broglia, D., Bonoldi, I., Mancini, L., et al. (2010). Increased dopamine DRD4 receptor mRNA expression in lymphocytes of musicians and autistic individuals: Bridging the music-autism connection. *Neuro Endocrinology Letters, 31*(1), 122–125.
146. Purcell, A. E., Jeon, O. H., Zimmerman, A. W., Blue, M. E., & Pevsner, J. (2001). Postmortem brain abnormalities of the glutamate neurotransmitter system in autism. *Neurology, 57*(9), 1618–1628.
147. Wong, C. C. Y., Meaburn, E. L., Ronald, A., Price, T. S., Jeffries, A. R., Schalkwyk, L. C., et al. (2013). Methylomic analysis of monozygotic twins discordant for autism spectrum disorder and related behavioural traits. *Molecular Psychiatry, 19*(4), 495.
148. James, S. J., Melnyk, S., Jernigan, S., Hubanks, A., Rose, S., & Gaylor, D. W. (2008). Abnormal transmethylation/transsulfuration metabolism and DNA hypomethylation among parents of children with autism. *Journal of Autism and Developmental Disorders, 38*(10), 1966–1975.
149. Nardone, S., Sams, D. S., Reuveni, E., Getselter, D., Oron, O., Karpuj, M., et al. (2014). DNA methylation analysis of the autistic brain reveals multiple dysregulated biological pathways. *Translational Psychiatry, 4*, e433.
150. Ladd-Acosta, C., Hansen, K. D., Briem, E., Fallin, M. D., Kaufmann, W. E., & Feinberg, A. P. (2013). Common DNA methylation alterations in multiple brain regions in autism. *Molecular Psychiatry, 19*(8), 862.
151. Ladd-Acosta, C., Hansen, K. D., Briem, E., Fallin, M. D., Kaufmann, W. E., & Feinberg, A. P. (2014). Common DNA methylation alterations in multiple brain regions in autism. *Molecular Psychiatry, 19*(8), 862–871.
152. Daniele Fallin, M., Feinberg, J. I., Bakulski, K. M., Brown, S. C., Tryggvadottir, R., Feinberg, A. P., et al. (2015). Paternal sperm DNA methylation associated with early signs of autism risk in an autism-enriched cohort. *International Journal of Epidemiology, 44*(4), 1199–1210.
153. Milekic, M. H., Xin, Y., O'Donnell, A., Kumar, K. K., Bradley-Moore, M., Malaspina, D., et al. (2014). Age-related sperm DNA methylation changes are transmitted to offspring and associated with abnormal behavior and dysregulated gene expression. *Molecular Psychiatry, 20*(8), 995.
154. Behnia, F., Parets, S. E., Kechichian, T., Yin, H., Dutta, E. H., Saade, G. R., et al. (2015). Fetal DNA methylation of autism spectrum disorders candidate genes: Association with spontaneous preterm birth. *American Journal of Obstetrics and Gynecology, 212*(4), 533.e531–533.e539.
155. James, S. J., Shpyleva, S., Melnyk, S., Pavliv, O., & Pogribny, I. P. (2013). Complex epigenetic regulation of engrailed-2 (EN-2) homeobox gene in the autism cerebellum. *Translational Psychiatry, 3*, e232.
156. Nguyen, A., Rauch, T. A., Pfeifer, G. P., & Hu, V. W. (2010). Global methylation profiling of lymphoblastoid cell lines reveals epigenetic contributions to autism spectrum disorders and a novel autism candidate gene, RORA, whose protein product is reduced in autistic brain. *The FASEB Journal, 24*(8), 3036.
157. Bremer, A., Giacobini, M., Nordenskjöld, M., Brøndum-Nielsen, K., Mansouri, M., Dahl, N., et al. (2010). Screening for copy number alterations in loci associated with autism spectrum disorders by two-color multiplex ligation-dependent probe amplification. *American Journal of Medical Genetics. Part B, Neuropsychiatric Genetics, 153B*(1), 280–285.
158. Chahrour, M., Jung, S. Y., Shaw, C., Zhou, X., Wong, S. T. C., Qin, J., et al. (2008). MeCP2, a key contributor to neurological disease, activates and represses transcription. *Science (New York, N.Y.), 320*(5880), 1224.

159. Vorstman, J. A. S., Staal, W. G., van Daalen, E., van Engeland, H., Hochstenbach, P. F. R., & Franke, L. (2006). Identification of novel autism candidate regions through analysis of reported cytogenetic abnormalities associated with autism. *Molecular Psychiatry, 11*(1), 1.
160. Liu, X., Solehdin, F., Cohen, I. L., Gonzalez, M. G., Jenkins, E. C., Lewis, M. E. S., et al. (2011). Population- and family-based studies associate the MTHFR gene with idiopathic autism in simplex families. *Journal of Autism and Developmental Disorders, 41*(7), 938–944.
161. Mohammad, N. S., Jain, J. M. N., Chintakindi, K. P., Singh, R. P., Naik, U., & Akella, R. R. D. (2009). Aberrations in folate metabolic pathway and altered susceptibility to autism. *Psychiatric Genetics, 19*(4), 171–176.
162. Schanen, N. C. (2006). Epigenetics of autism spectrum disorders. *Human Molecular Genetics, 15*(suppl_2), R138–R150.
163. Lasalle, J. M. (2013). Autism genes keep turning up chromatin. *OA Autism, 1*(2), 14.
164. Berko, E. R., Suzuki, M., Beren, F., Lemetre, C., Alaimo, C. M., Calder, R. B., et al. (2014). Mosaic epigenetic dysregulation of ectodermal cells in autism spectrum disorder. *PLoS Genetics, 10*(5), e1004402.
165. Irimia, M., Weatheritt, R. J., Ellis, J. D., Parikshak, N. N., Gonatopoulos-Pournatzis, T., Babor, M., et al. (2014). A highly conserved program of neuronal microexons is misregulated in autistic brains. *Cell, 159*(7), 1511–1523.
166. Vogel-Ciernia, A., & Wood, M. A. (2014). Neuron-specific chromatin remodeling: A missing link in epigenetic mechanisms underlying synaptic plasticity, memory, and intellectual disability disorders. *Neuropharmacology, 80*, 18–27.
167. LoParo, D., & Waldman, I. D. (2015). The oxytocin receptor gene (OXTR) is associated with autism spectrum disorder: A meta-analysis. *Molecular Psychiatry, 20*(5), 640–646.
168. Lopez, A. J., & Wood, M. A. (2015). Role of nucleosome remodeling in neurodevelopmental and intellectual disability disorders. *Frontiers in Behavioral Neuroscience, 9*, 100.
169. Kubota, T., & Mochizuki, K. (2016). Epigenetic effect of environmental factors on autism spectrum disorders. *International Journal of Environmental Research and Public Health, 13*(5), pii: E504.
170. Quesnel-Vallieres, M., Dargaei, Z., Irimia, M., Gonatopoulos-Pournatzis, T., Ip, J. Y., Wu, M., et al. (2016). Misregulation of an activity-dependent splicing network as a common mechanism underlying autism spectrum disorders. *Molecular Cell, 64*(6), 1023–1034.
171. Wang, Y., Fang, Y., Zhang, F., Xu, M., Zhang, J., Yan, J., et al. (2014). Hypermethylation of the enolase gene (ENO2) in autism. *European Journal of Pediatrics, 173*(9), 1233–1244.
172. Frye, R. E., & Rossignol, D. A. (2011). Mitochondrial dysfunction can connect the diverse medical symptoms associated with autism spectrum disorders. *Pediatric Research, 69*(5 Pt 2), 41R–47R.
173. Rossignol, D. A., & Frye, R. E. (2012). Mitochondrial dysfunction in autism spectrum disorders: A systematic review and meta-analysis. *Molecular Psychiatry, 17*(3), 290–314.
174. Lombard, J. (1998). Autism: A mitochondrial disorder? *Medical Hypotheses, 50*(6), 497–500.
175. Valenti, D., de Bari, L., De Filippis, B., Henrion-Caude, A., & Vacca, R. A. (2014). Mitochondrial dysfunction as a central actor in intellectual disability-related diseases: An overview of down syndrome, autism, fragile X and Rett syndrome. *Neuroscience and Biobehavioral Reviews, 46*(Pt), 2.
176. Giulivi, C., Zhang, Y.-F., Omanska-Klusek, A., Ross-Inta, C., Wong, S., Hertz-Picciotto, I., et al. (2010). Mitochondrial dysfunction in autism. *JAMA, 304*(21), 2389–2396.
177. Oliveira, G., Diogo, L., Grazina, M., Garcia, P., Ataíde, A., Marques, C., et al. (2005). Mitochondrial dysfunction in autism spectrum disorders: A population-based study. *Developmental Medicine and Child Neurology, 47*(3), 185–189.
178. Goldenthal, M. J., Damle, S., Sheth, S., Shah, N., Melvin, J., Jethva, R., et al. (2015). Mitochondrial enzyme dysfunction in autism spectrum disorders; a novel biomarker revealed from buccal swab analysis. *Biomarkers in Medicine, 9*(10), 957–965.
179. Parikshak, N. N., Gandal, M. J., & Geschwind, D. H. (2015). Systems biology and gene networks in neurodevelopmental and neurodegenerative disorders. *Nature Reviews Genetics, 16*(8), 441–458.

180. Napoli, E., Wong, S., Hertz-Picciotto, I., & Giulivi, C. (2014). Deficits in bioenergetics and impaired immune response in granulocytes from children with autism. *Pediatrics, 133*(5), 1405–1410.
181. Weissman, J. R., Kelley, R. I., Bauman, M. L., Cohen, B. H., Murray, K. F., Mitchell, R. L., et al. (2008). Mitochondrial disease in autism spectrum disorder patients: A cohort analysis. *PLoS One, 3*(11), e3815.
182. Palmieri, L., & Persico, A. M. (2010). Mitochondrial dysfunction in autism spectrum disorders: Cause or effect? *Biochimica et Biophysica Acta, 1797*(6–7), 1130–1137.
183. Fillano, J. J., Goldenthal, M. J., Rhodes, C. H., & Marín-García, J. (2002). Mitochondrial dysfunction in patients with hypotonia, epilepsy, autism, and developmental delay: HEADD syndrome. *Journal of Child Neurology, 17*(6), 435–439.
184. Gu, F., Chauhan, V., Kaur, K., Brown, W. T., LaFauci, G., Wegiel, J., et al. (2013). Alterations in mitochondrial DNA copy number and the activities of electron transport chain complexes and pyruvate dehydrogenase in the frontal cortex from subjects with autism. *Translational Psychiatry, 3*, e299.
185. Smith, M., Spence, M. A., & Flodman, P. (2009). Nuclear and mitochondrial genome defects in autisms. *Annals of the New York Academy of Sciences, 1151*, 102–132.
186. Pitceathly, R. D. S., Rahman, S., & Hanna, M. G. (2012). Single deletions in mitochondrial DNA--molecular mechanisms and disease phenotypes in clinical practice. *Neuromuscular Disorders, 22*(7), 577–586.
187. Haas, R. H., Parikh, S., Falk, M. J., Saneto, R. P., Wolf, N. I., Darin, N., et al. (2007). Mitochondrial disease: A practical approach for primary care physicians. *Pediatrics, 120*(6), 1326–1333.
188. Niyazov, D. M., Kahler, S. G., & Frye, R. E. (2016). Primary mitochondrial disease and secondary mitochondrial dysfunction: Importance of distinction for diagnosis and treatment. *Molecular Syndromology, 7*(3), 122–137.
189. Varga, N. Á., Pentelényi, K., Balicza, P., Gézsi, A., Reményi, V., Hársfalvi, V., et al. (2018). Mitochondrial dysfunction and autism: Comprehensive genetic analyses of children with autism and mtDNA deletion. *Behavioral and Brain Functions, 14*. https://doi.org/10.1186/s12993-018-0135-x
190. Coolen, M., & Bally-Cuif, L. (2009). MicroRNAs in brain development and physiology. *Current Opinion in Neurobiology, 19*(5), 461–470.
191. Hu, Y., Ehli, E. A., & Boomsma, D. I. (2017). MicroRNAs as biomarkers for psychiatric disorders with a focus on autism spectrum disorder: Current progress in genetic association studies, expression profiling, and translational research. *Autism Research, 10*(7), 1184–1203.
192. Stamova, B., Ander, B. P., Barger, N., Sharp, F. R., & Schumann, C. M. (2015). Specific regional and age-related small noncoding RNA expression patterns within superior temporal gyrus of typical human brains are less distinct in autism brains. *Journal of Child Neurology, 30*(14), 1930–1946.
193. Schumann, C. M., Sharp, F. R., Ander, B. P., & Stamova, B. (2017). Possible sexually dimorphic role of miRNA and other sncRNA in ASD brain. *Molecular Autism, 8*, 4.
194. Abu-Elneel, K., Liu, T., Gazzaniga, F. S., Nishimura, Y., Wall, D. P., Geschwind, D. H., et al. (2008). Heterogeneous dysregulation of microRNAs across the autism spectrum. *Neurogenetics, 9*(3), 153–161.
195. Buyske, S. (2009). Comment on the article "Heterogeneous dysregulation of microRNAs across the autism spectrum" by Abu-Elneel et al. *Neurogenetics, 10*(2), 167.
196. Constantin, L. (2017). The role of microRNAs in cerebellar development and autism spectrum disorder during embryogenesis. *Molecular Neurobiology, 54*(9), 6944–6959.
197. Sarachana, T., Zhou, R., Chen, G., Manji, H. K., & Hu, V. W. (2010). Investigation of post-transcriptional gene regulatory networks associated with autism spectrum disorders by microRNA expression profiling of lymphoblastoid cell lines. *Genome Medicine, 2*(4), 23.
198. Talebizadeh, Z., Butler, M. G., & Theodoro, M. F. (2008). Feasibility and relevance of examining lymphoblastoid cell lines to study role of microRNAs in autism. *Autism Research, 1*(4), 240–250.

199. Wu, Y. E., Parikshak, N. N., Belgard, T. G., & Geschwind, D. H. (2016). Genome-wide, integrative analysis implicates microRNA dysregulation in autism spectrum disorder. *Nature Neuroscience, 19*(11), 1463–1476.
200. Mundalil Vasu, M., Anitha, A., Thanseem, I., Suzuki, K., Yamada, K., Takahashi, T., et al. (2014). Serum microRNA profiles in children with autism. *Molecular Autism, 5*, 40.
201. Kichukova, T. M., Popov, N. T., Ivanov, I. S., & Vachev, T. I. (2017). Profiling of circulating serum microRNAs in children with autism spectrum disorder using stem-loop qRT-PCR assay. *Folia Medica (Plovdiv)., 59*(1), 43–52.
202. Weber, J. A., Baxter, D. H., Zhang, S., Huang, D. Y., Huang, K. H., Lee, M. J., et al. (2010). The microRNA spectrum in 12 body fluids. *Clinical Chemistry, 56*(11), 1733–1741.
203. Hicks, S. D., Ignacio, C., Gentile, K., & Middleton, F. A. (2016). Salivary miRNA profiles identify children with autism spectrum disorder, correlate with adaptive behavior, and implicate ASD candidate genes involved in neurodevelopment. *BMC Pediatrics, 16*, 52.
204. Nguyen, L. S., Lepleux, M., Makhlouf, M., Martin, C., Fregeac, J., Siquier-Pernet, K., et al. (2016). Profiling olfactory stem cells from living patients identifies miRNAs relevant for autism pathophysiology. *Molecular Autism, 7*, 1.
205. Mellios, N., & Sur, M. (2012). The emerging role of microRNAs in schizophrenia and autism spectrum disorders. *Frontiers in Psychiatry, 3*, 39.
206. Rogelj, B. (2006). Brain-specific small nucleolar RNAs. *Journal of Molecular Neuroscience, 28*(2), 103–109.
207. Cavaillé, J., Buiting, K., Kiefmann, M., Lalande, M., Brannan, C. I., Horsthemke, B., et al. (2000). Identification of brain-specific and imprinted small nucleolar RNA genes exhibiting an unusual genomic organization. *PNAS, 97*(26), 14311–14316.
208. Dykens, E. M., Lee, E., & Roof, E. (2011). Prader-Willi syndrome and autism spectrum disorders: An evolving story. *Journal of Neurodevelopmental Disorders, 3*(3), 225–237.
209. Gabriele, S., Sacco, R., & Persico, A. M. (2014). Blood serotonin levels in autism spectrum disorder: A systematic review and meta-analysis. *European Neuropsychopharmacology, 24*(6), 919–929.
210. Nakatani, J., Tamada, K., Hatanaka, F., Ise, S., Ohta, H., Inoue, K., et al. (2009). Abnormal behavior in a chromosome-engineered mouse model for human 15q11-13 duplication seen in autism. *Cell, 137*(7), 1235–1246.
211. Rajasethupathy, P., Antonov, I., Sheridan, R., Frey, S., Sander, C., Tuschl, T., et al. (2012). A role for neuronal piRNAs in the epigenetic control of memory-related synaptic plasticity. *Cell, 149*(3), 693–707.
212. Wilkinson, B., & Campbell, D. B. (2013). Contribution of long noncoding RNAs to autism spectrum disorder risk. *International Review of Neurobiology, 113*, 35–59.
213. Minshew, N. J., & Keller, T. A. (2010). The nature of brain dysfunction in autism: Functional brain imaging studies. *Current Opinion in Neurology, 23*(2), 124–130.
214. Kerin, T., Ramanathan, A., Rivas, K., Grepo, N., Coetzee, G. A., & Campbell, D. B. (2012). A noncoding RNA antisense to moesin at 5p14.1 in autism. *Science Translational Medicine, 4*(128), 128ra140.
215. DeWitt, J. J., Grepo, N., Wilkinson, B., Evgrafov, O. V., Knowles, J. A., & Campbell, D. B. (2016). Impact of the autism-associated long noncoding RNA MSNP1AS on neuronal architecture and gene expression in human neural progenitor cells. *Genes (Basel), 7*(10), pii: E76.
216. Vincent, J. B., Petek, E., Thevarkunnel, S., Kolozsvari, D., Cheung, J., Patel, M., et al. (2002). The RAY1/ST7 tumor-suppressor locus on chromosome 7q31 represents a complex multi-transcript system. *Genomics, 80*(3), 283–294.
217. Williams, J. M., Beck, T. F., Pearson, D. M., Proud, M. B., Cheung, S. W., & Scott, D. A. (2009). A 1q42 deletion involving DISC1, DISC2, and TSNAX in an autism spectrum disorder. *American Journal of Medical Genetics Part A, 149A*(8), 1758–1762.
218. Wang, Y., Zhao, X., Ju, W., Flory, M., Zhong, J., Jiang, S., et al. (2015). Genome-wide differential expression of synaptic long noncoding RNAs in autism spectrum disorder. *Translational Psychiatry, 5*, e660.

219. Parikshak, N. N., Swarup, V., Belgard, T. G., Irimia, M., Ramaswami, G., Gandal, M. J., et al. (2016). Genome-wide changes in lncRNA, splicing, and regional gene expression patterns in autism. *Nature, 540*(7633), 423–427.
220. Gudenas, B. L., Srivastava, A. K., & Wang, L. (2017). Integrative genomic analyses for identification and prioritization of long non-coding RNAs associated with autism. *PLoS One, 12*(5), e0178532.
221. Noor, A., Whibley, A., Marshall, C. R., Gianakopoulos, P. J., Piton, A., Carson, A. R., et al. (2010). Disruption at the PTCHD1 locus on Xp22.11 in autism spectrum disorder and intellectual disability. *Science Translational Medicine, 2*(49), 49ra68.
222. Velmeshev, D., Magistri, M., & Faghihi, M. A. (2013). Expression of non-protein-coding antisense RNAs in genomic regions related to autism spectrum disorders. *Molecular Autism, 4*(1), 32.
223. Fallah, H., Ganji, M., Arsang-Jang, S., Sayad, A., & Taheri, M. (2019). Consideration of the role of MALAT1 long noncoding RNA and catalytic component of RNA-induced silencing complex (Argonaute 2, AGO2) in autism spectrum disorders: Yes, or no? *Meta Gene, 19*, 193–198.
224. Bond, A. M., Vangompel, M. J., Sametsky, E. A., Clark, M. F., Savage, J. C., Disterhoft, J. F., et al. (2009). Balanced gene regulation by an embryonic brain ncRNA is critical for adult hippocampal GABA circuitry. *Nature Neuroscience, 12*(8), 1020–1027.
225. Lipovich, L., Dachet, F., Cai, J., Bagla, S., Balan, K., Jia, H., et al. (2012). Activity-dependent human brain coding/noncoding gene regulatory networks. *Genetics, 192*(3), 1133–1148.
226. Hammond, S. M. (2015). An overview of microRNAs. *Advanced Drug Delivery Reviews, 87*, 3–14.
227. Mohr, A. M., & Mott, J. L. (2015). Overview of microRNA biology. *Seminars in Liver Disease, 35*(1), 3–11.
228. Toma, C., Torrico, B., Hervas, A., Salgado, M., Rueda, I., Valdes-Mas, R., et al. (2015). Common and rare variants of microRNA genes in autism spectrum disorders. *The World Journal of Biological Psychiatry, 16*, 376–386.
229. DeWitt, J. J., Hecht, P. M., Grepo, N., Wilkinson, B., Evgrafov, O. V., Morris, K. V., et al. (2016). Transcriptional gene silencing of the autism-associated long noncoding RNA MSNP1AS in human neural progenitor cells. *Developmental Neuroscience, 38*(5), 375–383.
230. Tang, J., Yu, Y., & Yang, W. (2017). Long noncoding RNA and its contribution to autism spectrum disorders. *CNS Neuroscience & Therapeutics, 23*(8), 645–656.
231. Wang, P., Mokhtari, R., Pedrosa, E., Kirschenbaum, M., Bayrak, C., Zheng, D., et al. (2017). CRISPR/Cas9-mediated heterozygous knockout of the autism gene CHD8 and characterization of its transcriptional networks in cerebral organoids derived from iPS cells. *Molecular Autism, 8*, 11.

Genomics of Autism

Khalid A. Fakhro

Abstract Autism spectrum disorder (ASD) is a heterogeneous condition affecting >1% of all children, characterized by impaired social interactions, repetitive behavior and a widely variable spectrum of comorbidities. These comorbidities may include developmental delay, gastrointestinal problems, cardiac disorders, immune and autoimmune dysregulation, neurological manifestations (e.g., epilepsy, intellectual disability), and other clinical features. This wide phenotypic heterogeneity is difficult to predict and manifests across a wide range of ages and with a high degree of difference in severity, making disease management and prediction of a successful intervention very difficult. Recently, advances in genomics and other molecular technologies have enabled the study of ASD on a molecular level, illuminating genes and pathways whose perturbations help explain the clinical variability among patients, and whose impairments provide possible opportunities for better treatment options. In fact, there are now >1000 genes that have been linked to ASD through genetic studies of more than 10,000 patients and their families. This chapter discusses these discoveries and in the context of recent developments in genomics and bioinformatics, while also examining the trajectory of gene discovery efforts over the past few decades, as both better ascertainment and global attention have been given to this highly vulnerable patient population.

Keywords Autism · Genomics · Next-generation sequencing · NGS · Copy number variations · Chromosomal abnormalities · Exome sequencing · Genome sequencing · Bioinformatics

K. A. Fakhro (✉)
Sidra Medicine, Doha, Qatar
e-mail: kfakhro@sidra.org

M. M. Essa, M. W. Qoronfleh (eds.), *Personalized Food Intervention and Therapy for Autism Spectrum Disorder Management*, Advances in Neurobiology 24,
https://doi.org/10.1007/978-3-030-30402-7_3

1 Introduction

Autism spectrum disorders (ASD) are a group of highly heterogeneous disorders characterized by repetitive behaviors, impaired social interactions and a wide spectrum of neurodevelopmental and physical comorbidities. While the overall prevalence of ASD is estimated at 1 in 68 children [1], this may not represent an increase in incidence as much as it represents a widening of the scope of disorders that fit under the ASD umbrella in an era of improving clinical ascertainment. As a spectrum disorder, ASD may present as an isolated set of symptoms or with multiple comorbidities, including but not limited to intellectual disability, developmental delay, epilepsy, gastrointestinal complications, cardiac problems, immune disorders, etc. [2]. This heterogeneity is apparent even in the settings of identical genetic backgrounds (e.g., monozygotic twins discordant for co-morbidities), underscoring the complexity of understanding ASD on the molecular level.

While ASD disproportionately affects males (male: female ratio of 3.4:1), the reasons for this remain poorly understood. In fact, the search for a molecular etiology is further complicated by the interplay of both genetic and environmental factors which together contribute to pathogenesis. Importantly, high incidence despite the significant impairment of reproductive fitness means that the cause of ASD is likely different among most cases of ASD, i.e., unrelated patients will rarely share the same mutation or even the same gene. Despite that, it is clear that ASD has a major heritable component, with siblings of ASD patients usually having a one in five risk (ten-fold higher than population average) of developing ASD themselves [3]. Further, concordance between monozygotic twins ranges from 30% to 99%, and the overall heritability is estimated between 0.5 and 0.8 [4–8].

This complex landscape has made the search for and discovery of genetic factors using traditional methods very difficult in the general population, mainly due to studies being underpowered to detect causal variants in small sample sizes. As expected by the limitations imposed by previous technologies, the majority of loci identified were in the form of chromosomal abnormalities, with few individual genes identified. The subsequent introduction of high-throughput microarrays enabled the investigation of smaller chromosomal abnormalities termed copy number variations (CNVs), and study of associations between common variants and the trait of interest. Signals detected from these three approaches (linkage, karyotyping and microarrays) rarely produced single-candidate genes; usually narrowing the search space to several kilobases or megabases, in which the search for causative genes was iterative and time-consuming. Alternatively, some mutations could be found by resequencing genes known to cause similar phenotypes in model organisms in a larger patient cohort.

Subsequent technological improvements led to the advent of next-generation sequencing (NGS), which has transformed the field profoundly, allowing the discovery of different classes of variations (e.g., single nucleotide variants (SNVs) and insertions/deletions (indels)) genome-wide. This enabled making discoveries from nuclear families in the absence of multiple affected or large pedigrees to establish linkage. The proliferation and accessibility of NGS technologies mean the bottleneck is no longer the ability to detect variants in a cost-effective manner, but the ability to amass cohorts that are large enough to capture a significant proportion of the genetic and phenotypic heterogeneity underlying ASD in the general population.

This chapter summarizes gene discovery in ASD in the pre- and post-sequencing era, explaining the significance of these discoveries, and framing them in the larger context of the potential impact that genomics will have on ASD diagnosis and care in the future.

2 Pre-NGS Era

2.1 Introduction

Next-generation sequencing (NGS) refers to advancements in technology that have enabled large-scale sequencing of many DNA fragments at the same time. These advancements have allowed the interrogation of variation at many loci in the genome in parallel at reasonable speed and cost, thus increasing the efficiency of genetic research. While NGS technologies began to appear in academic environments over a decade ago [9], their proliferation and adoption into the mainstream was not until more than a decade later. Importantly, while several different technologies appeared initially to compete for adoption, it was Illumina's short-read sequencing technology that was able to capture the biggest market segment with a combination of price point, accuracy and speed. And while today the price of a single human genome is around $1000, the price was significantly higher up until just a few years ago, rendering large-scale studies still very costly. This section covers discoveries made in autism genetics prior to the introduction to NGS to study this condition, whereas the next chapter will cover discoveries made when large-scale genomic assessment became increasingly affordable, in what is known as the post-"genomic" or post-NGS era.

2.2 Linkage Studies

Due to the paucity of multiplex or extended pedigrees with ASD, linkage approaches were not a robust approach to gene discovery in ASD. Nevertheless, numerous studies were performed (reviewed in [10]), revealing few loci in total. Notably, of these, only two were ever replicated successfully in an independent study. These include linkage to chromosome 7q35, containing the CNTNAP2 gene [11], and to chromosome 20p13, containing the four genes [12].

2.3 Association Studies

The development of cheap high throughput microarray genotyping technologies with higher marker density empowered a flurry of genome-wide association studies (GWAS) in a wide variety of human diseases. The discovery of markers by GWAS has two main limitations. First, they indicate loci with small effect size on the trait, sometimes increasing odds ratio by as little as 0.05 [13]. Second, they require very large sample sizes to have sufficient power to discriminate alleles between cases and controls. For ASD, large cohorts were not possible to amass for sufficient power, and therefore while GWAS were attempted over the past 10 years, few were ever replicated [12–17]. This is in contrast to studies of other neurodevelopmental conditions such as Schizophrenia for which cohorts could be amassed in the tens of thousands to discover tens of loci that replicate in independent cohorts. For ASD, only two loci have been implicated using GWAS to date, including a locus on 5p14.1 (containing the CDH9 and CDH10 genes), and another on 20p12.1 (MACRO2 gene) [15, 16]. Importantly, consistent with the genetic heterogeneity and the need for very large numbers, neither of these loci has been replicated.

2.4 Chromosomal Abnormalities Studies

The association of ASD with other syndromic comorbidities such as Fragile X and intellectual disability was a first indicator that chromosomal-level events could be underlying a subset of the condition. The concurrent evolution of microarray technologies introduced the ability to rapidly detect structural copy number variations in human genomes at scale. Together, karyotyping and CNV analysis have uncovered tens of chromosomal segments involved in ASD, including duplication of 15q [18], deletion of 22q11.2 [19, p., 200], deletion of 16p11.2 [20] and deletion of Xp22.3 [21]. In addition, several recurrent hotspots of de novo CNVs with ASD include duplications on 7q11.2 and deletions of 16p11.2, the latter also associated with schizophrenia [22, 23].

A key feature of CNVs is that they range widely in size from single-gene deletions to large regions encompassing tens to hundreds of genes. Consistent with multi-genic contribution to other phenotypes, patients with multiple de novo CNVs or large chromosomal abnormalities usually have more severe, syndromic phenotypes [24, p. 2], [25].

As cohort sizes grow, it has also been shown that CNV-affected genes predominantly comprise candidates from three key pathways, including neuronal signaling, synaptic function, and chromatin remodeling [26], [27, p. 201]. Together, these studies not only identify novel loci, but demonstrate that de novo CNVs are strongly associated with ASD [28] and that recurrent CNVs point to shared architecture with other diseases.

2.5 *Candidate Gene Resequencing Studies*

In contrast to the paucity of discovery from linkage studies, work from both other syndromes and CNV studies identified several candidate human ASD genes that could be screened by resequencing in larger cohorts. By assessing larger cohorts for mutations in these genes, the following genes were all found to harbor damaging point mutations in ASD subjects: MECP2 (Rett syndrome), TSC1 and TSC2 (tuberos sclerosis), CACNA1C (Timothy syndrome), NLGN3 and NLGN4 (X-linked mental retardation), and CNTNAP2 (7q35 deletion), SLC9A9 and BCKDK (epilepsy), etc. [29–31]. Other genes also discovered to carry rare damaging variants by resequencing include SHANK1, SHANK2, SHANK3, NRXN1 and NRXN3 [32–36].

2.6 *Conclusion*

In conclusion, the pre-NGS era relied mainly on candidate gene resequencing and association studies to link genes and loci to Autism. Unlike other monogenic disorders, linkage analysis was not a very successful approach to finding genes linked to Autism primarily due to the requirements of large pedigrees or multiple kindreds segregating the same locus, which are difficult to find considering the genetic heterogeneity underlying Autism and the detrimental effect it has on reproductive fitness.

3 The NGS Era

3.1 *Next-Generation Sequencing as a Tool to Study Genetic Disease*

Over the past decade, there have been numerous tools developed for next-generation sequencing (NGS) (Table 1). At its core, NGS may be broadly classified into two categories: whole genome sequencing (WGS), and targeted NGS. While the former is concerned with reading the entire content of an organism's genetic material, targeted NGS methods focus on selectively sequencing a group of genes ("gene panels"), usually selected based on specific selection criteria, e.g., having been identified in smaller cohorts or in animal studies, or genes within the same pathway(s) as well-established candidate disease genes. These gene panels may be customized to include any number of genomic fragments of interest, including, for example, *all* coding regions—commonly known as whole exome sequencing (WES). Typical WES experiments also capture flanking regulatory regions, enabling discovery of variants affecting splice junctions and untranslated promoter and downstream sequences [37].

Table 1 Comparison of different sequencing technologies

Technology		Maximum reads per run	Read length per run (bp)	Accuracy (single read not consensus)	Time per run	Cost per 1 million bases (in US$)	Advantages	Disadvantages	References
Sequencing by synthesis (SBS) Illumina	MiSeq series	25 million	2 × 150	99.9%	4–55 h	0.05 to $0.15	Potential for high sequence yield depends on sequencer model and desired application	System can be very expensive. Requires high concentrations of DNA	[9]
	NextSeq series	400 million			12–30 h				
	HiSeq series	5 billion			<1–3.5 days (HiSeq 3000/4000) 7 h–6 days (HiSeq 2500)				
	HiSeq X series	6 billion			<3 days				
Sequencing by semiconductor Thermo Fisher Scientific	Ion torrent	up to 80 million	200–400	98%	2–4 h	$1	Less expensive equipment. Fast	Homopolymer errors	
	Ion proton		125						
Single-molecule real-time sequencing Pacific Biosciences[a]	Sequel system PacBio RS II	10,000–15,000 maximum read length 65,000 bases	50,000 per SMRT cell, or 500–1000 megabases	87% single-read accuracy	30 min to 6 h	$0.13–$0.60	Longest read length. Fast. Detects 4 mC, 5 mC, 6 mA	Moderate throughput. Equipment can be very expensive	
Oxford Nanopore	MinION GridION X5 PromethION SmidgION	Dependent on library preparation, not the device, so user chooses read length (up to 200 kb reported)	Dependent on read length selected by user	~92–97% single read (up to 99.96% consensus)	Data streamed in real time. Choose 1 min to 48 h	$500–999 per flow cell, base cost dependent on experiment	Very long reads. Portable (palm-sized)	Lower throughput than other machines. Single read accuracy in 90 s	

[a]On 1 November 2018, Illumina entered into a purchase agreement to buy PacBio for ~$1.2B in total

Whole genome sequencing (WGS), on the other hand, covers both WES regions as well as non-coding and inter-genic regions. It is usually faster and more uniform because it does not require target panel capture, and thus can be performed with minimal sample preparation, resulting in sequences that are evenly distributed across all chromosomes. This distribution of sequencing *coverage* means that variants can be confidently assigned at average depth of sequencing as low as 20X. Conversely, whole-exome and other panel sequencing requires target enrichment and PCR amplification, often resulting in highly variable coverage profiles with some regions (e.g., repetitive elements or GC-rich content) being missed due to the technical limitations. Another important advantage of WGS's even coverage is the ability to discover genome-wide structural variants (including copy number variants). Given the number of human disorders (including ASD) in which structural variants play a significant role, a single test that can assess both large and small genomic variation is often cited a reason to use WGS despite its slightly higher cost vis-à-vis using a combination of microarray and WES for each patient.

3.2 Bioinformatics and Variant Interpretation

One important aspect of the NGS approach is the generation of large quantities of data, often requiring sophisticated computational tools (bioinformatics) to interpret. Specifically, bioinformatics pipelines share three major steps in common, irrespective of the NGS technology used: read alignment to a reference genome, variant calling versus the reference, and variant interpretation to determine pathogenic from benign variation.

Genetic variants may belong to several different classes, including: single nucleotide variants (SNVs, including single nucleotide polymorphisms (SNPs)), multi-nucleotide variants (MNVs, including small insertions and deletions (indels)), and structural variations (SVs, including copy number variations (CNVs)). For all three variant classes, a number of statistical considerations need to be taken into account to sort out likely true positive variants from noise, including: depth of sequencing, sequencing quality, the number of times mutations are observed, and the likelihood that such a change is true rather than an artifact of sequencing [38]. Importantly, the joint steps of read alignment and variant calling may themselves introduce error into the experiment, e.g., for fragments coming from highly repetitive genomic segments [39, 40].

The most challenging aspect in bioinformatics pipeline is variant interpretation—the step where tens to hundreds of variants may require in-depth manual scrutiny to determine putative effect on disease. Robust variant interpretation requires a well-annotated reference genome (for both coding and non-coding elements) and a large number of control individuals to accurately discriminate putative disease causing variants from population-specific polymorphisms (that may rarely appear in public databases because inadequate numbers of population-matched controls are available) [41–45].

3.3 NGS Suitability in Routine Clinical Care

As NGS technologies become more widely adopted in academic hospital settings, there is a growing need to establish gold-standard pipelines to allow for genomics to enter routine clinical testing [46, 47]. While some guidelines do exist, especially for diagnostic laboratory settings, these guidelines vary widely and currently still require orthogonal validation before they are deemed actionable [46, 48]. The role of a clinical-grade pipeline is primarily to demonstrate processing and interpretation in a highly reproducible manner, thus ensuring disease management is not compromised from this approach [47]. These steps, however, are non-trivial—they would need to account for influences on data quality and sources of error, for example, sample prep using protocols, sequencing instruments and batch effects, ensuring all genes in a panel are adequately captured, errors in sequencing chemistry and noise from the sequence alignment and variant calling steps.

Further, these tasks scale in complexity with the number of samples being studied and the databases from which annotations are being drawn. Of key consideration, for example, is the large number of variant sites produced per NGS run (three to four million per genome). Amongst these, hundreds or thousands of variants would be considered variants of unknown significance (VUS) whose interpretation and relevance to health and disease is completely unknown [46, 49]. In many cases, the recruitment of parents and siblings could help with sorting through these variants, but still tens to hundreds remain "private" variants with unknown function. For NGS to be adopted in routine care, clinical platforms must deal with such cases systematically, bearing in mind not to discard these variants because they may have future value as the genome is better annotated in the academic literature. Moreover, clinical platforms should take into consideration the constantly evolving annotations of genes, e.g., >200 new genes and hundreds of variants are being linked to diseases each year [50–53], and thus variant sharing as part of consortia may mitigate the absence of variants in the publication record. Such considerations need to be taken into account when designing clinical NGS pipelines, to ensure that genetic testing of patients is accurate, reproducible and safe. Only by controlling for these factors in a statistically robust framework would it be possible to ensure reproducibility and standardization, thereby enabling precision in data interpretation in disease settings.

4 Successful Application of NGS to Autism Spectrum Disorder

4.1 Sample Size and Cohort Considerations

The evolution of NGS thus enables the assessment of single families and single cases at a rate not performed before. The biggest challenge lies in discriminating rare alleles from population-specific polymorphisms, a challenge that can only be adequately addressed by sequencing a large enough number of both patients and of ethnic/population-matched controls. This is especially important in the setting of high genetic heterogeneity, where it is unlikely to find individuals sharing mutations in the same gene, let alone the same pathogenic variant. In recent studies, for example, sample sizes of >2000 families were required to identify recurrent gene and copy number regions shared between individuals [20, 23, 27, 54].

Conversely, in settings with high consanguinity, the approach of finding recessive variants is boosted by the usual availability of affected siblings or additional cousins (in multiplex families) who share the same homozygous mutations in candidate genes. However, the identification of recessive genes causing ASD has been limited so far by the type of families studied—mostly outbred simplex families with unrelated parents. In rare cases where families with multiple affected siblings were identified, they were found to have two different de novo causative variants rather than the same recessive variant [55]. This is expected due to the high levels of genetic heterogeneity underlying ASD.

However, this presents an important opportunity for consanguineous populations attempting ASD studies, with some initial reports reporting promising results [56–59].

4.2 *Exome and Genome Sequencing*

Due to the paucity of studies in ASD families from areas of high consanguinity, recessive variants causing ASD have only been identified so far in the following genes: *AMT*, *BCKDK*, *CNTNAP2*, *PEX7*, *SLC9A9*, *SYNE1*, *VPS13B*, *PAH* and *POMGNT1* [60]. In contrast to the few recessive genes discovered, the vast majority of families studied to date have been outbred, in which single affecteds (simplex) are born to unaffected, unrelated parents. In these cases, the genetic architecture is usually driven by de novo mutations, or rare inherited variants; however, even when multiple siblings are found in the same family, they are sometimes found to harbor separate de novo variants, stressing the importance of this type of variation in ASD etiology. There are approximately 800 genes affected by de novo variants in ASD (not counting genes within de novo chromosomal abnormalities) [10]. Altogether, the contribution of de novo mutations in ASD is estimated to be between 15% and 25% [61].

In 2012, four groups published concurrent studies using exome-sequencing to identify de novo gene disrupting variants in ASD patients. Only approximately 20 of these genes were recurrently hit across the cohorts, including: ADNP, ANK2, ARID1B, BCL11A, CACNA2D3, CHD8, CUL3, DSCAM, DYRK1A, GRIN2B, KDM5B, KDM6B, KMT2C, KMT2E, KMT5B, NCKAP1, PHF2, RIMS1, SCN2A, SYNGAP1, TBR1, TCF7L2, TNRC6B, and WAC [22, 54, 62–67]. However, the majority of the other genes identified were singletons (only observed in a single patient without replication), but their potential role in ASD was supported by their impacting critical pathways in neurological development, such as cognition, synaptic formation, and regulation of transcription of brain-specific genes [54, 67, 68].

In addition to de novo variants affecting genes directly, more recent studies have found an enrichment of de novo and private disruptive mutations in DNAse I hypersensitivity sites in regions close to some of the genes that have been implicated in ASD [69]. This indicates that ASD genes disruption is not only through mutations that may alter function but also those that may alter gene regulation. Notably, one recurrent theme across most studies is that de novo point-mutations are predominantly paternal in origin, with the rate of de novo mutations increasing with paternal age.

Despite advances in WES, less than 10% of known patients receive a genetic diagnosis in ASD. This is far lower than the solve rate of neurodevelopmental disorders as a whole, with diagnoses above 30% of cases. Nevertheless, the utility of WES and WGS extends beyond simple diagnostic value, as it has allowed the identification of genes underlying more complex syndromes shared with ASD. For example, *de novo* mutations in the SWI-SNF-related gene *ADNP* causes a syndromic form of ASD with unique facial dysmorphism [70], whereas mutations in the NatA complex subunit *NAA15* cause a syndromic form of ASD with multiple congenital anomalies including craniofacial, neuromuscular, and cardiac complications [71]. Such families may not have been individually identified a priori to share similar genetic underpinnings prior to the advent of NGS technologies, which now enable patients to be stratified more precisely based on their genetic abnormality rather than phenotypic variability.

Importantly, accurate genetic diagnosis is critical for determining potential therapeutic approaches for patients with ASD. One area where the impact has been most recognizable is in ASD related to branched-chain amino acid deficiencies, for example, branched-chain keto-acid dehydrogenase kinase deficiency, in which mutations in *BCKDK* were identified. These mutations cause loss of function of BCKDK, itself a repressor of branched-chain amino acid degradation, and therefore patients have a concurrent deficiency of BCAAs. In murine models, supplementation of knockout mice with BCAAs significantly improved their neurologic phenotypes, suggesting that patients with BCKDK mutations may benefit from dietary supplementation of BCAAs to counteract the elevated degradation caused by the genetic mutation [31]. Similarly, ASD patients with a wide variety of comorbidities (e.g., sleep disorders, seizures and metabolic and immune abnormalities) have been found to have imbalances in compounds that could easily be rectified by dietary intervention, such as folate, carnitine, cobalamin, etc. [72]. More recently, one case–control randomized trial has demonstrated that supplementation with essential fatty acids, carnitine, digestive enzymes, and a hypoallergenic diet (e.g., gluten, soy, and casein-free) all improved ASD symptoms, including non-verbal IQ and nutritional status [73]. Therefore, as more cohorts of patients continue to be evaluated at the genomic and epidemiological levels, the future of ASD research can lead to novel tools and therapies that improve stratification and clinical management of patients based on their genomic information, ushering in an era of personalized medicine for ASD.

5 Conclusion

Autism spectrum disorders (ASD) are a heterogeneous group of disorders characterized by clinical comorbidities and extreme genetic heterogeneity. While a lot has been achieved to understand the molecular and genetic etiology, there is still a long way to go to understand how perturbations in genes ultimately lead to an ASD phenotype. Importantly, further studies may also reveal genetic markers of the development of different physical comorbidities, which can help in patient stratification and early intervention in cases predicted to become severe. Thus, as future studies are conceived, they ought not to only focus broadly on ASD patients across the entire spectrum, but also on important concepts such as data sharing and collaborations to aid in the interpretation, and eventually treatment of ASD across the globe.

References

1. Elsabbagh, M., et al. (2012). Global prevalence of autism and other pervasive developmental disorders. *Autism Research* - Wiley Online Library. [Online]. Retrieved January 28, 2019, from https://onlinelibrary.wiley.com/doi/full/10.1002/aur.239
2. Aldinger, K. A., et al. (2015). Patterns of risk for multiple co-occurring medical conditions replicate across distinct cohorts of children with autism spectrum disorder. *Autism Research* - Wiley Online Library. [Online]. Retrieved January 28, 2019, from https://onlinelibrary.wiley.com/doi/full/10.1002/aur.1492
3. Ozonoff, S., et al. (2011). Recurrence risk for autism spectrum disorders: A Baby Siblings Research Consortium study. *Pediatrics, 128*(3), e488–e495.
4. Bailey, A., et al. (1995). Autism as a strongly genetic disorder: Evidence from a British twin study. *Psychological Medicine, 25*(01), 63.
5. Rosenberg, R. E., Law, J. K., Yenokyan, G., McGready, J., Kaufmann, W. E., & Law, P. A. (2009). Characteristics and concordance of autism spectrum disorders among 277 twin pairs. *Archives of Pediatrics & Adolescent Medicine, 163*(10), 907.
6. Hallmayer, J. (2011). Genetic heritability and shared environmental factors among twin pairs with autism. *Archives of General Psychiatry, 68*(11), 1095.
7. Sandin, S., Lichtenstein, P., Kuja-Halkola, R., Larsson, H., Hultman, C. M., & Reichenberg, A. (2014). The familial risk of autism. *JAMA, 311*(17), 1770.
8. Colvert, E., et al. (2015). Heritability of autism spectrum disorder in a UK population-based twin sample. *JAMA Psychiatry, 72*(5), 415.
9. Goodwin, S., McPherson, J. D., & McCombie, W. R. (2016). Coming of age: Ten years of next-generation sequencing technologies. *Nature Reviews Genetics, 17*(6), 333–351.
10. Geschwind, D. H., & State, M. W. (2015). Gene hunting in autism spectrum disorder: On the path to precision medicine. *The Lancet Neurology, 14*(11), 1109–1120.
11. Alarcón, M., Cantor, R. M., Liu, J., Gilliam, T. C., & Geschwind, D. H. (2002). Evidence for a language quantitative trait locus on chromosome 7q in multiplex autism families. *The American Journal of Human Genetics, 70*(1), 60–71.
12. Weiss, L. A., et al. (2009). A genome-wide linkage and association scan reveals novel loci for autism. *Nature, 461*(7265), 802–808.
13. Anney, R., et al. (2012). Individual common variants exert weak effects on the risk for autism spectrum disorders. *Human Molecular Genetics, 21*(21), 4781–4792.
14. Ma, D., et al. (2009). A genome-wide association study of autism reveals a common novel risk locus at 5p141. *Annals of Human Genetics, 73*(3), 263–273.

15. Wang, Z., Gerstein, M., & Snyder, M. (2009). RNA-Seq: A revolutionary tool for transcriptomics. *Nature Reviews. Genetics, 10*(1), 57–63.
16. Anney, R., et al. (2010). A genome-wide scan for common alleles affecting risk for autism. *Human Molecular Genetics, 19*(20), 4072–4082.
17. Chaste, P., et al. (2015). A genome-wide association study of autism using the Simons simplex collection: Does reducing phenotypic heterogeneity in autism increase genetic homogeneity? *Biological Psychiatry, 77*(9), 775–784.
18. Bundey, S., Hardy, C., Vickers, S., Kilpatrick, M. W., & Corbett, J. A. (1994). Duplication of the 15q11-13 region in a patient with autism, epilepsy and ataxia. *Developmental Medicine and Child Neurology, 36*(8), 736–742.
19. Fine, S. E., et al. (2005). Autism spectrum disorders and symptoms in children with molecularly confirmed 22q11.2 deletion syndrome. *Journal of Autism and Developmental Disorders, 35*(4), 461–470.
20. Weiss, L. A., et al. (2008). Association between microdeletion and microduplication at 16p11.2 and autism. *New England Journal of Medicine, 358*(7), 667–675.
21. Thomas, N. S., et al. (1999). Xp deletions associated with autism in three females. SpringerLink. [Online]. Retrieved January 28, 2019, from https://link.springer.com/article/10.1007%2Fs004390050908
22. Sanders, S. J., et al. (2011). Multiple recurrent de novo CNVs, including duplications of the 7q11.23 Williams Syndrome Region, are strongly associated with autism. *Neuron, 70*(5), 863–885.
23. Kumar, R. A., et al. (2008). Recurrent 16p11.2 microdeletions in autism. *Human Molecular Genetics, 17*(4), 628–638.
24. Marshall, C. R., et al. (2008). Structural variation of chromosomes in autism spectrum disorder. *The American Journal of Human Genetics, 82*(2), 477–488.
25. Pinto, D., et al. (2010). Functional impact of global rare copy number variation in autism spectrum disorders. *Nature, 466*(7304), 368–372.
26. Pinto, D., et al. (2014). Convergence of genes and cellular pathways dysregulated in autism spectrum disorders. *The American Journal of Human Genetics, 94*(5), 677–694.
27. Sanders, S. J., et al. (2015). Insights into autism spectrum disorder genomic architecture and biology from 71 risk loci. *Neuron, 87*(6), 1215–1233.
28. Levy, D., et al. (2011). Rare de novo and transmitted copy-number variation in autistic spectrum disorders. *Neuron, 70*(5), 886–897.
29. Strauss, K. A., et al. (2006). Recessive symptomatic focal epilepsy and mutant contactin-associated protein-like 2. *The New England Journal of Medicine, 354*(13), 1370–1377.
30. Morrow, E. M., et al. (2008). Identifying autism loci and genes by tracing recent shared ancestry. *Science, 321*, 218.
31. Novarino, G., et al. (2012). Mutations in BCKD-kinase lead to a potentially treatable form of autism with epilepsy. *Science, 338*(6105), 394–397.
32. Moessner, R., et al. (2007). Contribution of SHANK3 Mutations to Autism Spectrum Disorder. *The American Journal of Human Genetics, 81*(6), 1289–1297.
33. Kim, H.-G., et al. (2008). Disruption of neurexin 1 associated with autism spectrum disorder. *The American Journal of Human Genetics, 82*(1), 199–207.
34. Berkel, S., et al. (2010). Mutations in the SHANK2 synaptic scaffolding gene in autism spectrum disorder and mental retardation. *Nature Genetics, 42*(6), 489–491.
35. Berkel, S., et al. (2012). Inherited and de novo SHANK2 variants associated with autism spectrum disorder impair neuronal morphogenesis and physiology. *Human Molecular Genetics, 21*(2), 344–357.
36. Vaags, A. K., et al. (2012). Rare deletions at the neurexin 3 locus in autism spectrum disorder. *The American Journal of Human Genetics, 90*(1), 133–141.
37. Bamshad, M. J., et al. (2011). Exome sequencing as a tool for Mendelian disease gene discovery. *Nature Reviews Genetics, 12*(11), 745–755.
38. Van der Auwera, G. A., et al. (2013). From FastQ data to high confidence variant calls: The Genome Analysis Toolkit best practices pipeline. *Current Protocols in Bioinformatics, 43*, 11.10.1–11.1033.

39. Pirooznia, M., et al. (2014). Validation and assessment of variant calling pipelines for next-generation sequencing. *Human Genomics, 8*(1), 14.
40. Pirooznia, M., Goes, F. S., & Zandi, P. P. (2015). Whole-genome CNV analysis: Advances in computational approaches. *Frontiers in Genetics, 06*, 138.
41. The 1000 Genomes Project Consortium, et al. (2015). A global reference for human genetic variation. *Nature, 526*(7571), 68–74.
42. The 1000 Genomes Project Consortium, et al. (2015). An integrated map of structural variation in 2,504 human genomes. *Nature, 526*(7571), 75–81.
43. Exome Aggregation Consortium, et al. (2016). Analysis of protein-coding genetic variation in 60,706 humans. *Nature, 536*(7616), 285–291.
44. Durbin, R. M., et al. (2010). A map of human genome variation from population-scale sequencing. *Nature, 467*(7319), 1061–1073.
45. Fakhro, K. A., et al. (2016). The Qatar genome: A population-specific tool for precision medicine in the Middle East. *Human Genome Variation, 3*, 16016.
46. For the Working Group of the American College of Medical Genetics and Genomics Laboratory Quality Assurance Committee, et al. (2013). ACMG clinical laboratory standards for next-generation sequencing. *Genetics in Medicine, 15*(9), 733–747.
47. Moorthie, S., Hall, A., & Wright, C. F. (2013). Informatics and clinical genome sequencing: Opening the black box. *Genetics in Medicine, 15*(3), 165–171.
48. Gargis, A. S., et al. (2012). Assuring the quality of next-generation sequencing in clinical laboratory practice. *Nature Biotechnology, 30*(11), 1033–1036.
49. Green, R. C., et al. (2013). ACMG recommendations for reporting of incidental findings in clinical exome and genome sequencing. *Genetics in Medicine, 15*(7), 565–574.
50. Bamshad, M. J., et al. (2012). The Centers for Mendelian Genomics: A new large-scale initiative to identify the genes underlying rare Mendelian conditions. *American Journal of Medical Genetics Part A, 158A*(7), 1523–1525.
51. Yang, Y., et al. (2013). Clinical whole-exome sequencing for the diagnosis of Mendelian disorders. *New England Journal of Medicine, 369*(16), 1502–1511.
52. Jurgens, J., et al. (2015). Assessment of incidental findings in 232 whole-exome sequences from the Baylor–Hopkins Center for Mendelian Genomics. *Genetics in Medicine, 17*(10), 782–788.
53. On behalf of the IRDiRC Consortium Assembly, et al. (2017). The International Rare Diseases Research Consortium: Policies and guidelines to maximize impact. *European Journal of Human Genetics, 25*(12), 1293–1302.
54. O'Roak, B. J., et al. (2014). Recurrent de novo mutations implicate novel genes underlying simplex autism risk. *Nature Communications, 5*(1), 5595.
55. Yuen, R. K. C., et al. (2017). Whole genome sequencing resource identifies 18 new candidate genes for autism spectrum disorder. *Nature Neuroscience, 20*(4), 602–611.
56. Rajab, A., et al. (2015). Recessive *DEAF1* mutation associates with autism, intellectual disability, basal ganglia dysfunction and epilepsy. *Journal of Medical Genetics, 52*(9), 607–611.
57. Al-Mubarak, B., et al. (2017). Whole exome sequencing reveals inherited and de novo variants in autism spectrum disorder: A trio study from Saudi families. *Scientific Reports, 7*(1), 5679.
58. Leblond, C. S., et al. (2019). Both rare and common genetic variants contribute to autism in the Faroe Islands. *NPJ Genomic Medicine, 4*, 1.
59. Jamra, R. (2018). Genetics of autosomal recessive intellectual disability. *Medizinische Genetik, 30*(3), 323–327.
60. Yu, T. W., et al. (2013). Using whole-exome sequencing to identify inherited causes of autism. *Neuron, 77*(2), 259–273.
61. Ronemus, M., Iossifov, I., Levy, D., & Wigler, M. (2014). The role of de novo mutations in the genetics of autism spectrum disorders. *Nature Reviews Genetics, 15*(2), 133–141.
62. Iossifov, I., et al. (2012). De novo gene disruptions in children on the autistic spectrum. *Neuron, 74*(2), 285–299.
63. Neale, B. M., et al. (2012). Patterns and rates of exonic de novo mutations in autism spectrum disorders. *Nature, 485*(7397), 242–245.

64. Sanders, S. J., et al. (2012). De novo mutations revealed by whole-exome sequencing are strongly associated with autism. *Nature, 485*(7397), 237–241.
65. Willsey, A. J., et al. (2013). Coexpression networks implicate human Midfetal deep cortical projection neurons in the pathogenesis of autism. *Cell, 155*(5), 997–1007.
66. Dong, S., et al. (2014). De novo insertions and deletions of predominantly paternal origin are associated with autism spectrum disorder. *Cell Reports, 9*(1), 16–23.
67. O'Roak, B. J., et al. (2012). Multiplex targeted sequencing identifies recurrently mutated genes in autism spectrum disorders. *Science, 338*(6114), 1619–1622.
68. The DDD Study, et al. (2014). Synaptic, transcriptional and chromatin genes disrupted in autism. *Nature, 515*(7526), 209–215.
69. Turner, T. N., et al. (2016). Genome sequencing of autism-affected families reveals disruption of putative noncoding regulatory DNA. *The American Journal of Human Genetics, 98*(1), 58–74.
70. Helsmoortel, C., et al. (2014). A SWI/SNF-related autism syndrome caused by de novo mutations in ADNP. *Nature Genetics, 46*(4), 380–384.
71. Cheng, H., et al. (2018). Truncating variants in NAA15 are associated with variable levels of intellectual disability, autism spectrum disorder, and congenital anomalies. *American Journal of Human Genetics, 102*(5), 985–994.
72. Frye, R. E., & Rossignol, D. A. (2016). Identification and treatment of pathophysiological comorbidities of autism spectrum disorder to achieve optimal outcomes. *Clinical Medicine Insights. Pediatrics, 10*, 43–56.
73. Adams, J. B., et al. (2018). Comprehensive nutritional and dietary intervention for autism spectrum disorder-a randomized, controlled 12-month trial. *Nutrients, 10*(3), pii: E369.

Neuropsychopathology of Autism Spectrum Disorder: Complex Interplay of Genetic, Epigenetic, and Environmental Factors

Ranjana Bhandari, Jyoti K. Paliwal, and Anurag Kuhad

Abstract *Autism spectrum disorder (ASD)* is a complex heterogeneous consortium of pervasive development disorders (PDD) which ranges from atypical autism, autism, and Asperger syndrome affecting brain in the developmental stage. This debilitating neurodevelopmental disorder results in both core as well as associated symptoms. Core symptoms observed in autistic patients are lack of social interaction, pervasive, stereotyped, and restricted behavior while the associated symptoms include irritability, anxiety, aggression, and several comorbid disorders.

ASD is a polygenic disorder and is multifactorial in origin. Copy number variations (CNVs) of several genes that regulate the synaptogenesis and signaling pathways are one of the major factors responsible for the pathogenesis of autism. The complex integration of various CNVs cause mutations in the genes which code for molecules involved in cell adhesion, voltage-gated ion-channels, scaffolding proteins as well as signaling pathways (PTEN and mTOR pathways). These mutated genes are responsible for affecting synaptic transmission by causing plasticity dysfunction responsible, in turn, for the expression of ASD.

Epigenetic modifications affecting DNA transcription and various pre-natal and post-natal exposure to a variety of environmental factors are also precipitating factors for the occurrence of ASD. All of these together cause dysregulation of glutamatergic signaling as well as imbalance in excitatory: inhibitory pathways resulting in glial cell activation and release of inflammatory mediators responsible for the aberrant social behavior which is observed in autistic patients.

In this chapter we review and provide insight into the intricate integration of various genetic, epigenetic, and environmental factors which play a major role in the pathogenesis of this disorder and the mechanistic approach behind this integration.

Keywords Autism · Genetic · Epigenetic · Environment · Glutamate · Microglial activation

---

R. Bhandari · J. K. Paliwal · A. Kuhad (✉)
Pharmacology Research Laboratory, University Institute of Pharmaceutical Sciences, UGC-Centre of Advanced Study, Panjab University, Chandigarh, India

M. M. Essa, M. W. Qoronfleh (eds.), *Personalized Food Intervention and Therapy for Autism Spectrum Disorder Management*, Advances in Neurobiology 24,
https://doi.org/10.1007/978-3-030-30402-7_4

1 Introduction

Autism spectrum disorder (ASD) is an intricate syndrome that has been characterized by a diverse category of neuropsychiatric disorders which affect the brain during its development. The symptoms usually become manifested by the time the child is three years old. The Diagnostic and Statistical Manual of Mental Disorders, 5th Edition (DSM-V), encompasses various disorders under the umbrella of ASD such as autistic disorder, Asperger's syndrome, pervasive developmental disorder not otherwise specified (PDD-NOS), childhood disintegrative disorder, and Rett syndrome [1]. According to reports by the US Center for Disease Control and Prevention, there has been an increase in the prevalence of autism in children from 1 in 88 in 2010 to 1 in 68 in 2014. World Health Organization (WHO) reports of 2017 show that one in 160 children are suffering from ASD worldwide [2]. In India approximately 23 out of every 10,000 children have autism, according to reports from India's first rigorous estimate of autism prevalence [3]. ASDs show significant skewness in occurrence in boys as compared to girls by a ratio of 4:1 [4–7]. ASD shows high heritability of approximately 80% [8]. It is a genetically heterogeneous disorder in which 10–25% of the cases suffer from a genetic disorder either occurring as a result of single gene mutation such as fragile X syndrome, tuberous sclerosis (TSC), Rett syndrome, Angelman syndrome or aberrations in chromosomes and imbalances in genome. These are also known as syndromic ASDs. Remaining cases for which the causes are unknown come under the category of idiopathic ASDs [9, 10]. Idiopathic autism is also sometimes referred to as non-syndromic ASD and includes those cases where autism is the primary diagnosis and is not secondary or part of an existing condition caused by a genetic syndromes like Rett syndrome, Fragile X syndrome, tuberous sclerosis, or Angelman syndrome.

ASD is characterized by core as well as associated symptoms. The primary symptoms associated with autism are lack of individual's ability to communicate and engage in social interaction. Such patients also show restriction, repetition, pervasiveness, and stereotypy in their behavior and activities. Along with the primary symptoms, ASD also results in symptoms such as aggressive behavior, irritable nature, and anxiety as well as some accompanied disorders like attention deficit hyperactivity disorder (ADHD), epilepsy, and disorders affecting the processing of sensory information [1, 11–14].

In this chapter we bring forth the role of the complex amalgamation of genetic factors and factors affecting environment in the pathogenesis of autism spectrum disorders (ASDs). This article has been divided into sections and sub-sections highlighting the following:

(a) Genetic factors involved in ASD
(b) Epigenetic modifications of genes involved in ASD
(c) Pre-natal and post-natal environmental factors affecting neurodevelopment and leading to ASD
(d) Mechanism of integration of genetic, epigenetic, and environmental factors affecting synaptic transmission and glial activation

2 Pathogenetic Mechanisms: Role of Genetic, Epigenetic, and Environmental Factors

2.1 Genetic Factors in ASD

ASD is a polygenic disorder. Its estimated heritability is approximately 80% [8, 15, 16]. Our focus here is on understanding the genetic mechanisms underlying the disease pathology.

The first section dealing with *genetic factors* involved in pathogenesis of ASD has a sub-section on monogenic autism, i.e., autism which results from mutation in a single gene. Under monogenic autism we explain copy number variations (CNVs), which are one of the reasons for the occurrence of monogenic autism. These can occur in the genes involved in synapse formation, such as genes for the cell adhesion molecule, scaffolding proteins, cytoskeletal proteins, signaling pathways, ion channels, and cell signaling molecules. Mutation in these genes disrupts regulatory or coding regions which contribute to the pathogenesis of ASD by affecting synapse formation, plasticity as well as transmission. All such genes have been discussed in the sub-section on *Genes regulating synaptic plasticity and transmission* covering mutations in several genes at the synapse which code for cell adhesion molecules (Neuroligin, Neurexins, Contactins, NrCAM, and Cadherins), ion-channels (sodium channel ($Na_v1.1$), calcium channel ($Ca_v1.2$), and potassium channel (Kir4.1 and BKCa)), scaffolding proteins (Shank, SHANK2, and SHANK3) and cytoskeletal proteins, and the abnormalities in the PI3K/AKT/mTOR pathway or the RAS pathway which lead to the aberrant synaptic protein synthesis. This in turn leads to the development of autism. Figure 1 depicts various genetic factors which have been described below.

2.1.1 Monogenic Autism

Autism which results from mutation in a single gene is known as monogenic autism. Monogenic mutation can be a part of either syndromic autism or part of non-syndromic autism which might involve many genes or environmental factors as likely causes of its etiology [17]. Monogenic mutations can be a result of mutations in genomic DNA or *de-novo* mutations as a result of copy number variations (CNVs) as well as single nucleotide polymorphisms (SNPs). These mutations, which occur as a result of CNVs or SNPs, can either affect synaptic transmission and plasticity by causing mutations in genes encoding for molecules causing cellular adhesion, voltage-gated ion-channels, scaffolding proteins, cytoskeletal proteins, and PTEN and mTOR signaling pathways [18] or occur in genetic syndromes such as Fragile X syndrome (FMR1 gene [19, 20], tuberous sclerosis (TSC1 and TSC2 genes) [21]) and Angelman syndrome (UBE3A gene [22, 23]) by affecting a particular gene resulting in the development of major intellectual disability disorders of which autism is a part.

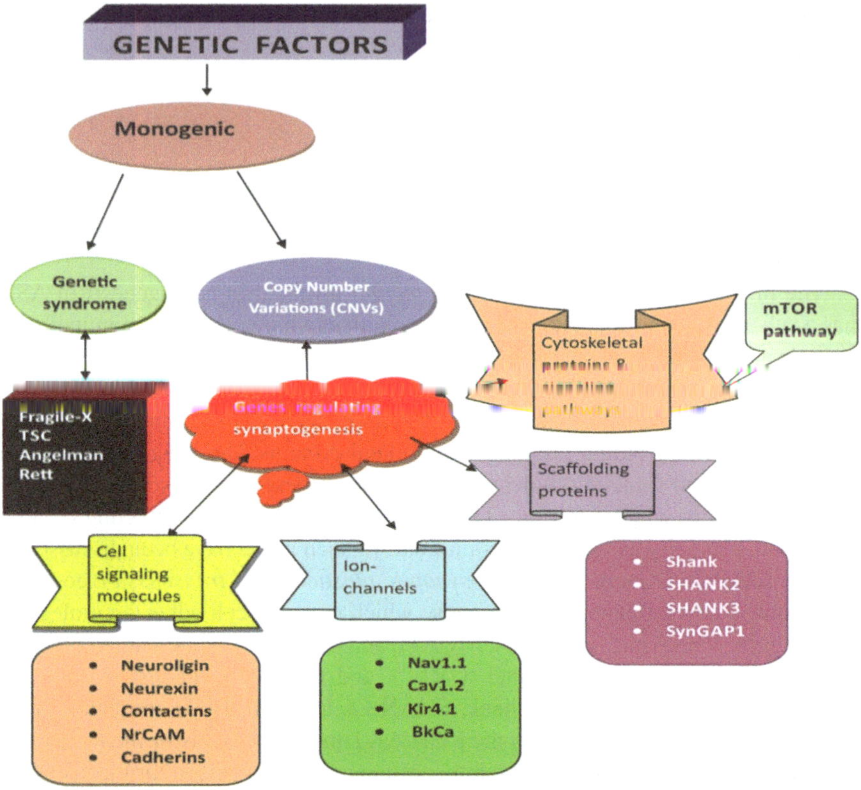

Fig. 1 Classification of the genetic factors associated with ASD. This figure illustrates the genes responsible for the development of ASD associated phenotype. $Na_v1.1$—sodium channel type 1, $Ca_v1.2$—voltage dependent L-type Ca^{2+} channel, Kir4.1, and $BKCa^{2+}$—potassium channels

Fragile X syndrome (FXS) occurs as a result of abnormal CGG repeats in the FMRP gene located at a specific locus, i.e., Xq27.3. The abnormal repeats result in transcription silencing of this gene by abnormal methylation and ultimately reduce cerebral proteins [20, 24]. Tuberous sclerosis (TSC) is a disorder affecting autosomes in which tumor like lesions occur in multiple organs. TSC occurs as a result of mutations in either of the two genes, i.e., TSC1 or TSC2 which are present at two different loci, i.e., 9q34 and 16p13.3, respectively [21, 25]. The incidence of ASD in FXS is 18–33%, while in TSC it is 25–60% [21]. Angelman syndrome is a neurodevelopmental disorder leading to severe intellectual disability, seizures, speech impairment, abnormal gait, cognitive impairment, and stereotypic mannerisms typical of autistic patients [26, 27]. Primary cause of this syndrome is the deletion or inactivation of UBE3A gene present on maternally inherited copy of chromosome 15 [22, 23].

Copy number variations are the structural variations in the DNA segments. The size range of these is from 50 base pairs to several megabases. The size of the affected DNA segment has to be larger than 1 kb to be classified as copy number variant. They occur as a result of mutation, i.e., deletion, duplication, insertion, inversion or complex recombination of the segments of chromosomes [28, 29]. They are a lesser or large number of copies of a particular sequence and occur with a frequency of 6–10% in ASD population. CNVs are known as de-novo CNVs if they are present spontaneously in offspring germ cells or somatic cells or are considered to be inherited if they arise in parental germ cells. In somatic cells, de-novo CNVs influence the degree of functional deficits during development [30]. These CNVs play an integral part in the development of ASD depending on the location at which they are present on the chromosomes, like duplication at loci 15q13 or microdeletion at loci 16p11.2, 15q11-13, 22q11.2, and 1q21.1 [31, 32]. CNVs are involved in monogenic forms of autism. However, there are also multigenic CNVs [33]. Menashe et al. [34] have indicated in their comprehensive analysis of CNV data from genetic database known as AutDB that there are 11 CNV loci either present on chromosome 16, 22, 15, 13, 9, 4, 3 or 1. These loci contain a total of 166 genes and cover 15,610 kb of genome. Out of these 11 CNV loci, there are seven multigenic loci and four monogenic loci. Sanders et al. [35] analyzed copy number variations via exome analysis in a cohort consisting of simplex ASD families known as Simons Simplex Collection. It consists of 2591 families. This exome analysis replicated the initial findings indicating strong connection with ASD and confirmed the identification of six specific loci considered as "risk loci." These risk loci are present on chromosome 1, 3, 7, 16, 15, 13, and 22. When data was added from prior published data from the Autism Genome Project (AGP), Simons Simplex Collection as well as Autism Sequencing Consortium (ASC), it indicated that genes within small de-novo deletions include high-effect ASD risk genes, whereas large de-novo CNVs include only medium-effect risk genes. Thus, Sanders et al. [35], through their exome analysis of de-novo CNVs in ASD, brought forth that 50% of de-novo CNVs arbitrate risk of ASD. There are 200 CNV loci and 800 genes considered as risk loci and risk genes, respectively, which are susceptible to de-novo mutations and 11% of these cases in the cohort are a result of de-novo mutations. They identified 71 independent ASD risk loci which comprised of 65 risk genes and six risk loci of ASD, many of which are either targeting chromatin biology or are involved in synaptogenesis. Iossifov et al. [36] performed exome analysis in 2500 simplex families each of which had a child suffering from ASD. They compared affected vs unaffected siblings. Exome analysis was done at Cold Spring Harbor Laboratory, Yale School of Medicine and University of Washington. The results indicated that 13% of de-novo missense mutations and 42% of de-novo likely gene disrupting mutations contributed to 12% and 9% of diagnoses, respectively. Their results estimated that 40% of the simplex families were at high-risk. There was no role of de-novo mutations in high-risk families. Hence, de-novo mutations contributed to 60% of Simon Simplex Collection families.

CNVs in the genes which code for scaffolding proteins, cytoskeletal proteins, signaling pathways, ion channels, and cell signaling molecules cause disruption of the regulatory as well as coding regions that are responsible for pathogenesis of disease [9].

Synaptic dysfunction occurs at multiple levels in ASD. There are a number of genes regulating various cell adhesion molecules, ion-channels, neurotransmitter receptors, scaffolding and cytoskeletal proteins and genes affecting the PTEN/TSC/mTOR signaling pathways. These mutations are part of idiopathic or non-syndromic autism where gene mutation is not part of the syndrome and can be a result of epigenetic changes occurring as a result of various environmental factors. Genetic mutations in the upstream targets of mTOR pathway, i.e., PTEN, TSC1/2, NF1 and the loss of fragile-X-mental retardation protein (FMRP) as a result of mutation of FMRP gene leads to hyperactivity of the mTORC1-eIF4E pathway. This is associated with the development of ASD as these genes are responsible for the integrity and function of the synapse [37, 38]. Mutation in this pathway, when affecting the TSC1/2 or FMRP, leads to syndromic ASD as it is a part of tuberous sclerosis or Fragile X syndrome. Figure 2 depicts various sites at the synapse which are involved in the mutation of the genes coding for synaptic proteins that regulate synaptogenesis and synaptic transmission.

2.1.2 Cell Adhesion Molecules in Synaptogenesis

The following cell adhesion molecules are involved in synaptogenesis. The mutation in the genes encoding for these molecules is one of the plausible causes of the pathogenesis of ASD:

a. *Neuroligins*: Neuroligins (type 1 membrane proteins required for cell adhesion) are found on the post-synaptic membrane of the glutamatergic or GABAergic synapse and are responsible for its formation [41, 42]. Neuroligins (NLGNs) act as ligands for pre-synaptic neurexins [43]. Neuroligins are involved in signaling by differentiation, maturation and stabilization of synaptic components [44] (Fig. 2).

Fig. 2 (continued) ion-channels like sodium, potassium, and calcium channels, cytoskeletal proteins (HOMER and CORTACTIN) and receptors such as NMDAR, TyrK, mGluR present post-synaptically as well as signaling pathways (such as mTOR pathway) responsible for the maintenance of synaptic plasticity and function. In the pre-synaptic assembly, the ion-channels like calcium channels are closely present to synaptic vesicles like glutamate and BDNF vesicles. This entire pre-synaptic assembly is in close proximity to the post-synaptic assembly with the help of cell adhesion molecules like neuroligin and neurexin located at post and pre-synaptic locations, respectively, for the maintenance of synapse function. At post-synaptic sites, neuroligin binds to the cytoplasmic scaffolding proteins like PSD-95, PSD-93, SAPAP which are enriched in post-synaptic densities for the maintenance of architecture of synapse. HOMER and SHANK along with PSD-95 serve as a link between the post-synaptic receptors like mGluR, TrkB and their downstream signaling components. Shanks are the post-synaptic scaffolding proteins which are enriched in post-synaptic densities (PSDs) and are responsible for stabilizing the PSD-95/SAPAP/SHANK/HOMER complex. It also interacts with NMDA receptors and actin regulatory proteins, Cortactin. Thus, Shank and HOMER stabilize the post-synaptic density as well as receptors like NMDAR, AMPAR, mGluR. TrKB (tyrosine kinase receptor) triggers the RAS/RAF/MAPK pathway and PI(3)K/AKT/mTORC1 pathway by the action of BDNF on them. These are responsible for the triggering of CAP-dependent translation and elongation of mRNA. Mutations in the genes coding for these synaptic components like cell adhesion molecules such as (Neuroligin, Neurexin, Cadherins, Contactins, NrCAM), cytoskeletal and scaffolding proteins like Shank, HOMER, PSD-95, ion-channels like calcium channel, sodium channel, and potassium channel, and the downstream and upstream signaling components of mTOR pathway like (mutations in PTEN, NF-1, TSC1/2, eIF4E, and 4E-BP) will lead to improper protein-protein interactions resulting in abnormal synaptic activity, loss of neuronal function precipitating ASD phenotype [39, 40]

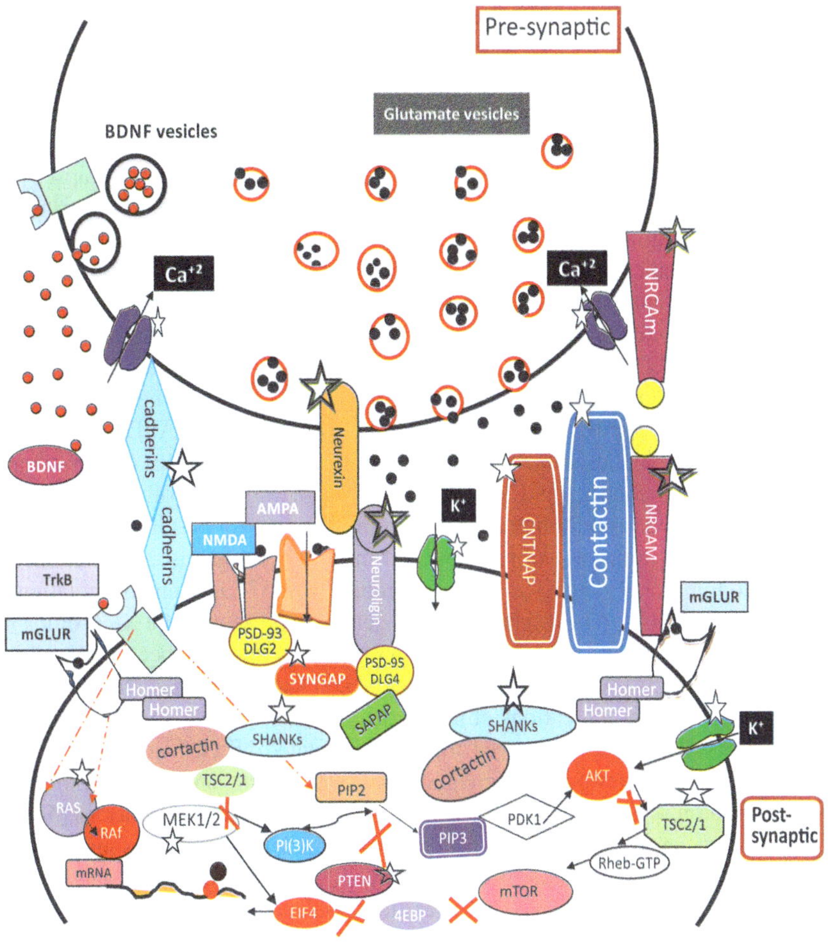

Genes which are mutated
Neuroligins: NlgN3, NLGN4X
Neurexins:
NRXN1
Contactins:
CNTN3, CNTN4
CNTNAP2
NRCAM

CADHERINS:
CDH18
PCDH9, PCDH10
Ion-channels:
SCN1A (Nav1.1)
CACNA1C (Cav1.2)
KCNMA1
KCNJ10(Kir4.1)

Scaffolding proteins:
Shank3, Shank2
SynGAP1
Signaling Pathways:
PTEN, TSC1 ,TSC2, RAS,RAF, MEK1/2

Fig. 2 Illustrates the various sites at the synapse which are involved in the genetic mutation of the genes coding for various synaptic proteins regulating synaptogenesis and synaptic transmission. Star marked indicates genes which are mutated. There are varieties of synaptic proteins which have been linked with ASD. This figure is a depiction of an assembly of various pre- as well as post-synaptic proteins present at the glutamatergic or GABAergic synapses. These synapses consist of both pre-synaptic and post-synaptic elements that work in partnership to maintain synaptic functionality and plasticity. Any disruption occurring in this synaptic assembly will result in neurodevelopmental disorder like ASD. This glutamatergic synapse is composed of various components such as neurexin, neuroligin, cadherins, and NrCAM, scaffolding proteins (such as PSD-95 and SHANK),

NLGNs consist of an extracellular domain which bears homology to acetylcholinesterases. But NLGNs lack esterase activity due to lack of critical residues in their active site. Neuroligins are present in humans, drosophila, C. *elegans*, and mice [45–48]. NLGN genes involved in humans include NLGN1, NLGN2, NLGN3, NLGN4X, and NLGN4Y [46, 49]. Neuroligin 1 is present at the glutamatergic synapses. Neuroligin 2 is present at the GABAergic and other inhibitory synapses [43]. Neuroligin 3 is expressed in CNS neurons in the excitatory and inhibitory synapses as well as in the glial cells in rats and mice. Neuroligin 4 is present at the postsynaptic site of glycinergic neurons [41, 42, 46, 50, 51]. Genes encoding NLGN3 and NLGN4 are present in humans on the X-chromosome. Humans also have a NLGN gene on Y chromosome complementary to NLGN4 gene which is referred to as NLGN4Y or NLGN5 gene [52]. The neurodevelopmental stage requires appropriate balance of excitatory and inhibitory (E/I) inputs known as the E/I ratio. This imbalance of the E/I ratio is responsible for the pathogenesis of ASD [53].

Experimental studies clearly indicate the involvement of NLGNs in synapse maturation, maintenance, and function [9, 42, 54, 55]. Neuroligin dysfunction occurs in ASD as a result of mutation in the genes coding for NLGN [31, 56]. These mutations can be point mutations, frameshift mutations, missense, or deletions and can lead to impaired synaptic cell adhesion molecules [57]. In humans, mutations of NLGN3 or NLGN4X gene can occur. These genes are present on the X-chromosome and autism occurring as a result of mutation in these genes has been referred to as X-linked autism [58]. Various studies conducted on knockout mice, Drosophila, and C. *elegans* gave clear indications of the involvement of NLGN3 and NLGN4 mutation in ASD. It has been observed that when mutated NLGN3 gene R451C was inserted in mice, it showed impairment in social interaction, enhancement of synaptic transmission across inhibitory synapses and spatial learning disabilities [58–60]. This mutation caused the mutant protein to be retained in the endoplasmic reticulum and thus its binding to neurexin was reduced. Similarly, NLGN3 and NLGN4 knockout mice showed behavioral alterations and vocalizations similar to those observed in ASD patients [9, 61]. NLGN1 and 2 also result in the reduction of inhibitory synapse transmission.

b. Neurexins: These are pre-synaptic cell adhesion molecules that help glue neurons together. Their extracellular domain interacts with neuroligins. They are encoded by three genes, i.e., Nrxn1, Nrxn2, and Nrxn3 in humans [62]. Each neurexin gene is regulated by α and β promoters resulting in α-Nrxn1-3 and β-Nrxn1-3 [63]. Intracellular domains are identical for α and β neurexins but have different extracellular domains. β-Neurexins act as receptors for neuroligin. C-terminus of the short intracellular section of both types of neurexins binds to synaptotagmins and PDZ domains CASK and MINT [*PDZ* is an acronym combining the first letters of three proteins—post-synaptic density protein (PSD95), Drosophila disc large tumor suppressor (Dlg1), and zonula occludens-1 protein (zo-1)]. These interactions are responsible for forming connections between intracellular synaptic vesicles and fusion proteins [64]. Nrxns are required for synaptic maintenance and functions and these interact with neuroligins to form a synapse between two neurons [65]. *De-novo* copy number variations occurring in Nrxn1 are also one of the plausible reasons for ASD phenotype [31, 66]. There have been several experimental studies describing phenotypes resulting from knocking out these genes. Some of these studies such as Dachtler et al. [67] have reported that as a result of mutation in NRXN2 gene, (Nrxn2α) protein is deficient and knockout experiments in mice have shown the development of core symptoms of autism like deficits in social interaction, social novelty preferences and social preference over novel object. These mice also showed anxiety-like and pervasive behavior as indicated by their experiments on elevated plus-maze and open-field tests. They also showed less tendency to explore novel objects. As the Nrxn2α protein is necessary for synaptogenesis, it affected the pre-synaptic vesicle release, as indicated by the results showing a decrease in protein expression in excitatory and inhibitory transmission across synapse. Rabaneda et al. [68] have reported in *Cell Reports* that transgenic mice, which they produced by causing mutation in neurexin-1β gene, resulted in phenotype similar to that of autism patients, as indicated by various specific behavioral tests like tests for sociability, repetitive self-grooming and olfactory habituation–dishabituation. These mice also showed alteration in transmission across the glutamatergic synapse, indicating the important role of neurexin in synaptogenesis and showing how impairment of circuits was responsible for the development of autistic phenotype. Etherton et al. [69] have also indicated that knock out of Neurexin-1α gene resulted in significant reduction of the frequency of occurrence of miniature excitatory post-synaptic currents (mEPSCs) in mice. These mice also showed significant behavioral alterations specific to the autistic phenotype like repetitive self-grooming, pervasive behavior and decrease in prepulse inhibition. Wang et al. [70] have indicated through their study in Chinese population that variants of neurexin gene family are linked to ASD. Thus, neurexins are involved in synaptogenesis and synaptic transmission, and any mutation in the genes encoding neurexins could lead to the development of autistic phenotype.

c. Contactins: Contactins (CNTN) are proteins which belong to the super family of immunoglobulin (Ig) which are widely and exclusively expressed in the CNS. They do not have transmembrane or an intracellular domain but are attached to the cell membrane with the help of a glycophosphatidylinositol (GPI)-anchor [71]. They are involved in synaptogenesis, plasticity and myelination [10, 72]. There is no report in the literature about the role of CNTN1 and CNTN2 in ASD. There are reports, e.g., [73] about the deletion in the gene CNTN3. Human BIG-2/contactin-4 and contactin-3 gene present at loci 3p25-p26 are involved in ASD. Mutations in CNTN4 and CNTN3 genes alter the formation and function of synapse [74, 75]. There are reports about the occurrence of mutations in genes CNTN4, CNTN5, and CNTN6 and the involvement of CNTN4 in ASD [31, 76, 77]. Mutations concerning CNTN4 are either a result of removal of the distal end of one of the arms of chromosome 3 or are a result of copy number variations (CNVs) of the gene, as reported by Pinto et al. [78] using results of patient data of the Autism Genome Project. Results from the Utrecht patient cohort study have also revealed the occurrence of *de-novo* CNVs and gene deletions in CNTN5 as well as CNTN6 genes [78, 79].

d. Neuronal cell adhesion molecule: NrCAM is a protein encoded by the NrCAM gene which is important for cellular interactions occurring during the development of the brain and is known to be expressed in the cerebral cortex, striatum, hippocampus, and cerebellum in rodents [80–82]. It is capable of interacting with Contactin-1, Contactin-2, and Neurofascin. Patients with mutations like deletions in the gene located on chromosome 7q which encode for NrCAM or having SNPs are presented with autism [83, 84] and show obsessive-compulsive behavior [85]. There have been studies on mice which are deficient in NrCAM. Such mice have shown reduction in the size of cerebellum [86] as well as anxiety like behavior [87]. NrCAM-null mice have been used by Moy et al. [88] to study the role of these neuronal cell adhesion molecules in the pathogenesis of autism. They observed that the loss of NrCAM resulted in the lack of sociability, reversal learning and sensorimotor gating deficit in male mice while female mice, in addition to the above, showed anxiety, motor in-coordination and deficit in acquisition task of the Morris water maze. Another study by Mohan et al. [89] has indicated the involvement of the neuronal cell adhesion molecule in regulating density of dendritic spines of pyramidal neurons in the medial and visual frontal cortex. These results provide strong indication that these neuronal cell adhesion molecules have a role in autism. NrCAM has been also linked to the disruption of visual acuity in autism spectrum disorders (ASDs) as its deletion has been known to alter the thalamocortical connections to the visual cortex [90].

e. Cadherins: Cadherins (CDH) are Ca^{2+} dependent glycosylated transmembrane proteins which are involved in the cell adhesion [91]. These are required for the proper synapse function and formation [92]. CDHs are involved in the intracellular signaling pathways [93, 94] and are associated with various neuropsychiatric disorders [94, 95]. Cadherin dysfunction has been associated with ASD [96]. The genes involved in ASD belong to the classes of both classic cadherins as well protocadherins [97]. The genes belonging to the class of classic cadherins associated with ASD are CDH15 [98], CDH5 [99], CDH8, CDH9, and CDH10 [100, 101] which are present on chromosome 5; also CDH13, which is present at 16q23 loci, is involved in the pathogenesis of ASD because of the deletion of the gene occurring at this loci [32, 102]. The genes encoding protocadherins, which are involved with ASD as well as some of its associated symptoms such as epilepsy, are PCDH10 [73], PCDH19 [103, 104], and PCDHb [99]. They have been identified after several gene mapping and linkage studies in cohorts. Thus, this class of cell adhesion molecules (CAMs) are involved in the pathogenesis of autism.

Having provided an overview of the mutations in the genes encoding for cell adhesion molecules, we will now describe, in brief, the mutations involving voltage gated sodium, potassium, and calcium channels. These present post- and presynaptically, respectively, and are responsible for the conduction of action potential.

Ion Channels Associated with the Development of ASD Phenotype: Ion-channels are the transmembrane proteins essential for regulation and maintenance of neuronal excitation by conduction of action potentials. The three major ion-channels associated with the development of ASD phenotypes are sodium channel type 1 ($Na_v1.1$), voltage-dependent L-type Ca^{2+} channel ($Ca_v1.2$), and potassium channels (Kir4.1

and BKCa^{2+}). The genes associated with these channels are SCN1A (Na$_v$1.1), CACNA1C (Ca$_v$1.2), KCNMA1 (BKCa^{2+}), and KCNJ10 (Kir4.1) [9, 105, 106]. A brief description of their association with ASDs is given below:

a. Sodium channel: Studies have identified the presence of SCN1A haplosufficiency to be responsible for the development of autism [107]. Mutation in SCN1A gene results in haplosufficiency resulting in altered Na$^+$ channel activity, impaired GABA signaling, concurrent occurrence of Dravet syndrome (myoclonic epilepsy in infants) and behavioral abnormalities associated with ASD like impaired social interaction, ADHD, and cognitive deficits [108, 109]. SCN2A gene is also strongly associated with ASD and encodes for Na$_v$1.2 sodium channel. This sodium channel is expressed at the site in neurons where the action potential is initiated and at the nodes of Ranvier during the early developmental stage. Ben-Shalom et al. [110] have characterized the effect of 11 de-novo mutations in SCN2A gene occurring in ASD patients through exome sequencing. They have indicated that SCN2A gene mutations such as nonsense, missense, and frameshift mutations (11 de-novo mutations) result in inhibition of the function of Na$_v$1.2 sodium channel. Mutations in the SCN2A gene in the cortical pyramidal excitatory neurons result in deficit of neuronal excitability in the early stage of development of the human brain. Sanders et al. [111] have also indicated two de-novo loss-of-function (LoF) variants in SCN2A gene among 200 ASD families from the Simons Simplex Collection. De Rubeis et al. [112] used exome sequencing analysis to show rare coding variation in 3871 patients suffering from ASD and 9937 ancestry-matched or paternal controls from the Autism Sequencing Consortium (ASC) and have indicated that SCN2A gene had a 99% chance of being a true autism gene.

b. L-Type voltage gated Ca^{2+} channels are responsible for intracellular signaling and the activation of various transcription factors. CACNA1C and CACNA2D4 are two genes coding for this channel. CACNA1C codes for the alpha-1 subunit of voltage gated Ca^{2+} channel and CACNA2D4 is a member of the alpha-2/delta subunit. Mutations in the genes coding for these channels result in neurological complications, developmental deficits, and autism [113, 114]. Copy number variations (CNVs) of CACNA1C and CACNA2D4 result in 2p:12p chromosomal translocation resulting in the deletion of 12p. This leads to the removal of one copy of CACNA1C and CACNA2D4 genes as indicated by FISH analysis. CNVs for these genes occur in the 12pter-p25.2 region of the chromosome and involves 27 genes causing ASD like phenotype in such patients [115]. The G406R codon is affected by mutation of CACNA1C resulting in Timothy syndrome associated with autism [116]. There are other genes which code for this channel and have been implicated in ASD like CACNB2, CACNA1H, CACNA1G. Missense mutations in CACNB2 have been identified in families affected by ASD causing calcium channel dysfunction [117]. Some of the other mutations are associated with gene CACNA1H and CACNA1G. Splawski et al. [118] have reported that mutant allele of gene CACNA1H is present in patients affected by ASD. On the other hand, the CACNA1G gene contains single nucleotide polymorphisms (SNPs) at the locus 17q11-q21, a susceptible region for ASD [119].

c. Kir4.1 and BKC^{a2+}: Potassium channels Kir4.1 and BKC^{a2+} are found in neurons and astrocytes. Mutations in these channels are involved in pathogenesis of ASD. Kir4.1 is the inward rectifier potassium channel extensively found in the brain, encoded by the gene KCNJ10 present on chromosome 1q22 and its missense mutation results in the upregulation of Kir4.1. It has been found in locus coeruleus neurons of MECP knockout mice and is involved in pathogenesis of Rett syndrome with concurrent occurrence of autism and epilepsy. Mutations of KCNJ10 gene occur in the regions of pR18Q and pV84M and are responsible for the replacement of amino acid residues [120]. Another potassium channel associated with ASD is BKC^{a2+} encoded by gene KCNMA1. Mutations in both these genes result in the development of poor social interaction, intellectual disability, haploinsufficiency, and other phenotypic characteristics associated with ASD due to excessive ion-channel activity as these channels play an important role in the regulation of neuronal function and synaptic plasticity [105, 121].

Mutations such as copy number variations (CNVs) or single nucleotide polymorphisms (SNPs) can occur in the genes encoding for scaffolding proteins and are responsible for the development of autistic phenotype. These have been discussed below.

Scaffolding Proteins in the Pathology of ASD: Scaffolding proteins are post-synaptic proteins like SHANK. These create a link between the post-synaptic receptors and the downstream components of signaling pathways along with the regulation of actin cytoskeleton [122]. There are varieties of scaffolding proteins which are involved in the regulation of various signal transduction pathways like KSR in RAS/MAPK pathway, HOMER (calcium signaling), BCl-10 (MAPK pathway), and SHANKs (Fig. 2).

Shanks are a family of scaffolding proteins which are a major component of PSD (post-synaptic density) of all excitatory glutamatergic synapses in the brain [123]. Shanks have three genes encoding for their proteins, i.e., SHANK1, SHANK2, and SHANK3. These are expressed at the brain and the products of these genes are found at the synapse [124]. Shank proteins are rich in PSDs and they interact with other proteins like PSD-95, HOMER, and SAPAP via proline rich regions, PDZ domain, ankyrin repeats, SH3 domains and alpha motif domains forming PSD-95/SAPAP/SHANK/HOMER complex [123]. Shanks also interacts with NMDA and metabotropic glutamate receptors (mGluRs) forming and stabilizing NMDA/PSD-95/GKAP complex besides its interaction with Cortactin [124]. Shank proteins are responsible for shaping the morphology of dendritic spines using HOMER. The role played by mutant SHANK genes in the pathology of ASD is described below:

(a) SHANK1: It was first reported by Hung et al. [125] that SHANK1 has a role in the development of the ASD phenotype. It was observed that SHANK1 mutations resulted in reduced spine density and smaller and thinner PSD by targeting exons 14-15/PDZ in mutant mice. It was also observed that AMPA receptor-mediated transmission was reduced. SHANK1 mutation resulted in reduced social sniffing, reduction in reciprocal social interaction, increased repetitive behavior, reduction in scent-marking behavior and enhanced special learning and memory task and aber-

rations in cognitive function and alteration in expression of BDNF in hippocampus [40, 125–127].

(b) SHANK2: SHANK2 gene mutations resulted in reduction in NMDA receptor function and basal synaptic transmission. Mutations in SHANK2 gene also resulted in reduction in spine density and social behavior, reduction in ultrasonic vocalization, increased anxiety-like behavior, impaired nesting behavior, reduction in spatial learning/memory and occurrence of repetitive behavior, hyperactivity and enhanced motivation leading to dysfunction of striatal neurons in mutant mice [40, 128, 129].

(c) SHANK3: Mutations of SHANK3 were the first to be described and are considered to be the best-characterized SHANK mutation in humans for ASD [40, 124, 130, 131]. De-novo point mutations occurring in SHANK3 gene are of the type, missense, frameshift, deletion, and splice site mutations. They are involved in synaptogenesis and synaptic plasticity. The SHANK3 gene is also involved in 22q13.3 deletion syndrome along with other genes and is responsible for the development of autism-like behavior in this syndrome [132]. Several studies conducted on SHANK3 mutant mice targeting several exons like exons 4-9, 4-7, 13-16, and 11 indicated reduction in synaptic protein like GKAP, GluA1, GluN2A, SAPAP3, HOMER1, PSD-93, and increase in GluN2B. It has been observed that spine volume, thickness of PSD and dendritic spines are reduced [133, 134]. Synaptic physiology indicated a reduction in synaptic transmissions and long-term potentiation (LTP) [135–137]. Social behaviors in knockout mice indicated reduced interest in social sniffing, impairment of social interaction, reduced interest in novel mice, repetitive behavior, reduced ultrasonic vocalizations (USVs) and impairment of acquisition, reversal and novel object recognition [40]. Duplications have also been observed in SHANK3 resulting in hyperactivity suggesting that SHANK3 is also responsible for maintenance of excitatory and inhibitory balance in neurons [138].

The pathogenetic mechanisms of ASD also involve mutations in the genes coding for the downstream or upstream components of the signaling cascades such as mTOR pathway or the RAS pathway. We will now discuss the various components of these signaling cascades, mutation of which can lead to the development of autistic phenotype.

2.1.3 Signaling Pathways mTOR, PI3K/AKT, and RAS/MAPK

Genes coding GTPase-activating proteins (GAPs), guanosine exchange factors (GEFs), and tuberous sclerosis (TSC1/TSC2) are responsible for the structural and functional regulation of actin filaments. Alteration in the structure and function of the cytoskeleton, specifically microtubules and actin filaments, is a hall-mark of autism. Mutation in TSC1 or TSC2 results in the alteration of the structure of microtubules, actin filaments and dendritic spine structure (Fig. 3). KATNAL2 is another microtubule-associated protein that is considered to play an important role in the development of ASD [9, 140].

A description of the abnormalities in the components of the mTOR/PI3K/AKT/RAS/MAPK pathways and their role in ASD has been detailed below:

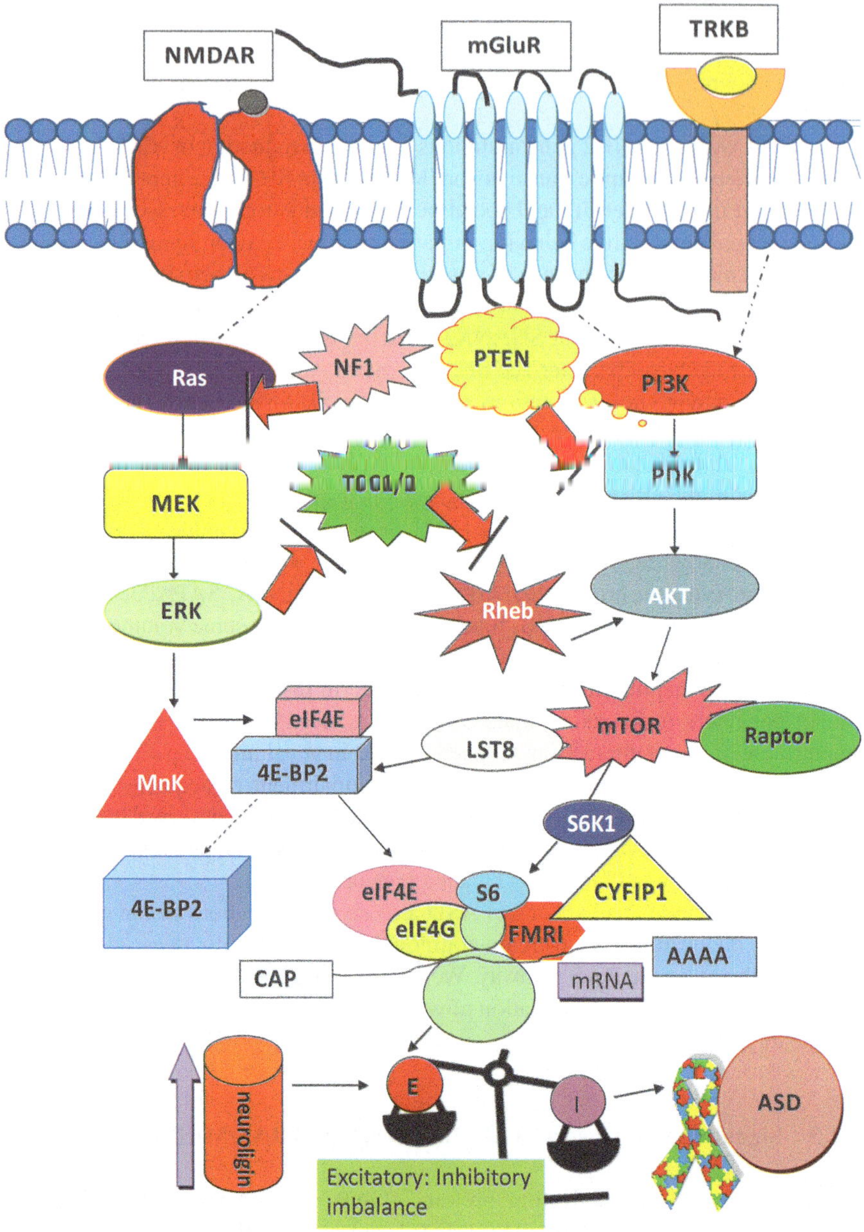

Fig. 3 Pathogenesis of mTOR signaling in autism spectrum disorders [139]: Akt, also known as PKB, protein kinase B; ASD: autism spectrum disorder; 4E-BP2, eIF4E-binding protein2; E/I: excitation/inhibition; ERK: extracellular signal regulated kinase; FMRP: fragile X mental retardation protein; CYFIP1: FMRP interacting protein1; MEK: mitogen-activated protein/ERK kinase; MnK: transcription factor activated by ERK/MAPK; mGluR: metabotropic glutamate receptor; mTOR: mammalian target of rapamycin ; mTORC1: mTORcomplex1; NF1: neurofibromatosis1; NLGN: neuroligin; NMDAR: NMDA receptor; PDK: phosphoinositide dependent kinase; PI3K: phosphoinositide-3kinase; PTEN: Phosphatase and tensin homolog; Raptor: regulatory associated protein of mTOR; LST8: mammalian lethal with SEC13 protein 8 Rheb: Ras homologen enriched in brain; TRKB: receptor tyrosine kinase; S6K1: p70 ribosomal S6 kinase1: TSC1/2: tuberous sclerosis complex 1 and 2. Star marked indicates genes which are mutated

The mammalian target of rapamycin (mTOR) is a serine/threonine protein kinase which plays an important role in synaptogenesis as many components of signaling cascade are present at the synapse [37, 141–143]. The normal physiological role of mTOR signaling cascade is that it gets inputs from different receptors such as NMDAR, AMPA-R, mGluR, and tyrosine receptor kinase B (TRKB) which are necessary for the maintenance of synaptic plasticity [144–148]. Various ligands of these receptors trigger two major signaling cascades like RAS-MAPK pathway and PI(3)K pathway, both of which converge to regulate the activity of mTORC1 complex consisting of a catalytic subunit mTOR, a regulatory associated protein of mTOR (Raptor), mammalian lethal with SEC13 protein 8 (LST8) which is an mTOR complex subunit and some non-core components [141, 149–151]. The downstream signaling cascade of mTOR performs phosphorylation of 4E-BP and p70S6 kinases by mTOR. The disassociation of 4E-BP, which are initiation factor binding proteins, from the initiation factor, i.e., eIF4E, results in CAP-dependent translation and mRNA elongation [152, 153].

PI3K converts PIP_2 to PIP_3, for which phosphatase and tensin homolog (PTEN) are the negative regulators [154]. PDK1 activates AKT or protein kinase B which inhibits TSC1/2, an mTOR protein synthesis regulator. TSC1/2 is a GTPase activating protein for GTPase Rheb and is responsible for the conversion of Rheb-GTP to inactive Rheb-GDP for the regulation of protein synthesis during extreme conditions like oxidative stress and DNA damage. Rheb-GTP activates mTOR to induce mRNA translation [155].

MAPK/ERK also have an inhibitory role on TSC1/2. The RAS/MAPK cascade is a protein kinase activated by RAS. RAS activates RAF which further phosphorylates MEK and MAPK [139]. MAPK is also known by the name of extracellular signal regulated kinases (ERK) and it phosphorylates MNK which is a transcription factor. It also phosphorylates CREB, responsible for phosphorylation of eIF4E [139]. MAPK is responsible for phosphorylation of 40S ribosomal protein S6 kinase (S6K1) which phosphorylates ribosomal protein S6 [139, 156].

Abnormalities in the PI3K/AKT/mTOR pathway or the RAS pathway lead to aberrant synaptic protein synthesis and the development of autism [38, 157–159]. Approximately 8–10% of the cases of ASD are due to a faulty mTOR signaling pathway [37, 160]. In ASD, these downstream and upstream signaling cascades are affected by various gene mutations. Mutations in the genes coding for proteins which are negative regulators of mTORC1 activated by PI3K pathway are *TSC1/TSC2, NF1, PTEN*. It leads to the development of syndromic ASD [37, 160–163]. Phosphatase and the tensin homolog (PTEN) gene are present on chromosome 10 and have a significant role in brain development [154, 155]. The rate of ASD occurrence in patients who suffer from PTEN gene mutation is 1–17% [164], while 1–2% of ASD cases are a result of mutations in the gene coding for PTEN and TSC1/TSC2 [164, 165]. PTEN gene mutation results in the overgrowth of synapse and has been associated with the development of macrocephaly, social behavioral deficits, anxiety, and learning deficits [166, 167]. TSC1 or TSC2 gene mutations result in tuberous sclerosis complex (TSC) and the rate of prevalence of autism is 25–50% [168–170]. These precipitate hypertrophy of the synapse resulting in macrocephaly, seizures, and learning deficits [171–173].

mTORC1, a principle regulator of downstream translation, promotes translation of mRNA by the recognition of initiation factor eIF4E. This 4E-BPs (eIF4E-binding protein 2) is the initiation factor-binding protein which inhibits the initiation of translation. It is the phosphorylation of binding proteins by mTORC1 which promotes eIF4E release and initiates mRNA translation [37, 142, 174, 175]. If a single nucleotide polymorphism (SNP) occurs in the promoter region of the initiation factor, i.e., eIF4E, it enhances the promoter activity of eIF4E resulting in the increased translation of neuroligins and disruption of excitatory/inhibitory balance leading to the development of ASD phenotype [176–179]. The protein 4E-BP2 also competes with eIF4G for eIF4E binding and inhibits translation. Its removal can also lead to enhanced CAP dependent translation [158, 159].

It was also observed by Gkogkas et al. [158] in their study on Eif4ebp2 knockout mice that if the gene which codes for 4E-BP2 is deleted, it leads to the development of the autistic phenotype. This occurs as a result of increased translation of NLGNs due to over expression of eIF4E leading to altered synaptic excitation/inhibition ratio resulting in ASD. The treatment of these mice with eIF4E inhibitor resulted in reduction in the levels of NLGNs.

Mutations in the FMR1 gene led to the lack of production of fragile X mental retardation protein (FMRP) in turn leading to fragile X syndrome. This could be a result of hyperactivity of mTORC1–eIF4E pathway [161, 180–182] suggesting that downstream mTOR signaling might be one of the probable mechanisms in ASD. Approximately, 2–8% of the cases of autism are a result of fragile X Syndrome [183]. FMR1 gene contains CGG trinucleotide repeats which are subjected to mutations resulting in abnormal repeats ranging from 200 to 1000 times. This causes gene silencing and loss of FMRP protein [184]. CYFIP1 is a cytoplasmic and functional protein that works in partnership with FMRP. CYFIP1 is also known as the FMRP interacting protein 1. FMRP directly binds to CYFIP1, prohibiting eIF4E-dependent initiation and hence, affecting translation [185]. FMRP controls translation of target mRNAs at synapses and is responsible for impaired synaptic plasticity. Along with FMRP, this causes repression of protein synthesis in neurons. When TRKB receptors or mGluRs are activated, CYFIP1 is released from eIF4E and the translation begins. Thus, CYFIP1 inhibits local protein synthesis and favors actin remodeling [185, 186]. CYFIP1 is located at loci chr15q11.2, which is considered a hot spot for ASD. Any deletions or duplications in this region result in ASD [187]. The connection between FMRP, mTOR, and translation either involves CYFIP1 or S6K, which is an important substrate of this cascade, and can phosphorylate and regulate the mRNA binding activity of FMRP [188]. Animal studies have also indicated dysregulation of mTOR signaling in *Fmr1* KO [189, 190] suggesting that genetic changes in both downstream and upstream signaling cascades might be one of the probable reasons for the development of the ASD phenotype (Fig. 3).

In the next section, epigenetic modifications of genes involved in ASD will be discussed.

2.2 *Epigenetic Modifications of Genes Involved in ASD*

Epigenetic modifications share a deep relationship with ASD [191, 192]. Several genes involved in autism spectrum disorders undergo epigenetic modifications under environmental and other pathogenic influences, i.e., DNA methylation or post-translational modifications of histones [191, 193–195].

2.2.1 Environment and Epigenetics

Epigenetic modifications are deeply influenced by the environment in which the gene is present. Thus, epigenetics refers to the link between the susceptible genes and the environmental influences [192, 196]. Positive correlation has been established between the development of autism and pre-natal as well as post-natal exposure to air emissions like nitrogen dioxide, particulate matters, heavy metals like lead, nickel, arsenic, chromium, cobalt, cadmium, benzene, radiation, peroxisome proliferators, and tobacco smoke which lead to epigenetic modifications [197]. Pre-natal exposure to sodium valproate and other anti-epileptic medications and bisphenol A [198, 199] affect genes of various receptors such as oxytocin, estrogen, and vasopressin by causing epigenetic regulation of these genes resulting in the development of ASD phenotype [200–203]. There are other chemicals like PBDEs (polybrominated diphenyl ethers), exposure to which affects brain-derived neurotropic factor (BDNF), calcium-calmodulin kinase II (CAMKII), and GAP-43 (growth associated protein 43) which are essential for the function and survival of neurons and efficient synapse formation [204]. Such epigenetic modifications result in the development of ASD by altering the levels of serotonin, axon-thinning, and immune activation. Epigenetic modifications also occur as a result of post-natal exposure to valproic acid, estradiol, and citalopram [32]. Even maternal vitamin D deficiency during gestation is associated with an increased risk of development of ASD in children [205, 206]. It has been found that mothers having mid-gestational deficiency of vitamin D had more than a two-fold increase in the risk of developing ASD [207]. Histone modifications occurring via enzymes like histone acetyltransferases and deacetylases of the VDR (vitamin D receptor) have also been reported in the pathogenesis of ASD [205, 206, 208–210].

Epigenetic modifications such as DNA methylation or histone modifications can take place in the genes responsible for regulation of synaptic plasticity and transmission. These can either code for cell adhesion molecules, ion-channels, scaffolding proteins or are involved in signaling pathways.

2.2.2 DNA Methylation

Epigenetic modifications of genes such as methylation of DNA take place at several loci on the chromosomes. These epigenetic changes can potentially play a significant role in the development of ASD [191]. Several genes of interest are as follows:

OXTR, AFF2, NLGN3, NRXN1, SHANK3, AUTS2, SLC6A4, GABRB3, BCL2, Ubiquitin protein ligase E3A, and Reelin [195, 211–213]. DNA methylation of the promoter region of oxytocin receptor gene leads to the inhibition of this gene expression by blocking transcription [214, 215]. Epigenetic changes in oxytocin receptor lead to the development of ASD phenotypes as oxytocin regulates various social behaviors [200]. The genes responsible for synaptogenesis and regulation of synaptic plasticity also undergo epigenetic modifications. It has been observed that various genes that code for cell adhesion molecules like SHANK3 and neuroligins (NLGN3 and NLGN4) undergo epigenetic modifications like DNA methylation. SHANK3 specifically undergoes epigenetic regulation. SHANK3 gene comprises of five CpG islands which undergo methylation [213, 216]. This DNA methylation at the CpG dinucleotides is essential for the proper functioning of genome. It has been observed that expression of SHANK3 has been epigenetically regulated by DNA methylation of CpG-island2, resulting in a specific expression of SHANK3 [216]. Methylation rate of CpG-island2 peaks at two weeks after birth and the expression of SHANK3 changes as the synapse matures [213, 217, 218]. SHANK3 is a scaffolding protein and is associated with neuroligin, necessary for synapse formation and maintenance. NLGNs, especially NLGN3 and NLGN4, also undergo epigenetic regulation [58, 219]. Other genes which undergo epigenetic regulation are Bcl-2 genes and RORA genes [220, 221]. It is the hypermethylation of particular CpG sites in the promoter regions of these genes that is responsible for their involvement in ASD. Bcl-2 is necessary for the regulation of anti-apoptotic processes, and its alteration leads to the development of decreased cognitive functions. RORA gene, a nuclear receptor, is involved in neuronal oxidative stress [214, 221–223]. Nardone et al. [224] performed DNA methylation analysis in the cortical region of ASD individuals and found that DNA methylation and expression of gene are inversely correlated. Genes which are hypomethylated are overexpressed.

2.2.3 Histone Modifications

There are several kinds of post-translational modifications in histone proteins such as acetylation of lysine residues, ubiquitinylation, sumoylation, phosphorylation, and methylation which occur in the amino and carboxy tail of histone proteins [225]. These modifications basically occur in either lysine, arginine, serine, or proline. It has been observed that the methylation of lysine residues of H3 histone protein (H3K4) results in altered social interaction, stereotype, and repetitive behavior which are characteristic of autism phenotype [226, 227]. This histone protein is encoded by gene SMCX which also regulates other genes associated with this disorder such as genes coding for type-1 sodium channel ($Na_v1.1$), L-Ca^{2+} channel (CACNA1H), and BDNF gene [228]. Acetylation of histone proteins of the genes coding for oxytocin and vasopressin has also been associated with ASD like behavior, as indicated by the increase in the up-regulation of these receptors after the administration of histone deacetylase inhibitors [192, 229]. Sun et al. [230] studied effect of histone modification in brain of patients suffering from autism spectrum disorders. They performed the immunoprecipitation sequencing technique of H3K27 chromatin protein on post-mortem brain samples of 275 ASD patients and compared this with control brain samples. Their results showed that acetylation of the histone protein is widespread in the cerebral cortex and cerebellum of ASD patients. James et al. [231] also indicated that trimethylation levels were reduced in H3K27 in the cerebellum of patients suffering from autism as compared to control subjects. Autism spectrum disorders are associated with changes in neurons of the pre-frontal cortex and these are responsible for behavioral and cognitive deficits occurring as part of autism. Shulha et al. [227] studied the structure of chromatin of pre-frontal cortex neurons of ASD patients in order to check for the presence of any epigenetic modifications in histone proteins which might be responsible for the pathogenesis. They collected 15 billion base pairs of tri-methylated H3K4 histone protein enriched sequences from 32 brains (16 = autistic brain and 16 = brain of control subjects). They observed changes in the H3K4 histone proteins indicating that autism results in changes in structure of chromatin.

The forthcoming section of the chapter is about environmental factors which bring changes at the genetic and epigenetic levels leading to the development of the autism phenotype and several behavioral abnormalities associated with this disorder.

2.3 Pre-natal and Post-natal Environmental Factors Affecting Neurodevelopment and Leading to ASD

We have discussed in a previous section about mutations in genes encoding for various synaptic proteins regulating synaptogenesis and synaptic transmission. Epigenetic changes caused by various environmental factors in these genes have also been discussed. In this section, we will discuss various pre-natal and post-natal environmental risk factors and how they can affect the genes encoding for proteins regulating synaptogenesis and synaptic transmission. There are several environmental factors which have become a major cause of concern for a neurodevelopmental disorder like autism. ASD is largely a genetic disorder involving multiple genes. But factors affecting environment also contribute to the alteration of genes associated with ASD. Environmental risk factors responsible for the development of ASD can be divided into pre-natal and post-natal risk factors (Fig. 4). Pre-natal exposure of the developing fetus to these environmental risk factors leads to hampered neurodevelopment, becoming one of the major causes of ASD in children [232].

In utero exposure of fetus to pre-natal infections like influenza, rubella, and CMV put the new-born at a risk for developing ASD. Exposure of the mother to these infections during pregnancy affects the immune system of fetus as placenta is the most important stem cell source for the fetus [233]. Altered immune system of fetus, circulating cytokines, and HLA antigens have been linked to the development of autism in the developing fetus [234]. The likelihood of pre-natal infection

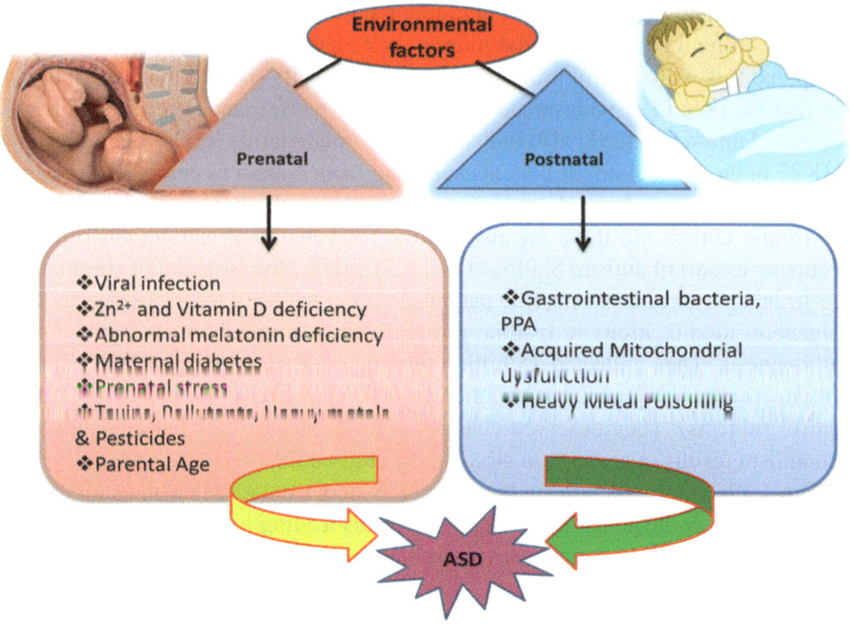

Fig. 4 Various environmental risk factors that might be responsible for development of autism spectrum disorders by affecting various genetic hits

being a risk factor for autism depends on the maternal immune system and that of the fetus. The plausible reason for this is placental barrier between mother and fetus which acts as a source of hematopoietic stem cells and might be responsible for deregulation of immune system of fetus [233, 235]. Deficiencies of Zn^{2+}, vitamin D, and abnormal levels of melatonin in the mother have also been linked to autism. Deficiency of various metal ions during pregnancy can lead to cognitive diseases and neurological deficits in the fetus as these are essential for the proper functioning of the brain. Zn^{2+} deficiency affects maturation of neurons. Enzymes requiring Zn^{2+} as a co-factor are impaired, and this can lead to impairment of learning and memory as well as dysfunction of brain [236]. There are various clinical reports indicating Zn^{2+} deficiency in the autistic children. It has been found that excess of Cu^{2+} can lead to Zn^{2+} deficiency. Many autistic children suffer from this deficiency [237]. Yasuda et al. [238] reported that hair samples of autistic children indicated zinc deficiency in the infants up to 3 years of age. This in turn indicated a connection between zinc deficiency and autism. Zinc deficiency can lead to cognitive impairment, behavioral deficit as well as glutamate excitotoxicity in autistic children [136, 239]. Zinc deficiency has been linked to autism as it affects the PSD scaffold by affecting SHANK2 and SHANK3 gene, thus affecting the regulation of synaptogenesis. It also affects the immune system of the fetus and glutamatergic synapses by affecting Nrxn-Nlgn-Shank pathway [239, 240]. Apart from this, there are other genes which are affected by zinc deficiency such as COMMD1 (COMM domain-containing protein 1), ERK1 (extracellular signal-regulated kinase 1),

TrkB (tyrosine-related kinase B), MTF1 (metal regulatory transcription factor 1), and metallothioneins (MTs) [99, 111, 241, 242]. Hence, zinc might act as a fulcrum for gene-environment interactions and its deficiency in mother during gestation might act as a risk factor for autism. Melatonin, a neurohormone, is essential for the regulation of the biological clock and for maintenance of synaptic plasticity. Melatonin levels were decreased in children suffering from autism. Apart from this, these children also have sleep problems such as prolonged latency to onset of sleep, increase in the frequency of waking up during night and reduction in the duration of sleep [243]. Abnormalities in circadian rhythm in autism can occur as a result of mutations in genes coding for enzymes involved in melatonin synthesis. Hence, it has been observed that deletion of gene encoding for acetylserotonin O-methyltransferase) has been found in autistic individuals and might be the result of abnormalities in circadian rhythm [244]. Environmental factors such as Zn^{2+} deficiency or maternal stress can also affect synthesis of melatonin leading to melatonin deficiency during pregnancy and are considered to be risk factors for the development of ASD [245, 246].

Maternal diabetes, prenatal stress and parental age are other risk factors in the pathogenesis of ASD [247]. Prenatal stress can have a detrimental effect on the developing fetus and increased exposure of mother to prenatal stress can increase the risk of developing ASD in the child [248]. The plausible mechanism behind this is abnormal activation of HPA axis, causing a reduction in the volume of hippocampus. When the volume of the hippocampus is reduced, there is increased secretion of cortisol [249]. Cortisol levels are normally elevated during pregnancy, but abnormal activation as a result of stressful conditions alters the development of the fetus leading to reduction in volume of the hippocampus and amygdala. During excessive stress, excessive release of corticotrophin-releasing factor (CRF) from placenta affects the fetus by crossing the blood–brain barrier [249]. Advanced parental age is another risk factor associated with the development of autism [250]. Advanced age of either mother or father is considered to be a risk factor, but the relative risk of developing autism with advanced age is more with paternal age as compared to maternal age. Risk of developing ASD was 5.75 times higher in child born to fathers who are more than 40 years of age as compared to fathers of 30 years or younger [251]. There are increased chances of developing de-novo mutations with advanced parental age especially in the case of fathers as sperm production continues throughout lifetime and these small de novo mutations can accumulate over time in comparison to ova [252]. Advanced maternal age is a risk factor as there can be certain complications during pregnancy or as a result of maternal activation of immune system and development of autoimmune disorders, these risks increase with advanced age. Immune system activation can lead to increased release of pro-inflammatory cytokines and chemokines affecting the developing brain of the fetus [247]. Exposure of the fetus to various toxins, pollutants, pesticides and heavy metals, known as teratogens, can lead to the development of ASD. The exposure of fetus to drugs like valproic acid causes neurodevelopmental deficits like non-social behavior, anxiety, deficits in motor performance, repetitive and perseverative behavior specific to autism. Valproic acid is used as an anticonvulsant for the treatment of epilepsy and is a histone deacetylase inhibitor causing changes in epigenetic regulation resulting in altered gene expression [253–256]. Valproic acid exposure has been found to significantly decrease expression of mRNA of Neuroligin3, in the CA1 region of hippocampus and somatosensory cortex and dentate gyrus. Neuroligin3 is a cell-adhesion molecule found on the post-synaptic membrane of glutamatergic or GABAergic synapse and interacts with neurexins. We have described NLGN3 and neurexin gene mutations previously in the article [257]. Thalidomide, used as an anti-emetic during pregnancy, was banned because of its tendency to cause birth defects [258]. It has immunomodulatory action and modulates the cytokine levels affecting the activation of T-cell and NK cells. Hence, maternal immune stimulation increases the levels and expression of fetal cytokines, thus affecting microarchitecture of fetus brain [259]. Prenatal exposure to organophosphate pesticides like chlorpyrifos, diazinon and organochlorine pesticides like endosulfan is neurotoxic and may lead to development of autism. Roberts et al. [260] indicated that women who were exposed to these organochlorine pesticides in their second trimester of pregnancy were several times more likely to give birth to children suffering from autism. The post-natal exposure to heavy metals like cadmium, nickel, mercury, and ethanol not only results in epigenetic modifications in DNA methylation and acetylation but also causes post-translational histone modifications. There is a large body of evidence suggesting that even pre-conceptual exposure to mercury, cadmium, nickel, vinyl chloride, and trichloroethylene can result in de novo point mutations as these are mutagenic and thus, increase the risk for development of autism. Exposure to these chemicals results in production of reactive oxygen species (ROS) responsible for oxidative damage of DNA. These chemicals also inhibit pathways responsible for repair of damaged DNA. They cause depletion of endogenous antioxidants such as glutathione and superoxide dismutase which normally protect cells from damage. Hence, DNA is damaged as a result of oxidative damage by ROS. They increased point mutations, and it resulted in chromosomal aberrations [261, 262]. Increased exposure to synthetic chemicals and fragrances that disrupt endocrine functions are emerging as risk factors for autism [263]. Glyphosate is a widely used herbicide as it is considered to be non-toxic, but it can indirectly affect human population as it inhibits the shikimate pathway of gut-microbes. This pathway is responsible for production of aromatic amino acids which are precursors for synthesis of neurotransmitters such as serotonin and dopamine. It kills good bacteria in the gut and results in overgrowth of pathogenic bacteria of the Clostridia and Bacteroides species responsible for leaky gut phenomenon in autistic patients. These gut bacteria also produce excess amount of short-chain fatty acids such as propanoic acid (PPA) which can further cause biochemical, behavioral, and neurochemical alterations in the autistic individuals [264, 265]. Exposure to aluminum has also been linked to ASD. Aluminum has been used as an adjuvant in vaccines. Mold et al. [266] have found through post-mortem brain analysis of autistic brain that aluminum content was very high in the neurons as well as non-neuronal cells such as microglia and astrocytes. Microglia cells were loaded with aluminum and this results in their dysfunction, hence affecting synaptic pruning. Aluminum was also found in other cells such as meninges and grey and white matter. Thus, excess exposure to aluminum pre- or post-natally might be a risk factor for development of ASD. Exposure to these environmental toxins, pesticides, and chemicals not only results in de novo mutations by affecting various genes involved in synaptogenesis but also results in activation of microglia and astrocytes resulting in the release of pro-inflammatory cytokines. This can have a detrimental effect on the fetus during development as it is during gestation that neurodevelopment and immune system development takes place [267].

Post-natal exposure to gastrointestinal bacteria and acquired mitochondrial dysfunction also results in ASD. Propanoic acid (PPA) is present in various foods like cheese and is also present as preservative in processed foods. Propanoic acid (PPA) and other short-chain fatty acids are also produced as a result of fermentation by gut bacteria such as Clostridia, Bacteroides as well as Desulfovibrio and are associated with ASD. Short-chain fatty acids can affect gut, brain, and behavior. The gut-brain cross talk in ASD is emerging as a new concept in autism. There is a bidirectional communication between the gut and brain as the excitotoxicity, free radical generation, and neuroinflammation can alter gut microflora composition on the one hand. While, on the other hand, the products of these altered microbiota can also affect the function of brain. This is the cause of "leaky gut phenomenon" occurring in children with autism. As a result, they have increased gastrointestinal permeability resulting in the production of microbiota products such as short-chain fatty acids (SCFAs) to cross the blood–brain barrier and reach brain after entering into blood stream. These will then stimulate the immune system and worsen the symptoms of autism [268, 269]. There have been reports that when PPA is administered by ICV to the brain, it results in various neurochemical changes similar to those associated with autism. It causes neuroinflammation, ROS production, glutathione depletion, and changes in acylcarnitine, resulting in acquired mitochondrial dysfunction [270–274].

2.3.1 How Environmental Factors Affect the Genetic Hits?

There is a strong interconnection and cross-talk between several environmental and genetic factors. We had also explained earlier in the chapter under section Genetic Factors in ASD that synaptogenesis and synaptic transmission might be altered as a result of copy number variations (CNVs) or single nucleotide polymorphisms (SNPs) in the genes encoding for cell adhesion molecules. Examples include Neuroligins, Neurexins, Contactins, NrCAM, Cadherins or ion-channels such as sodium channel (SCN1A/SCN2A gene), L-type voltage-gated Ca^{2+} channel, and potassium channel or scaffolding proteins of the ProSAP/Shank family and signaling pathways such as mTOR-PI3K/Akt pathway or RAS/MAPK pathway. These can lead to changes in a number of receptors or their composition, affecting mGluR/NMDAR/AMPAR. Environmental risk factors affecting ASD are interconnected with each other. Immune system abnormalities can occur as a result of pre-natal stress, pre-natal viral infection, advanced parental age, environmental toxins, melatonin deficiency, and Zn^{2+} deficiency. Zn^{2+} deficiency can occur as a result of malnutrition or copper overload or melatonin deficiency [264]. Hence, immune system deregulation and zinc deficiency may act as central environmental risk factors (other factors converge into them). Zinc deficiency, toxins, and parental age (discussed earlier) may act on genetic factors and are involved in the cross-talk between genes and environment. Zinc deficiency causes reduction in SHANK3 and levels of NMDA receptors. It affects the PSD scaffold by affecting SHANK2 and SHANK3 gene [239, 275]. This in turn will affect synaptic transmission via Neuroligin-Neurexin complex and mGlu5 signaling. Valproic acid exposure has also been found to significantly decrease expression of mRNA of Neuroligin3, in the CA1 region of the hippocampus and somatosensory cortex and dentate gyrus [257]. Exposure to various environmental toxins as well as parental age can also cause de-novo point mutations in the genes. Regulation of synaptic components occur via signaling cascades such as p38MAPK and ERK pathway. The ERK pathway is responsible for synaptic delivery of AMPA receptor and modifications of the dendritic spine. Zinc deficiency affects ERK kinases, and it is ERK2 which affects social behaviors [242]. Zinc deficiency also affects GPR39 signaling and alters the composition of glutamatergic synapses [276]. Moreover, immune system deregulation results in release of pro-inflammatory cytokines, chemokines, and NO release leading to oxidative stress. Immune cells infiltrate the CNS activating the microglial cells and resulting in upregulation of pro-inflammatory cytokines such as IL-18, IL-6, IL-β, and TNF-α which in turn act on NMDA, AMPA, and mGlu receptors of excitatory glutamatergic synapses, thus modulating p38MAPK and ERK signaling [277, 278]. Hence, immune system dysfunction can cause alteration of NMDA receptor mediated synaptic plasticity and transmission. Therefore, either genetic mutations in the various components of excitatory synapses or immune system deregulation or zinc deficiency can lead to disruption of balance of excitatory and inhibitory pathway which is already known to be disrupted in ASD.

In the concluding section of this chapter we have discussed the complex mechanisms of integration of genetic, epigenetic, and environmental factors and their subsequent consequences. As a result of this integration there is glutamate excitotoxicity, mitochondrial dysfunction as well as activation of microglia, astrocytes, as well as oligodendrocytes in ASD.

3 Plausible Complex Mechanism of Integration of Genetic, Epigenetic, and Environmental Factors Affecting Synaptic Transmission and Glial Activation

There is a complicated integration of genetic, epigenetic, and environmental factors associated with the pathogenesis of ASD. A predisposing genetic tendency might amalgamate with environmental factors and the epigenetic changes brought by them might manifest in ASD, a complex neurodevelopmental disorder. A predisposing genetic tendency as a result of complex genetic syndromes and de-novo copy number variations (CNVs) might interact with various environmental factors. Environmental factors may precipitate epigenetic regulations like DNA methylation

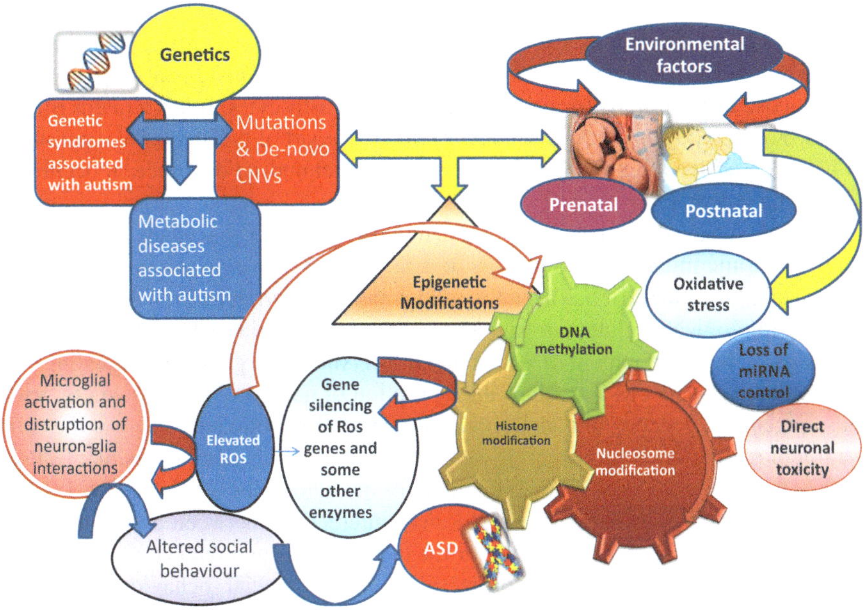

Fig. 5 Diagrammatic illustration of the complex amalgamation of environment, genetics, and epigenetics in the development of phenotype of ASD

and post-translational histone modifications, which could cause gene silencing of various enzymes which scavenge free-radicals. This results in elevation of reactive oxygen species (ROS). Environmental factors could also result in oxidative stress and neural toxicity by triggering glutamate toxicity and microglial activation and dysregulation of miRNA control. These epigenetic alterations, DNA damage, and gene silencing of antioxidant enzymes could result in excessive elevation of ROS triggering microglial activation and damage neuron-glia interactions. This aggravates the existing neurobiology of ASD. Excessive production of reactive oxygen species might lead to the impairment of DNA methylation resulting in a positive feedback mechanism. Hence, patients suffering from ASD are more vulnerable to oxidative stress and neuronal toxicity (Fig. 5) [191, 279, 280].

We now describe various consequences of the complex integration of genetic, epigenetic, and environmental factors.

3.1 Glutamatergic Excitotoxicity & Mitochondrial Dysfunction

Glutamate is an excitatory neurotransmitter. Its concentration is maintained by various uptake transporters like EAAT1, EAAT2, EAAT3, EAAT4, and EAAT5 in normal physiological conditions. Excess glutamate is converted to glutamine after its uptake by astrocytes or is involved in GSH synthesis. Glutamine is stored in vesicles after its conversion to glutamate. During glutamate excitotoxicity, the release of GSH is significantly enhanced. Glutamate is responsible for learning and memory in the region of cerebellum and hippocampus [281]. Glutamate is also involved in the glucose utilization pathway via GLUT-1 in the astrocytes. It has a major role in the GSK-3 (glycogen synthase kinase3), which is a part of β-Catenin/Wnt pathway. It has been noticed that GSK-3 is involved in mitochondrial dysfunction occurring in autistic individuals.

Mitochondrial dysfunction has been involved in ASD that leads to oxidative stress. Chronic mitochondrial dysfunction, associated with electron transport chain (ETC) complex I and III, has been found in patients suffering from ASD [282, 283]. Thus, pathologically glutamate signaling and mitochondrial dysfunction converge and lead to the development of ASD phenotype. Abnormal glutamate signaling and excitotoxicity occurs as a result of the disruption of neuronal transmission by various epigenetic, genetic, and environmental factors. Excitotoxicity triggers the activation of resting microglia that releases various neuroprotective factors like BDNF, free radicals like ROS, NO, pro-inflammatory mediators like TNF-α, NF-kB, IL-β, IL-6, IL-12, and excess glutamate. All these interactions between astrocytes, neurons, and glial cells are responsible for the disruption of neuronal connectivity and thus are implicated in the abnormal and aberrant social behavior associated with ASD.

Glutamate excitotoxicity discussed above leads to neuroinflammation via subsequent involvement of activation of microglia and release of inflammatory mediators. In the next subsection we discuss, in more detail, the involvement of the activation of microglia, astrocytes, and oligodendrocytes in the pathogenesis of ASD.

3.2 The Role of Microglial Dysfunction and Activation of Microglia, Astrocytes, and Oligodendrocytes in ASD

Microglial cells are the myeloid progenitor cells that have significant physiological functions like synaptogenesis, secretion of various neuroprotective factors like BDNF, TNF-α, secretion of synaptic transmission, and neurogenesis. The microglial pathway has been widely implicated in ASD. Microglial activation results in the secretion of inflammatory mediators and cytokines like IL-6, IL-8, TNF-α, INF-ϒ, NF-kB, GM-CSF as a result of neuronal excitotoxicity [280, 284]. Deficient microglial function has been involved in various syndromes associated with ASD and the behavior abnormalities occurring in ASD like repetitive behavior, stereotypy, abnormal social interaction, etc. Rett syndrome involves mutations in MeCP2 gene that codes for methyl-CpG–binding protein 2 in microglia resulting in glutamatergic toxicity [279, 285].

Activated microglia also results in the activation of TLR-4 and TLR-3 receptors causing the production of inflammatory cytokines which have a significant impact on the cerebellum, white matter, and cortex of ASD patients. Cytokines affect neurogenesis, synaptogenesis, and ASD-associated behavioral phenotypes. This immune system abnormality is clearly observed during PET scanning of the brains of ASD patients where there is enlargement of lateral ventricle and increase in microglial cell density in grey matter. Microglial activation also results in the significant production of ROS and proinflammatory cytokines affecting mitochondrial function. Thus, oxidative stress and ROS production are key features of neurodevelopmental disorders [280, 286]. Abnormal fatty acid metabolism, which increases peroxisomal-β-oxidation and CD-38 activation, has been linked to the pathogenesis of ASD [287]. These are responsible for affecting glutamatergic pathways and hence, activation of microglia cells resulting in neurodegeneration. This neuroimmune dysfunction in the brains of autistic patients results in neuroinflammation which triggers the development of various abnormalities associated with this disorder.

Astrocytes are star-shaped neuroglial cells accompanying neurons that are responsible for neurogenesis, synaptogenesis, maintenance of synaptic transmission, involvement of glutamate-glutamine and ROS generation [288–290]. As a result of neuroinflammation induced by pathology of autism, these astrocytes undergo anisomorphic and isomorphic astrogliosis resulting in the formation of glial scar. There are various environmental factors which cause epigenetic changes and have an effect on astrocytes resulting in down-regulation of GLAST, reducing astrocytes' uptake of glutamate and the blockage of AMPA receptors. Due to ASD, there is inactivation of GSK-3. This results in the lack of the ability to synthesize glucose for neurons and induction of apoptosis by hippocampal neurons causing the release of TNF-α from astrocytes leading to down-regulation of GLAST. Other processes which are controlled by astrocytes are their role in cholesterol metabolism by stabilizing OTR receptors, NAD+ metabolism, and the production of inflammatory cytokines. All of these abnormalities form part of ASD and result in disturbed neuron-glia interactions regulating synaptic plasticity [291].

Oligodendrocytes are the neuroglia cells responsible for myelination and mechanical and metabolic support for the axon. Pathogenesis of ASD involving oligodendrocytes has been observed as a result of impaired metabolism of N-acetyl aspartate and increased myelination in the cortex of ASD patients and NG-2 immunoreactivity [292–295].

4 Conclusions

ASD is a complicated diverse neurodevelopmental disease that integrates genetic, epigenetic, and environmental mechanisms to form a complex phenotype. Genetic predisposition of an individual as a result of single gene disorder, chromosomal aberrations and copy number variations (CNVs) disrupt the process of synaptogenesis and affect synaptic plasticity. There are various environmental factors which bring about epigenetic modifications like DNA methylation and post-translational histone modifications resulting in gene silencing and affected transcription. This complex integration results in the interplay between the immune processes and synaptic function. Immune cells play an important role in neuroinflammation. This entire integration causes immune system activation and starts a cascade of events which triggers the development of behavioral phenotype of ASD. Microglia and astrocytes have been found to be activated in the brains of ASD patients and play a significant role in synaptic pruning. Increased microglia is present in the brains of ASD patients [296–298]. Microglia and astrocytes have a very important role in synapse formation and whenever mutations occur in cell adhesion molecules of neurons. It clearly indicates that neurons involved in autism are prone to immune cell dysfunction. Microglia activation triggers the release of various inflammatory cytokines and chemokines like IL-6, IL-12, IL-β, TNF-α, resulting in the excess production of iNOS, ROS and altering synaptic plasticity and causing behavioral changes associated with ASD. The role of the complex interplay of the immune system could provide vital insight into the development of novel pharmacotherapeutic targeting of the core symptoms of autism.

Acknowledgements Research grants sanctioned by SERB, Department of Science & Technology (grant no SB/FT/LS-284/2012), All India Council of Technical Education (11-25/RIFD/CAYT/POL-II/2013-14) and University Grants Commission (20-29(12)/2012(BSR), New Delhi to Dr Anurag Kuhad are gratefully acknowledged. Senior Research Fellowship sanctioned by Indian Council of Medical Research, New Delhi to Ms. Ranjana Bhandari is also gratefully acknowledged.

Conflict of Interest The authors declare no conflicts of interest.

References

1. American Psychiatric Association. (2013). *Diagnostic and statistical manual of mental disorders*. American Psychiatric Association.
2. WHO. (2017). *Autism spectrum disorders*. WHO.
3. Rudra, A., Belmonte, M.K., Soni, P. K., Banerjee, S., Mukerji, S., Chakrabarti, B. (2017). Prevalence of autism spectrum disorder and autistic symptoms in a school-based cohort of children in Kolkata, India. *Autism Research, 10*, 1597–1605. https://doi.org/10.1002/aur.1812
4. Fombonne, E. (2005). Epidemiology of autistic disorder and other pervasive developmental disorders. *The Journal of Clinical Psychiatry*, 3–8.
5. Holt, R., & Monaco, A. P. (2011). Links between genetics and pathophysiology in the autism spectrum disorders. *EMBO Molecular Medicine, 3*, 438–450. https://doi.org/10.1002/emmm.201100157
6. Santangelo, S. L., & Tsatsanis, K. (2005). What is known about autism: Genes, brain, and behavior. *American Journal of Pharmacogenomics, 5*, 71–92.
7. Werling, D. M., & Geschwind, D. H. (2013). Sex differences in autism spectrum disorders. *Current Opinion in Neurology, 26*, 146–153. https://doi.org/10.1097/WCO.0b013e32835ee548
8. Sandin, S., Lichtenstein, P., Kuja-Halkola, R., Hultman, C., Larsson, H., Reichenberg, A. (2017). The heritability of autism spectrum disorder. *JAMA, 318*, 1182. https://doi.org/10.1001/jama.2017.12141
9. Banerjee, S., Riordan, M., & Bhat, M. A. (2014). Genetic aspects of autism spectrum disorders: Insights from animal models. *Frontiers in Cellular Neuroscience, 8*, 58. https://doi.org/10.3389/fncel.2014.00058

10. Betancur, C., Sakurai, T., & Buxbaum, J. D. (2009). The emerging role of synaptic cell-adhesion pathways in the pathogenesis of autism spectrum disorders. *Trends in Neurosciences, 32*, 402–412. https://doi.org/10.1016/j.tins.2009.04.003
11. Antshel, K.M., Zhang-James, Y., Wagner, K. E., Ledesma, A., Faraone, SV. (2016). An update on the comorbidity of ADHD and ASD: A focus on clinical management. *Expert Review of Neurotherapeutics, 16*, 279–293. https://doi.org/10.1586/14737175.2016.1146591
12. Estabillo, J. A., Matson, J. L., & Cervantes, P. E. (2018). Autism symptoms and problem behaviors in children with and without developmental regression. *Journal of Developmental and Physical Disabilities, 30*, 17–26. https://doi.org/10.1007/s10882-017-9573-x
13. Frye, R. E., & Rossignol, D. A. (2016). Identification and treatment of pathophysiological comorbidities of autism spectrum disorder to achieve optimal outcomes. *Clinical Medicine Insights: Pediatrics, 10*, 43–56. https://doi.org/10.4137/CMPed.S38337
14. Helverschou, S. B., Bakken, T. L., & Martinsen, H. (2011). Psychiatric disorders in people with autism spectrum disorders: Phenomenology and recognition. In *International handbook of autism and pervasive developmental disorders* (pp. 53–74). New York: Springer.
15. El-Fishawy, P., & State, M. W. (2010). The genetics of autism: Key issues, recent findings, and clinical implications. *The Psychiatric Clinics of North America, 33*, 83–105. https://doi.org/10.1016/j.psc.2009.12.002
16. Liu, J., Nyholt, D. R., Magnussen, P., Parano, E., Pavone, P., Geschwind, D., et al. (2001). A genome-wide screen for autism susceptibility loci. *American Journal of Human Genetics, 69*, 327–340.
17. Caglayan, A. O. (2010). Genetic causes of syndromic and non-syndromic autism. *Developmental Medicine and Child Neurology, 52*, 130–138. https://doi.org/10.1111/j.1469-8749.2009.03523.x
18. Persico, A. M., & Sacco, R. (2014). Endophenotypes in autism spectrum disorders. In *Comprehensive guide to autism* (pp. 77–95). New York: Springer.
19. Pieretti, M., Zhang, F.P., Fu, Y.H., Warren, ST., Oostra, BA., Caskey, CT., et al. (1991). Absence of expression of the FMR-1 gene in fragile X syndrome. *Cell, 66*, 817–822.
20. Sitzmann, A.F., Hagelstrom, R.T., Tassone, F., Hagerman, R.J., Butler, M.G. (2018). Rare *FMR1* gene mutations causing fragile X syndrome: A review. *American Journal of Medical Genetics - Part A, 176*, 11–18. https://doi.org/10.1002/ajmg.a.38504
21. Persico, A. M., & Napolioni, V. (2013). Autism genetics. *Behavioural Brain Research, 251*, 95–112. https://doi.org/10.1016/j.bbr.2013.06.012
22. Clayton-Smith, J., & Laan, L. (2003). Angelman syndrome: A review of the clinical and genetic aspects. *Journal of Medical Genetics, 40*, 87–95.
23. Vijayakumar, P. (2018). Identification imprinting gene expression at 15q11-q13 region in Angelman syndrome. *Parkinsonism & Related Disorders, 46*, e53–e54. https://doi.org/10.1016/j.parkreldis.2017.11.181
24. Kotulska, K., & Jóźwiak, S. (2011). Autism in monogenic disorders. *European Journal of Paediatric Neurology, 15*, 177–180. https://doi.org/10.1016/j.ejpn.2010.08.007
25. Napolioni, V., & Curatolo, P. (2008). Genetics and molecular biology of tuberous sclerosis complex. *Current Genomics, 9*, 475–487. https://doi.org/10.2174/138920208786241243
26. Peters, S., Beaudet, A., Madduri, N., & Bacino, C. (2004). Autism in Angelman syndrome: Implications for autism research. *Clinical Genetics, 66*, 530–536. https://doi.org/10.1111/j.1399-0004.2004.00362.x
27. Tan, W.-H., Bacino, C. A., Skinner, S. A., Irina, Anselm., Rene, Barbieri-Welge., Astrid, Bauer-Carlin., et al. (2011). Angelman syndrome: Mutations influence features in early childhood. *American Journal of Medical Genetics - Part A, 155*, 81–90. https://doi.org/10.1002/ajmg.a.33775
28. McCarroll, S. A., & Altshuler, D. M. (2007). Copy-number variation and association studies of human disease. *Nature Genetics, 39*, S37–S42. https://doi.org/10.1038/ng2080
29. Sharp, A. J., Locke, D. P., McGrath, S. D., Cheng, Z., Bailey, J. A., Vallente, R. U., et al. (2005). Segmental duplications and copy-number variation in the human genome. *American Journal of Human Genetics, 77*, 78–88. https://doi.org/10.1086/431652

30. Marshall, C. R., & Scherer, S. W. (2012). Detection and characterization of copy number variation in autism spectrum disorder. *Methods in Molecular Biology (Clifton, N.J.)*, 115–135.
31. Glessner, J. T., Wang, K., Cai, G., Korvatska, O, Kim, C. E,, Wood, S., et al. (2009). Autism genome-wide copy number variation reveals ubiquitin and neuronal genes. *Nature, 459*, 569–573. https://doi.org/10.1038/nature07953
32. Sanders, S. J., Ercan-Sencicek, A. G., Hus, V., Luo, R., Murtha, M. T., Moreno-De-Luca, D., et al. (2011). Multiple recurrent de novo CNVs, including duplications of the 7q11.23 Williams syndrome region, are strongly associated with autism. *Neuron, 70*, 863–885. https://doi.org/10.1016/j.neuron.2011.05.002
33. Sener, E. F. (2014). Association of copy number variations in autism spectrum disorders: A systematic review. *Chinese Journal of Biology, 2014*, 1–9. https://doi.org/10.1155/2014/713109
34. Menashe, I., Larsen, E. C., & Banerjee-Basu, S. (2013). Prioritization of copy number variation loci associated with autism from AutDB-An integrative multi-study genetic database. *PLoS One*, 8. https://doi.org/10.1371/journal.pone.0066707
35. Sanders, S. J., He, X., Willsey, A. J., Ercan-Sencicek, A. G., Samocha, K. E., Cicek, A. E., et al. (2015). Insights into autism spectrum disorder genomic architecture and biology from 71 risk loci. *Neuron, 87*, 1215–1233. https://doi.org/10.1016/j.neuron.2015.09.016
36. Iossifov, I., O'Roak, B. J., Sanders, S. J., Ronemus, M., Krumm, N., Levy, D., et al. (2014). The contribution of de novo coding mutations to autism spectrum disorder. *Nature, 515*, 216–221. https://doi.org/10.1038/nature13908
37. Hoeffer, C. A., & Klann, E. (2010). mTOR signaling: At the crossroads of plasticity, memory and disease. *Trends in Neurosciences, 33*, 67–75. https://doi.org/10.1016/j.tins.2009.11.003
38. Yeung, K. S., Tso, W. W. Y., Ip, J. J. K., Mak, C. C. Y., Leung, G. K. C., Tsang, M. H. Y., et al. (2017). Identification of mutations in the PI3K-AKT-mTOR signaling pathway in patients with macrocephaly and developmental delay and/or autism. *Molecular Autism, 8*, 66. https://doi.org/10.1186/s13229-017-0182-4
39. Ebert, D. H., & Greenberg, M. E. (2013). Activity-dependent neuronal signaling and autism spectrum disorder. *Nature, 493*, 327–337. https://doi.org/10.1038/nature11860
40. Jiang, Y. H., & Ehlers, M. D. (2013). Modeling autism by SHANK gene mutations in mice. *Neuron, 78*, 8–27. https://doi.org/10.1016/j.neuron.2013.03.016
41. Song, J. Y., Ichtchenko, K., Südhof, T. C., & Brose, N. (1999). Neuroligin 1 is a postsynaptic cell-adhesion molecule of excitatory synapses. *Proceedings of the National Academy of Sciences of the United States of America, 96*, 1100–1105.
42. Varoqueaux, F., Jamain, S., & Brose, N. (2004). Neuroligin 2 is exclusively localized to inhibitory synapses. *European Journal of Cell Biology, 83*, 449–456. https://doi.org/10.1078/0171-9335-00410
43. Südhof, T. C. (2017). Synaptic neurexin complexes: A molecular code for the logic of neural circuits. *Cell, 171*, 745–769. https://doi.org/10.1016/j.cell.2017.10.024
44. Fabrichny, I. P., Leone, P., Sulzenbacher, G., Comoletti, D., Miller, M. T., Taylor, P., et al. (2007). Structural analysis of the synaptic protein neuroligin and its β-neurexin complex: Determinants for folding and cell adhesion. *Neuron, 56*, 979–991. https://doi.org/10.1016/j.neuron.2007.11.013
45. Banovic, D., Khorramshahi, O., Owald, D., Wichmann, C., Riedt, T., Fouquet, W., et al. (2010). Drosophila neuroligin 1 promotes growth and postsynaptic differentiation at glutamatergic neuromuscular junctions. *Neuron, 66*, 724–738. https://doi.org/10.1016/j.neuron.2010.05.020
46. Budreck, E. C., & Scheiffele, P. (2007). Neuroligin-3 is a neuronal adhesion protein at GABAergic and glutamatergic synapses. *The European Journal of Neuroscience, 26*, 1738–1748. https://doi.org/10.1111/j.1460-9568.2007.05842.x
47. Hunter, J. W., Mullen, G. P., McManus, J. R., Heatherly, J. M., Duke, A., Rand, J. B. (2010). Neuroligin-deficient mutants of C. elegans have sensory processing deficits and are hypersensitive to oxidative stress and mercury toxicity. *Disease Models & Mechanisms, 3*, 366–376. https://doi.org/10.1242/dmm.003442

48. Tabuchi, K., & Südhof, T. C. (2002). Structure and evolution of neurexin genes: Insight into the mechanism of alternative splicing. *Genomics, 79*, 849–859. https://doi.org/10.1006/geno.2002.6780
49. Ichtchenko, K., Nguyen, T., & Südhof, T. C. (1996). Structures, alternative splicing, and neurexin binding of multiple neuroligins. *The Journal of Biological Chemistry, 271*, 2676–2682. https://doi.org/10.1074/JBC.271.5.2676
50. Hoon, M., Bauer, G., Fritschy, J.-M., Moser, T., Falkenburger, B. H., Varoqueauxet, F. (2009). Neuroligin 2 controls the maturation of GABAergic synapses and information processing in the retina. *The Journal of Neuroscience, 29*, 8039–8050. https://doi.org/10.1523/JNEUROSCI.0534-09.2009
51. Hoon, M., Soykan, T., Falkenburger, B., Matthieu, H., Annarita, P., Karl-Friedrich, S., et al. (2011). Neuroligin-4 is localized to glycinergic postsynapses and regulates inhibition in the retina. *Proceedings of the National Academy of Sciences, 108*, 3053–3058. https://doi.org/10.1073/pnas.1006946108
52. Bolliger, M. F., Frei, K., Winterhalter, K. H., & Gloor, S. M. (2001). Identification of a novel neuroligin in humans which binds to PSD-95 and has a widespread expression. *Biochemical Journal, 356*, 581–588.
53. Levinson, J. N., & El-Husseini, A. (2005). Building excitatory and inhibitory synapses: Balancing neuroligin partnerships. *Neuron, 48*, 171–174. https://doi.org/10.1016/j.neuron.2005.09.017
54. Chen, Y.-C., Lin, Y. Q., Banerjee, S., Venken, K., Li, J., Ismat, A., et al. (2012). Drosophila neuroligin 2 is required presynaptically and postsynaptically for proper synaptic differentiation and synaptic transmission. *The Journal of Neuroscience, 32*, 16018–16030. https://doi.org/10.1523/JNEUROSCI.1685-12.2012
55. Chih, B., Engelman, H., & Scheiffele, P. (2005). Control of excitatory and inhibitory synapse formation by neuroligins. *Science, 307*, 1324–1328. https://doi.org/10.1126/science.1107470
56. Parente, D. J., Garriga, C., Baskin, B., Douglas, G., Cho, M. T., Araujo, G. C. et al. (2017). Neuroligin 2 nonsense variant associated with anxiety, autism, intellectual disability, hyperphagia, and obesity. *American Journal of Medical Genetics - Part A, 173*, 213–216. https://doi.org/10.1002/ajmg.a.37977
57. Bottos, A., Rissone, A., Bussolino, F., & Arese, M. (2011). Neurexins and neuroligins: Synapses look out of the nervous system. *Cellular and Molecular Life Sciences, 68*, 2655–2666. https://doi.org/10.1007/s00018-011-0664-z
58. Jamain, S., Quach, H., Betancur, C., Råstam, M., Colineaux, C., Gillberg, I. C., et al. (2003). Mutations of the X-linked genes encoding neuroligins NLGN3 and NLGN4 are associated with autism. *Nature Genetics, 34*, 27–29. https://doi.org/10.1038/ng1136
59. Martella, G., Meringolo, M., Trobiani, L., De Jaco, A., Pisani, A., Bonsi, P., et al. (2018). The neurobiological bases of autism spectrum disorders: The R451C-neuroligin 3 mutation hampers the expression of long-term synaptic depression in the dorsal striatum. *The European Journal of Neuroscience, 47*, 701–708. https://doi.org/10.1111/ejn.13705
60. Tabuchi, K., Blundell, J., Etherton, M. R., Hammer, R. E., Liu, X., Powell, C. M., et al. (2007). A neuroligin-3 mutation implicated in autism increases inhibitory synaptic transmission in mice. *Science, 318*, 71–76. https://doi.org/10.1126/science.1146221
61. Jamain, S., Radyushkin, K., Hammerschmidt, K., Sylvie, G., Susann, B., Frederique V. et al. (2008). Reduced social interaction and ultrasonic communication in a mouse model of monogenic heritable autism. *Proceedings of the National Academy of Sciences, 105*, 1710–1715. https://doi.org/10.1073/pnas.0711555105
62. Taniguchi, H., Gollan, L., Scholl, F. G., Veeravan, M., Elizabeth, D., Nicolas, L., et al. (2007). Silencing of neuroligin function by postsynaptic neurexins. *The Journal of Neuroscience, 27*, 2815–2824. https://doi.org/10.1523/JNEUROSCI.0032-07.2007
63. Baudouin, S., & Scheiffele, P. (2010). SnapShot: neuroligin-neurexin complexes. *Cell, 141*, 908–908.e1. https://doi.org/10.1016/J.CELL.2010.05.024
64. Craig, A. M., & Kang, Y. (2007). Neurexin–neuroligin signaling in synapse development. *Current Opinion in Neurobiology, 17*, 43–52. https://doi.org/10.1016/j.conb.2007.01.011

65. Scheiffele, P., Fan, J., Choih, J., Fetter, R., Serafini, T. (2000). Neuroligin expressed in non-neuronal cells triggers presynaptic development in contacting axons. *Cell, 101*, 657–669.
66. Zahir, F. R., Baross, A., Delaney, A. D., Eydoux. P., Fernandes, N. D., Pugh., T. et al. (2007). A patient with vertebral, cognitive and behavioural abnormalities and a de novo deletion of NRXN1. *Journal of Medical Genetics, 45*, 239–243. https://doi.org/10.1136/jmg.2007.054437
67. Dachtler, J., Glasper, J., Cohen, R. N., Ivorra, J. L., Swiffen, D. J., Jackson, A. J., et al. (2014). Deletion of α-neurexin II results in autism-related behaviors in mice. *Translational Psychiatry, 4*, e484. https://doi.org/10.1038/tp.2014.123
68. Rabaneda, L. G., Robles-Lanuza, E., Nieto-González, J. L., & Scholl, F. G. (2014). Neurexin dysfunction in adult neurons results in autistic-like behavior in mice. *Cell Reports, 8*, 338–346. https://doi.org/10.1016/j.celrep.2014.06.022
69. Etherton, M. R., Blaiss, C. A., Powell, C. M., & Sudhof, T. C. (2009). Mouse neurexin-1 deletion causes correlated electrophysiological and behavioral changes consistent with cognitive impairments. *Proceedings of the National Academy of Sciences, 106*, 17998–18003. https://doi.org/10.1073/pnas.0910297106
70. Wang, J., Gong, J., Li, L., Chen, Y., Liu, L., Gu, H., et al. (2018). Neurexin gene family variants as risk factors for autism spectrum disorder. *Autism Research, 11*, 37–43. https://doi.org/10.1002/aur.1881
71. Zuko, A., Kleijer, K. T. E., Oguro-Ando, A., Kas, M. J. H., van Daalen, E., van der Zwaag, B., et al. (2013). Contactins in the neurobiology of autism. *European Journal of Pharmacology, 719*, 63–74. https://doi.org/10.1016/j.ejphar.2013.07.016
72. Berglund, E. O., Murai, K. K., Fredette, B., Sekerková, G., Marturano, B., Weber, L., et al. (1999). Ataxia and abnormal cerebellar microorganization in mice with ablated contactin gene expression. *Neuron, 24*, 739–750.
73. Morrow, E. M., Yoo, S.-Y., Flavell, S. W., Kim, T. K., Lin, Y., Hill, R. S., et al. (2008). Identifying autism loci and genes by tracing recent shared ancestry. *Science, 321*, 218–223. https://doi.org/10.1126/science.1157657
74. Shimoda, Y., & Watanabe, K. (2009). Contactins: Emerging key roles in the development and function of the nervous system. *Cell Adhesion & Migration, 3*, 64–70.
75. Tong, D., Chen, R., Lu, Y., Li, W., Zhang, Y. F., Lin, J. K., et al. (2018). The critical role of ASD-related gene CNTNAP3 in regulating synaptic development and social behavior in mice. bioRxiv 260083. https://doi.org/10.1101/260083
76. Cottrell, C. E., Bir, N., Varga, E., Alvarez, C. E., Bouyain, S., Zernzach, R., et al. (2011). Contactin 4 as an autism susceptibility locus. *Autism Research, 4*, 189–199. https://doi.org/10.1002/aur.184
77. Guo, H., Xun, G., Peng, Y., Li, X., et al. (2012). Disruption of Contactin 4 in two subjects with autism in Chinese population. *Gene, 505*, 201–205. https://doi.org/10.1016/J.GENE.2012.06.051
78. Pinto, D., Pagnamenta, A. T., Klei, L., Anney, R., Merico, D., Regan, R., et al. (2010). Functional impact of global rare copy number variation in autism spectrum disorders. *Nature, 466*, 368–372. https://doi.org/10.1038/nature09146
79. van Daalen, E., Kemner, C., Verbeek, N. E., Zwaag, B. V., Dijkhuizen, T., Rump P., et al. (2011). Social responsiveness scale-aided analysis of the clinical impact of copy number variations in autism. *Neurogenetics, 12*, 315–323. https://doi.org/10.1007/s10048-011-0297-2
80. Heyden, A., Angenstein, F., Sallaz, M., Seidenbecher, C., Montag, D. (2008). Abnormal axonal guidance and brain anatomy in mouse mutants for the cell recognition molecules close homolog of L1 and NgCAM-related cell adhesion molecule. *Neuroscience, 155*, 221–233. https://doi.org/10.1016/j.neuroscience.2008.04.080
81. Ishiguro, H., Liu, Q. R., Gong, J.-P., Hall, F. S., Ujike, H., Morales, M., et al. (2006). NrCAM in addiction vulnerability: Positional cloning, drug-regulation, haplotype-specific expression and altered drug reward in knockout mice. *Neuropsychopharmacology, 31*, 572–584. https://doi.org/10.1038/sj.npp.1300855

82. Sytnyk, V., Leshchyns'ka, I., & Schachner, M. (2017). Neural cell adhesion molecules of the immunoglobulin superfamily regulate synapse formation, maintenance, and function. *Trends in Neurosciences, 40*, 295–308. https://doi.org/10.1016/J.TINS.2017.03.003
83. Bonora, E., Lamb, J. A., Barnby, G., Nuala, Sykes., Moberly, T., Beyer, K. S., et al. (2005). Mutation screening and association analysis of six candidate genes for autism on chromosome 7q. *European Journal of Human Genetics, 13*, 198–207. https://doi.org/10.1038/sj.ejhg.5201315
84. Marui, T., Funatogawa, I., Koishi, S., et al. (2009). Association of the neuronal cell adhesion molecule (NrCAM) gene variants with autism. *The International Journal of Neuropsychopharmacology, 12*, 1. https://doi.org/10.1017/S1461145708009127
85. Sakurai, T., Ramoz, N., Reichert, J. G., Corwin, T. E., Kryzak, L., Smith., C. J., et al. (2006). Association analysis of the NrCAM gene in autism and in subsets of families with severe obsessive-compulsive or self-stimulatory behaviors. *Psychiatric Genetics, 16*, 251–257. https://doi.org/10.1097/01.ypg.0000242196.81891.c9
86. Sakurai, T., Lustig, M., Babiarz, J., Furley, A. J. W., Tait, S., Brophy, P. J., et al. (2001). Overlapping functions of the cell adhesion molecules Nr-CAM and L1 in cerebellar granule cell development. *The Journal of Cell Biology, 154*, 1259–1273. https://doi.org/10.1083/jcb.200104122
87. Matzel, L. D., Babiarz, J., Townsend, D. A., Grossman H. C., Grumet, M., et al. (2008). Neuronal cell adhesion molecule deletion induces a cognitive and behavioral phenotype reflective of impulsivity. *Genes, Brain and Behavior, 7*, 470–480. https://doi.org/10.1111/j.1601-183X.2007.00382.x
88. Moy, S. S., Nonneman, R. J., Young, N. B., et al. (2009). Impaired sociability and cognitive function in NrCAM-null mice. *Behavioural Brain Research, 205*, 123–131. https://doi.org/10.1016/j.bbr.2009.06.021
89. Mohan, V., Sullivan, C. S., Guo, J., Wade, S. D., Majumder, S., Agarwal, A., et al. (2018). Temporal regulation of dendritic spines through NrCAM-Semaphorin3F receptor signaling in developing cortical pyramidal neurons. *Cerebral Cortex*. https://doi.org/10.1093/cercor/bhy004
90. Demyanenko, G. P., Riday, T. T., Tran, T. S., Dalal, J., Darnell, E. P., Brennaman, L. H., et al. (2011). NrCAM deletion causes topographic mistargeting of thalamocortical axons to the visual cortex and disrupts visual acuity. *Journal of Neuroscience, 31*(4), 1545–1558. https://doi.org/10.1523/JNEUROSCI.4467-10.2011.NrCAM
91. Shapiro, L., Love, J., & Colman, D. R. (2007). Adhesion molecules in the nervous system: Structural insights into function and diversity. *Annual Review of Neuroscience, 30*, 451–474. https://doi.org/10.1146/annurev.neuro.29.051605.113034
92. Dalva, M. B., McClelland, A. C., & Kayser, M. S. (2007). Cell adhesion molecules: Signalling functions at the synapse. *Nature Reviews Neuroscience, 8*, 206–220. https://doi.org/10.1038/nrn2075
93. Hirano, S., Suzuki, S. T., & Redies, C. (2003). The cadherin superfamily in neural development: Diversity, function and interaction with other molecules. *Frontiers in Bioscience, 8*, d306–d355.
94. Hirano, S., & Takeichi, M. (2012). Cadherins in brain morphogenesis and wiring. *Physiological Reviews, 92*, 597–634. https://doi.org/10.1152/physrev.00014.2011
95. El-Amraoui, A., & Petit, C. (2010). Cadherins as targets for genetic diseases. *Cold Spring Harbor Perspectives in Biology, 2*, a003095–a003095. https://doi.org/10.1101/cshperspect.a003095
96. Hawi, Z., Tong, J., Dark, C., Yates, H., Johnson, B., Bellgrove, M.A., et al. (2018). The role of cadherin genes in five major psychiatric disorders: A literature update. *American Journal of Medical Genetics Part B: Neuropsychiatric Genetics, 177*, 168–180. https://doi.org/10.1002/ajmg.b.32592
97. Redies, C., Hertel, N., & Hübner, C. A. (2012). Cadherins and neuropsychiatric disorders. *Brain Research, 1470*, 130–144. https://doi.org/10.1016/j.brainres.2012.06.020

98. Willemsen, M. H., Fernandez, B. A., Bacino, C. A., Gerkes, E., de Brouwer, A. P., Pfundt, R., et al. (2010). Identification of ANKRD11 and ZNF778 as candidate genes for autism and variable cognitive impairment in the novel 16q24.3 microdeletion syndrome. *European Journal of Human Genetics, 18*, 429–435. https://doi.org/10.1038/ejhg.2009.192
99. O'Roak, B. J., Vives, L., Girirajan, S., Karakoc, E., Krumm, N., Coe, B. P., et al. (2012). Sporadic autism exomes reveal a highly interconnected protein network of de novo mutations. *Nature, 485*, 246–250. https://doi.org/10.1038/nature10989
100. Pagnamenta, A. T., Khan, H., Walker, S., Gerrelli, D., Wing, K., Bonaglia, M. C., et al. (2011). Rare familial 16q21 microdeletions under a linkage peak implicate cadherin 8 (CDH8) in susceptibility to autism and learning disability. *Journal of Medical Genetics, 48*, 48–54. https://doi.org/10.1136/jmg.2010.079426
101. Wang, K., Zhang, H., Ma, D., Bucan, M., Glessner, J. T., Abrahams, B. S., et al. (2009). Common genetic variants on 5p14.1 associate with autism spectrum disorders. *Nature, 459*, 528–533. https://doi.org/10.1038/nature07999
102. Chapman, N. H., Estes, A., Munson, J., Bernier, R., Webb, S. J., Rothstein, J. H., et al. (2011). Genome scan for IQ discrepancy in autism: Evidence for loci on chromosomes 10 and 16. *Human Genetics, 129*, 59–70. https://doi.org/10.1007/s00439-010-0899-z
103. Camacho, A., Simón, R., Sanz, R., Viñuela, A., Martínez-Salio, A., Mateos, F., et al. (2012). Cognitive and behavioral profile in females with epilepsy with PDCH19 mutation: Two novel mutations and review of the literature. *Epilepsy & Behavior, 24*, 134–137. https://doi.org/10.1016/J.YEBEH.2012.02.023
104. Depienne, C., Trouillard, O., Saint-Martin, C., Gourfinkel-An, I., Bouteiller, D., Carpentier, W., et al. (2009). Spectrum of SCN1A gene mutations associated with Dravet syndrome: Analysis of 333 patients. *Journal of Medical Genetics, 46*, 183–191. https://doi.org/10.1136/jmg.2008.062323
105. Ji, L., Chauhan, A., Brown, W. T., & Chauhan, V. (2009). Increased activities of Na+/K+-ATPase and Ca2+/Mg2+-ATPase in the frontal cortex and cerebellum of autistic individuals. *Life Sciences, 85*, 788–793. https://doi.org/10.1016/j.lfs.2009.10.008
106. Li, B.-M., Liu, X.-R., Yi, Y.-H., Deng, Y. H., Su, T., Zou, X., et al. (2011). Autism in Dravet syndrome: Prevalence, features, and relationship to the clinical characteristics of epilepsy and mental retardation. *Epilepsy & Behavior, 21*, 291–295. https://doi.org/10.1016/j.yebeh.2011.04.060
107. Weiss, L. A., Escayg, A., Kearney, J. A., Trudeau, M., MacDonald, B.T., Mori, M., et al. (2003). Sodium channels SCN1A, SCN2A and SCN3A in familial autism. *Molecular Psychiatry, 8*, 186–194. https://doi.org/10.1038/sj.mp.4001241
108. Han, S., Tai, C., Westenbroek, R. E., Yu, F. H., Cheah, C. S., Potter, G. B., et al. (2012). Autistic-like behaviour in Scn1a+/− mice and rescue by enhanced GABA-mediated neurotransmission. *Nature, 489*, 385–390. https://doi.org/10.1038/nature11356
109. Harkin, L. A., McMahon, J. M., Iona, X., Dibbens, L., Pelekanos, J. L., Zuberi, S. M., et al. (2007). The spectrum of SCN1A-related infantile epileptic encephalopathies. *Brain, 130*, 843–852. https://doi.org/10.1093/brain/awm002
110. Ben-Shalom, R., Keeshen, C. M., Berrios, K. N., An, J. Y., Sanders, S., & Bender, K. (2017). Opposing effects on NaV 1.2 function underlie differences between SCN2A variants observed in individuals with autism spectrum disorder or infantile seizures. *Biological Psychiatry, 82*, 224–232. https://doi.org/10.1016/j.biopsych.2017.01.009.Opposing
111. Sanders, S. J., Murtha, M. T., Gupta, A. R., Murdoch, J. D., Raubeson, M. J., Willsey A. J. et al. (2012). De novo mutations revealed by whole-exome sequencing are strongly associated with autism. *Nature, 485*, 237–241. https://doi.org/10.1038/nature10945
112. De Rubeis, S., He, X., Goldberg, A. P., Poultney, C. S., Samocha, K., Cicek, A. E., et al. (2014). Synaptic, transcriptional and chromatin genes disrupted in autism. *Nature, 515*, 209–215. https://doi.org/10.1038/nature13772
113. Kabir, Z. D., Martínez-Rivera, A., & Rajadhyaksha, A. M. (2017). From gene to behavior: L-type calcium channel mechanisms underlying neuropsychiatric symptoms. *Neurotherapeutics, 14*, 588–613. https://doi.org/10.1007/s13311-017-0532-0

114. Liao, P., & Soong, T. W. (2010). CaV1.2 channelopathies: From arrhythmias to autism, bipolar disorder, and immunodeficiency. *Pflügers Archiv - European Journal of Physiology, 460*, 353–359. https://doi.org/10.1007/s00424-009-0753-0
115. Smith, M., Flodman, P. L., Gargus, J. J., Simon, M. T., Verrell, K., Haas, R., et al. (2012). Mitochondrial and ion channel gene alterations in autism. *Biochimica et Biophysica Acta - Bioenergetics, 1817*, 1796–1802. https://doi.org/10.1016/j.bbabio.2012.04.004
116. Splawski, I., Yoo, D. S., Stotz, S. C., Cherry, A., Clapham, D. E., Keating, M. T. (2006). *CACNA1H* mutations in autism spectrum disorders. *The Journal of Biological Chemistry, 281*, 22085–22091. https://doi.org/10.1074/jbc.M603316200
117. Breitenkamp, A. F. S., Matthes, J., Nass, R. D., Sinzig, J., Lehmkuhl, G., Nürnberg, P., et al. (2014). Rare mutations of CACNB2 found in autism spectrum disease-affected families alter calcium channel function. *PLoS One, 9*. https://doi.org/10.1371/journal.pone.0095579
118. Splawski, I., Timothy, K. W., Sharpe, L. M., Decher, N., Kumar, P., Bloiseet R., et al. (2004). CaV1.2 calcium channel dysfunction causes a multisystem disorder including arrhythmia and autism. *Cell, 119*, 19–31. https://doi.org/10.1016/j.cell.2004.09.011
119. Strom, S. P., Stone, J. L., Ten Bosch, J. R., Merriman, B., Cantor, R. M., Geschwind D. H., et al. (2010). High-density SNP association study of the 17q21 chromosomal region linked to autism identifies CACNA1G as a novel candidate gene. *Molecular Psychiatry, 15*, 996–1005. https://doi.org/10.1038/mp.2009.41
120. D'Adamo, M. C., Catacuzzeno, L., Di Giovanni, G., Franciolini., F & Pessia M., et al. (2013). K+ channelepsy: Progress in the neurobiology of potassium channels and epilepsy. *Frontiers in Cellular Neuroscience, 7*, 134. https://doi.org/10.3389/fncel.2013.00134
121. Sicca, F., Imbrici, P., D'Adamo, M. C., Moro, F., Bonatti, F., Brovedani, P., et al. (2011). Autism with seizures and intellectual disability: Possible causative role of gain-of-function of the inwardly-rectifying K+ channel Kir4.1. *Neurobiology of Disease, 43*, 239–247. https://doi.org/10.1016/J.NBD.2011.03.016
122. Verpelli, C., Schmeisser, M. J., Sala, C., & Boeckers, T. M. (2012). Scaffold proteins at the postsynaptic density. *Advances in Experimental Medicine and Biology*, 29–61.
123. Sheng, M., & Kim, E. (2000). The Shank family of scaffold proteins. *Journal of Cell Science, 113*.
124. Monteiro, P., & Feng, G. (2017). SHANK proteins: Roles at the synapse and in autism spectrum disorder. *Nature Reviews Neuroscience, 18*, 147–157. https://doi.org/10.1038/nrn.2016.183
125. Hung, A. Y., Futai, K., Sala, C., Valtschanoff, J. G., Ryu, J., Woodworth, M. A., et al. (2008). Smaller dendritic spines, weaker synaptic transmission, but enhanced spatial learning in mice lacking Shank1. *The Journal of Neuroscience, 28*, 1697–1708. https://doi.org/10.1523/JNEUROSCI.3032-07.2008
126. Silverman, J. L., Yang, M., Lord, C., & Crawley, J. N. (2010). Behavioural phenotyping assays for mouse models of autism. *Nature Reviews Neuroscience, 11*, 490–502. https://doi.org/10.1038/nrn2851
127. Sungur, A. Ö., Jochner, M. C. E., Harb, H., Rust, M. B. (2017). Aberrant cognitive phenotypes and altered hippocampal BDNF expression related to epigenetic modifications in mice lacking the post-synaptic scaffolding protein SHANK1: Implications for autism spectrum disorder. *Hippocampus, 27*, 906–919. https://doi.org/10.1002/hipo.22741
128. Modi, M. E., Brooks, J. M., Guilmette, E. R., Mercedes, B., Radka G., Dominik R., Hyperactivity and hypermotivation associated with increased striatal mGluR1 signaling in a Shank2 rat model of autism. *Frontiers in Molecular Neuroscience, 11*, 107. https://doi.org/10.3389/FNMOL.2018.00107
129. Won, H., Lee, H.-R., Gee, H. Y., Mah, W., Kim, J. I., Lee, J., et al. (2012). Autistic-like social behaviour in Shank2-mutant mice improved by restoring NMDA receptor function. *Nature, 486*, 261–265. https://doi.org/10.1038/nature11208
130. Boccuto, L., Lauri, M., Sarasua, S. M., Skinner, C. D., Buccella, D., Dwivedi, A., et al. (2013). Prevalence of SHANK3 variants in patients with different subtypes of autism spec-

trum disorders. *European Journal of Human Genetics, 21*, 310–316. https://doi.org/10.1038/ejhg.2012.175

131. Qin, L., Ma, K., Wang, Z. J., Hu, Z., Matas, E., Wei, J., et al. (2018). Social deficits in Shank3-deficient mouse models of autism are rescued by histone deacetylase (HDAC) inhibition. *Nature Neuroscience, 21*, 564–575. https://doi.org/10.1038/s41593-018-0110-8
132. Sarasua, S. M., Dwivedi, A., Boccuto, L., Rollins, J. D., Chen, C. F., Rogers, R. C., et al. (2011). Association between deletion size and important phenotypes expands the genomic region of interest in Phelan-McDermid syndrome (22q13 deletion syndrome). *Journal of Medical Genetics, 48*, 761–766. https://doi.org/10.1136/jmedgenet-2011-100225
133. Sala, C., Piëch, V., Wilson, N. R., Passafaro, M., Liu, G., Sheng, M., et al. (2001). Regulation of dendritic spine morphology and synaptic function by Shank and Homer. *Neuron, 31*, 115–130.
134. Tu, J. C., Xiao, B., Naisbitt, S., Tu, J. C., Xiao, B., Naisbitt, S., et al. (1999). Coupling of mGluR/Homer and PSD-95 complexes by the Shank family of postsynaptic density proteins. *Neuron, 23*, 583–592.
135. Peça, J., Feliciano, C., Ting, J. T., Wang, W., Wells, M. F., Venkatraman, T. N., et al. (2011). Shank3 mutant mice display autistic-like behaviours and striatal dysfunction. *Nature, 472*, 437–442. https://doi.org/10.1038/nature09965
136. Schmeisser, M. J., Ey, E., Wegener, S., Bockmann, J., Stempel, A. V., Kuebler, A., et al. (2012). Autistic-like behaviours and hyperactivity in mice lacking ProSAP1/Shank2. *Nature, 486*, 256–260. https://doi.org/10.1038/nature11015
137. Wang, X., McCoy, P. A., Rodriguiz, R. M., Pan, Y., Je, H. S., Roberts, A. C., et al. (2011). Synaptic dysfunction and abnormal behaviors in mice lacking major isoforms of Shank3. *Human Molecular Genetics, 20*, 3093–3108. https://doi.org/10.1093/hmg/ddr212
138. Han, K., Holder Jr., J. L., Schaaf, C. P., Lu, H., Chen, H., Kang, H., et al. (2013). SHANK3 overexpression causes manic-like behaviour with unique pharmacogenetic properties. *Nature, 503*, 72–77. https://doi.org/10.1038/nature12630
139. Avruch, J., Khokhlatchev, A., Kyriakis, J. M., Luo, Z., Tzivion, G., Vavvas, D., et al. (2001). Ras activation of the Raf kinase: Tyrosine kinase recruitment of the MAP kinase cascade. *Recent Progress in Hormone Research, 56*, 127–155.
140. Neale, B. M., Kou, Y., Liu, L., Samocha, K. E., Sabo, A., Lin, C. F., et al. (2012). Patterns and rates of exonic de novo mutations in autism spectrum disorders. *Nature, 485*, 242–245. https://doi.org/10.1038/nature11011
141. Hay, N., & Sonenberg, N. (2004). Upstream and downstream of mTOR. *Genes & Development, 18*, 1926–1945. https://doi.org/10.1101/gad.1212704
142. Hershey, J. W. B., Sonenberg, N., & Mathews, M. B. (2012). Principles of translational control: An overview. *Cold Spring Harbor Perspectives in Biology, 4*, a011528–a011528. https://doi.org/10.1101/cshperspect.a011528
143. Kumar, V., Zhang, M. X., Swank, M. W., Kunz, J., Wu, G. Y. (2005). Regulation of dendritic morphogenesis by Ras-PI3K-Akt-mTOR and Ras-MAPK signaling pathways. *The Journal of Neuroscience, 25*, 11288–11299. https://doi.org/10.1523/JNEUROSCI.2284-05.2005
144. Banko, J. L., Poulin, F., Hou, L., DeMaria, C. T., Sonenberg, N., Klann E. (2005). The translation repressor 4E-BP2 is critical for eIF4F complex formation, synaptic plasticity, and memory in the hippocampus. *The Journal of Neuroscience, 25*, 9581–9590. https://doi.org/10.1523/JNEUROSCI.2423-05.2005
145. Hou, L., Antion, M. D., Hu, D., Spencer, C. M., Paylor, R., Klann, E. (2006). Dynamic translational and proteasomal regulation of fragile X mental retardation protein controls mGluR-dependent long-term depression. *Neuron, 51*, 441–454. https://doi.org/10.1016/j.neuron.2006.07.005
146. Hou, L., & Klann, E. (2004). Activation of the phosphoinositide 3-kinase-Akt-mammalian target of rapamycin signaling pathway is required for metabotropic glutamate receptor-dependent long-term depression. *The Journal of Neuroscience, 24*, 6352–6361. https://doi.org/10.1523/JNEUROSCI.0995-04.2004

147. Huber, K. M., Roder, J. C., & Bear, M. F. (2001). Chemical induction of mGluR5- and protein synthesis--dependent long-term depression in hippocampal area CA1. *Journal of Neurophysiology, 86*, 321–325.
148. Zheng, F., & Gallagher, J. P. (1992). Metabotropic glutamate receptors are required for the induction of long-term potentiation. *Neuron, 9*, 163–172. https://doi.org/10.1016/0896-6273(92)90231-2
149. Sarbassov, D. D., Ali, S. M., Kim, D. H., Guertin, D. A., Latek, R. R., Erdjument-Bromage, H., et al. (2004). Rictor, a novel binding partner of mTOR, defines a rapamycin-insensitive and raptor-independent pathway that regulates the cytoskeleton. *Current Biology, 14*, 1296–1302. https://doi.org/10.1016/j.cub.2004.06.054
150. Jacinto, E. (2008). What controls TOR? *IUBMB Life, 60*, 483–496. https://doi.org/10.1002/iub.56
151. Kim, D. H., Sarbassov, D. D., Ali, S. M., King, J. E., Latek, R. R., Erdjument-Bromage, H., et al. (2002). mTOR interacts with raptor to form a nutrient-sensitive complex that signals to the cell growth machinery. *Cell, 110*, 163–175.
152. Costa-Mattioli, M., Sossin, W. S., Klann, E., & Sonenberg, N. (2009). Translational control of long-lasting synaptic plasticity and memory. *Neuron, 61*, 10–26. https://doi.org/10.1016/j.neuron.2008.10.055
153. Proud, C. G. (2009). mTORC1 signaling and mRNA translation: Figure 1. *Biochemical Society Transactions, 37*, 227–231. https://doi.org/10.1042/BST0370227
154. Li, J., Yen, C., Liaw, D., Podsypanina, K., Bose, S., Wang, S. I., et al. (1997). PTEN, a putative protein tyrosine phosphatase gene mutated in human brain, breast, and prostate cancer. *Science, 275*, 1943–1947.
155. Endersby, R., & Baker, S. J. (2008). PTEN signaling in brain: Neuropathology and tumorigenesis. *Oncogene, 27*, 5416–5430. https://doi.org/10.1038/onc.2008.239
156. Pende, M., Um, S. H., Mieulet, V., Sticker, M., Goss, V. L., Mestan, J., et al. (2004). S6K1(-/-)/S6K2(-/-) mice exhibit perinatal lethality and rapamycin-sensitive 5′-terminal oligopyrimidine mRNA translation and reveal a mitogen-activated protein kinase-dependent S6 kinase pathway. *Molecular and Cellular Biology, 24*, 3112–3124.
157. Chen, J., Alberts, I., & Li, X. (2014). Dysregulation of the IGF-I/PI3K/AKT/mTOR signaling pathway in autism spectrum disorders. *International Journal of Developmental Neuroscience, 35*, 35–41. https://doi.org/10.1016/j.ijdevneu.2014.03.006
158. Gkogkas, C. G., Khoutorsky, A., Ran, I., Rampakakis, E., Nevarko, T., Weatherill, D. B., et al. (2012). Autism-related deficits via dysregulated eIF4E-dependent translational control. *Nature, 493*, 371–377. https://doi.org/10.1038/nature11628
159. Wang, H., & Doering, L. C. (2013). Reversing autism by targeting downstream mTOR signaling. *Frontiers in Cellular Neuroscience, 7*, 28. https://doi.org/10.3389/fncel.2013.00028
160. Kelleher, R. J., & Bear, M. F. (2008). The autistic neuron: Troubled translation? *Cell, 135*, 401–406. https://doi.org/10.1016/j.cell.2008.10.017
161. Auerbach, B. D., Osterweil, E. K., & Bear, M. F. (2011). Mutations causing syndromic autism define an axis of synaptic pathophysiology. *Nature, 480*, 63–68. https://doi.org/10.1038/nature10658
162. Bourgeron, T. (2009). A synaptic trek to autism. *Current Opinion in Neurobiology, 19*, 231–234. https://doi.org/10.1016/j.conb.2009.06.003
163. Sawicka, K., & Zukin, R. S. (2012). Dysregulation of mTOR signaling in neuropsychiatric disorders: Therapeutic implications. *Neuropsychopharmacology, 37*, 305–306. https://doi.org/10.1038/npp.2011.210
164. Buxbaum, J. D., Cai, G., Chaste, P., Nygren, G., Goldsmith, J., Reichert, J., et al. (2007). Mutation screening of the PTEN gene in patients with autism spectrum disorders and macrocephaly. *American Journal of Medical Genetics Part B: Neuropsychiatric Genetics, 144B*, 484–491. https://doi.org/10.1002/ajmg.b.30493
165. Baker, P., Piven, J., & Sato, Y. (1998). Autism and tuberous sclerosis complex: Prevalence and clinical features. *Journal of Autism and Developmental Disorders, 28*, 279–285.

166. Kwon, C. H., Luikart, B. W., Powell, C. M., Zhou, J., Matheny, S. A., Zhang W., et al. (2006). Pten regulates neuronal arborization and social interaction in mice. *Neuron, 50*, 377–388. https://doi.org/10.1016/j.neuron.2006.03.023
167. Wong, C. W., Or, P. M. Y., Wang, Y., Li, L., Li, J., Yan, M., et al. (2018). Identification of a PTEN mutation with reduced protein stability, phosphatase activity, and nuclear localization in Hong Kong patients with autistic features, neurodevelopmental delays, and macrocephaly. *Autism Research*. https://doi.org/10.1002/aur.1950
168. Smalley, S. L. (1998). Autism and tuberous sclerosis. *Journal of Autism and Developmental Disorders, 28*, 407–414.
169. Smalley, S. L., Tanguay, P. E., Smith, M., & Gutierrez, G. (1992). Autism and tuberous sclerosis. *Journal of Autism and Developmental Disorders, 22*, 339–355.
170. Wiznitzer, M. (2004). Autism and tuberous sclerosis. *Journal of Child Neurology, 19*, 675–679. https://doi.org/10.1177/08830738040190090701
171. Lv, J. W., Cheng, T. L., Qiu, Z. L., & Zhou, W. H. (2013). Role of the PTEN signaling pathway in autism spectrum disorder. *Neuroscience Bulletin, 29*, 773–778. https://doi.org/10.1007/s12264-013-1382-3
172. Meikle, L., Talos, D. M., Onda, H., Pollizzi, K., Rotenberg, A., Sahin, M., et al. (2007). A mouse model of tuberous sclerosis: Neuronal loss of Tsc1 causes dysplastic and ectopic neurons, reduced myelination, seizure activity, and limited survival. *The Journal of Neuroscience, 27*, 5546–5558. https://doi.org/10.1523/JNEUROSCI.5540-06.2007
173. Zeng, L.-H., Rensing, N. R., Zhang, B., Gutmann, D. H., Gambello, M. J., Wong, M., et al. (2011). Tsc2 gene inactivation causes a more severe epilepsy phenotype than Tsc1 inactivation in a mouse model of tuberous sclerosis complex. *Human Molecular Genetics, 20*, 445–454. https://doi.org/10.1093/hmg/ddq491
174. Richter, J. D., & Klann, E. (2009). Making synaptic plasticity and memory last: Mechanisms of translational regulation. *Genes & Development, 23*, 1–11. https://doi.org/10.1101/gad.1735809
175. Richter, J. D., & Sonenberg, N. (2005). Regulation of cap-dependent translation by eIF4E inhibitory proteins. *Nature, 433*, 477–480. https://doi.org/10.1038/nature03205
176. Bourgeron, T. (2007). The possible interplay of synaptic and clock genes in autism spectrum disorders. *Cold Spring Harbor Symposia on Quantitative Biology, 72*, 645–654. https://doi.org/10.1101/sqb.2007.72.020
177. Neves-Pereira, M., Muller, B., Massie, D., Williams, J. H., O'Brien, P. C., Hughes, A., et al. (2009). Deregulation of EIF4E: A novel mechanism for autism. *Journal of Medical Genetics, 46*, 759–765. https://doi.org/10.1136/jmg.2009.066852
178. Rubenstein, J. L. R., & Merzenich, M. M. (2003). Model of autism: Increased ratio of excitation/inhibition in key neural systems. *Genes, Brain, and Behavior, 2*, 255–267.
179. Uhlhaas, P. J., & Singer, W. (2012). Neuronal dynamics and neuropsychiatric disorders: Toward a translational paradigm for dysfunctional large-scale networks. *Neuron, 75*, 963–980. https://doi.org/10.1016/j.neuron.2012.09.004
180. Santoro, M. R., Bray, S. M., & Warren, S. T. (2012). Molecular mechanisms of fragile X syndrome: A twenty-year perspective. *Annual Review of Pathology: Mechanisms of Disease, 7*, 219–245. https://doi.org/10.1146/annurev-pathol-011811-132457
181. Wang, H., Kim, S. S., & Zhuo, M. (2010). Roles of fragile X mental retardation protein in dopaminergic stimulation-induced synapse-associated protein synthesis and subsequent α-amino-3-hydroxyl-5-methyl-4-isoxazole-4-propionate (AMPA) receptor internalization. *The Journal of Biological Chemistry, 285*, 21888–21901. https://doi.org/10.1074/jbc.M110.116293
182. Wang, T., Bray, S. M., & Warren, S. T. (2012). New perspectives on the biology of fragile X syndrome. *Current Opinion in Genetics & Development, 22*, 256–263. https://doi.org/10.1016/j.gde.2012.02.002
183. Jacquemont, S., Hagerman, R. J., Hagerman, P. J., & Leehey, M. A. (2007). Fragile-X syndrome and fragile X-associated tremor/ataxia syndrome: Two faces of FMR1. *Lancet Neurology, 6*, 45–55. https://doi.org/10.1016/S1474-4422(06)70676-7

184. O'Donnell, W. T., & Warren, S. T. (2002). A decade of molecular studies of fragile X syndrome. *Annual Review of Neuroscience, 25*, 315–338. https://doi.org/10.1146/annurev.neuro.25.112701.142909
185. Napoli, I., Mercaldo, V., Boyl, P. P., Eleuteri, B., Zalfa, F., De Rubeis, S., et al. (2008). The fragile X syndrome protein represses activity-dependent translation through CYFIP1, a new 4E-BP. *Cell, 134*, 1042–1054. https://doi.org/10.1016/j.cell.2008.07.031
186. Schenck, A., Bardoni, B., Langmann, C., Harden, N., Mandel, J. L., Giangrande, A. (2003). CYFIP/Sra-1 controls neuronal connectivity in Drosophila and links the Rac1 GTPase pathway to the fragile X protein. *Neuron, 38*, 887–898.
187. Oguro-Ando, A., Rosensweig, C., Herman, E., Nishimura, Y., Werling, D., Bill, B. R., et al. (2015). Increased CYFIP1 dosage alters cellular and dendritic morphology and dysregulates mTOR. *Molecular Psychiatry, 20*, 1069–1078. https://doi.org/10.1038/mp.2014.124
188. Narayanan, U., Nalavadi, V., Nakamoto, M., Thomas, G., Ceman, S., Bassell, G. J., et al. (2008). S6K1 phosphorylates and regulates Fragile X Mental Retardation Protein (FMRP) with the neuronal protein synthesis-dependent Mammalian Target of Rapamycin (mTOR) signaling cascade. *The Journal of Biological Chemistry, 283*, 18478–18482. https://doi.org/10.1074/jbc.C800055200
189. Magdalon, J., Sánchez-Sánchez, S., Griesi-Oliveira, K., & Sertié, A. (2017). Dysfunctional mTORC1 signaling: A convergent mechanism between syndromic and nonsyndromic forms of autism spectrum disorder? *International Journal of Molecular Sciences, 18*, 659. https://doi.org/10.3390/ijms18030659
190. Sharma, A., Hoeffer, C. A., Takayasu, Y., Miyawaki, T., McBride, S. M., Klann, E. (2010). Dysregulation of mTOR signaling in fragile X syndrome. *The Journal of Neuroscience, 30*, 694–702. https://doi.org/10.1523/JNEUROSCI.3696-09.2010
191. Eshraghi, A. A., Liu, G., Kay, S.-I. S., Eshraghi, R. S., Mittal, J., Moshiree, B., et al. (2018). Epigenetics and autism spectrum disorder: Is there a correlation? *Frontiers in Cellular Neuroscience, 12*, 78. https://doi.org/10.3389/fncel.2018.00078
192. LaSalle, J. M. (2013). Epigenomic strategies at the interface of genetic and environmental risk factors for autism. *Journal of Human Genetics, 58*, 396–401. https://doi.org/10.1038/jhg.2013.49
193. Miyake, K., Hirasawa, T., Koide, T., & Kubota, T. (2012). Epigenetics in autism and other neurodevelopmental diseases. *Advances in Experimental Medicine and Biology*, 91–98.
194. Rangasamy, S., D'mello, S. R., & Narayanan, V. (2013). Epigenetics, autism spectrum, and neurodevelopmental disorders. *Neurotherapeutics, 10*(4), 742–756. https://doi.org/10.1007/s13311-013-0227-0
195. Zhubi, A., Cook, E. H., Guidotti, A., & Grayson, D. R. (2014). Epigenetic mechanisms in autism spectrum disorder. *International Review of Neurobiology*, 203–244.
196. Schaevitz, L. R., & Berger-Sweeney, J. E. (2012). Gene-environment interactions and epigenetic pathways in autism: The importance of one-carbon metabolism. *ILAR Journal, 53*, 322–340. https://doi.org/10.1093/ilar.53.3-4.322
197. Volk, H. E., Lurmann, F., Penfold, B., Hertz-Picciotto, I., McConnell, R. (2013). Traffic-related air pollution, particulate matter, and autism. *JAMA Psychiatry, 70*, 71. https://doi.org/10.1001/jamapsychiatry.2013.266
198. Rasalam, A. D., Hailey, H., Williams, J. H. G., Moore, S. J., Turnpenny, P. D., et al. (2005). Characteristics of fetal anticonvulsant syndrome associated autistic disorder. *Developmental Medicine and Child Neurology, 47*, 551–555.
199. Wolstenholme, J. T., Edwards, M., Shetty, S. R. J., Gatewood, J. D., Taylor, J. A., Rissman, E. F., et al. (2012). Gestational exposure to bisphenol A produces transgenerational changes in behaviors and gene expression. *Endocrinology, 153*, 3828–3838. https://doi.org/10.1210/en.2012-1195
200. Kumsta, R., Hummel, E., Chen, F. S., & Heinrichs, M. (2013). Epigenetic regulation of the oxytocin receptor gene: Implications for behavioral neuroscience. *Frontiers in Neuroscience, 7*, 83. https://doi.org/10.3389/fnins.2013.00083

201. Lukas, M., & Neumann, I. D. (2013). Oxytocin and vasopressin in rodent behaviors related to social dysfunctions in autism spectrum disorders. *Behavioural Brain Research, 251*, 85–94. https://doi.org/10.1016/j.bbr.2012.08.011
202. Murakami, G., Hunter, R. G., Fontaine, C., Ribeiro, A., Pfaff, D. (2011). Relationships among estrogen receptor, oxytocin and vasopressin gene expression and social interaction in male mice. *The European Journal of Neuroscience, 34*, 469–477. https://doi.org/10.1111/j.1460-9568.2011.07761.x
203. Veenema, A., & Neumann, I. (2008). Central vasopressin and oxytocin release: Regulation of complex social behaviours. In *Advances in vasopressin and oxytocin - From genes to behaviour to disease* (pp. 261–276). Elsevier.
204. Viberg, H., Mundy, W., & Eriksson, P. (2008). Neonatal exposure to decabrominated diphenyl ether (PBDE 209) results in changes in BDNF, CaMKII and GAP-43, biochemical substrates of neuronal survival, growth, and synaptogenesis. *Neurotoxicology, 29*, 152–159. https://doi.org/10.1016/j.neuro.2007.10.007
205. Ali, A., Cui, X., & Eyles, D. (2018). Developmental vitamin D deficiency and autism: Putative pathogenic mechanisms. *The Journal of Steroid Biochemistry and Molecular Biology, 175*, 108–118. https://doi.org/10.1016/j.jsbmb.2016.12.018
206. Whitehouse, A. J. O., Holt, B. J., Serralha, M., Holt, P. G., Hart, P. H., Kusel, M. M. (2013). Maternal vitamin D levels and the autism phenotype among offspring. *Journal of Autism and Developmental Disorders, 43*, 1495–1504. https://doi.org/10.1007/s10803-012-1676-8
207. Vinkhuyzen, A. A. E., Eyles, D. W., Burne, T. H. J., Blanken, L. M. E., Kruithof C. J., Verhulst, F., et al. (2017). Gestational vitamin D deficiency and autism spectrum disorder. *British Journal of Psychiatry Open, 3*, 85–90. https://doi.org/10.1192/bjpo.bp.116.004077
208. Baccarelli, A., & Bollati, V. (2009). Epigenetics and environmental chemicals. *Current Opinion in Pediatrics, 21*, 243–251.
209. Bahrami, A., Sadeghnia, H. R., Tabatabaeizadeh, S.-A., Bahrami-Taghanaki, H., Behboodi, N., Esmaeili, H., et al. (2018). Genetic and epigenetic factors influencing vitamin D status. *Journal of Cellular Physiology, 233*, 4033–4043. https://doi.org/10.1002/jcp.26216
210. LaSalle, J. M. (2011). A genomic point-of-view on environmental factors influencing the human brain methylome. *Epigenetics, 6*, 862–869. https://doi.org/10.4161/EPI.6.7.16353
211. Fatemi, S. H., Snow, A. V., Stary, J. M., Araghi-Niknam, M., Reutiman, T.J., Suzanne Lee, et al. (2005). Reelin signaling is impaired in autism. *Biological Psychiatry, 57*, 777–787. https://doi.org/10.1016/j.biopsych.2004.12.018
212. Siniscalco, D., Cirillo, A., Bradstreet, J. J., & Antonucci, N. (2013). Epigenetic findings in autism: New perspectives for therapy. *International Journal of Environmental Research and Public Health, 10*, 4261–4273. https://doi.org/10.3390/ijerph10094261
213. Zhu, L., Wang, X., Li, X.-L., Towers, A., Cao, X., Wang, P., et al. (2014). Epigenetic dysregulation of SHANK3 in brain tissues from individuals with autism spectrum disorders. *Human Molecular Genetics, 23*, 1563–1578. https://doi.org/10.1093/hmg/ddt547
214. Behnia, F., Parets, S. E., Kechichian, T., Yin, H., Dutta, E. H., Saade, G. R., et al. (2015). Fetal DNA methylation of autism spectrum disorders candidate genes: Association with spontaneous preterm birth. *American Journal of Obstetrics and Gynecology, 212*, 533.e1–533.e9. https://doi.org/10.1016/j.ajog.2015.02.011
215. Jack, A., Connelly, J. J., & Morris, J. P. (2012). DNA methylation of the oxytocin receptor gene predicts neural response to ambiguous social stimuli. *Frontiers in Human Neuroscience, 6*, 280. https://doi.org/10.3389/fnhum.2012.00280
216. Beri, S., Tonna, N., Menozzi, G., Clara Bonaglia, M. C., Sala, C., Giorda, R., et al. (2007). DNA methylation regulates tissue-specific expression of Shank3. *Journal of Neurochemistry, 101*, 1380–1391. https://doi.org/10.1111/j.1471-4159.2007.04539.x
217. Durand, C. M., Betancur, C., Boeckers, T. M., Bockmann, J., Chaste, P., Fauchereau, F., et al. (2007). Mutations in the gene encoding the synaptic scaffolding protein SHANK3 are associated with autism spectrum disorders. *Nature Genetics, 39*, 25–27. https://doi.org/10.1038/ng1933

218. Uchino, S., & Waga, C. (2013). SHANK3 as an autism spectrum disorder-associated gene. *Brain Dev, 35*, 106–110. https://doi.org/10.1016/j.braindev.2012.05.013
219. Ramaswami, G. (2018). Genetics of autism spectrum disorder. *Handbook of Clinical Neurology, 147*, 321–329. https://doi.org/10.1016/B978-0-444-63233-3.00021-X
220. García-Sáez, A. J. (2012). The secrets of the Bcl-2 family. *Cell Death and Differentiation, 19*, 1733–1740. https://doi.org/10.1038/cdd.2012.105
221. Nguyen, A., Rauch, T. A., Pfeifer, G. P., & Hu, V. W. (2010). Global methylation profiling of lymphoblastoid cell lines reveals epigenetic contributions to autism spectrum disorders and a novel autism candidate gene, RORA, whose protein product is reduced in autistic brain. *The FASEB Journal, 24*, 3036–3051. https://doi.org/10.1096/fj.10-154484
222. Boukhtouche, F., Vodjdani, G., Jarvis, C. I., Bakouche, J., Staels B., Mallet J., et al. (2006). Human retinoic acid receptor-related orphan receptor α1 overexpression protects neurones against oxidative stress-induced apoptosis. *Journal of Neurochemistry, 96*, 1778–1789. https://doi.org/10.1111/j.1471-4159.2006.03708.x
223. Hu, V. W. (2012). Is retinoic acid-related orphan receptor-alpha (RORA) a target for gene–environment interactions contributing to autism? *Neurotoxicology, 33*, 1434–1435. https://doi.org/10.1016/J.NEURO.2012.07.009
224. Nardone, S., Sharan Sams, D., Reuveni, E., Getselter, D., Oron, O., Karpuj, M., et al. (2014). DNA methylation analysis of the autistic brain reveals multiple dysregulated biological pathways. *Translational Psychiatry, 4*, e433. https://doi.org/10.1038/tp.2014.70
225. Berger, S. L. (2007). The complex language of chromatin regulation during transcription. *Nature, 447*, 407–412. https://doi.org/10.1038/nature05915
226. Akbarian, S., & Huang, H.-S. (2009). Epigenetic regulation in human brain—Focus on histone lysine methylation. *Biological Psychiatry, 65*, 198–203. https://doi.org/10.1016/j.biopsych.2008.08.015
227. Shulha, H. P., Cheung, I., Whittle, C., Wang, J., Virgil, D., Lin, C. L., et al. (2012). Epigenetic signatures of autism. *Archives of General Psychiatry, 69*, 314. https://doi.org/10.1001/archgenpsychiatry.2011.151
228. Adegbola, A., Gao, H., Sommer, S., & Browning, M. (2008). A novel mutation inJARID1C/SMCX in a patient with autism spectrum disorder (ASD). *American Journal of Medical Genetics - Part A, 146A*, 505–511. https://doi.org/10.1002/ajmg.a.32142
229. Wang, H., Duclot, F., Liu, Y., Wang, Z., & Mohamed, Kabbaj. (2013). Histone deacetylase inhibitors facilitate partner preference formation in female prairie voles. *Nature Neuroscience, 16*, 919–924. https://doi.org/10.1038/nn.3420
230. Sun, W., Poschmann, J., Cruz-Herrera del Rosario, R., Parikshak, N. N., Hajan H. S., Kumar, V., et al. (2016). Histone acetylome-wide association study of autism spectrum disorder. *Cell, 167*, 1385–1397.e11. https://doi.org/10.1016/j.cell.2016.10.031
231. James, S. J., Shpyleva, S., Melnyk, S., Pavliv, O., Pogribny, I. P., (2013). Complex epigenetic regulation of Engrailed-2 (EN-2) homeobox gene in the autism cerebellum. *Translational Psychiatry, 3*, e232–e238. https://doi.org/10.1038/tp.2013.8
232. Kalkbrenner, A. E., Windham, G. C., Serre, M. L., Akita, Y., Wang, X., Hoffman K., et al. (2015). Particulate matter exposure, prenatal and postnatal windows of susceptibility, and autism spectrum disorders. *Epidemiology, 26*, 30–42. https://doi.org/10.1097/EDE.0000000000000173
233. Gekas, C., Dieterlen-Lièvre, F., Orkin, S. H., & Mikkola, H. K. A. (2005). The placenta is a niche for hematopoietic stem cells. *Developmental Cell, 8*, 365–375. https://doi.org/10.1016/j.devcel.2004.12.016
234. Patterson, P. H. (2009). Immune involvement in schizophrenia and autism: Etiology, pathology and animal models. *Behavioural Brain Research, 204*, 313–321. https://doi.org/10.1016/j.bbr.2008.12.016
235. Rhodes, K. E., Gekas, C., Wang, Y., Lux, C. T., Francis, C. S., Chan, D. N., et al. (2008). The emergence of hematopoietic stem cells is initiated in the placental vasculature in the absence of circulation. *Cell Stem Cell, 2*, 252–263. https://doi.org/10.1016/j.stem.2008.01.001

236. Takeda, A. (2001). Zinc homeostasis and functions of zinc in the brain. *BioMetals, 14*, 343–351. https://doi.org/10.1023/A:1012982123386
237. Russo, A. J. (2011). Increased copper in individuals with autism normalizes post zinc therapy more efficiently in individuals with concurrent GI disease. *Nutrition and Metabolic Insights, 4*, 49–54. https://doi.org/10.4137/NMI.S6827
238. Yasuda, H., Yoshida, K., Yasuda, Y., & Tsutsui, T. (2011). Infantile zinc deficiency: Association with autism spectrum disorders. *Scientific Reports, 1*, 129. https://doi.org/10.1038/srep00129
239. Grabrucker, A. M., Knight, M. J., Proepper, C., Bockmann, J., Joubert, M., Rowan, M., et al. (2011). Concerted action of zinc and ProSAP/Shank in synaptogenesis and synapse maturation. *The EMBO Journal, 30*, 569–581. https://doi.org/10.1038/emboj.2010.336
240. Cezar, L. C., Kirsten, T. B., da Fonseca, C. C. N., de Lima, A. P. N., Bernardi, M. M., Felicio, L. F. (2018). Zinc as a therapy in a rat model of autism prenatally induced by valproic acid. *Progress in Neuro-Psychopharmacology and Biological Psychiatry, 84*, 173–180. https://doi.org/10.1016/J.PNPBP.2018.02.008
241. Huang, Y. Z., Pan, E., Xiong, Z.-Q., & McNamara, J. O. (2008). Zinc-mediated transactivation of TrkB potentiates the hippocampal mossy fiber-CA3 pyramid synapse. *Neuron, 57*, 546–558. https://doi.org/10.1016/J.NEURON.2007.11.026
242. Nuttall, J. R., & Oteiza, P. I. (2012). Zinc and the ERK kinases in the developing brain. *Neurotoxicity Research, 21*, 128–141. https://doi.org/10.1007/s12640-011-9291-6
243. Rossignol, D. A., & Frye, R. E. (2011). Melatonin in autism spectrum disorders: A systematic review and meta-analysis. *Developmental Medicine and Child Neurology, 53*, 783–792. https://doi.org/10.1111/j.1469-8749.2011.03980.x
244. Veatch, O. J., Goldman, S. E., Adkins, K. W., & Malow, B. A. (2015). Melatonin in children with autism spectrum disorders: How does the evidence fit together? *Journal of Nature and Science, 1*, e125.
245. Rossignol, D. A., & Frye, R. E. (2012). A review of research trends in physiological abnormalities in autism spectrum disorders: Immune dysregulation, inflammation, oxidative stress, mitochondrial dysfunction and environmental toxicant exposures. *Molecular Psychiatry, 17*, 389–401. https://doi.org/10.1038/mp.2011.165
246. Won, J., Jin, Y., Lee, T. H., Park, S., Lee, T. H., Lee S. R., et al. (2017). Melatonin as an interventional novel candidate for the individual with autistic fragile X syndrome in human
247. Gardener, H., Spiegelman, D., & Buka, S. L. (2009). Prenatal risk factors for autism: Comprehensive meta-analysis. *The British Journal of Psychiatry, 195*, 7–14. https://doi.org/10.1192/bjp.bp.108.051672
248. Kinney, D. K., Munir, K. M., Crowley, D. J., & Miller, A. M. (2008). Prenatal stress and risk for autism. *Neuroscience and Biobehavioral Reviews, 32*, 1519–1532. https://doi.org/10.1016/j.neubiorev.2008.06.004
249. Charil, A., Laplante, D. P., Vaillancourt, C., & King, S. (2010). Prenatal stress and brain development. *Brain Research Reviews, 65*, 56–79. https://doi.org/10.1016/J.BRAINRESREV.2010.06.002
250. Durkin, M. S., Maenner, M. J., Newschaffer, C. J., Lee, L. C., Cunniff, C. M., Daniels, J. L., et al. (2008). Advanced parental age and the risk of autism spectrum disorder. *American Journal of Epidemiology, 168*, 1268–1276. https://doi.org/10.1093/aje/kwn250
251. Reichenberg, A., Gross, R., Weiser, M., Bresnahan, M., Silverman, J., Harlap, S., et al. (2006). Advancing paternal age and autism. *Archives of General Psychiatry, 63*, 1026. https://doi.org/10.1001/archpsyc.63.9.1026
252. Chandley, A. C. (1991). On the parental origin of de novo mutation in man. *Journal of Medical Genetics, 28*, 217–223. https://doi.org/10.1136/JMG.28.4.217
253. Bennett, G. D., Wlodarczyk, B., Calvin, J. A., Craig, J. C., Finnell, R. H. (2000). Valproic acid-induced alterations in growth and neurotrophic factor gene expression in murine embryos [corrected]. *Reproductive Toxicology, 14*, 1–11. https://doi.org/10.1016/S0890-6238(99)00064-7
254. Du, L., Zhao, G., Duan, Z., & Li, F. (2017). Behavioral improvements in a valproic acid rat model of autism following vitamin D supplementation. *Psychiatry Research, 253*, 28–32. https://doi.org/10.1016/j.psychres.2017.03.003

255. Kumar, H., & Sharma, B. (2016). Memantine ameliorates autistic behavior, biochemistry & blood brain barrier impairments in rats. *Brain Research Bulletin, 124*, 27–39. https://doi.org/10.1016/j.brainresbull.2016.03.013
256. Ornoy, A., Weinstein-Fudim, L., Tfilin, M., Ergaz, Z., Yanai, J., Szyf, M., et al. (2018). S-adenosyl methionine prevents ASD like behaviors triggered by early postnatal valproic acid exposure in very young mice. *Neurotoxicology and Teratology*. https://doi.org/10.1016/J.NTT.2018.01.005
257. Kolozsi, E., Mackenzie, R. N., Roullet, F. I., de Catanzaro, D., Foster, J. A. (2009). Prenatal exposure to valproic acid leads to reduced expression of synaptic adhesion molecule neuroligin 3 in mice. *Neuroscience, 163*, 1201–1210. https://doi.org/10.1016/J.NEUROSCIENCE.2009.07.021
258. Smithells, R. W., & Newman, C. G. (1992). Recognition of thalidomide defects. *Journal of Medical Genetics, 29*, 716–723.
259. Kumar, V., Harjai, K., & Chhibber, S. (2010). Thalidomide treatment modulates macrophage pro-inflammatory function and cytokine levels in Klebsiella pneumoniae B5055 induced pneumonia in BALB/c mice. *International Immunopharmacology, 10*, 777–783. https://doi.org/10.1016/J.INTIMP.2010.04.008
260. Roberts, E. M., English, P. B., Grether, J. K., Windham, G. C., Somberg, L., & Wolff, C. (2007). Maternal residence near agricultural pesticide applications and autism spectrum disorders among children in the California Central Valley. *Environmental Health Perspectives, 115*, 1482–1489. https://doi.org/10.1289/ehp.10168
261. Kinney, D. K., Barch, D. H., Chayka, B., Napoleon, S., Munir, K. M. (2010). Environmental risk factors for autism: Do they help cause de novo genetic mutations that contribute to the disorder? *Medical Hypotheses, 74*, 102–106. https://doi.org/10.1016/j.mehy.2009.07.052
262. Windham, G. C., Zhang, L., Gunier, R., Croen, L. A., Grether, J. K. (2006). Autism spectrum disorders in relation to the distribution of hazardous air pollutants in the San Francisco Bay Area. *Environmental Health Perspectives, 114*, 1438–1444. https://doi.org/10.1289/ehp.9120
263. Bagasra, O., Golkar, Z., Garcia, M., Rice, L. N., Pace, D. G. (2013). Role of perfumes in pathogenesis of Autism. *Medical Hypotheses, 80*, 795–803. https://doi.org/10.1016/J.MEHY.2013.03.014
264. Sealey, L. A., Hughes, B. W., Sriskanda, A. N., Guest J. R., Gibson, A. D., Johnson-Williams L., et al. (2016). Environmental factors in the development of autism spectrum disorders. *Environment International, 88*, 288–298. https://doi.org/10.1016/J.ENVINT.2015.12.021
265. Shaw, W. (2017). Elevated urinary glyphosate and clostridia metabolites with altered dopamine metabolism in triplets with autistic spectrum disorder or suspected seizure disorder: A case study. *Integrative Medicine (Encinitas), 16*, 50–57.
266. Mold, M., Umar, D., King, A., & Exley, C. (2018). Aluminium in brain tissue in autism. *Journal of Trace Elements in Medicine and Biology, 46*, 76–82. https://doi.org/10.1016/j.jtemb.2017.11.012
267. Kim, S. M., Han, D. H., Lyoo, H. S., Min, K. J., Kim, K. H., Renshaw, P. (2010). Exposure to environmental toxins in mothers of children with autism spectrum disorder. *Psychiatry Investigation, 7*, 122–127. https://doi.org/10.4306/pi.2010.7.2.122
268. Cristiano, C., Lama, A., Lembo, F., Mollica, M. P., Calignano, A., & Raso G. M. (2018). Interplay between peripheral and central inflammation in autism spectrum disorders: Possible nutritional and therapeutic strategies. *Frontiers in Physiology, 9*, 184. https://doi.org/10.3389/fphys.2018.00184
269. Vuong, H. E., & Hsiao, E. Y. (2017). Emerging roles for the gut microbiome in autism spectrum disorder. *Biological Psychiatry, 81*, 411–423. https://doi.org/10.1016/J.BIOPSYCH.2016.08.024
270. Bhandari, R., & Kuhad, A. (2015). Neuropsychopharmacotherapeutic efficacy of curcumin in experimental paradigm of autism spectrum disorders. *Life Sciences, 141*, 156–169. https://doi.org/10.1016/j.lfs.2015.09.012
271. Bhandari, R., & Kuhad, A. (2017). Resveratrol suppresses neuroinflammation in the experimental paradigm of autism spectrum disorders. *Neurochemistry International, 103*, 8–23. https://doi.org/10.1016/j.neuint.2016.12.012

272. Choi, J., Lee, S., Won, J., Jin, Y., Hong, Y., Hur, T. Y., et al. (2018). Pathophysiological and neurobehavioral characteristics of a propionic acid-mediated autism-like rat model. *PLoS One, 13*, e0192925. https://doi.org/10.1371/journal.pone.0192925
273. Frye, R. E., Melnyk, S., & MacFabe, D. F. (2013). Unique acyl-carnitine profiles are potential biomarkers for acquired mitochondrial disease in autism spectrum disorder. *Translational Psychiatry, 3*, e220. https://doi.org/10.1038/tp.2012.143
274. Macfabe, D. F. (2012). Short-chain fatty acid fermentation products of the gut microbiome: Implications in autism spectrum disorders. *Microbial Ecology in Health and Disease, 23*. https://doi.org/10.3402/mehd.v23i0.19260
275. Ha, H. T. T., Leal-Ortiz, S., Lalwani, K., Kiyonaka, S., Hamachi, I., Mysore, S. P., Montgomery, J. M., et al. (2018). Shank and zinc mediate an AMPA receptor subunit switch in developing neurons. *Frontiers in Molecular Neuroscience, 11*, 405. https://doi.org/10.3389/fnmol.2018.00405
276. Hershfinkel, M. (2014). The zinc-sensing receptor, ZnR/GPR39: Signaling and significance. In *Zinc signals in cellular functions and disorders* (pp. 111–133). Tokyo: Springer.
277. Goines, P. E., & Ashwood, P. (2013). Cytokine dysregulation in autism spectrum disorders (ASD): Possible role of the environment. *Neurotoxicology and Teratology, 36*, 67–81. https://doi.org/10.1016/j.ntt.2012.07.006.Cytokine
278. Pickering, M., Cumiskey, D., & O'Connor, J. J. (2005). Actions of TNF-α on glutamatergic synaptic transmission in the central nervous system. *Experimental Physiology, 90*, 663–670. https://doi.org/10.1113/expphysiol.2005.030734
279. Ohja, K., Gozal, E., Fahnestock, M., Cai, L., Cai, J., Freedmanet J. H., et al. (2017). Neuroimmunologic and neurotrophic interactions in autism spectrum disorders: Relationship to neuroinflammation. *Neuromolecular Medicine, 1*. https://doi.org/10.1007/s12017-018-8488-8
280. Zeidán-Chuliá, F., Salmina, A. B., Malinovskaya, N. A., Noda, M., Verkhratsky, A., Moreira, J. C. (2014). The glial perspective of autism spectrum disorders. *Neuroscience and Biobehavioral Reviews, 38*, 160–172. https://doi.org/10.1016/j.neubiorev.2013.11.008
281. Ozawa, S., Kamiya, H., & Tsuzuki, K. (1998). Glutamate receptors in the mammalian central nervous system. *Progress in Neurobiology, 54*, 581–618. https://doi.org/10.1016/S0301-0082(97)00085-3
282. Bauman, M. L. (2010). Medical comorbidities in autism: Challenges to diagnosis and treatment. *Neurotherapeutics, 7*, 320–327. https://doi.org/10.1016/j.nurt.2010.06.001
283. Valiente-Pallejà, A., Torrell, H., Muntané, G., Cortés, M. J., Martínez-Leal, R., Abasolo, N., et al. (2018). Genetic and clinical evidence of mitochondrial dysfunction in autism spectrum disorder and intellectual disability. *Human Molecular Genetics, 27*, 891–900. https://doi.org/10.1093/hmg/ddy009
284. Li, X., Chauhan, A., Sheikh, A. M., Patil, S., Chauhan, V., Li, X. M., et al. (2009). Elevated immune response in the brain of autistic patients. *Journal of Neuroimmunology, 207*, 111–116. https://doi.org/10.1016/j.jneuroim.2008.12.002
285. Ballas, N., Lioy, D. T., Grunseich, C., & Mandel, G. (2009). Non-cell autonomous influence of MeCP2-deficient glia on neuronal dendritic morphology. *Nature Neuroscience, 12*, 311–317. https://doi.org/10.1038/nn.2275
286. Lull, M. E., & Block, M. L. (2010). Microglial activation and chronic neurodegeneration. *Neurotherapeutics, 7*, 354–365. https://doi.org/10.1016/j.nurt.2010.05.014
287. Mazahery, H., Stonehouse, W., Delshad, M., Kruger, M. C., Conlon, C. A., Beck, K. L., et al. (2017). Relationship between long chain n-3 polyunsaturated fatty acids and autism spectrum disorder: Systematic review and meta-analysis of case-control and randomised controlled trials. *Nutrients, 9*, 155. https://doi.org/10.3390/nu9020155
288. Hertz, L., Dringen, R., Schousboe, A., & Robinson, S. R. (1999). Astrocytes: Glutamate producers for neurons. *Journal of Neuroscience Research, 57*, 417–428.
289. Iadecola, C., & Nedergaard, M. (2007). Glial regulation of the cerebral microvasculature. *Nature Neuroscience, 10*, 1369–1376. https://doi.org/10.1038/nn2003

290. Wang, D., & Bordey, A. (2008). The astrocyte odyssey. *Progress in Neurobiology, 86*, 342–367. https://doi.org/10.1016/j.pneurobio.2008.09.015
291. Li, X., Bijur, G. N., & Jope, R. S. (2002). Glycogen synthase kinase-3beta, mood stabilizers, and neuroprotection. *Bipolar Disorders, 4*, 137–144.
292. Carmody, D. P., & Lewis, M. (2010). Regional white matter development in children with autism spectrum disorders. *Developmental Psychobiology, 52*, 755–763. https://doi.org/10.1002/dev.20471
293. Corrigan, N. M., Shaw, D. W. W., Estes, A. M., Todd L, R., Jeff, M., Friedman, S. D., et al. (2013). Atypical developmental patterns of brain chemistry in children with autism spectrum disorder. *JAMA Psychiatry, 70*, 964. https://doi.org/10.1001/jamapsychiatry.2013.1388
294. Horder, J., Lavender, T., Mendez, M. A., O'Gorman, R., Daly, E., Craig, M. C., et al. (2013). Reduced subcortical glutamate/glutamine in adults with autism spectrum disorders: A [1H] MRS study. *Translational Psychiatry, 3*, e279. https://doi.org/10.1038/tp.2013.53
295. Stephenson, D. T., O'Neill, S. M., Narayan, S., Tiwari, A., Arnold, E., Samaroo, H. D., et al. (2011). Histopathologic characterization of the BTBR mouse model of autistic-like behavior reveals selective changes in neurodevelopmental proteins and adult hippocampal neurogenesis. *Molecular Autism, 2*, 7. https://doi.org/10.1186/2040-2392-2-7
296. Morgan, J. T., Chana, G., Pardo, C. A., Achim, C., Semendeferi, K., Buckwalter, J., et al. (2010). Microglial activation and increased microglial density observed in the dorsolateral prefrontal cortex in autism. *Biological Psychiatry, 68*, 368–376. https://doi.org/10.1016/j.biopsych.2010.05.024
297. Suzuki, K., Sugihara, G., Ouchi, Y., Nakamura, K., Futatsubashi, M., Takebayashi, K., et al. (2013). Microglial activation in young adults with autism spectrum disorder. *JAMA Psychiatry, 70*, 49. https://doi.org/10.1001/jamapsychiatry.2013.272
298. Vargas, D. L., Nascimbene, C., Krishnan, C., Zimmerman, A. W., Pardo, C. A. (2005). Neuroglial activation and neuroinflammation in the brain of patients with autism. *Annals of Neurology, 57*, 67–81. https://doi.org/10.1002/ana.20315

Maternal Prenatal Exposures in Pregnancy and Autism Spectrum Disorder: An Insight into the Epigenetics of Drugs and Diet as Key Environmental Influences

Kholoud N. Bastaki, Sura Alwan, and Farah R. Zahir

Abstract Autism spectrum disorder (ASD) is a rapidly growing global pandemic that affects an estimated 1 in 59–68 children. It is a complex disease with both genetic and environmental etiologies. Due to the rapid increase in the incidence of ASD, environmental causes for ASD are gaining attention. Efforts to probe several environmental exposures that could contribute to causing ASD are underway. In this regard, this chapter is directed towards understanding prenatal exposure to key environmental factors i.e., drugs and dietary nutrients that may act via the same molecular pathway - epigenetics as a potential etiological factor for ASD. Epigenetic regulation is a molecular mechanism known to be a significant contributor to neurodevelopmental disorders. It also offers a means to explain how environmental exposures can impact genetics. We discuss the impact of maternal exposures to certain drugs, and dietary intake, on the developing fetus during pregnancy. Maternal Exposure to some drugs during gestation are associated with a higher risk of ASD, while exposure to other dietary compounds may offer promise to rescue epigenetic regulatory insults related to ASD. However, more work in this important area is still required, nevertheless preliminary research already has important implications in the understanding, prevention and treatment of ASD.

Keywords Autism spectrum disorder · Epigenetic regulation · Prenatal drugs · Epigenetic diet · Epigenetic drugs

---

K. N. Bastaki
Hamad Bin Khalifa University, Doha, Qatar
e-mail: khbastaki@mail.hbku.edu.qa

S. Alwan · F. R. Zahir (✉)
University of British Columbia, Vancouver, BC, Canada
e-mail: alwans@bcchr.ca; farahz@bcchr.ca

M. M. Essa, M. W. Qoronfleh (eds.), *Personalized Food Intervention and Therapy for Autism Spectrum Disorder Management*, Advances in Neurobiology 24,
https://doi.org/10.1007/978-3-030-30402-7_5

1 Introduction to NDs in General and ASD in Particular

Autism spectrum disorder (ASD) is an umbrella term used to describe an expanse of neurodevelopmental disorders (NDs), mainly characterized by deficits in social interaction and communication and repetitive patterns of behavior (see Box 1 for a summarized definition of ASD from the most recent edition of the American Psychiatric Association handbook, The Diagnostic and Statistical Manual of Mental Disorders (DSM-5), a globally accepted standard [1]). In this context it is important to note that ASD is one of several types of NDs. Symptoms found in ASD can also be found in other types of ND, notably in intellectual disability (ID) [2], as each is an umbrella term encompassing a wide spectrum of disease presentations.

Both ASD and ID are not considered single disorders, rather they are on a spectrum defined by a set of common criteria that are broad in nature. ASD often appears as a co-morbidity in children primarily diagnosed with ID and *vice versa*, with reports of up to 70% of ASD patients' having ID [3, 4]. In addition, several other co-morbidities, such as; major congenital anomalies, blindness, deafness, motor dysfunction, cerebral palsy, and epilepsy for example, are found among patients with ID/ASD or both [4]. This is important as it implies that evidence for causation of other ND conditions and co-morbidities, when discussing prenatal exposure to possible ASD environmental risk factors, may also play a role in the development of ASD. Therefore in this chapter, while we will focus on ASD, we will not restrict ourselves to it alone, and consider what is known about the role of epigenetics in prenatal exposures to the etiology of ND in general, when necessary.

Currently, NDs are among the most commonly diagnosed conditions, globally. According to a parental survey, around 15% of children aged between 3 and 17 years were affected by NDs, in the USA alone. These include ASD, attention deficit hyperactivity disorder (ADHD), learning disabilities, ID, cerebral palsy, seizures, stuttering or stammering, moderate to profound hearing loss, blindness, and other developmental delays [5, 6]. A study in 2016 found that an estimated 1 in every 68 children in the USA had ASD [7], and it is currently recognized as one of the most common disorders worldwide [8]. However, an update of the estimated prevalence of ASD among children in the USA released by the Centers for Disease Control and Prevention (CDC) reported a 15% increase from 2012 to 2014 (1 in 68 children in 2012 to 1 in 59 children in 2014) [9]. Estimated ASD prevalence was at 2.47% among US children and adolescents in 2014–2016 (95% confidence intervals, 2.20–2.73%) [9, 10]. Alarmingly these data do not stand alone, as ASD is currently recognized as being a burgeoning global pandemic [11, 12].

Therefore taken together, NDs are among the most prevalent disorders globally, and as they appear in childhood, they present an extreme burden of cost of care over lifespan, accounting for costs greater than that combined for heart disease, cancer and stroke [13]. This underscores the urgent need to find effective prophylactic and therapeutic strategies. The first step toward combating any disease condition is to understand what causes it. We will now provide an overview of what is known about causes and other risk factors for ASD below.

Box 1 Autism Spectrum Disorder Definition as given by DMS V [1]
ASD is diagnosed when all five of the following major symptoms are present in a child.

A. Persistent deficits in social communication and social interaction across multiple contexts.
B. Restricted, repetitive patterns of behavior, interests, or activities, as manifested by at least two of the following, currently or by history (examples are illustrative).
    1. Stereotyped or repetitive motor movements, use of objects, or speech (e.g., simple motor stereotypies, lining up toys or flipping objects, echolalia, idiosyncratic phrases).
    2. Insistence on sameness, inflexible adherence to routines, or ritualized patterns of verbal or nonverbal behavior (e.g., extreme distress at small changes, difficulties with transitions, rigid thinking patterns, greeting rituals, need to take same route or eat same food every day).
    3. Highly restricted, fixated interests that are abnormal in intensity or focus (e.g., strong attachment to or preoccupation with unusual objects, excessively circumscribed or perseverative interest).
    4. Hyper- or hypo-reactivity to sensory input or unusual interests in sensory aspects of the environment (e.g., apparent indifference to pain/temperature, adverse response to specific sounds or textures, excessive smelling or touching of objects, visual fascination with lights or movement).
C. Symptoms must be present in the early developmental period (but may not fully manifest until social demands exceed limited capacities or may be masked by learned strategies in later stages of life).
D. Symptoms cause clinically significant impairment in social, occupational, or other important areas of current functioning.
E. These disturbances are not better explained by ID (intellectual developmental disorder) or global developmental delay. ID and ASD frequently co-occur; to make comorbid diagnoses of ASD and ID, social communication should be below that expected for general developmental level.

Note: Individuals with a well-established DSM-IV diagnosis of autistic disorder, Asperger's disorder, or pervasive developmental disorder not otherwise specified should be given the diagnosis of ASD. Individuals who have marked deficits in social communication, but whose symptoms do not otherwise meet criteria for autism spectrum disorder, should be evaluated for social (pragmatic) communication disorder.

2 ASD Etiology

ASD is considered a complex disorder; one caused by the interaction of genetics with the environment. Although ASD is known to be highly heritable [14, 15], only 10–20% of patients are diagnosed with a definitive genetic cause [15]. Very low diagnostic yield despite comprehensive genetic screens has led to a significant "missing heritability" problem in ASD research.

The missing heritability conundrum has shifted attention to the environmental component as a likely explanation for the increased rates of ASD diagnoses. Several environmental risk factors have been suggested to contribute to the development of ASD in existing literature, including air pollution, pesticide exposure via food and otherwise, plastics, psychosocial and socio economic factors that influence lifestyle and family, maternal obesity and metabolic conditions such as diabetes during pregnancy, maternal mental illness, prenatal and delivery complications and the use of certain supplements and medications during pregnancy [16, 17]. Importantly, the breadth of possible environmental risk factors, taken together with the extremely low percentage of ASD caused by 100% penetrant genetic factors (i.e., only a small fraction of ASD patients are found to have disease due to purely genetic causes, despite it being known to be a highly heritable disorder), emphasizes that ASD is truly a complex disorder. Multifactorial "causal pies" [16] comprising of more than one environmental and/or genetic factor that act in concert, may often be what causes this disorder to manifest.

It is possible that some environmental factors exert their adverse developmental effects via mechanisms that influence the genome, in which context a genome regulatory mechanism termed *epigenetics* is of particular interest (see Box 2 for a brief primer on Epigenetics). Epigenetics refers to control of the genome by external factors by a process termed epigenetic regulation. These factors are of two main types: (a) enzymes that catalyze epigenetic regulatory reactions, which themselves are encoded by genes, such as DNA methyltransferases, which add methyl groups onto the DNA strand, and (b) chemical moieties which are substrates (e.g., methyl groups, ethyl groups, etc.), that are supplied by the cellular environment. Physiological disturbances of either via maternal environmental exposures during prenatal developmental, may have the potential to contribute to ND/ASD development.

3 Epigenetics and ASD

The initial interest in epigenetic deregulation as a mechanism important in the development of ASD was fueled by the observation that several single gene disorders, that include ASD in their presentation, are caused by perturbation of the genes that encode enzymes involved in epigenetic regulation [18–20]. The subsequent recognition that genes having a role in epigenetic programming are among the most frequently mutated genes related to ASD [21, 22], further shone the spotlight on epigenetic deregulation as a possible prime etiological mechanism for ASD/ND. Recently, epigenetic deregulation was recognized to be as significant a cause for ND, as the long-time lead molecular causative mechanism—defects in synaptogenesis [23]. As genome-wide epigenome screens started becoming more affordable, efforts got underway to profile the epigenome (i.e., the DNA methylation profile, and histone modification profiles for example) in an attempt to search for etiological clues, with the focus shifting to characterization of epigenomic changes, regardless of the underlying genetics, as causative profiles for ND in general, and ASD in particular. Epigenetic profile alterations such as DNA methylation for example, have been suggested as key contributing factors for ASD development at the genome- [24] and gene-level [25]. Eshraghi et al. [26], provide a comprehensive summary of epigenomic profile changes associated with ASD. These include DNA methylation changes that are associated with maternal health conditions [27], DNA methylation changes in the placenta of subjects with ASD [28], histone acetylation changes in syndromic and non-syndromic ASD cases [29], and even RNAi signatures associated with ASD [30].

Box 2 Epigenetics Primer

The term Epigenetics comes from the Greek word "epi" for "over/above/on top of," thus, epigenetics refer to heritable traits that are not encoded in the DNA sequence, rather they are formed by changes to factors that sit "on top" of the genome and thereby regulate gene expression. Epigenetic processes include three main components;

1. DNA methylation
2. Histone modification
3. Chromatin remodeling

Together, all three processes cause genes to be either turned on, or shut down, as well as control fine-tuning of gene expression via non-coding RNA dependent mechanisms [31]. Particularly, histone modification and chromatin remodeling act in concert. DNA methylation at gene promoter sites which often contain CpG islands, is a driver of both histone modification and chromatin remodeling [20]. Figure 1 below is a reproduction of Figure 1 from the Zahir and Brown review paper [20] of the impact of epigenetic processes on neurodevelopment, and shows the cross-talk between the three processes above.

In addition a fourth component, RNAi (RNA interference) is a mechanism by which non-coding RNA interacts with some epigenetic components, or with gene transcripts directly in order to regulate genic expression, and is gaining prominence in regulation for neurodevelopmental and neuro-functional processes. It is often included as a fourth epigenetic process; RNA interference based gene expression regulation.

Box 2 (continued)

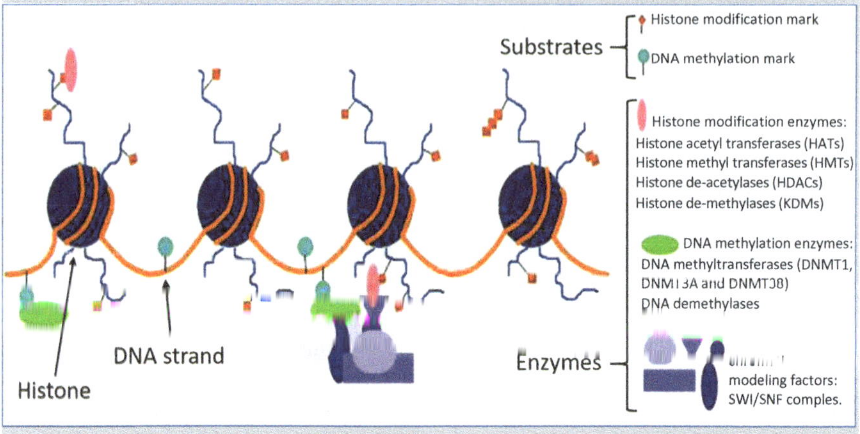

Fig. 1 Modified reproduced Figure 1 from Zahir and Brown. *Ped. Research*. 2011. Caption for the image – Interactions between DNA methylation, histone modification and chromatin remodeling. The DNA strand is wrapped around histone protein cores to form repeating nucleosomes that make up chromatin. Histone tail modifications, are attached to histone tails, and DNA methylation marks are attached to the DNA strand. Epigenetic regulators that have DNA methylation, histone modification or chromatin remodeling interact with each other and display cross-recruitment

4 Environmentally Controlled Epigenetic Regulatory Impacts in utero and its Role in ASD

As noted previously, epigenetic regulation can be disrupted during the neurodevelopmental phase by an environmental stimulus either by altering the enzymes that lay down epigenetic marks, or by altering the substrate concentrations. In this regard, maternal ingestion of certain substances during pregnancy, such as drugs and nutritional supplements, that are suggested to induce epigenetic changes, is of prime importance in impacting embryonic and fetal neurodevelopment. We discuss the evidence of their influence below.

4.1 The Impact of Drugs Taken in utero on Epigenetic Regulation During Development

Epigenetic regulation is impacted by drugs primarily via disruption of the enzymatic processes involved, while nutritional supplements on the other hand, influence the concentration of substrate available (see Box 1). However, both drugs and

Table 1 Compounds and drugs that act via epigenetic mechanisms

Epigenetic mechanism involved	Example compounds/drugs
DNA methyltransferase (DNMT) inhibitors Reviewed by Yang et al. [34], Gnyszka et al. [35], and Ahuja et al. [36]	5-Azacytidine Decitabine Zebularine Non-nucleoside DNMT inhibitors (hydralazine and procainamide)
Histone deacetylase (HDAC) inhibitors Reviewed by Eckschlager et al. [37], Ahuja et al. [36], and Goey et al. [38]	Short chain fatty acids such as valproic acid Benzamides such as entinostat and tacedinaline Hydroxamic acids such as resminostat, abexinostat, quisinostat, rocilinostat Cyclic tetrapeptides such as romidepsin
Histone acetyltransferases (HAT) inhibitors Reviewed by Dekker et al. [39]	Anacardic acid and its derivatives
Exact mechanism unclear—Informative studies are given next to drug/compound	Methotrexate [40] Selective serotonin reuptake inhibitors (SSRIs) [41] Thalidomide [42] Sirtuin modulators such as resveratrol and polyphenol [43]

supplements have been identified and used traditionally, based not on the mechanism of action, but on observed clinical or health outcomes. This is no longer the case, especially with respect to drugs. In the past few decades as synthetic drug development, fueled by the large and powerful pharmaceutical industry (so called "big pharma"), has grown to dominate drug provision world-wide, the focus has shifted to mechanism of drug action. This is primarily due to big pharma-led large scale efforts to develop advanced targeted synthetic therapeutics [32, 33]. During the process of synthetic drug development, the mechanism of action of traditionally prescribed drugs was studied and many well-known drugs were found to act via an epigenetic mechanism. This, in turn, fueled interest in a new wave of "epigenetic drugs" [32, 33]. In Table 1, we present an overview of well-known classes and examples for drugs that have been shown to act via epigenetic mechanisms.

As epigenetic deregulation has been highly implicated in the development of ND/ASD, it could be hypothesized that maternal intake of drugs that act via epigenetic mechanisms during pregnancy may be associated with adverse neurodevelopmental outcomes. While information is presently scarce due to the novelty of the field, some examples of maternal drug treatments during pregnancy that are associated with ASD have been suggested in the literature and are discussed below. However, we note that with the exception of valproate, there is very limited information on the role of possible epigenetic mechanisms that could be induced by the drugs discusse below and their association with ASD.

4.2 Associations of Maternal Drug Treatment in Pregnancy with the Diagnosis of ASD in the Prenatally Exposed Children

4.2.1 Thalidomide

In a report of 100 thalidomide embryopathy patients, four cases met the criteria of autism diagnosis (DSM 3), which was a high prevalence of the disorder at the time compared to the general population [44] (Table 2). However, with thalidomide not prevalently used during pregnancy, it would probably account for very few cases of ASD.

4.2.2 Anti-Epileptic Drugs

Animal studies on rats have demonstrated that exposure to AEDs in utero may carry an increased risk of the development of autism [55]. In a study of 260 children exposed to an anti-epileptic medication (AED) during pregnancy, an increased risk of ASD was reported among mothers prenatally exposed to valproate alone, or in combination with other AEDs as well as among mothers exposed to carbamazepine alone or in combination with other AEDs [56]. However, the strongest association in this study was found with valproate exposure. Similarly, 6.3% of children prenatally exposed to valproate monotherapy had a diagnosis of ASD, compared to 0.9%

Table 2 Drug treatments in pregnancy and risk of ASD in the infants

Medications	Summary of positive associations	Influential studies and reviews (*R*)
Thalidomide	Reports of a 100 thalidomide embryopathy patients with elevated risk to autistic disorder diagnosis (DSM 3)	[44]
Antiepileptic drugs (AEDs)	Associations of several AEDs, including valproic acid, with increased risk of ASD and other neurodevelopmental outcomes.	[45, 46]
Selective Serotonin Reuptake inhibitors (SSRIs)	Associations of SSRI use during pregnancy and risk of ASD in the infants have been reported in several large cohort and case–control studies. However, confounding by indication could not be excluded, and most studies that have adjusted for indication of use found conflicting results.	[47–52]
Acetaminophen	An increased risk of ASD is suggested with maternal acetaminophen use in pregnancy, but empirical data is lacking.	[53]
B-2 adrenergic receptor agonists	An increased risk of ASD is suggested with maternal use in pregnancy, but large empirical data is lacking.	[54]

of children of mothers who did not take antiepileptic medications during pregnancy [57]. In a larger, population-based Danish study of mothers prescribed with valproate monotherapy during pregnancy, elevated risks of ASD among school-aged children have also been demonstrated compared to children of mothers unexposed to valproate or another AED during pregnancy [58]. Recently, a large systematic review and meta-analysis of 29 cohort studies of maternal exposure to AED during pregnancy and neurological development of their children concluded that valproate monotherapy, or valproate in combination with other AEDs, showed the strongest association with adverse neurodevelopmental outcomes. On the other hand, oxcarbazepine and lamotrigine in utero exposure were mostly associated with risk to ASD in children compared to unexposed healthy mothers [46] (Table 2). In utero exposure to valproate is also a recognized risk factor for several developmental abnormalities, such as spina bifida, cardiac, skeletal, and craniofacial defects. The proposed mechanisms for valproate teratogenicity that have an epigenetic component include interference with folate metabolism and inhibition of histone deacetylases [59].

4.2.3 Selective Serotonin-Reuptake Inhibitors

Various studies on animals and humans have suggested that increased serotonergic activity during fetal brain development may be one of the causal pathways leading to the development of ASD [60–62]. As a result, the hypothesis that maternal treatment with a specific type of antidepressant termed selective serotonin reuptake inhibitors (SSRIs), during gestation may increase the risk of having a child with ASD, has emerged [48]. As SSRIs are the most commonly prescribed antidepressants, the literature has since been expanding with regard to studies suggesting that such an association probably exists [50, 63], specifically when mothers take an SSRI in the first trimester of pregnancy [49, 64, 65]. A recent, large systematic review of the literature and meta-analyses assessing such an association from preconception and across all trimesters of pregnancy included 10 studies, and concluded that a positive association between SSRI exposure and ASD risk is consistent across all trimesters [51]. However, the calculated odds ratio was 1.8, which is still considered small at the population level. When partially adjusted by controlling for the underlying maternal condition, the association remained significant but with an even lower odds ratio of 1.5. In conclusion, and in light of the current evidence, it remains difficult to separate the effect of the underlying disease and/or related comorbidities, lifestyle and other risk factors from the effect of the medication used [52]. Therefore, it is hard to attribute risk to one causal factor when it is more likely the interaction of the psychological status, pharmacological treatment, genetic factors, and other associated factors present in each individual case that determines the risk of an adverse outcome [66].

4.2.4 Other Suspected Drugs Associated with ASD

Acetaminophen is an over-the-counter drug that is currently recommended as a safe pain and fever treatment during pregnancy. However, in recent years, studies have suggested a possible association between maternal acetaminophen use during pregnancy and ASD, but evidence remains controversial [53, 67].

A few reports have indicated a possible association of maternal exposure to β-2 adrenergic receptor agonists, used for treatment of asthma, during pregnancy and risk of ASD [47, 54], but such results need to be replicated in future studies to establish a definite link.

4.3 *Epigenetic Diet: The Impact of Nutritional Supplements on Epigenetic Regulation During Development*

Nutritional supplements are categorized as part of complementary and alternative medicine (CAM) practices: a diverse group of medical and health care systems, practices, and products that are not generally considered part of conventional medicine or standard medical care. They include interventions such as massage, acupuncture and dietary supplements [68]. Currently, however, very little information is available about mechanisms of action for most CAM, precluding an in-depth discussion of possible links to ASD etiology.

Nutritional supplements that have epigenetic impacts (termed epigenetic diet nutrients) are however, gaining ground as an emerging area of research [69]. Epigenetic diets may influence epigenetic regulation by changing the concentration of available substrate for epigenetic regulatory reactions. Contrary to the scenario with drugs taken by pregnant women and the possible adverse effects upon the fetus discussed above, the effects on the developing fetus due to maternal epigenetic diet are predominantly positive [69, 70]. While a comprehensive review of an epigenetic diet is beyond the scope of this chapter, we shall highlight research foci showing associations of neurodevelopmental outcomes, especially with respect to ASD/ND, and maternal epigenetic diet.

4.3.1 Gestational Intake of Methyl Donors

The earliest report of maternal intake of methyl donors' ability to alter fetal outcome was reported in 1998 by Wolff et al., in a seminal paper that paved the way for a new area of research. Wolff et al. [71], showed that feeding mice dams a methyl supplemented diet was able to alter the coat color of their pups. Albeit the experiment was conducted on a carefully controlled genetic background, it showed that maternal diet can impact fetal development via epigenetic mechanisms.

Subsequently several studies have shown that gestational intake of various forms of methyl donors are able to influence fetal development. A recent report from the

Maternal Nutrition and Offsprings' Epigenome (MANOE) study of 115 mother–infant pairs, found that maternal intake of methyl donor groups, via diet (in this case methionine, betaine, choline, folate) and via supplementation (folic acid), both before and during pregnancy, was able to significantly alter DNA methylation in cord-blood [72]. However, Boeke et al. [73] conducted a similar study in a folate-replete population, estimating maternal intake of methyl donors nutrients via vitamin B12, betaine, choline, folate, cadmium, zinc, and iron among mother–infant pairs, and found a negative association to DNA methylation levels in male offspring. This study is important for two reasons: firstly it shows that for a healthy population, there is likely no potentially adverse effect due to intake of methyl donor nutrients, and secondly, it also highlights a possible sex-specific signature that requires further study.

Additionally, work in animal models has shown that methyl donor supplementation exerts a protective effect; in chick embryos, it was found that maternal dietary zinc addition was able to protect the growing embryo against negative impacts of maternal heat-shock, by increasing antioxidant activity in the embryo [74]. While in rats, it has been shown that late pregnancy supplementation with folate is able to rescue structural and functional defects of the brain [75, 76].

4.3.2 Gestational Intake of Epigenetic Diet Nutrients of Unknown Mechanism

As epigenomic profiling techniques become more accessible, researchers are discovering that dietary elements are able to exert DNA methylation and histone acetylation profile changes, though the exact mechanism of doing so is unknown. While the reader is referred to more comprehensive reviews [77, 78], we highlight selected dietary nutrients for which epigenetic profile alterations important in neural development and function have been documented, following prenatal exposure.

Fish Oils

Fish oils have long been consumed during pregnancy as a dietary supplement. They contain polyunsaturated fatty acids (PUFA) whose role as epigenetic diet nutrients important for brain development is increasingly gaining attention [79]. In a large scale randomized control trial, Van Dijk et al. [76] investigated epigenetic profile changes upon the child following high dose maternal gestational supplementation with docosahexaenoic acid (DHA), a long-chain PUFA. They found a sex-specific increase in differentially methylated regions genome-wide in children of high-dose exposure mothers, interestingly with a higher impact on boys than girls [76]. Another group probed changes in DNA methylation due to prenatal DHA exposure, using a targeted approach. In two separate publications, they screened for DNA methylation differences in a cohort of pregnant Mexican women who were given dietary supplements of 400 mg of DHA daily. They showed small changes in DNA methylation of imprinted loci and repetitive elements [80, 81].

Curcumin

Curcumin (diferuloylmethane), is a component of turmeric (Curcuma longa), one of the most common Asian spices. Turmeric has been recognized and used for its many medicinal properties in both Ayurvedic and Traditional Chinese Medicine for several centuries, if not millennia. In the past decade, interest has grown exponentially in exploring the pharmacological benefits and pharmacoepigenomical effects of curcumin [82, 83]. Currently, epigenetic regulatory roles as a modulator of DNA methylation, histone acetylation and epigenetic programming via RNAi have been identified [84]. Curcumin health benefits are most documented for cancer. However, it also shows therapeutic potential for neurological and inflammatory disorders [84, 85]. Salehi et al. [86] have presented the most up to date review of clinical trials of curcumin, while Lopresti [85] reviewed evidence for clinical and animal trials of curcumin in neuropsychiatric disease. The interested reader is referred to these papers. However, while we could find no specific information related to ASD, we note that the well-documented effects of curcumin as an anti-oxidative and anti-inflammatory agent, coupled with its ability to influence epigenomic programming broadly, suggests the hypothesis that curcumin may have an impact on ASD etiology, is plausible.

Others

Other important epigenetic dietary components that may impact fetal brain development via epigenetic mechanisms include polyphenols such as those found in green tea; which was shown to have a potential protective effect. In a mouse model of fetal alcohol spectrum disorder, pregnant dams fed on epigallocatechin-3-gallate (EGCG), the major anti-oxidative component of green tea [87], showed rescue of embryo size back to normal. Trans-resveratrol, another polyphenol, has been shown to be able to rescue aberrant epigenetic programming in rats following perinatal asphyxia [88], an important finding as perinatal asphyxia may cause ND [89].

5 Prevalence of Pregnant Women's Exposure to Drugs with Potential for Adverse Fetal Outcomes, and Epigenetic Diet

Given the importance of maternal intake of drugs, that may exert an adverse effect on the developing embryo or fetus, and dietary supplements that induce epigenetic mechanisms influencing neurodevelopment, we discuss exposures below.

5.1 *Drug Intake in Pregnant Women and Potential for Adverse Developmental Effects*

Current evidence has suggested that 65–94% of women take at least one prescription drug during pregnancy [90–92]. Lupattelli et al. [93] showed that more than 80% of the pregnant women in the USA, Europe, and Australia use at least one prescribed drug during pregnancy. Importantly, the rate of pregnant women's exposure to drugs has seen a rapid increase in the past few decades [94]. Furthermore, approximately 70% of women have been reported to be taking medication in the first trimester (encompassing the period of organogenesis, when the fetus's important organs are developing) [92, 95], including both over the counter drugs and herbal medications [90, 92].

Unfortunately there has not been a concomitant increase in the amount of information available with respect to the potentially adverse effects of drugs on the developing embryo or fetus. This is because the effect of drugs on development cannot be studied in humans through clinical trials [91] becuase such studies are ethically unacceptable. According to the Automated Teratogen Information System (TERIS), the teratogenic risk in human pregnancy is undetermined for 92% of the drug treatments approved by the US FDA system between 1980 and 2000 [96]. In another review of the safety of 172 drugs approved by the US FDA between 2000 and 2010, it was found that the teratogenic risk in human pregnancy was undetermined for 98% of drugs, and for 74% there were no available data about the risk in pregnancy [97]. However, while clear clinical trials of drug safety in pregnant women is often unavailable, information on potentially adverse outcomes is usually gathered via a collection of animal and model organism studies and epidemiological investigations that usually occur several years after a drug has been put on the market.

Nevertheless, given the growing interest in drugs that act via epigenetic mechanisms (Table 1), and already emerging associations of certain drugs that may act via epigenetic mechanisms with ASD (Table 2), we raise a note of caution that further studies specific to possible teratogenic effects of drugs with epigenetic mechanisms of action are warranted.

5.2 *Complementary and Alternative Medicine (CAM) Intake During Pregnancy*

Parallel to the rise of drug intake among pregnant women, there is also a rising trend of CAM use, though it is less well documented. A literature review of CAM usage in industrialized countries in 2011, found that 1–87% of pregnant women use CAM [98]. On the other hand, a representative survey of pregnant women in the USA, published in 2008, found that over half of the respondents used CAM [99]. Furthermore, a European study found that approximately 60% of pregnant women use a dietary supplement [100].

Pregnant women use CAM to relieve specific pregnancy related problems such as nausea, vomiting, tiredness, and back pain [98, 101]. Examples include the use of ginger root, shown to be a safe and effective non-pharmacological option for nausea and vomiting during early pregnancy [102], and supplementing with Vitamin B6, also considered a safe alternative pharmacological treatment for nausea and vomiting [102, 103].

Thus, the widespread use of CAM during pregnancy, makes it clinically relevant [68, 104].

6 Conclusion

Epigenetic mechanisms are a molecular mechanism by which the environment is able to impact the genome. Recognizing ASD as a complex condition with a likely substantial causative environmental component [16], epigenetic modalities, by which the environment may cause disease, is receiving increased research attention. However, here we only discuss efforts in this area which focus on a key environment: that which the developing fetus is exposed to in utero. There are two main exposures in this regard: fetal exposure to drugs that may have adverse neurodevelopmental potential, and fetal exposure to an epigenetic diet, both of which are emerging areas of research.

Additionally, understanding epigenetic modes of action in disease causation accurately, is important as there the possibility of correcting them [59, 105]. Indeed, the entire field of epigenetic drugs is a direct result of efforts probing how epigenetic deregulation can be corrected back to the normal. Such precision-based medicine efforts, where treatment is based on directly addressing the molecular cause of disease, have seen success, especially in cancer research [106]. However, in this chapter, we drew attention to an overlooked area of epigenetic drugs: those that when injested by a pregnant mother, may impact ASD risk for their unborn child. We emphasize that more research is needed to specifically understand both their epigenetic mechanism of action and their potential to cause harm to the developing fetus.

On the other hand, encouragingly, the emerging area of epigenetic diet is gaining attention due to its potential for prevention and rescue from adverse effects for the developing fetus. Studies in epigenetic diet have focused on elucidating molecular mechanisms by which nutritional supplements act, and evidence is growing for possible epigenetic deregulation rescue outcomes of these "nutraceuticals." Current reports, while few in number, predominantly detail protective and beneficial effects on neurodevelopment. Especially important in context of this discussion is the ability of epigenetic diet components to ameliorate known harmful environmental exposures for pregnant women; an area of research we highlight here and for which future studies hold promise.

In summary, the significant role epigenetics plays in ASD, as a molecular mechanism translating environment into genomic or genetic control, and the recognition that there are both drugs and diet that a pregnant woman can be exposed to, which act via epigenetic mechanisms, highlight the importance of more focused research on what such exposures in utero could mean for the baby. Thus, we end this chapter calling for further research to understand the epigenetic mechanisms underlying gestational exposures to drugs and diet, and highlighting the remarkable potential of epigenetic regulatory compounds to serve as therapeutics.

References

1. American Psychiatric Association. (2013). *Diagnostic and statistical manual of mental disorders (DSM-5®)*. Washington, DC: Author.
2. Homberg, J. R., Kyzar, E. J., Scattoni, M. L., Norton, W. H., Pittman, J., Gaikwad, S., et al. (2016). Genetic and environmental modulation of neurodevelopmental disorders: Translational insights from labs to beds. *Brain Research Bulletin, 125*, 79–91.
3. Mpaka, D. M., Okitundu, D. L. E. A., Ndjukendi, A. O., N'situ, A. M., Kinsala, S. Y., Mukau, J. E., et al. (2016). Prevalence and comorbidities of autism among children referred to the outpatient clinics for neurodevelopmental disorders. *The Pan African Medical Journal, 25*, 82–82.
4. Vissers, L. E. L. M., Gilissen, C., & Veltman, J. A. (2015). Genetic studies in intellectual disability and related disorders. *Nature Reviews Genetics, 17*, 9.
5. Boyle, C. A., Boulet, S., Schieve, L. A., Cohen, R. A., Blumberg, S. J., Yeargin-Allsopp, M., et al. (2011). Trends in the prevalence of developmental disabilities in US Children, 1997–2008. *Pediatrics, 127*, 1034–1042.
6. Gupta, S., Venkatesan, S. P., Goswami, S., & Kumar, R. (2018). Emerging trends in the diagnosis and intervention of neurodevelopmental disorders. IGI Global.
7. Christensen, D. L., Baio, J., Van Naarden Braun, K., Bilder, D., Charles, J., Constantino, J. N., et al. (2016). Prevalence and characteristics of autism spectrum disorder among children aged 8 Years--Autism and Developmental Disabilities Monitoring Network, 11 Sites, United States, 2012. *MMWR Surveillance Summaries, 65*, 1–23.
8. Elsabbagh, M., Divan, G., Koh, Y.-J., Kim, Y. S., Kauchali, S., Marcín, C., et al. (2012). Global prevalence of autism and other pervasive developmental disorders. *Autism Research: Official Journal of the International Society for Autism Research, 5*, 160–179.
9. Baio, J., Wiggins, L., Christensen, D. L., Maenner, M. J., Daniels, J., Warren, Z., et al. (2018). Prevalence of autism spectrum disorder among children aged 8 years - autism and developmental disabilities monitoring network, 11 Sites, United States, 2014. *Morbidity and Mortality Weekly Report. Surveillance Summaries (Washington, D.C. : 2002), 67*, 1–23.
10. Xu, G., Strathearn, L., Liu, B., & Bao, W. (2018). Prevalence of autism spectrum disorder among us children and adolescents, 2014–2016. *JAMA, 319*, 81–82.
11. Bilbo, S. D., Jones, J. P., & Parker, W. (2012). Is autism a member of a family of diseases resulting from genetic/cultural mismatches? Implications for treatment and prevention. *Autism Research and Treatment, 2012*, 910946.
12. Bilbo, S. D., Nevison, C. D., & Parker, W. (2015). A model for the induction of autism in the ecosystem of the human body: The anatomy of a modern pandemic? *Microbial Ecology in Health and Disease, 26*, 26253.
13. Meerding, W. J., Bonneux, L., Polder, J. J., Koopmanschap, M. A., & Van Der Maas, P. J. (1998). Demographic and epidemiological determinants of healthcare costs in Netherlands: Cost of illness study. *BMJ (Clinical research ed.), 317*, 111–115.
14. El-Fishawy, P., & State, M. W. (2010). The genetics of autism: Key issues, recent findings, and clinical implications. *The Psychiatric Clinics of North America, 33*, 83–105.

15. Geschwind, D. H. (2011). Genetics of autism spectrum disorders. *Trends in Cognitive Sciences, 15*, 409–416.
16. Hertz-Picciotto, I., Schmidt, R. J., & Krakowiak, P. (2018). Understanding environmental contributions to autism: Causal concepts and the state of science. *Autism Research, 11*, 554–586.
17. Lyall, K., Schmidt, R. J., & Hertz-Picciotto, I. (2014). Maternal lifestyle and environmental risk factors for autism spectrum disorders. *International Journal of Epidemiology, 43*, 443–464.
18. Grayson, D. R., & Guidotti, A. (2016). Merging data from genetic and epigenetic approaches to better understand autistic spectrum disorder. *Epigenomics, 8*, 85–104.
19. Loke, Y. J., Hannan, A. J., & Craig, J. M. (2015). The role of epigenetic change in autism spectrum disorders. *Frontiers in Neurology, 6*, 107.
20. Zahir, F. R., & Brown, C. J. (2011). Epigenetic impacts on neurodevelopment: Pathophysiological mechanisms and genetic modes of action. *Pediatric Research, 69*, 92R.
21. Bernier, R., Golzio, C., Xiong, B., Stessman, H. A., Coe, B. P., Penn, O., et al. (2014). Disruptive CHD8 mutations define a subtype of autism early in development. *Cell, 158*, 263–276.
22. O'Roak, B. J., Vives, L., Girirajan, S., Karakoc, E., Krumm, N., Coe, B. P., et al. (2012). Sporadic autism exomes reveal a highly interconnected protein network of de novo mutations. *Nature, 485*, 246–250.
23. Zahir, F. R., Tucker, T., Mayo, S., Brown, C. J., Lim, E. L., Taylor, J., et al. (2016). Intragenic CNVs for epigenetic regulatory genes in intellectual disability: Survey identifies pathogenic and benign single exon changes. *American Journal of Medical Genetics. Part A, 170*, 2916–2926.
24. Keil, K. P., & Lein, P. J. (2016). DNA methylation: A mechanism linking environmental chemical exposures to risk of autism spectrum disorders? *Environmental Epigenetics, 2*, dvv012.
25. Elagoz Yuksel, M., Yuceturk, B., Karatas, O. F., Ozen, M., & Dogangun, B. (2016). The altered promoter methylation of oxytocin receptor gene in autism. *Journal of Neurogenetics, 30*, 280–284.
26. Eshraghi, A. A., Liu, G., Kay, S.-I. S., Eshraghi, R. S., Mittal, J., Moshiree, B., et al. (2018). Epigenetics and autism spectrum disorder: Is there a correlation? *Frontiers in Cellular Neuroscience, 12*, 78–78.
27. Gunawardhana, L. P., Baines, K. J., Mattes, J., Murphy, V. E., Simpson, J. L., & Gibson, P. G. (2014). Differential DNA methylation profiles of infants exposed to maternal asthma during pregnancy. *Pediatric Pulmonology, 49*, 852–862.
28. Ladd-Acosta, C., Hansen, K. D., Briem, E., Fallin, M. D., Kaufmann, W. E., & Feinberg, A. P. (2014). Common DNA methylation alterations in multiple brain regions in autism. *Molecular Psychiatry, 19*, 862–871.
29. Sun, W., Poschmann, J., Cruz-Herrera Del Rosario, R., Parikshak, N. N., Hajan, H. S., Kumar, V., et al. (2016). Histone acetylome-wide association study of autism spectrum disorder. *Cell, 167*, 1385–1397.e11.
30. Wu, Y. E., Parikshak, N. N., Belgard, T. G., & Geschwind, D. H. (2016). Genome-wide, integrative analysis implicates microRNA dysregulation in autism spectrum disorder. *Nature Neuroscience, 19*, 1463–1476.
31. Rosikiewicz, W., & Makalowska, I. (2016). Biological functions of natural antisense transcripts. *Acta Biochimica Polonica, 63*, 665–673.
32. Altucci, L., & Rots, M. G. (2016). Epigenetic drugs: From chemistry via biology to medicine and back. *Clinical Epigenetics, 8*, 56–56.
33. Heerboth, S., Lapinska, K., Snyder, N., Leary, M., Rollinson, S., & Sarkar, S. (2014). Use of epigenetic drugs in disease: An overview. *Genetics & Epigenetics, 6*, 9–19.
34. Yang, X., Lay, F., Han, H., & Jones, P. A. (2010). Targeting DNA methylation for epigenetic therapy. *Trends in Pharmacological Sciences, 31*, 536–546.

35. Gnyszka, A., Jastrzębski, Z., & Flis, S. (2013). DNA methyltransferase inhibitors and their emerging role in epigenetic therapy of cancer. *Anticancer Research, 33*, 2989–2996.
36. Ahuja, N., Sharma, A. R., & Baylin, S. B. (2016). Epigenetic therapeutics: A new weapon in the war against cancer. *Annual Review of Medicine, 67*, 73–89.
37. Eckschlager, T., Plch, J., Stiborova, M., & Hrabeta, J. (2017). Histone deacetylase inhibitors as anticancer drugs. *International Journal of Molecular Sciences, 18*, 1414.
38. Goey, A. K., Sissung, T. M., Peer, C. J., & Figg, W. D. (2016). Pharmacogenomics and histone deacetylase inhibitors. *Pharmacogenomics, 17*, 1807–1815.
39. Dekker, F. J., Van Den Bosch, T., & Martin, N. I. (2014). Small molecule inhibitors of histone acetyltransferases and deacetylases are potential drugs for inflammatory diseases. *Drug Discovery Today, 19*, 654–660.
40. Forster, V. J., Mcdonnell, A., Theobald, R., & Mckay, J. A. (2017). Effect of methotrexate/vitamin B(12) on DNA methylation as a potential factor in leukemia treatment-related neurotoxicity. *Epigenomics, 9*, 1205–1218.
41. Williams, K., Brignell, A., Randall, M., Silove, N., & Hazell, P. (2013). Selective serotonin reuptake inhibitors (SSRIs) for autism spectrum disorders (ASD). *Cochrane Database of Systematic Reviews*, Cd004677.
42. Ahmadvand, M., Noruzinia, M., Fard, A. D., Zohour, M. M., Tabatabaiefar, M. A., Soleimani, M., et al. (2014). The role of epigenetics in the induction of fetal hemoglobin: A combination therapy approach. *International Journal of Hematology-Oncology and Stem Cell Research, 8*, 9–14.
43. Mahajan, S. S., Leko, V., Simon, J. A., & Bedalov, A. (2011). Sirtuin modulators. *Handbook of Experimental Pharmacology, 206*, 241–255.
44. Stromland, K., Nordin, V., Miller, M., Akerstrom, B., & Gillberg, C. (1994). Autism in thalidomide embryopathy: A population study. *Developmental Medicine and Child Neurology, 36*, 351–356.
45. Christensen, J., Grønborg, T. K., Sørensen, M. J., Schendel, D., Parner, E. T., Pedersen, L. H., et al. (2013). Prenatal valproate exposure and risk of autism spectrum disorders and childhood autism. *JAMA, 309*, 1696–1703.
46. Veroniki, A. A., Rios, P., Cogo, E., Straus, S. E., Finkelstein, Y., Kealey, R., et al. (2017). Comparative safety of antiepileptic drugs for neurological development in children exposed during pregnancy and breast feeding: A systematic review and network meta-analysis. *BMJ Open, 7*, e017248.
47. Croen, L. A., Connors, S. L., Matevia, M., Qian, Y., Newschaffer, C., & Zimmerman, A. W. (2011). Prenatal exposure to beta2-adrenergic receptor agonists and risk of autism spectrum disorders. *Journal of Neurodevelopmental Disorders, 3*, 307–315.
48. Harrington, R. A., Lee, L. C., Crum, R. M., Zimmerman, A. W., & Hertz-Picciotto, I. (2013). Serotonin hypothesis of autism: Implications for selective serotonin reuptake inhibitor use during pregnancy. *Autism Research, 6*, 149–168.
49. Harrington, R. A., Lee, L.-C., Crum, R. M., Zimmerman, A. W., & Hertz-Picciotto, I. (2014). Prenatal SSRI use and offspring with autism spectrum disorder or developmental delay. *Pediatrics, 133*, e1241–e1248.
50. Gidaya, N. B., Lee, B. K., Burstyn, I., Yudell, M., Mortensen, E. L., & Newschaffer, C. J. (2014). In utero exposure to selective serotonin reuptake inhibitors and risk for autism spectrum disorder. *Journal of Autism and Developmental Disorders, 44*, 2558–2567.
51. Mezzacappa, A., Lasica, P. A., Gianfagna, F., Cazas, O., Hardy, P., Falissard, B., et al. (2017). Risk for autism spectrum disorders according to period of prenatal antidepressant exposure: A systematic review and meta-analysis. *JAMA Pediatrics, 171*, 555–563.
52. Morales, D. R., Slattery, J., Evans, S., & Kurz, X. (2018). Antidepressant use during pregnancy and risk of autism spectrum disorder and attention deficit hyperactivity disorder: Systematic review of observational studies and methodological considerations. *BMC Medicine, 16*, 6.
53. Bauer, A. Z., Kriebel, D., Herbert, M. R., Bornehag, C. G., & Swan, S. H. (2018). Prenatal paracetamol exposure and child neurodevelopment: A review. *Hormones and Behavior, 101*, 125–147.

54. Gidaya, N. B., Lee, B. K., Burstyn, I., Michael, Y., Newschaffer, C. J., & Mortensen, E. L. (2016). In utero exposure to beta-2-adrenergic receptor agonist drugs and risk for autism spectrum disorders. *Pediatrics, 137*, e20151316.
55. Ingram, J. L., Peckham, S. M., Tisdale, B., & Rodier, P. M. (2000). Prenatal exposure of rats to valproic acid reproduces the cerebellar anomalies associated with autism. *Neurotoxicology and Teratology, 22*, 319–324.
56. Rasalam, A. D., Hailey, H., Williams, J. H., Moore, S. J., Turnpenny, P. D., Lloyd, D. J., et al. (2005). Characteristics of fetal anticonvulsant syndrome associated autistic disorder. *Developmental Medicine and Child Neurology, 47*, 551–555.
57. Bromley, R. L., Mawer, G., Clayton-Smith, J., & Baker, G. A. (2008). Autism spectrum disorders following in utero exposure to antiepileptic drugs. *Neurology, 71*, 1923–1924.
58. Christensen, J., Gronborg, T. K., Sorensen, M. J., Schendel, D., Parner, E. T., Pedersen, L. H., et al. (2013). Prenatal valproate exposure and risk of autism spectrum disorders and childhood autism. *JAMA, 309*, 1696–1703.
59. Grafodatskaya, D., Chung, B., Szatmari, P., & Weksberg, R. (2010). Autism spectrum disorders and epigenetics. *Journal of the American Academy of Child and Adolescent Psychiatry, 49*, 794–809.
60. Anderson, G. M., Freedman, D. X., Cohen, D. J., Volkmar, F. R., Hoder, E. L., Mcphedran, P., et al. (1987). Whole blood serotonin in autistic and normal subjects. *Journal of Child Psychology and Psychiatry, 28*, 885–900.
61. Cook Jr., E. H., Leventhal, B. L., & Freedman, D. X. (1988). Free serotonin in plasma: Autistic children and their first-degree relatives. *Biological Psychiatry, 24*, 488–491.
62. Vorhees, C. V., Acuff-Smith, K. D., Schilling, M. A., Fisher, J. E., Moran, M. S., & Buelke-Sam, J. (1994). A developmental neurotoxicity evaluation of the effects of prenatal exposure to fluoxetine in rats. *Fundamental and Applied Toxicology, 23*, 194–205.
63. Rai, D., Lee, B. K., Dalman, C., Golding, J., Lewis, G., & Magnusson, C. (2013). Parental depression, maternal antidepressant use during pregnancy, and risk of autism spectrum disorders: Population based case-control study. *BMJ, 346*, f2059.
64. Croen, L. A., Grether, J. K., Yoshida, C. K., Odouli, R., & Hendrick, V. (2011). Antidepressant use during pregnancy and childhood autism spectrum disorders. *Archives of General Psychiatry, 68*, 1104–1112.
65. Sorensen, M. J., Gronborg, T. K., Christensen, J., Parner, E. T., Vestergaard, M., Schendel, D., et al. (2013). Antidepressant exposure in pregnancy and risk of autism spectrum disorders. *Clinical Epidemiology, 5*, 449–459.
66. Alwan, S., Friedman, J. M., & Chambers, C. (2016). Safety of selective serotonin reuptake inhibitors in pregnancy: A review of current evidence. *CNS Drugs, 30*, 499–515.
67. Andrade, C. (2016). Use of acetaminophen (paracetamol) during pregnancy and the risk of autism spectrum disorder in the offspring. *The Journal of Clinical Psychiatry, 77*, e152–e154.
68. Steel, A., Adams, J., Sibbritt, D., & Broom, A. (2015). The outcomes of complementary and alternative medicine use among pregnant and birthing women: Current trends and future directions. *Women's Health, 11*, 309–323.
69. Li, Y., Saldanha, S. N., & Tollefsbol, T. O. (2013). Impact of epigenetic dietary compounds on transgenerational prevention of human diseases. *The AAPS Journal, 16*, 27–36.
70. Bianco-Miotto, T., Craig, J. M., Gasser, Y. P., Van Dijk, S. J., & Ozanne, S. E. (2017). Epigenetics and DOHaD: From basics to birth and beyond. *Journal of Developmental Origins of Health and Disease, 8*, 513–519.
71. Wolff, G. L., Kodell, R. L., Moore, S. R., & Cooney, C. A. (1998). Maternal epigenetics and methyl supplements affect agouti gene expression in Avy/a mice. *The FASEB Journal, 12*, 949–957.
72. Pauwels, S., Ghosh, M., Duca, R. C., Bekaert, B., Freson, K., Huybrechts, I. A. S., et al. (2016). Dietary and supplemental maternal methyl-group donor intake and cord blood DNA methylation. *Epigenetics, 12*, 1–10.
73. Boeke, C. E., Baccarelli, A., Kleinman, K. P., Burris, H. H., Litonjua, A. A., Rifas-Shiman, S. L., et al. (2012). Gestational intake of methyl donors and global LINE-1 DNA methylation

in maternal and cord blood: Prospective results from a folate-replete population. *Epigenetics, 7*, 253–260.
74. Zhu, Y., Liao, X., Lu, L., Li, W., Zhang, L., Ji, C., et al. (2017). Maternal dietary zinc supplementation enhances the epigenetic-activated antioxidant ability of chick embryos from maternal normal and high temperatures. *Oncotarget, 8*, 19814–19824.
75. Geoffroy, A., Kerek, R., Pourié, G., Helle, D., Guéant, J.-L., Daval, J.-L., et al. (2017). Late maternal folate supplementation rescues from methyl donor deficiency-associated brain defects by restoring let-7 and miR-34 pathways. *Molecular Neurobiology, 54*, 5017–5033.
76. Van Dijk, S. J., Zhou, J., Peters, T. J., Buckley, M., Sutcliffe, B., Oytam, Y., et al. (2016). Effect of prenatal DHA supplementation on the infant epigenome: Results from a randomized controlled trial. *Clinical Epigenetics, 8*, 114–114.
77. Hardy, T. M., & Tollefsbol, T. O. (2011). Epigenetic diet: Impact on the epigenome and cancer. *Epigenomics, 3*, 503–518.
78. Meeran, S. M., Ahmed, A., & Tollefsbol, T. O. (2010). Epigenetic targets of bioactive dietary components for cancer prevention and therapy. *Clinical Epigenetics, 1*, 101–116.
79. Schuchardt, J. P., Huss, M., Stauss-Grabo, M., & Hahn, A. (2010). Significance of long-chain polyunsaturated fatty acids (PUFAs) for the development and behaviour of children. *European Journal of Pediatrics, 169*, 149–164.
80. Lee, H.-S., Barraza-Villarreal, A., Biessy, C., Duarte-Salles, T., Sly, P. D., Ramakrishnan, U., et al. (2014). Dietary supplementation with polyunsaturated fatty acid during pregnancy modulates DNA methylation at IGF2/H19 imprinted genes and growth of infants. *Physiological Genomics, 46*, 851–857.
81. Lee, H.-S., Barraza-Villarreal, A., Hernandez-Vargas, H., Sly, P. D., Biessy, C., Ramakrishnan, U., et al. (2013). Modulation of DNA methylation states and infant immune system by dietary supplementation with ω-3 PUFA during pregnancy in an intervention study. *The American Journal of Clinical Nutrition, 98*, 480–487.
82. Aggarwal, B. B., & Harikumar, K. B. (2009). Potential therapeutic effects of curcumin, the anti-inflammatory agent, against neurodegenerative, cardiovascular, pulmonary, metabolic, autoimmune and neoplastic diseases. *The International Journal of Biochemistry & Cell Biology, 41*, 40–59.
83. Lopresti, A. L. (2017). Curcumin for neuropsychiatric disorders: A review of in vitro, animal and human studies. *Journal of Psychopharmacology, 31*, 287–302.
84. Boyanapalli, S. S. S., & Kong, A.-N. T. (2015). “Curcumin, the King of Spices”: Epigenetic regulatory mechanisms in the prevention of cancer, neurological, and inflammatory diseases. *Current Pharmacology Reports, 1*, 129–139.
85. Zhu, L.-N., Mei, X., Zhang, Z.-G., Xie, Y.-P., & Lang, F. (2019). Curcumin intervention for cognitive function in different types of people: A systematic review and meta-analysis. *Phytotherapy Research, 33*(3), 524–533.
86. Salehi, B., Stojanovic-Radic, Z., Matejic, J., Sharifi-Rad, M., Anil Kumar, N. V., Martins, N., et al. (2018). The therapeutic potential of curcumin: A review of clinical trials. *European Journal of Medicinal Chemistry, 163*, 527–545.
87. Long, L., Li, Y., Wang, Y. D., He, Q. Y., Li, M., Cai, X. D., et al. (2010). The preventive effect of oral EGCG in a Fetal Alcohol Spectrum Disorder Mouse Model. *Alcoholism: Clinical and Experimental Research, 34*, 1929–1936.
88. Isac, S., Panaitescu, A. M., Spataru, A., Iesanu, M., Totan, A., Udriste, A., et al. (2017). Trans-resveratrol enriched maternal diet protects the immature hippocampus from perinatal asphyxia in rats. *Neuroscience Letters, 653*, 308–313.
89. Van Handel, M., Swaab, H., De Vries, L. S., & Jongmans, M. J. (2007). Long-term cognitive and behavioral consequences of neonatal encephalopathy following perinatal asphyxia: A review. *European Journal of Pediatrics, 166*, 645–654.
90. Ayad, M., & Costantine, M. M. (2015). Epidemiology of medications use in pregnancy. *Seminars in Perinatology, 39*, 508–511.
91. Mosley 2nd, J. F., Smith, L. L., & Dezan, M. D. (2015). An overview of upcoming changes in pregnancy and lactation labeling information. *Pharmacy Practice (Granada), 13*, 605.

92. Temming, L. A., Cahill, A. G., & Riley, L. E. (2016). Clinical management of medications in pregnancy and lactation. *American Journal of Obstetrics and Gynecology, 214*, 698–702.
93. Lupattelli, A., Spigset, O., Twigg, M. J., Zagorodnikova, K., Mårdby, A. C., Moretti, M. E., et al. (2014). Medication use in pregnancy: A cross-sectional, multinational web-based study. *BMJ Open, 4*, e004365.
94. Mitchell, A. A., Gilboa, S. M., Werler, M. M., Kelley, K. E., Louik, C., & Hernandez-Diaz, S. (2011). Medication use during pregnancy, with particular focus on prescription drugs: 1976-2008. *American Journal of Obstetrics and Gynecology, 205*, 51.e1–51.e8.
95. Mitchell, A. A., Gilboa, S. M., Werler, M. M., Kelley, K. E., Louik, C., Hernández-Díaz, S., et al. (2011). Medication use during pregnancy, with particular focus on prescription drugs: 1976-2008. *American Journal of Obstetrics and Gynecology, 205*, 51.e1–51.e518.
96. Lo, W., & Friedman, J. (2002). Teratogenicity of recently introduced medications in human pregnancy. *Obstetrics & Gynecology, 100*, 465–473.
97. Adam, M. P., Polifka, J. E., & Friedman, J. M. (2011). Evolving knowledge of the teratogenicity of medications in human pregnancy. *American Journal of Medical Genetics Part C: Seminars in Medical Genetics, 157*, 175–182.
98. Hall, H. G., Griffiths, D. L., & Mckenna, L. G. (2011). The use of complementary and alternative medicine by pregnant women: A literature review. *Midwifery, 27*, 817–824.
99. Wade, C., Chao, M., Kronenberg, F., Cushman, L., & Kalmuss, D. (2008). Medical pluralism among American women: Results of a national survey. *Journal of Women's Health (2002), 17*, 829–840.
100. Holst, L., Wright, D., Haavik, S., & Nordeng, H. (2011). Safety and efficacy of herbal remedies in obstetrics-review and clinical implications. *Midwifery, 27*, 80–86.
101. Steel, A., Adams, J., Sibbritt, D., Broom, A., Gallois, C., & Frawley, J. (2012). Utilisation of complementary and alternative medicine (CAM) practitioners within maternity care provision: Results from a nationally representative cohort study of 1,835 pregnant women. *BMC Pregnancy and Childbirth, 12*, 146.
102. Thomson, M., Corbin, R., & Leung, L. (2014). Effects of ginger for nausea and vomiting in early pregnancy: A meta-analysis. *Journal of American Board of Family Medicine, 27*, 115–122.
103. Firouzbakht, M., Nikpour, M., Jamali, B., & Omidvar, S. (2014). Comparison of ginger with vitamin B6 in relieving nausea and vomiting during pregnancy. *Ayu, 35*, 289–293.
104. Birdee, G. S., Kemper, K. J., Rothman, R., & Gardiner, P. (2014). Use of complementary and alternative medicine during pregnancy and the postpartum period: An analysis of the National Health Interview Survey. *Journal of Women's Health (2002), 23*, 824–829.
105. Siu, M. T., & Weksberg, R. (2017). Epigenetics of autism spectrum disorder. *Advances in Experimental Medicine and Biology, 978*, 63–90.
106. Moran, S., Martinez-Cardus, A., Boussios, S., & Esteller, M. (2017). Precision medicine based on epigenomics: The paradigm of carcinoma of unknown primary. *Nature Reviews. Clinical Oncology, 14*, 682–694.

Psychological Comorbidities in Autism Spectrum Disorder

Eman Shaltout, Nader Al-Dewik, Muthanna Samara, Hisham Morsi, and Azhar Khattab

Abstract Autism spectrum disorder (ASD) is characterized by impairment in behavior, communication, and social interaction. Thus, accurate identification, regular behavioral and other nonmedical interventions would improve the diagnosis, management, and treatment of this condition.

In this chapter, we investigate the importance of diagnosing and identifying comorbid psychiatric disorders that occur with ASD as these conditions can often complicate treatment, and failure to recognize them can result in deficits that can persist into adolescence and adulthood. In addition, we explore the impact of comprehensive psychological intervention in ASD patients with comorbid psychiatric disorders with the ultimate goal of improving overall quality of life.

Keywords Autism spectrum disorders · Psychiatric comorbidities · Cognitive behavior therapy · Psychological interventions

---

E. Shaltout
Medical Research Center, Hamad Medical Corporation (HMC), Doha, Qatar

Department of Psychology, Kingston University London, Kingston upon Thames, UK

N. Al-Dewik
Clinical and Metabolic Genetics, Pediatrics Department, Hamad General Hospital (HGH), Hamad Medical Corporation (HMC), Doha, Qatar

College of Health and Life Sciences, Hamad Bin Khalifa University (HBKU), Doha, Qatar

M. Samara (✉)
Department of Psychology, Kingston University London, Kingston upon Thames, UK
e-mail: M.Samara@kingston.ac.uk

H. Morsi
Department of Psychology, Kingston University London, Kingston upon Thames, UK

Quality of Life Unit, National Center for Cancer Care and Research, (NCCCR), Hamad Medical Corporation (HMC), Doha, Qatar

A. Khattab
Qatar Rehabilitation Institute, Pediatric Rehabilitation, Hamad Medical Corporation (HMC), Doha, Qatar

M. M. Essa, M. W. Qoronfleh (eds.), *Personalized Food Intervention and Therapy for Autism Spectrum Disorder Management*, Advances in Neurobiology 24,
https://doi.org/10.1007/978-3-030-30402-7_6

Highlights

- Substantial overlapping occurs between autism spectrum disorders (ASDs) and psychological disorders.
- Mood disorders, anxiety disorders, and ADHD are among the psychological disorders most frequently related with ASD.
- Symptom presentation is similar whether ASD occurs alone or with other conditions.
- Numerous assessments after initial diagnosis of ASD are commonly required.
- The majority of ASD patients had poor QoL.

1 Introduction

Central autism features like behavior, social, and communication impairments are well-documented lifetime functional deficits [1].

The role of psychology in ASDs is classically to provide a comprehensive roadmap to evaluate patients' weaknesses and strengths and provide a guide for treatment in these areas. Subsequent recommendations are based on afflicted patients' cognitive, behavioral, emotional, and academic or vocational needs. The overall aim is to improve functioning by identifying and adjusting maladaptive behaviors associated with the diagnosis along with helping patients and their families succeed at key transition points such as starting school, entering adolescence, and moving into adulthood [2].

Each individual with ASD is unique and has a range of strengths and challenges. Some individuals with ASD are able to succeed in their traditional schools, hold jobs, and perform functions of daily living with varying levels of support. Others have substantial intellectual impairments, need to be integrated into special schools, and need extensive support and assistance throughout their lives.

The reality of this disorder as a wide spectrum of symptom severity shed light on the importance of a dynamic and holistic approach to diagnosis and treatment.

2 Diagnostic Criteria of ASD

One of the biggest changes in the DSM 5 [1] was the introduction of ASD. Previously, in the DSM IV [3], autistic symptoms were categorized into four groups: autistic disorder, Asperger's disorder, childhood disintegrative disorder or the broader diagnosis of pervasive developmental disorder not otherwise specified. The main reason for this shift in diagnostic criteria was to limit the inconsistency in diagnosis across medical centers and practitioners, ultimately creating a comprehensive unified structure for assessing autism that would allow for greater efficacy in developing treatment plans [1]. Table 1 highlights the DSM 5 diagnostic criteria for ASD [1].

Table 1 DSM 5 diagnostic criteria for ASD

A. Persistent deficits in social communication and social interaction across multiple contexts, as manifested by the following, currently or by history (examples are illustrative, not exhaustive, see text)
Deficits in social-emotional reciprocity, ranging, for example, from abnormal social approach and failure of normal back-and-forth conversation; to reduced sharing of interests, emotions, or affect; to failure to initiate or respond to social interactions
Deficits in nonverbal communicative behaviors used for social interaction, ranging, for example, from poorly integrated verbal and nonverbal communication; to abnormalities in eye contact and body language or deficits in understanding and use of gestures; to a total lack of facial expressions and nonverbal communication
Deficits in developing, maintaining, and understanding relationships, ranging, for example, from difficulties adjusting behavior to suit various social contexts; to difficulties in sharing imaginative play or in making friends; to absence of interest in peers
B. Restricted, repetitive patterns of behavior, interests, or activities, as manifested by at least two of the following, currently or by history (examples are illustrative, not exhaustive; see text)
Stereotyped or repetitive motor movements, use of objects, or speech (e.g., simple motor stereotypies, lining up toys or flipping objects, echolalia, idiosyncratic phrases).
Insistence on sameness, inflexible adherence to routines, or ritualized patterns or verbal nonverbal behavior (e.g., extreme distress at small changes, difficulties with transitions, rigid thinking patterns, greeting rituals, need to take same route or eat food every day)
Highly restricted, fixated interests that are abnormal in intensity or focus (e.g., strong attachment to or preoccupation with unusual objects, excessively circumscribed or perseverative interest)
Hyper- or hypo-reactivity to sensory input or unusual interests in sensory aspects of the environment (e.g., apparent indifference to pain/temperature, adverse response to specific sounds or textures, excessive smelling or touching of objects, visual fascination with lights or movement)
C. Symptoms must be present in the early developmental period (but may not become fully manifest until social demands exceed limited capacities or may be masked by learned strategies in later life)
D. Symptoms cause clinically significant impairment in social, occupational, or other important areas of current functioning
E. These disturbances are not better explained by intellectual disability (intellectual developmental disorder) or global developmental delay. Intellectual disability and autism spectrum disorder frequently co-occur; to make comorbid diagnoses of autism spectrum disorder and intellectual disability, social communication should be below that expected for general developmental level

For diagnostic clusters A and B, it is necessary to specify and categorize severity of symptoms into three levels based on social communication impairments and restricted, repetitive patterns of behavior (requiring support, requiring substantial support, requiring very substantial support). Considering the effect of the treatment plan, practitioners should also specify if the disorder is:

- With or without accompanying intellectual impairment
- With or without accompanying language impairment
- Associated with a known medical or genetic condition or environmental factor
- Associated with another neurodevelopmental, mental or behavioral disorder
- With catatonia

3 Comorbid Psychological Conditions in ASDs

While the DSM 5 goes some distance to standardize the method for assessing impairments or medical and neurodevelopmental disorders that co-occur with autism, it fails to do the same for psychological comorbidities. In fact, the DSM 5 remains dependent on categorical definitions of psychological disorders rather than dimensional classifications [4].

This limitation in the DSM 5, i.e., lack of standardized assessment of comorbidities, generates a major gap in the ability to create an effective treatment plan that adequately meets the individual needs of each patient, and subsequently improve functioning. A burgeoning area of research has attempted to document the importance of identifying comorbidities in ASD. In a twin study in Sweden, Lundstrom et al. [5] found that half of the 272 ASD patients he studied had four or more coexisting disorders and that only 4% did not have a comorbid diagnosis. Talisa et al. [6] found that some neuropsychiatric and behavioral conditions were related to anxiety and not autism, indicating that failure to diagnose this would result in an inability to adequately improve function. Practitioners should become attuned to spotting signs of existing comorbidities like severe and incapacitating problem behavior, worsening of symptoms or abrupt changes from baseline and not responding to treatment as expected. Should these issues arise, a thorough assessment of psychological comorbidities should be undertaken using standardized assessment tools like:

- Young Mania Rating Scale (YMRS)
- Inventory of Depressive Symptomatology (IDS)
- Structured Clinical Interview for DSM IV for personality disorders (SCID-II)
- Structured Clinical Interview for DSM IV Childhood Diagnoses (Kid SCID)

Psychological conditions that commonly occur with ASDs are diverse, comprising of mood disorders (depression and bipolar), anxiety disorders, obsessive-compulsive disorder, and attention-deficit/hyperactivity disorder (ADHD). These conditions were found to be biologically based and situationally induced. In the following sections, each of these disorders will be discussed and will also be preceded by their DSM 5 diagnostic criteria.

4 Depression and Bipolar Disorder

In the DSM IV, depressive disorders and bipolar disorders were grouped under the category of "mood disorders"; however, in the DSM 5 these were reclassified into separate categories. Despite this, the diagnostic criteria for major depressive disorder (MDD) and bipolar I and II have remained more or less the same and changes were mostly conceptual in nature. Tables 2, 3, and 4 four outline the DSM 5 diagnostic criteria for MDD and bipolar disorder I and II, respectively.

Table 2 DSM 5 criteria for major depressive disorder

Major depressive disorder
The individual must experience five or more symptoms during the same 2-week period and at least one of the symptoms should be either (1) depressed mood or (2) loss of interest or pleasure
1. Depressed mood most of the day, nearly every day
2. Markedly diminished interest or pleasure in all, or almost all, activities most of the day, nearly every day
3. Significant weight loss when not dieting or weight gain or decrease or increase in appetite nearly every day
4. A slowing down of thought and a reduction of physical movement (observable by others, not merely subjective feelings of restlessness or being slowed down)
5. Fatigue or loss of energy nearly every day
6. Feelings of worthlessness or excessive or inappropriate guilt nearly every day
7. Diminished ability to think or concentrate, or indecisiveness, nearly every day
8. Recurrent thoughts of death, recurrent suicidal ideation without a specific plan, or a suicide attempt or a specific plan for committing suicide

Table 3 DSM 5 criteria for bipolar disorder I

Bipolar disorder I
A. Characterized by the occurrence of one or more manic or mixed episodes (the manic episode may have been preceded by and may be followed by hypomanic or major depressive episodes, but these are not required for diagnosis)
B. Distinct period of abnormally and persistently elevated, expansive, or irritable mood, and increased goal-directed activity or energy lasting ≥1 week (any duration if hospitalized), present most of the day, nearly every day
C. During the mood disturbance and increased energy or activity, at least three (or four if irritable mood only) of the following – Inflated self-esteem or grandiosity – Decreased need for sleep – Pressured speech – Racing thoughts or flight of ideas – Distractibility – Increased activity – Excess pleasurable or risky activity
D. Marked impairment not due to a substance or medical condition. In addition, these symptoms a. Do not meet criteria for a mixed episode b. Cause functional impairment, necessitate hospitalization, or there are psychotic features c. Are not related to substance misuse d. Are not due to a general medical condition e. Are not caused by somatic antidepressant therapy

Postorino et al. [7] reported the prevalence of co-occurrence of mood disorders (such as bipolar and depression) in ASDs to be between 1.4% and 70% [8–38]. Previous studies used different criteria and different assessment tools, both self-report and clinician administered which can greatly alter diagnosis, especially when

Table 4 DSM 5 criteria for bipolar disorder II

Bipolar disorder II
A. Never had a full manic episode; at least one hypomanic episode and at least one major depressive episode
B. Distinct period of abnormally and persistently elevated, expansive, or irritable mood, and increased goal-directed activity or energy lasting ≥4 but <7 days, and clearly different from usual nondepressed mood, present most of the day, nearly every day
C. During the hypomanic episode, at least three (or four if irritable mood only) of the following – Inflated self-esteem or grandiosity – Decreased need for sleep – Pressured speech – Racing thoughts or flight of ideas Distractibility Increased activity – Excess pleasurable or risky activity
D. Episode is unequivocal change in functioning, uncharacteristic of person, and observable by others
E. Not severe enough to cause marked impairment, not due to substance or medical condition, and no psychosis (if present, then this is mania by definition)
F. During the major depressive episode, at least five of the following symptoms are present during the same 2-week period and represent a change from previous functioning. At least one of the symptoms is either depressed mood or loss of interest or pleasure – Depressed mood most of the day, nearly every day – Markedly diminished interest or pleasure, nearly every day – Significant weight loss when not dieting or weight gain, or decrease or increase in appetite, nearly every day – Insomnia or hypersomnia, nearly every day – Psychomotor agitation or retardation, nearly every day – Fatigue or loss of energy, nearly every day Feelings of worthlessness or excessive or inappropriate guilt (which may be delusional), nearly every day – Diminished ability to think or concentrate, or indecisiveness, nearly every day – Recurrent thoughts of death (not just fear of dying), recurrent suicidal ideation with or without a specific plan
G. In addition, these depressive symptoms – Cause functional impairment (e.g., social, occupational) – Are not better explained by substance misuse, medication side effects, or other psychiatric or somatic medical conditions

taking into account the capabilities of the child [39]. The wide variance of this prevalence highlights the importance of a single standardized diagnostic method and assessment for psychiatric disorders.

There is evidence to support that mood disorders are associated with greater adaptability in ASD. Several studies found that symptoms of depression and mania are directly correlated with higher levels of functioning and adaptation, more insight or self-awareness of own impairments, and a higher cognitive level of functioning [24, 40]. Similarly, Vickerstaff et al. [20] found that there are significant associations between self-perception of social competence and depressive symptoms. In addition, high-functioning autism (HFA) patients were found to be predominantly afflicted with mood disorders [11, 14, 17, 20, 25, 30, 31, 41]. The rates of these comorbid mood disorders were found to be even higher in adolescent and adult HFA patients [19, 22, 24, 28, 35].

Conversely, other reports have indicated that with more severe symptoms of autism, patients are more vulnerable to stressors as well as to the development of depression [41, 42]. This is compounded further by Mazurek et al.'s [43] findings that poorer quality of friendship is correlated with higher levels of anxiety and depression indicating the importance of protective factors against general stressors. Depression-induced regression was found to be noticeably present mainly in low-functioning autism patients who are characterized by loss of language, social withdrawal, loss of eye contact, moodiness, tantrums, fearfulness, obsessiveness, stereotypies, hyperactivity, and occasionally self-injurious behaviors [40, 44].

Age was also found to be a significant predictor of comorbid depression as symptoms were found to increase with age, with emotional age being a more reliable predictor of the development of depression than chronological age [20]. Several studies also showed that the age of onset of co-occurrence of depression is predominantly around pre-adolescence and adolescence. This could be attributed to the transition period of ASD patients becoming more conscious of their own social skills as well as awareness of lower self-perceived social competence [9, 20, 26, 31, 40].

The diagnosis of depression is substantially based on self-reported feelings and how those feelings impact daily functioning. This is often difficult to obtain in the ASD population due to inherent impairments in social interaction and verbal communication.

From a clinical point of view, the diagnosis of depression in ASD remains a challenge despite characteristic symptoms like depressed mood, irritability, anhedonia, sleep or appetite disturbances, cognitive problems like impaired concentration, indecision, feelings of hopelessness, morbid thoughts, and somatic complaints being recognized. Other symptoms like aggression, mood lability, hyperactivity, decreased self-care, decreased level of functioning, regression, changes in core symptoms, increased compulsions, self-injurious behavior, and catatonia, and overall changes in adaptive functioning are often neglected in the observation of ASD patients [45]. The failure to identify these symptoms as depression and assuming them to be an extension of the ASD diagnosis can lead to a loss of the patient's ability to learn new skills that might greatly improve their ability to live with ASD and may lead to missing suitable interventions that could help tackle these problems.

5 Anxiety Disorders

Anxiety disorders in the DSM 5 include separation anxiety disorder, selective mutism, specific phobia, social phobia, panic disorder, agoraphobia, and generalized anxiety disorder (GAD). The common symptoms across each of these diagnoses are best explained by the diagnostic criteria for GAD in Table 5.

Table 5 DSM 5 diagnostic criteria for generalized anxiety disorder

Generalized anxiety disorder		
A.	The presence of excessive anxiety and worry about a variety of topics, events, or activities. Worry occurs more often than not for at least 6 months and is clearly excessive	
B.	–	The worry is experienced as very challenging to control. The worry in both adults and children may easily shift from one topic to another
C.	–	The anxiety and worry are accompanied with at least three of the following physical or cognitive symptoms. (In children, only one symptom is necessary for a diagnosis of GAD)
	–	Edginess or restlessness
	–	Tiring easily; more fatigued than usual
	–	Impaired concentration or feeling as though the mind goes blank
	–	Irritability (which may or may not be observable to others)
	–	Increased muscle aches or soreness
	–	Difficulty sleeping (due to trouble falling asleep or staying asleep, restlessness at night or unsatisfying sleep)

Due to the nature of ASD heavily impairing social and communication skills, school-age children and adolescents are often commonly affected by anxiety-related concerns [40]. Simonoff et al. [18] supported this further with findings showing that 41.9% of 112 ASD children aged from 10 to 14 years met the criteria for at least one anxiety disorder.

Reported prevalence of anxiety in ASD varies widely, with estimates ranging from 13.6% to 84.1% [14, 46–49]. A recent systematic review obtained from 31 studies [50] identified that clinically significant levels of anxiety were present in 39.6% of a pooled sample of 2121 individuals under the age of 18 with ASD. Although findings are inconsistent, the most frequent anxiety disorders in ASD appear to be specific phobias, generalized anxiety disorder, separation anxiety disorder, and social phobia with social anxiety being the most prevalent in ASDs (29.2%) [10, 49, 51–55].

Sukhodolsky et al. [54] similarly found that 43% of 171 children with ASD aged 5–14 years met the criteria for at least one anxiety disorder. They also reported that increased anxiety was associated with higher IQ and less ASD severity, which could be attributed to more self-awareness of social dysfunction.

Children with ASD presented a distinctive set of fears when compared to chronological- and mental-age matched peers, reporting more frequent situation phobias and medical fears, less often related to fears of being harmed or injured [51].

In conclusion, anxiety seems to be more common in ASD than in both the general population and several clinical groups with probably up to 40% of ASDs patients presenting with at least one anxiety subtype.

6 Obsessive-Compulsive Disorder

Obsessive-compulsive disorder (OCD) is characterized by recurrent disturbing thoughts or images, and repetitive behaviors. In the DSM IV, OCD was previously categorized as an anxiety disorder; however, in the DSM 5 it was reclassified as a distinct disorder due to the focus on the behavioral component. Table 6 describes the DSM 5 criteria for diagnosing OCD.

OCD often begins in childhood and adolescence. Several studies show an increased incidence of OCD in the ASD population, as well as increased ASD among those diagnosed with OCD [56, 57]. Postorino et al. [58] reported that the prevalence of OCD in ASD cases ranged between 2.6% and 37.2%.

It can be difficult to determine OCD diagnosis in an autistic child as there are overlapping rituals common across both such as repetitive behavior and rigid adherence to routines [35, 59, 60]. However, the compulsions are characterized by their

Table 6 DSM 5 diagnostic criteria for obsessive-compulsive disorder

Obsessive-compulsive disorder (OCD)
A. Presence of obsessions, compulsions, or both – Obsessions are defined by (1) and (2) 1. Recurrent and persistent thoughts, urges, or impulses that are experienced, at some time during the disturbance, as intrusive and unwanted, and that in most individuals cause marked anxiety or distress 2. The individual attempts to ignore or suppress such thoughts, urges, or images, or to neutralize them with some other thought or action (i.e., by performing a compulsion) – Compulsions are defined by (1) and (2) 1. Repetitive behaviors (e.g., hand washing, ordering, checking) or mental acts (e.g., praying, counting, repeating words silently) that the individual feels driven to perform in response to an obsession or according to rules that must be applied rigidly 2. The behaviors or mental acts are aimed at preventing or reducing anxiety or distress, or preventing some dreaded event or situation; however, these behaviors or mental acts are not connected in a realistic way with what they are designed to neutralize or prevent or are clearly excessive
B. The obsessions or compulsions are time-consuming (e.g., take more than 1 h/day) or cause clinically significant distress or impairment in social, occupational, or other important areas of functioning
C. The obsessive-compulsive symptoms are not attributable to the physiological effects of a substance (e.g., a drug of abuse, a medication) or another medical condition
D. The disturbance is not better explained by the symptoms of another mental disorder (e.g., excessive worries, as in generalized anxiety disorder; preoccupation with appearance, as in body dysmorphic disorder; difficulty discarding or parting with possessions, as in hoarding disorder; hair pulling, as in trichotillomania [hair-pulling disorder]; skin picking, as in excoriation [skin-picking] disorder; stereotypies, as in stereotypic movement disorder; ritualized eating behavior, as in eating disorders; preoccupation with substances or gambling, as in substance-related and addictive disorders; preoccupation with having an illness, as in illness anxiety disorder; sexual urges or fantasies, as in paraphilic disorders; impulses, as in disruptive, impulse-control, and conduct disorders; guilty ruminations, as in major depressive disorder; thought insertion or delusional preoccupations, as in schizophrenia spectrum and other psychotic disorders; or repetitive patterns of behavior, as in autism spectrum disorder)

distressing effect on the individual and anxiety peaks as a result of the attempt to resist carrying out the compulsive behavior. Rituals of autistic patients, on the other hand, are not characterized by any preceding anxiety or distress and are often a rewarding and pleasant experience for the child.

Ruta et al. [61] summarized the differences between children who received a diagnosis for OCD only, ASD only, and those with a comorbidity of OCD and ASD. They found that OCD groups and ASD groups reported different types of obsessive behaviors, with OCD children reporting higher frequencies of aggressive obsessions and checking compulsions, while ASD children displaying higher frequencies of saving/hoarding behaviors. However, they found that groups with comorbid diagnoses, ASD with OCD or Tourette syndrome, had comparable levels of symptom severity and impairment.

Anholt et al. [62] reported that adults with OCD show increased frequency of ADHD and autism symptoms and speculated common etiological factors to ASD, ADHD, and OCD.

7 Attention-Deficit/Hyperactivity Disorder

Attention-deficit/hyperactivity disorder (ADHD) is characterized by symptoms of inattention, hyperactivity, and impulsivity across multiple settings. Table 7 specifies the diagnostic criteria for ADHD according to the DSM 5.

So far, no meta-analyses have been conducted on the prevalence of ADHD in ASDs. However, ADHD was found to co-occur in as many as 30–80% of the ASD cases, while the presence of ASD is estimated to be between 20% and 50% of the ADHD children [63–66].

For instance, van der Meer et al. [64] conducted a study on three groups of patients ((1) ADHD plus ASD; (2) predominant ASD plus ADHD; and (3) ADHD only) and found a significantly slower identification of facial emotions in the ASD + ADHD, and ADHD + ASD groups when compared with the ADHD-alone group. Significant differences were also found in visual spatial attention, verbal attention, and working memory amongst the groups but no significant differences in inhibition and cognitive flexibility was noticed [64]. The ADHD plus ASD and ADHD-alone groups performed significantly worse in detail-focused processing [64].

8 Personality Disorders (PD)

In the DSM 5, the 10 PD outlined in the DSM IV were retained and they are: paranoid personality disorder, schizoid personality disorder, schizotypal personality disorder, antisocial personality disorder, borderline personality disorder, histrionic personality, narcissistic personality disorder, avoidant personality disorder, dependent personality disorder, and obsessive-compulsive personality disorder.

Table 7 DSM 5 diagnostic criteria for attention-deficit/hyperactivity disorder

Attention-deficit/hyperactivity disorder (ADHD)
A. Persistent pattern of inattention and/or hyperactivity–impulsivity that interferes with functioning or development, as characterized by (1) and/or (2)
1. Inattention: Six (or more) of the following symptoms have persisted for at least 6 months to a degree that is inconsistent with developmental level and that negatively impacts directly on social and academic/occupational activities
– Often fails to give close attention to details or makes careless mistakes in schoolwork, at work, or during other activities (e.g., overlooks or misses details, work is inaccurate)
– Often has difficulty sustaining attention in tasks or play activities (e.g., has difficulty remaining focused during lectures, conversations, or lengthy reading)
– Often does not seem to listen when spoken to directly (e.g., mind seems elsewhere, even in the absence of any obvious distraction)
– Often does not follow through on instructions and fails to finish schoolwork, chores, or duties in the workplace (e.g., starts tasks but quickly loses focus and is easily sidetracked)
– Often has difficulty organizing tasks and activities (e.g., difficulty managing sequential tasks; difficulty keeping materials and belongings in order; messy, disorganized work; has poor time management; fails to meet deadlines)
– Often avoids, dislikes, or is reluctant to engage in tasks that require sustained mental effort (e.g., schoolwork or homework; for older adolescents and adults, preparing reports, completing forms, reviewing lengthy papers)
– Often loses things necessary for tasks or activities (e.g., school materials, pencils, books, tools, wallets, keys, paperwork, eyeglasses, mobile telephones)
– Is often easily distracted by extraneous stimuli (for older adolescents and adults, may include unrelated thoughts)
– Is often forgetful in daily activities (e.g., doing chores, running errands; for older adolescents and adults, returning calls, paying bills, keeping appointments)
2. Hyperactivity and impulsivity: Six (or more) of the following symptoms have persisted for at least 6 months to a degree that is inconsistent with developmental level and that negatively impacts directly on social and academic/occupational activities
– Often fidgets with or taps hands or feet or squirms in seat
– Often leaves seat in situations when remaining seated is expected (e.g., leaves his or her place in the classroom, in the office or other workplace, or in other situations that require remaining in place)
– Often runs about or climbs in situations where it is inappropriate. (Note: In adolescents or adults, may be limited to feeling restless)
– Often unable to play or engage in leisure activities quietly
– Is often "on the go," acting as if "driven by a motor" (e.g., is unable to be or uncomfortable being still for extended time, as in restaurants, meetings; may be experienced by others as being restless or difficult to keep up with)
– Often talks excessively
– Often blurts out an answer before a question has been completed (e.g., completes people's sentences; cannot wait for turn in conversation)
– Often has difficulty waiting his or her turn (e.g., while waiting in line)
– Often interrupts or intrudes on others (e.g., butts into conversations, games, or activities; may start using other people's things without asking or receiving permission; for adolescents and adults, may intrude into or take over what others are doing)

(continued)

Table 7 (continued)

Attention-deficit/hyperactivity disorder (ADHD)	
B.	Several inattentive or hyperactive-impulsive symptoms were present prior to age 12 years
C.	Several inattentive or hyperactive-impulsive symptoms are present in two or more settings (e.g., at home, school, or work; with friends or relatives; in other activities)
D.	There is clear evidence that the symptoms interfere with, or reduce the quality of, social, academic, or occupational functioning
E.	The symptoms do not occur exclusively during the course of schizophrenia or another psychotic disorder and are not better explained by another mental disorder (e.g., mood disorder, anxiety disorder, dissociative disorder, personality disorder, substance intoxication or withdrawal)

However, much like autism and as opposed to schizophrenia or posttraumatic stress disorder, personality disorders are not categorical and do exist on a continuum. For this reason, the DSM 5 has put forward proposed changes for further study in a separate section. The proposed model would evaluate impairments in personality functioning and assess five broad areas of pathological personality traits. This model includes only six PD as evidenced by research: antisocial, avoidant, borderline, narcissistic, obsessive-compulsive, and schizotypal.

In this vein, distinct PD are not as commonly found to be diagnosed as comorbidities of ASD. However, certain traits like aggression and self-injurious behavior that are symptomatic of PD, like borderline personality disorder and antisocial personality disorder, were prevalent ASD comorbidities [43, 67]. While there is a correlation between aggression, self-injurious behavior, and ASD, we cannot infer a causal relationship. In fact, it is difficult to even determine whether these variables affect each other distinctly or if they are manifestations of the same problem. However, as it is not possible to diagnose any PD before the age of 18 as personality is still in its formative stage, problematic traits should be monitored using functional analysis to identify factors that might perpetuate or reinforce the trait or behavior [68].

9 Interventional Models

Children with ASD generally require a combination of therapies and interventions to address their individual constellation of symptoms. Approaches can be broadly categorized according to conceptual models. However, there is no uniformly agreed upon classification system. The availability of programs varies by region and access to interventions may affect the choice of programming. A systematic review found insufficient evidence to suggest that any interventional model is superior to another [69]. However, there is moderate evidence that greater intensity (in hours per week) and greater duration (in months) of treatment lead to better outcomes [70].

Table 8 summarizes five interventions commonly used to treat ASD and the strengths of each therapy.

Table 8 Summary of interventional models for ASD

Intervention	Description	Type	Objective	Strengths
Developmental behavioral interventions	This therapy is applied in the client's natural setting or in a structured environment, and includes behavioral modification, structured teaching, and is relationship based. Generally, it works by reinforcing productive behaviors and discouraging maladaptive behaviors Examples of developmental behavioral interventions include: [insert names]	Behavioral and developmental therapy	Focuses on using a variety of behavioral strategies to teach necessary skills relevant to the development stage	– Targets specific domains (e.g., social, language, cognitive) – May occur in various settings (e.g., naturalistic versus structured) – Involves the parents, particularly when interventions are provided in the home
Treatment and Education of Autistic and Related Communication Handicapped Children (TEACCH)	The TEACCH method uses structured teaching to help individuals overcome areas of weakness	Generalist (does not identify with one developmental discipline)	The goal is to modify the environment and improve skills	– Understands culture of autism – Uses an individualized person and family-centered plan – Organizes the physical environment – Uses a predictable sequence of activities – Utilizes visual schedules and visually structured activities – Implements routines with flexibility – Structured work/activity systems
Occupational therapy	Occupational therapy is often used to address deficits in adaptive functioning and fine motor skills that affect academic and everyday functioning	Occupational	To enhance functioning inhibited by a specific deficit and encourage self-sufficiency	In young children with ASD, occupational therapy focuses on enhancing – sensory processing, sensorimotor, and social-behavioral performance – self-care (e.g., dressing, hygiene) – participation in play In older children, the focus of occupational therapy may include: – social and behavioral performance – transition to work and independence in the community

(continued)

Table 8 (continued)

Intervention	Description	Type	Objective	Strengths
CBT (self-management)	CBT focuses on replacing negative or ineffective patterns of thought and behavior with structured strategies that are effective in improving mood and adaptive functioning	Cognitive-behavioral therapy	Helping those with ASD to learn to independently regulate their own behaviors and act appropriately in a variety of home, school, and community-based situations	Those with ASD are taught to – discriminate between appropriate and inappropriate behaviors – accurately monitor and record their own behaviors – reward themselves for behaving appropriately – eventually take on greater responsibility in their own self-care

10 Treating Psychological Comorbidities in ASD

Once a diagnosis of a comorbidity has been ascertained, an individualized treatment plan that compliments the interventions he or she is already receiving needs to be determined.

Comprehensive integrative models address multiple domains of function. For example, the Early Start Denver Model (ESDM) uses a combination of behavioral programming and developmental- and relationship-based approaches and includes parents as therapists. These types of comprehensive therapies are often beneficial with comorbid psychiatric disorders and tend to directly and indirectly target symptoms that often complicate the treatment of ASD. A randomized trial comparing the ESDM program with interventions commonly available in the community demonstrated significant language, cognitive, and adaptive functioning gains in 48 toddlers over a 2-year period [71]. The Agency for Healthcare Research and Quality (AHRQ) published a systematic review [72] suggesting the utility of parent training for improving behavioral outcomes in general and of adding parent training to medication interventions for children with challenging behaviors. However, the studies were small, relied on parent report, and used varying intervention models.

Nevertheless, the National Autism Center's National Standards Reports [73] considers targeted behavioral interventions to be the general standard of treatment. Historically, behavioral interventions have also been found to be beneficial. A systematic review of 251 studies conducted between 1980 and 1996 of targeted behavioral interventions found that focal behavioral interventions consistently result in positive behavioral outcomes across a wide range of targets, including aberrant behaviors (e.g., self-injury, aggression), language skills, daily living skills, social skills, etc. [74].

On the other hand, a 2014 systematic review of studies published after 2000 suggested the efficacy of CBT interventions in reducing anxiety symptoms in individuals with ASD and IQ scores above 70 [2]. Moreover, a systematic review published by the US Massachusetts National Standards projects classified CBT as an established intervention for children and adolescents [73]. Similarly, a meta-analysis of 12 studies for anxiety comorbidity involving 511 youth with high functioning ASD found statistically significant pooled treatment effect for CBT with significant IQ heterogeneity [75].

A systematic review [76] evaluating the efficacy of CBT on ASD and OCD comorbidities found that although CBT with various modifications has been shown to be beneficial, the research includes small populations and a variety of nonstandard modifications; the lack of standardization in applying CBT limits the generalizability of the findings. Nevertheless, all the studies did show at least some treatment gains despite the variation in age and severity of diagnosis. The methods involved in the studies, while varied, generally included mapping, cognitive restructuring, fear hierarchy development, (exposure and response prevention) and relapse prevention.

With a comorbid diagnosis of ASD and ADHD, nonpharmacological treatments found to be moderately effective include dietary interventions (restricted elimination diets, artificial food color exclusions, and free fatty acid supplementation), behavioral interventions, cognitive training, and neurofeedback [77].

It is likely that comorbid emotional or behavioral problems would influence outcomes of social skills interventions. In a study observing the effect of a social skills training program, it was reported that social skills improved for children with ASD, and children with ASD and comorbid anxiety, but that there was no improvement among children with ASD and comorbid ADHD, highlighting the importance of individualizing treatment plans for different comorbid diagnoses [78].

11 Quality of Life

In the simplest terms, Quality of Life (QoL) is defined as inner subjective personal satisfaction across four basic domains: physical, emotional, social, and vocational [79]. QoL interventions from a positive psychology point of view aim at promoting a life satisfaction in which humans identify, pursue, and fulfill their most cherished goals, desires, and wishes across all valued areas of life [80]. In the context of ASD, QoL Clinical Practice (QoLCP) normalizes the life of patients and their families so that it does not fall below a predetermined cut-off threshold [81].

With this definition, QoL Clinical Practice could be a precise, patient and family cantered care method for measuring improvement, monitoring ASD symptoms, optimizing interventions, and personalizing medicosocial care amongst individuals with ASD.

11.1 Key Features in ASD Conventional QoLCP

11.1.1 CASIO Rubric

QoL provides a rubric model for life satisfaction (Change in *C*ircumstances, *A*ttitude, *S*tandards, *I*mportance, and *O*ther aspects; CASIO) as a blueprint for positive psychological intervention. The model presented in Fig. 1 offers a strategy for

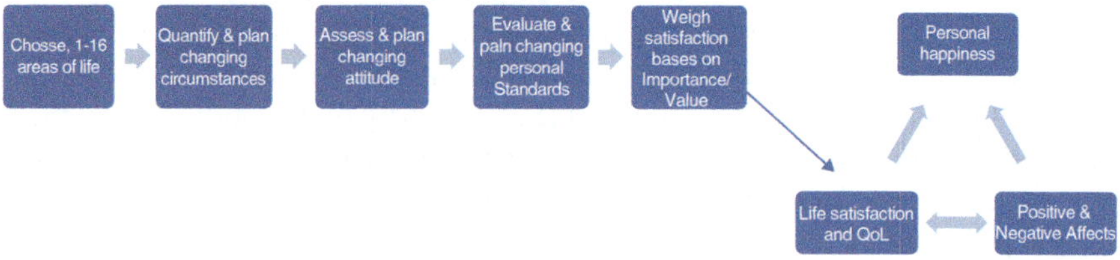

Fig. 1 CASIO model for life satisfaction

Table 9 CASIO eight-session program for improving QoL

Session 1 Introduce participants		Session 5	
Review	Goals QoL interventions and rationale	Review	CASIO model in values Homework
Discuss	16 areas of life satisfaction Difficult areas	Discuss	Relationships and its role in life satisfaction
Homework	Think how to improve QoL	Homework	Everyday life skills: Enhance relationships using writing a letter and basket of eggs techniques [79]
Session 2		Session 6	
Review	QoL progress Homework	Review	CASIO model in relationships Homework
Discuss	Role of self-esteem in happiness increasing and present skills in these areas	Discuss	The role of play and leisure in increasing happiness
Homework	Everyday life skills: Improve strengths and gratitude through BAT (Blessings, Accomplishments, Talents, and Traits) technique [79]	Homework	Everyday life skills: Increase play and family recreation time [79]
Session 3		Session 7	
Review	CASIO model in self-esteem Homework	Review	CASIO model of play Homework
Discuss	Health topics and concerns	Discuss	Learning and skills
Homework	Everyday life skills: Report on frequent health concerns using Trigger, Actions, and Consequences (TAC) technique [79]	Homework	Everyday life skills: Boost learning satisfaction using problem solving technique [79]
Session 4		Session 8	
Review	CASIO model in health concerns Homework	Review	CASIO model of learning Homework All treatment sessions
Discuss	Goals and important values	Discuss	Transition to being own QoL therapist and using relapse prevention techniques
Homework	Everyday life skills: Tweak goals and values using Daily Action Plan (DAP) and Life Script techniques [79]	Homework	Further study and work in QoL

management of 16 areas of life over 8 therapeutic sessions [79, 80]. The program (illustrated in Table 9) starts with introducing clinical participants, and each session consists of reviews, discussions, and assigning homework steps.

12 Constructive Mode Activation for ASD Comorbidities

The QoLCP also provides patients access to constructive cognitive creation of life satisfaction and happiness through the above CASIO model. Individual differences in relation to life satisfaction is accommodated via recognition of interaction between external life conditions and patients' own circumstances, personal values attached to life goals, and personal standards for reaching goals in 16 areas of life [80]. Table 10 highlights definitions of the 16 areas of life focused on in QoLCP.

Table 10 16 areas of life for QoLCP

1.	*Health* is being physically fit, not sick, and without pain or disability
2.	*Self-Esteem* means liking and respecting yourself in light of your strengths and weaknesses, successes and failures, and ability to handle problems
3.	*Goals-and-Values* ± Spiritual Life: are beliefs about what matters most in life and how you should live, both now and in the future
4.	*Money* (or Standard of Living) is made of the money you earn, the things you own (like a car or furniture) and believing that you will have the money and things that you need in the future
5.	*Work* means your career or how you spend most of your time
6.	*Play* (or Recreation) means what you do in your free time to relax, have fun, or improve yourself. This could include watching movies, visiting friends, or pursuing a hobby like sports or gardening
7.	*Learning* means gaining new skills or information about things that interest you. Learning can come from reading books or taking classes on subjects like history, car repair, or using a computer
8.	*Creativity* is using your imagination to come up with new and clever ways to solve every day problems or to pursue a hobby like painting, photography, or needlework. This can include decorating your home, playing the guitar, or finding a new way to solve a problem at work
9.	*Helping* (Social Service and Civic Action) means helping others (not just friends or relatives) in need or helping to make your community a better place to live
10.	*Love* (or Love Relationship) is a very close romantic relationship with another person. Love usually includes sexual feelings and feeling loved, cared for, and understood
11.	*Friends* (or Friendships) are people (not relatives) you know well and care about who have interests and opinions like yours
12.	*Children* includes a measure of how you get along with your child (or children). Think of how you get along as you care for, visit, or play with your child (or children)
13.	*Relatives* means how you get along with your parents, grandparents, brothers, sisters, aunts, uncles, and in-laws
14.	*Home* is where you live. It is your house or apartment and the yard around it
15.	*Neighborhood* is the area around your home
16.	*Community* is the whole city, town, or rural area where you live (not just your neighborhood). Community includes how nice the area looks, the amount of crime, and how well you like the people. It also includes places to go for fun like parks, concerts, sporting events, and restaurants

Adapted with minimal modifications from Frisch [79]

12.1 Innovative Key Features of ASD QoLCP

As the QoL of autistic patients and their families is lower than that of the general population [82], it requires innovative practice in addition to these two key conventional features. The traditional QoL/psychological diagnosis of autistic patients and associated comorbidities involves medical and psychological history taking, mental state examination, and psychological screening. The end result of such a process is a subjective diagnosis of the case and the difficulties that families might be going through as a result of the disorder. Recently, these subjective projections of health care practitioners are being challenged, and objective nonbiased assessment tools are being pursued [83]. This represents a key requirement in personalizing QoL management of patients and families and optimizing their well-being in several domains of the 16 areas of life of the CASIO model.

12.2 Assessment of QoL in ASD Patients

If the assessment is carried out by a QoL practitioner who is not a physician, he or she interacts with the primary physician to get a medical report and a green light to carry out the QoL interventions [79]. However, if a physician is carrying out the intervention, then QoL assessment followed by a psychiatric ASD assessment should be performed starting with comprehensive history taking and physical and mental state examinations. Screening tools are then applied as a baseline and a follow up investigation.

12.2.1 Screening Tools for Adults

The Research Autism of the National Autistic Society of UK validated the use of Autism Specific QoL survey (ASQoL) to be used alongside the World Health Organization Quality of Life-Brief (WHOQoL-BREF) and World Health Organization Quality of Life (WHOQoL) disabilities modules. It is used with adults to evaluate total ASQoL score (eight items), and a score for the global item (item 9) about "autistic identity" [84].

12.2.2 Screening Tools for Children and Adolescents

The most commonly used instruments are the QoL Battery of Varni [85, 86]. The battery contains the Pediatric Quality of Life Inventory ™ (PedsQL) and other instruments to assess a wide variety of domains related to QoL, family satisfaction, and burden of diseases [87].

In conclusion, ASD patients experience a specific and unique form of QoL, the normalization of which is an endpoint medical care and requires a multidisciplinary team effort that includes a QoL therapist. This normalization takes place for ASD and all its associated comorbidities through the conventional and innovative QoLCP key features, and it encompasses all aspects of patient's life and his/her family.

13 Pharmacotherapy

While nonpharmacological treatments have been shown to be effective in treating comorbidities of ASD, a valid treatment option is medication. Pharmacotherapy should be considered when symptoms of comorbidities are extremely severe (e.g. depression or OCD), if there is severe functional impairment secondary to disruptive behavior or if there is no response to behavioral interventions. Moreover, as patients with ASD often undergo several hours of weekly interventions to improve general functioning, it can be overwhelming to recommend further interventions for their comorbidities.

Interventions should be guided by evidence and appropriate treatment guidelines [57]. Below is a summary of medications found to be effective in treating comorbidities in ASD:

- Depression: The efficacy of selective serotonin reuptake inhibitor (SSRIs) and serotonin norepinephrine reuptake inhibitor (SNRIs) in the treatment of depression and ASD has not been sufficiently validated through randomized controlled trials; nonetheless, empirical data support their use as indicated in neurotypical children [88].
- Anxiety: The treatment of anxiety in children with ASD and neurotypical children is similar. A multimodal approach is recommended, including modified cognitive behavioral therapy, with some evidence that supports its efficacy in high functioning ASD. Pharmacological data in this population is limited [89]. Behavioral interventions should also be considered in addressing sensory and special education needs [90].
- OCD: Similarities between OCD and the repetitive behaviors of ASD led researchers to investigate the use of SSRIs in the autism core domain [91]. In a randomized placebo-control crossover study of 44 children with ASD, SSRI (fluoxetine) was found to be beneficial in reducing repetitive behaviors in ASDs patients. The strength of evidence for the effect of other SSRIs (e.g., citalopram and escitalopram) is insufficient [92]. The evidence indicating that medication is effective in treating similar symptoms common in both OCD and ASD [93, 94].
- ADHD: Medications could be considered in the treatment of ADHD in the context of ASD [95, 96]. Methylphenidate (Ritalin) is the most commonly used drug and is effective in reducing symptoms of inattention and hyperactivity in children with ASD although response rates may be lower than that of children with typical ADHD. Randomized control trials suggest less benefit and more side

effects for ADHD + ASD as compared with ADHD alone [97]. Methylphenidate was found to significantly improve joint attention and emotional self-regulation as well as improvement in hyperactive and impulsive behaviors. However, the results on the efficacy of amphetamines are less conclusive. Alpha-2 adrenergic agonists were also effective when dealing with ADHD/ASD comorbidities and were found to significantly improve behavioral symptoms in 62 children when compared with a placebo. Alternatively, norepinephrine reuptake inhibitor (NERI), namely atomoxetine, was found to improve ADHD symptoms in two randomized controlled trials.

- Aggression: Haloperidol, a typical neuroleptic, is commonly used to treat severe aggression in autistic children; however, these have been found to significantly impair movement in recipients [98]. In addition risperidone was found to reduce irritability, aggression, self-injurious behaviors, and severe tantrums in ASD [92, 99, 100]. For younger ASD cases aged between 6 and 17 years, aripiprazole is recommended to treat aggression, and in a longitudinal study both risperidone and aripiprazole were found to adequately treat aggression and irritability in ASD patients, especially when combined with parent training in behavioral management [101].

14 Conclusions and Future Directions

Psychological comorbidities are relatively recently recognized phenomena in ASD although the majority of ASDs have at least one comorbid psychological disorder.

The high level of comorbidities could be attributed to similar or associated risk factors, i.e., the occurrence of one disorder increases the risk of another disorder. In addition, limitations could include misdiagnosis and inadequacy of the diagnostic systems to reflect the factual nature of psychiatric disorders that co-occur with an ASD diagnosis.

These comorbid conditions persist from childhood to adolescence to adulthood and are associated with more impaired social functioning [102, 103].

The current understanding of the processes that contribute to the high rates of comorbidities in ASD remains incomplete. Furthermore, there has been nearly no research on interventions involving comorbid presentations in ASD with other psychological and psychiatric disorders.

Thus, research studies in this field that may provide important clues about the underlying mechanisms and potential risk and protective factors involved in ASD are highly required.

Targeting two comprehensive modules of processes likely involved in high rates of comorbidities in ASD may be mainly useful. The first class is central developmental processes directly linked to the etiology of ASD, while the second module includes wider, transdiagnostic risk processes. It is possible that as developing social neural systems increasingly advance from "normal" trajectories in ASD children, other processes related to mental health may be affected as well.

In this vein, we can consider core processes such as social detachment and atypical social information processing in the possible pathogenesis of comorbid conditions. As an example, decreased hedonic responses to the social-emotional bids of others may be involved in the development of oppositional problems or aggression.

The second class is transdiagnostic processes that are not necessarily causally linked to the core impairments of ASD. Rather, they are "fundamental" in the sense that they are central to many forms of psychopathology. There are many transdiagnostic processes such as attentional avoidance, persistent negative affect, and rumination. Poor emotion regulation, for example, is a transdiagnostic process that has been linked theoretically to the high rates of anxiety disorders seen in people with ASD [104, 105]. These processes occur over the course of development, and thus, it will be important for future research to consider the longitudinal course of comorbidity and the possibility of sequential comorbidities over the course of a lifetime [106].

The early identification and treatment of the psychological comorbidities are useful for symptom relief, quality of life and daily adaptive functioning.

However, it is also equally important to remember that comorbid conditions are not meant to take clinical attention away from core/primary ASD symptoms in need of immediate intervention.

Previous studies on ASD and their comorbidities used different criteria and assessment tools completed by different informants (e.g., parents, teachers, practitioners/clinicians or self-report), resulting in different diagnoses and comorbidity results. Thus, future research and intervention should concentrate on comprehensive standardized diagnostic methods and assessments for ASD and psychiatric and psychological comorbidities. Furthermore, some ASD impairments overlap with some of the features of comorbid disorders making it difficult to differentiate between them. For example, OCD diagnosis and ASD impairments have overlapping rituals in common. These include repetitive behavior and rigid adherence to routines. However, whilst OCD compulsions are characterized by distress and anxiety, similar rituals of autistic patients are often a rewarding and pleasant experience for the child and free of such anxiety and distress. Thus, assessment tools should have the ability to distinguish clearly between pure ASD and ASD and its comorbidities. In some cases, it may be difficult to extract information for self-report from ASD patients (e.g., self-report of feelings and how those feelings impact daily functioning) due to inherent impairments in social interaction and verbal communication. This can result in a diagnosis being missed and the patient not receiving helpful intervention.

Cognitive behavioral therapy (CBT) has been shown to be effective in treating ASD and some of its comorbidities. However, research in this particular area come with a significant set of limitations. These limitations included small population size, a lack of standardization in applying CBT, and the neglect of ASD comorbidities and/or different outcomes for different comorbidities of ASD (e.g., improvement in some but not for others). This highlights the importance of individualizing treatment plans for different comorbid diagnoses.

In conclusion, improved, comprehensive, diagnostic assessment tools, taking into account various comorbidities and how they relate to ASD, are needed. Once accurate diagnoses have been made, better individualized and comprehensive interventions should be constructed to yield optimum outcomes for patients.

References

1. American Psychiatric Association. (2013). *Diagnostic and statistical manual of mental disorders* (5th ed.). Arlington, VA: American Psychiatric Association.
2. Weitlauf, A. S., McPheeters, M. L., Peters, B., Sathe, N., Travis, R., Aiello, R., et al. (2014). Therapies for children with autism spectrum disorder: Behavioral interventions update. Report No.: 14-EHC036-EF. *AHRQ Comparative Effectiveness Reviews*. Rockville, MD: Agency for Healthcare Research and Quality (US).
3. American Psychiatric Association (APA). (1994). *Diagnostic and statistical manual of mental disorders (DSM-IV)* (4th ed.). Washington DC: American Psychiatric Association.
4. Frenz, D. (2016). *Diagnostic comorbidity in DSM-5: Origins, current status, and potential solutions*. Retrieved from https://pro.psychcentral.com/diagnostic-comorbidity-in-dsm-5-origins-current-status-and-potential-solutions/
5. Lundstrom, S., Reichenberg, A., Melke, J., Rastam, M., Kerekes, N., Lichtenstein, P., et al. (2015). Autism spectrum disorders and coexisting disorders in a nationwide Swedish twin study. *Journal of Child Psychology and Psychiatry, 56*(6), 702–710. https://doi.org/10.1111/jcpp.12329
6. Talisa, V. B., Boyle, L., Crafa, D., & Kaufmann, W. E. (2014). Autism and anxiety in males with fragile X syndrome: An exploratory analysis of neurobehavioral profiles from a parent survey. *American Journal of Medical Genetics. Part A, 164a*(5), 1198–1203. https://doi.org/10.1002/ajmg.a.36468
7. Postorino, V., & Mazzone, L. (2016). *Mood disorders and autism spectrum disorder*. Cham: Springer. https://doi.org/10.1007/978-3-319-29695-1_1
8. Barnhill, G. P., & Myles, B. S. (2001). Attributional style and depression in adolescents with Asperger syndrome. https://doi.org/10.1177/109830070100300305
9. Brereton, A. V., Tonge, B. J., & Einfeld, S. L. (2006). Psychopathology in children and adolescents with autism compared to young people with intellectual disability. *Journal of Autism and Developmental Disorders, 36*(7), 863–870. https://doi.org/10.1007/s10803-006-0125-y
10. de Bruin, E. I., Ferdinand, R. F., Meester, S., de Nijs, P. F., & Verheij, F. (2007). High rates of psychiatric co-morbidity in PDD-NOS. *Journal of Autism and Developmental Disorders, 37*(5), 877–886. https://doi.org/10.1007/s10803-006-0215-x
11. Ghaziuddin, M., & Greden, J. (1998). Depression in children with autism/pervasive developmental disorders: A case-control family history study. *Journal of Autism and Developmental Disorders, 28*(2), 111–115.
12. Green, J., Gilchrist, A., Burton, D., & Cox, A. (2000). Social and psychiatric functioning in adolescents with Asperger syndrome compared with conduct disorder. *Journal of Autism and Developmental Disorders, 30*(4), 279–293.
13. Hedley, D., & Young, R. (2006). Social comparison processes and depressive symptoms in children and adolescents with Asperger syndrome. *Autism, 10*(2), 139–153.
14. Kim, J. A., Szatmari, P., Bryson, S. E., Streiner, D. L., & Wilson, F. J. (2000). The prevalence of anxiety and mood problems among children with autism and Asperger syndrome. *Autism, 4*(2), 117–132.
15. Lainhart, J. E., & Folstein, S. E. (1994). Affective disorders in people with autism: A review of published cases. *Journal of Autism and Developmental Disorders, 24*(5), 587–601.

16. Leyfer, O. T., Folstein, S. E., Bacalman, S., Davis, N. O., Dinh, E., Morgan, J., et al. (2006). Comorbid psychiatric disorders in children with autism: Interview development and rates of disorders. *Journal of Autism and Developmental Disorders, 36*(7), 849–861. https://doi.org/10.1007/s10803-006-0123-0
17. Munesue, T., Ono, Y., Mutoh, K., Shimoda, K., Nakatani, H., & Kikuchi, M. (2008). High prevalence of bipolar disorder comorbidity in adolescents and young adults with high-functioning autism spectrum disorder: A preliminary study of 44 outpatients. *Journal of Affective Disorders, 111*(2–3), 170–175. https://doi.org/10.1016/j.jad.2008.02.015
18. Simonoff, E., Pickles, A., Charman, T., Chandler, S., Loucas, T., & Baird, G. (2008). Psychiatric disorders in children with autism spectrum disorders: Prevalence, comorbidity, and associated factors in a population-derived sample. *Journal of the American Academy of Child and Adolescent Psychiatry, 47*(8), 921–929. https://doi.org/10.1097/CHI.0b013e318179964f
19. Stahlberg, O., Soderstrom, H., Rastam, M., & Gillberg, C. (2004). Bipolar disorder, schizophrenia, and other psychotic disorders in adults with childhood onset AD/HD and/or autism spectrum disorders. *Journal of Neural Transmission (Vienna), 111*(7), 891–902. https://doi.org/10.1007/s00702-004-0115-1
20. Vickerstaff, S., Heriot, S., Wong, M., Lopes, A., & Dossetor, D. (2007). Intellectual ability, self-perceived social competence, and depressive symptomatology in children with high-functioning autistic spectrum disorders. *Journal of Autism and Developmental Disorders, 37*(9), 1647–1664. https://doi.org/10.1007/s10803-006-0292-x
21. Wozniak, J., Biederman, J., Faraone, S. V., Frazier, J., Kim, J., Millstein, R., et al. (1997). Mania in children with pervasive developmental disorder revisited. *Journal of the American Academy of Child and Adolescent Psychiatry, 36*(11), 1552–1559. https://doi.org/10.1016/s0890-8567(09)66564-3
22. Hofvander, B., Delorme, R., Chaste, P., Nyden, A., Wentz, E., Stahlberg, O., et al. (2009). Psychiatric and psychosocial problems in adults with normal-intelligence autism spectrum disorders. *BMC Psychiatry, 9*, 35. https://doi.org/10.1186/1471-244x-9-35
23. Mattila, M. L., Hurtig, T., Haapsamo, H., Jussila, K., Kuusikko-Gauffin, S., Kielinen, M., et al. (2010). Comorbid psychiatric disorders associated with Asperger syndrome/high-functioning autism: A community- and clinic-based study. *Journal of Autism and Developmental Disorders, 40*(9), 1080–1093. https://doi.org/10.1007/s10803-010-0958-2
24. Sterling, L., Dawson, G., Estes, A., & Greenson, J. (2008). Characteristics associated with presence of depressive symptoms in adults with autism spectrum disorder. *Journal of Autism and Developmental Disorders, 38*(6), 1011–1018. https://doi.org/10.1007/s10803-007-0477-y
25. Whitehouse, A. J. O., Watt, H. J., Line, E. A., & Bishop, D. V. M. (2009). Adult psychosocial outcomes of children with specific language impairment, pragmatic language impairment and autism. *International Journal of Language & Communication Disorders, 44*(4), 511–528. https://doi.org/10.1080/13682820802708098
26. Williamson, S., Craig, J., & Slinger, R. (2008). Exploring the relationship between measures of self-esteem and psychological adjustment among adolescents with Asperger syndrome. *Autism, 12*(4), 391–402. https://doi.org/10.1177/1362361308091652
27. Amr, M., Bu Ali, W., Hablas, H., Raddad, D., El-Mehesh, F., El-Gilany, A. H., et al. (2012). Sociodemographic factors in Arab children with autism Spectrum disorders. *The Pan African Medical Journal, 13*, 65.
28. Cassidy, S., Bradley, P., Robinson, J., Allison, C., McHugh, M., & Baron-Cohen, S. (2014). Suicidal ideation and suicide plans or attempts in adults with Asperger's syndrome attending a specialist diagnostic clinic: A clinical cohort study. *Lancet Psychiatry, 1*(2), 142–147. https://doi.org/10.1016/s2215-0366(14)70248-2
29. Gotham, K., Unruh, K., & Lord, C. (2015). Depression and its measurement in verbal adolescents and adults with autism spectrum disorder. *Autism, 19*(4), 491–504. https://doi.org/10.1177/1362361314536625
30. Joshi, G., Wozniak, J., Petty, C., Martelon, M. K., Fried, R., Bolfek, A., et al. (2013). Psychiatric comorbidity and functioning in a clinically referred population of adults with

autism spectrum disorders: A comparative study. *Journal of Autism and Developmental Disorders, 43*(6), 1314–1325. https://doi.org/10.1007/s10803-012-1679-5
31. Mazzone, L., Postorino, V., De Peppo, L., Fatta, L., Lucarelli, V., Reale, L., et al. (2013). Mood symptoms in children and adolescents with autism spectrum disorders. *Research in Developmental Disabilities, 34*(11), 3699–3708. https://doi.org/10.1016/j.ridd.2013.07.034
32. Pouw, L., Rieffe, C., Stockmann, L., & Gadow, K. (2013). The link between emotion regulation, social functioning, and depression in boys with ASD. *Research in Autism Spectrum Disorders, 7*, 549–556. https://doi.org/10.1016/j.rasd.2013.01.002
33. Rosenberg, R., Kaufmann, W., Law, K., & Law, P. (2011). Parent report of community psychiatric comorbid diagnoses in autism spectrum disorders. *Autism Research and Treatment, 2011*, 1–10. https://doi.org/10.1155/2011/405849
34. Strang, J. F., Kenworthy, L., Daniolos, P., Case, L., Wills, M. C., Martin, A., et al. (2012). Depression and anxiety symptoms in children and adolescents with autism spectrum disorders without intellectual disability. *Research in Autism Spectrum Disorder, 6*(1), 406–412. https://doi.org/10.1016/j.rasd.2011.06.015
35. Lugnegard, T., Hallerback, M. U., & Gillberg, C. (2011). Psychiatric comorbidity in young adults with a clinical diagnosis of Asperger syndrome. *Research in Developmental Disabilities, 32*(5), 1910–1917. https://doi.org/10.1016/j.ridd.2011.03.025
36. Simonoff, E. (2012). Autism spectrum disorder: Prevalence and cause may be bound together. *The British Journal of Psychiatry, 201*, 88–89. https://doi.org/10.1192/bjp.bp.111.104703
37. Mazefsky, C. A., Kao, J., & Oswald, D. P. (2011). Preliminary evidence suggesting caution in the use of psychiatric self-report measures with adolescents with high-functioning autism spectrum disorders. *Research in Autism Spectrum Disorders, 5*(1), 164–174.
38. Henry, J. D., Terrett, G., Altgassen, M., Raponi-Saunders, S., Ballhausen, N., Schnitzspahn, K. M., et al. (2014). A Virtual Week study of prospective memory function in autism spectrum disorders. *Journal of Experimental Child Psychology, 127*, 110–125.
39. Chandrasekhar, T., & Sikich, L. (2015). Challenges in the diagnosis and treatment of depression in autism spectrum disorders across the lifespan. *Dialogues in Clinical Neuroscience, 17*(2), 219–227.
40. Ghaziuddin, M., Ghaziuddin, N., & Greden, J. (2002). Depression in persons with autism: Implications for research and clinical care. *Journal of Autism and Developmental Disorders, 32*(4), 299–306.
41. Stewart, M. E., Barnard, L., Pearson, J., Hasan, R., & O'Brien, G. (2006). Presentation of depression in autism and Asperger syndrome: A review. *Autism, 10*(1), 103–116. https://doi.org/10.1177/1362361306062013
42. Ghaziuddin, M., Alessi, N., & Greden, J. F. (1995). Life events and depression in children with pervasive developmental disorders. *Journal of Autism and Developmental Disorders, 25*(5), 495–502.
43. Mazurek, M. O., & Kanne, S. M. (2010). Friendship and internalizing symptoms among children and adolescents with ASD. *Journal of Autism and Developmental Disorders, 40*(12), 1512–1520. https://doi.org/10.1007/s10803-010-1014-y
44. Myers, K., & Winters, N. C. (2002). Ten-year review of rating scales. II: Scales for internalizing disorders. *Journal of the American Academy of Child and Adolescent Psychiatry, 41*(6), 634–659. https://doi.org/10.1097/00004583-200206000-00004
45. Magnuson, K. M., & Constantino, J. N. (2011). Characterization of depression in children with autism spectrum disorders. *Journal of Developmental and Behavioral Pediatrics, 32*(4), 332–340. https://doi.org/10.1097/DBP.0b013e318213f56c
46. Bellini, S. (2004). Social skill deficits and anxiety in high-functioning adolescents with autism Spectrum disorders. https://doi.org/10.1177/10883576040190020201
47. Bradley, E. A., Summers, J. A., Wood, H. L., & Bryson, S. E. (2004). Comparing rates of psychiatric and behavior disorders in adolescents and young adults with severe intellectual disability with and without autism. *Journal of Autism and Developmental Disorders, 34*(2), 151–161.

48. Lidstone, J., Uljarević, M., Sullivan, J., Rodgers, J., McConachie, H., Freeston, M., et al. (2014). Relations among restricted and repetitive behaviors, anxiety and sensory features in children with autism spectrum disorders. *Research in Autism Spectrum Disorders, 8*, 82–92. https://doi.org/10.1016/j.rasd.2013.10.001
49. Muris, P., Steerneman, P., Merckelbach, H., Holdrinet, I., & Meesters, C. (1998). Comorbid anxiety symptoms in children with pervasive developmental disorders. *Journal of Anxiety Disorders, 12*(4), 387–393.
50. van Steensel, F. J., Bogels, S. M., & Perrin, S. (2011). Anxiety disorders in children and adolescents with autistic spectrum disorders: A meta-analysis. *Clinical Child and Family Psychology Review, 14*(3), 302–317. https://doi.org/10.1007/s10567-011-0097-0
51. Evans, D. W., Canavera, K., Kleinpeter, F. L., Maccubbin, E., & Taga, K. (2005). The fears, phobias and anxieties of children with autism spectrum disorders and down syndrome: Comparisons with developmentally and chronologically age matched children, *Child Psychiatry and Human Development, 36*(1), 3–26. https://doi.org/10.1007/s10578-004-3619-x
52. Gadow, K. D., Devincent, C. J., Pomeroy, J., & Azizian, A. (2005). Comparison of DSM-IV symptoms in elementary school-age children with PDD versus clinic and community samples. *Autism, 9*(4), 392–415. https://doi.org/10.1177/1362361305056079
53. Gillott, A., & Standen, P. J. (2007). Levels of anxiety and sources of stress in adults with autism. *Journal of Intellectual Disabilities, 11*(4), 359–370. https://doi.org/10.1177/1744629507083585
54. Sukhodolsky, D. G., Scahill, L., Gadow, K. D., Arnold, L. E., Aman, M. G., McDougle, C. J., et al. (2008). Parent-rated anxiety symptoms in children with pervasive developmental disorders: Frequency and association with core autism symptoms and cognitive functioning. *Journal of Abnormal Child Psychology, 36*(1), 117–128. https://doi.org/10.1007/s10802-007-9165-9
55. Weisbrot, D. M., Gadow, K. D., DeVincent, C. J., & Pomeroy, J. (2005). The presentation of anxiety in children with pervasive developmental disorders. *Journal of Child and Adolescent Psychopharmacology, 15*(3), 477–496. https://doi.org/10.1089/cap.2005.15.477
56. Kumar, B., Prakash, A., Sewal, R. K., Medhi, B., & Modi, M. (2012). Drug therapy in autism: A present and future perspective. *Pharmacological Reports, 64*(6), 1291–1304.
57. West, L., Waldrop, J., & Brunssen, S. (2009). Pharmacologic treatment for the core deficits and associated symptoms of autism in children. *Journal of Pediatric Health Care, 23*(2), 75–89. https://doi.org/10.1016/j.pedhc.2008.12.001
58. Postorino, V., Sharp, W. G., McCracken, C. E., Bearss, K., Burrell, T. L., Evans, A. N., et al. (2017). A systematic review and meta-analysis of parent training for disruptive behavior in children with autism spectrum disorder. *Clinical Child and Family Psychology Review, 20*(4), 391–402. https://doi.org/10.1007/s10567-017-0237-2
59. Mack, H., Fullana, M. A., Russell, A. J., Mataix-Cols, D., Nakatani, E., & Heyman, I. (2010). Obsessions and compulsions in children with Asperger's syndrome or high-functioning autism: A case-control study. *The Australian and New Zealand Journal of Psychiatry, 44*(12), 1082–1088. https://doi.org/10.3109/00048674.2010.515561
60. South, M., Ozonoff, S., & McMahon, W. M. (2005). Repetitive behavior profiles in Asperger syndrome and high-functioning autism. *Journal of Autism and Developmental Disorders, 35*(2), 145–158.
61. Ruta, L., Mugno, D., D'Arrigo, V. G., Vitiello, B., & Mazzone, L. (2010). Obsessive-compulsive traits in children and adolescents with Asperger syndrome. *European Child & Adolescent Psychiatry, 19*(1), 17–24. https://doi.org/10.1007/s00787-009-0035-6
62. Anholt, G. E., Cath, D. C., van Oppen, P., Eikelenboom, M., Smit, J. H., van Megen, H., et al. (2010). Autism and ADHD symptoms in patients with OCD: Are they associated with specific OC symptom dimensions or OC symptom severity? *Journal of Autism and Developmental Disorders, 40*(5), 580–589. https://doi.org/10.1007/s10803-009-0922-1
63. Rommelse, N. N. J., Franke, B., Geurts, H. M., Hartman, C. A., & Buitelaar, J. K. (2010). Shared heritability of attention-deficit/hyperactivity disorder and autism spectrum disor-

der. *European Child & Adolescent Psychiatry, 19*(3), 281–295. https://doi.org/10.1007/s00787-010-0092-x

64. van der Meer, J. M., Oerlemans, A. M., van Steijn, D. J., Lappenschaar, M. G., de Sonneville, L. M., Buitelaar, J. K., et al. (2012). Are autism spectrum disorder and attention-deficit/hyperactivity disorder different manifestations of one overarching disorder? Cognitive and symptom evidence from a clinical and population-based sample. *Journal of the American Academy of Child and Adolescent Psychiatry, 51*(11), 1160–1172. https://doi.org/10.1016/j.jaac.2012.08.024
65. Grzadzinski, R., Di Martino, A., Brady, E., Mairena, M. A., O'Neale, M., Petkova, E., et al. (2011). Examining autistic traits in children with ADHD: Does the autism spectrum extend to ADHD? *Journal of Autism and Developmental Disorders, 41*(9), 1178–1191. https://doi.org/10.1007/s10803-010-1135-3
66. Mahajan, R., Bernal, M. P., Panzer, R., Whitaker, A., Roberts, W., Handen, B., et al. (2012). Clinical practice pathways for evaluation and medication choice for attention-deficit/hyperactivity disorder symptoms in autism spectrum disorders. *Pediatrics, 130*(Suppl 2), S125–S138. https://doi.org/10.1542/peds.2012-0900J
67. Wallace, S., Fein, D., Rosanoff, M., Dawson, G., Hossain, S., Brennan, L., et al. (2012). A global public health strategy for autism spectrum disorders. *Autism Research, 5*(3), 211–217. https://doi.org/10.1002/aur.1236
68. Belardinelli, C., Raza, M., & Taneli, T. (2016). Comorbid behavioral problems and psychiatric disorders in autism Spectrum disorders. *Journal of Childhood & Developmental Disorders*, 2–11. https://doi.org/10.4172/2472-1786.100019
69. Maglione, M. A., Gans, D., Das, L., Timbie, J., & Kasari, C. (2012). Nonmedical interventions for children with ASD: Recommended guidelines and further research needs. *Pediatrics, 130*(Suppl 2), S169–S178. https://doi.org/10.1542/peds.2012-0900O
70. Linstead, E., Dixon, D. R., Hong, E., Burns, C. O., French, R., Novack, M. N., et al. (2017). An evaluation of the effects of intensity and duration on outcomes across treatment domains for children with autism spectrum disorder. *Translational Psychiatry, 7*(9), e1234. https://doi.org/10.1038/tp.2017.207
71. Touzet, S., Occelli, P., Schroder, C., Manificat, S., Gicquel, L., Stanciu, R., et al. (2017). Impact of the early start Denver model on the cognitive level of children with autism spectrum disorder: Study protocol for a randomised controlled trial using a two-stage Zelen design. *BMJ Open, 7*(3), e014730. https://doi.org/10.1136/bmjopen-2016-014730
72. Agency for Healthcare Research and Quality. (2014). *Therapies for children with autism spectrum disorder: Behavioral interventions update*. Rockville, MD: U.S. Department of Health & Human Services. https://effectivehealthcare.ahrq.gov/products/autism-update/research
73. National Autism Center. (2015). *Findings and conclusions: National standards project, Phase 2*. Randolph, MA: Author.
74. Matson, J. L., Benavidez, D. A., Compton, L. S., Paclawskyj, T., & Baglio, C. (1996). Behavioral treatment of autistic persons: A review of research from 1980 to the present. *Research in Developmental Disabilities, 17*(6), 433–465.
75. Ung, D., Selles, R., Small, B. J., & Storch, E. A. (2015). A systematic review and meta-analysis of cognitive-behavioral therapy for anxiety in youth with high-functioning autism spectrum disorders. *Child Psychiatry and Human Development, 46*(4), 533–547. https://doi.org/10.1007/s10578-014-0494-y
76. Kose, L. K., Fox, L., & Storch, E. A. (2018). Effectiveness of cognitive behavioral therapy for individuals with autism spectrum disorders and comorbid obsessive-compulsive disorder: A review of the research. *Journal of Developmental and Physical Disabilities, 30*(1), 69–87. https://doi.org/10.1007/s10882-017-9559-8
77. Daley, D., van der Oord, S., Ferrin, M., Danckaerts, M., Doepfner, M., Cortese, S., et al. (2014). Behavioral interventions in attention-deficit/hyperactivity disorder: A meta-analysis of randomized controlled trials across multiple outcome domains. *Journal of the American Academy of Child and Adolescent Psychiatry, 53*(8), 835–847. https://doi.org/10.1016/j.jaac.2014.05.013

78. Antshel, K. M., Polacek, C., McMahon, M., Dygert, K., Spenceley, L., Dygert, L., et al. Comorbid ADHD and anxiety affect social skills group intervention treatment efficacy in children with autism spectrum disorders. *Journal of Developmental and Behavioral Pediatrics, 32*(6), 439–446.
79. Frisch, M. (2006). *Quality of life therapy: Applying a life satisfaction approach to positive psychology and cognitive therapy*. New York: Wiley.
80. Toghyani, M., Kalantari, M., Amiri, S., & Molavi, H. (2011). The effectiveness of quality of life therapy on subjective Well-being of male adolescents. *Procedia – Social and Behavioral Sciences, 30*, 1752–1757. https://doi.org/10.1016/j.sbspro.2011.10.338
81. Morsi, H., Perkins, J. D., Alsaied, A., Hassan, A., Langford, C., Alemayehu, E., et al. (2016). Understanding the quality of life (QoL) and quality adjusted survival (QAS) in children, teens, and young adults with cancer. *International Journal of Advanced Biomedicine, 1*, 29–31.
82. Mason, D., McConachie, H., Garland, D., Petrou, A., Rodgers, J., & Parr, J. R. (2018). Predictors of quality of life for autistic adults. *Autism Research, 11*(8), 1138–1147. https://doi.org/10.1002/aur.1965
83. Morsi, H., Clark, H., & Todorovic, N. (2018). *Advances in developmental psychology and subjective vs objective patient and family centred QoL care "DIFI"*. Retrieved from https://www.difi.org.qa/presentations/advances-in-developmental-psychology-and-subjective-vs-objective-patient-and-family-centred-qol-care/
84. McConachie, H., Mason, D., Parr, J. R., Garland, D., Wilson, C., & Rodgers, J. (2018). Enhancing the validity of a quality of life measure for autistic people. *Journal of Autism and Developmental Disorders, 48*(5), 1596–1611. https://doi.org/10.1007/s10803-017-3402-z
85. Varni, J. W., Limbers, C., & Burwinkle, T. M. (2007). Literature review: Health-related quality of life measurement in pediatric oncology: Hearing the voices of the children. *Journal of Pediatric Psychology, 32*(9), 1151–1163. https://doi.org/10.1093/jpepsy/jsm008
86. Varni, J. W., Seid, M., & Kurtin, P. S. (2001). PedsQL 4.0: Reliability and validity of the pediatric quality of life inventory version 4.0 generic core scales in healthy and patient populations. *Medical Care, 39*(8), 800–812.
87. Ikeda, E., Hinckson, E., & Krageloh, C. (2014). Assessment of quality of life in children and youth with autism spectrum disorder: A critical review. *Quality of Life Research, 23*(4), 1069–1085. https://doi.org/10.1007/s11136-013-0591-6
88. Posey, D. J., Erickson, C. A., Stigler, K. A., & McDougle, C. J. (2006). The use of selective serotonin reuptake inhibitors in autism and related disorders. *Journal of Child and Adolescent Psychopharmacology, 16*(1–2), 181–186. https://doi.org/10.1089/cap.2006.16.181
89. Vasa, R. A., Carroll, L. M., Nozzolillo, A. A., Mahajan, R., Mazurek, M. O., Bennett, A. E., et al. (2014). A systematic review of treatments for anxiety in youth with autism spectrum disorders. *Journal of Autism and Developmental Disorders, 44*(12), 3215–3229. https://doi.org/10.1007/s10803-014-2184-9
90. White, S., Oswald, D., Ollendick, T., & Scahill, L. (2009). Anxiety in children and adolescents with autism spectrum disorders. *Clinical Psychology Review, 29*, 216–229.
91. West, L., Brunssen, S. H., & Waldrop, J. (2009). Review of the evidence for treatment of children with autism with selective serotonin reuptake inhibitors. *Journal for Specialists in Pediatric Nursing, 14*(3), 183–191. https://doi.org/10.1111/j.1744-6155.2009.00196.x
92. McPheeters, M. L., Warren, Z., Sathe, N., Bruzek, J. L., Krishnaswami, S., Jerome, R. N., et al. (2011). A systematic review of medical treatments for children with autism spectrum disorders. *Pediatrics, 127*(5), e1312–e1321. https://doi.org/10.1542/peds.2011-0427
93. Chez, M. G., Burton, Q., Dowling, T., Chang, M., Khanna, P., & Kramer, C. (2007). Memantine as adjunctive therapy in children diagnosed with autistic spectrum disorders: An observation of initial clinical response and maintenance tolerability. *Journal of Child Neurology, 22*(5), 574–579. https://doi.org/10.1177/0883073807302611
94. Wink, L. K., Erickson, C. A., Stigler, K. A., & McDougle, C. J. (2011). Riluzole in autistic disorder. *Journal of Child and Adolescent Psychopharmacology, 21*(4), 375–379. https://doi.org/10.1089/cap.2010.0154

95. Abramson, R. K., Ravan, S. A., Wright, H. H., Wieduwilt, K., Wolpert, C. M., Donnelly, S. A., et al. (2005). The relationship between restrictive and repetitive behaviors in individuals with autism and obsessive compulsive symptoms in parents. *Child Psychiatry and Human Development, 36*(2), 155–165. https://doi.org/10.1007/s10578-005-2973-7
96. Goel, R., Hong, J. S., Findling, R. L., & Ji, N. Y. (2018). An update on pharmacotherapy of autism spectrum disorder in children and adolescents. *International Review of Psychiatry, 30*(1), 78–95. https://doi.org/10.1080/09540261.2018.1458706
97. Research Units on Pediatric Psychopharmacology Autism Network. (2005). Randomized, controlled, crossover trial of methylphenidate in pervasive developmental disorders with hyperactivity. *Archives of General Psychiatry, 62*(11), 1266–1274. https://doi.org/10.1001/archpsyc.62.11.1266
98. Miral, S., Gencer, O., Inal-Emiroglu, F. N., Baykara, B., Baykara, A., & Dirik, E. (2008). Risperidone versus haloperidol in children and adolescents with AD: A randomized, controlled, double-blind trial. *European Child & Adolescent Psychiatry, 17*(1), 1–8. https://doi.org/10.1007/s00787-007-0620-5
99. Huffman, L. C., Sutcliffe, T. L., Tanner, I. S., & Feldman, H. M. (2011). Management of symptoms in children with autism spectrum disorders: A comprehensive review of pharmacologic and complementary-alternative medicine treatments. *Journal of Developmental and Behavioral Pediatrics, 32*(1), 56–68. https://doi.org/10.1097/DBP.0b013e3182040acf
100. McVoy, M., & Findling, R. (2009). Child and adolescent psychopharmacology update. *The Psychiatric Clinics of North America, 32*(1), 111–133. https://doi.org/10.1016/j.psc.2008.11.002
101. Arnold, L. E., Aman, M. G., Li, X., Butter, E., Humphries, K., Scahill, L., et al. (2012). Rupp autism NETWORK randomized clinical trial of parent training and medication: One-year follow-up. *Journal of the American Academy of Child and Adolescent Psychiatry, 51*(11), 1173–1184. https://doi.org/10.1016/j.jaac.2012.08.028
102. Chang, Y.-C., Quan, J., & Wood, J. J. (2012). Effects of anxiety disorder severity on social functioning in children with autism spectrum disorders. *Journal of Developmental and Physical Disabilities, 24*(3), 235–245.
103. Simonoff, E., Jones, C. R. G., Baird, G., Pickles, A., Happé, F., & Charman, T. (2013). The persistence and stability of psychiatric problems in adolescents with autism spectrum disorders. *Journal of Child Psychology and Psychiatry, 54*(2), 186–194.
104. Mazefsky, C., Herrington, J., Siegel, M., Scarpa, A., Maddox, B., Scahill, L., et al. (2013). The role of emotion regulation in autism spectrum disorder. *Journal of the American Academy of Child and Adolescent Psychiatry, 52*, 679–688.
105. White, S. W., Mazefsky, C. A., Dichter, G. S., Chiu, P. H., Richey, J. A., & Ollendick, T. H. (2014). Social-cognitive, physiological, and neural mechanisms underlying emotion regulation impairments: Understanding anxiety in autism spectrum disorder. *International Journal of Developmental Neuroscience, 39*, 22–36.
106. Rutter, M., Kim-Cohen, J., & Maughan, B. (2006). Continuities and discontinuities in psychopathology between childhood and adult life. *Journal of Child Psychology and Psychiatry, 47*(3-4), 276–295.

Role of Oxidative Stress and Antioxidants in Autism

Thamilarasan Manivasagam, Selvaraj Arunadevi, Mustafa Mohamed Essa, Chidambaram SaravanaBabu, Anupom Borah, Arokiasamy Justin Thenmozhi, and M. Walid Qoronfleh

Abstract Autism spectrum disorder (ASD) is a heterogeneous group of neurodevelopmental disorders with poorly understood etiology that are defined exclusively on the basis of behavioral observations. This disorder has been linked to increased levels of oxidative stress and lower antioxidant capacity. Oxidative stress in autism has been studied at the membrane level and also by measuring products of lipid peroxidation, detoxifying agents (such as glutathione), and antioxidants involved in the defense system against reactive oxygen species (ROS). Several studies have suggested alterations in the activities of antioxidant enzymes such as superoxide dismutase, glutathione peroxidase, and catalase in autism. Additionally, altered glutathione levels and homocysteine/methionine metabolism, increased inflammation, excitotoxicity, as well as mitochondrial and immune dysfunction have been suggested in autism. Moreover, environmental and genetic risk factors may intensify vulnerability to oxidative stress in autism. Collectively, these studies suggest increased oxidative stress in autism that may contribute to the development of this disease both in terms of pathogenesis and clinical symptoms. Antioxidant

T. Manivasagam · S. Arunadevi · A. Justin Thenmozhi (✉)
Department of Biochemistry and Biotechnology, Faculty of Science, Annamalai University, Chidambaram, Tamil Nadu, India

M. M. Essa
Department of Food Science and Nutrition, CAMS, Sultan Qaboos University, Muscat, Oman

Ageing and Dementia Research Group, Sultan Qaboos University, Muscat, Oman

Food and Brain Research Foundation, Chennai, Tamil Nadu, India

C. SaravanaBabu
Department of Pharmacology, JSS College of Pharmacy, JSS Academy of Higher Education & Research (JSSAHER), Mysuru, India

A. Borah
Cellular and Molecular Neurobiology Laboratory, Department of Life Science and Bioinformatics, Assam University, Silchar, Assam, India

M. W. Qoronfleh
Research & Policy Department, World Innovation Summit for Health (WISH), Qatar Foundation, Doha, Qatar

M. M. Essa, M. W. Qoronfleh (eds.), *Personalized Food Intervention and Therapy for Autism Spectrum Disorder Management*, Advances in Neurobiology 24,
https://doi.org/10.1007/978-3-030-30402-7_7

supplementation, or ways to improve the altered metabolite levels in the interconnected transmethylation and transsulfuration pathways, has been associated with decreased autistic behaviors and severity. This chapter provides a conceptual framework on oxidative stress and antioxidants utility. These types of interventions should be further studied in order to determine their effectiveness at improving metabolic imbalances.

Keywords ASD · Autism · Oxidative stress · Reactive oxygen species · Free radicals · Antioxidants

1 The Concept of Oxidative Stress

Oxidative stress is a condition that occurs due to the imbalance between synthesis of reactive oxygen/nitrogen species (ROS/RNS) and the organism's ability to reduce their deleterious effect by antioxidative protection systems. It arises due to enhanced ROS/RNS formation or from a reduced synthesis or functional antioxidant protective ability, being resulting in diminished combating against oxidative attack towards target biomolecules. ROS or RNS is not just considered a species able to damage biomolecules but is also involved in chemical means of defense or detoxification and for cell signaling and biosynthetic reactions. Free radical-induced oxidative damage has been confirmed as a key contributor to the occurrence, progression and severity of over a 100 pathogenic disease conditions such as Alzheimer's, Huntington's, Parkinson's disease, autism and amyotrophic lateral sclerosis, diabetes mellitus, cardiovascular and inflammatory diseases, emphysema, cataracts, and cancer [1, 2].

2 Occurrence, Characterization and Activity of Reactive Oxygen Species

Free radicals are reactive chemical substances capable of independent existence with an unpaired electron in the external orbit. ROS are represented by free radical (superoxide (O_2•), singlet oxygen ($_{1/2}$ O_2), and the hydroxyl radical (•OH) and non-free radical oxygenated molecules (hydrogen peroxide (H_2O_2). Both endogenous and exogenous free radical generation cannot be hindered, owing to continuous occurrence of metabolic processes and the action of environmental oxidants. They are synthesized in the phagocyte during exposure to microbial infections in cells during aerobic processes such respiration or during extensive physical activity or by the action of exogenous pollutants/toxins such as alcohol, pesticides, cigarette smoke, ozone ionizing, and UV radiations. In low concentrations, ROS act as the signaling molecules which are reported in the regulation of apoptosis, cell proliferation and gene expression by regulating various transcription factors. Their phagocytic generation is essential for defense mechanisms against few bacterial or fungal strains [3]. During the aerobic process, oxygen is employed to oxidize hydrogen and carbon-containing biomolecules to generate chemical energy and heat, which is reduced stepwise to a series of intermediate compounds such as hydroperoxyl radical, superoxide radical anion, hydrogen peroxide, hydroxyl anion, and hydroxyl radical.

$$O_2 + e^- + H^+ \rightarrow HO_2$$

$$HO_2 \rightarrow H^+ + O_2^{\bullet -}$$

$$O_2^{\bullet -} + 2H^+ + e^- \rightarrow H_2O_2$$

$$H_2O_2 + e^- \rightarrow HO^- + HO$$

$$HO + H^+ + e^- \rightarrow H_2O$$

A superoxide radical anion is formed by various oxidases such as dihydronicotinamide adenine dinucleotide phosphate oxidase, cyclooxygenase, xanthine oxidase, etc., when an electron enters the Π* 2p orbitals of oxygen. Increased concentrations of superoxide radical anions are due to processes that occur during normal activities like oxidative phosphorylation, which yields ATP in the mitochondrial electron transport chain (ETC). In the ETC, the electrons are transferred by four membrane bound complexes from NADH and FADH2 to molecular oxygen by yielding water [4]. Electrons may leak from the inner membrane and are able to reduce molecular oxygen to superoxide radical anions ($O_2\bullet$). Although superoxide radical is considered a stronger reducing agent than iron complexes like cytochrome C and ferric-EDTA, it proved to be weak against ascorbic acid and thiols in an aqueous solution. Hydroperoxyl radical in its protonated form is reported to have strong oxidant and reductant properties with much less stability at physiological pH (7.4). Superoxide anion is an active nucleophile which can attack positively charged substances and react with hydrogen donors such as tocopherol and ascorbate. Superoxide anion form spontaneously or by the enzyme superoxide dismutase to generate molecular oxygen and hydrogen peroxide [5].

$$2O_2^{\bullet -} + 2H^+ \rightarrow O_2 + H_2O_2$$

Hydrogen peroxide can be formed by superoxide and/or by direct transfer of two electrons to molecular oxygen with the help of urate oxidase, D-amino acid oxidase, and glucose oxidase. After interaction with metal ions, it forms highly reactive radicals. H_2O_2 attacks the hemeprotein and releases iron, inactivates enzymes and oxidizes lipids, DNA, -thiol groups and keto-acids [5]. It is depleted via conversion to water by catalase (CAT), the ferriheme-containing enzyme. The OH· radical is a potent radical species formed from Fenton-type reactions or by the radiolysis of water [6]. It can interact at its site of generation with proteins, DNA, lipids, sugars, amino acids, and metals. Myeloperoxidase present in macrophages and neutrophils is responsible for the formation of hypochlorous acid from hydrogen peroxide in the presence of chloride anion. Hypochlorous acid induces oxidative chlorination on proteoglycans, lipids, amino acids, and other membrane components. Molecular oxygen is not considered a free radical, but has high reactivity. This happens if the spin restriction is removed enhancing its oxidative power [6]. Ozone oxidizes lung proteins, lipids and DNA.

3 ROS and Biomolecular Impairment

By targeting all substances, ROS modulates the function of all these bio-molecules in the cell. Lipids are the most susceptible to oxidative processes especially the polyunsaturated fatty acids (PUFA) like arachidonic acid and docosahexaenoic acid that form malondialdehyde and 4-hydroxynonenal, which are the indicators of lipid oxidative decay. The oxidation of PUFA induces cell membrane damage, as they are rich in PUFA. Peroxidation of PUFA leads to the formation of isoprostanes and results in generation of reactive aldehydes, like 4-hydroxynonenal and malondialdehyde. The 4-hydroxynonenal binds to proteins and impairs their function. ROS oxidize the amino acids present in the backbone and side chain of proteins resulting in unfolding and misfolding leading to inactivity. Oxidation of thiol groups and carbonylation of amino acids lead to the generation of advanced glycation end products. Like other amino acids, cysteines and methionines readily undergo oxidization, reversible due to the activity of disulfide reductases. Oxidation of proline, arginine, lysine, and threonine forms carbonyl derivatives which are markers of ROS-mediated protein oxidation [7]. Aromatic amino acids are prone to oxidation reaction and forms different oxygenated products—tyrosine reacts with OH radical to form dityrosine, with nitrogenated species to form 3-nitrotyrosine and with hypochlorite oxide to form 3-chlorotyrosine. It also reacts with nucleic acids and causes DNA strand breaking, DNA-protein crosslinking and modification of purine and pyridine-base structures resulting in DNA mutations. Currently marker referred to for DNA oxidation is 8-hydroxydeoxyguanosine which arises from the oxidation process of guanosine by OH•. Oxidation of RNA bases leads to the breakage of the nucleotide strand and by ribosomal dysfunction and forms a homologue of 8-hydroxydeoxyguanosine and 8-hydroxyguanosine [8]. As compared to DNA, RNA can undergo oxidation easily, as it located near ROS occurrence sites in the cell.

An interesting aspect to oxidative stress is that the damage of the mitochondrial membranes and the protein structure itself further enhances reactive oxygenated species generation, leading to DNA damage and cell death by apoptosis. Apoptosis signal regulating kinase 1 is considered an important marker of apoptosis initiated by oxidative stress. Its activity is primarily controlled by hioredoxin-1, the redox sensitive oxidoreductase that binds the reduced form of apoptosis signal regulating kinase 1. When thioredoxin-1 is oxidized, the binding to apoptosis signal-regulating kinase 1 is hindered resulting in activation of the subsequent c-Jun N-terminal kinase apoptosis pathway. The activity of apoptosis signal-regulating kinase 1 is also tuned by other redox proteins including glutaredoxin, heat-shock proteins, and glutathione S-transferase. Another regulator of oxidative-mediated apoptosis is p53 which, after its translocation in the nucleus, is capable of triggering proapoptotic genes [9].

4 Antioxidant Defense Mechanisms

Excessive production of free radicals or impaired antioxidant mechanism may cause oxidative stress which may induce several pathophysiological processes. The two main roles of cellular antioxidant defense mechanism are to prevent the generation of free radicals and to inactivate them after generation. Impaired antioxidant defense mechanism can result in cell membrane damage, alteration in membrane fluidity and permeability and oxidative changes in proteins, lipids and DNA. Several enzymatic and non-enzymatic defense molecules are evolved to inhibit the oxidant attack by transferring the excess electrons during the detoxification process.

5 Enzymatic Antioxidants

Antioxidant enzymes such as superoxide dismutase, catalase and glutathione peroxidase play a key role against ROS. As superoxide is the main ROS formed by various sources, its dismutation to H_2O_2 by SOD is part of the key protection for each cell. SOD exists in three forms: Cu-ZnSOD found mainly in the cytoplasm while MnSOD is a key enzyme of mitochondria. H_2O_2 is reduced to water by the activities of catalase and/or glutathione peroxidase (GSH-Px). Catalase consists of four identical monomers (tetramer), each of them having a heme group at its active site. H_2O_2 is degraded via the conversion between two forms of catalase-ferricatalase and compound I. Catalase utilizes NADPH as a reducing equivalent to avert oxidative inactivation of the enzyme by H_2O_2 as it is reduced to water. Like catalase, GSH-Pxs is a family of tetrameric enzymes which consist of a selenocysteine in their active sites and utilize GSH to reduce H_2O_2 and lipid peroxides also to their alcohols. Four forms of GSH-Pxs exist and are encoded by different genes: GSH-Px-1 is ubiquitous and reduces fatty acid peroxides and H_2O_2; GSH-Px-2 is present only in gastrointestinal epithelial cells that reduce dietary peroxides. GSH-Px-3 is found in the extracellular compartment which is the most important mammalian extracellular antioxidant enzyme. GSH-Px-4 is a membrane bound enzyme that reduces esterified lipids by using several low-molecular-weight thiols as reducing equivalents [10].

6 Non-enzymatic Antioxidants

The examples for non-enzymatic antioxidants are low-molecular-weight compounds, such as vitamins (vitamins C and E), β-carotene, uric acid and GSH, a tripeptide (L-γ-glutamyl-L-cysteinyl-L-glycine). Water-soluble vitamin C (ascorbic acid) acts both as intracellular and extracellular aqueous-phase antioxidant that mainly scavenges oxygen free radicals. It helps revert vitamin E free radicals into vitamin E. Lipid soluble vitamin E is found primarily inside the cell membrane and acts against oxidant-induced membrane injury. GSH is present in abundance in all cell compartments and is the key antioxidant. It helps detoxify H_2O_2 and lipid peroxides by GPx. GSH reduces H_2O_2 to H_2O and O_2 by donating an electron. Oxidized glutathione is reduced again into GSH by glutathione reductase that utilizes NAD(P)H as the electron donor. GSH donates the protons to membrane lipids and is involved in protection from oxidant attacks. It acts as cofactor for various detoxifying enzymes GPx and GSH transferase. It helps in converting oxidized vitamin C and E into their active reduced forms. In addition, thiol compounds, such as thioredoxin, are capable of detoxifying hydrogen peroxide, but, in turn require conversion back to their reduced form by thioredoxin reductase. Ceruloplasmin and transferrin also play important roles by sequestering free iron ions, thereby inhibiting the Fenton reaction and production of OH•. Carotenoids are plant pigments that are able to react with eroxyl (ROO), hydroxyl (OH), and superoxide (O_2) radicals [11]. They showed high antioxidant effects in low oxygen concentration and also inhibited the oxidant-induced NF-kB activation and inflammation.

7 Brain and Autism

The nervous system is vulnerable to oxidative stress-mediated injury because of the following: (1) its high energy needs, the brain consumes more oxygen leading to the excessive production of ROS; (2) neuronal membranes contain more polyunsaturated fatty acids susceptible to free radical attack; (3) high membrane surface area to cytoplasmic volume ratio; (4) specialized neuronal activity and synaptic transmission requiring competent membrane function; (5) axons are prone to peripheral injury; (6) neuronal activities are also affected by disruptions; (7) the excitotoxic nature of glutamate, a major neurotransmitter also causes oxidative stress; (8) enhanced Ca^{2+} flow across the neuronal membranes and intrusion of ion transport maximizing intracellular Ca^{2+} often leading to OS; (9) oxidation of neurotransmitters can produce O_2 and quinones that reduce glutathione; (10) formation of Iron throughout the entire brain which releases iron ions capable of catalyzing free radical reactions; (11) modest antioxidant defense mechanisms, in particular, low levels of catalase, glutathione peroxidase and vitamin E; (12) ROS directly downregulates proteins of tight junctions and indirectly activates matrix metalloproteinases (MMP) that contribute to open the blood–brain barrier (BBB); (13) activated microglia produce ROS and cytokines in a perpetual process; (14) cytochrome P450 produces ROS; (15) loss of trophic support can activate NADPH oxidase, which increases ROS; (16) the presence of hemoglobin within the neural tissues secondary to spontaneous, iatrogenic, or traumatic causes is neurotoxic, heme and iron are released and promote ROS, neuronal mitochondria generate O_2, and the interaction of NO with superoxide can be implicated also in neuronal degeneration; and (17) Neuronal cells are nonreplicating and thus are sensitive to ROS. In comparison with other organs, the neuronal network may be especially vulnerable to ROS-mediated injury because of the anatomic, physiological, and biochemical properties of the brain.

8 Autism and Oxidative Stress

Numerous studies have indicated the presence of oxidative stress in individuals with ASD [12] and their parents [13]. Direct markers of lipid peroxidation such as serum lipid peroxides and thiobarbituric acid reactive substances, urinary isoprostanes and their indirect markers like phospholipase A2 and loss of membrane lipoprotein asymmetry are higher in autism. Moreover, the levels of pro-oxidants such as organic toxins including perchlorethylene, hexane and pentane and heavy metals like mercury, lead, and arsenic were accumulated in the ASD patients. Enhanced levels of cytokines and xanthine oxidase were found in the blood circulation of autistic patients, and both can generate free radicals. Few viruses and bacteria induce excess local production of $NO^{\bullet}$ in the gut, which can affect the brain by their circulation. Higher levels of nitrite in autism may link chronic gut and brain injury. Previous studies have suggested the presence of a leaky BBB in autism, relatively sensitive to oxidative damage. Overstimulation of excitatory receptors in autistic patients leads to oxidative injury in neurons and enhanced oxidative stress enhances the release of glutamate and stimulation of excitatory receptors. Muscarinic impairment found in autism may exaggerate oxidative stress as muscarinic signals shield the neurons from oxidative stress and apoptosis.

Abnormal metabolism of glutathione particularly with low reduced glutathione, elevated oxidized glutathione, and diminished glutathione redox ratio was found in the temporal cortex and cerebellum of autistic patients, and increased heme oxygenase-1 was reported in the parietal and frontal lobes and the cerebellum. SOD activity has been shown to be decreased, increased, or unchanged in plasma and erythrocyte. Erythrocyte SOD activity was found to be higher in ASD children than in the control group in this study. An increased SOD level is considered a compensatory response as protection against the cell damage caused by oxidative stress. The catalase enzyme is directly involved in ROS elimination. Previous studies have reported that catalase activity is reduced in erythrocytes, but remains unchanged in plasma and erythrocyte. Consistent with these increased oxidative stress biomarkers in children with autism, reduced endogenous antioxidant capacity, specifically the total GSH levels, altered GPx, SOD and CAT activities, were found in individuals with autism as compared to controls.

9 Clinical Implication

Several double-blind, placebo-controlled therapeutic trials in autism are being conducted using potent antioxidants such as vitamin C, carnosine, zinc, reduced glutathione, fish oil (rich in EPA, Eicosapentaenoic acid is an omega-3 fatty acid), melatonin, and vitamin B-6 in combination with magnesium. In some clinical trials treatment with high dose vitamin C or carnosine or combined vitamin B-6 and magnesium improved the behavior of individuals with autism. Additionally, melatonin has been reported to be useful in the treatment of sleep disorders in autism (https://clinicaltrials.gov/). Overall, oxidative stress-related metabolites could also have potential use as biomarkers for diagnosis and help determine future intervention/treatments.

Developing new therapeutic strategies targeting the mitochondria may shed a new light onto autism treatment. Antioxidants such as CoQ10, NADH, α-lipoic acid (LA), glutathione and Mito Q, Szeto Schiller peptide all have some potential therapeutic value in the treatment of certain neurodegenerative diseases where mitochondrial dysfunction is implicated through catabolizing H_2O_2. Preventive antioxidants leading to neuronal protection against many oxidative damages can be used to treat behavioral and cognitive symptoms of ASD. Examples include enzymes like GSH-Px, MnSOD and Cu-ZnSOD besides repair enzymes such as lipases and DNA repair enzymes.

The reader is referred to Table 1 for the therapeutic implications of antioxidants in experimental models of autism, and Table 2 for the therapeutic role of antioxidants in clinical trials of autistic patients. Glutathione is present in pools within mitochondria and freely in the cytosol. Reduction of mitochondrial glutathione levels has been associated with neuronal susceptibility to oxidative stress. Glutathione deficiencies increase vulnerability to oxidative stress in children with autism. Excessive ROS and depleted antioxidants/antioxidant enzymes can create a negative cycle within mitochondria, which has been linked to mitochondrial dysfunction in autism. Figure 1 shows the interconnected pathways of GSH biosynthesis, and transmethylation and transsulfuration reactions for possible diet/supplementation interventions. Metabolic abnormalities have been noted in autism and are related to the interconnected pathways of folate, methionine and glutathione metabolism. Metabolic differences between those with autism and controls exist in transmethylation and transsulfuration pathways. Children with autism exhibit lower levels of adenosine deaminase (ADA), which leads to increase levels of adenosine or homocysteine. This accumulation inactivates S-adenosylhomocysteine hydrolase (SAHH)

Table 1 The therapeutic implications of antioxidants in experimental models of autism

Antioxidant	Neurotoxin	Animal model	Mechanism	References
Resveratrol	Norethindrone	Sprague–Dawley rats	Enhanced estrogen receptor β, which regulates the basal expression of superoxide dismutase Upregulated estrogen-related receptor α (ERRα), which regulates mitochondrial function and lipid metabolism	[14]
Leptin and camel milk	Valproic acid	Sprague–Dawley rats	Ameliorated the oxidative stress (significant reduction in the MDA level and enhancement of activities of SOD, GPx and catalase) and its related inflammation and apoptosis	[15]
Hesperetin and nano-hesperetin	Valproic acid	Wistar–Albino rats	Attenuated the oxidative stress (significant reduction in the MDA level and enhancement of the expression of SOD, GPx and catalase) and its related inflammation and apoptosis	[16]
Flavonoid	–	–	Regulation of the glyoxalase pathway is an antioxidant defense mechanism (Review)	[17]
Resveratrol	–	BTBR T+ Itpr3tf/J (BTBR) mice	Attenuated the oxidative stress mediated inflammation and apoptosis related signaling markers	[18, 19]
Minocycline and doxycycline	Terbutaline	Albino Wistar rats	Ameliorated the oxidative stress mediated inflammation	[20]
N-acetylcysteine (NAC)	Valproic acid	Sprague–Dawley rats	Regulation of the canonical Wnt signaling pathway has been implicated in oxidative processes	[21]
Selol, an organic selenium donor	Lipopolysaccharide (LPS)	Wistar rats	Nullified the neuroinflammation by inhibiting pro-inflammatory cytokine release, by boosting antioxidant systems and BDNF level	[22]

(continued)

Table 1 (continued)

Antioxidant	Neurotoxin	Animal model	Mechanism	References
Resveratrol	Propanoic acid	Wistar rats	Suppressed the oxidative-nitrosative stress, mitochondrial dysfunction, inflammation and also amelioration of neurobehavioral and biochemical deficits	[23]
Docosahexaenoic acid	Valproic acid	Male and female Wistar rats	Alleviated oxidative stress mediated apoptosis	[24]
Laser acupuncture	Valproic acid	Wistar rats	Repaired brain damage and reduced autism-like behaviors and decreased oxidative stress in the cortex, striatum and hippocampus	[25]
Tagara	Methyl mercury	Adult Wistar rats	Ameliorated the oxidative stress mediated mitochondrial dysfunction	[26]
Sulindac	Valproic acid	Wistar rats	Ameliorates autism-like behavioral abnormalities and inhibition of the oxidative mediated activation of the canonical Wnt pathway	[27]
N-acetyl-cysteine	Propionic acid	Male Western Albino rats	Ameliorated the impaired biochemical parameters representing neurochemical, inflammatory, detoxification and DNA damage processes	[28]
Bacopa monnieri (L.) Wettst	Sodium valproate	Female pregnant rats	Improved behavioral alterations in two developmental periods, ameliorated oxidative stress markers and histopathological findings	[29]

therefore increasing S-adenosylhomocysteine (SAH) and inactivating methyltransferase. Methylation is hindered in children with autism. Protein synthesis is also reduced due to insufficient methionine levels, which has major downstream effects. Cysteine levels are diminished and are the ultimate cause of decreased glutathione production in this disorder. Various ways to ameliorate the abnormalities in the transmethylation and transsulfuration pathways and their consequences have been studied in cardiovascular, cancer, autoimmune conditions and neurodegenerative diseases. Increased antioxidant support is needed to maintain proper health in these conditions and could be a novel pharmacologic intervention in autism.

Table 2 The therapeutic role of antioxidants in clinical trials of autistic patients

Antioxidant	Patient	Outcome	Treatment	References
Dark chocolate 70% cacao and 30% organic cane sugar	Children with ASD	Improved social communication, unusual behaviors, and self-regulation behaviors of children with ASD	4-week	[30]
Sulforaphane, a supplement with indirect antioxidant effects that are derived from broccoli sprouts and seeds	Children with ASD	Affected pathways of oxidative stress, amino acid/gut microbiome, neurotransmitters, hormones, and sphingomyelin metabolism	12-week	[31]
Coenzyme Q10 supplementation	Children with ASD	Reduced the MDA levels and enhanced total antioxidant status (TAS) assay, and antioxidant enzymes (superoxide dismutase or SOD and glutathione peroxidase or GPx) activity	12-weeks	[32]
N-acetylcysteine (NAC)	Children with ASD	Restored GSH levels and scavenges oxidants such as hydroxyl radical and hydrogen peroxide	10-weeks	[33]
Camel milk	Children with ASD	Decreased oxidative stress by alteration of antioxidant enzymes and nonenzymatic antioxidant molecules levels and improvement of autistic behavior	2-weeks	[34]

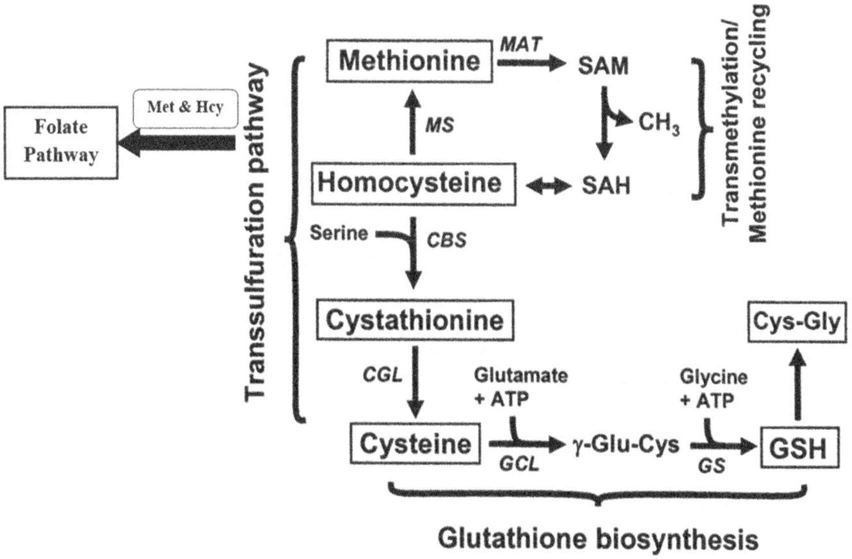

Fig. 1 GSH biosynthesis and folate and transmethylation/transsulfuration pathways. Adapted from Toroser and Sohal [35]

10 Conclusion

The notion of oxidative stress involvement in autism has been derived from several lines of evidences: elevated nitric oxide concentration, thiobarbituric acid reactive substance levels and xanthine oxidase activity have been detected in the red blood cells of autistic individuals. Consistent with these increased oxidative stress biomarkers in children with autism, a reduced endogenous antioxidant capacity, specifically the total GSH levels, altered GPx, SOD and CAT activities were found in autistic individuals compared to controls. Further indications of oxidative stress role in autism are derived from evidence of impaired energy metabolism. Reduced synthesis of adenosine triphosphate (ATP) and higher lactate and pyruvate levels may suggest mitochondrial dysfunction in autism. The most critical function of mitochondria is producing ATP, the primary energy currency in the brain and in the body. Increased ROS metabolism induced by dysfunctional mitochondria could elicit chronic oxidative stress. The understanding of oxidative stress alterations and mechanisms in autism may lead to new diagnostic testing and therapeutic intervention strategies in individuals with ASD.

References

1. Lopez-Alarcona, C., & Denicola, A. (2013). Evaluating the antioxidant capacity of natural products: A review on chemical and cellular-based assays. *Analytica Chimica Acta, 763*, 1–10.
2. Sies, H. (1985). *Oxidative stress: Introductory remarks*. London: Academic Press.
3. Poljsak, B., Suput, D., & Milisav, I. (2013). Achieving the balance between ROS and antioxidants: When to use the synthetic antioxidants. *Oxidative Medicine and Cellular Longevity, 2013*, 956792.
4. Koopman, W. J., Nijtmans, L. G., Dieteren, C. E., Roestenberg, P., Valsecchi, F., Smeitink, J. A. M., et al. (2010). Mammalian mitochondrial complex I: Biogenesis, regulation, and reactive oxygen species generation. *Antioxidants & Redox Signaling, 12*, 1431–1470.
5. Kohen, R., & Nyska, A. (2002). Oxidation of biological systems: Oxidative stress phenomena, antioxidants, redox reactions, and methods for their quantification. *Toxicologic Pathology, 30*, 620–650.
6. Gutteridge, J. M. C. (1994). Biological origin of free radicals, and mechanisms of antioxidant protection. *Chemico-Biological Interactions, 91*, 133–140.
7. Berlett, B. S., & Stadtman, E. R. (1997). Protein oxidation in aging, disease, and oxidative stress. *The Journal of Biological Chemistry, 272*, 20313–20316.
8. Poulsen, H. E., Specht, E., Broedbaek, K., Henriksen, T., Ellervik, C., MandrupPoulsen, T., et al. (2012). RNA modifications by oxidation: A novel disease mechanism. *Free Radical Biology & Medicine, 52*, 1353–1361.
9. Yamamoto, H., Ozaki, T., Nakanishi, M., Kikuchi, H., Yoshida, K., Horie, H., et al. (2007). Oxidative stress induces p53-dependent apoptosis in hepatoblastoma cell through its nuclear translocation. *Genes to Cells, 12*, 461–471.
10. Chu, F. F., Doroshow, J. H., & Esworthy, R. S. (1993). Expression, characterization, and tissue distribution of a new cellular selenium-dependent glutathione peroxidase, GSHPx-GI. *The Journal of Biological Chemistry, 268*, 2571–2576.
11. El-Agamey, A., Lowe, G. M., McGarvey, D. J., Mortensen, A., Phillip, D. M., Truscott, T. G., et al. (2004). Carotenoid radical chemistry and antioxidant/pro-oxidant properties. *Arch Biochem Biophys, 430*(1), 37–48.

12. Rose, S., Melnyk, S., Pavliv, O., Bai, S., Nick, T. G., Frye, R. E., et al. (2012). Evidence of oxidative damage and inflammation associated with low glutathione redox status in the autism brain. *Translational Psychiatry, 2*(7), 134.
13. James, S. J., Melnyk, S., Jernigan, S., Pavliv, O., Trusty, T., Lehman, S., et al. (2010). A functional polymorphism in the reduced folate carrier gene and DNA hypomethylation in mothers of children with autism. *American Journal of Medical Genetics Part B: Neuropsychiatric Genetics, 153*(6), 1209–1220.
14. Xie, W., Ge, X., Li, L., Yao, A., Wang, X., Li, M., et al. (2018). Resveratrol ameliorates prenatal progestin exposure-induced autism-like behavior through ERß activation. *Molecular Autism, 2*, 9–43.
15. Hamzawy, M. A., El-Ghandour, Y. B., Abdel-Aziem, S. H., & Ali, Z. H. (2018). Leptin and camel milk abate oxidative stress status, genotoxicity induced in valproic acid rat model of autism. *International Journal of Immunopathology and Pharmacology, 32*, 2058738418785514.
16. Khalaj, R., Hajizadeh Moghaddam, A., & Zare, M. (2018). Hesperetin and it nanocrystals ameliorate social behavior deficits and oxido-inflammatory stress in rat model of autism. *International Journal of Developmental Neuroscience, 69*, 80–87.
17. Frandsen, J. R., & Narayanasamy, P. (2018). Neuroprotection through flavonoid: Enhancement of the glyoxalase pathway. *Redox Biology, 14*, 465–473.
18. Ahmad, S. F., Ansari, M. A., Nadeem, A., Bakheet, S. A., Alzahrani, M. Z., Alshammari, M. A., et al. (2018). Resveratrol attenuates pro-inflammatory cytokines and activation of JAK1-STAT3 in BTBR T+ Itpr3tf/J autistic mice. *European Journal of Pharmacology, 829*, 70–78.
19. Ahmad, S. F., Ansari, M. A., Nadeem, A., Alzahrani, M. Z., Bakheet, S. A., & Attia, S. M. (2018). Resveratrol improves neuroimmune dysregulation through the inhibition of neuronal toll-like receptors and COX-2 signaling in BTBR T+ Itpr3tf/J mice. *Neuromolecular Medicine, 20*(1), 133–146.
20. Rani, V., Gautam, S., Rawat, J. K., Singh, M., Devi, U., Yadav, R. K., et al. (2018). Effects of minocycline and doxycycline against terbutaline induced early postnatal autistic changes in albino rats. *Physiology & Behavior, 183*, 49–56.
21. Zhang, Y., Cui, W., Zhai, Q., Zhang, T., & Wen, X. (2017). N-acetylcysteine ameliorates repetitive/stereotypic behavior due to its antioxidant properties without activation of the canonical Wnt pathway in a valproic acid-induced rat model of autism. *Molecular Medicine Reports, 16*(2), 2233–2240.
22. Dominiak, A., Wilkaniec, A., Jesko, H., Czapski, G. A., Lenkiewicz, A. M., Kurek, E., et al. (2017). Selol, an organic selenium donor, prevents lipopolysaccharide-induced oxidative stress and inflammatory reaction in the rat brain. *Neurochemistry International, 108*, 66–77.
23. Bhandari, R., & Kuhad, A. (2017). Resveratrol suppresses neuroinflammation in the experimental paradigm of autism spectrum disorders. *Neurochemistry International, 103*, 8–23.
24. Gao, J., Wang, X., Sun, H., Cao, Y., Liang, S., Wang, H., et al. (2016). Neuroprotective effects of docosahexaenoic acid on hippocampal cell death and learning and memory impairments in a valproic acid-induced rat autism model. *International Journal of Developmental Neuroscience, 49*, 67–78.
25. Khongrum, J., & Wattanathorn, J. (2015). Laser acupuncture improves behavioral disorders and brain oxidative stress status in the valproic acid rat model of autism. *Journal of Acupuncture and Meridian Studies, 8*(4), 183–191.
26. Ayyathan, D. M., Chandrasekaran, R., & Thiagarajan, K. (2015). Neuroprotective effect of Tagara, an Ayurvedic drug against methyl mercury induced oxidative stress using rat brain mitochondrial fractions. *BMC Complementary and Alternative Medicine, 15*, 268.
27. Zhang, Y., Yang, C., Yuan, G., Wang, Z., Cui, W., & Li, R. (2015). Sulindac attenuates valproic acid-induced oxidative stress levels in primary cultured cortical neurons and ameliorates repetitive/stereotypic-like movement disorders in Wistar rats prenatally exposed to valproic acid. *International Journal of Molecular Medicine, 35*(1), 263–270.

28. Aldbass, A. M., Bhat, R. S., & El-Ansary, A. (2013). Protective and therapeutic potency of N-acetyl-cysteine on propionic acid-induced biochemical autistic features in rats. *Journal of Neuroinflammation, 27*, 10–42.
29. Sandhya, T., Sowjanya, J., & Veeresh, B. (2012). Bacopa monnieri (L.) Wettst ameliorates behavioral alterations and oxidative markers in sodium valproate induced autism in rats. *Neurochemical Research, 37*, 1121–1131.
30. Sadek, A., Berk, L. S., Mainess, K., & Daher, N. S. (2018). Antioxidants and autism: Teachers' perceptions of behavioral changes. *Advances in Mind-Body Medicine, 32*(3), 12–17.
31. Bent, S., Lawton, B., Warren, T., Widjaja, F., Dang, K., Fahey, J. W., et al. (2018). Identification of urinary metabolites that correlate with clinical improvements in children with autism treated with sulforaphane from broccoli. *Molecular Autism, 9*, 35.
32. Mousavinejad, E., Ghaffari, M. A., Riahi, F., Hajmohammadi, M., Tiznobeyk, Z., & Mousavinejad, M. (2018). Coenzyme Q10 supplementation reduces oxidative stress and decreases antioxidant enzyme activity in children with autism spectrum disorders. *Psychiatry Research, 265*, 62–69.
33. Nikoo, M., Radnia, H., Farokhnia, M., Mohammadi, M. R., & Akhondzadeh, S. (2015). N-acetylcysteine as an adjunctive therapy to risperidone for treatment of irritability in autism: A randomized, double-blind, placebo-controlled clinical trial of efficacy and safety. *Clinical Neuropharmacology, 38*(1), 11–17.
34. Al-Ayadhi, L. Y., & Elamin, N. E. (2013). Camel milk as a potential therapy as an antioxidant in autism spectrum disorder (ASD). *Evidence-based Complementary and Alternative Medicine, 2013*, 602834.
35. Toroser, D., & Sohal, R. S. (2007). Age-associated perturbations in glutathione synthesis in mouse liver. *The Biochemical Journal, 405*(3), 583–589.

The Regulation of Reactive Neuroblastosis, Neuroplasticity, and Nutraceuticals for Effective Management of Autism Spectrum Disorder

G. P. Poornimai Abirami, Risna Kanjirassery Radhakrishnan, Esther Johnson, Syed Aasish Roshan, Ajisha Yesudhas, Suhadha Parveen, Abir Biswas, Vijaya Roobini Ravichandran, Anusuyadevi Muthuswamy, and Mahesh Kandasamy

Abstract Autism spectrum disorder (ASD) encompasses a cluster of neurodevelopmental and genetic disorders that has been characterized mainly by social withdrawal, repetitive behavior, restricted interests, and deficits in language processing mainly in children. ASD has been known to severely impair behavioral patterns and cognitive functions including learning and memory due to defects in neuroplasticity. The biology of the ASD appears to be highly complex and heterogeneous, and thus, finding a therapeutic target for autism remains obscure. There

G. P. Poornimai Abirami and Risna Kanjirassery Radhakrishnan share the first authorship.

G. P. Poornimai Abirami
School of Life Sciences, Bharathidasan University, Tiruchirappalli, Tamil Nadu, India

R. K. Radhakrishnan · E. Johnson · A. Yesudhas · V. R. Ravichandran
Laboratory of Stem Cells and Neuroregeneration, Department of Animal Science, School of Life Sciences, Bharathidasan University, Tiruchirappalli, Tamil Nadu, India

S. A. Roshan · S. Parveen · A. Biswas
Molecular Gerontology Laboratory, Department of Biochemistry, School of Life Sciences, Bharathidasan University, Tiruchirappalli, Tamil Nadu, India

A. Muthuswamy (✉)
School of Life Sciences, Bharathidasan University, Tiruchirappalli, Tamil Nadu, India

Molecular Gerontology Laboratory, Department of Biochemistry, School of Life Sciences, Bharathidasan University, Tiruchirappalli, Tamil Nadu, India
e-mail: janushyas@bdu.ac.in

M. Kandasamy (✉)
School of Life Sciences, Bharathidasan University, Tiruchirappalli, Tamil Nadu, India

Laboratory of Stem Cells and Neuroregeneration, Department of Animal Science, School of Life Sciences, Bharathidasan University, Tiruchirappalli, Tamil Nadu, India

Faculty Recharge Programme, University Grants Commission (UGC-FRP), New Delhi, India
e-mail: mahesh.kandasamy@bdu.ac.in

M. M. Essa, M. W. Qoronfleh (eds.), *Personalized Food Intervention and Therapy for Autism Spectrum Disorder Management*, Advances in Neurobiology 24,
https://doi.org/10.1007/978-3-030-30402-7_8

has been no complete prevention or disease-modifying cure for this disorder. Recently, individuals with autism have been characterized by reactive neurogenesis, obstructions in axonal growth, heterotopia, resulting from dysplasia of neuroblasts in different brain regions. Therefore, it can be assumed that the aforementioned neuropathological correlates seen in the autistic individuals might originate from the defects mainly in the regulation of neuroblasts in the developing as well as adult brain. Nutrient deficiencies during early brain development and intake of certain allergic foods have been proposed as main reasons for the development of ASD. However, the integrated understanding of neurodevelopment and functional aspects of neuroplasticity working through neurogenesis in ASD is highly limited. Moreover, neurogenesis at the level of neuroblasts can be regulated by nutrition. Hence, defects in neuroblastosis underlying the severity of autism potentially could be rectified by appropriate implementation of nutraceuticals.

Keywords Autism · Neuroplasticity · Neuroblasts · Neurogenesis · Neurotransmission · Food and nutrition · Nutraceuticals

1 Introduction

The term Autism, derived from the Latin word "Autismus" by Paul Eugen Bleuler in 1911, refers to the mental state of individuals with schizoaffective disorders in early adolescence and adulthood [1]. In 1943, Leo Kanner proposed the term "autistic disturbances of affective contact," to describe a similar mental state in children [1, 2]. In 1944, Asperger used the term "autistic psychopathy" in his case report examining several similarities and some differences between symptoms of autism [1]. In 1981, Lorna Wing introduced the term Asperger's syndrome and proposed the concept of autism spectrum disorder (ASD) [3]. As of today, ASD has been understood as encompassing a range of complex neurodevelopmental syndromes that cause language impairment and abnormal social behavior among children [4]. In accordance with the *Diagnostic and Statistical Manual of Mental Disorders, Fifth Edition (DSM-5)*, autistic disorder, Asperger's syndrome, Rett syndrome, childhood disintegrative disorder, and pervasive developmental disorder-not otherwise specified (PDD-NOS) have been categorized under ASD [5]. Among individuals with ASD, the symptoms can be identified within 3 years of age [6]. ASD has been characterized by impairments in social interaction, speech pathology, attention deficits, and stereotypical and repetitive behaviors [4]. Other symptoms include difficulty in carrying out daily routines and poor self-care [4]. Depending on the severity of the disorder, the behavioral symptoms may vary from hyperactivity, aggression, obsessiveness, depression, anxiety, and sensory abnormalities to self-injury [4]. The clinical symptoms have been known to overlap with seizures, gastrointestinal (GI) abnormalities, dysregulation of the immune system, and disruption in circadian rhythm [7]. ASD has been estimated to be increasing exponentially worldwide [8]. About 1% of the global population has been affected by ASD [9]. The risk of developing ASD appears to be higher in males than in females [10]. Recent studies suggest that an increased level of ubiquitin-protein ligase E3A (UBE3A) gene product, its copy number variations and duplication in the allele, in association with the regulatory activity of protein kinase A (PKA) and impaired synaptic transmission, contribute to the development of ASD [11]. Though there exist some controversial reports, chronic inflammation and elevated levels of the circulating pro-inflammatory cytokines such as TNF-α and IL-12 have been known to be associated with ASD [12]. Moreover, an immunological response mediated by certain foods followed by the elicitation of allergic reactions through the activation of T cells have been shown to be linked with ASD [13]. Some associative studies revealed that the intake of certain foods like cow's milk with immunogenic proteins appears to predispose the individuals to ASD [4, 14]. Additionally, recent data gathered from genome-wide association studies (GWAS) suggest that mutations, copy number variations, and SNPs in many genes appear to be linked with ASD [15]. Thus, ASDs appear to be complex and multifaceted disorders with prenatal exposure of drugs, malnutrition, endocrine dysfunctions, genetic, epigenetic, and nongenetic risk factors, and their reciprocal interactions being considered as the potential risk factors for the onset of the disease. However, the precise etiology of ASD remains obscure due to the abovementioned comorbidities. Therefore, the biology of underlying mechanisms of ASD and the therapeutic target remains unclear, making the availability of a complete cure much harder.

Notably, the behavioral symptoms of ASD appear to emerge in toddlers when brain development occurs in response to nutrition, learning, and environmental stimuli [4, 6]. The development of the cortex, cerebellum, hippocampus, and amygdala has been reported to be different in the brain during the early postnatal period among individuals affected by ASD [16, 17]. An increasing body of evidence from brain-imaging and postmortem studies point to unusual cerebral growth followed by growth arrest coupled with aberrant neuroplasticity as the potential underlying pathophysiology of ASD [18, 19]. Specifically, the neurogenesis and synaptogenesis responsible for neuroplasticity of the brain have been known to be determined by the generation of neuroblasts and their migration, followed by layering and their integration into the neural circuit [20, 21]. Thus, proper development of the brain is determined by highly regulated neural stem cell (NSC)-derived neurogenesis through an intermediate process called neuroblastosis [21]. Eventually, this neurogenic process in the developing adult brains can be regulated by ingested food, environmental stimuli, learning paradigms, and physical activities through a wide range of genetic, epigenetic, and signaling pathways [20, 22]. Considering the causative nature of ASD, malnutrition responsible for defects in the metabolic process may play a critical role in the establishment of cytoarchitecture and neuroplasticity through neuroblasts [21, 23]. Thus, abnormal regulation of neurogenesis at the level of neuroblastosis can be posited as a key cellular mechanism for the development of ASD [22].

1.1 *Abnormal Brain Development and Reactive Neuroblastosis in Autism Spectrum Disorder*

It has been well-established that abnormal neurogenesis is linked to the development of neuropsychiatric, neurodegenerative, and neurodevelopmental disorders, including ASD [20, 22, 23]. Autistic children have been characterized by macrocephaly due to the abnormal production of neurons in their developing brains [24, 25]. In ASD, abnormality in neurogenesis has been known to cause unusual growth of different regions of the cerebrum soon after birth [24–26]. This is mainly due to the overproduction of neurons primarily affecting the structures of the limbic system [27]. The abnormal overgrowth occurs mainly in the regions responsible for emotional learning and memory, social behavior, and language processing [16, 20, 22, 24]. Among them, abnormal neuroanatomical changes in areas of the cerebral cortex like the frontal and temporal cortices and limbic system (mainly the amygdala and hippocampus) are affected in individuals with ASD [22, 25, 28]. Besides, abnormal development of the cerebellum resulting from substantial overproliferation of neuroblasts has been reported in subjects with ASD [8, 26]. In particular, abnormal development of the cortex has been studied extensively, and it has been predicted that the generation of a surplus amount of neuroblasts may be the underlying cause for the development of ASD [25, 29, 30]. In order to ensure the development of the cortex, radial glial cells (RGCs) need to undergo mitotic divisions in the developing brain in a highly regulated manner [19, 25]. Asymmetric divisions in glial cells can result in the generation of intermediate progenitors followed by neuroblastosis leading to the generation of the postmitotic neurons in the brain [25, 30, 31]. Subsequently, symmetric divisions of intermediate progenitor cells in the subventricular zone (SVZ) can give rise to a subset of neuroblasts through transit-amplifying cells which migrate in an "inside–out" fashion to establish the six layers of the cortex [19, 25, 31]. RGCs display a characteristic bipolar structure with a short apical end foot at the ventricular zone and a long radial glial filament that spans the neocortex of the brain to support the migration of neuroblasts from the ventricular zone to constitute the pyramidal neurons of the cortical layers [19, 30, 31]. This is done in such a way that the migration of neuroblasts leads to the cellular arrangement of the deeper layer first, followed by the systematic formation of the superficial layers of the cortex along with the guidance and support of RGCs [19, 25, 31]. Therefore, an alteration in the generation and unusual migratory pattern of neuroblasts and abnormalities in their differentiation fate can lead to an unpredicted number and phenotype of neurons, thereby contributing to the abnormal laminar changes and aberrant synaptic plasticity of the cortex. Interestingly, ASD has been characterized by abnormal neurogenesis and aberrant migratory pattern of neuroblasts in the cortex [19, 25, 27]. In 1998, a neuropathological study by Bailey et al. indicated abnormal mitosis in the cortical regions, limbic system, and cerebellum in the brain tissues of subjects with ASD. In 2007, a histology-based postmortem study by Hutsler et al. indicated the presence of abnormal neuronal populations in cortical regions of ASD subjects [32]. In 2010, Wegiel and colleagues demonstrated dysregulation of neurogenesis, abnormal neuronal migration, and impaired neuronal maturation in the cortex, hippocampus, and cerebellum in the brain of individuals with ASD [27]. These findings have been validated by subsequent studies, and therefore, the concept of altered neurogenesis resulting from the altered cell cycle event of neural progenitors and abnormal migration of neuroblasts has been considered a key event in neuropathological outcomes of ASD. Interestingly, it has been well-established that the Wnt signaling pathway and phosphatase and tensin homolog (PTEN) gene have a key role in the development of the central nervous system (CNS) and cell cycle regulation as they are important for the regulation of neurogenesis at the level of stem cell proliferation [33, 34]. Thus, mutations in the PTEN gene and the genes that are involved in Wnt signaling have been linked to the occurrence of abnormal neurogenesis-mediated macrocephaly in ASD. Besides, mutations in some genes participating in the migration of neuroblasts in the developing brain have also been linked to ASD. For example, the T-brain-1 (TBR1) gene that encodes for a brain-specific T-box transcription factor has been known to play a decisive role in the regulation of neuronal migration and differentiation of intermediate progenitors to post neurons, thereby contributing to brain development [35]. It has been noted that missense mutations and de novo truncation in the TBR1 gene, responsible for the disruption of cortical neurogenesis, appear to be linked with the development of ASD [35]. Reelin (RELN)-mediated signaling has been a well-established regulator of neuronal migration during brain development. Several lines of evidence have ascertained that variants and functional loss of the RELN gene have been a potential risk factor for the development of ASD [36, 37]. Genetic loss, mutation, and defects in contactin-associated protein-like 2 gene (CNTNAP2) have also been found to be associated with ectopic migration of neuroblasts in many neurological disorders [38]. Thus, CNTNAP2 has also been proposed as another potent candidate gene for the development of ASD [38]. While the role of distal-less homeobox (Dlx) genes has been well-established in the regulation of migration of neuroblasts with the commitment of GABAergic phenotype, a subset of ASD subjects have been characterized for the presence of Dlx gene variants [39]. In addition to this, genetic abnormalities and variation in genes that are essential for neuronal migration such as astrotactin 1 (ASTN1) [40], autism susceptibility candidate 2 (AUTS2) [41], WD repeat and FYVE domain - containing 3 (WDFY3) [42], and NudE neurodevelopment protein 1 (NDE1) [43] have been implicated as the risk factors of ASD. Notably, reactive neuroblastosis and ectopic migration of neuroblasts have recently been identified as central cellular traits responsible for abnormal neurogenesis in many neurological syndromes including Huntington's disease (HD) [23]. Interestingly, reactive neuroblasts have been proposed to have immunogenic properties during brain pathological state [21]. However, the role of reactive neuroblastosis in neuropathogenesis has been less explored in the context of ASD. Considering the immunogenic and neurogenic nature of reactive neuroblasts, reactive neuroblastosis might be an ultimate mechanism responsible for ASD. While neuroblasts represent the functional unit of the neuroplasticity, future investigation into the regulation of reactive neuroblastosis in subjects with ASD may provide a potential therapeutic target for the effective management of the neurological, behavioral, and cognitive symptoms.

2 Changes in Synaptic Components and Neurotransmitters in Autism Spectrum Disorder

Synapses are highly specialized structures, and their function through neurotransmitters is essential for neuroplasticity [44]. Synaptic proteins that are present in the terminal ends of the axon and dendrites ensure the storage, trafficking, and release of neurotransmitters and their subsequent receptor-mediated neurotransmission [21, 44]. Mutiple genetic factors have been proposed to be linked to the development of ASD, Presently, more than 40 genetic loci and 100 genes have been linked to ASD [45, 46]. Specifically, there are some mutations in synaptic proteins that have been linked to the development of ASD. For example, neurexin (NRXN) and synapsin (SYN) come under the group of presynaptic proteins. NRXNs are synaptic adhesion proteins that consist of three isoforms, namely NRXN1, NRXN2, and NRXN3 [47, 48]. In ASD, multiple mutations or copy number variations have been identified in the NRXN genes [49]. The SYN is involved in vesicle-mediated neurotransmitter release and neurite outgrowth. While the family of SYN proteins also comprises three isoforms (SYN1, SYN2, and SYN3), mutations in the SYN1 gene have been identified as a risk factor for ASD and epileptic phenotypes [50].

Apart from NRXNs, neuroligins (NLGNs) and SH3 and multiple ankyrin repeat domains (SHANK) are part of the family of postsynaptic proteins linked with ASD [51]. The interaction between the single-pass transmembrane molecules, NRXNs and NLGNs, have been known to regulate the synaptogenesis in the brain thoughout the life [51]. Notably, dysregulation in the interaction between NRXN and NLGN has been proposed to be involved in the development of abnormal neuroplasticity in ASD [51]. The SHANK proteins are required for the proper formation and function of N-methyl-D-aspartate (NMDA) receptors and are attached to NLGNs with the help of postsynaptic density-95 (PSD-95) protein and involved in the funtions of postsynaptic signaling machinery [52]. Many studies have indicated that the defects in the isoforms of SHANK, namely, SHANK2 and SHANK3 can also be associated with the development of ASD [52]. Gephyrin, a key postsynaptic scaffolding protein, have also been interlinked to ASD in asscociation with NLGNs and NRXNs [53]. Therefore, the dysregulation of NLGN, NRXN and SHANK proteins have been known to be associated with behavioral abnormalities noticed in ASD [47, 51]. Proteins are, as 2 has been found to be localized. Taken together, dysregulation or mutation in pre- and postsynaptic proteins appears to contribute to the dysregulation of synaptic plasticity in ASD. SHANK3 strengthens the glutamatergic α-amino-3-hydroxy-5-methyl-4-isoxazolepropionic acid receptor (AMPAR) and *NMDA* receptor mediated synaptic transmission and increases glutamate release via the formation of transsynaptic signaling complexes with NRXN and NLGN. Several point mutations, deletions, and truncations in *SHANK*2, 3 genes have been identified in subjects with ASD [54]. Notably, *SHANK*3 mutations have been correlated with moderate-to-severe intellectual deficits [52]. The function of SHANK3 on the synaptic stability is known to be modulated by zinc, in order to enhance the strength of the excitatory synaptic properties [52]. While ASD-associated SHANK3 mutations have been proposed to retain responsiveness to zinc, dietary zinc supplementation appears to promote the recruitment of zinc-sensitive SHANK2 protein to synapses and alter the synaptic transmission via NMDA-type glutamate receptors [52]. Mouse models expressing mutant *SHANK* genes have been characterised by the defects in glutamatergic synaptic function and display increased anxiety and psychiatric distrubances [55, 56]. Therefore, glutamatergic synapses have gained major research focus for developing therapeutic target for the behavioral deficits in ASD. Studies have revealed that there are many other neurotransmitter abnormalities in glutamate, γ-aminobutyric acid (GABA), dopamine (DA), 5-hydroxytryptamine (5-HT), and oxytocin that can also lead to ASD. The imbalance of excitatory glutamatergic neurotransmitters and inhibitory GABAergic neurotransmitters has been found to be closely associated with the pathogenesis of ASD [30, 57–59]. Deficiency in some dietary-derived essential compounds, such as the amino acid tryptophan, appear to be prominent in patients with ASD [60].

Neurotransmitter system dysfunction has been known to affect neuronal cell migration, differentiation, and synaptogenesis, eventually affecting the neuroplasticity of the brain [21, 61]. The function of neurotransmitters in the brain is linked not only to synaptic remodeling but also to other roles including brain development and cortical and cerebellar organization. Thus, defects in neurotransmitter systems including serotonin, dopamine, noradrenaline (norepinephrine), GABA, glutamate, and neuropeptides and their subsequent receptor-mediated signal pathways and gene regulation have been implicated in the development of ASD [61]. Serotonin (5-hydroxytryptamine [5-HT]) is an important biogenic amine that has wide-ranging effects on numerous physiological processes such as circadian rhythm, appetite, mood, sleep, anxiety, motor activity, and cognition [57, 62]. The levels of serotonin in the circulatory and the digestive system have been found to be high among those with ASD [57, 62]. Tryptophan depletion has previously been shown to result in an increase in autistic behaviors. Nonetheless, it has been proposed that there exists a normal developmental process characterized by high serotonin levels typically in children up to 5 years of age which is disrupted in autistic subjects [57, 63]. Fatemi et al. [66] reported around 50% decrease in GABA-synthesizing enzymes, GAD65 and GAD67, from different parts of the cerebellum of subjects with ASD. Individuals with autism experience a proportionately high prevalence of seizures, with temporal lobe epilepsy developing in one-fifth to one-third of affected subjects, suggesting an abnormality in the GABA system [67]. The change in levels and funtions of neurotransmitters including GABA, glutamate, DA, and 5-HT have also been related to exposure of some neurotoxins like organochlorines, organophosphates, and phthalates [68]. Hence, the effect of neurotoxic compounds on the functioning of neurotransmitters might also play a key role in the development of ASD. Taken together, aberrant neural transmission has been one of the major underlying mechanisms for ASD.

3 A Potential Link Between Deficiency in Dietary Supplements, Foodborne Factors, and Autism Spectrum Disorder

Exponentially increasing experimental data and population-based studies have signified that the pathological changes of ASD appear to originate during fetal development. Neurological and behavioral correlates of ASD in the fetus have been suggested to be acquired from the metabolic abnormalities of maternal origin. An earlier study by Krakowiak et al. [69] indicated that metabolic disorders including obesity and diabetes during gestation may predispose the risk for development of ASD in the fetus. Proper intake of nutrients during pregnancy is crucial for brain development and maturation [22, 70]. Dietary deficiency or intake of certain allergic foods in association with genetic, biochemical factors or immunological reactions can be transmitted from mother to fetus and elicit some adverse effects on the developing brain of the fetus [71]. Some associative measures of dietary and nutritional status in children have advocated that the deficiency of certain supplements, vitamins and minerals including omega-3 fatty acid, pyridoxine (vitamin B6), folic acid (vitamin B9), cholecalciferol (vitamin D), tocopherols (vitamin E), magnesium, calcium, potassium, iron, and zinc can be the potential risk factors of ASD [72–74]. For example, the deficiency of zinc has been associated with ASD [75, 76]. Moreover, the presence of abnormal levels of copper, calcium, folic acid, and iron has been known to interfere with the absorption of zinc [76, 77]. Pregnant women consuming excessive amounts of calcium- and iron-rich supplements may be encountered with a deficiency in the absorption of zinc [73]. In the prenatal stage, expression of the genes that are inventible for neuroplasticity and neurogenesis such as BDNF, SDF-1, CamKIIa, and PSD-95 appears to be affected by zinc and iron deficiencies. Some pharmacological agents like angiotensin-converting enzyme (ACE) inhibitors, used to treat high blood pressure, may also decrease blood zinc levels [78, 79]. Taken together, deficiency of zinc has been known to influence embryonic development and alters the neuroplasticity of the developing brain and poses as a potential risk factor of ASD [30, 80, 81]. Next, folic acid plays a very important role in erythropoiesis and development of the brain and spinal cord from the neural tube [82, 83]. While some studies have suggested that folic acid intake has been beneficial in treating ASD, abnormalities in folate metabolism and an overdose of folic acid during pregnancy have been linked with the development of the symptoms of ASD in progenies [84]. Though a trace amount of vitamin D can be found in food, it is generally obtained from the exposure of the skin to sunlight [85]. Deficiency in vitamin D synthesis due to some environmental factors, including the weather, has been associated with ASD [85, 86]. Vitamin E has been considered a potent antioxidant that prevents oxidative stress in the body. Some reports indicate that children with vitamin E deficiency tend to display autism-like behavioral changes [87]. While ASD has been characterized by elevated levels of free radicals, lipid peroxidation and mitochondrial deformity, lower levels of antioxidants, including serum proteins, namely transferrin (iron-binding protein), ceruloplasmin (copper-binding protein), metallothionein (zinc-buffering protein), and glutathione, have been evident in ASD [75, 88]. Electrolytes like magnesium, calcium, and potassium are key cofactors responsible for signal transduction and many enzyme-based biochemical and cellular events. ASD has been characterized by a deficiency in these minerals, thereby indicating that electrolyte imbalance can also play a role in the development of ASD [88, 89]. Polyunsaturated fatty acids (omega-3 and omega-6 fatty acids) appear to play a decisive role in brain development and regulation of neuroplasticity. Due to lifestyle modification, dietary intake of omega fatty acids seems to be reduced. As a result, deficiency in omega fatty acids has been recognized as a risk factor of ASD [90].

Gastrointestinal (GI) disorders have been a common hallmark in subjects with ASD [91]. On the one hand, abnormalities in the gut microbiome and GI disorders could interfere with the assimilation of dietary supplements leading to the deficiency of vitamins, minerals, and other essential nutrients. However, on the other hand, it may lead to the dysregulation of the gut–brain axis [92]. The gut–brain axis has been known to influence brain development and behaviors through the modulation of neurogenesis, neuroplasticity, and neuroendocrine and neuroimmune functions [93]. Dairy and gluten-rich foods tend to greatly affect the homeostasis of gut microbe environment, and hence, such foods will impair the gut–brain axis and further impair neuronal functions [92–95]. Thus, dysregulation of the gut–brain axis, noticed in the individuals with abnormal behavioral patterns, can also be highly relevant for the onset of ASD [13]. Moreover, food that cause allergy and also the accumulation of some toxic elements such as cadmium, arsenic, and mercury have been reported as candidate factors for the development of ASD [96]. Taken together, both malnutrition and overnutrition could severely impair neuroplasticity in prenatal and postnatal stages. The aforementioned difficulties, encountered due to substandard regulation of neural plasticity, could be alleviated or managed in part by nutrition. Hence, a distinct approach for improving neuroplasticity and neurogenesis through nutraceuticals needs to be promoted for the management of ASD.

4 Nutraceuticals for the Effective Management of Autism Spectrum Disorder

Nutraceuticals are part of a modern, scientific, and medical attempt to facilitate tissue regeneration or prevention, management, and cure of diseases through the implementation of nutrition and dietary supplements in a regulated manner. In accordance with the US Foundation for Innovation in Medicine (FIM), any edible substance or a constituent of food which ensures medical or health benefits can be considered as nutraceuticals. Nutraceuticals include products or modified traditional foods with a dietary supplement of vitamins, minerals, amino acids, herbal derivatives, and probiotic microbes that claim to have beneficial effects in health care and medical systems [74, 97]. Recently, ASD has been subjected to the extensive intervention of nutraceuticals, as dietary supplements are considered to elicit no or minimal side effects when compared to synthetic and pharmacological agents [14, 95]. Data from several lines of evidence advocate that the risk and symptoms of ASD could be minimized by supplementation of multivitamins and minerals [75, 88]. In general, vitamins and minerals function as coenzymes and cofactors responsible for many metabolic and immunological pathways and the development and maintenance of organs, including the brain. Hence, the implementation of multivitamin and mineral has been proposed to rectify key, if not all, susceptible metabolic and molecular pathways found in ASD. Vitamin B6 is an essential cofactor for many neurotransmitter systems. Magnesium forms part of the essential macrominerals for the majority of enzyme-catalyzed reactions. Studies suggest that the combination of vitamin B6 and magnesium tends to reduce the symptoms of ASD to a great extent [98]. Naturally occurring neurotransmitter reuptake inhibitors and precursors of neurotransmitters such as amino acids have been shown to manage abnormal neurotransmission in the brain [63]. Likewise, some dietary fatty acids, antioxidants, herbal products, and probiotics have been proposed to yield beneficial effects among individuals with ASD. Precursors of L-tryptophan can be found in eggs, meat, cereal, milk, bananas, fish, seafood, and plums, though amino acid is also synthesized by gut microflora, which contribute to altered tryptophan metabolism, yielding increased levels of indole-3-acetic acid and indolyl lactate [99]. In the eliminative approaches, a gluten-free and casein-free (GFCF) diet has been shown to minimize the risk and symptoms of ASD [100]. Gluten is an ingredient in many widely used food products such as wheat, barley, and grains such as oats. Wheat remains the major source of caloric intake (50%) in the majority of developed and developing countries. Gluten in the flour provides suitable viscosity and elasticity in the food. Glutens are resistant to digestion in the human GI tract, and thus, they confer to the permeability of the enterocytes in the small intestine. To reduce the effects of casein and gluten by eliciting an immune response, GFCF diet should be provided to children with autism. The GFCF diet has been known to give promising results in ASD [100]. Elevated positive results include coordination in the motor area, social interaction, eye contact, and ritualistic behavior and language. Moreover, dietary factors and gut–brain axis have been known to regulate neurogenesis in the brain depending on the associative elements in adulthood [101]. However, the roles of these factors specifically acting on the regulation of reactive neuroblastosis and ectopic migration of neuroblasts have not been specifically addressed in the developing brain. Thus, future studies are needed for further understanding of the modulation of neuroblastosis by nutrition in subjects with ASD.

5 Summary and Conclusion

Though ASD, comprising a wide range of neurodevelopmental disorders, is caused primarily by inherited or acquired defects in numerous genes, malnutrition, especially during the prenatal stage and several environmental factors including air pollution and pesticides could also play a major role in either causing or aggravating the disease condition. As there is no complete cure for the disorder, current treatment procedures aim at alleviating the symptoms, and one of the best ways to do so is by monitoring and controlling dietary intake. This, in turn, can be facilitated by the intake of nutraceuticals. Though the reason for inducing intolerance and the symptom-aggravating nature of certain diets has been proposed by a number of theories, the actual mechanism remains obscure. Nutraceuticals appear to ameliorate the pathogenic state by exerting control over the neurotransmitters and neuroblastosis; hence, the disease can be prevented, or proper functioning of the brain can be restored in ASD by the effective employment of nutraceuticals.

Acknowledgments MK has been supported by the Faculty Recharge Programme, University Grants Commission (UGC-FRP), New Delhi, India. MK would like to acknowledge a start-up grant from UGC-FRP, a research grant (EEQ/2016/000639) and an Early Career Research Award (ECR/2016/000741) from DST-SERB, New Delhi, India. AY has been supported as JRF from DST SERB-EEQ/2016/000639. RKR has been supported as JRF from DST SERB-ECR/2016/000741. SAR has been supported as JRF from Department of Biotechnology (DBT), New Delhi, India. The authors acknowledge UGC-SAP, DST-FIST, and PURSE for the infrastructure of the Department of Animal Science and Department of Biochemistry, Bharathidasan University.

References

1. Evans, B. (2013). How autism became autism. *History of the Human Sciences, 26*(3), 3–31.
2. Olmsted, D., & Blaxill, M. (2016). Leo Kanner's mention of 1938 in his report on autism refers to his first patient. *Journal of Autism and Developmental Disorders, 46*(1), 340–341.
3. N. C. C. for M. Health (UK). (2012). *Introduction to autism in adults*. Leicester: British Psychological Society.
4. Lai, M.-C., Lombardo, M. V., & Baron-Cohen, S. (2014). Autism. *Lancet, 383*(9920), 896–910.
5. Yaylaci, F., & Miral, S. (2017). A comparison of DSM-IV-TR and DSM-5 diagnostic classifications in the clinical diagnosis of autistic spectrum disorder. *Journal of Autism and Developmental Disorders, 47*(1), 101–109.
6. Ozonoff, S., Heung, K., Byrd, R., Hansen, R., & Hertz-Picciotto, I. (2008). The onset of autism: Patterns of symptom emergence in the first years of life. *Autism Research, 1*(6), 320–328.
7. Frye, R. E., & Rossignol, D. A. (2016). Identification and treatment of pathophysiological comorbidities of autism spectrum disorder to achieve optimal outcomes. *Clinical Medicine Insights. Pediatrics, 10*, 43–56.
8. Park, H. R., Lee, J. M., Moon, H. E., Lee, D. S., Kim, B. N., Kim, J., et al. (2016). A short review on the current understanding of autism spectrum disorders. *Experimental Neurobiology, 25*(1), 1–13.
9. Boat, T. F., Wu, J. T., eds (2015). *Prevalence of autism spectrum disorder*. Washington, DC: National Academies Press (US).
10. Halladay, A. K., Bishop, S., Constantino, J. N., Daniels, A. M., Koenig, K., Palmer, K., et al. (2015). Sex and gender differences in autism spectrum disorder: Summarizing evidence gaps and identifying emerging areas of priority. *Molecular Autism, 6*, 36.
11. Smith, S. E. P., Zhou, Y.-D., Zhang, G., Jin, Z., Stoppel, D. C., & Anderson, M. P. (2011). Increased gene dosage of Ube3a results in autism traits and decreased glutamate synaptic transmission in mice. *Science Translational Medicine, 3*(103), 103ra97.

12. Xu, N., Li, X., & Zhong, Y. (2015). Inflammatory cytokines: Potential biomarkers of immunologic dysfunction in autism spectrum disorders. *Mediators of Inflammation, 2015*, 531518.
13. Jyonouchi, H. (2009). Food allergy and autism spectrum disorders: Is there a link? *Current Allergy and Asthma Reports, 9*(3), 194–201.
14. Li, Y.-J., Ou, J.-J., Li, Y.-M., & Xiang, D.-X. (2017). Dietary supplement for core symptoms of autism spectrum disorder: Where are we now and where should we go? *Frontiers in Psychiatry, 8*, 155.
15. Autism Spectrum Disorders Working Group of The Psychiatric Genomics Consortium. (2017). Meta-analysis of GWAS of over 16,000 individuals with autism spectrum disorder highlights a novel locus at 10q24.32 and a significant overlap with schizophrenia. *Molecular Autism, 8*, 21.
16. Blatt, G. J. (2012). The neuropathology of autism. *Scientifica (Cairo), 2012*, 703675.
17. Pickett, J., & London, E. (2005). The neuropathology of autism: A review. *Journal of Neuropathology and Experimental Neurology, 64*(11), 925–935.
18. Ha, S., Sohn, I.-J., Kim, N., Sim, H. J., & Cheon, K. A. (2015). Characteristics of brains in autism spectrum disorder: Structure, function and connectivity across the lifespan. *Experimental Neurobiology, 24*(4), 273–284.
19. Courchesne, E. (2004). Brain development in autism: Early overgrowth followed by premature arrest of growth. *Mental Retardation and Developmental Disabilities Research Reviews, 10*(2), 106–111.
20. Ming, G.-L., & Song, H. (2011). Adult neurogenesis in the mammalian brain: Significant answers and significant questions. *Neuron, 70*(4), 687–702.
21. Kandasamy, M., & Aigner, L. (2018). Neuroplasticity, limbic neuroblastosis and neuro-regenerative disorders. *Neural Regeneration Research, 13*(8), 1322–1326.
22. Kaushik, G., & Zarbalis, K. S. (2016). Prenatal neurogenesis in autism spectrum disorders. *Frontiers in Chemistry, 4*, 12.
23. Kandasamy, M., & Aigner, L. (2018). Reactive neuroblastosis in Huntington's disease: A putative therapeutic target for striatal regeneration in the adult brain. *Frontiers in Cellular Neuroscience, 12*, 37.
24. Packer, A. (2016). Neocortical neurogenesis and the etiology of autism spectrum disorder. *Neuroscience and Biobehavioral Reviews, 64*, 185–195.
25. Courchesne, E., Mouton, P. R., Calhoun, M. E., Semendeferi, K., Ahrens-Barbeau, C., Hallet, M. J., et al. (2011). Neuron number and size in prefrontal cortex of children with autism. *JAMA, 306*(18), 2001–2010.
26. Wang, S. S.-H., Kloth, A. D., & Badura, A. (2014). The cerebellum, sensitive periods, and autism. *Neuron, 83*(3), 518–532.
27. Wegiel, J., Kuchna, I., Nowicki, K., Imaki, H., Wegiel, J., Marchi, E., et al. (2010). The neuropathology of autism: Defects of neurogenesis and neuronal migration and dysplastic changes. *Acta Neuropathologica, 119*(6), 755–770.
28. Mercadante, M. T., Cysneiros, R. M., Schwartzman, J. S., Arida, R. M., Cavalheiro, E. A., & Scorza, F. A. (2008). Neurogenesis in the amygdala: A new etiologic hypothesis of autism? *Medical Hypotheses, 70*(2), 352–357.
29. Casanova, E. L., & Casanova, M. F. (2014). Genetics studies indicate that neural induction and early neuronal maturation are disturbed in autism. *Frontiers in Cellular Neuroscience, 8*, 397.
30. Gilbert, J., & Man, H.-Y. (2017). Fundamental elements in autism: From neurogenesis and neurite growth to synaptic plasticity. *Frontiers in Cellular Neuroscience, 11*, 359.
31. Beattie, R., & Hippenmeyer, S. (2017). Mechanisms of radial glia progenitor cell lineage progression. *FEBS Letters, 591*(24), 3993–4008.
32. Hutsler, J. J., Love, T., & Zhang, H. (2007). Histological and magnetic resonance imaging assessment of cortical layering and thickness in autism spectrum disorders. *Biological Psychiatry, 61*(4), 449–457.

33. Frazier, T. W., Embacher, R., Tilot, A. K., Koenig, K., Mester, J., & Eng, C. (2015). Molecular and phenotypic abnormalities in individuals with germline heterozygous PTEN mutations and autism. *Molecular Psychiatry, 20*(9), 1132–1138.
34. Kalkman, H. O. (2012). A review of the evidence for the canonical Wnt pathway in autism spectrum disorders. *Molecular Autism, 3*, 10.
35. Chuang, H.-C., Huang, T.-N., & Hsueh, Y.-P. (2015). T-Brain-1—A potential master regulator in autism spectrum disorders. *Autism Research, 8*(4), 412–426.
36. Wang, Z., Hong, Y., Zou, L., Zhong, R., Zhu, B., Shen, N. et al. (2014). Reelin gene variants and risk of autism spectrum disorders: An integrated meta-analysis. *American Journal of Medical Genetics. Part B, Neuropsychiatric Genetics, 165B*(2), 192–200.
37. Lammert, D. B., & Howell, B. W. (2016). RELN mutations in autism spectrum disorder. *Frontiers in Cellular Neuroscience, 10*, 84.
38. Liska, A., Bertero, A., Gomolka, R., Sabbioni, M., Galbusera, A., Barsotti, N., et al. (2018). Homozygous loss of autism-risk gene CNTNAP2 results in reduced local and long-range prefrontal functional connectivity. *Cerebral Cortex, 28*(4), 1141–1153.
39. Liu, X., Novosedlik, N., Wang, A., Hudson, M. L., Cohen, I. L., Chudley, A. E., et al. (2009). The DLX1and DLX2 genes and susceptibility to autism spectrum disorders. *European Journal of Human Genetics, 17*(2), 228–235.
40. Lionel, A. C., Tammimies, K., Vaags, A. K., Rosenfeld, J. A., Ahn, J. W., Merico, D., et al. (2014). Disruption of the ASTN2/TRIM32 locus at 9q33.1 is a risk factor in males for autism spectrum disorders, ADHD and other neurodevelopmental phenotypes. *Human Molecular Genetics, 23*(10), 2752–2768.
41. Oksenberg, N., Stevison, L., Wall, J. D., & Ahituv, N. (2013). Function and regulation of AUTS2, a gene implicated in autism and human evolution. *PLoS Genetics, 9*(1), e1003221.
42. Napoli, E., Song, G., Panoutsopoulos, A., Riyadh, M. A., Kaushik, G., Halmai, J., et al. (2018). Beyond autophagy: A novel role for autism-linked Wdfy3 in brain mitophagy. *Scientific Reports, 8*, 11348.
43. Paciorkowski, A. R., Keppler-Noreuil, K., Robinson, L., Sullivan, C., Sajan, S., Christian, S. L., et al. (2013). Deletion 16p13.11 uncovers NDE1 mutations on the non-deleted homolog and extends the spectrum of severe microcephaly to include fetal brain disruption. *American Journal of Medical Genetics. Part A, 161*(7), 1523–1530.
44. Cramer, S. C., Sur, M., Dobkin, B. H., O'Brien, C., Sanger, T. D., Trojanowski, J. Q., et al. (2011). Harnessing neuroplasticity for clinical applications. *Brain, 134*(6), 1591–1609.
45. Yuen, R. K. C., Merico, D., Bookman, M., Howe, L. J., Thiruvahindrapuram, B., Patel, R. V., et al. (2017). Whole genome sequencing resource identifies 18 new candidate genes for autism spectrum disorder. *Nature Neuroscience, 20*(4), 602–611.
46. Buxbaum, J. D. (2009). Multiple rare variants in the etiology of autism spectrum disorders. *Dialogues in Clinical Neuroscience, 11*(1), 35–43.
47. Wang, J., Gong, J., Li, L., Chen, Y., Liu, L., Gu, H., et al. (2018). Neurexin gene family variants as risk factors for autism spectrum disorder. *Autism Research, 11*(1), 37–43.
48. Greco, B., Managò, F., Tucci, V., Kao, H.-T., Valtorta, F., & Benfenati, F. (2013). Autism-related behavioral abnormalities in synapsin knockout mice. *Behavioural Brain Research, 251*, 65–74.
49. Gauthier, J., Siddiqui, T. J., Huashan, P., Yokomaku, D., Hamdan, F. F., Champagne, N., et al. (2011). Truncating mutations in NRXN2 and NRXN1 in autism spectrum disorders and schizophrenia. *Human Genetics, 130*(4), 563–573.
50. Fassio, A., Patry, L., Congia, S., Onofri, F., Piton, A., Gauthier, J., et al. (2011). SYN1 loss-of-function mutations in autism and partial epilepsy cause impaired synaptic function. *Human Molecular Genetics, 20*(12), 2297–2307.
51. Südhof, T. C. (2008). Neuroligins and neurexins link synaptic function to cognitive disease. *Nature, 455*(7215), 903–911.
52. Leblond, C. S., Nava, C., Polge, A., Gauthier, J., Huguet, G., Lumbroso, S., et al. (2014). Meta-analysis of SHANK mutations in autism spectrum disorders: A gradient of severity in cognitive impairments. *PLoS Genetics, 10*(9), e1004580.

53. Chen, J., Yu, S., Fu, Y., & Li, X. (2014). Synaptic proteins and receptors defects in autism spectrum disorders. *Frontiers in Cellular Neuroscience, 8*, 276.
54. Moessner, R., Marshall, C. R., Sutcliffe, J.S., Skaug, J., Pinto, D., Vincent, J., et al. (2007). Contribution of SHANK3 mutations to autism spectrum disorder. *American Journal of Human Genetics, 81*(6), 1289–1297.
55. Mei, Y., Monteiro, P., Zhou, Y., Kim, J. A., Gao, X., Fu, Z., et al. (2016). Adult restoration of Shank3 expression rescues selective autistic-like phenotypes. *Nature, 530*(7591), 481–484.
56. Schmeisser, M. J., Ey, E., Wegener, S., Bockmann, J., Stempel, A. V., Kuebler, A., et al. (2012). Autistic-like behaviours and hyperactivity in mice lacking ProSAP1/Shank2. *Nature, 486*(7402), 256–260.
57. Muller, C. L., Anacker, A. M. J., & Veenstra-VanderWeele, J. (2016). The serotonin system in autism spectrum disorder: From biomarker to animal models. *Neuroscience, 321*, 24–41.
58. Polšek, D., Jagatic, T., Cepanec, M., Hof, P. R., & Šimić, G. (2011). Recent developments in neuropathology of autism spectrum disorders. *Translational Neuroscience, 2*(3), 256–264.
59. Rojas, D. C. (2014). The role of glutamate and its receptors in autism and the use of glutamate receptor antagonists in treatment. *Journal of Neural Transmission, 121*(8), 891–903.
60. Zheng, H.-F., Wang, W.-Q., Li, X.-M., Rauw, G., & Baker, G. B. (2017). Body fluid levels of neuroactive amino acids in autism spectrum disorders: A review of the literature. *Amino Acids, 49*(1), 57–65.
61. Gerhard, S. D., Evelien, M. B., Albert, P. A., Tamar, M. V. V., Nicolaas, P., Richard, A. E. E., et al. (2016). Altered neurotransmitter metabolism in adolescents with high-functioning autism. *Psychiatry Research, 256*, 44–49.
62. Zafeiriou, D., Ververi, A., & Vargiami, E. (2009). The serotonergic system: Its role in pathogenesis and early developmental treatment of autism. *Current Neuropharmacology, 7*(2), 150–157.
63. Kałużna-Czaplińska, J., Jóźwik-Pruska, J., Chirumbolo, S., & Bjørklund, G. (2017). Tryptophan status in autism spectrum disorder and the influence of supplementation on its level. *Metabolic Brain Disease, 32*(5), 1585–1593.
64. Schubert, D., Martens, G. J. M., & Kolk, S. M. (2015). Molecular underpinnings of prefrontal cortex development in rodents provide insights into the etiology of neurodevelopmental disorders. *Molecular Psychiatry, 20*(7), 795–809.
65. Rogers, T. D., Dickson, P. E., McKimm, E., Heck, D. H., Goldowitz, D., Blaha, C. D., et al. (2013). Reorganization of circuits underlying cerebellar modulation of prefrontal cortical dopamine in mouse models of autism spectrum disorder. *Cerebellum, 12*(4), 547–556.
66. Fatemi, S. H., Aldinger, K. A., Ashwood, P., Bauman, M. L., Blaha, C. D., Blatt, G. J., et al. (2012). Consensus paper: Pathological role of the cerebellum in autism. *Cerebellum, 11*(3), 777–807.
67. Coghlan, S., Horder, J., Inkster, B., Mendez, M. A., Murphy, D. G., & Nutt, D. J. (2012). GABA system dysfunction in autism and related disorders: From synapse to symptoms. *Neuroscience and Biobehavioral Reviews, 36*(9), 2044–2055.
68. Quaak, I., Brouns, M. R., & de Bor, M. V. (2013). The dynamics of autism spectrum disorders: How neurotoxic compounds and neurotransmitters interact. *International Journal of Environmental Research and Public Health, 10*(8), 3384–3408.
69. Krakowiak, P., Walker, C. K., Bremer, A. A., Baker, A. S., Ozonoff, S., Hansen, R. L., et al. (2012). Maternal metabolic conditions and risk for autism and other neurodevelopmental disorders. *Pediatrics, 129*(5), e1121–e1128.
70. Kawicka, A., & Regulska-Ilow, B. (2013). How nutritional status, diet and dietary supplements can affect autism. A review. *Roczniki Państwowego Zakładu Higieny, 64*(1), 1–12.
71. Hyman, S. L., Stewart, P. A., Schmidt, B., Cain, U., Lemcke, N., Foley, J. T., et al. (2012). Nutrient intake from food in children with autism. *Pediatrics, 130*(Suppl 2), S145–S153.
72. Mazahery, H., Camargo, C. A., Conlon, C., Beck, K. L., Kruger, M. C., & von Hurst, P. R. (2016). Vitamin D and autism spectrum disorder: A literature review. *Nutrients, 8*(4), 236.
73. Ranjan, S., & Nasser, J. A. (2015). Nutritional status of individuals with autism spectrum disorders: Do we know enough? *Advances in Nutrition, 6*(4), 397–407.

74. Fujiwara, T., Morisaki, N., Honda, Y., Sampei, M., & Tani, Y. (2016). Chemicals, nutrition, and autism spectrum disorder: A mini-review. *Frontiers in Neuroscience, 10*, 174.
75. Adams, J. B., Audhya, T., McDonough-Means, S., Rubin, R. A., Quig, D., Geis, E., et al. (2011). Nutritional and metabolic status of children with autism vs. neurotypical children, and the association with autism severity. *Nutrition & Metabolism (London), 8*, 34.
76. Yasuda, H., Yoshida, K., Yasuda, Y., & Tsutsui, T. (2011). Infantile zinc deficiency: Association with autism spectrum disorders. *Scientific Reports, 1*, 129.
77. Babaknejad, N., Sayehmiri, F., Sayehmiri, K., Mohamadkhani, A., & Bahrami, S. (2016). The relationship between zinc levels and autism: A systematic review and meta-analysis. *Iranian Journal of Child Neurology, 10*(4), 1–9.
78. Hagmeyer, S., Sauer, A. K., & Grabrucker, A. M. (2018). Prospects of zinc supplementation in autism spectrum disorders and Shankopathies such as Phelan McDermid Syndrome. *Frontiers in Synaptic Neuroscience, 10*, 11.
79. Velusamy, T., Archana, S. P., Purushottam, M., Anusuyadevi, M., Pal, K. P., Jain, S., et al. (2017). Protective effect of antioxidants on neuronal dysfunction and plasticity in Huntington's Disease. *Oxidative Medicine and Cellular Longevity, 2017*, 3279061.
80. Grabrucker, A. M. (2013). Environmental factors in autism. *Frontiers in Psychiatry, 3*, 118.
81. Grabrucker, S., Jannetti, L., Eckert, M., Gaub, S., Chhabra, R., Pfaender, S., et al. (2014). Zinc deficiency dysregulates the synaptic ProSAP/Shank scaffold and might contribute to autism spectrum disorders. *Brain, 137*(Pt 1), 137–152.
82. Sun, C., Zou, M., Zhao, D., Xia, W., & Wu, L. (2016). Efficacy of folic acid supplementation in autistic children participating in structured teaching: An open-label trial. *Nutrients, 8*(6), E337.
83. Castro, K., Klein, L. d. S., Baronio, D., Gottfried, C., Riesgo, R., & Perry, I. S. (2016). Folic acid and autism: What do we know? *Nutritional Neuroscience, 19*(7), 310–317.
84. Wiens, D., & DeSoto, M. C. (2017). Is high folic acid intake a risk factor for autism? A review. *Brain Sciences, 7*(11), pii: E149.
85. Saad, K., Abdel-Rahman, A. A., Elserogy, Y. M. , Al-Atram, A. A. , Cannell, J. J., Bjørklund, G., et al. (2016). Vitamin D status in autism spectrum disorders and the efficacy of vitamin D supplementation in autistic children. *Nutritional Neuroscience, 19*(8), 346–351.
86. Cannell, J. J. (2017). Vitamin D and autism, what's new? *Reviews in Endocrine & Metabolic Disorders, 18*(2), 183–193.
87. Krajcovicova-Kudlackova, M., Valachovicova, M., Mislanova, C., Hudecova, Z., Sustrova, M., & Ostatnikova, D. (2009). Plasma concentrations of selected antioxidants in autistic children and adolescents. *Bratislavské Lekárske Listy, 110*(4), 247–250.
88. Adams, J. B., Audhya, T., McDonough-Means, S., Rubin, R. A., Quig, D., Geis, E., et al. (2011). Effect of a vitamin/mineral supplement on children and adults with autism. *BMC Pediatrics, 11*, 111.
89. Skalny, A. V., Simashkova N. V., Klyushnik, T. P., Grabeklis, A. R., Radysh, I. V., Skalnaya, M. G., et al. (2017). Assessment of serum trace elements and electrolytes in children with childhood and atypical autism. *Journal of Trace Elements in Medicine and Biology, 43*, 9–14.
90. Parletta, N., Niyonsenga, T., & Duff, J. (2016). Omega-3 and Omega-6 polyunsaturated fatty acid levels and correlations with symptoms in children with attention deficit hyperactivity disorder, autistic spectrum disorder and typically developing controls. *PLoS One, 11*(5), e0156432.
91. Chaidez, V., Hansen, R. L., & Hertz-Picciotto, I. (2014). Gastrointestinal problems in children with autism, developmental delays or typical development. *Journal of Autism and Developmental Disorders, 44*(5), 1117–1127.
92. van De Sande, M. M. H., van Buul, V. J., & Brouns, F. J. P. H. (2014). Autism and nutrition: The role of the gut-brain axis. *Nutrition Research Reviews, 27*(2), 199–214.
93. Tognini, P. (2017). Gut microbiota: A potential regulator of neurodevelopment. *Frontiers in Cellular Neuroscience, 11*, 25.
94. Berding, K., & Donovan, S. M. (2018). Diet can impact microbiota composition in children with autism spectrum disorder. *Frontiers in Neuroscience, 12*, 515.

95. Sanctuary, M. R., Kain, J. N., Angkustsiri, K., & German, J. B. (2018). Dietary considerations in autism spectrum disorders: The potential role of protein digestion and microbial putrefaction in the gut-brain axis. *Frontiers in Nutrition, 5*, 40.
96. Rossignol, D. A., Genuis, S. J., & Frye, R. E. (2014). Environmental toxicants and autism spectrum disorders: A systematic review. *Translational Psychiatry, 4*(2), e360.
97. Alanazi, A. S. (2013). The role of nutraceuticals in the management of autism. *Saudi Pharmaceutical Journal, 21*(3), 233–243.
98. Nye, C., & Brice, A. (2005). Combined vitamin B6-magnesium treatment in autism spectrum disorder. *Cochrane Database of Systematic Reviews, 4*, CD003497.
99. Richard, D. M., Dawes, M. A., Mathias, C. W., Acheson, A., Hill-Kapturczak, N., & Dougherty, D. M. (2009). L-Tryptophan: Basic metabolic functions, behavioral research and therapeutic indications. *International Journal of Tryptophan Research, 2*, 45–60.
100. Marí-Bauset, S., Zazpe, I., Mari-Sanchis, A., Llopis-González, A., & Morales-Suárez-Varela, M. (2014). Evidence of the gluten-free and casein-free diet in autism spectrum disorders: A systematic review. *Journal of Child Neurology, 29*(12), 1718–1727.
101. Stangl, D., & Thuret, S. (2009). Impact of diet on adult hippocampal neurogenesis. *Genes & Nutrition, 4*(4), 271–282.

Part II
Specific Foods and Nutrient Qualities in Autism

"Let food be thy medicine and medicine be thy food." – Hippocrates, 400 BC

1.1 Overview and Reflection

This section of the book discusses various kinds of specific foods and their nutrient qualities. Proper diet and nutrient intake form the foundations of good health. In an era of processed food, artificial and/or synthetic additives or preservatives the likes of food flavors, coloring, and dyes, appropriate nourishment along with a comprehensive understanding of nutritional value is exceedingly important to ensure good health. It is not possible to cover all food items in this chapter. However, few representative types from different food families have been selected. The choice of food variety is predicated on ordinarily available, edible foods present in grocery stores in the West and the East.

Since the beginning of this millennium interest in and awareness of natural medicine, often referred to as complementary and alternative therapies, have significantly increased. While not part of conventional medicine, the practice has taken hold in our society and this wide acceptance emphasizes an integrative health approach that addresses body, mind, and spirit. For instance, the interaction of food products with the immune system leading to sensitivities, intolerance, or allergies is quite widespread where complementary and alternative medicine (CAM) is being used as a form of treatment or combined with conventional therapy. It is not the intent of this section to provide nutritional and dietary intervention or therapeutic approaches to food consumption as this will be covered in other chapters of this book. Indeed, personalized nutrition and emerging dietary management of various health conditions including autism have seen sharp increase as of late. Other parts of this book will address the use of food and diet as therapy for individuals with autism. These include probiotic intervention or the *gluten*-free/*casein*-free (GFCF) diet where some parents reported improvements in *autism* symptoms with this dietary regimen.

Of importance is the "gate" role that the gut plays. Nowadays, brain–gut interaction and immune activation due to gut microflora are well established. Microbiota modification takes place as a result of the kind of food intake leading to microbiome changes. Research indicates there is a strong link between microbiome alteration and psychiatric disorders, mood changes and even ASD.

In the following chapters, the collection of vegetables, fruits, grains, nuts, seeds, spices, and other edible natural products gives scope for further research besides providing clues for useful food and their benefits. Our predisposition here is holistic in nature. Nutritional facts and values will be provided for each item discussed, and the general health benefits (antioxidant, antiinflammatory, anticancer, neuroprotective, immunostimulant, or other advantages) will also be mentioned and properly referenced. Families often turn to CAM when they have a long lasting issue that conventional medicine has not addressed. CAM is often perceived as "natural" without the side effects of conventional medical treatments. We completely realize and comprehend that the role of certain active pharmacological ingredients or compounds in food is neither substantiated nor fully understood and that additional research and rigorous clinical trials are needed to support some of the claims. We will make links to autism or other neurodevelopmental disorders where it is appropriate and pertinent. No less important, drawing attention to valuable characteristics permitting the management of comorbid health conditions like diabetes, cancer, and cardiovascular diseases. We believe this global, comprehensive perspective will benefit not only the scientists but also the community in its entirety.

Finally, health information access is at the fingertips of many families with many actively participate in their health management. Therefore, continued growth of interest in CAM can be anticipated. Clinicians must remember that parents may have different beliefs regarding the effectiveness of therapy/intervention methods for ASD management and different tolerance for treatment risks. Medical practitioners must keep avenues of communication open with families and remain open-minded regarding medical care of their patients. Moreover, the insightfulness and empathetic understanding that prescribing dietary/nutritional recommendations is highly personalized to the autistic individual. It is vital to discuss alternative therapies for autism openly and compassionately as some CAM interventions or therapies are supported by scientific evidence. Physicians need access to balanced education that will inform their own recommendations for specific CAM interventions or therapies and adequate information to care for families who elect their use.

Vegetables

Sawsan G. Mohammed and M. Walid Qoronfleh

Abstract Vegetables come in varied colors, forms, and tastes. Health specialists highly encourage inclusion of vegetables in one's diet due to their inherent nutritional worth. This chapter will cover selected vegetables from wide-ranging families like roots, stems, leafy greens, and cruciferous varieties. The broad choice of the designated vegetables is predicated on popular household preferences, nourishing value, and health benefits. These vegetables are also in common use and are obtainable at the market. The health features of the vegetables are covered in such a way that they provide the distinctive phytonutrient quality coupled with prominent health findings.

Keywords Vegetables · Fiber · Antioxidants · Anthocyanins · Xanthophylls · Chlorophyll · Carotenoids · Flavonoids · Glycolipids · Glucosinolates · Isothiocyanates · Sulforaphane · Allicin · Polyphenols · Phytoestrogens · Phytosterols · Terpenoids · Phthalides

S. G. Mohammed (✉)
Qatar Research Leadership Program (QRLP), Qatar Foundation, Doha, Qatar
e-mail: sgmohammed@qf.org.qa

M. W. Qoronfleh (✉)
Research & Policy Department, World Innovation Summit for Health (WISH), Qatar Foundation, Doha, Qatar
e-mail: wqoronfleh@qf.org.qa

M. M. Essa, M. W. Qoronfleh (eds.), *Personalized Food Intervention and Therapy for Autism Spectrum Disorder Management*, Advances in Neurobiology 24,
https://doi.org/10.1007/978-3-030-30402-7_9

1 Asparagus

Family: Asparagaceae
Genus: *Asparagus*
Common name: Asparagus, garden asparagus, or sparrow grass

Asparagus is a spring vegetable that is full of essential vitamins like vitamins A, B-1 (thiamin), and C and minerals including trace mineral chromium and antioxidants (Table 1). It is low caloric with no fat or sodium. There are three varieties of asparagus. The most common type of asparagus is green, but the white asparagus, which is more delicate and difficult to harvest and the purple, which is smaller with fruitier flavor are also available. In ancient times, it was used fresh when in season and dried or even frozen in winter.

Asparagus is rich in antioxidants like vitamins E and C and glutathione, as well as different flavonoids and polyphenols typically associated with decreased risk of cancers. Glutathione is a free radical scavenger capable of keeping several cancers in check including breast [1], bone [2], colon [3], prostate [4], larynx [5], and lung [6] cancers. Finally, asparagus is a potential source for prevention and treatment of liver cancer [7] as the asparagus polysaccharides can selectively inhibit carcinoma cell proliferation through induction of G_2/M phase arrest and apoptosis via modulation of Bax, Bcl-2, and capase-3. Moreover, asparanin A, a steroidal saponin or an amphipathic glycoside, has displayed antiproliferative activities against many cancers such as esophageal cancer, gastric cancer, lung cancer, and leukemia [7]. They also improved immune response by increasing IgG production and IL-12-specific response—cellular immunity—while inhibiting pro-inflammatory cytokines IL-6 and tumor necrosis factor (TNF) with low allergic response and cytotoxicity at the same time [8]. As a rich source of dietary fiber, it promotes digestion and reduces blood pressure and cholesterol, thus lowering the risk of heart diseases, diabetes, as well as colorectal cancer [9–11]. The antidiabetic effect is attributed to chromium that enhances the ability of insulin to transport glucose, B vitamins like B-6 (pyridoxine) that regulates blood glucose levels, and antioxidants by improving overall insulin secretion and β-cell function, as well as the oxidative stress status [12]. The presence of folic acid is vital for DNA replication and reduces the risk of pregnancy-related complications and neural tube defects [13, 14]. High vitamin K content helps maintain healthy bones by absorbing calcium accordingly reducing the

Table 1 Asparagus nutrition facts of 100 g, from the United States Department of Agriculture (USDA SR-21)

Serving size of 100 g of raw asparagus	
Per serving	% Daily value[a]
Calories 20	
Total fat 0.1 g	0
Saturated fat 0.0 g	0
Polyunsaturated fat 0.1 g	
Monounsaturated fat 0 g	
Total omega-3 fatty acids 10.0 mg	
Total omega-6 fatty acids 40.0 mg	
Cholesterol 0 mg	0
Phytosterols 24.0 mg	
Total carbohydrates 3.88 g	1
Dietary fiber 2.1 g	8
Sugars 1.88 g	
Protein 2.2 g	4
Vitamins	
Vitamin A	15
Vitamin E	6
Vitamin K	52
Vitamin C	9
Vitamin B-6	5
Folic acid	13
Minerals	
Sodium	0
Calcium	2
Magnesium	3
Copper	9
Potassium	6
Iron	12
Manganese	8
Zinc	4

National Nutrient Database
[a]Based on a 2000-calorie diet

risk of fractures and is crucial for proper blood coagulation [15]. It also contains high levels of the amino acid asparagine which acts as a natural diuretic, thereby helping control mild hypertension [16] and some urinary tract conditions [17], and is necessary for the development and function of the brain [18]. Potassium in asparagus helps lower blood pressure resulting in healthy blood vessels and heart [19]. Asparagus contains different B vitamins including B-6, B-9, and B-12 that are vital for controlling homocysteine levels via conversion to cysteine. Homocysteine is a definite trigger of inflammation leading to blood vessel damage, vascular dysfunction, and eventually a possible risk for atherosclerosis (hardening of the arteries) and blood clots [20], while vitamin B3 (niacin), on the other hand, was found to reduce joint inflammation and associated swelling including pain [21] perhaps due to interference in neutrophil migration and inhibition of the protein kinase C pathway.

There are no known conditions associated with asparagus consumption, but some people may experience a number of uncomfortable effects, especially when consumed excessively. Excessive consumption of asparagus can cause dry mouth and sudden weight loss. As a natural diuretic, it causes fluid loss from the body and dehydration. For this reason, it might also lead to dramatic drop in blood pressure when used with antihypertensive agents and could enhance the effect of diuretics too. On the other hand, the US NIH recommends that those who suffer from uric acid kidney stones should avoid asparagus and advises caution for those with low blood sugar. Its high fiber content may negatively affect the small intestine and cause constipation and abdominal cramps. High-complex carbohydrate in asparagus is difficult to breakdown; hence, it will be fermented by bacteria triggering excessive gas formation. In some people, asparagus is associated with causing urine smell because of the presence of the chemical asparagusic acid which upon degradation causes the sulfur-containing compounds to give rise to an unpleasant scent [22]. In others, runny or blocked nose, throat irritation, skin rashes, and itching are common allergic reactions. On rare occasions, few might develop nausea, headaches, and dizziness. Fewer people may exhibit allergic conjunctivitis with itching, redness, and swelling of the eyes.

2 Beets

Family: Amaranthaceae
Genus: *Beta*
Common names: Beet, table beet, garden beet, red beet, or golden beet

Beet is the taproot portion of the beet plant. It is a seasonal, ancient, prehistoric vegetable, frequently used as a natural coloring agent. Beet is a rich source of sucrose and is often used to make refined sugar. Beets are highly nutritious root vegetables, rich in vitamins and minerals (Table 2). It is a high source of folic acid, vitamin C, vitamin B complex, manganese, iron, copper, potassium, and antioxidants.

They are also a very rich source of phytochemical compounds, e.g., anthocyanins, carotenoids (lutein and zeaxanthin), glycine, and betaine including an important class of water-soluble, red and yellow indole-derived pigments called betalains [23]. The betalain red pigment betacyanins of the beets (include betanin, isobetanin, probetanin, and neobetanin) are considered a potent antioxidant with anti-inflammatory and anticancer agents [24]. Betanin is considered a food additive used as a coloring agent with E-number E162 (see Chap. 15). Beets are low caloric, low fat and cholesterol, and a great source of dietary fiber. However, beets have the highest sugar content of all vegetables and are relatively high in carbohydrates.

A number of benefits are associated with beetroot consumption. High iron content of the red beetroots helps prevent anemia and enhances the regeneration of red blood cells [25]. Rich levels of vitamin C augment iron absorption. β-Carotene—a form of vitamin A with antioxidant properties—was found to effectively reduce or slow down macular degeneration and protect the eyes against damaging free radicals [26] and age-related cataract [27]. Raw beet greens have *carotenoids zeaxanthin* and *lutein* which were reported to protect the retina of the eye [28]. It has been suggested that phytochemicals in beetroots can protect against skin, lung, and colon cancers in multiple animal models [23]. It has been shown that the betacyanin pigments in beets counteract cancerous cell growth [29]. Red beetroot extract has been demonstrated to have synergistic cytotoxicity effects with the anticancer drug, doxorubicin, in some human cancer cell lines [30]. Beet consumption also helps prevent cardiovascular diseases in several ways; this includes the effect of fiber which can decrease the level of triglycerides and low-density lipoprotein (LDL) in the body and increase the "good" high-density lipoprotein (HDL) cholesterol [9, 10, 31]. Betaine is a water-soluble trimethylglycine amino acid. It is a potent bioactive compound originally discovered in sugar beets. Biologically, betaine serves as an osmolyte (osmotic stress protectant), a methyl donor in biochemical pathways, and a detoxification agent. Both betaine (amino acid) and betanin (pigment) enhance the level of detoxifying enzymes (glutathione peroxidase and superoxide dismutase) in the liver stimulating glutathione production (reduces hepatic toxicity) [32], inhibiting chronic inflammation [33], and decreasing homocysteine levels in the body particularly in combination with vitamin B-9 (folic acid) [34]. Elevated homocysteine levels can lead to atherosclerosis, heart attacks, and strokes [20]. Beets are rich in dietary nitrates which will be converted into nitric oxide in the body. Nitric oxide helps dilate blood vessels and lower blood pressure [35]. Nitrate and carbohydrate-rich constituents in beet juice and whole beets provide energy and so enhance athletic performance. It has been found to increase runners' oxygen uptake [36]. Due to nitrate effect on oxygenation, beetroot juice could also improve brain neuroplasticity and prevent cognition deterioration in early dementia [37]. Beets' content of fiber in addition to magnesium and potassium helps the body to get rid of excess water, preventing bloating and aiding weight loss. Beet is a rich source of folic acid that helps prevent neural tube defects [14, 38].

Like any other vegetable, if taken excessively, beetroots and their juice can lead to some health problems. Beetroot-rich oxalate content can increase the risk of kidney stone formation [39] and gouty arthritis. Drinking beet juice may lead to the

Table 2 Beet nutrition facts of 100 g, from the United States Department of Agriculture (USDA SR-21)

Serving size of 100 g of raw beetroots	
Per serving	% Daily value[a]
Calories 43.0	
Total fat 0.2 g	0
Saturated fat 0.0 g	0
Polyunsaturated fat 0.1 g	
Monounsaturated fat 0.0 g	
Total omega-3 fatty acids 5.0 mg	
Total omega-3 fatty acids 55.0 mg	
Cholesterol 0.0 mg	0
Phytosterols 25.00 mg	
Total carbohydrates 9.6 g	3
Dietary fiber 2.8 g	11
Starch 0.0 g	
Sugars 6.8 g	
Protein 1.6 g	3
Vitamins	
Vitamin A	1
Vitamin E	0
Vitamin K	0
Vitamin C	8
Vitamin B-6	3
Folic acid	27
Minerals	
Sodium	3
Calcium	2
Magnesium	6
Copper	4
Potassium	9
Iron	4
Manganese	16
Zinc	2

National Nutrient Database
[a]Based on a 2000-calorie diet

accumulation of metals like copper, magnesium, phosphorous, and iron in liver and pancreas tissues. Nitrates in the beetroot juice may cause sudden drop in the blood pressure [40]. Nitrates are reported to be a cause of sudden gastrointestinal symptoms associated with raw beet consumption [41]. Abdominal pain, skin rashes, hives, and fever can be signs and symptoms of allergy to beets. High glycemic index beets might increase the blood sugar. Beets contain betanin pigment which can change the color of the stool as well as the color of the urine in a harmless condition known as beeturia.

3 Butternut Squash

Family: Cucurbitaceae
Genus: *Cucurbita*
Common names: Butternut squash, butternut pumpkin, or gramma

Butternut squash is the most common type of winter squash fruit though used as a vegetable. It is a low caloric and has a sweet, nutty taste similar to pumpkin. It has a yellow, hard inedible skin and orange firm bulb with the seed part in the lowest portion. Ripe squash has a deep orange color and usually has a sweeter taste. It is a good source of fiber, vitamin C, manganese, magnesium, and potassium. Butternut squash is a very rich source of vitamin A and also is a good source of vitamins E, B-1 (thiamin), B-3 (niacin), B-5 (pantothenic acid), B-6 (pyridoxine), and B-9 (folic acid) (Table 3).

The rich amount of vitamin A in butternut squash makes it a potent antioxidant fruit used to treat oxidative stress [42]. Vitamin A was also observed to improve the immune response and decrease inflammation [43]. It has been revealed that it can decrease the risk of asthma in children too [44, 45]. Zeaxanthin and lutein are powerful antioxidants known to preserve eye health and decrease the risk of cataract [46]. Vitamin A in butternut squash seeds was discovered to inhibit the growth of melanoma (skin cancer) [47]. As a consequence, it is believed that high content of vitamin A could cause butternut squash to serve as a good treatment for other types of cancers like lung cancer [48], ovarian cancer [49], and colon cancer [50], in addition to its potentiation effect on some chemotherapeutic agents. Both vitamins A and C support healthy skin and hair maintenance [51, 52]. Butternut squash contains manganese and other microelements that act as cofactors for antioxidant enzymes required for optimum catalytic activity, and it also boosts the different antioxidant enzyme reactions [53]. Manganese and potassium contribute to the role of butternut squash in decreasing the risk of osteoporosis and having strong bones, specifically in postmenopausal women and old men [54]. Butternut squash is thought to have anti-obesity potential [55]. Both manganese and potassium have been suggested not only to reduce weight but also to reduce premenstrual symptoms [56] and muscle cramps in general. High potassium content helps lowering the blood pressure and prevents heart diseases and stroke as well [19]. In animal models, butternut squash was discovered to decrease fatigue and increase physical

Table 3 Butternut winter squash nutrition facts of 100 g, from the United States Department of Agriculture (USDA SR-21)

Serving size of 100 g of raw butternut winter squash	
Per serving	% Daily value[a]
Calories 45	
Total fat 0.1 g	0
Saturated fat 0.0 g	0
Polyunsaturated fat 0.0 g	
Monounsaturated fat 0 g	
Total omega-3 fatty acids 26 mg	
Total omega-6 fatty acids 16 mg	
Cholesterol 0 mg	0
Phytosterols	
Total carbohydrates 12 g	4
Dietary fiber 2 g	8
Sugars 2.2 g	
Protein 1 g	2
Vitamins	
Vitamin A	213
Vitamin E	7
Vitamin K	1
Vitamin C	35
Vitamin B-6	8
Folic acid	7
Minerals	
Sodium	0
Calcium	5
Magnesium	8
Copper	4
Potassium	10
Iron	4
Manganese	10
Zinc	1

National Nutrient Database
[a]Based on a 2000-calorie diet

activity tolerance [57], whereas its high vitamin C content helps improve the physical performance. The high fiber content of this vegetable decreases inflammation and prevents constipation to sustain a healthy digestive system [9, 10] and helps cultivate a healthy cardiovascular system [9, 10, 31].

Side effects of eating butternut squash are uncommon, but like any other food item, butternut squash may be associated with minimal allergic reaction, mainly in the form of contact dermatitis and swelling of the hands and around the mouth. As a protective measure, peeling the butternut squash before it becomes fully ripe can cause dryness of the skin on the hand, though this is not an allergic reaction.

4 Carrots

Family: Apiaceae
Genus: *Daucus*
Common name: Carrot

Carrots are a root vegetable. Originally purple, black, red, white, or yellow in color, they used to be bitter with a woody core. In the seventeenth century, orange-colored, sweeter carrots were developed in the Netherlands. The taproot is the most commonly eaten part, although the stems and leaves are also edible. The roots are very rich source of antioxidants, vitamins A, C, and K and B vitamins like B-5 (pantothenic acid), B-6 (pyridoxine), and B-9 (folic acid), and the minerals potassium, iron, copper, and manganese (Table 4).

The health benefits of carrots are numerous. Most of the benefits of carrots can be attributed to their β-carotene and fiber content. This includes its positive effect on the heart and the rest of the cardiovascular system. Carrot consumption reduces the total cholesterol. It has been noticed that people consuming more carrots are at a lower risk of developing heart attacks compared to people who consume less carrots [58]. The potassium content of carrots helps reduce arterial blood pressure, prevents atherosclerosis in normotensive animals [59], and reduces stroke risk by 68% [60]. High fiber content prevents constipation and keeps a healthy digestive system, and it also eliminates cholesterol and LDL from the body keeping the heart and the cardiovascular system healthy [9, 10, 31]. It is generally accepted that carotenoids and their metabolites reduce the risk of developing several chronic diseases, such as type 2 diabetes, atherosclerosis, and cancer since they modulate inflammatory and oxidative stress pathways [61]. Carrot's high level of β-carotene is linked to risk reduction of different cancers [62] like lung cancer [48] and breast cancer [63]. In addition, vitamin A, its derivatives, and other retinoids affect cell differentiation, proliferation, and apoptosis, play an important physiologic role in a wide range of biological processes [64], and are regarded as a promising chemotherapeutic or chemopreventive agents [65, 66]. The FDA has approved synthetic retinoid for dermatological purposes with demonstrated antitumor activity. For example, bexarotene has been approved for lymphoma treatment, while tazarotene clinically showed good efficacy in carcinoma therapy [64]. Fiber content of carrots was found to reduce the risk of colon and breast cancers [9, 67, 68]. Vitamin C content in carrots boosts the immune system [69], and the high level of β-carotene helps reduce the

Table 4 Carrot nutrition facts of 100 g, from the United States Department of Agriculture (USDA SR-21)

Serving size of 100 g of raw carrots	
Per serving	% Daily value[a]
Calories 41	
Total fat 0.2 g	0
Saturated fat 0.0 g	0
Polyunsaturated fat 0.1 g	
Monounsaturated fat 0 g	
Total omega-3 fatty acids 2.0 mg	
Total omega-3 fatty acids 115.0 mg	
Cholesterol 0 mg	0
Phytosterols	
Total carbohydrates 10 g	3
Dietary fiber 3 g	11
Sugars 5 g	
Protein 0.9 g	2
Vitamins	
Vitamin A	334
Vitamin E	3
Vitamin K	16
Vitamin C	10
Vitamin B-6	7
Folic acid	5
Minerals	
Sodium	3
Calcium	3
Magnesium	3
Copper	2
Potassium	9
Iron	2
Manganese	7
Zinc	2

National Nutrient Database
[a]Based on a 2000-calorie diet

risk of macular degeneration [70]. The high levels of vitamin A improve eye sight and reduce risk of night blindness and glaucoma [71]. Carotenoids in carrots regulate insulin, glucose metabolism, and blood glucose level [72].

Overconsumption of carrot is generally not harmful to human health; it may only cause carotenemia and color the skin orange yellow. A limited number of people are allergic to carrots. It is not common to develop overdose of vitamin A only from eating carrots, but people taking vitamin A-derived medications or vitamin A supplements should avoid consuming large quantities of carrots to prevent toxic hypervitaminosis A.

5 Celery

Family: Apiaceae
Genus: *Apium*
Common name: Celery

Celery is a green vegetable with a long fibrous stalk tapering into leaves. All parts of celery including the seeds, roots, and leaves are edible. Celery is an excellent source of antioxidants and beneficial enzymes. Rich in vitamins K, C, and B-6 (pyridoxine), it is a very good source of dietary fiber, folic acid (B-9), potassium, manganese, and pantothenic acid (B-5). Celery is also high in sodium, copper, calcium, phosphorus, magnesium, vitamin A, and riboflavin (B-2) (Table 5).

There are a lot of health benefits associated with consuming celery. Celery possesses a unique bioactive chemical compound called phthalides [73]. The phthalide core chemical structure is a lactone. Phthalides are recognized for their distinctive medicinal properties. They have been used as a herbal remedy to treat several conditions such as cancer and inflammation [74]. Phthalides were found to reduce the levels of chemical messengers known as stress hormones. They act as smooth muscle relaxants most likely through influencing calcium and potassium flow inside the cells causing blood vessel expansion and as a result lowering blood pressure. Several studies have investigated the effect of celery seed extract on blood pressure and blood lipid profiles. The studies concluded that the extract has clinically relevant blood pressure-lowering effects in mild-to-moderate hypertensive patients along with lowering cholesterol levels [75]. In animal studies, they led to a significant reduction in serum total cholesterol, triglyceride, and low-density lipoprotein (LDL) cholesterol levels [76]. In due course, this manifests as reduction in the risk of atherosclerosis and coronary heart diseases. Correspondingly, high potassium levels in celery, a vasodilator, decrease blood pressure and reduce the risk of developing atherosclerosis to avert heart attacks and strokes [77]. Indeed, vitamin C presence also enhances this protection. In addition, celery's high fiber content plays a role in improving intestinal function and boosting cardiovascular health [9, 10]. With dietary fiber, blood glucose levels tended to decrease along with LDL and total cholesterol [78]. The high content of potassium and sodium in this vegetable helps regulate body fluid balance [77]. The anti-inflammatory and diuretic effect of celery seeds and celery sticks help eliminate uric acid, thereby reducing rheumatism and gouty arthritis. The combined effect of diuretic action with its antibacterial activity prevents urinary tract infection [77]. Celery juice increases urination and, as a result,

Table 5 Celery nutrition facts of 100 g, from the United States Department of Agriculture (USDA SR-21)

Serving size of 100 g of raw celery	
Per serving	% Daily value[a]
Calories 16	
Total fat 0.17 g	0
Saturated fat 0.0 g	0
Polyunsaturated fat 0.0 g	
Monounsaturated fat 0.0 g	
Total omega-3 fatty acids	
Total omega-6 fatty acids 79.0 mg	
Cholesterol 0 mg	
Phytosterols 6.0 mg	
Total carbohydrates 3.4 g	1
Dietary fiber 1.6 g	6
Sugars 1.8 g	
Protein 0.7 g	1
Vitamins	
Vitamin A	9
Vitamin E	1
Vitamin K	37
Vitamin C	5
Vitamin B-6	4
Folic acid	9
Minerals	
Sodium	3
Calcium	4
Magnesium	3
Copper	2
Potassium	7
Iron	1
Manganese	5
Zinc	1

National Nutrient Database
[a]Based on a 2000-calorie diet

aids in the flush of toxins, salt, and fat out of the body, thus improving liver health and status [79, 80]. For instance, diet supplementation with celery is effective in decreasing serum level of liver enzymes aspartate aminotransferase, alanine aminotransferase, and alkaline phosphatase. Other phyto-substances in celery may improve liver enzyme function and detoxification, control gene expression, limit DNA damage, and facilitate DNA repair [81]. Celery is a rich source of varied antioxidants [82] such as phenolic acids, flavones (apigenin and luteolin), flavonols (quercetin and kaempferol), phytosterols, dihydrostilbenoids, and furanocoumarins. Celery flavonoids (flavones and flavonols) have reportedly anti-inflammatory activity which have also been shown in human studies [83, 84]. Flavonoids display several anti-inflammatory mechanisms. They act as preferential inhibitors of COX-2, and it is thought that they modulate inflammation by reducing TNF-α and nuclear factor-kappa B (NF-κB), decreasing levels of the pro-inflammatory cytokines interleukin 1B (IL-1β) and interleukin 8 (IL-8) [85] and/or microRNA (miR) expression [86]. Antioxidants in celery such as phthalides, flavonoids, and polyacetylenes activate certain white blood cells and coumarins that enhance their activity; in other words, they effectively fend off certain types of cancer [83, 87, 88]. In addition, the antioxidant vitamin C stimulates the immune system [89] and has anti-inflammatory properties along with other antioxidants which enable celery to prevent conditions associated with severe inflammation such as asthma [90]. Celery is rich with the flavones luteolin and apigenin [91], which are linked to anticancer properties [92, 93]. Luteolin causes induction of apoptosis and inhibition of cell proliferation, metastasis, and angiogenesis. Furthermore, luteolin sensitizes cancer cells to therapeutic-induced cytotoxicity through suppression of cell survival pathways such as phosphatidylinositol 3-kinase (PI3K)/Akt, NF-κB, and X-linked inhibitor of apoptosis protein (XIAP) and stimulation of apoptosis pathways including those that induce the tumor suppressor p53. Apigenin has the ability to promote cell cycle arrest and induce apoptosis through the p53-related pathway. These flavones have been found to reduce tumor size and mitigate metastasis to other organs [94]. Phenolic acids and flavonoids in celery juice were found to reduce age spots and skin wrinkles [95]. Furocoumarins and multiple vitamins present in celery juice can treat or provide relief from a number of skin conditions such as psoriasis, eczema, acne, and rosacea [96]. Coumarins were thought to alleviate migraine headaches mostly via nitric oxide release suppression in the brain [97].

Celery seeds and oil are generally safe. Overconsumption of celery is associated with some undesirable effects. High fiber content of celery usually leads to abdominal pain, bloating, and diarrhea. Some natural products found in celery can prevent the body's ability to use iodine appropriately. Excessive consumption of mainly raw celery for a long time may lead to iodine deficiency and goiter. Celery contains multiple forms of furanocoumarins that may lead to increased light sensitivity after eating or even handling. Some essential oils found in celery may cause skin irritation. Celery is rich in natural anticoagulants and can enhance the effect of anticoagulant medications. Overconsumption of celery and celery seeds may induce bleeding and uterine contraction, so it is advisable to avoid during pregnancy. The natural diuretics in celery may increase urination leading to loss of minerals such as potassium, calcium, and magnesium. Allergic reactions are very common with celery. They range from rashes, itching, and stomach upset to severe life-threatening anaphylactic shock. Celery oil contains sedative properties; even a moderate consumption of celery may lead to sleepiness and drowsiness. Celery is one of the foods that contains the most residual pesticides. Sustained exposure to contaminated celery may lead to brain and nervous system toxicity and to skin, eye, and lung irritation as well.

6 Coriander (Cilantro)

Family: Apiaceae
Genus: *Coriandrum*
Common names: Coriander, cilantro, or Chinese parsley

Coriander is an herb that is widely used to give flavor or to garnish dishes. Its leaves and fruits have a distinguishable and pleasant scent and are usually used either raw or dried. Coriander is packed with vitamins and minerals. Its leaves are a rich source of *dietary fiber*; vitamins A, C, K, and E; and folic acid (B-9), as well as manganese, iron, and magnesium. It also contains appreciable quantities of potassium, calcium, and copper (Table 6). Coriander contains a number of essential oils that have numerous health benefits [98, 99].

Cineole is a natural monoterpene also known as eucalyptol. An essential oil, chemically, it is a monoterpenoid cyclic ether. It is also found in other herbs such as common sage, sweet basil, and rosemary. It is a major constituent in coriander that exhibits multiple therapeutic properties. Taxol is a very famous terpenoid. It is an approved anticancer drug and a chemotherapeutic medicine used to treat a number of types of cancer. Cineole is reported to have anti-inflammatory, antimicrobial, and antioxidant activities where several clinical trials have established potent anti-inflammatory mode of action [100]. In particular, cineole possesses antirheumatic and anti-arthritic properties and inhibits the pro-inflammatory cytokines and TNF-α [101]. It offers relief through its analgesic effect and reduces swelling associated with these conditions [102]. Cineole was also shown to be effective against a range of respiratory conditions, including chronic obstructive pulmonary disease (COPD), asthma, bronchitis, sinusitis, common cold, cough, or flu [103, 104]. Coriander's essential oils have also been demonstrated to be beneficial in reducing skin inflammation and improve its appearance [105]. They stimulate the immune response and have an immunomodulatory effect [106, 107]. For instance, cineole antihistamine properties help treat allergic seasonal rhinitis and urticaria and reduce other allergic reactions due to contact with plant, food, and insects [108–111]. Coriander essential oils boast of wide-ranging antiseptic, disinfectant, insecticidal biological activities. Numerous studies have shown that cineole has a broad spectrum of antimicrobial action against viruses [112], bacteria [113], and fungi [114]. Few examples are cited

Table 6 Coriander nutrition facts of 100 g, from the United States Department of Agriculture (USDA SR-21)

Serving size of 100 g of raw coriander leaves	
Per serving	% Daily value[a]
Calories 23	
Total fat 0.5 g	1
Saturated fat 0.0 g	0
Polyunsaturated fat 0.0 g	
Monounsaturated fat 0.3 g	
Total omega-3 fatty acids	
Total omega-6 fatty acids 40.0 mg	
Cholesterol 0.0 mg	0
Phytosterols 5.0 mg	
Total carbohydrates 3.7 g	1
Dietary fiber 2.8 g	11
Starch 0.0 g	
Sugars 0.9 g	
Protein 2.1 g	4
Vitamins	
Vitamin A	135
Vitamin E	13
Vitamin K	388
Vitamin C	45
Vitamin B-6	7
Folic acid	16
Minerals	
Sodium	2
Calcium	7
Magnesium	6
Copper	11
Potassium	15
Iron	10
Manganese	21
Zinc	3

National Nutrient Database
[a]Based on a 2000-calorie diet

below proving its utility in various infections. The antibacterial activity against dermatological skin pathogen is well studied. However, it is also a great cure for skin conditions like eczema or irritations such as acne, boils, cysts, wounds, cuts, burns, sores, and even insect bites [115]. It displayed inhibitory effects against respiratory bacteria and viruses [116]. Finally, it showed a considerable inhibitory effect on *Candida albicans* [117] improving candidiasis treatment in normal and diabetic rats [118] and working as an antifungal topical treatment for infected toenails (onychomycosis) [119]. Additionally, as a home remedy, it demonstrated efficacy against pink eye (conjunctivitis), healing the eye from infection. Cineole coriander contains a number of other essential oil components such as citronellol, a good mosquito repellent that also possess antimicrobial and healing effects. The essential oils prevent mouth wound complications, enhance mouth ulcer healing, and thwart bad breath [120, 121]. Other essential oil components like borneol, limonene, alpha-pinene, and beta-phellandrene have antibacterial effects as well, thereby promoting digestive system health since eating coriander can help indigestion (dyspepsia) [122]. They aid in treating diarrhea that are fungal (*Candida* spp.) or microbial in origin (*Salmonella*, *Shigella*, and enterotoxigenic *E. coli*) [123]. They also prevent nausea, vomiting, and stomach aches [99]. Coriander contains dodecanal, a very powerful natural bactericidal compound that protects one from *Salmonella*-based illnesses [124]. As a leafy green, coriander leaves have a high content of calcium, other essential minerals, and vitamin K (promotes calcium absorption and helps support blood coagulation processes as well as cardiovascular health), which are important components for bone regrowth, durability and limiting demineralization, the principal cause of osteoporosis [125–127]. High iron levels in coriander preclude iron deficiency and anemia [99]. Low iron content in the blood can result in shortness of breath, heart palpitations, extreme fatigue, and a decrease in cognitive functions. Minerals such as calcium and potassium help reduce the blood pressure and aid in keeping a healthy cardiovascular system [122]. Findings were reported regarding the role of cineole in acting as a calcium channel blocker [128]. Furthermore, coriander and its seeds contain vitamin C and other acids, e.g., linoleic acid, oleic acid, palmitic acid, and stearic acid, which are effective in reducing the bad LDL cholesterol levels and increasing the healthy HDL cholesterol levels in the blood offering protection against cardiovascular complications and stroke [129]. Coriander possesses antioxidant and detoxifying properties [130]. The antioxidant feature of coriander is largely attributed to the presence of polyphenols, particularly flavonoids [131] that are well known for their numerous health benefits and protection against oxidative stress [132]. The antioxidants vitamins A and C in coriander help protect the eyes, reduce vision disorders, and lower the risk of macular degeneration. In animal studies, coriander helps stimulate insulin secretion and reduces glucose level in the blood [133]. Coriander promotes neuroprotection [134] and stimulates the memory [135].

On a more precautionary note, cineole is toxic if ingested at higher than normal doses. Like any other food, consuming coriander may cause allergic skin reaction in susceptible people. Overconsumption of coriander leaves may increase light sensitivity potentially leading to sunburns and subsequently to skin cancer [136]. Coriander may lower blood glucose and blood pressure and therefore has to be consumed with caution by people suffering from diabetes or taking antihypertensive agents and before any surgical procedures. Overconsumption of coriander may lead to severe diarrhea, stomach pain, and dehydration.

7 Cruciferous

Family: Brassicaceae
Genus: *Brassica* (broccoli, cauliflower, and cabbage)

7.1 Broccoli

Broccoli is an edible vegetable in the cabbage family whose large green flowering head is eaten raw or cooked. Boiling broccoli leads to the loss of some of its antineoplastic substances, while steaming, microwaving, or stir-frying for few minutes preserves such nutrients to a great extent. It is considered one of the healthiest vegetables for its low calories, low fat, and rich content of vitamins such as vitamins C and K and other antioxidants. In addition, broccoli is a good source of dietary fiber, folic acid, vitamin A, and minerals like potassium and calcium (Table 7).

Broccoli's dietary content of sulforaphane (SFN), a sulfur-containing compound belonging to the glucosinolate family, is associated with lower risk of cancer [137]. This dietary component of broccoli and broccoli sprout preparations has been shown to have an effect on different types of cancers [138]. For instance, in vitro, it inhibits breast cancer stem cells [139] and bladder cancer [140]. Sulforaphane has been studied in multiple clinical trials to test its ability to delay or slow the growth of cancer cells. It has been shown to be effective against pancreatic cancer [141] as well as recurrent prostate cancer [142]. Broccoli contains folic acid which has been linked to decreased risk of breast cancer [143], colorectal cancer [144], and prostate cancer [145]. The in vitro and in vivo studies of brassica-derived phytochemicals including sulforaphane and isothiocyanates suggest chemopreventive activity through redox-sensitive transcription factor Nrf2. The effect of SFN on Nrf2 pathways has been intensively investigated. It appears that SFN induces Nrf2 [146, 147]. There is evidence too that overexpression of Nrf2 can modulate NF-κB expression as well [148]. Vitamin K in broccoli reduces the risk of osteoporosis and bone fractures by improving calcium absorption and reducing calcium excreted from the body [15, 126], in addition to vitamin K's well-documented important role in blood coagulation and cardiovascular health [127]. Broccoli has different vitamins like folic acid and vitamins A and E that exhibit antiaging effects and are essential in

Table 7 Broccoli nutrition facts of 100 g, from the United States Department of Agriculture (USDA SR-21)

Serving size of 100 g of raw broccoli	
Per serving	% Daily value[a]
Calories 34.0	
Total fat 0.4 g	1
Saturated fat 0.0 g	0
Polyunsaturated fat 0.0 g	
Monounsaturated fat 0.0 g	
Total omega-3 fatty acids 21.0 mg	
Total omega-6 fatty acids 17.0 mg	
Cholesterol 0.0 mg	0
Phytosterols	
Total carbohydrates 6.6 g	2
Dietary fiber 2.6 g	10
Starch 0.0 g	
Sugars 1.7 g	
Protein 2.8 g	6
Vitamins	
Vitamin A	12
Vitamin E	4
Vitamin K	127
Vitamin C	149
Vitamin B-6	9
Folic acid	16
Minerals	
Sodium	1
Calcium	5
Magnesium	5
Copper	2
Potassium	9
Iron	4
Manganese	10
Zinc	3

National Nutrient Database
[a]Based on a 2000-calorie diet

maintaining healthy skin. Moreover, the antioxidant vitamin C contributes to collagen formation and alleviates the impacts of pollution and ultraviolet sunlight on the skin [51, 52]. Dietary fiber in broccoli enhances digestive system health, prevents constipation, and reduces the risk of colorectal cancer. It is thought that fiber consumption is associated with lower risk of developing chronic diseases [9, 11] and helps to detoxify the body by excreting toxins through the digestive system. Phytochemicals and natural isothiocyanates (a glucosinolate major hydrolysis by-product) from broccoli help combat inflammation and detoxify the body at the epigenetic level. It has been shown that isothiocyanates may inhibit histone deacetylase transferases and DNA methyltransferases in cultured cells [147]. In this context, NF-κB is a central player in inflammatory processes. A variety of naturally occurring NF-κB inhibitors have been described including brassica-derived phytochemicals like sulforaphane [147]. Fresh broccoli juice mixed with other fresh vegetable and fruit juices was found to reduce the total cholesterol and therefore the risk of stroke and coronary heart diseases [149]. Kaempferol, a small antioxidant flavonol molecule, was shown to improve insulin sensitivity to avert diabetes [150]. In animal studies, both fiber and potassium in broccoli help control the blood pressure [151] and indirectly encourage weight loss in obese people. Inflammation has been linked to many modern chronic diseases including obesity, cancer, and atherosclerosis [152–154]. Omega-3 fatty acids are powerful anti-inflammatory substances that reduce the production of immunomodulatory molecules such as cytokines [155, 156]. Thus, higher omega-3 intake has been consistently shown to reduce inflammation [157–159]. It also has been shown to reduce the gastric inflammation caused by *H. pylori* [160]. Docosahexaenoic (DHA) acid is an omega-3 fatty type that is a major structural component of the brain and retina of the eye. Omega-3 has been linked to a reduced risk of macular degeneration [161, 162]. Broccoli also contains vitamins A, B complex, C, and E that are all good for eye health preventing macular degeneration and cataract.

Generally, broccoli is safe to consume, and any side effects are not serious. It is common to have bowel irritation and gas formation when eating cruciferous including broccoli. It is advisable to monitor the consumption of broccoli when taking anticoagulant medications as broccoli's high content of vitamin K might interact with it.

7.2 Cauliflower

Cauliflower is another member of the same family. Normally, the cauliflower head is the only edible part. There are multiple types of cauliflowers, e.g., white, which is the most common. Similarly, nutritious yellow, purple, and green types are also available. Cauliflower is a low-caloric vegetable with a lot of nutrients. It is a rich source of vitamins C, K, and B complex and folic acid. It also contains minerals such as potassium, manganese, and calcium (Table 8).

There are many health benefits associated with cauliflower consumption. First, it is quite high in fiber. Previous sections delineated the importance of dietary fiber in health, disease, and digestive conditions [9, 163, 164]. Cauliflower features as a good source of fiber, being low in carbohydrates and having high-water content, and generally, low calories permit body weight management [165, 166] and aid in the excretion of toxins from the body with the help of its high content of antioxidants and glucosinolate such as sulforaphane, glucoraphanin, glucobrassicin, and gluconasturtiin [130, 167]. Second, cauliflower is extremely rich in antioxidants and anti-inflammatory compounds that control the oxidative stress and damage caused by the

Table 8 Cauliflower nutrition facts of 100 g, from the United States Department of Agriculture (USDA SR-21)

Serving size of 100 g of raw cauliflower	
Per serving	% Daily value[a]
Calories 25.0	
Total fat 0.1 g	0
Saturated fat 0.0 g	0
Polyunsaturated fat 0.0 g	
Monounsaturated fat 0.0 g	
Total omega-3 fatty acids 37.0 mg	
Total omega-6 fatty acids 11.0 mg	
Cholesterol 0.0 mg	0
Phytosterols 18.0 mg	
Total carbohydrates 5.3 g	2
Dietary fiber 2.5 g	10
Starch	
Sugars 2.4 g	
Protein 2.0 g	4
Vitamins	
Vitamin A	0
Vitamin E	0
Vitamin K	20
Vitamin C	77
Vitamin B-6	11
Folic acid	14
Minerals	
Sodium	1
Calcium	2
Magnesium	4
Copper	2
Potassium	9
Iron	2
Manganese	8
Zinc	2

National Nutrient Database
[a]Based on a 2000-calorie diet

free radicals [132, 168–170]. Several of these carotenoid and flavonoid phytochemicals also possess anticarcinogenic properties. Research results indicate that there is a positive correlation between cancer prevention and cruciferous vegetable consumption [171, 172]. Examples of antioxidants available in cauliflower include beta-cryptoxanthin, quercetin, rutin, kaempferol, cinnamic acid, caffeic acid, and ferulic acid, in addition to vitamin C, which helps fight inflammation [173, 174]. Sulforaphane in cauliflower has been shown to protect the body from cancer and chronic inflammation-related diseases. It has been shown that combining cauliflower with curcumin, the active compound in turmeric spice, may help control prostate cancer cell growth [175]. Sulforaphane is also found to reduce the risk of chemically induced breast cancer in animals [176] and cause cell death of certain human breast cancer cell lines [177]. Isothiocyanates, another chemical group found naturally in cruciferous vegetables, exhibit antioxidant properties [178]. Several reports provide evidence for its protective effects against cancer where they were found to inhibit many cancers such as multiple gastrointestinal, bladder, breast, and lung cancer [179–181]. The oxidative stress effect of sulforaphane can ensure good eye health as it prevents macular degeneration and cataract and reduces the risk of loss of vision [182]. Sulforaphane also improves kidney functions and controls blood pressure [183]. Antioxidants' role in cardiovascular health and reduced risk of heart diseases is well supported by numerous findings [168, 184, 185]. In addition, the cardiovascular benefits of potassium, vitamin K, and omega-3 fatty acids are well supported by research as they help to contribute to blood pressure regulation, total cholesterol reduction, and blood coagulation [127, 186–192] and ensure overall bone health preventing osteoporosis [126, 193]. Choline is a water-soluble, B vitamin-like lipotropic substance [194, 195]. It is an essential nutrient involved in lipid metabolism. It is a precursor to the neurotransmitter acetylcholine and phospholipids that are indispensable to cell membrane integrity and cell signaling. It also plays a role in lipid transport as well as methyl-group transfer [196]. Choline is a neuroprotective agent [197] shown to improve memory, cognitive functions, and brain aging [197–199]. The nutrient is found in cauliflower, and it decreases the signs of dementia [196, 197]. Choline intake during pregnancy has been found to boost the cognitive function and memory of animals [200], and it is thought to continue its effect till later in life. In addition to these components, cauliflower is a good source of vitamin B complex which is known to enhance brain development [201].

Like other high-fiber-contained foods, overconsumption of cauliflower may cause bloating and accumulation of gas. However, it can mostly be tolerated when consumed in moderate amounts. High content of vitamin K in cauliflower may interact with blood anticoagulants and lead to bleeding. Therefore, it should be consumed with caution.

7.3 *Cabbage*

Cabbage is a leafy vegetable grown for its dense-leaved heads. It has multiple varieties of different shapes and colors, including red, purple, white (light green), and green. Its leaves can be crinkled or smoothed. The most common type is the green cabbage that has smooth leaves with a firm head. Cabbage can be eaten raw, cooked, or pickled. It is an excellent source of vitamins K and C and dietary fiber. It also contains folic acid and potassium (Table 9).

Cabbage and other cruciferous naturally contain glucosinolates, which impart the characteristic bitter flavor to the vegetables. Glucosinolates are a secondary metabolite derived from glucose and either aliphatic or aromatic amino acids.

Table 9 Cabbage nutrition facts of 100 g, from the United States Department of Agriculture (USDA SR-21)

Serving size of 100 g of raw cabbage	
Per serving	% Daily value[a]
Calories 25.0	
Total fat 0.1 g	0
Saturated fat 0.0 g	0
Polyunsaturated fat 0.0 g	
Monounsaturated fat 0.0 g	
Total omega-3 fatty acids	
Total omega-6 fatty acids 17.0 mg	
Cholesterol 0.0 mg	0
Phytosterols 11.0 mg	
Total carbohydrates 5.8 g	2
Dietary fiber 2.5 g	10
Starch 0.0 g	
Sugars 3.2 g	
Protein 1.3 g	3
Vitamins	
Vitamin A	2
Vitamin E	1
Vitamin K	95
Vitamin C	61
Vitamin B-6	6
Folic acid	11
Minerals	
Sodium	1
Calcium	4
Magnesium	3
Copper	1
Potassium	5
Iron	3
Manganese	8
Zinc	1

National Nutrient Database
[a]Based on a 2000-calorie diet

Indoles and isothiocyanates are the hydrolysis by-products' active substances [202]. Glucosinolates have been studied in vivo for their potential to affect human health, in particular their anticancer properties [203]. As a rich source for glucosinolates, high intake of cruciferous is associated with lower risk of several cancers in humans such as lung and colorectal cancers considering the evidence for their anti-oxidative stress/inflammation properties [181]. As anticancer agents, their chemopreventative role is linked to the induction of cellular defense detoxifying/antioxidant enzymes and their epigenetic mechanisms [203]. One breakdown product of the glucosinolate glucobrassicin is a small molecule known as 3,3′-diindolylmethane (DIM). This compound possesses anticarcinogenic qualities against breast or prostate cancer in humans and human papilloma virus infection. The antitumor activity is thought to be due to its action as a histone deacetylase inhibitor [204]. This chemical has also been shown to reduce the toxic effect of radiotherapy without causing DNA damage in normal tissue cells [205]. Another group of predominant phytochemicals is anthocyanins. They belong to the flavonoid phenolic class. The primary compounds in this parent class include the anthocyanins (e.g., cyanidin, pelargonidin, delphinidin, malvidin, peonidin, petunidin), flavonols (quercetin, kaempferol), flavones (luteolin, apigenin), flavanones (myricetin, naringin, hesperetin, naringenin), flavan-3-ols (catechin, epicatechin, gallocatechin), and isoflavones (genistein, daidzein) [206]. Anthocyanins are water-soluble, glycosylated pigments with wide-ranging colors dependent on the pH environment (see Chap. 15). They are found mostly in flowers, fruits, and vegetables; for instance, they are present in red cabbage and give it its color [207]. Anthocyanins and anthocyanidins (the sugar-free counterpart of anthocyanins) are very potent, efficient antioxidants that often interact with other phytochemicals to potentiate bioactivity. They have been investigated for their various health and therapeutic effects. The potential health benefits of anthocyanins have been summarized in a recent review article [207]. These include cardiovascular, anticancer, and antidiabetic benefits, visual health, anti-obesity improvements, neuroprotection, and antimicrobial effects. Different mechanisms and pathways are involved in enabling these protective effects, including free radical scavenging pathway, cyclooxygenase pathway, mitogen-activated protein kinase pathway, and inflammatory cytokine signaling [207]. We provide few examples to illustrate these points. Anthocyanin compounds have been found to slow the proliferation of cancer cells and prevent the formation of new ones [208]. Anthocyanins were found also to impede inflammation and therefore decrease the risk of chronic inflammation including obesity and related disorders such as cardiovascular diseases [209]. Anthocyanins have been shown to improve night vision and overall sight. It has been suggested that anthocyanins protect the eyes through different mechanisms including shielding the eyes from damage by free radicals. In a mouse model, anthocyanin-rich extract had a protective vision effect during retinal inflammation through suppression of STAT3, IL-6 expression, and NF-κB [210]. Studies revealed that an anthocyanin-rich diet enhances the cognitive function and memory in older adults with mild-to-moderate dementia [211] and augments both acute and long-term outcomes [212]. The anti-inflammatory effect of cabbage is thought to be due to other antioxidants like sulforaphane [213] and kaempferol [214]. The rich content of polyphenol in cabbage is thought to lower the risk of cardiovascular diseases by preventing platelet aggregation and reducing blood pressure [215]. Red cabbage is a very good source of potassium and anthocyanins which are known for their role in lowering blood pressure and boosting the cardiovascular system and heart health [216–218]. Cabbage and other vegetable juices also lower the total cholesterol [149], which enhances cardiovascular system health and prevents coronary heart diseases [219]. Cabbage contains phytosterols, natural compounds structurally similar to cholesterol, and prevents its absorption from the gut leading to reduction in LDL cholesterol levels [220]. Another health benefit of consuming cabbage comes from its high content of vitamin C. It is another powerful antioxidant that decreases the risk of cancer [221] and protects the body from the free radical-related damage linked to number of chronic conditions including cancer [222]. In addition, vitamin C helps absorb iron better and evade anemia as a result of iron deficiency [223]. It also helps in building collagen, thus maintaining a healthy-looking skin [52] and ensuring healthy bones, blood vessels, and muscle. Cabbage also enhances the digestive system. Its fiber content promotes regular bowel movement and prevents constipation [224]. Fiber in cabbage acts as a fuel for the normal flora in the gut [225]. This boosts the immune system and produces nutrients like vitamins K and B-12.

There is no evidence of serious side effects related to cabbage consumption though glucosinolates have been shown to have toxic effects in both humans and animals, when ingested at high doses. Like other cruciferous vegetables, overconsumption of cabbage causes bloating and gas accumulation [226]. It might also affect breast-fed babies if their mother has a lot of cabbage in her diet. Consuming a lot of raw cabbage might interfere with thyroid hormone functions [227]. It might prevent normal uptake and processing of iodine by the thyroid gland leading to iodine deficiency. High vitamin K content might affect anticoagulant medications and should be consumed with caution. Cabbage normally reduces blood glucose, and hence, overeating it might lead to hypoglycemia. It should be avoided when planning for surgical procedures [228].

8 Garlic and Onion

Family: Amaryllidaceae
Genus: *Allium*
Common names: Garlic and onion

Onions and garlic are vegetables belonging to the same family; they can be eaten either cooked or raw or just be used to add a flavor to the food. The sulfur compound (allyl propyl disulfide) is responsible for their strong odor and flavor. It delivers few health benefits when eaten, as reported by the Linus Pauling Institute.

Different types of onions exist each with a distinct flavor, such as white, red, and yellow onions. Green onions are the immature ones that have not formed the bulb yet. Onions are a rich source of allicin, an organosulfur compound, copper, and

Table 10 Onion nutrition facts of 100 g, from the United States Department of Agriculture (USDA SR-21)

Serving size of 100 g of raw onions	
Per serving	% Daily value[a]
Calories 40.0	
Total fat 0.1 g	0
Saturated fat 0.0 g	0
Polyunsaturated fat 0.0 g	
Monounsaturated fat 0.0 g	
Total omega-3 fatty acids 4.0 mg	
Total omega-6 fatty acids 13.0 mg	
Cholesterol 0.0 mg	0
Phytosterols 15.0 mg	
Total carbohydrates 9.3 g	3
Dietary fiber 1.7 g	7
Starch 0.0 g	
Sugars 4.2 g	
Protein 1.1 g	2
Vitamins	
Vitamin A	0
Vitamin E	0
Vitamin K	0
Vitamin C	12
Vitamin B-6	6
Folic acid	5
Minerals	
Sodium	0
Calcium	2
Magnesium	2
Copper	2
Potassium	4
Iron	1
Manganese	6
Zinc	1

National Nutrient Database
[a]Based on a 2000-calorie diet

selenium. It is also rich in a number of vitamins, e.g., B-6 and C (Table 10). Onions stimulate the body's production of glutathione, a potent antioxidant.

Garlic is the most common herb used around the world. The bulb of the plant is the most commonly used part. Garlic bulbs are divided into a number of sections called cloves. There are various health benefits associated with garlic consumption. It is a very rich source of manganese, selenium, and vitamins B-6 and C. In addition, garlic is a good source of some minerals, including phosphorous, calcium, potassium, iron, zinc, and copper (Table 11). Many of garlics' therapeutic effects are thought to be due to its active ingredient, allicin.

Table 11 Garlic nutrition facts of 100 g, from the United States Department of Agriculture (USDA SR-21)

Serving size of 100 g of raw garlic	
Per serving	% Daily value[a]
Calories 149	
Total fat 0.5 g	1
Saturated fat 0.1 g	0
Polyunsaturated fat 0.2 g	
Monounsaturated fat 0.0 g	
Total omega-3 fatty acids 20.0 mg	
Total omega-6 fatty acids 229.0 mg	
Cholesterol 0.0 mg	0
Phytosterols	
Total carbohydrates 33.1 g	11
Dietary fiber 2.1 g	8
Starch	
Sugars 1.0 g	
Protein 6.4 g	13
Vitamins	
Vitamin A	0
Vitamin E	0
Vitamin K	2
Vitamin C	52
Vitamin B-6	62
Folic acid	1
Minerals	
Sodium	1
Calcium	18
Magnesium	6
Copper	15
Potassium	11
Iron	9
Manganese	84
Zinc	8

National Nutrient Database
[a]Based on a 2000-calorie diet

The antimicrobial activity of both garlic and onion extracts has been investigated heavily and well documented in the literature. By way of example, efficacy studies included bacterial clinical isolates [229], other human bacterial pathogens [230], and fungi species [231]. Furthermore, it has been demonstrated that onion and garlic extracts potentiate the efficacy of conventional antibiotics against standard and clinical bacterial isolates [232].

Onion and garlic phytochemicals such as allicin, polyphenols, vitamin C, and the mineral selenium naturally stimulate immune functions or prevent excessive immune response and impede viral, bacterial, and fungal infections. A recent review article has detailed the in vitro and in vivo immunomodulator activities of *Allium*

[233] where various chemical bioactive compounds, in particular, organosulfur ones, maintain immune system homeostasis and exhibit beneficial effects on immune cells, especially through activation, regulation of proliferation, and cytokine gene expression. To illustrate, few studies have found that daily garlic supplements can combat common cold as it reduced the number of times of colds and duration of cold symptoms [234–236]. However, more studies are needed to validate these findings. Vitamin C and sulfur contain compounds that stimulate collagen production which indirectly aids the formation of skin cells and hair growth, thereby retaining healthy skin and hair [52]. The juices of these vegetables have even been used as a topical treatment for skin infections and hair loss condition [237]. The main constituents of onions and garlic are sulfur compounds particularly allicin, and its metabolites have been ascribed anti-osteoporotic activity as it can inhibit bone resorption and their pseudo-estrogen-like action also has an effect on bone density as seen in postmenopausal women [238, 239]. A review article focusing on bone health attributed other benefits pertinent to connective tissues and preservation of healthy bones (metabolism, growth, and remodeling) [240].

Garlic and onions appear to possess anticancer properties [241, 242]. Epidemiological studies indicate some associations of *Allium* vegetable consumption with decreased risk of cancer, particularly cancers of the gastrointestinal tract. A recent review article elegantly presented *Allium*-derived bioactive sulfur compounds along with strong epidemiological and intervention evidence of the protective effects of several types of digestive tract cancers [243]. These include stomach, colorectal, esophageal, and prostate cancers and cancer of the oral cavity/pharynx, larynx, kidney, breast, ovary, and endometrium. Antioxidants, especially the phytoestrogen type like isoflavones, possess antitumor activity of hormone-responsive cancers [242]. For instance, red onion's chemical-free extract which is rich in the antioxidants flavonol quercetin and anthocyanin both have anti-inflammatory effect and anti-stomach cancer and anti-colorectal cancer properties through activation mechanism of apoptosis [244]. In addition, quercetin [245] and organosulfur compounds have been found to promote cardiovascular health [246]. Garlic and onions have numerous other cardiovascular benefits, for example, lowering blood pressure by onions [247]. Garlic supplements were found to have a significant impact on reducing blood pressure [248, 249] where aged garlic extract was just as effective as the drug atenolol at reducing blood pressure over a 24-week period [250]. Another example is the decrease in total and/or LDL cholesterol levels by onions [245] and garlic [251–253]. A third example is they were also found to reduce the risk of developing blood clots. It appears that rutin (a flavonol) displays antithrombic activity as it inhibits a key enzyme called protein disulfide isomerase (PDI) involved in thrombosis [254, 255]. Rutin has been demonstrated to possess other favorable pharmacological activities [256]. Besides flavonols, other antioxidants like antioxidant vitamins A, C, and E protect against harmful UV rays as well as free radical damage fighting the skin's ageing process [51, 52]. Moreover, polyphenols in onion and garlic improve glucose homeostasis. Glycemic control is thought to be through multiple mechanisms of action in the intestine, liver, muscle adipocytes, and pancreatic β-cells, as well as through prebiotic effects in the digestive tract [257, 258].

Garlic and onion modulate oxidative stress [259, 260]. Both are rich in allicin and selenium with detoxification and activation of antioxidant liver enzyme properties [261–263]. For instance, when garlic is consumed at high doses, the sulfur compounds have been shown to ameliorate organ damage from lead toxicity [264]. Onion consumption was effective in nonalcoholic fatty liver disease (NAFLD) management [265]. Allicin, its metabolites [266–268], and polyphenols [269, 270] present in garlic and onion are well investigated for their anti-inflammatory properties. Neuroinflammation is known as a risk factor for cognitive deficits and dementia, and its incidence only increases with aging. Extracts or compounds from onion and garlic help to buffer free radical action and reduce inflammation and hence related diseases such as Alzheimer's, Parkinson's, and dementia [271]. For example, S-allyl cysteine (SAC) is the main active component of aged garlic extract with anti-inflammatory, neuroprotective, and nootropic potential [272]. Finally, onions act as a natural laxative and relieves stomach aches because of their high fructan fiber and inulin, a group of naturally occurring polysaccharides, content [164]. In short, combining garlic with onions has greater health benefits. Potentially considering their various pharmacological properties, they act synergistically as a potent antidepressant, effective painkiller, anticoagulant, and anti-inflammatory.

Consuming onions and garlic may have some undesirable effects. They might lead to heartburn and reflux esophagitis. Overconsumption of onions and garlic can cause bad breath, stomach ache, vomiting, and diarrhea. They have the potential to increase the risk of bleeding and therefore should be used with caution after surgical procedures and along with anticoagulant medications. Onions also may lower blood sugar. Overconsumption of garlic can cause liver toxicity [273]. Garlic has blood pressure-lowering properties and should be eaten with caution in case of blood pressure medication use [274]. Direct prolonged contact with garlic may cause skin irritation, rashes, and an eczema-like condition. Excessive garlic consumption may lead to hyphemia, which is a bleeding inside the eye, and may cause permanent loss of vision. Onion and garlic may also trigger migraine headache [275].

9 Leafy Greens

Family: Brassicaceae
Genus: *Brassica* (kale, mustard greens)
Genus: *Eruca* (arugula)

9.1 Arugula

Arugula, also known as salad rocket, roquette, or rucola, is a popular low-caloric, low-fat, leafy green vegetable and a type of cruciferous that contains tremendous amounts of nutrients. Arugula leaves are tender and small with strong, peppery flavor that is usually consumed raw and can also be cooked. Arugula is a very rich source of nitrate, calcium, magnesium, manganese, and vitamins such as vitamins A, K, and C and folic acid. It is also a good source of dietary fiber; protein; vitamins B-1 (thiamin), B-2 (riboflavin), B-5 (pantothenic acid), and B-6 (pyridoxine); and the minerals zinc and copper (Table 12). Consuming arugula is thought to decrease the risk of obesity, diabetes, cardiovascular diseases, and cancer and help in body weight control or weight reduction.

Arugula is rich in glucosinolates, sulfur-containing compounds which break down to biologically active substances such as indoles, and isothiocyanates that enable carcinogen clearance from the body before they initiate DNA damage [137]. One family member, sulforaphane, that gives cruciferous vegetables, including arugula, their special taste has been proposed to inhibit some enzymes associated with cancer cell progression [276]. In the aforementioned broccoli section, we have detailed sulforaphane cancer chemoprevention action and proposed mechanisms. Another major compound "erucin," metabolically and structurally related to sulforaphane, is present in large quantities in arugula and cabbage and inhibits tumors via microtubule dynamics suppression [277]. Other mechanisms have been described for isothiocyanates effectiveness and the remarkably broad anticancer-type activities [278]. Arugula also contains chlorophyll which has been shown to block the effects of mutagenic carcinogenic substances during grilling of food at a high temperature [279]. Additionally, arugula is rich in flavonols, an antioxidant flavonoid subclass. Some in vitro studies have shown that they demonstrate anticancer, anti-inflammatory, and antidiabetic characteristics [270, 280–282]. Arugula, therefore, is thought to decrease risk of several cancers, e.g., skin, esophageal, prostate, lung, and pancreatic cancers. Another antioxidant, alpha-lipoic acid, is a vitamin-like antioxidant and is sometimes referred to as the "universal antioxidant" because it is soluble in both fat and water. It was shown to decrease blood glucose, increase insulin sensitivity, and decrease peripheral nerve damage secondary to diabetes [283–286]. In Germany, alpha-lipoic acid is approved as a drug for the treatment of diabetic neuropathy since 1966 and is available by prescription [287]. Arugula is a low-caloric, high-nutrient vegetable (it contains fiber, antioxidants, vitamin K, and the minerals potassium, magnesium, and calcium). As a cruciferous vegetable, their intake is linked with better blood pressure, improved circulation, and a lower risk for heart disease and overall mortality [184, 288]. Vitamin K and calcium are key players in developing a strong skeletal system, contributing to improvement in bone health [126] and osteoporosis prevention [15, 193]. Arugula's content of folic acid is vital during pregnancy to prevent neural tube defects [14, 289] and is essential for amino acid metabolism [290, 291]. Disturbed homocysteine and folate metabolism is implicated in many different heart diseases [292–294].

Table 12 Arugula nutrition facts of 100 g, from the United States Department of Agriculture (USDA SR-21)

Serving size of 100 g of raw arugula	
Per serving	% Daily value[a]
Calories 25.0	
Total fat 0.7 g	1
Saturated fat 0.1 g	0
Polyunsaturated fat 0.3 g	
Monounsaturated fat 0.0 g	
Total omega-3 fatty acids 170.0 mg	
Total omega-6 fatty acids 130.0 mg	
Cholesterol 0.0 mg	0
Phytosterols	
Total carbohydrates 3.7 g	1
Dietary fiber 1.6 g	6
Starch 0.0 g	
Sugars 2.1 g	
Protein 2.6 g	5
Vitamins	
Vitamin A	47
Vitamin E	2
Vitamin K	136
Vitamin C	25
Vitamin B-6	4
Folic acid	24
Minerals	
Sodium	1
Calcium	16
Magnesium	12
Copper	4
Potassium	11
Iron	8
Manganese	16
Zinc	3

National Nutrient Database

[a]Based on a 2000-calorie diet

Arugula is a top source of nitrate which has been suggested to increase exercise tolerance during long-term strenuous exercise. Nitrate is converted in the body to nitrite and stored and circulated in the blood. In conditions of low oxygen availability, nitrite can be converted into nitric oxide, which is known to play a number of important roles in vascular and metabolic control. Increase in plasma nitrite concentration reduces resting blood pressure [295]. Arugula is a source of number of nutrients that support eye health. These include vitamin A and two other carotenoids like beta-carotene lutein and zeaxanthin that are known to protect the retina, cornea, and other delicate parts of the eyes from UV damage and other effects [296]. These can also protect the eyes from age-related macular degeneration and decrease the risk of loss of vision [28].

Arugula, like any other vegetables, might have undesirable effects when overconsumed. Eating large quantity of arugula may cause abdominal pain and flatulence due to its high sulforaphane content. Arugula is rich in vitamin K and should be consumed with caution if the person has been prescribed anticoagulant agents such as warfarin. It is important to store nitrate-containing vegetable juices such as arugula in proper conditions to maintain their nutritional benefits. Nitrate in arugula could be reduced into nitrite, which in high levels could harm the cardiovascular system. Nitrite may also interact with some nitrite medications used for angina [297, 298].

9.2 Spinach

Family: Amaranthaceae
Genus: *Spinacia* (spinach)

Spinach is a flowering vegetable, and its leaves can be eaten either raw or cooked. Despite being cooked, spinach generally improves nutrient absorption. Spinach leaves differ in shape, from oval to triangular, and are variable in size. Typically, the larger leaves are at the base of the plant, while the smaller ones are higher at the top of the flowering stem. Raw and cooked spinach have a number of nutrients like vitamins A, K, and C, folic acid (B-9), and minerals like potassium, magnesium, manganese, iron, and copper (Table 13). It also contains pyridoxine (B-6), niacin (B-3), riboflavin (B-2), and thiamin (B-1). In addition, spinach is packed with nitrates and glycolipids, which may act as anti-inflammatory substances.

Spinach is rich in nitrate which can augment nitric oxide status and improve endothelial function in healthy individuals [299]. Nitrates may improve skeletal muscle blood flow and function, thus enhancing the performance of athletes, and may therefore improve the quality of life of older people with muscle weakness and exercise intolerance [300]. Nitrates also have the ability to considerably decrease the level of serum triglycerides, total cholesterol, and unhealthy cholesterol LDL and increase the level of HDL, thus improving lipid homeostasis [301]. Nitrates were also shown to decrease the blood glucose level. As a dietary supplement, it showed promise in managing insulin resistance and endothelial dysfunction [301]. Glycolipids in spinach enhance the production of nitric oxide in the body as well, which helps relax the blood vessels, lower the blood pressure, and reduce the risk of atherosclerosis [35]. Glycolipids also exhibit anticancer activity. The in vitro and ex vivo antiangiogenesis effects are mediated via the inhibition of replicative DNA polymerase [302, 303]. Additionally, N-oxalylglycine (NOG), a natural product in spinach, is a substance that was found to have anticancer characteristics. NOG is an inhibitor of 2-oxoglutarate and ferrous iron-dependent oxygenases. 2-Oxoglutarate

Table 13 Spinach nutrition facts of 100 g, from the United States Department of Agriculture (USDA SR-21)

Serving size of 100 g of raw spinach	
Per serving	% Daily value[a]
Calories 23.0	
Total fat 0.4 g	1
Saturated fat 0.1 g	0
Polyunsaturated fat 0.2 g	
Monounsaturated fat 0.0 g	
Total omega-3 fatty acids 138 mg	
Total omega-6 fatty acids 26.0 mg	
Cholesterol 0.0 mg	0
Phytosterols 9.0 mg	
Total carbohydrates 3.6 g	1
Dietary fiber 2.2 g	9
Starch	
Sugars 0.4 g	
Protein 2.9 g	6
Vitamins	
Vitamin A	188
Vitamin E	10
Vitamin K	604
Vitamin C	47
Vitamin B-6	10
Folic acid	49
Minerals	
Sodium	3
Calcium	10
Magnesium	20
Copper	6
Potassium	16
Iron	15
Manganese	45
Zinc	4

National Nutrient Database
[a]Based on a 2000-calorie diet

oxygenase is a presumed cancer target of the tricarboxylic acid cycle for its role in epigenetic regulation [304].

Antioxidants [305, 306] and carotenoids [61, 307] in spinach contribute immensely to its anticancer and anti-inflammatory effectiveness as they limit oxidative stress and DNA damage. The antioxidant flavonols kaempferol is linked to lowering the risk of cancer [308], and quercetin also has potent anti-inflammatory properties [309] that were suggested to protect the body against certain forms of cancer. Studies have revealed that the carotenoid xanthophylls, which are natural fat-soluble pigments, improve inflammation status, serum triglyceride levels, blood pressure levels, and liver function test values. On the other hand, recent investigation has shown that xanthophylls possess high anticancer, antidiabetic, anti-obesity, and antioxidant properties [310]. Both neoxanthin and violaxanthin have been demonstrated to possess antiproliferative activity and induce apoptosis in PC-3 human prostate cancer cells [311, 312] and obesity-associated cancers [310]. Other valuable carotenoid plant compounds that boost health include the xanthophylls lutein and zeaxanthin that reduce age-related macular degeneration and improve eye health [313].

The insoluble fiber in spinach maintains a healthy digestive system and prevents constipation [163]. High-fiber diet works to reduce high cholesterol levels and slow the absorption of sugar into the bloodstream [314, 315]. In addition to countless health benefits, fiber contributes to lowering the risk of various noncommunicable diseases [9, 11] such as diabetes [316] and its secondary complications [317] and cardiovascular diseases [10]. Spinach is an extremely rich source of fiber and vitamin K, referred to earlier for their collective cardiovascular health benefits and role in bone health [10, 126, 127]. Spinach is a very rich source of iron which is better absorbed in the presence of vitamin C. The presence of both in spinach prevents anemia and reduces its prevalence [223].

Consuming spinach is not free of undesirable health effects. Raw spinach contains oxalic acid which binds to calcium and interferes with its absorption. Spinach is high in calcium and oxalates, and both can lead to kidney stone formation [318]. Spinach is a rich source of vitamin K, which interferes with blood clotting. Therefore, spinach should be consumed with caution if the person takes warfarin or other anticoagulant medication [319].

9.3 Lettuce

Family: Asteraceae
Genus: *Lactuca* (romaine lettuce)

Lettuce is a leafy green vegetable with many different types, e.g., iceberg (low in fiber yet has a high water content than other types), green leaf (similar to spinach), and red leaf lettuce (has higher carotenoids and phenolic compounds). The vegetable is mostly grown for its leaves though sometimes also for its stem and seeds. Lettuce is often eaten raw in salads and sandwiches. It can also be grilled or prepared as a soup. Most lettuce types are low sodium and low caloric (1 calorie per leaf) and contain high dietary fiber (cellulose), protein, and sugars. Considering these features, lettuce can help in body weight control. It contains quite few minerals and vitamins including magnesium, manganese, phosphorus, potassium, sodium, and zinc along with vitamins like B-1 (thiamin), B-2 (riboflavin), B-3 (niacin), B-6 (pyridoxine), B-9 (folic acid), A, C, and K. It provides significant amounts of vitamins A and K (Table 14). Baby green romaine is especially high in vitamin C.

Table 14 Lettuce nutrition facts of 100 g, from the United States Department of Agriculture (USDA SR-21)

Serving size of 100 g of raw lettuce, green leaf	
Per serving	% Daily value[a]
Calories 15.0	
Total fat 0.2 g	0
Saturated fat 0.0 g	0
Polyunsaturated fat 0.1 g	
Monounsaturated fat 0.0 g	
Total omega-3 fatty acids 58.0 mg	
Total omega-6 fatty acids 24.0 mg	
Cholesterol 0.0 mg	0
Phytosterols 38.0 mg	
Total carbohydrates 2.8 g	1
Dietary fiber 1.3 g	5
Starch 0.0 g	
Sugars 0.8 g	
Protein 1.4 g	3
Vitamins	
Vitamin A	148
Vitamin E	1
Vitamin K	217
Vitamin C	30
Vitamin B-6	4
Folic acid	10
Minerals	
Sodium	1
Calcium	4
Magnesium	3
Copper	1
Potassium	6
Iron	5
Manganese	13
Zinc	1

National Nutrient Database
[a]Based on a 2000-calorie diet

Lettuce is believed to have a number of health benefits. The bioactive compounds and their corresponding health benefits were reviewed recently [320]. In vitro, in vivo, and clinical studies have shown anti-inflammatory, cholesterol-lowering, and antidiabetic activities attributed to the bioactive compounds in lettuce, largely focusing on vitamins (B-9, C, and E), carotenoids (β-carotene and xanthophylls), phenolic compounds (phenolic acids and flavonoids), and key minerals. Lettuce possesses antifungal and antibacterial properties. Lettuce leaf extracts and latex saps of lettuce were effective against different types of bacterial species such as

Staphylococcus, *Proteus*, and *Klebsiella*, yeasts including *Candida albicans*, and fungi like *Aspergillus* [321, 322]. Antiviral activity has also been reported [323]. Additionally, lettuce is a powerful anti-inflammatory. Its extracts contain triterpene lactones that significantly inhibit lipoxygenase and carrageenan-induced edema [324]. Lettuce leaf extracts were also found to be a rich source of antioxidants. The antioxidant capacity of lettuce has been detailed in recent reviews for different lettuce types [325, 326]. The antioxidant properties of vitamins C and E and carotenoids in reducing risk of oxidative stress-related diseases are well documented. The carotenoid and flavonoid role in inflammation and chronic disease prevention have been described earlier [61, 84]. For instance, as stated above, the pigments xanthophylls and anthocyanins possess high anticancer, antidiabetic, anti-obesity, and antioxidant properties [207, 310]. Vitamin C has also been evaluated for its role in disease prevention [173]. Potentiation or synergistic effect has been observed among lettuce antioxidants [327]. Dark-green leafy vegetables may decrease colon cancer risk because the water-insoluble green pigment chlorophyll prevents the detrimental, cytotoxic, and hyper-proliferative colonic effects of dietary heme (eating red meat) [328]. Antioxidants decrease the overall risk of cancer. For example, consumption of either lettuce or the leave extract was found to control some types of cancer such as breast cancer [329, 330] and leukemia. The antileukaemic effects correlate with Chk2 kinase activation, p21 tumor suppressor induction, proto-oncogene cyclin D1 sever downregulation, and acetylation of alpha-tubulin [331]. Raw lettuce displays anxiolytic and minor tranquilizing properties [332, 333] and, along with other leafy greens, was shown to significantly improve mental health outcomes including cognition [334, 335]. Lettuce displays an opium-like effect as well. Lactucarium, a milky fluid with sedative and analgesic properties (depressant chemical), can be isolated from the base of the wild-lettuce stem. The chemical constituent lactucin and its derivatives along with other extract components were found to decrease the heart rate and induce sleep in animal experiments [336, 337]. Several components of lettuce such as phenolic antioxidants, folic acid, and potassium have been demonstrated to have neurological protective effects and can be used to manage ischemia-induced neuronal damage and decrease the risk of Alzheimer's disease [333, 334, 338]. Lettuce's anti-neurotoxicity mechanism appears to suppress Bax and caspase-3 proteins and decrease Bcl-2 [339]. Studies have also found that lettuce impacts cardiovascular health by decreasing total cholesterol level and improving the antioxidant status [10, 340]. Leafy greens, including lettuce, are a good source of omega-3 fatty acids [341]. Hence, lettuce omega-3 content boosts mental and physical health and makes it a good alternative source for vegetarians. Vitamin K in lettuce improves bone health [126], supports blood coagulation, and prevents atherosclerosis by decreasing the accumulation of calcium in the blood vessel wall, boosting cardiovascular system health [127]. Vitamin A and zinc in lettuce can promote eye health too. They prevent age-related macular degeneration and decrease the formation of cataract [342]. Vitamin A also stimulates immune functions and minimizes infections [343]. It is essential for skin health as it can diminish signs of aging [51]. Finally, the role of folic acid in pregnancy and fetus development has been denoted above [14, 289].

In general, consuming lettuce is safe. Some people might develop an allergic reaction when eating lettuce. Adverse symptoms like diarrhea, often bloody, abdominal pain, and vomiting a few days following lettuce consumption could be signs of ingestion of contaminated lettuce with *E. coli*. It is mostly a self-limited condition but might lead to serious complications such as hemolytic uremic syndrome and kidney failure. It is advisable to wash lettuce leaves before consuming it to avoid the ingestion of pesticides as well. Lettuce has high vitamin A content, and overconsumption might cause temporary, harmless, yellowish discoloration of the skin in a case known as carotenodermia. Lettuce is rich in vitamin K and therefore has to be consumed with caution, just like other leafy greens when warfarin or other anticoagulants are used.

References

1. Li, S., Lang, G. T., Zhang, Y. Z., Yu, K. D., Shao, Z. M., & Zhang, Q. (2018). Interaction between glutathione S-transferase M1-null/present polymorphism and adjuvant chemotherapy influences the survival of breast cancer. *Cancer Medicine, 7*(9), 4202–4207.
2. Kageyama, S., Ii, H., Taniguchi, K., Kubota, S., Yoshida, T., Isono, T., et al. (2018). Mechanisms of tumor growth inhibition by depletion of gamma-Glutamylcyclotransferase (GGCT): A novel molecular target for anticancer therapy. *International Journal of Molecular Sciences, 19*(7), E2054.
3. Mikešová, L., Mikeš, J., Koval', J., Gyurászová, K., Culka, L., Vargová, J., et al. (2013). Conjunction of glutathione level, NAD(P)H/FAD redox status and hypericin content as a potential factor affecting colon cancer cell resistance to photodynamic therapy with hypericin. *Photodiagnosis and Photodynamic Therapy, 10*(4), 470–483.
4. Cheteh, E. H., Augsten, M., Rundqvist, H., Bianchi, J., Sarne, V., Egevad, L., et al. (2017). Human cancer-associated fibroblasts enhance glutathione levels and antagonize drug-induced prostate cancer cell death. *Cell Death & Disease, 8*(6), e2848.
5. Li, Q., & Liu, M. (2014). Glutathione S-transferase T1 null genotype and laryngeal cancer risk: A meta-analysis. *Tumour Biology, 35*(9), 8781–8785.
6. Tang, S. C., Wu, C. H., Lai, C. H., Sung, W. W., Yang, W. J., Tang, L. C., et al. (2013). Glutathione S-transferase mu2 suppresses cancer cell metastasis in non-small cell lung cancer. *Molecular Cancer Research, 11*(5), 518–529.
7. Zhou, Y., Li, Y., Zhou, T., Zheng, J., Li, S., & Li, H. B. (2016). Dietary natural products for prevention and treatment of liver cancer. *Nutrients, 8*(3), 156.
8. Tiwari, N., Gupta, V. K., Pandey, P., Patel, D. K., Banerjee, S., Darokar, M. P., et al. (2017). Adjuvant effect of Asparagus racemosus Willd. Derived saponins in antibody production, allergic response and pro-inflammatory cytokine modulation. *Biomedicine & Pharmacotherapy, 86*, 555–561.
9. Anderson, J. W., Baird, P., Davis Jr., R. H., Ferreri, S., Knudtson, M., Koraym, A., et al. (2009). Health benefits of dietary fiber. *Nutrition Reviews, 67*(4), 188–205.
10. Tang, G. Y., Meng, X., Li, Y., Zhao, C. N., Liu, Q., & Li, H. B. (2017). Effects of vegetables on cardiovascular diseases and related mechanisms. *Nutrients, 9*(8), E857.
11. Timm, D. A., & Slavin, J. L. (2008). Dietary fiber and the relationship to chronic diseases. *American Journal of Lifestyle Medicine, 2*(3), 233–240.
12. Hafizur, R. M., Kabir, N., & Chishti, S. (2012). Asparagus officinalis extract controls blood glucose by improving insulin secretion and beta-cell function in streptozotocin-induced type 2 diabetic rats. *The British Journal of Nutrition, 108*(9), 1586–1595.

13. Greenberg, J. A., Bell, S. J., Guan, Y., & Yu, Y. H. (2011). Folic acid supplementation and pregnancy: More than just neural tube defect prevention. *Reviews in Obstetrics & Gynecology, 4*(2), 52–59.
14. Wilson, R. D., Davies, G., Désilets, V., Reid, G. J., Summers, A., Wyatt, P., et al. (2003). The use of folic acid for the prevention of neural tube defects and other congenital anomalies. *Journal of Obstetrics and Gynaecology Canada, 25*(11), 959–973.
15. Pearson, D. A. (2007). Bone health and osteoporosis: The role of vitamin K and potential antagonism by anticoagulants. *Nutrition in Clinical Practice, 22*(5), 517–544.
16. Teymoori, F., Asghari, G., Mirmiran, P., & Azizi, F. (2017). Dietary amino acids and incidence of hypertension: A principle component analysis approach. *Scientific Reports, 7*(1), 16838.
17. Flores-Mireles, A. L., Walker, J. N., Caparon, M., & Hultgren, S. J. (2015). Urinary tract infections: Epidemiology, mechanisms of infection and treatment options. *Nature Reviews. Microbiology, 13*(5), 269–284.
18. Ruzzo, E. K., Capo-Chichi, J. M., Ben-Zeev, B., Chitayat, D., Mao, H., Pappas, A. L., et al. (2013). Deficiency of asparagine synthetase causes congenital microcephaly and a progressive form of encephalopathy. *Neuron, 80*(2), 429–441.
19. Aaron, K. J., & Sanders, P. W. (2013). Role of dietary salt and potassium intake in cardiovascular health and disease: A review of the evidence. *Mayo Clinic Proceedings, 88*(9), 987–995.
20. Iuliano, M., De Tommaso, G., & Ragone, R. (2009). Homocysteine disulphides and vascular disease. *Disease Markers, 27*(2), 55–61.
21. Freitas, C. S., Roveda Jr., A. C., Truzzi, D. R., Garcia, A. C., Cunha, T. M., Cunha, F. Q., et al. (2015). Anti-inflammatory and anti-nociceptive activity of ruthenium complexes with isonicotinic and nicotinic acids (Niacin) as ligands. *Journal of Medicinal Chemistry, 58*(11), 4439–4448.
22. Markt, S. C., Nuttall, E., Turman, C., Sinnott, J., Rimm, E. B., Ecsedy, E., et al. (2016). Sniffing out significant "Pee values": Genome wide association study of asparagus anosmia. *BMJ, 355*, i6071.
23. Clifford, T., Howatson, G., West, D. J., & Stevenson, E. J. (2015). The potential benefits of red beetroot supplementation in health and disease. *Nutrients, 7*(4), 2801–2822.
24. Georgiev, V. G., Weber, J., Kneschke, E. M., Denev, P. N., Bley, T., & Pavlov, A. I. (2010). Antioxidant activity and phenolic content of betalain extracts from intact plants and hairy root cultures of the red beetroot Beta vulgaris cv. Detroit dark red. *Plant Foods for Human Nutrition, 65*(2), 105–111.
25. Lakshimi, E. (2016). Food fortification and iron deficiency anaemia – School model approach. *International Journal of Pharma and Bio Sciences, 7*(4), (B) 831–(B) 835.
26. Raju, M., Sadineni, V., Lakshminarayana, R., Krishnakantha, T. P., & Baskaran, V. (2007). Carotenoid composition and vitamin A activity of medicinally important green leafy vegetables. *Food Chemistry, 101*(4), 1598–1605.
27. Vardavas, C., Majchrzak, D., Wagner, K.-H., Elmadfa, I., & Kafatos, A. (2006). The antioxidant and phylloquinone content of wildly grown greens in Crete. *Food Chemistry, 99*, 813–821.
28. Wu, J., Cho, E., Willett, W. C., Sastry, S. M., & Schaumberg, D. A. (2015). Intakes of lutein, zeaxanthin, and other carotenoids and age-related macular degeneration during 2 decades of prospective follow-up. *JAMA Ophthalmology, 133*(12), 1415–1424.
29. Kapadia, G. J., Tokuda, H., Konoshima, T., & Nishino, H. (1996). Chemoprevention of lung and skin cancer by Beta vulgaris (beet) root extract. *Cancer Letters, 100*(1–2), 211–214.
30. Kapadia, G. J., Rao, G. S., Ramachandran, C., Iida, A., Suzuki, N., & Tokuda, H. (2013). Synergistic cytotoxicity of red beetroot (Beta vulgaris L.) extract with doxorubicin in human pancreatic, breast and prostate cancer cell lines. *Journal of Complementary and Integrative Medicine, 10.*

31. Threapleton, D. E., Greenwood, D. C., Evans, C. E., Cleghorn, C. L., Nykjaer, C., Woodhead, C., et al. (2013). Dietary fibre intake and risk of cardiovascular disease: Systematic review and meta-analysis. *BMJ, 347*, f6879.
32. Vali, L., Stefanovits-Bányai, E., Szentmihályi, K., Fébel, H., Sárdi, E., Lugasi, A., et al. (2007). Liver-protecting effects of table beet (Beta vulgaris var. rubra) during ischemia-reperfusion. *Nutrition, 23*(2), 172–178.
33. Kanner, J., Harel, S., & Granit, R. (2001). Betalains—A new class of dietary cationized antioxidants. *Journal of Agricultural and Food Chemistry, 49*(11), 5178–5185.
34. Craig, S. A. (2004). Betaine in human nutrition. *The American Journal of Clinical Nutrition, 80*(3), 539–549.
35. Kapil, V., Khambata, R. S., Robertson, A., Caulfield, M. J., & Ahluwalia, A. (2015). Dietary nitrate provides sustained blood pressure lowering in hypertensive patients: A randomized, phase 2, double-blind, placebo controlled study. *Hypertension, 65*(2), 320–327.
36. Bailey, S. J., Winyard, P., Vanhatalo, A., Blackwell, J. R., Dimenna, F. J., Wilkerson, D. P., et al. (2009). Dietary nitrate supplementation reduces the O2 cost of low-intensity exercise and enhances tolerance to high-intensity exercise in humans. *Journal of Applied Physiology (Bethesda, MD: 1985), 107*(4), 1144–1155.
37. Petrie, M., Rejeski, W. J., Basu, S., Laurienti, P. J., Marsh, A. P., Norris, J. L., et al. (2017). Beet root juice: An ergogenic aid for exercise and the aging brain. *The Journals of Gerontology. Series A, Biological Sciences and Medical Sciences, 72*(9), 1284–1289.
38. Wang, M., & Goldman, I. (1997). Transgressive segregation and reciprocal effect for free folic acid content in a red beet (Beta vulgaris L.) population. *Euphytica, 96*, 317–321.
39. Han, H., et al. (2015). Nutritional management of kidney stones (Nephrolithiasis). *Clinical Nutrition Research, 4*(3), 137–152.
40. Coles, L. T., & Clifton, P. M. (2012). Effect of beetroot juice on lowering blood pressure in free-living, disease-free adults: A randomized, placebo-controlled trial. *Nutrition Journal, 11*, 106.
41. Jacks, A., Toikkanen, S., Pihlajasaari, A., Johansson, T., Hakkinen, M., Hemminki, K., et al. (2013). Raw grated beetroot linked to several outbreaks of sudden-onset gastrointestinal illness, Finland 2010. *Epidemiology and Infection, 141*(8), 1640–1646.
42. Wu, H., Zhu, J., Diao, W., & Wang, C. (2014). Ultrasound-assisted enzymatic extraction and antioxidant activity of polysaccharides from pumpkin (Cucurbita moschata). *Carbohydrate Polymers, 113*, 314–324.
43. Kim, H. Y., Nam, S. Y., Yang, S. Y., Kim, H. M., & Jeong, H. J. (2016). Cucurbita moschata Duch. and its active component, beta-carotene effectively promote the immune responses through the activation of splenocytes and macrophages. *Immunopharmacology and Immunotoxicology, 38*(5), 319–326.
44. Riccioni, G., Barbara, M., Bucciarelli, T., di Ilio, C., & D'Orazio, N. (2007). Antioxidant vitamin supplementation in asthma. *Annals of Clinical and Laboratory Science, 37*(1), 96–101.
45. Chen, F., Shao, F., Hinds, A., Yao, S., Ram-Mohan, S., Norman, T. A., et al. (2018). Retinoic acid signaling is essential for airway smooth muscle homeostasis. *JCI Insight, 3*(16), e120398.
46. Delcourt, C., Carrière, I., Delage, M., Barberger-Gateau, P., Schalch, W., & POLA Study Group. (2006). Plasma lutein and zeaxanthin and other carotenoids as modifiable risk factors for age-related maculopathy and cataract: The POLA study. *Investigative Ophthalmology & Visual Science, 47*(6), 2329–2335.
47. Xia, H. C., Li, F., Li, Z., & Zhang, Z. C. (2003). Purification and characterization of Moschatin, a novel type I ribosome-inactivating protein from the mature seeds of pumpkin (Cucurbita moschata), and preparation of its immunotoxin against human melanoma cells. *Cell Research, 13*(5), 369–374.
48. Yu, N., Su, X., Wang, Z., Dai, B., & Kang, J. (2015). Association of dietary vitamin A and beta-carotene intake with the risk of lung cancer: A meta-analysis of 19 publications. *Nutrients, 7*(11), 9309–9324.

49. Williams, S. J., Cvetkovic, D., & Hamilton, T. C. (2009). Vitamin A metabolism is impaired in human ovarian cancer. *Gynecologic Oncology, 112*(3), 637–645.
50. Nkondjock, A., & Ghadirian, P. (2004). Dietary carotenoids and risk of colon cancer: Case-control study. *International Journal of Cancer, 110*(1), 110–116.
51. Kafi, R., Kwak, H. S., Schumacher, W. E., Cho, S., Hanft, V. N., Hamilton, T. A., et al. (2007). Improvement of naturally aged skin with vitamin A(retinol). *Archives of Dermatology, 143*(5), 606–612.
52. Telang, P. S. (2013). Vitamin C in dermatology. *Indian Dermatology Online Journal, 4*(2), 143–146.
53. Krishnamurthy, P., & Wadhwani, A. (2012). Antioxidant enzymes and human health. In M. A. El-Missiry (Ed.), *Antioxidant enzyme* (pp. 3–18). London, UK: IntechOpen.
54. Zhu, K., Devine, A., & Prince, R. L. (2009). The effects of high potassium consumption on bone mineral density in a prospective cohort study of elderly postmenopausal women. *Osteoporosis International, 20*(2), 335–340.
55. Lee, J., Kim, D., Choi, J., Choi, H., Ryu, J. H., Jeong, J., et al. (2012). Dehydrodiconiferyl alcohol isolated from Cucurbita moschata shows anti-adipogenic and anti-lipogenic effects in 3T3-L1 cells and primary mouse embryonic fibroblasts. *The Journal of Biological Chemistry, 287*(12), 8839–8851.
56. Penland, J. G., & Johnson, P. E. (1993). Dietary calcium and manganese effects on menstrual cycle symptoms. *American Journal of Obstetrics and Gynecology, 168*(5), 1417–1423.
57. Wang, S. Y., Huang, W. C., Liu, C. C., Wang, M. F., Ho, C. S., Huang, W. P., et al. (2012). Pumpkin (Cucurbita moschata) fruit extract improves physical fatigue and exercise performance in mice. *Molecules, 17*(10), 11864–11876.
58. Kidmose, U., Hansen, S. L., Christensen, L. P., Edelenbos, M., Larsen, E., & Nørbæk, R. (2004). Effects of genotype, root size, storage, and processing on bioactive compounds in organically grown carrots (Daucus carota L.). *Journal of Food Science, 69*(9), S388–S394.
59. Gilani, A.-H., Shaheen, F., Saeed, S. A., Bibi, S., Sadiq, M., & Faizi, S. (2000). Hypotensive action of coumarin glycosides from Daucus carota. *Phytomedicine, 7*, 423–426.
60. Vandekinderen, I., Van Camp, J., Devlieghere, F., Veramme, K., Denon, Q., Ragaert, P., et al. (2008). Effect of decontamination agents on the microbial population, sensorial quality, and nutrient content of grated carrots (Daucus carota L.). *Journal of Agricultural and Food Chemistry, 56*(14), 5723–5731.
61. Kaulmann, A., & Bohn, T. (2014). Carotenoids, inflammation, and oxidative stress—Implications of cellular signaling pathways and relation to chronic disease prevention. *Nutrition Research, 34*(11), 907–929.
62. Surles, R. L., Weng, N., Simon, P. W., & Tanumihardjo, S. A. (2004). Carotenoid profiles and consumer sensory evaluation of specialty carrots (Daucus carota, L.) of various colors. *Journal of Agricultural and Food Chemistry, 52*(11), 3417–3421.
63. He, J., Gu, Y., & Zhang, S. (2018). Vitamin A and breast cancer survival: A systematic review and meta-analysis. *Clinical Breast Cancer, 18*(6), e1389–e1400.
64. Doldo, E., Costanza, G., Agostinelli, S., Tarquini, C., Ferlosio, A., Arcuri, G., et al. (2015). Vitamin A, cancer treatment and prevention: The new role of cellular retinol binding proteins. *BioMed Research International, 2015*, 14.
65. Jiang, L., Dong, R., Ying, M., He, Q., Cao, J., & Yang, B. (2018). Immune cells in the tumour: New routes of retinoids for chemoprevention and chemotherapeutics. *British Journal of Pharmacology, 175*(23), 4285–4294.
66. Ni, X., Hu, G., & Cai, X. (2019). The success and the challenge of all-trans retinoic acid in the treatment of cancer. *Critical Reviews in Food Science and Nutrition, 59*, S71–S80.
67. Shankar, S., & Lanza, E. (1991). Dietary fiber and cancer prevention. *Hematology/Oncology Clinics of North America, 5*(1), 25–41.
68. Moore, M. A., Park, C. B., & Tsuda, H. (1998). Soluble and insoluble fiber influences on cancer development. *Critical Reviews in Oncology/Hematology, 27*(3), 229–242.
69. Carr, A. C., & Maggini, S. (2017). Vitamin C and immune function. *Nutrients, 9*(11), E1211.

70. Sun, T., Simon, P. W., & Tanumihardjo, S. A. (2009). Antioxidant phytochemicals and antioxidant capacity of biofortified carrots (Daucus carota L.) of various colors. *Journal of Agricultural and Food Chemistry, 57*(10), 4142–4147.
71. Horng, C. T., Tsai, M. L., Shiang, J. C., Chien, S. T., Chien, C. H., Chang, T. H., et al. (2011). Glaucoma treatment with the extract of astragalus membranaceus in rats experimental model. *Life Science Journal, 8*, 124–132.
72. Coyne, T., Ibiebele, T. I., Baade, P. D., Dobson, A., McClintock, C., Dunn, S., et al. (2005). Diabetes mellitus and serum carotenoids: Findings of a population-based study in Queensland, Australia. *The American Journal of Clinical Nutrition, 82*(3), 685–693.
73. Leon, A., Del-Ángel, M., Ávila, J. L., & Delgado, G. (2017). Phthalides: Distribution in nature, chemical reactivity, synthesis, and biological activity. *Progress in the Chemistry of Organic Natural Products, 104*, 127–246.
74. Wassenhove, F. V., Dirinck, P., Vulsteke, G., & Schamp, N. (1990). Aromatic volatile composition of celery and celeriac cultivars. *HortScience, 25*, 556–559.
75. Madhavi, D., Kagan, D., Rao, V., & Murray, M. T. (2013). A pilot study to evaluate the antihypertensive effect of a celery extract in mild to moderate hypertensive patients. *Natural Medicine Journal, 5*(4), 1–5.
76. Mansi, K., Abushoffa, A. M., Disi, A., & Aburjai, T. (2009). Hypolipidemic effects of seed extract of celery (Apium graveolens) in rats. *Pharmacognosy Magazine, 5*(20), 301–305.
77. Fazal, S., & Singla, R. K. (2012). Review on the pharmacognostical & pharmacological characterization of Apium Graveolens Linn. *Indo Global Journal of Pharmaceutical Sciences, 2*, 36–42.
78. Mongeau, R., Siddiqui, I. R., Emery, J., & Brassard, R. (1990). Effect of dietary fiber concentrated from celery, parsnip, and rutabaga on intestinal function, serum cholesterol, and blood glucose response in rats. *Journal of Agricultural and Food Chemistry, 38*, 185–200.
79. Mahran, G. H., Kadry, H. A., Isaac, Z. G., Thabet, C. K., Al-Azizi, M. M., & El-Olemy, M. M. (1991). Investigation of diuretic drug plants. 1. Phytochemical screening and pharmacological evaluation of Anethum graveolens L., Apium graveolens L., Daucus carota L. and Eruca sativa mill. *Phytotherapy Research, 5*(4), 169–172.
80. Abd El-Mageed, N. M. (2011). Hepatoprotective effect of feeding celery leaves mixed with chicory leaves and barley grains to hypercholesterolemic rats. *Pharmacognosy Magazine, 7*(26), 151–156.
81. Chang, J. L., Chen, G., Ulrich, C. M., Bigler, J., King, I. B., Schwarz, Y., et al. (2010). DNA damage and repair: Fruit and vegetable effects in a feeding trial. *Nutrition and Cancer, 62*(3), 329–335.
82. Kooti, W., & Daraei, N. (2017). A review of the antioxidant activity of celery (Apium graveolens L). *Journal of Evidence Based Complementary Alternative Medicine, 22*(4), 1029–1034.
83. Li, M. Y., Hou, X. L., Wang, F., Tan, G. F., Xu, Z. S., & Xiong, A. S. (2018). Advances in the research of celery, an important Apiaceae vegetable crop. *Critical Reviews in Biotechnology, 38*(2), 172–183.
84. Panche, A. N., Diwan, A. D., & Chandra, S. R. (2016). Flavonoids: An overview. *Journal of Nutrition Science, 5*, e47.
85. Hostetler, G., Riedl, K., Cardenas, H., Diosa-Toro, M., Arango, D., Schwartz, S., et al. (2012). Flavone deglycosylation increases their anti-inflammatory activity and absorption. *Molecular Nutrition & Food Research, 56*(4), 558–569.
86. Arango, D., Diosa-Toro, M., Rojas-Hernandez, L. S., Cooperstone, J. L., Schwartz, S. J., Mo, X., et al. (2015). Dietary apigenin reduces LPS-induced expression of miR-155 restoring immune balance during inflammation. *Molecular Nutrition & Food Research, 59*(4), 763–772.
87. Christensen, L. P., & Brandt, K. (2006). Bioactive polyacetylenes in food plants of the Apiaceae family: Occurrence, bioactivity and analysis. *Journal of Pharmaceutical and Biomedical Analysis, 41*(3), 683–693.

88. Purup, S., Larsen, E., & Christensen, L. P. (2009). Differential effects of falcarinol and related aliphatic C(17)-polyacetylenes on intestinal cell proliferation. *Journal of Agricultural and Food Chemistry, 57*(18), 8290–8296.
89. Wills, R. B. H., Wimalasiri, P., & Greenfield, H. (1984). Dehydroascorbic acid levels in fresh fruit and vegetables in relation to total vitamin C activity. *Journal of Agricultural and Food Chemistry, 32*(4), 836–838.
90. Tharib, S. M., & Veitch, G. B. A. (1985). The anti-inflammatory activity of celery Apium graveolens L. (Fam. Umbelliferae) AU - Lewis, David A. *International Journal of Crude Drug Research, 23*(1), 27–32.
91. Crozier, A., Lean, M. E., McDonald, M. S., & Black, C. (1997). Quantitative analysis of the flavonoid content of commercial tomatoes, onions, lettuce, and celery. *Journal of Agricultural and Food Chemistry, 45*(3), 590–595.
92. Lin, Y., Shi, R., Wang, X., & Shen, H. M. (2008). Luteolin, a flavonoid with potential for cancer prevention and therapy. *Current Cancer Drug Targets, 8*(7), 634–646.
93. Sung, B., Chung, H. Y., & Kim, N. D. (2016). Role of apigenin in cancer prevention via the induction of apoptosis and autophagy. *Journal of Cancer Prevention, 21*(4), 216–226.
94. Patel, D., Shukla, S., & Gupta, S. (2007). Apigenin and cancer chemoprevention: Progress, potential and promise (review). *International Journal of Oncology, 30*(1), 233–245.
95. Schagen, S. K., Zampeli, V. A., Makrantonaki, E., & Zouboulis, C. C. (2012). Discovering the link between nutrition and skin aging. *Dermatoendocrinology, 4*(3), 298–307.
96. Beier, R. C., Ivie, G. W., Oertli, E. H., & Holt, D. L. (1983). HPLC analysis of linear furocoumarins (psoralens) in healthy celery (Apium graveolens). *Food and Chemical Toxicology, 21*(2), 163–165.
97. Morales-Asin, F., Iñiguez, C., Cornudella, R., Mauri, J. A., Espada, F., & Mostacero, E. E. (2000). Patients with acenocoumarol treatment and migraine. *Headache, 40*(1), 45–47.
98. Mandal, S., & Mandal, M. (2015). Coriander (Coriandrum sativum L.) essential oil: Chemistry and biological activity. *Journal of Tropical Biomedicine, 4*, 1.
99. Rajeshwari, U., & Andallu, B. (2011). Medicinal benefits of coriander(Coriandrum Sativum L). *Spatula DD, 1*(1), 51–58.
100. Brown, K., Garver, S. W., & Orlando, R. (2017). 1,8-cineole: An underappreciated anti-inflammatory therapeutic. *Journal of Biomolecular Research & Therapeutics, 6*, 1.
101. Choudhary, M., Kumar, V., Malhotra, H., & Singh, S. (2015). Medicinal plants with potential anti-arthritic activity. *Journal of Intercultural Ethnopharmacology, 4*(2), 147–179.
102. Silva, J., Abebe, W., Sousa, S. M., Duarte, V. G., Machado, M. I., & Matos, F. J. (2003). Analgesic and anti-inflammatory effects of essential oils of eucalyptus. *Journal of Ethnopharmacology, 89*(2–3), 277–283.
103. Juergens, U. R. (2014). Anti-inflammatory properties of the monoterpene 1.8-cineole: Current evidence for co-medication in inflammatory airway diseases. *Drug Research (Stuttg), 64*(12), 638–646.
104. Ben-Arye, E., Dudai, N., Eini, A., Torem, M., Schiff, E., & Rakover, Y. (2011). Treatment of upper respiratory tract infections in primary care: A randomized study using aromatic herbs. *Evidence-based Complementary and Alternative Medicine, 2011*, 690346.
105. Reuter, J., Huyke, C., Casetti, F., Theek, C., Frank, U., Augustin, M., et al. (2008). Anti-inflammatory potential of a lipolotion containing coriander oil in the ultraviolet erythema test. *Journal der Deutschen Dermatologischen Gesellschaft, 6*(10), 847–851.
106. Sadlon, A. E., & Lamson, D. W. (2010). Immune-modifying and antimicrobial effects of eucalyptus oil and simple inhalation devices. *Alternative Medicine Review, 15*(1), 33–47.
107. Serafino, A., Sinibaldi Vallebona, P., Andreola, F., Zonfrillo, M., Mercuri, L., Federici, M., et al. (2008). Stimulatory effect of eucalyptus essential oil on innate cell-mediated immune response. *BMC Immunology, 9*, 17.
108. Wang, X. Y., Lim-Jurado, M., Prepageran, N., Tantilipikorn, P., & Wang de, Y. (2016). Treatment of allergic rhinitis and urticaria: A review of the newest antihistamine drug bilastine. *Therapeutics and Clinical Risk Management, 12*, 585–597.

109. Choi, S. Y., & Park, K. (2016). Effect of inhalation of aromatherapy oil on patients with perennial allergic rhinitis: A randomized controlled trial. *Journal of Evidence-Based Complementary and Alternative Medicine, 2016*, 7.
110. Leal-Cardoso, J. H., Lahlou, S., Weinreich, D., & Caldas Magalhães, P. J. (2010). The essential oil of croton nepetaefolius selectively blocks histamine-augmented neuronal excitability in Guinea-pig celiac ganglion. *The Journal of Pharmacy and Pharmacology, 62*(8), 1045–1053.
111. Kehrl, W., Sonnemann, U., & Dethlefsen, U. (2004). Therapy for acute nonpurulent rhinosinusitis with cineole: Results of a double-blind, randomized, placebo-controlled trial. *Laryngoscope, 114*(4), 738–742.
112. Schnitzler, P., Schon, K., & Reichling, J. (2001). Antiviral activity of Australian tea tree oil and eucalyptus oil against herpes simplex virus in cell culture. *Pharmazie, 56*(4), 343–347.
113. Bachir, R. G., & Benali, M. (2012). Antibacterial activity of the essential oils from the leaves of Eucalyptus globulus against Escherichia coli and Staphylococcus aureus. *Asian Pacific Journal of Tropical Biomedicine, 2*(9), 739–742.
114. Musyimi, D. M., & Ogur, J. A. (2008). Comparative assessment of antifungal activity of extracts from Eucalyptus globulus and Eucalyptus citriodora. *Research Journal of Phytochemistry, 2*, 35–43.
115. Orchard, A., & van Vuuren, S. (2017). Commercial essential oils as potential antimicrobials to treat skin diseases. *Evidence-based Complementary and Alternative Medicine, 2017*, 4517971.
116. Cermelli, C., Fabio, A., Fabio, G., & Quaglio, P. (2008). Effect of eucalyptus essential oil on respiratory bacteria and viruses. *Current Microbiology, 56*(1), 89–92.
117. Noumi, E., Mejd, S., & Bakhrouf, A. (2010). In vitro effect of Melaleuca alternifolia and Eucalyptus globulus essential oils on mycelia formation by oral Candida albicans strains. *African Journal of Microbiology Research, 4*(12), 1332–1336.
118. Bokaeian, M., Nakhaee, A., Moodi, B., & Ali Khazaei, H. (2010). Eucalyptus globulus (eucalyptus) treatment of candidiasis in normal and diabetic rats. *Iranian Biomedical Journal, 14*(3), 121–126.
119. Bramston, C., & Robinson, C. (2015). Is eucalyptus oil an effective antifungal treatment for onychomycosis with and without nail matrix infection? *Journal of Foot and Ankle Research, 8*(Suppl 2), P1.
120. Farhana, K., Islam, H., Emran, E. H., & Islam, N. (2006). Toxicity and repellant activity of three spice materials on Tribolium castaneum (Herbst) adults. *Journal of Bio-Science, 14*, 127–130.
121. Vlachojannis, C., Winsauer, H., & Chrubasik, S. (2013). Effectiveness and safety of a mouthwash containing essential oil ingredients. *Phytotherapy Research, 27*(5), 685–691.
122. Jabeen, Q., Bashir, S., Lyoussi, B., & Gilani, A. H. (2009). Coriander fruit exhibits gut modulatory, blood pressure lowering and diuretic activities. *Journal of Ethnopharmacology, 122*(1), 123–130.
123. Uma, B., Prabhakar, K., Rajendran, S., & Sarayu, Y. L. (2009). Antimicrobial activity and phytochemical analysis of coriander sativum against infectious diarrhea. *Ethnobotanical Leaflets, 2009*, 4.
124. Kubo, I., Fujita, K., Kubo, A., Nihei, K., & Ogura, T. (2004). Antibacterial activity of coriander volatile compounds against salmonella choleraesuis. *Journal of Agricultural and Food Chemistry, 52*(11), 3329–3332.
125. Rios, J. J., Lochlainn, S. O., Devonshire, J., Graham, N. S., Hammond, J. P., King, G. J., et al. (2012). Distribution of calcium (Ca) and magnesium (mg) in the leaves of Brassica rapa under varying exogenous Ca and mg supply. *Annals of Botany, 109*(6), 1081–1089.
126. Weber, P. (2001). Vitamin K and bone health. *Nutrition, 17*(10), 880–887.
127. van Ballegooijen, A. J., & Beulens, J. W. (2017). The role of vitamin K status in cardiovascular health: Evidence from observational and clinical studies. *Current Nutrition Reports, 6*(3), 197–205.

128. Soares, M. C., Damiani, C. E., Moreira, C. M., Stefanon, I., & Vassallo, D. V. (2005). Eucalyptol, an essential oil, reduces contractile activity in rat cardiac muscle. *Brazilian Journal of Medical and Biological Research, 38*(3), 453–461.
129. Dhanapakiam, P., Joseph, J. M., Ramaswamy, V. K., Moorthi, M., & Kumar, A. S. (2008). The cholesterol lowering property of coriander seeds (Coriandrum sativum): Mechanism of action. *Journal of Environmental Biology, 29*(1), 53–56.
130. Guan, Y.-S., & He, Q. (2015). Plants consumption and liver health. *Journal of Evidence-Based Complementary and Alternative Medicine, 2015*, 10.
131. Barros, L. (2012). Phenolic profiles of in vivo and in vitro grown Coriandrum sativum L. *Food Chemistry, 132*(2), 841–848.
132. Kozlowska, A., & Szostak-Wegierek, D. (2014). Flavonoids—Food sources and health benefits. *Roczniki Państwowego Zakładu Higieny, 65*(2), 79–85.
133. Chithra, V., & Leelamma, S. (1999). Coriandrum sativum - Mechanism of hypoglycemic action. *Food Chemistry, 67*(3), 229–231.
134. Kannappan, R., Gupta, S. C., Kim, J. H., Reuter, S., & Aggarwal, B. B. (2011). Neuroprotection by spice-derived nutraceuticals: You are what you eat! *Molecular Neurobiology, 44*(2), 142–159.
135. Akram, M., & Nawaz, A. (2017). Effects of medicinal plants on Alzheimer's disease and memory deficits. *Neural Regeneration Research, 12*(4), 660–670.
136. Hwang, E., Lee, D. G., Park, S. H., Oh, M. S., & Kim, S. Y. (2014). Coriander leaf extract exerts antioxidant activity and protects against UVB-induced photoaging of skin by regulation of procollagen type I and MMP-1 expression. *Journal of Medicinal Food, 17*(9), 985–995.
137. Hayes, J. D., Kelleher, M. O., & Eggleston, I. M. (2008). The cancer chemopreventive actions of phytochemicals derived from glucosinolates. *European Journal of Nutrition, 47*(Suppl 2), 73–88.
138. Li, Y., & Zhang, T. (2013). Targeting cancer stem cells with sulforaphane, a dietary component from broccoli and broccoli sprouts. *Future Oncology, 9*(8), 1097–1103.
139. Li, Y., Zhang, T., Korkaya, H., Liu, S., Lee, H. F., Newman, B., et al. (2010). Sulforaphane, a dietary component of broccoli/broccoli sprouts, inhibits breast cancer stem cells. *Clinical Cancer Research, 16*(9), 2580–2590.
140. Abbaoui, B., Riedl, K. M., Ralston, R. A., Thomas-Ahner, J. M., Schwartz, S. J., Clinton, S. K., et al. (2012). Inhibition of bladder cancer by broccoli isothiocyanates sulforaphane and erucin: Characterization, metabolism, and interconversion. *Molecular Nutrition & Food Research, 56*(11), 1675–1687.
141. Lozanovski, V. J., Houben, P., Hinz, U., Hackert, T., Herr, I., & Schemmer, P. (2014). Pilot study evaluating broccoli sprouts in advanced pancreatic cancer (POUDER trial) - study protocol for a randomized controlled trial. *Trials, 15*, 204.
142. Alumkal, J. J., Slottke, R., Schwartzman, J., Cherala, G., Munar, M., Graff, J. N., et al. (2015). A phase II study of sulforaphane-rich broccoli sprout extracts in men with recurrent prostate cancer. *Investigational New Drugs, 33*(2), 480–489.
143. Chen, P., Li, C., Li, X., Li, J., Chu, R., & Wang, H. (2014). Higher dietary folate intake reduces the breast cancer risk: A systematic review and meta-analysis. *British Journal of Cancer, 110*(9), 2327–2338.
144. Figueiredo, J. C., Mott, L. A., Giovannucci, E., Wu, K., Cole, B., Grainge, M. J., et al. (2011). Folic acid and prevention of colorectal adenomas: A combined analysis of randomized clinical trials. *International Journal of Cancer, 129*(1), 192–203.
145. Figueiredo, J. C., Grau, M. V., Haile, R. W., Sandler, R. S., Summers, R. W., Bresalier, R. S., et al. (2009). Folic acid and risk of prostate cancer: Results from a randomized clinical trial. *Journal of the National Cancer Institute, 101*(6), 432–435.
146. Keum, Y. S. (2011). Regulation of the Keap1/Nrf2 system by chemopreventive sulforaphane: Implications of posttranslational modifications. *Annals of the New York Academy of Sciences, 1229*, 184–189.

147. Wagner, A. E., Terschluesen, A. M., & Rimbach, G. (2013). Health promoting effects of brassica-derived phytochemicals: From chemopreventive and anti-inflammatory activities to epigenetic regulation. *Oxidative Medicine and Cellular Longevity, 2013*, 964539.
148. Liu, G. H., Qu, J., & Shen, X. (2008). NF-kappaB/p65 antagonizes Nrf2-ARE pathway by depriving CBP from Nrf2 and facilitating recruitment of HDAC3 to MafK. *Biochimica et Biophysica Acta, 1783*(5), 713–727.
149. Aiso, I., Inoue, H., Seiyama, Y., & Kuwano, T. (2014). Compared with the intake of commercial vegetable juice, the intake of fresh fruit and komatsuna (Brassica rapa L. var. perviridis) juice mixture reduces serum cholesterol in middle-aged men: A randomized controlled pilot study. *Lipids in Health and Disease, 13*, 102.
150. Alkhalidy, H., Moore, W., Zhang, Y., McMillan, R., Wang, A., Ali, M., et al. (2015). Small molecule kaempferol promotes insulin sensitivity and preserved pancreatic beta-cell mass in middle-aged obese diabetic mice. *Journal Diabetes Research, 2015*, 532984.
151. Obika, L. F. O. (2006). *Kallikrein, potassium, and blood pressure: What has the kidney got to do with them?* Benin City: University of Benin Press.
152. Lumeng, C. N., & Saltiel, A. R. (2011). Inflammatory links between obesity and metabolic disease. *The Journal of Clinical Investigation, 121*(6), 2111–2117.
153. Coussens, L. M., & Werb, Z. (2002). Inflammation and cancer. *Nature, 420*(6917), 860–867.
154. Libby, P. (2002). Inflammation in atherosclerosis. *Nature, 420*(6917), 868–874.
155. Li, H., Ruan, X. Z., Powis, S. H., Fernando, R., Mon, W. Y., Wheeler, D. C., et al. (2005). EPA and DHA reduce LPS-induced inflammation responses in HK-2 cells: Evidence for a PPAR-gamma-dependent mechanism. *Kidney International, 67*(3), 867–874.
156. Calder, P. C. (2006). N-3 polyunsaturated fatty acids, inflammation, and inflammatory diseases. *The American Journal of Clinical Nutrition, 83*(6 Suppl), 1505s–1519s.
157. Simopoulos, A. P. (2002). Omega-3 fatty acids in inflammation and autoimmune diseases. *Journal of the American College of Nutrition, 21*(6), 495–505.
158. Li, K., Huang, T., Zheng, J., Wu, K., & Li, D. (2014). Effect of marine-derived n-3 polyunsaturated fatty acids on C-reactive protein, interleukin 6 and tumor necrosis factor alpha: A meta-analysis. *PLoS One, 9*(2), e88103.
159. Kiecolt-Glaser, J. K., Belury, M. A., Andridge, R., Malarkey, W. B., & Glaser, R. (2011). Omega-3 supplementation lowers inflammation and anxiety in medical students: A randomized controlled trial. *Brain, Behavior, and Immunity, 25*(8), 1725–1734.
160. Keenan, J. I., Salm, N., Wallace, A. J., & Hampton, M. B. (2012). Using food to reduce H. pylori-associated inflammation. *Phytotherapy Research, 26*(11), 1620–1625.
161. Lim, L. S., Mitchell, P., Seddon, J. M., Holz, F. G., & Wong, T. Y. (2012). Age-related macular degeneration. *Lancet, 379*(9827), 1728–1738.
162. Merle, B. M., Benlian, P., Puche, N., Bassols, A., Delcourt, C., Souied, E. H., et al. (2014). Circulating omega-3 fatty acids and neovascular age-related macular degeneration. *Investigative Ophthalmology & Visual Science, 55*(3), 2010–2019.
163. Otles, S., & Ozgoz, S. (2014). Health effects of dietary fiber. *Acta Scientiarum Polonorum. Technologia Alimentaria, 13*(2), 191–202.
164. Slavin, J. (2013). Fiber and prebiotics: Mechanisms and health benefits. *Nutrients, 5*(4), 1417–1435.
165. Lattimer, J. M., & Haub, M. D. (2010). Effects of dietary fiber and its components on metabolic health. *Nutrients, 2*(12), 1266–1289.
166. Stelmach-Mardas, M., Rodacki, T., Dobrowolska-Iwanek, J., Brzozowska, A., Walkowiak, J., Wojtanowska-Krosniak, A., et al. (2016). Link between food energy density and body weight changes in obese adults. *Nutrients, 8*(4), 229.
167. Barba, F. J., Nikmaram, N., Roohinejad, S., Khelfa, A., Zhu, Z., & Koubaa, M. (2016). Bioavailability of glucosinolates and their breakdown products: Impact of processing. *Frontiers in Nutrition, 3*, 24.
168. Fiedor, J., & Burda, K. (2014). Potential role of carotenoids as antioxidants in human health and disease. *Nutrients, 6*(2), 466–488.

169. Larocca, M., Perna, A. M., Simonetti, A., Gambacorta, E., Iannuzzi, A., Perucatti, A., et al. (2017). Antioxidant and anti-inflammatory effects of cauliflower leaf powder-enriched diet against LPS induced toxicity in rabbits. *Food & Function, 8*(9), 3288–3296.
170. Ahmed, F. A., & Ali, R. F. (2013). Bioactive compounds and antioxidant activity of fresh and processed white cauliflower. *BioMed Research International, 2013*, 367819.
171. Abdull Razis, A. F., & Noor, N. M. (2013). Cruciferous vegetables: Dietary phytochemicals for cancer prevention. *Asian Pacific Journal of Cancer Prevention, 14*(3), 1565–1570.
172. Murillo, G., & Mehta, R. G. (2001). Cruciferous vegetables and cancer prevention. *Nutrition and Cancer, 41*(1–2), 17–28.
173. Padayatty, S. J., Katz, A., Wang, Y., Eck, P., Kwon, O., Lee, J. H., et al. (2003). Vitamin C as an antioxidant: Evaluation of its role in disease prevention. *Journal of the American College of Nutrition, 22*(1), 18–35.
174. Chambial, S., Dwivedi, S., Shukla, K. K., John, P. J., & Sharma, P. (2013). Vitamin C in disease prevention and cure: An overview. *Indian Journal of Clinical Biochemistry, 28*(4), 314–328.
175. Khor, T. O., Keum, Y. S., Lin, W., Kim, J. H., Hu, R., Shen, G., et al. (2006). Combined inhibitory effects of curcumin and phenethyl isothiocyanate on the growth of human PC-3 prostate xenografts in immunodeficient mice. *Cancer Research, 66*(2), 613–621.
176. Azarenko, O., Okouneva, T., Singletary, K. W., Jordan, M. A., & Wilson, L. (2008). Suppression of microtubule dynamic instability and turnover in MCF7 breast cancer cells by sulforaphane. *Carcinogenesis, 29*(12), 2360–2368.
177. Pledgie-Tracy, A., Sobolewski, M. D., & Davidson, N. E. (2007). Sulforaphane induces cell type-specific apoptosis in human breast cancer cell lines. *Molecular Cancer Therapeutics, 6*(3), 1013–1021.
178. de Figueiredo, S. M., Filho, S. A., Nogueira-Machado, J. A., & Caligiorne, R. B. (2013). The anti-oxidant properties of isothiocyanates: A review. *Recent Patents on Endocrine Metabolic & Immune Drug Discovery, 7*(3), 213–225.
179. Keck, A. S., & Finley, J. W. (2004). Cruciferous vegetables: Cancer protective mechanisms of glucosinolate hydrolysis products and selenium. *Integrative Cancer Therapies, 3*(1), 5–12.
180. Bianchini, F., & Vainio, H. (2004). Isothiocyanates in cancer prevention. *Drug Metabolism Reviews, 36*(3–4), 655–667.
181. Higdon, J. V., Delage, B., Williams, D. E., & Dashwood, R. H. (2007). Cruciferous vegetables and human cancer risk: Epidemiologic evidence and mechanistic basis. *Pharmacological Research, 55*(3), 224–236.
182. Gao, X., & Talalay, P. (2004). Induction of phase 2 genes by sulforaphane protects retinal pigment epithelial cells against photooxidative damage. *Proceedings of the National Academy of Sciences of the United States of America, 101*(28), 10446–10451.
183. Senanayake, G. V., Banigesh, A., Wu, L., Lee, P., & Juurlink, B. H. (2012). The dietary phase 2 protein inducer sulforaphane can normalize the kidney epigenome and improve blood pressure in hypertensive rats. *American Journal of Hypertension, 25*(2), 229–235.
184. Zhang, X., Shu, X. O., Xiang, Y. B., Yang, G., Li, H., Gao, J., et al. (2011). Cruciferous vegetable consumption is associated with a reduced risk of total and cardiovascular disease mortality. *The American Journal of Clinical Nutrition, 94*(1), 240–246.
185. Wang, C. Z., Mehendale, S. R., Calway, T., & Yuan, C. S. (2011). Botanical flavonoids on coronary heart disease. *The American Journal of Chinese Medicine, 39*(4), 661–671.
186. Haddy, F. J., Vanhoutte, P. M., & Feletou, M. (2006). Role of potassium in regulating blood flow and blood pressure. *American Journal of Physiology. Regulatory, Integrative and Comparative Physiology, 290*(3), R546–R552.
187. Houston, M. C. (2011). The importance of potassium in managing hypertension. *Current Hypertension Reports, 13*(4), 309–317.
188. Verma, H., & Garg, R. (2019). Effect of vitamin K supplementation on cardiometabolic risk factors: A systematic review and meta-analysis. *Endocrine, Metabolic & Immune Disorders Drug Targets, 19*(1), 13–25.

189. He, F. J., & MacGregor, G. A. (2008). Beneficial effects of potassium on human health. *Physiologia Plantarum, 133*(4), 725–735.
190. Weaver, C. M. (2013). Potassium and health. *Advances in Nutrition, 4*(3), 368s–377s.
191. Bowen, K. J., Harris, W. S., & Kris-Etherton, P. M. (2016). Omega-3 fatty acids and cardiovascular disease: Are there benefits? *Current Treatment Options in Cardiovascular Medicine, 18*(11), 69.
192. Hartley, L., Clar, C., Ghannam, O., Flowers, N., Stranges, S., & Rees, K. (2015). Vitamin K for the primary prevention of cardiovascular disease. *Cochrane Database of Systematic Reviews, 9*, Cd011148.
193. Lanham-New, S. A. (2008). Importance of calcium, vitamin D and vitamin K for osteoporosis prevention and treatment. *The Proceedings of the Nutrition Society, 67*(2), 163–176.
194. The National Academies. (1998). *The B vitamins and choline: Overview and methods, in dietary reference intakes for thiamin, riboflavin, niacin, vitamin B6, folate, vitamin B12, pantothenic acid, biotin, and choline*. Washington, DC: The National Academies Press.
195. Cornatzer, W. E. (1960). Lipotropic agents and lipid transport. *The American Journal of Clinical Nutrition, 8*(3), 306–309.
196. Zeisel, S. H., & da Costa, K. A. (2009). Choline: An essential nutrient for public health. *Nutrition Reviews, 67*(11), 615–623.
197. Blusztajn, J. K., Slack, B. E., & Mellott, T. J. (2017). Neuroprotective actions of dietary choline. *Nutrients, 9*(8), E815.
198. Poly, C., Massaro, J. M., Seshadri, S., Wolf, P. A., Cho, E., Krall, E., et al. (2011). The relation of dietary choline to cognitive performance and white-matter hyperintensity in the Framingham offspring cohort. *The American Journal of Clinical Nutrition, 94*(6), 1584–1591.
199. Bonetti, F., Brombo, G., & Zuliani, G. (2017). Chapter 15: The role of B group vitamins and choline in cognition and brain aging. In R. R. Watson (Ed.), *Nutrition and functional foods for healthy aging* (pp. 139–158). Cambridge, MA: Academic Press.
200. Li, Q., Guo-Ross, S., Lewis, D. V., Turner, D., White, A. M., Wilson, W. A., et al. (2004). Dietary prenatal choline supplementation alters postnatal hippocampal structure and function. *Journal of Neurophysiology, 91*(4), 1545–1555.
201. Kennedy, D. O. (2016). B vitamins and the brain: Mechanisms, dose and efficacy—A review. *Nutrients, 8*(2), 68.
202. Carlson, D. G. (1987). Glucosinolates in crucifer vegetables: Broccoli, Brussels sprouts, cauliflower, collards, kale, mustard greens, and kohlrabi. *Journal of the American Society for Horticultural Science, 112*(1), 173.
203. Fuentes, F., Paredes-Gonzalez, X., & Kong, A.-N. T. (2015). Dietary glucosinolates sulforaphane, phenethyl isothiocyanate, Indole-3-Carbinol/3,3′-Diindolylmethane: Antioxidative stress/inflammation, Nrf2, epigenetics/epigenomics and in vivo cancer chemopreventive efficacy. *Current Pharmacology Reports, 1*(3), 179–196.
204. Rajendran, P., Ho, E., Williams, D. E., & Dashwood, R. H. (2011). Dietary phytochemicals, HDAC inhibition, and DNA damage/repair defects in cancer cells. *Clinical Epigenetics, 3*(1), 4.
205. Fan, S., Meng, Q., Xu, J., Jiao, Y., Zhao, L., Zhang, X., et al. (2013). DIM (3,3′-diindolylmethane) confers protection against ionizing radiation by a unique mechanism. *Proceedings of the National Academy of Sciences of the United States of America, 110*(46), 18650–18655.
206. Lila, M. A. (2004). Anthocyanins and human health: An in vitro investigative approach. *Journal of Biomedicine & Biotechnology, 2004*(5), 306–313.
207. Khoo, H. E., Azlan, A., Tang, S. T., & Lim, S. M. (2017). Anthocyanidins and anthocyanins: Colored pigments as food, pharmaceutical ingredients, and the potential health benefits. *Food & Nutrition Research, 61*(1), 1361779.
208. Hagiwara, A., Yoshino, H., Ichihara, T., Kawabe, M., Tamano, S., Aoki, H., et al. (2002). Prevention by natural food anthocyanins, purple sweet potato color and red cabbage color, of 2-amino-1-methyl-6-phenylimidazo[4,5-b]pyridine (PhIP)-associated colorectal carcinogenesis in rats initiated with 1,2-dimethylhydrazine. *The Journal of Toxicological Sciences, 27*(1), 57–68.

209. Lee, Y. M., Yoon, Y., Yoon, H., Park, H. M., Song, S., & Yeum, K. J. (2017). Dietary anthocyanins against obesity and inflammation. *Nutrients, 9*(10), 1089.
210. Miyake, S., Takahashi, N., Sasaki, M., Kobayashi, S., Tsubota, K., & Ozawa, Y. (2012). Vision preservation during retinal inflammation by anthocyanin-rich bilberry extract: Cellular and molecular mechanism. *Laboratory Investigation, 92*(1), 102–109.
211. Kent, K., Charlton, K., Roodenrys, S., Batterham, M., Potter, J., Traynor, V., et al. (2017). Consumption of anthocyanin-rich cherry juice for 12 weeks improves memory and cognition in older adults with mild-to-moderate dementia. *European Journal of Nutrition, 56*(1), 333–341.
212. Kent, K., Charlton, K. E., Netzel, M., & Fanning, K. (2017). Food-based anthocyanin intake and cognitive outcomes in human intervention trials: A systematic review. *Journal of Human Nutrition and Dietetics, 30*(3), 260–274.
213. Durham, A., Jazrawi, E., Rhodes, J. A., Williams, C., Kilty, I., Barnes, P., et al. (2014). The anti-inflammatory effects of sulforaphane are not mediated by the Nrf2 pathway. *European Respiratory Journal, 44*(Suppl 58), P3332.
214. Rajendran, P., Rengarajan, T., Nandakumar, N., Palaniswami, R., Nishigaki, Y., & Nishigaki, I. (2014). Kaempferol, a potential cytostatic and cure for inflammatory disorders. *European Journal of Medicinal Chemistry, 86*, 103–112.
215. Chong, M. F., Macdonald, R., & Lovegrove, J. A. (2010). Fruit polyphenols and CVD risk: A review of human intervention studies. *The British Journal of Nutrition, 104*(Suppl 3), S28–S39.
216. Ahmadiani, N., Robbins, R. J., Collins, T. M., & Giusti, M. M. (2014). Anthocyanins contents, profiles, and color characteristics of red cabbage extracts from different cultivars and maturity stages. *Journal of Agricultural and Food Chemistry, 62*(30), 7524–7531.
217. Kolodziejczyk, J., Saluk, J., Posmyk, M. M., Janas, K. M., & Wachowicz, B. (2011). Red cabbage anthocyanins may protect blood plasma proteins and lipids. *Central European Journal of Biology, 6*(4), 565.
218. McDonough, A. A., Veiras, L. C., Guevara, C. A., & Ralph, D. L. (2017). Cardiovascular benefits associated with higher dietary K(+) vs. lower dietary Na(+): Evidence from population and mechanistic studies. *American Journal of Physiology. Endocrinology and Metabolism, 312*(4), E348–E356.
219. Ren, J., Grundyb, S. M., Liua, J., Wanga, W., Wanga, M., Suna, J., et al. (2010). Long-term coronary heart disease risk associated with very-low-density lipoprotein cholesterol in Chinese: The results of a 15-year Chinese Multi-Provincial Cohort Study (CMCS). *Atherosclerosis, 211*(1), 327–332.
220. Ras, R. T., Geleijnse, J. M., & Trautwein, E. A. (2014). LDL-cholesterol-lowering effect of plant sterols and stanols across different dose ranges: A meta-analysis of randomised controlled studies. *The British Journal of Nutrition, 112*(2), 214–219.
221. Shareck, M., Rousseau, M. C., Koushik, A., Siemiatycki, J., & Parent, M. E. (2017). Inverse association between dietary intake of selected carotenoids and vitamin C and risk of lung cancer. *Frontiers in Oncology, 7*, 23.
222. Lobo, V., Patil, A., Phatak, A., & Chandra, N. (2010). Free radicals, antioxidants and functional foods: Impact on human health. *Pharmacognosy Reviews, 4*(8), 118–126.
223. Teucher, B., Olivares, M., & Cori, H. (2004). Enhancers of iron absorption: Ascorbic acid and other organic acids. *International Journal for Vitamin and Nutrition Research, 74*(6), 403–419.
224. Yang, J., Wang, H. P., Zhou, L., & Xu, C. F. (2012). Effect of dietary fiber on constipation: A meta analysis. *World Journal of Gastroenterology, 18*(48), 7378–7383.
225. Ehle, F. R., Robertson, J. B., & Van Soest, P. J. (1982). Influence of dietary fibers on fermentation in the human large intestine. *The Journal of Nutrition, 112*(1), 158–166.
226. Larijani, B., Esfahani, M. M., Moghimi, M., Shams Ardakani, M. R., Keshavarz, M., Kordafshari, G., et al. (2016). Prevention and treatment of flatulence from a traditional persian medicine perspective. *Iranian Red Crescent Medical Journal, 18*(4), e23664.

227. Weng, H. X., Hong, C., Yan, A. L., Pan, L. H., Qin, Y. C., Bao, L. T., et al. (2008). Mechanism of iodine uptake by cabbage: Effects of iodine species and where it is stored. *Biological Trace Element Research, 125*(1), 59–71.
228. Platel, K., & Srinivasan, K. (1997). Plant foods in the management of diabetes mellitus: Vegetables as potential hypoglycaemic agents. *Nahrung, 41*(2), 68–74.
229. Lekshmi, P., Viveka, S., Jeeva, S., & Raja Brindha, J. (2015). Efficacy of crude extracts of Allium sativum and Allium cepa against human pathogens. *Advances in Applied Science Research, 6*(1), 72–78.
230. Hamza, H. J. (2014). In vitro antimicrobial activity of garlic, onion, garlic-onion combination (aquatic and oil) extract on some microbial pathogens in Babylon province, Iraq. *World Journal of Pharmacy and Pharmaceutical Science, 3*(8), 65–78.
231. Mariana, L., Coprean, D., Dinica, R. M., Lupoae, P., Gurău, G., & Bahrim, G. (2013). Antimicrobial activity of extracts of wild garlic (Allium ursinum) from Romanian spontaneous flora. *Scientific Study and Research: Chemistry and Chemical Engineering, 14*(4), 221–227.
232. Mahomoodally, F., Ramcharun, S., & Zengin, G. (2018). Onion and garlic extracts potentiate the efficacy of conventional antibiotics against standard and clinical bacterial isolates. *Current Topics in Medicinal Chemistry, 18*(9), 787–796.
233. Moutia, M., Habti, N., & Badou, A. (2018). In vitro and in vivo immunomodulator activities of Allium sativum L. *Journal of Evidence-Based Complementary and Alternative Medicine, 2018*, 10.
234. Josling, P. (2001). Preventing the common cold with a garlic supplement: A double-blind, placebo-controlled survey. *Advances in Therapy, 18*(4), 189–193.
235. Nantz, M. P., Rowe, C. A., Muller, C. E., Creasy, R. A., Stanilka, J. M., & Percival, S. S. (2012). Supplementation with aged garlic extract improves both NK and gammadelta-T cell function and reduces the severity of cold and flu symptoms: A randomized, double-blind, placebo-controlled nutrition intervention. *Clinical Nutrition, 31*(3), 337–344.
236. Lissiman, E., Bhasale, A. L., & Cohen, M. (2014). Garlic for the common cold. *Cochrane Database of Systematic Reviews, 11*, Cd006206.
237. Sharquie, K. E., & Al-Obaidi, H. K. (2002). Onion juice (Allium cepa L.), a new topical treatment for alopecia areata. *The Journal of Dermatology, 29*(6), 343–346.
238. Matheson, E. M., Mainous 3rd, A. G., & Carnemolla, M. A. (2009). The association between onion consumption and bone density in perimenopausal and postmenopausal non-Hispanic white women 50 years and older. *Menopause, 16*(4), 756–759.
239. Mozaffari-Khosravi, H., Hesabgar, H. A., Owlia, M. B., Hadinedoushan, H., Barzegar, K., & Fllahzadeh, M. H. (2012). The effect of garlic tablet on pro-inflammatory cytokines in postmenopausal osteoporotic women: A randomized controlled clinical trial. *Journal of Dietary Supplements, 9*(4), 262–271.
240. Putnam, S. E., Scutt, A. M., Bicknell, K., Priestley, C. M., & Williamson, E. M. (2007). Natural products as alternative treatments for metabolic bone disorders and for maintenance of bone health. *Phytotherapy Research, 21*(2), 99–112.
241. Schafer, G., & Kaschula, C. H. (2014). The immunomodulation and anti-inflammatory effects of garlic organosulfur compounds in cancer chemoprevention. *Anti-Cancer Agents in Medicinal Chemistry, 14*(2), 233–240.
242. Bacciottini, L., Falchetti, A., Pampaloni, B., Bartolini, E., Carossino, A. M., & Brandi, M. L. (2007). Phytoestrogens: food or drug? *Clinical Cases in Mineral and Bone Metabolism: The Official Journal of the Italian Society of Osteoporosis, Mineral Metabolism, and Skeletal Diseases, 4*(2), 123–130.
243. Nicastro, H. L., Ross, S. A., & Milner, J. A. (2015). Garlic and onions: Their cancer prevention properties. *Cancer Prevention Research (Philadelphia, Pa.), 8*(3), 181–189.
244. Murayyan, A. I., Manohar, C. M., Hayward, G., & Neethirajan, S. (2017). Antiproliferative activity of Ontario grown onions against colorectal adenocarcinoma cells. *Food Research International, 96*, 12–18.

245. Egert, S., Bosy-Westphal, A., Seiberl, J., Kürbitz, C., Settler, U., Plachta-Danielzik, S., et al. (2009). Quercetin reduces systolic blood pressure and plasma oxidised low-density lipoprotein concentrations in overweight subjects with a high-cardiovascular disease risk phenotype: A double-blinded, placebo-controlled cross-over study. *The British Journal of Nutrition, 102*(7), 1065–1074.
246. Vazquez-Prieto, M. A., & Miatello, R. M. (2010). Organosulfur compounds and cardiovascular disease. *Molecular Aspects of Medicine, 31*(6), 540–545.
247. Brull, V., Burak, C., Stoffel-Wagner, B., Wolffram, S., Nickenig, G., Müller, C., et al. (2015). Effects of a quercetin-rich onion skin extract on 24 h ambulatory blood pressure and endothelial function in overweight-to-obese patients with (pre-)hypertension: A randomised double-blinded placebo-controlled cross-over trial. *The British Journal of Nutrition, 114*(8), 1263–1277.
248. Ried, K., Frank, O. R., & Stocks, N. P. (2010). Aged garlic extract lowers blood pressure in patients with treated but uncontrolled hypertension: A randomised controlled trial. *Maturitas, 67*(2), 144–150.
249. Dhawan, V., & Jain, S. (2005). Garlic supplementation prevents oxidative DNA damage in essential hypertension. *Molecular and Cellular Biochemistry, 275*(1–2), 85–94.
250. Ashraf, R., Khan, R. A., Ashraf, I., & Qureshi, A. A. (2013). Effects of Allium sativum (garlic) on systolic and diastolic blood pressure in patients with essential hypertension. *Pakistan Journal of Pharmaceutical Sciences, 26*(5), 859–863.
251. Stevinson, C., Pittler, M. H., & Ernst, E. (2000). Garlic for treating hypercholesterolemia. A meta-analysis of randomized clinical trials. *Annals of Internal Medicine, 133*(6), 420–429.
252. Silagy, C., & Neil, A. (1994). Garlic as a lipid lowering agent—A meta-analysis. *Journal of the Royal College of Physicians of London, 28*(1), 39–45.
253. Ried, K., Toben, C., & Fakler, P. (2013). Effect of garlic on serum lipids: An updated meta-analysis. *Nutrition Reviews, 71*(5), 282–299.
254. Bekendam, R. H., & Flaumenhaft, R. (2016). Inhibition of protein disulfide isomerase in thrombosis. *Basic & Clinical Pharmacology & Toxicology, 119*(Suppl 3), 42–48.
255. Flaumenhaft, R. (2017). Advances in vascular thiol isomerase function. *Current Opinion in Hematology, 24*(5), 439–445.
256. Ganeshpurkar, A., & Saluja, A. K. (2017). The pharmacological potential of rutin. *Saudi Pharmaceutical Journal, 25*(2), 149–164.
257. Kim, Y., Keogh, J. B., & Clifton, P. M. (2016). Polyphenols and glycemic control. *Nutrients, 8*(1), E17.
258. The Endocrine Society. (2015, March). *Onion extract may improve high blood sugar and cholesterol*. Retrieved from https://www.sciencedaily.com/releases/2015/03/150306102513.htm
259. Ademiluyi, A. O., Oboh, G., Owoloye, T. R., & Agbebi, O. J. (2013). Modulatory effects of dietary inclusion of garlic (Allium sativum) on gentamycin-induced hepatotoxicity and oxidative stress in rats. *Asian Pacific Journal of Tropical Biomedicine, 3*(6), 470–475.
260. Borek, C. (2001). Antioxidant health effects of aged garlic extract. *The Journal of Nutrition, 131*(3s), 1010s–1015s.
261. Amagase, H., Petesch, B. L., Matsuura, H., Kasuga, S., & Itakura, Y. (2001). Intake of garlic and its bioactive components. *The Journal of Nutrition, 131*(3s), 955s–962s.
262. El-Barbary, M. I. (2016). Detoxification and antioxidant effects of garlic and curcumin in Oreochromis niloticus injected with aflatoxin B(1) with reference to gene expression of glutathione peroxidase (GPx) by RT-PCR. *Fish Physiology and Biochemistry, 42*(2), 617–629.
263. Avci, A., Ergüder, I. B., Varli, M., Devrim, E., Aras, S., et al. (2008). Effects of garlic consumption on plasma and erythrocyte antioxidant parameters in elderly subjects. *Gerontology, 54*(3), 173–176.
264. Kianoush, S., Balali-Mood, M., Mousavi, S. R., Moradi, V., Sadeghi, M., Dadpour, B., et al. (2012). Comparison of therapeutic effects of garlic and d-Penicillamine in patients with chronic occupational lead poisoning. *Basic & Clinical Pharmacology & Toxicology, 110*(5), 476–481.

265. Emamat, H., Foroughi, F., Eini-Zinab, H., Taghizadeh, M., Rismanchi, M., & Hekmatdoost, A. (2016). The effects of onion consumption on treatment of metabolic, histologic, and inflammatory features of nonalcoholic fatty liver disease. *Journal of Diabetes and Metabolic Disorders, 15*, 25.
266. Chu, C. C., Wu, W. S., Shieh, J. P., Chu, H. L., Lee, C. P., & Duh, P. D. (2017). The anti-inflammatory and vasodilating effects of three selected dietary organic sulfur compounds from allium species. *Journal of Functional Biomaterials, 8*(1), 5.
267. Lee, D. Y., Li, H., Lim, H. J., Lee, H. J., Jeon, R., & Ryu, J. H. (2012). Anti-inflammatory activity of sulfur-containing compounds from garlic. *Journal of Medicinal Food, 15*(11), 992–999.
268. Azab, A., Nassar, A., & Azab, A. N. (2016). Anti-inflammatory activity of natural products. *Molecules, 21*(10), E1321.
269. Hussain, T., Tan, B., Yin, Y., Blachier, F., Tossou, M. C., & Rahu, N. (2016). Oxidative stress and inflammation: What polyphenols can do for us? *Oxidative Medicine and Cellular Longevity, 2016*, 7432797.
270. Gonzalez, R., Ballester, I., López-Posadas, R., Suárez, M. D., Zarzuelo, A., Martínez-Augustin, O., et al. (2011). Effects of flavonoids and other polyphenols on inflammation. *Critical Reviews in Food Science and Nutrition, 51*(4), 331–362.
271. Borek, C. (2006). Garlic reduces dementia and heart-disease risk. *The Journal of Nutrition, 136*(3 Suppl), 810s–812s.
272. Zarezadeh, M., Baluchnejadmojarad, T., Kiasalari, Z., Afshin-Majd, S., & Roghani, M. (2017). Garlic active constituent s-allyl cysteine protects against lipopolysaccharide-induced cognitive deficits in the rat: Possible involved mechanisms. *European Journal of Pharmacology, 795*, 13–21.
273. Banerjee, S. K., Mukherjee, P. K., & Maulik, S. K. (2003). Garlic as an antioxidant: The good, the bad and the ugly. *Phytotherapy Research, 17*(2), 97–106.
274. Ried, K., Frank, O. R., & Stocks, N. P. (2013). Aged garlic extract reduces blood pressure in hypertensives: A dose-response trial. *European Journal of Clinical Nutrition, 67*(1), 64–70.
275. Mitra, J., Shrivastava, S. L., & Rao, P. S. (2012). Onion dehydration: A review. *Journal of Food Science and Technology, 49*(3), 267–277.
276. Tortorella, S. M., Royce, S. G., Licciardi, P. V., & Karagiannis, T. C. (2015). Dietary sulforaphane in cancer chemoprevention: The role of epigenetic regulation and HDAC inhibition. *Antioxidants & Redox Signaling, 22*(16), 1382–1424.
277. Azarenko, O., Jordan, M. A., & Wilson, L. (2014). Erucin, the major isothiocyanate in arugula (Eruca sativa), inhibits proliferation of MCF7 tumor cells by suppressing microtubule dynamics. *PLoS One, 9*(6), e100599.
278. Zhang, Y. (2004). Cancer-preventive isothiocyanates: Measurement of human exposure and mechanism of action. *Mutation Research, 555*(1–2), 173–190.
279. Shaughnessy, D. T., Gangarosa, L. M., Schliebe, B., Umbach, D. M., Xu, Z., MacIntosh, B., et al. (2011). Inhibition of fried meat-induced colorectal DNA damage and altered systemic genotoxicity in humans by crucifera, chlorophyllin, and yogurt. *PLoS One, 6*(4), e18707.
280. Lea, M. A. (2015). Flavonol regulation in tumor cells. *Journal of Cellular Biochemistry, 116*(7), 1190–1194.
281. Romagnolo, D. F., & Selmin, O. I. (2012). Flavonoids and cancer prevention: A review of the evidence. *Journal of Nutrition in Gerontology and Geriatrics, 31*(3), 206–238.
282. Batra, P., & Sharma, A. K. (2013). Anti-cancer potential of flavonoids: Recent trends and future perspectives. *3 Biotech, 3*(6), 439–459.
283. Mijnhout, G. S., Alkhalaf, A., Kleefstra, N., & Bilo, H. J. (2010). Alpha lipoic acid: A new treatment for neuropathic pain in patients with diabetes? *The Netherlands Journal of Medicine, 68*(4), 158–162.
284. El-Missiry, M. A., & El Gindy, A. M. (2000). Amelioration of alloxan induced diabetes mellitus and oxidative stress in rats by oil of Eruca sativa seeds. *Annals of Nutrition & Metabolism, 44*(3), 97–100.

285. Gomes, M. B., & Negrato, C. A. (2014). Alpha-lipoic acid as a pleiotropic compound with potential therapeutic use in diabetes and other chronic diseases. *Diabetology and Metabolic Syndrome, 6*(1), 80.
286. Golbidi, S., Badran, M., & Laher, I. (2011). Diabetes and alpha lipoic acid. *Frontiers in Pharmacology, 2*, 69–69.
287. Ziegler, D., Reljanovic, M., Mehnert, H., & Gries, F. A. (1999). Alpha-lipoic acid in the treatment of diabetic polyneuropathy in Germany: Current evidence from clinical trials. *Experimental and Clinical Endocrinology & Diabetes, 107*(7), 421–430.
288. Zusman, R. (2014). Key minerals to help control blood pressure. It's usually best to get calcium, magnesium, and potassium from food. Are you getting enough. *Harvard Health Letter, 39*(10), 5.
289. Pitkin, R. M. (2007). Folate and neural tube defects. *The American Journal of Clinical Nutrition, 85*(1), 285s–288s.
290. Brosnan, J. T., & Brosnan, M. E. (2006). The sulfur-containing amino acids: An overview. *The Journal of Nutrition, 136*(6 Suppl), 1636s–1640s.
291. Mahmood, L. (2014). The metabolic processes of folic acid and vitamin B12 deficiency. *Journal of Health Research and Review, 1*(1), 5–9.
292. Ganguly, P., & Alam, S. F. (2015). Role of homocysteine in the development of cardiovascular disease. *Nutrition Journal, 14*, 6.
293. Blom, H. J., & Smulders, Y. (2011). Overview of homocysteine and folate metabolism. With special references to cardiovascular disease and neural tube defects. *Journal of Inherited Metabolic Disease, 34*(1), 75–81.
294. Aleman, G., Tovar, A. R., & Torres, N. (2001). Homocysteine metabolism and risk of cardiovascular diseases: Importance of the nutritional status on folic acid, vitamins B6 and B12. *Revista de Investigación Clínica, 53*(2), 141–151.
295. Jones, A. M. (2014). Dietary nitrate supplementation and exercise performance. *Sports Medicine, 44*(Suppl 1), S35–S45.
296. Johnson, E. J. (2002). The role of carotenoids in human health. *Nutrition in Clinical Care, 5*(2), 56–65.
297. Nossaman, V. E., Nossaman, B. D., & Kadowitz, P. J. (2010). Nitrates and nitrites in the treatment of ischemic cardiac disease. *Cardiology in Review, 18*(4), 190–197.
298. Corleto, K. A., Singh, J., Jayaprakasha, G. K., & Patil, B. S. (2018). Storage stability of dietary nitrate and phenolic compounds in beetroot (Beta vulgaris) and arugula (Eruca sativa) juices. *Journal of Food Science, 83*(5), 1237–1248.
299. Bondonno, C. P., Yang, X., Croft, K. D., Considine, M. J., Ward, N. C., Rich, L., et al. (2012). Flavonoid-rich apples and nitrate-rich spinach augment nitric oxide status and improve endothelial function in healthy men and women: A randomized controlled trial. *Free Radical Biology & Medicine, 52*(1), 95–102.
300. Affourtit, C., Bailey, S. J., Jones, A. M., Smallwood, M. J., & Winyard, P. G. (2015). On the mechanism by which dietary nitrate improves human skeletal muscle function. *Frontiers in Physiology, 6*, 211.
301. Li, T., Lu, X., Sun, Y., & Yang, X. (2016). Effects of spinach nitrate on insulin resistance, endothelial dysfunction markers and inflammation in mice with high-fat and high-fructose consumption. *Food & Nutrition Research, 60*, 32010.
302. Maeda, N., Matsubara, K., Yoshida, H., & Mizushina, Y. (2011). Anti-cancer effect of spinach glycoglycerolipids as angiogenesis inhibitors based on the selective inhibition of DNA polymerase activity. *Mini Reviews in Medicinal Chemistry, 11*(1), 32–38.
303. Matsubara, K., Matsumoto, H., Mizushina, Y., Mori, M., Nakajima, N., Fuchigami, M., et al. (2005). Inhibitory effect of glycolipids from spinach on in vitro and ex vivo angiogenesis. *Oncology Reports, 14*(1), 157–160.
304. Al-Qahtani, K., Jabeen, B., Sekirnik, R., Riaz, N., Claridge, T. D. W., Schofield, C. J., et al. (2015). The broad spectrum 2-oxoglutarate oxygenase inhibitor N-oxalylglycine is present in rhubarb and spinach leaves. *Phytochemistry, 117*, 456–461.
305. Castenmiller, J. (2000). *Spinach as a source of carotenoids, folate and antioxidant activity* (PhD Thesis, Wageningen University, p. 183).

306. Ko, S. H., Park, J. H., Kim, S. Y., Lee, S. W., Chun, S. S., & Park, E. (2014). Antioxidant effects of spinach (Spinacia oleracea L.) supplementation in hyperlipidemic rats. *Preventive Nutrition and Food Science, 19*(1), 19–26.
307. Porrini, M., Riso, P., & Oriani, G. (2002). Spinach and tomato consumption increases lymphocyte DNA resistance to oxidative stress but this is not related to cell carotenoid concentrations. *European Journal of Nutrition, 41*(3), 95–100.
308. Chen, A. Y., & Chen, Y. C. (2013). A review of the dietary flavonoid, kaempferol on human health and cancer chemoprevention. *Food Chemistry, 138*(4), 2099–2107.
309. Boots, A. W., Haenen, G. R., & Bast, A. (2008). Health effects of quercetin: From antioxidant to nutraceutical. *European Journal of Pharmacology, 585*(2–3), 325–337.
310. Terasaki, M., Mutoh, M., Fujii, G., Takahashi, M., Ishigamori, R., & Masuda, S. (2014). Potential ability of xanthophylls to prevent obesity-associated cancer. *World Journal of Pharmacology, 3*(4), 140–152.
311. Pasquet, V., Morisset, P., Ihammouine, S., Chepied, A., Aumailley, L., & Berard, J. B. (2011). Antiproliferative activity of violaxanthin isolated from bioguided fractionation of Dunaliella tertiolecta extracts. *Marine Drugs, 9*(5), 819–831.
312. Kotake-Nara, E., Asai, A., & Nagao, A. (2005). Neoxanthin and fucoxanthin induce apoptosis in PC-3 human prostate cancer cells. *Cancer Letters, 220*(1), 75–84.
313. Krinsky, N. I., Landrum, J. T., & Bone, R. A. (2003). Biologic mechanisms of the protective role of lutein and zeaxanthin in the eye. *Annual Review of Nutrition, 23*, 171–201.
314. Hagander, B., Asp, N. G., Efendić, S., Nilsson-Ehle, P., & Scherstén, B. (1988). Dietary fiber decreases fasting blood glucose levels and plasma LDL concentration in noninsulin-dependent diabetes mellitus patients. *The American Journal of Clinical Nutrition, 47*(5), 852–858.
315. Brown, L., Rosner, B., Willett, W. W., & Sacks, F. M. (1999). Cholesterol-lowering effects of dietary fiber: A meta-analysis. *The American Journal of Clinical Nutrition, 69*(1), 30–42.
316. Post, R. E., Mainous 3rd, A. G., King, D. E., & Simpson, K. N. (2012). Dietary fiber for the treatment of type 2 diabetes mellitus: A meta-analysis. *Journal of American Board of Family Medicine, 25*(1), 16–23.
317. Saraswat, M., Muthenna, P., Suryanarayana, P., Petrash, J. M., & Reddy, G. B. (2008). Dietary sources of aldose reductase inhibitors: Prospects for alleviating diabetic complications. *Asia Pacific Journal of Clinical Nutrition, 17*(4), 558–565.
318. Massey, L. K., Roman-Smith, H., & Sutton, R. A. (1993). Effect of dietary oxalate and calcium on urinary oxalate and risk of formation of calcium oxalate kidney stones. *Journal of the American Dietetic Association, 93*(8), 901–906.
319. Chang, C. H., Wang, Y. W., Yeh Liu, P. Y., & Kao Yang, Y. H. (2014). A practical approach to minimize the interaction of dietary vitamin K with warfarin. *Journal of Clinical Pharmacy and Therapeutics, 39*(1), 56–60.
320. Kim, M. J., Moon, Y., Tou, J. C., Mou, B., & Waterland, N. L. (2016). Nutritional value, bioactive compounds and health benefits of lettuce (Lactuca sativa L.). *Journal of Food Composition and Analysis, 49*, 19–34.
321. Zdravković, J., Pavlovic, N. V., Pavlovic, R., Maskovic, P., Mladenović, J., Durić, M., et al. (2012). Antimicrobial activity of lettuce (Lactuca sativa L.) extract grown in plastic and glasshouses. *Acta Horticulturae, 960*, 299–303.
322. Moulin-Traffort, J., Giordani, R., & Regli, P. (1990). Antifungal action of latex saps from Lactuca sativa L. and Asclepias curassavica L. *Mycoses, 33*(7–8), 383–392.
323. Edziri, H. L., Smach, M. A., Ammar, S., Mahjoub, M. A., Gannoun, S., Aouni, M., et al. (2011). Antioxidant, antibacterial and antiviral activities of Lactuca Sativa extracts. *Industrial Crops and Products, 34*(1), 1182–1185.
324. Araruna, K., & Carlos, B. (2010). Anti-inflammatory activities of triterpene lactones from Lactuca sativa. *Phytopharmacology, 1*(1), 1–6.
325. Kim, D.-E., Shang, X., Assefa, A. D., Keum, Y. S., & Saini, R. K. (2018). Metabolite profiling of green, green/red, and red lettuce cultivars: Variation in health beneficial compounds and antioxidant potential. *Journal of Food Research International, 105*, 361–370.

326. Lopez, A., Javier, G. A., Fenoll, J., Hellín, P., & Flores, P. (2014). Chemical composition and antioxidant capacity of lettuce: Comparative study of regular-sized (Romaine) and baby-sized (Little Gem and Mini Romaine) types. *Journal of Food Composition and Analysis, 33*(1), 39–84.
327. Altunkaya, A., Arzu, B., Eleonora, M. G., Vural, S., & Leif, H. (2009). Antioxidant activity of lettuce extract (Lactuca sativa) and synergism with added phenolic antioxidants. *Food Chemistry, 115*(1), 163–168.
328. de Vogel, J., Jonker-Termont, D. S., van Lieshout, E. M., Katan, M. B., & van der Meer, R. (2005). Green vegetables, red meat and colon cancer: Chlorophyll prevents the cytotoxic and hyperproliferative effects of haem in rat colon. *Carcinogenesis, 26*(2), 387–393.
329. Mourouti, N., Papavagelis, C., Plytzanopoulou, P., Kontogianni, M., Vassilakou, T., Malamos, N., et al. (2015). Dietary patterns and breast cancer: A case-control study in women. *European Journal of Nutrition, 54*(4), 609–617.
330. Katsouyanni, K., Trichopoulos, D., Boyle, P., Xirouchaki, E., Trichopoulou, A., Lisseos, B., et al. (1986). Diet and breast cancer: A case-control study in Greece. *International Journal of Cancer, 38*(6), 815–820.
331. Gridling, M., Popescu, R., Kopp, B., Wagner, K. H., Krenn, L., & Krupitza, G. (2010). Anti-leukaemic effects of two extract types of Lactuca sativa correlate with the activation of Chk2, induction of p21, downregulation of cyclin D1 and acetylation of alpha-tubulin. *Oncology Reports, 23*(4), 1145–1151.
332. Harsha, S. N., & Anilakumar, K. R. (2013). Anxiolytic property of Lactuca sativa, effect on anxiety behaviour induced by novel food and height. *Asian Pacific Journal of Tropical Medicine, 6*(7), 532–536.
333. Anilakumar, K. R., Harsha, S. N., & Sharma, R. K. (2017). Lettuce: A promising leafy vegetable with functional properties. *Defence Life Science Journal, 2*(2), 178–185.
334. Morris, M. C., Wang, Y., Barnes, L. L., Bennett, D. A., Dawson-Hughes, B., & Booth, S. L. (2018). Nutrients and bioactives in green leafy vegetables and cognitive decline: Prospective study. *Neurology, 90*(3), e214–e222.
335. Brookie, K. L., Best, G. I., & Conner, T. S. (2018). Intake of raw fruits and vegetables is associated with better mental health than intake of processed fruits and vegetables. *Frontiers in Psychology, 9*, 487.
336. Kim, H. D., Hong, K. B., Noh, D. O., & Suh, H. J. (2017). Sleep-inducing effect of lettuce (Lactuca sativa) varieties on pentobarbital-induced sleep. *Food Science and Biotechnology, 26*(3), 807–814.
337. Ghorbani, A., Rakhshandeh, H., & Sadeghnia, H. R. (2013). Potentiating effects of Lactuca sativa on pentobarbital-induced sleep. *Iranian Journal of Pharmaceuticals and Research, 12*(2), 401–406.
338. Im, S. E., Yoon, H., Nam, T. G., Heo, H. J., Lee, C. Y., & Kim, D. O. (2010). Antineurodegenerative effect of phenolic extracts and caffeic acid derivatives in romaine lettuce on neuron-like PC-12 cells. *Journal of Medicinal Food, 13*(4), 779–784.
339. Ghorbani, A., Sadeghnia, H. R., & Asadpour, E. (2015). Mechanism of protective effect of lettuce against glucose/serum deprivation-induced neurotoxicity. *Nutritional Neuroscience, 18*(3), 103–109.
340. Nicolle, C., Cardinault, N., Gueux, E., Jaffrelo, L., Rock, E., Mazur, A., et al. (2004). Health effect of vegetable-based diet: Lettuce consumption improves cholesterol metabolism and antioxidant status in the rat. *Clinical Nutrition, 23*(4), 605–614.
341. Kim, M. J., Moon, Y., Kopsell, D. A., Park, S., Tou, J. C., & Waterland, N. L. (2016). Nutritional value of crisphead 'iceberg' and romaine lettuces (Lactuca sativa L.). *Journal of Agricultural Science, 8*(11), 1–10.
342. Schleicher, M., Weikel, K., Garber, C., & Taylor, A. (2013). Diminishing risk for age-related macular degeneration with nutrition: A current view. *Nutrients, 5*(7), 2405–2456.
343. Stephensen, C. B. (2001). Vitamin A, infection, and immune function. *Annual Review of Nutrition, 21*, 167–192.

Fruits

Sawsan G. Mohammed and M. Walid Qoronfleh

Abstract Fruits come in a wide variety of colors, shapes, and flavors. This chapter will cover selected fruits that are known to be healthy and highly nutritious. These fruits were chosen due to their common usage and availability. Since it is not possible to cover all health benefits or essential nutrients and important phytochemicals of the fruit composition, this chapter will focus on the key valuable constituents and their potential health effects.

Keywords Fruits · Fiber · Antioxidants · Monounsaturated fatty acids · Phenolic acids · Polyphenols · Catechins · Flavonoids · Anthocyanins · Xanthophylls · Chlorophyll · Carotenoids · Capsaicins · Triterpenoids · Tannins · Limonoids · Cucurbitacins · Resveratrol

S. G. Mohammed (✉)
Qatar Research Leadership Program (QRLP), Qatar Foundation, Doha, Qatar
e-mail: sgmohammed@qf.org.qa

M. W. Qoronfleh (✉)
Research & Policy Department, World Innovation Summit for Health (WISH), Qatar Foundation, Doha, Qatar
e-mail: wqoronfleh@qf.org.qa

M. M. Essa, M. W. Qoronfleh (eds.), *Personalized Food Intervention and Therapy for Autism Spectrum Disorder Management*, Advances in Neurobiology 24,
https://doi.org/10.1007/978-3-030-30402-7_10

1 Avocados

Family: Lauraceae
Genus: *Persea*
Common names: Avocado or Alligator Pear

Avocado is a unique stone fruit with a creamy texture and single large seed and color ranging from green to almost black when ripe. In shape, they vary from pear or egg shape to round shape. The edible portion, i.e., the green-yellow flesh, is incredibly nutritious and healthy [1]. It contains low saturated fat but a high content of mono- and polyunsaturated fatty acids in addition to about twenty different vitamins and minerals. Avocado contains high dietary fiber content and is particularly rich in vitamins B, C, E, and K along with important minerals like potassium, copper, and magnesium (Table 1). It also contains appreciable amounts of the common carotenoids, *lutein* and *zeaxanthin*.

Avocado is high in monounsaturated oleic acid (as seen in olive oil): a fatty acid that is believed to be a significant source of its anti-inflammatory [2, 3] and anti-carcinogenic [4] effects. For instance, there is evidence that avocado extracts have a dampening effect on osteoarthritis [5] and that an avocado- and soybean-based nutritional supplement appears to be beneficial for osteoarthritis patients [6]. Existing evidence supports the notion that monounsaturated fatty acids (MFA) like oleic acid may influence breast cancer risk [7, 8], though it appears that this is population dependent. Avocado extract can selectively induce cell cycle arrest, inhibit growth, and induce apoptosis in precancerous and cancer cell lines. Examples include human oral cancer cell lines [9] or specific constituents like aliphatic acetogenins which inhibit human oral cancer cell proliferation by targeting the EGFR/RAS/RAF/MEK/ERK1/2 pathway [10]. Some in vitro studies have shown that carotenoids and vitamin E in avocado extract may have a role in preventing prostate cancer as the extract inhibits the growth of both androgen-dependent (LNCaP) and androgen-independent (PC-3) prostate cancer cell lines [11]. It has been also demonstrated that it displays chemoprotective effects by lowering the side effects of cyclophosphamide chemotherapy in human lymphocyte cell culture [12]. β-Sitosterol is a ubiquitous plant sterol, and avocado is a rich source for the same [13]. The component β-sitosterol is currently being investigated for its potential to reduce benign prostatic hyperplasia (BPH) [14].

Table 1 Avocado nutrition facts of 100 g, from the United States Department of Agriculture (USDA SR-21)

Serving size of 100 g of raw avocado	
Per serving	% Daily value[a]
Calories 160	
Total fat 14.7 g	23
Saturated fat 2.1 g	11
Polyunsaturated fat 1.8 g	
Monounsaturated fat 9.8 g	
Total omega-3 fatty acids 110.0 mg	
Total omega-6 fatty acids 1689.0 mg	
Cholesterol 0 mg	0
Phytosterols	
Total carbohydrates 8.5 g	3
Dietary fiber 6.7 g	27
Starch 0.1 g	
Sugars 0.7 g	
Protein 2 g	4
Vitamins	
Vitamin A	3
Vitamin C	17
Vitamin E	10
Vitamin B-6	13
Vitamin K	26
Folic acid	20
Minerals	
Sodium	0
Potassium	14
Calcium	1
Iron	3
Magnesium	7
Manganese	7
Copper	9
Zinc	4

National Nutrient Database
[a]Based on a 2000 calorie diet

Avocado is high in fiber (25% of which is soluble fiber and the 75% remainder is insoluble fiber) [15]; it is an important component in maintaining body weight and controlling energy level [16]. In a clinical trial among overweight adults, avocado intake increased satisfaction, lowered desire to eat over the next 5 h, and influenced glucose and insulin response [17]. A restricted avocado diet, a source rich in MFA, did not compromise weight loss or adversely impact serum lipids [18, 19]. A vegetarian diet enriched with avocado has been shown to decrease the risk of heart diseases due to dramatic decrease in total cholesterol, 20% reduction in triglysrides and lowered LDL by 22% [1, 20, 21]. Positive effects on lipid profile have largely been attributed to avocado (as a source for MFA) [21, 22]. A study of patients with non-insulin-dependent diabetes showed that an enriched avocado diet (high MFA) improved glycemic control and resulted in favorable lipid profile [23]. In all, avocado consumption has been found to be associated with better diet quality and nutrient intake and lower metabolic syndrome risk in US adults [24]. The fruit also provides a substantial amount of fatty acids, a structural component of cell membranes, thus playing an important role in brain development and function and, ultimately, mental health [25–27].

The fatty content of avocado helps the body absorb fat-soluble nutrients such as vitamins A, K, D, and E as well as carotenoid antioxidants [28]. Potassium intake by the general population is low [29, 30]. Avocado is a rich source of potassium (actually more than bananas). It is linked to lower blood pressure [31] and along with other phytonutrient factors to a healthy cardiovascular system [1]. Evidence has supported that fact, that is, the role of the appreciable content of vitamin K in cardiovascular health [32] and keeping healthy bones [33, 34]. The high levels of folic acid in this fruit is important in maintaining healthy pregnancy and reducing the risk of miscarriage, neural tube defects, and other congenital anomalies [35–37]. Folic acid may also help prevent the buildup of homocysteine, thereby improving one's mood and reducing depression, the foremost reason being that homocysteine buildup depletes production of dopamine and serotonin neurotransmitters promoting a sense of well-being. Other reasons include avocados' high tyrosine (an amino acid) content that is a precursor to dopamine and MFAs that support the production of acetylcholine. Furthermore, homocysteine is a definite trigger for inflammation. Elevated homocysteine levels can lead to various heart diseases [38–41].

Avocado is rich in the antioxidants lutein and zeaxanthin, found to decrease the risk of macular degeneration and cataracts and improve overall eye health [42]. Moreover, these antioxidants are capable of lessening signs of aging by protecting the skin from damage by both UV rays and radiation [43]. On the other hand, other vitamins such as C, E, and K stimulate collagen synthesis [44–46] and scavenge radicals [47]. Lutein from avocado has also been found to improve memory in older adults [48].

Overconsumption of avocado is associated with migraine headache, nausea, vomiting, fever, and sensitivity to light. People with allergy to latex may exhibit some allergic symptoms such as itching, skin rash, skin redness, or eczema after consuming avocado. People suffering from a compromised liver function must avoid avocado as it contains estragole and anethole. It is advisable to avoid avocado during pregnancy and lactation too as it may lead to damage of mammary glands and reduction in milk production. Avocado is rich in β-sitosterol that absorbs necessary cholesterol from the body. Hence, consuming large amounts of avocado may lead to a harmful decrease of total cholesterol.

2 Bananas

Family: Musaceae
Genus: *Musa*
Common name: Banana

Bananas are among the most popular fruits in the world. Many kinds of bananas exist in various sizes and shapes. The skin color usually varies from green to yellow. They possess a fair amount of fiber, antioxidants, and minerals [49] (Table 2). Botanically speaking, banana is technically classified as a berry.

Bananas are a great source of both fiber and potassium [50, 51]. Potassium regulates blood flow [52], lowers the blood pressure [53–55], and prevents cardiovascular diseases [56–58]. They are also a source abundant in pyridoxine (vitamin B-6), vitamin C, antioxidants, and other phytonutrients [59, 60]. They contain other essential minerals such as magnesium [61–63] and manganese [64–67]. Their role in health and disease has been well-investigated. While bananas are low in protein, they are an extraordinary source of energy yet almost no fat.

Intake of dietary fiber provides numerous health benefits [68, 69]. In terms of fiber, a medium-sized banana contains roughly 3 g of fiber that is almost 10% of the daily intake [49]. Dietary benefits of banana differ depending on their ripeness. Green, unripe bananas have lower sugar content and more resistant starch which functions as a fiber and keeps the person feeling full for longer periods of time [70]. It also has prebiotics effect which is essential for gut health, especially for the colon, and can help absorb nutrients such as calcium [71, 72]. On the other hand, green bananas are low in antioxidants. Banana fiber helps in heart health, moderates blood sugar level, aids in weight management, and improves digestive health. Largely, unripen banana has two forms of fiber: pectin and resistant starch [73, 74]. The benefit of fiber-rich diet and cardiovascular risk reduction is well documented [75–77]. Banana's main fiber types contribute to fullness feeling and appetite reduction [78–81] and help with weight loss [78, 81–83] which in turn cause insulin sensitivity improvement [84, 85] and blood sugar control [23, 86, 87]. The impact of fiber on digestive health is well-known [88, 89]. Other recognized value of these fibers is lowered risk of kidney [90, 91] and colon [92, 93] cancers. The presence of potassium and magnesium in sufficient concentrations has been correlated with lower kidney stone risk incidence [94–96].

Yellow, ripe bananas are sweeter and have higher levels of antioxidants [60]. Yellow bananas with brown spots are particularly rich in the antioxidant dopamine

Table 2 Banana nutrition facts of 100 g, from the United States Department of Agriculture (USDA SR-21)

Serving size of 100 g of raw banana	
Per serving	% Daily value[a]
Calories 89	
Total fat 0.3 g	1
Saturated fat 0.1 g	1
Polyunsaturated fat 0.1 g	
Monounsaturated fat 0.0 g	
Total omega-3 fatty acids 27.0 mg	
Total omega-6 fatty acids 46.0 mg	
Cholesterol 0.0 mg	0
Phytosterols 16.0 mg	
Total carbohydrates 22.8 g	8
Dietary fiber 2.6 g	10
Starch 5.4 g	
Sugars 12.2 g	
Protein 1.1 g	2
Vitamins	
Vitamin A	1
Vitamin C	15
Vitamin E	1
Vitamin B-6	18
Vitamin K	1
Folic acid	5
Minerals	
Sodium	0
Potassium	10
Calcium	1
Iron	1
Magnesium	7
Manganese	13
Copper	4
Zinc	1

National Nutrient Database
[a]Based on a 2000 calorie diet

[59] and flavan-3-ols such as catechins [60], though they may increase blood glucose levels. Catechins belong to the antioxidant flavonoid polyphenolic class. They are linked to lower incidence of chronic diseases, such as certain cancers and cardiovascular and neurodegenerative diseases [97–100]. These and other antioxidant substances in ripe bananas are immunostimulants [101]. Research has shown that consumption of overripe bananas enhances the immune system's ability to produce TNF-α [101, 102]. Tumor necrosis factor (TNF-α), a cytokine, plays many pivotal biological roles [103] including tumorigenesis inhibition, i.e., anticancerous [102, 103].

Bananas are postulated to reduce depression and anxiety and stabilize sleep. They contain tryptophan which helps the body to manufacture serotonin, a neurotransmitter known as "happiness" hormone, i.e., hormone preventing mood disorders [102, 104]. They also have a high content of dopamine [59]. Though it is an important neurotransmitter, dopamine from banana does not appear to cross the blood-brain barrier to affect mood. Rather it seems to behave as a potent antioxidant [59].

Bananas as a rich source of carbohydrates and minerals can help to increase one's energy and prevent fatigue. More importantly, they are a great snack pre- and post-exercise [105, 106], although there are other studies that have found contrary evidence [107]. Despite this one study, research supports the notion that bananas help reduce exercise-related muscle cramps and soreness resulting from dehydration and electrolyte imbalance, thus enhancing performance and endurance of athletes [105, 108–110].

Despite the various health benefits, there are certain side effects associated mostly with overconsumption of this fruit. High starch content in banana dissolves slowly in the mouth leading to tooth decay. Resistant starch in unripe banana is associated with abdominal pain, nausea, vomiting, and constipation. Soluble fiber and fructose in banana can cause indigestion and gas formation. Conversely, ripe banana has been found to reduce constipation. Banana's high content of B-6 can be rarely associated with nerve damage when consuming large quantities. High consumption of overripe banana might increase the weight and increase blood glucose [111]. In comparison to other fruits, bananas are somewhat high in sugar and lower in fiber. Therefore, individuals with blood sugar and weight management concerns like diabetic patients should consume banana in moderation [111]. Banana allergies are relatively uncommon; however, due to certain banana protein similarity with latex, people with latex allergy have the possibility of being allergic to banana and some other fruits as well. Allergy to banana manifests as lips and tongue swelling, and throat irritation is well identified in some people when consuming banana or even handling it. The allergic reaction of muscle cramping, skin rashes, and wheezing can be complicated by severe immune reaction and serious anaphylaxis shock [112].

Bananas contain the amino acid tyrosine which is converted into tyramine in the body. Tyramine may trigger migraine headaches in some people. A number of medications such as monoamine oxidase inhibitors (MAOIs) were found to interact with tyramine prompting a dietary restriction. Oxazolidinone antibiotics, being weak MAOIs, might be associated with increased blood pressure when administered for a long period at high dose; thus, it is recommended to avoid tyramine-rich food [113, 114]. Bananas are a good source for potassium and magnesium which help muscle relaxation; they also contain the amino acid tryptophan which is converted into serotonin and therefore can be most helpful in promoting sleep [115, 116]. Banana is a highly rich source of potassium. Hyperkalemia (high level of potassium in the blood) [117] may affect the heart rate and can even lead to heart attack especially in people with impaired renal function [118]. Excessive consumption of banana might alter the action of some blood pressure-reducing agents such as beta-blockers, diuretics, and angiotensin-converting enzyme inhibitors (ACEIs).

3 Berries

Family: Rosaceae
Genus: *Rubus* (Blackberry, Black Raspberry, Raspberry, Tayberry, Boysenberry, Loganberry, Dewberry, and Cloudberry)
Genus: *Fragaria* (Strawberry)
Genus: *Aronia* (Chokeberry)

Berries are a widely popular variety of fruit worldwide and often are used in various kitchen products and recipes. Interestingly, the term "berry" is an encompassing term. The scientific usage of the name "berry" radically differs from the common terminology usage where many fruits are excluded by the botanical definition. These include strawberries, raspberries, blackberries, and mulberries. Here we will apply the commonly used fruits termed as berries by the masses and select few for further discussion. Berries provide a number of impressive, extraordinary health benefits. Strong scientific evidence suggests that eating various types of berries affects different body systems positively. Not only are berries delicious and nutritious, but they are also packed with phytochemicals that have medicinal properties. While most berries are edible, they are few that are poisonous to humans.

The health value of berries is largely derived from their antioxidant constituents which are the highest among commonly consumed fruits, next only to pomegranates [119]. More specifically, the impact is due the presence of phenolic compounds [120]. Berries' phenolics and polyphenols effects have been reviewed in many studies recently [121–127]. Within the selected group of berries, it is not possible to cover espoused health benefits comprehensively. Hence, we will choose key, critical components to delineate their disease prevention qualities including any crucial vitamins and minerals. By way of illustration, the emerging health impact comprises the following: cardiovascular [128–132] and metabolic syndromes [133–136], inflammation [137–140], digestive system [141], cancer [142–145], skin aging [146], and neuroprotection and cognition [147–151]. Many of these reviews are evidence-based and cite intervention clinical studies.

3.1 Blackberry

Blackberry is an edible black fruit, which botanically speaking is an "aggregate fruit." It has a group of over 375 of closely related species. Like other berries, blackberries are commonly eaten raw. They also can be eaten in baked goods, added to salads, or made into jams, jellies, or sauces. Interestingly, ancient cultures considered blackberry as a wild plant weed despite its traditional, medicinal usage of the fruit, leaf, bark, and roots for healing several health conditions including infections and poisonous bites. Blackberries are a good source of fiber, essential amino acids, vitamins C and K, and minerals (Table 3). Blackberries also retain high antioxidant content. They contain the most quantity of vitamin K and manganese and have the lowest calories when compared to other berries.

Blackberries contain numerous amounts of potent antioxidant compounds [152–154] such as phenolic acids and polyphenols like tannins and flavonoids (subclasses: anthocyanins, flavonols, and flavanols, i.e., flavon-3-ols). Among the anthocyanins (anthocyanosides), the glycosylated form of anthocyanidine being mostly 3-glucosides, the pigment cyanidin-3-glucoside (C3G) is the major anthocyanin found in most of the berries. They are present in the different tissues of the plant [155] and are responsible for the blackish color of fresh blackberries. They boost the immune system [156] and counteract the actions of free radicles while protecting the body from diseases caused by oxidative damage and inflammation [157–161]. In addition, blackberries have been found to prevent DNA damage in cells, therefore diminishing the risk of mutation-related maladies like cancer [152, 162–165]; examples include colorectal [166, 167], breast [168], and lung [169] cancers. They also play an important role in preventing cardiovascular diseases in many ways [170–172] and as demonstrated in separate intervention studies [173–176]. It has been reported that anthocyanins [173, 177] or C3G delay and can even halt the development of cardiovascular diseases [178] as evident in biomarker profile improvement. Blackberry anthocyanin-rich extract has been shown to protect the skin from the damage caused by ultraviolet light [179]. Interestingly, blackberry leaves are also used to treat some skin conditions like eczema, acne, different skin rashes, and itching due to insect bites. The combination of anthocyanins and vitamins improves vision and protects the eyes from age-related macular degeneration, cataract, and night blindness [180–182]. Blackberry leaves, bark, and root are a rich source of tannins, a class of polyphenols with astringent properties, and have been used as herbal medications for a number of conditions including digestive tract complaints, such as diarrhea, dysentery, and gastroenteritis. The antioxidant and anti-inflammatory activity of blackberry polyphenols may also limit the decline in age-related cognitive and motor abilities [183–185], which is further enforced by the availability of vitamins C and E [186]. The high concentration of manganese, an essential trace metal, is thought to affect brain functions and temperate some neurological conditions similar to Parkinson's disease [66].

The additive effect of the high vitamin K content in blackberries is thought to contribute to their anticancerous properties. It has been shown that blackberries play

Table 3 Blackberry nutrition facts of 100 g, from the United States Department of Agriculture (USDA SR-21)

Serving size of 100 g of raw blackberry	
Per serving	% Daily value[a]
Calories 43	
Total fat 0.5 g	1
Saturated fat 0.0 g	0
Polyunsaturated fat 0.3 g	
Monounsaturated fat 0.0 g	
Total omega-3 fatty acids 94.0 mg	
Total omega-6 fatty acids 186.0 mg	
Cholesterol 0.0 mg	0
Phytosterols	
Total carbohydrates 10.2 g	3
Dietary fiber 5.3 g	21
Starch 0.0 g	
Sugars 4.9 g	
Protein 1.4 g	3
Vitamins	
Vitamin A	4
Vitamin C	35
Vitamin E	6
Vitamin B-6	1
Vitamin K	25
Folic acid	6
Minerals	
Sodium	0
Potassium	5
Calcium	3
Iron	3
Magnesium	5
Manganese	32
Copper	8
Zinc	4

National Nutrient Database Nutrition
[a]Based on a 2000 calorie diet

a role in fighting and preventing the spread of many cancers including prostate [187], stomach [188], liver [189], and colorectal cancers [190]. Furthermore, vitamin K in combination with vitamin C could be a promising treatment for leukemia [191] and bladder cancer [192]. Vitamin K also plays a role in the blood's normal clotting process inhibiting excessive bleeding and promoting wound healing [193]. In addition, it boosts bone health by preventing osteoporosis and reducing the risk of fractures [33]. Vitamin K also reduces blood pressure as it decreases inflammation in cells lining the blood vessels and reduces the risk of heart attacks.

The phytoestrogens, vitamins, and minerals in blackberries and their extract were found to have natural antibacterial effects against some infections such as oral infections [194]. Also, blackberry extract has been suggested to have antiviral activity; they were found to inhibit the early stages of oral epithelial cells replication of HSV-1 [195].

The consumption of reasonable amounts of ripe blackberries has shown no reported side effects, but overconsumption may lead to diarrhea, while the relatively higher content of tannins in unripe blackberries can cause constipation, nausea, and/or vomiting in some people.

3.2 Raspberry

Raspberry is a palatable, sweet, soft berry fruit of multiple species. They occur in a multitude of colors like black, golden or yellow, pink, white, purple and even bright blue, and red which is most popular. Red raspberries can be eaten raw, dried, included in some cooking recipes, or made into jam; its leaves are used to make teas and to give flavor. Red raspberry leaves and fruit are also used for their medicinal value. Raspberries are excellent sources of vitamin C, dietary fiber, and manganese. They are also rich in different B vitamins, copper, and iron (Table 4). They have high concentration of antioxidants such as simple phenols (gallic acid and salicylic acid); ellagic acid, an ellagitannin by-product; and the flavonoids quercetin, kaempferol, catechins, cyanidin, and pelargonidin. However, light-colored raspberries, like the yellow ones, have a much lower concentration of anthocyanins.

Raspberry consumption has been associated with many health benefits [196, 197]. Both anthocyanins and ellagitannins are major pharmacological contributors [198]. Ellagic acid, a tannin hydrolysis-derived product, has been reported to have multiple bioactivities, including anti-inflammatory, antioxidative, anticancerous as well as antiviral capabilities [199, 200]. Ellagic acid has displayed anti-inflammatory responses and been shown to reduce collagen destruction caused by ultraviolet light damage in an in vitro model [201]. Ellagic acid has been shown to inhibit certain types of cancer, including lung [202], bladder [203], breast [204], and skin [205] cancers. Anthocyanins in black raspberries have been demonstrated to have anticancerous properties as well. They reduce oxidative stress, affect apoptosis, and restore tumor suppressive activity [206], especially in esophageal [207] and oral cavity cancers [208] and colorectal cancer [209]. Anthocyanins were also found to prevent transformation of cells, decrease inflammatory reactions, and induce normal functions in cancer cells [210]. The polyphenols anthocyanins and ellagitannins in red raspberries and their extract were also proven to lower the risk of heart conditions [131, 196, 211–213]. The same was found for black raspberry [200]. Black raspberry seed oil, rich in α-linolenic acid (omega-3 fatty acid), has been found to significantly lower the plasma triglyceride level [214]. Whereas red raspberry seed oil is rich in vitamin E, which is a potent antioxidant, along with the fatty

Table 4 Raspberry nutrition facts of 100 g, from the United States Department of Agriculture (USDA SR-21)

Serving size of 100 g of raw raspberry	
Per serving	% Daily value[a]
Calories 52	
Total fat 0.7 g	1
Saturated fat 0.0 g	0
Polyunsaturated fat 0.4 g	
Monounsaturated fat 0.1 g	
Total omega-3 fatty acids 126 mg	
Total omega-6 fatty acids 249 mg	
Cholesterol 0.0 mg	0
Phytosterols	
Total carbohydrates 11.9 g	4
Dietary fiber 6.5 g	26
Starch 0.0 g	
Sugars 4.4 g	
Protein 1.2 g	2
Vitamins	
Vitamin A	1
Vitamin C	44
Vitamin E	4
Vitamin B-6	3
Vitamin K	10
Folic acid	5
Minerals	
Sodium	0
Potassium	4
Calcium	2
Iron	4
Magnesium	5
Manganese	34
Copper	4
Zinc	3

National Nutrient Database Nutrition
[a]Based on a 2000 calorie diet

acids' omega-3 and omega-6, plus vitamin A, they limit skin aging and temperate ultraviolet light harm [201, 215, 216].

There are no reported side effects associated with moderate intake of raspberries. Red raspberries act like the hormone estrogen (isoflavones are natural endocrine active phytoestrogens) [217, 218]. Therefore, red raspberry should be consumed with caution during pregnancy, and it is better to be avoided in conditions sensitive to hormones such as breast, uterine, and ovarian cancers [219, 220], endometriosis, and uterine fibroids, as they might be exacerbated by estrogen like action of raspberries.

3.3 Strawberry

A strawberry is a sweet, heart-shaped fruit, growing in bushes, known for its distinct aroma, bright red color, and juicy consistency. The fruit can be consumed raw as a whole fruit or as a juice. Strawberries can be added to recipes or used to make jam, pies, ice creams, milkshakes, and chocolates or simply to add as a flavor to the food or to body care items such as lip balms, lotions, shampoos, and toothpastes. Strawberry leaves are also edible: they can be consumed either raw or cooked and used to prepare tea. Strawberries are low-calorie, fiber-rich fruits, with very high antioxidant and polyphenol content. They are also rich in vitamins C and K, folic acid, and minerals such as manganese, magnesium, and potassium (Table 5).

Strawberry consumption has been linked to many health benefits [221–225]. Strawberries harbor several active pharmacological ingredients against chronic diseases [226, 227]. Strawberries and its leaves have large quantities of antioxidants such as vitamin C, phenolic acids, and flavonoids, mainly as flavonols and anthocyanins, that neutralize free radicle action and offer protection against many incurable chronic conditions [157, 228, 229]. Strawberries generally exhibit chemopreventive activities against several cancer types [230–236] and have been shown to inhibit oral tumor formation [237]. Phenolic compounds in strawberries are thought to enhance apoptosis in cervical cancer [238]. Polyphenols from strawberry extract have shown protective properties against breast cancer as well [235], while the specific polyphenol compound kaempferol, a natural flavonol, induces cell cycle arrest and is thought to have an effect on colorectal cancer [239]. It appears that the anticancer features of these extracts or compounds exert their effect by targeting multiple signaling pathways. The flavonol compound fisetin, present at the highest concentration only in strawberry [240], was also observed to affect prostate cancer [241] and triple-negative breast cancer [242] plus other notable digestive cancers in cell cultures [243]. Besides being an anticancerous agent, fisetin has multiple other health benefits including neuroprotection [244, 245]. Fisetin seems to possess senolytic quality, i.e., extending lifespan. In laboratory animals it has been shown it is the most potent senolytic agent reducing senescence biomarkers in multiple tissues [246]. It is currently undergoing clinical trials in the USA to show efficacy in humans.

In various clinical intervention trials, strawberry consumption has been associated with better cardio-profile biomarkers and risk reduction of cardiovascular diseases [247–252]. Anthocyanin flavonoids in strawberries reduce the risk of myocardial infarctions [253]. Furthermore, the flavonoid quercetin in strawberries has anti-inflammatory features that helps lower the risk of atherosclerosis [254]. Strawberries also contain significant amounts of potassium and magnesium [255], both of which are efficient in lowering high blood pressure. Potassium is a vasodilator that reduces high blood pressure as well as the rigidity of blood vessels, thereby decreasing the risk of cardiovascular diseases [256].

Strawberries are rich in vitamin C and phytochemicals that help boost the nervous system and avert age-related cognitive deterioration and Alzheimer's disease

Table 5 Strawberry nutrition facts of 100 g, from the United States Department of Agriculture (USDA SR-21)

Serving size of 100 g of raw strawberry	
Per serving	% Daily value[a]
Calories 32	
Total fat 0.3 g	0
Saturated fat 0.0 g	0
Polyunsaturated fat 0.2 g	
Monounsaturated fat 0.0 g	
Total omega-3 fatty acids 65.0 mg	
Total omega-6 fatty acids 90.0 mg	
Cholesterol 0.0 mg	0
Phytosterols 12.0 mg	
Total carbohydrates 7.7 g	3
Dietary fiber 2.0 g	8
Starch 0.0 g	
Sugars 4.9 g	
Protein 0.7 g	1
Vitamins	
Vitamin A	0
Vitamin C	98
Vitamin E	1
Vitamin B-6	2
Vitamin K	3
Folic acid	6
Minerals	
Sodium	0
Potassium	4
Calcium	2
Iron	2
Magnesium	3
Manganese	19
Copper	2
Zinc	1

National Nutrient Database Nutrition
[a]Based on a 2000 calorie diet

[257]. Moreover, strawberries are also a very rich source of iodine that is useful for optimal brain functioning and healthy nervous system [258]. Potassium content has been linked to improved cognitive function by improving blood flow to the brain [259]. Likewise, potassium and flavonols together improve blood flow to the brain and improve the memory, thereby minimizing the risk of Alzheimer's disease [260]. Actually, it has been suggested that high intake of potassium and fisetin, the flavonol that gives the strawberry its distinct red color, improves memory and may help reduce the risk of Alzheimer's disease [261] besides other anthocyanins [184, 185] and vitamins [186].

It has also been shown that strawberry polyphenol-enriched anthocyanin extract protects against ultraviolet radiation, thusly limiting skin damage [262]. Anthocyanins in strawberries improve vision and protect the eyes from age-related macular degeneration, cataract, and night blindness [180].

These beneficial fruits are not, however, free of some unwanted health effects, especially when consumed excessively. Strawberries contain histamine which causes itching, dizziness, and nausea and may complicate food intolerance and cause digestive problems [263]. People allergic to strawberry should avoid eating it. High fiber content in raw strawberries might prevent nutrient absorption and may lead to diarrhea when overconsumed. With its high potassium content, eating lots of strawberries may cause harm to the kidneys and the heart especially during the use of heart medications. Eating unripe strawberries may lead to mouth irritation and burning sensation.

Family: Ericaceae
Genus: *Vaccinium* (Blueberry, Cranberry, Huckleberry, and Lingonberry)
Genus: *Arctostaphylos* (Bearberry)
Genus: *Empetrum* (Crowberry)

3.4 Blueberry

Blueberries are perennial flowering shrubs with blue- or purple-colored fruits that grow in clusters. The fruit is a berry shape with a wide crown at the end; they have pale greenish color at the top that then changes to reddish-purple, and when ripe it turns dark purple. Ripe blueberries have light green flesh and a sweet taste with variable levels of acidity. There are two main types of blueberry bushes: the low-bush or wild blueberries, which are typically smaller in size with fruits richer in certain antioxidants, and the highbush blueberries which are the most commonly grown type. Blueberries can be eaten raw, had as a juice, used in cooking recipes, or made into jam. Blueberry leaves are also used to make tea. Blueberries are low caloric fruit, high in carbohydrates and fiber. They are rich in vitamins C and K and minerals like manganese (Table 6). They are believed to be the richest source of antioxidants, including the important polyphenol flavonoid family, among all common vegetables and fruits.

Blueberries have been reported to have wide-ranging and generous health benefits including impact on the biology of aging [186, 264] with their high antioxidant capacity [119, 126, 265, 266] and ability to improve DNA damage protection in humans [267, 268]. A significant portion of these health attributes is the result of flavonoids particularly the antioxidants subgroup anthocyanins [269] as they can be detected in the serum of human subjects upon ingestion [270, 271]. Anthocyanins are present in high concentrations in blueberries, are responsible for its blue color, and can partly explain the generous health benefits of this fruit.

Blueberries' exceedingly rich content of flavonoids such as anthocyanin, catechin, quercetin, and kaempferol along with vitamin C makes them a potent anticancerous

Table 6 Blueberry nutrition facts of 100 g, from the United States Department of Agriculture (USDA SR-21)

Serving size of 100 g of raw blueberry	
Per serving	% Daily value[a]
Calories 57	
Total fat 0.3 g	1
Saturated fat 0.0 g	0
Polyunsaturated fat 0.1 g	
Monounsaturated fat 0.0 g	
Total omega-3 fatty acids 58.0 mg	
Total omega-6 fatty acids 88.0 mg	
Cholesterol 0.0 mg	0
Phytosterols	
Total carbohydrates 14.5 g	5
Dietary fiber 2.4 g	10
Starch 0.0 g	
Sugars 10.0 g	
Protein 0.7 g	1
Vitamins	
Vitamin A	1
Vitamin C	16
Vitamin E	3
Vitamin B-6	3
Vitamin K	24
Folic acid	1
Minerals	
Sodium	0
Potassium	2
Calcium	1
Iron	2
Magnesium	1
Manganese	17
Copper	3
Zinc	1

National Nutrient Database Nutrition
[a]Based on a 2000 calorie diet

fruit [272–274], fighting against breast cancer [275–278], colon and small intestine adenomas [279, 280], melanoma [281], and prostate cancer [282].

Another benefit of polyphenols in blueberries is their impact on risk factors of stroke and cardiovascular diseases [283]. High anthocyanin intake has been found to lower the risk of heart attacks in middle-aged women [253], and series of studies concluded that blueberries reduce LDL and that phenolic compounds in blueberries are thought to decrease the oxidation of the unhealthy LDL cholesterol [284–286]. Blueberries contain potassium, magnesium, and calcium which are known to lower blood pressure with several investigations showing that blueberries can lower the blood pressure in obese people [286–288]. The dietary fiber in blueberries amplifies the health of the digestive and cardiovascular systems, reduces cholesterol, prevents constipation, and helps control body weight [69].

Paradoxically, consuming the whole berry fruit is thought to decrease the risk of developing type 2 diabetes, while blueberry juice has been associated with a higher diabetes risk [289]. Anthocyanin compounds have also been thought to give blueberries their anti-diabetic properties observed in pancreatic cells and peripheral tissues. Few studies, including ones with clinical intervention, have established that blueberries improve fasting glucose and insulin sensitivity [290, 291]. Extracts from the stem, leaf, and fruit have been shown to protect against glucose toxicity [292] and mend glycemic profile via digestion and absorption reduction of starch, thereby controlling glycemia [293]. Findings from animal studies support the contention of blueberries capability to promote weight loss [294, 295].

Blueberry juice antioxidants and anti-inflammatory polyphenolic compounds, especially anthocyanins, are thought to be neuroprotective [184, 185, 296]. They have been shown to improve cognitive functions and memory among older people, mitigating Alzheimer's disease symptoms [297–299] as well as children cognition/memory [300]. Blueberries contain vitamins such as A, C, and E and minerals like selenium, copper, magnesium, and phosphorus that also help thwart cognitive damage and reduce mood swings/neurotic conditions [186, 301]. Consuming blueberries is also thought to slow down degenerative processes in the eye, preventing age-related macular degeneration and blindness. Vitamin A content of blueberries has been shown to prevent retinal oxidative damage [302]. Blueberry and its products are thought to reduce the risk of urinary tract infections (UTI) [303] via its anti-adhesin bioactivity [304].

Blueberries' high content of delphinidin, the major anthocyanin compound, was found to decrease bone loss, improve bone density, and induce bone formation [305]. Their richness in calcium, magnesium, phosphorus, manganese, iron, zinc, and vitamin K collectively augments bone health further. Zinc and iron were found to improve elasticity of bone and decrease the risk of osteoporosis [306], while vitamin K also decreases the risk of pathologic fractures [33].

Prolonged strenuous exercise may result in transient inflammation and muscle damage [307]. Blueberry intake prior to intensive exercising enhances performance, while supplementation after exercising was accompanied with decreased inflammation and oxidative stress biomarkers and increased total antioxidant status that extended 36 h post-exercise and accelerated muscle recovery [308–310]. These findings are comparable to those with cherry and watermelon consumption.

Ingesting blueberries in moderate amounts as part of a balanced diet is unlikely to cause serious health effects. However, individuals allergic to blueberries might experience diarrhea and vomiting once they eat it. On rare occasions, few allergic people might also develop an immune reaction in the form of rash, asthma, and/or nasal congestion. Those who use blood anti-coagulants, such as warfarin, should be extra cautious when consuming blueberries since the high vitamin K content may affect the blood clotting process. Furthermore, blueberries naturally contain excessive quantities of salicylates—the active component in aspirin—which can lead to some health side effects in people who cannot tolerate salicylates. For such individuals, blueberry juice might cause skin rash, headaches, or gastrointestinal symptoms such as nausea, vomiting, diarrhea or constipation, reflux, bloating, and gas formation. The blueberry fruit is expected to be safe for diabetics when eaten in moderate amounts, but blueberry leaves supplement was found to cause dangerous drop in the blood glucose.

Family: Moraceae
Genus: *Morus* (Red and White)

3.5 *Mulberry*

Mulberry is a berry from a wild flowering tree of few related species. Immature mulberries are white, green, or light yellow. Across most types, as they ripen, they become pink and then red, after which they turn dark purple or black. Mulberries can be eaten raw or dried and as part of different cooked sweets. Mulberries can be used to make jams, juice, jellies, smoothies, tea, pancakes, sauces, or canned food. Mulberry essential oil has a pleasant scent and can be added to give fragrance to goods such as lotions, shampoos, soaps, and candles. Leaves of white mulberries are the favorite or probably even the only food of silkworms. Mulberries were used in Chinese medicine to treat countless conditions including heart diseases, diabetes, anemia, and arthritis. They contain carbohydrates, fiber, fat, and proteins. They are a wealthy source of vitamins C, K, and B vitamins, calcium, and iron (Table 7). All parts of the plant are rich in antioxidants: the fruit, bark, stem, and leaves. Mulberries are antioxidants rich in simple phenols like chlorogenic acid (caffeic acid derivative); the flavonoids cyanidin, myricetin, and rutin, along with others like the stilbene resveratrol; and the carotenoids lutein and zeaxanthin.

Mulberries have few health benefits which have been reviewed recently along with its bioactives [311–313]. Mulberry anthocyanin (C3G) and its extract have been found to reduce oxidative stress and inflammation [314, 315]. They reduce the total cholesterol level and inhibit the oxidation of unhealthy LDL, thus lowering the risk of coronary heart diseases and atherosclerosis [316, 317]. Leaves extract was equally promising for modulating cardiometabolic risks and atherosclerosis [283, 318, 319]. Mulberry extract was demonstrated to treat a number of diseases including cancer [320–323], obesity [324], and type 2 diabetes [325–327]. The mulberry leaves contain 1-deoxynojirimycin, a poly-hydroxylated piperidine alkaloid secondary product that slows down carbohydrate degradation in the gut, thereby reducing the release of glucose after meals [328–331]. Mulberries and their root bark extract possess the phytochemicals anthocyanins and resveratrol. Their antioxidants and anti-inflammatory properties enable battling various cancers [332]. For example, resveratrol has been demonstrated to inhibit different cancer types including breast [333], prostate [334], thyroid [335], and colorectal cancers [336]. Resveratrol and potassium do

Table 7 Mulberry nutrition facts of 100 g, from the United States Department of Agriculture (USDA SR-21)

Serving size of 100 g of raw mulberry	
Per serving	% Daily value[a]
Calories 43	
Total fat 0.4 g	1
Saturated fat 0.0 g	0
Polyunsaturated fat 0.2 g	
Monounsaturated fat 0.0 g	
Total omega-3 fatty acids 1.0 mg	
Total omega-6 fatty acids 206 mg	
Cholesterol 0.0 mg	0
Phytosterols	
Total carbohydrates 9.8 g	3
Dietary fiber 1.7 g	7
Starch 0.0 g	
Sugars 8.1 g	
Protein 1.4 g	3
Vitamins	
Vitamin A	0
Vitamin C	61
Vitamin E	4
Vitamin B-6	3
Vitamin K	10
Folic acid	1
Minerals	
Sodium	0
Potassium	6
Calcium	4
Iron	10
Magnesium	5
Manganese	
Copper	3
Zinc	1

National Nutrient Database Nutrition
[a]Based on a 2000 calorie diet

help lower the blood pressure and relax the blood vessels which reduces the risk of cardiovascular diseases [337]. Furthermore, resveratrol and zeaxanthin protect the eye from free radicals that cause macular degeneration and loss of vision and cataract [338]. Extract from black mulberry—rich in antioxidants vitamins C and E—has been found to enhance cognitive functions in older people ameliorating Alzheimer's disease and age-related memory impairment [339, 340]. White mulberry has been shown to confer protection against nephrotoxicity [341].

Allergy to mulberries is not common, but pollen from its trees has been found to cause allergic reactions in sensitive individuals. The high potassium content may complicate kidney disorders. Mulberry extract may also cause sudden drop in blood glucose (hypoglycemia). Mulberries contain arbutin (a chalcone flavonoid), chemically a hydroquinone compound, which helps make the skin fairer. Since it prevents melanin release, it potentially raises the risk of skin cancer [342, 343].

4 Cherries

Family: Rosaceae
Genus: *Prunus*
Common name: Cherry

Cherries are small, round, deep red fruits with a pit in the middle which must be removed before eating or cooking. There are limited varieties that differ in their size and flavor. The two main types are sweet usually eaten fresh or sour which are used for cooking. Cherries are a good source of fiber, omega-3 and omega-6 fatty acids, vitamins A and C and folic acid, and minerals such as potassium, manganese, and copper (Table 8). They also contain very powerful antioxidants, e.g., quercetin, anthocyanins, and cyanidin, that also have anti-inflammatory and anticancerous properties [344, 345].

Cherries have high total antioxidant content [346]. Antioxidants along with other minerals and vitamins improve immune system status and help prevent infections. Researchers have found that consumption of cherry juice by marathon runners showed a significant reduction in upper respiratory tract symptoms [347]. Antioxidants, particularly anthocyanins, and the high fiber content in cherries aid in risk reduction of cancers like colon cancer [348]. Cherry antioxidants have cardiovascular benefits as well. Anthocyanin which gives the cherries its red color may contribute to its ability in regulating blood sugar and cholesterol levels, thereby lowering the risk of developing stroke via PPAR (peroxisome proliferator-activated receptors) isoforms activation that is involved in fat and glucose metabolism [349].

Table 8 Cherry nutrition facts of 100 g, from the United States Department of Agriculture (USDA SR-21)

Serving size of 100 g of red, raw cherry	
Per serving	% Daily value[a]
Calories 50	
Total fat 0.3 g	0
Saturated fat 0.1 g	0
Polyunsaturated fat 0.1 g	
Monounsaturated fat 0.1 g	
Total omega-3 fatty acids 44.0 mg	
Total omega-6 fatty acids 46.0 mg	
Cholesterol 0.0 mg	
Phytosterols	0
Total carbohydrates 12.2 g	4
Dietary fiber 2.0 g	6
Sugars 8.0 g	
Protein 1.0 g	2
Vitamins	
Vitamin A	26
Vitamin C	17
Vitamin E	0
Vitamin B-6	2
Vitamin K	3
Folic acid	2
Minerals	
Sodium	0
Potassium	5
Calcium	2
Iron	2
Magnesium	2
Manganese	6
Copper	5
Zinc	1

National Nutrient Database
[a]Based on a 2000 calorie diet

The high polyphenol content in cherries' juice including anthocyanin helps reduce blood pressure [350, 351]. Cherry also has the capacity to reduce the signs and symptoms of inflammation associated with different types of arteritis [352, 353]. The high level of anthocyanin in cherry juice has been shown to be neuroprotective [354] and improve memory and cognitive functions in older adults with dementia [355–357].

Consuming cherries may act as a natural pain reliever. Cherries have a long history as a treatment for gout [358] and joint pain [359, 360]. It reduces uric acid serum level precluding gouty arthritis [361, 362] and is an effective treatment for peripheral polyneuropathies [363]. It has been shown that eating or drinking the juice of cherries before and during strenuous exercises like running leads to less muscle inflammation and pain and less soreness and enhances faster recovery after vigorous workout [364, 365] by increasing total antioxidative capacity, reducing inflammation and lipid peroxidation, and hence aiding the recovery of muscle function. Cherries are significantly associated with weight management, specially reducing abdominal fat [366]. Cherries' rich content of melatonin may improve sleeping quality and duration for individuals suffering from insomnia or disturbed sleep disorder [216] including older adults [367, 368].

Consuming cherries is fairly safe. If a person is not allergic to cherries, their consumption is unlikely to have grave effects on health. Among people sensitive to them, allergic reaction can lead to shortness of breath, difficulty in swallowing, hives, nausea, or diarrhea. Overconsumption of cherries may cause abdominal cramps and bloating due to its high dietary fiber content and high amounts of sorbitol.

5 Chili Peppers

Family: Solanaceae
Genus: *Capsicum*
Common name: Chili

Chili pepper is actually a berry fruit with a distinct burning, hot flavor. It can be eaten fresh or dried, raw or cooked or simply added to dishes as a spice or as the main ingredient in different sauces. The mildly bitter leaves are also edible and almost as hot as the fruit itself. Multiple pepper varieties exist and they differ in color and the degree of hotness the likes of habanero, jalapeño, cayenne, piri, fresno, etc. The hot spicy taste of chili depends on the amount of the active alkaloid capsaicin it contains: the more the capsaicin, the hotter it will be. The intensity of capsaicin reflects the pepper type and its growing conditions. Chili pepper contains multiple vitamins and minerals. It is a very rich source of vitamins B-6, C, and K. Red pepper in particular is rich in vitamin A. Chili is also a good source of copper, manganese, and potassium (Table 9). Mature chili peppers have high levels of carotenoid antioxidants [369, 370], which protect against multiple chronic diseases,

Table 9 Chili pepper nutrition facts of 100 g, from the United States Department of Agriculture (USDA SR-21)

Serving size of 100 g of raw, red, hot, chili pepper	
Per serving	% Daily value[a]
Calories 40	
Total fat 0.4 g	1
Saturated fat 0.0 g	0
Polyunsaturated fat 0.2 g	
Monounsaturated fat 0.0 g	
Total omega-3 fatty acids 11.0 mg	
Total omega-6 fatty acids 228 mg	
Total cholesterol 0.0 mg	
Phytosterols	0
Carbohydrates 8.8 g	3
Dietary fiber 1.5 g	6
Starch	
Sugars 5.3 g	
Protein 1.9 g	4
Vitamins	
Vitamin A	19
Vitamin C	239
Vitamin E	3
Vitamin B-6	25
Vitamin K	17
Folic acid	6
Minerals	
Sodium	0
Potassium	9
Calcium	1
Iron	6
Magnesium	6
Manganese	9
Copper	6
Zinc	2

National Nutrient Database
[a]Based on a 2000 calorie diet

like the capsanthin in red peppers [371, 372], violaxanthin in yellow peppers [373, 374], lutein in green or immature peppers [375, 376], and phenolic compounds in the seeds such as hydroxycinnamic acids like sinapic acid and ferulic acid. In addition to other flavonoids and bioactive compounds [377, 378].

Chili pepper is the source of a unique set of bioactive compounds that have been associated with quite a few health benefits along with effectiveness against some chronic conditions [379, 380]. Chili peppers are also used as food preservatives [381]. Among these key bioactive chemical families are alkaloids (capsaicin and other capsaicinoids), the antioxidant carotenoids (capsanthin and xanthophylls), and phenolic acids (*sinapic and ferulic acids*). Capsaicin and other capsaicinoids are well-studied compounds and responsible for chili peppers' properties of pungent hot flavor as well as irritant quality. Capsaicin binds to a receptor called the vanilloid receptor subtype 1 (TRPV1) on sensory neurons leading to the induction of the painful and burning sensations without causing real burns [382]. This desensitization is temporary. By its very nature, capsaicins have been explored for pain management therapy [383] and neuropathic pain treatment [384]. Capsaicins are mainly utilized as an analgesic in topical ointments and dermal patches at low concentrations to relieve pain; however, it appears that topical formulations containing high capsaicin concentration at 8% are more effective and possess powerful pain-relieving properties [383, 384]. Capsaicin use in pain research and its mechanisms of action and use in pre-clinical as well as clinical utility have been reviewed recently [385–388]. Noteworthy are the initial findings among a small group of patients where capsaicin was used in migraine relief [389]. This work has since been extended, and the research area has been reviewed recently for the development of anti-migraine therapeutics [390–392].

Capsaicins desensitize sensory receptors [393], yet retain anti-inflammatory effects [394, 395]. Therefore, it can be used topically to treat arthritis and osteoarthritis [396]. This research area and the safety profile and the efficacy of capsaicin in reducing arthritis/osteoarthritis pain have been reviewed extensively using clinical trials findings that all indicate it is an effective sole or adjunct treatment [380, 397–400]. It can also be used topically to relieve pain caused by diabetic neuropathy [401–404]. Besides capsaicin's pain-relieving properties, it causes vasoconstriction of the blood vessels so it relieves nasal congestion [405, 406] or non-allergic rhinitis [407, 408] as well. Additionally, it can be used in cases of *Helicobacter pylori*-induced gastritis and/or gastric ulcers by either eliminating the bacteria or minimizing the inflammatory disease [409–412]. It also reverses dyspepsia pain commonly known as acid reflux or heartburn [413, 414]. However, by affecting other pain sensation capabilities, the person could be rendered insensitive to the heartburn caused by acid reflux.

There is an extensive body of literature that has explored capsaicins' role in weight management and as an anti-obesity potency. Selected reviews are referenced here [415–420]. However, the exact mechanism of action of capsaicin is not understood fully yet [386]. Several lines of evidence suggest that the desired outcome is through a combined effect of appetite suppression and gastrointestinal satiety [421–423], enhanced fat oxidation [419, 424–426], thermogenesis and increase in energy expenditure [419, 427–430], and alteration in gut microbiota [431–433]. Other highlighted beneficial effects on glucose and insulin homeostasis and diabetes, cardiovascular system (regulating blood pressure, atherosclerosis reduction, platelet aggregation inhibition, and cardioprotection) [386, 420, 434, 435], and cancer have been revealed. In terms of cancer, there is mixed and conflicting evidence showing capsaicin's chemopreventive and chemotherapeutic effects and that it may also act as carcinogenic or co-carcinogenic agent [436–441].

The valuable health benefits of antioxidant carotenoids have been discussed in previous vegetables and fruit sections. Among chili pepper, lutein in green peppers and xanthophylls in orange/yellow peppers were found to improve eye health [442–

447]. Several other types of antioxidant phenolic compounds have been isolated and identified in chili peppers [378, 448]. Prominent among these are phenolic acids [378, 449–452]. The therapeutic potential of phenolic acids has also been reviewed [448, 453–455]. Chief among these phenolic acids in chili peppers are sinapic acid and ferulic acid. Their strong antioxidant/oxidative stress capacity has been described in animal studies and clinical trials. Hence, they have the potential to protect or attenuate cellular stress-induced diseases and aging [456] [ferulic acid [457–459] and sinapic acid [460–462]]. The therapeutic potential of ferulic acid has been expounded including its use in cosmetics, as a food preservative, and as a precursor for vanillin for the flavor market [463, 464]. The medicinal potential of sinapic acid has been elucidated as well [462, 465]. A particular role for these acids is being ascribed in neurotoxicity protection and Alzheimer's disease [466–471].

Chili pepper consumption is not free of certain, minor side effects. Overconsumption of peppers (capsaicin) may lead to damage in pain receptors and—over a period of time—to the loss of the burning flavor of chili. Oleoresin capsicum is an extract of chili peppers and the main component of self-defense pepper sprays [472]. The exposure to high concentration of oleoresin capsicum leads to temporary acute irritation of the eyes, excess secretion of tears, inflammation of the conjunctiva, and involuntary closure of eyelids (blepharospasm), in addition to mild to moderate respiratory distress [473]. Asthma patients using theophylline—a bronchodilator—should consume chili peppers moderately due to drug interaction and potentiation effect. Eating chili could cause gastrointestinal symptom issues in some people in the form of abdominal pain, painful diarrhea, and burning sensation in the gut. It can also temporarily cause hypersensitivity and worsen the symptoms of irritable bowel disease [474]. Some studies have associated chili consumption with different gastrointestinal cancers [475–477]. Excessive use of capsaicin cream products may cause skin irritation, burning, and itching.

6 Citrus Fruits

Family: Rutaceae
Genus: *Citrus*
Common names: Orange, Lemon, Grapefruit, Tangerine, Clementine, Mandarin, Pomelo, and Lime

Owing to the sheer availability of variety and their distinct scent and flavorful taste, citrus fruits have become an integrated part of our daily diet. Interestingly, virtually a third of all citrus fruits are consumed as juice. They are known for their low protein and fat content and supplying different sugars (fructose, sucrose, and glucose) yet possessing low glycemic index [478] because of the attenuating influence of polyphenols and fiber [479]. Additionally, they are a good source of dietary fiber, vitamins C and B, and phytochemicals such as alkaloids, flavonoids, and carotenoids. The health benefits of citrus fruits have largely been attributed to the bioactivity of these natural, secondary metabolites [480]. Since citrus fruits' popular varieties, nutritious content, and contribution to health are decidedly comparable, they are collectively covered below with a particular focus on key biological active constituents. In ripe citrus fruits, the emphasis will be on fiber content; the antioxidants alkaloid triterpenoids (limonoids), flavonoids (flavanones), and carotenoids (β,β-xanthophylls); and vitamin C.

Citrus fruits may supply up to 18% of the daily recommended dietary fiber intake. The main soluble fibers present are pectin and some hemicellulose, while the main insoluble fibers are cellulose, hemicellulose, and lignin [481], exclusively enjoying a higher ratio of soluble to insoluble fiber compared to other fruits or vegetables [482]. The role of dietary fiber in the prevention or management of chronic diseases [69, 483, 484] including metabolic disorders [485] has been discussed extensively in the vegetable section and key fruits in this section. By way of reference, we provide here a brief narrative that is applicable to citrus fruits [486]. Most studies have focused on the usual risk reduction of chronic diseases such as diabetes, cardiovascular diseases, and colon cancer [487, 488]. Meta-analyses have summarized findings with regard to metabolic syndromes [485, 489, 490] with recent investigations focusing on type 2 diabetes [491–493]. In addition, results from these reviews or research articles show significant metabolic profile improvement. Numerous meta-analyses and research investigations have established that dietary fiber intake is associated with a lower risk of both cardiovascular and coronary heart disease [75–77, 494, 495]. There seems to be a universal agreement on their ability of lowering blood pressure [496], reducing cholesterol [497, 498], decreasing glucose levels [499], and contributing to weight loss [500] regardless of population age, i.e., young [501] or old [499, 502]; gender, i.e., male [503] or female [504]; and ethnicity [505–508]. Furthermore, a systematic review and meta-analysis implicated the protective role of dietary fiber intake on colorectal cancer [509]. Other studies indicated better odds ratio with fruit fiber intake [510–512]. Indeed, fiber also aids in maintaining a healthy digestive system. Fiber was found to improve constipation condition [513–515], promote satiety [83, 516], and permit body weight management [500, 517, 518]. Finally, dietary fiber influences the gut microbiome [519]. As a result, the gastrointestinal microbiota indeed affects human health and disease states [520]. Several articles have reviewed the role of prebiotic dietary fiber [521–524].

The antioxidant properties, biological functions, potential applications, and contribution to prevention/therapy of chronic diseases have been reviewed for terpenoids [525–530], flavonoids [99, 100, 531–534], and carotenoids [535–538]. The antioxidant capacity of citrus fruits is enormous [539, 540]. They contain a variety of important compounds that contribute to their attributed health benefits [539, 541–543].

Limonoids are phytochemicals classified as tetranortriterpenes. They exist in large quantities in sweet oranges and give the citrus juice its special taste [541]. The accumulation/distribution [544], chemistry and pharmacology [545], and bioactivity and biomedical prospects of limonoids have been reviewed [546, 547]. To illustrate, they have been shown to play an efficacious role in diabetes [548], atherosclerosis [549], cancer [550], and neurological diseases [551]. Citrus is the major food source for flavanones. They comprise approximately 95% of the total available flavonoids with many unique to the fruit. The nutraceutical value of citrus flavanones has been reviewed recently [552–561]. A subclass of citrus-derived flavanones occur naturally as glycosides (i.e., the aglycone is coupled to a sugar moiety). The most predominant and widely studied of the flavanones in citrus are hesperidin (hesperetin) in oranges, naringin (naringenin) in grapefruit, eriodictyol in lemon, and tangeretin (exist as aglycone only) in tangerines, though it is a flavone. For example, hesperidin's health benefits have been reviewed [562–564]. Its anti-inflammatory properties [562, 565], role in prevention of cancer and cardiovascular diseases [566, 567], and neuroprotective effects [568] have been reviewed. Likewise, naringin therapeutic potential has been reviewed [569, 570] including its effect on metabolic disorders [571] and cognitive dysfunction [572]. Citrus carotenoids have been studied widely, and their biological roles have been elucidated [573, 574]. They are responsible for the external and internal fruit color. The accumulation pattern of carotenoids among the different citrus fruits is similar. However, the content and composition profile diverge with variety, growth condition, and season [575, 576]. For oranges, the most abundant carotenoids are β,β-xanthophylls contributing up to 98% of the total amount with violaxanthin, an epoxy carotenoid, being the predominant constituent. On the other hand, in mandarins β-cryptoxanthin, a bicyclic carotenoid considered a provitamin A, is the major component. In lemons and grapefruits, phytoenes, lineal carotenoids, can be the most accumulated compound. Carotenoids' role in human health and diseases has been expansively covered in the literature [535, 537, 577–581]. Here we select few references that pertain to anti-inflammatory activity [582–584], cardiovascular care [585–588], cancer prevention [589–593], and diabetes care [1, 594, 595]. Vitamin A in citrus fruits along with other valuable carotenoids such as α- and β-carotenes, β-cryptoxanthin, zeaxanthin, and lutein helps in maintaining eye health and prevents macular degeneration [596]. Several reviews have covered this topic extensively [442, 445–447, 597].

Countless research articles have demonstrated the chronic disease prevention involvement and anticancer properties of these individual chemical constituents of citrus fruits [598, 599]. In our holistic view, these numerous, valuable antioxidants, for instance, were jointly found to contribute immensely to chemoprevention and chemoprotection against many types of cancers along with their anti-inflammatory, cardiovascular, and neuroprotective effects [480]. In the case of cancer, more importantly, citrus fruit intake specifically was inversely associated with the risk of several types of cancers: esophageal [600], breast [601], pancreatic [602], stomach [603], and prostate [604].

Citrus fruits retain high amounts of citric acid (citrate derivatives—salt, ester, or metal complex forms), which contributes to the sour taste. Lemons and limes possess the highest concentration where it can constitute nearly 8% of the fruits' dry weight. Consumption of citrus fruits and/or their juices correlates well with lower risk incidence of renal stones [1, 605–609]. Moreover, citrus fruits are an excellent source of vitamin C. Vitamin C has also been covered in the vegetable section and other key fruits in this section. In short, vitamin C seems to play a critical role in disease prevention and cure [610, 611]. As a strong antioxidant, it enhances immune functions to combat infections [612, 613], and it contributes to immune defenses as it is implicated in preventing chronic inflammation [614, 615], cardiovascular diseases [616, 617], and cancer [598, 618–620]. Vitamin C is vital for the development and maintenance of connective tissues including wound healing; it also plays an essential role in bone formation [621] and musculoskeletal injuries recovery [622]. Vitamin C plays an important role in skin health since it is a potent antioxidant [44, 623, 624]. Moreover, accompanied by collagen peptide, it diminishes age-related skin atrophy [625] and is highly efficient as a rejuvenation therapy [626]; along with folic acid, it also ensures better hair growth and prevents hair loss [627]. Both vitamin C and citric acid can increase iron absorption in the digestive tract alleviating anemia caused by iron deficiency [628, 629].

6.1 Orange

Orange is the most common among the citrus fruits. It has two main varieties, sweet, such as Persian and blood oranges, and bitter oranges which can be used as part of different recipes. Oranges are low caloric and low-fat fruits with low glycemic index. Oranges are a rich source of vitamin C and are a good source of fiber, antioxidants, polyphenols, folic acid (B-9), and potassium. They also contain flavonoids; vitamins A, B-1 (thiamin), B-2 (riboflavin), B-3 (niacin), B-5 (pantothenic acid), and B-6 (pyridoxine); and calcium (Table 10). Orange peels, although not commonly eaten, are rich in fiber and vitamin C, even more than the flesh of the fruit.

Orange juice consumption has been associated with cognitive improvements [630, 631] and neurodegenerative disease protection [632, 633]. Orange peels are rich in many of the same compounds of the fruit flesh and as such provide similar health benefits [634, 635]. Eating oranges or applying it topically helps reverse skin damage caused by the ultraviolet light from the sun, reduces wrinkles, and improves the skin texture [636–638]. The peel has also displayed antimicrobial properties against periodontal pathogens [639]. Various citrus species including oranges produce essential oils with varied biological activities with few carrying low risk of irritation or phototoxicity [640].

It is rare to develop allergy to oranges or citrus fruits in general. Some people may experience heartburn, or it may aggravate acid reflux symptoms due to the high content of organic acids. Overeating oranges may lead to abdominal cramps and

Table 10 Orange nutrition facts of 100 g, from the United States Department of Agriculture (USDA SR-21)

Serving size of 100 g of raw oranges	
Per serving	% Daily value[a]
Calories 47	
Total fat 0.1 g	0
Saturated fat 0.0 g	0
Polyunsaturated fat 0.0 g	
Monounsaturated fat 0.0 g	
Total omega-3 fatty acids 7.0 mg	
Total omega-6 fatty acids 18.0 mg	
Cholesterol 0.0 mg	
Phytosterols	
Total carbohydrates 11.7 g	4
Dietary fiber 2.4 g	10
Starch 0.0 mg	
Sugars 9.4 g	
Protein 0.9 g	2
Vitamins	
Vitamin A	4
Vitamin C	89
Vitamin E	1
Vitamin B-6	3
Vitamin K	0
Folic acid	8
Minerals	
Sodium	0
Potassium	5
Calcium	4
Iron	1
Magnesium	2
Manganese	1
Copper	2
Zinc	0

National Nutrient Database
[a]Based on a 2000 calorie diet

diarrhea due to their high fiber content. The overconsumption of orange- or citrus-type beverages may also increase the risk of dental caries [641] or dental enamel erosion [642]. Citrus fruits, including oranges, are rich in furocoumarins, photocarcinogenic and/or phototoxic agents, due to interaction with cell DNA [643]; therefore, it is advisable to avoid prolonged sun exposure when consuming oranges in order to lower the chance of phytophotodermatitis condition or the risk of melanoma [644, 645], although this is more of an issue with grapefruit consumption including certain drug interactions.

6.2 Mandarins

Mandarins are smaller-sized citrus fruit and appear less "rounded" than common oranges. Mandarins are of assorted hybrid types, such as clementine and tangerine oranges. Since clementine and tangerines are specific varieties of mandarin oranges, the names are used interchangeably. Tangerines are mostly seedless with few exceptions. It has a less sour, much sweeter, and stronger taste than an orange. A ripe tangerine is heavy for its size, firm to a little soft, and pebbly skinned with no deep grooves. It has thin reddish-orange color peel, with a thin layer of bitter-tasting white mesocarp, which makes it easier to peel and to split into segments. Tangerines are low caloric, rich in dietary fiber, antioxidants, vitamins C and A, and folic acid (Table 11).

There are a number of health benefits for mandarin orange peels as well. Tangerine peel is used to treat bronchial asthma, pain, and indigestion [646]. The peel also can be utilized in anti-wrinkle skin care formulations [647]. The phytonutrients from the pulp or peel of mandarin oranges provide a protective effect against age-related cognitive dysfunction decreasing chronic inflammation and limiting cell damage. For instance, cell culture studies showed that hesperidin, nobiletin, and tangeretin are collectively responsible for the anti-neuroinflammatory capacity of tangerine peel [648], while β-cryptoxanthin, a potent antioxidant abundant in mandarins, was found to suppress DNA oxidative damage and enhance cognitive abilities in mice [649]. Tangerine juice contains vitamin A that stimulates the immune system [650] and has antimicrobial activity [540] to combat infections. Tangerine essential oil is a potent antiseptic and bactericide that combats *Staphylococcus aureus* sepsis, the most common cause in the pediatric age group [651–655]. In addition, tangerine essential oils promote cell division, stimulate cell growth, and help tissue repair [640]. They also have a sedative effect and are able to regulate stress symptoms and insomnia [656–658]. On the other hand, vitamin B-12 in tangerine helps hair growth and reduces hair loss [659].

Tangerine has no reported serious side effects. However, anaphylactic reaction of tangerine seeds—but not the fruit pulp itself—has been described [660]. Overconsumption of tangerine and the high fiber it contains may lead to abdominal pain and diarrhea.

6.3 Grapefruit

Grapefruit is a citrus fruit known for its sour to semi-sweet, slightly bitter taste. It is yellow-orange skinned and generally has a spherical shape. It grows in clusters, similar to grapes, hence the derivation of their name. The flesh is segmented and acidic and variable in color and sweetness depending on the variety. Among the common types of the fruit are the white, yellow, pink, and red pulp grapefruits. Grapefruits are low caloric, low fat, and full of nutrients. It is an excellent source of

Table 11 Tangerine (mandarin orange) nutrition facts of 100 g, from the United States Department of Agriculture (USDA SR-21)

Serving size of 100 g of raw tangerine (mandarin orange)	
Per serving	% Daily value[a]
Calories 53	
Total fat 0.3 g	0
Saturated fat 0.0 g	0
Polyunsaturated fat 0.1 g	
Monounsaturated fat 0.1 g	
Total omega-3 fatty acids 18.0 mg	
Total omega-6 fatty acids 48.0 mg	
Cholesterol 0.0 mg	
Phytosterols	
Carbohydrates 13.3 g	4
Dietary fiber 1.8 g	7
Sugars 10.6 g	
Protein 0.8 g	2
Vitamins	
Vitamin A	14
Vitamin C	44
Vitamin E	1
Vitamin B-6	4
Vitamin K	0
Folic acid	4
Minerals	
Sodium	0
Potassium	5
Calcium	4·
Iron	1
Magnesium	3
Manganese	2
Copper	2
Zinc	0

National Nutrient Database
[a]Based on a 2000 calorie diet

vitamins C and A and fiber (Table 12). The bitter taste pith, the white flesh inside the peel which is often thrown out, is rich in antioxidants and fibers.

Key attributes ascribed to grapefruit have been its quality to control appetite and manage weight loss and diabetes condition. It is reasonable to conclude that features like high water content, low caloric input, and high fiber content all contribute to appetite control and promote fullness leading to low calorie intake [69, 83, 500, 661–663]. These findings have been substantiated by clinical trials [661, 664, 665] even in the case of overweight/obese adults [666, 667], prominently assuring better health [482, 668]. The two primary risk factors in diabetes are insulin resistance and high blood sugar levels [669, 670]. With grapefruit's low glycemic

Table 12 Grapefruit nutrition facts of 100 g, from the United States Department of Agriculture (USDA SR-21)

Serving size of 100 g of raw, pink, and red and white grapefruit	
Per serving	% Daily value[a]
Calories 32	
Total fat 0.1 g	0
Saturated fat 0.0 g	0
Polyunsaturated fat 0.0 g	
Monounsaturated fat 0.0 g	
Total omega-3 fatty acids 5.0 mg	
Total omega-6 fatty acids 19.0 mg	
Cholesterol 0.0 mg	
Phytosterols	
Carbohydrates 8.1 g	3
Dietary fiber 1.1 g	4
Sugars 7.0 g	
Protein 0.6 g	1
Vitamins	
Vitamin A	19
Vitamin C	57
Vitamin E	1
Vitamin B-6	2
Vitamin K	0
Folic acid	2
Minerals	
Sodium	0
Potassium	4
Calcium	1
Iron	0
Magnesium	2
Manganese	1
Copper	2
Zinc	0

National Nutrient Database
[a]Based on a 2000 calorie diet

index, it appears that eating a fresh grapefruit before meals controls glucose levels and reduces both insulin levels and insulin resistance [661] and, consequently, the risk of type 2 diabetes [671, 672]. The actual mechanism of this is not known. Fresh red grapefruit was found to reduce the total cholesterol and triglycerides, as well as the unhealthy LDL cholesterol, which reduces the risk of coronary artery diseases and stroke [673]. Grapefruit flavonoids were found to reduce the risk of ischemic stroke in women [674]. Another value of the red and pink grapefruits is that they possess higher content of the antioxidants β-carotene and lycopene adding to its health benefits. The combination of vitamin A and C in grapefruits not only boosts the immune system but also plays an important role in wound healing

[675] and managing bronchial asthma and its symptoms [676]. Grapefruits have high water content of about 88% of its total weight, which keeps the body well hydrated.

Grapefruit consumption is not risk-free. Grapefruits are rich in furanocoumarin, a natural phototoxic chemical that affects the liver's drug metabolism [677]. This compound inhibits a cytochrome P450 enzyme called CYP3A4, which plays a vital role in the metabolism of about 50% of all drugs in the body. This may lead to certain adverse effects [678] and potentially dangerous drug interaction [679]. This has prompted the US FDA to put out a consumer advisory note on some prescription and over-the-counter (OTC) medication use and simultaneous grapefruit consumption [680]. Therefore, it is not advisable to eat grapefruit or drink its juice prior to some medications. Examples from this list include anti-retrovirals like Indinavir, allergic drugs like loratadine (Claritin) or fexofenadine (Allegra), cholesterol-lowering agents like some statins (Lipitor and Zocor), anti-hypertensive Ca^{+2}-channel blockers (nifedipines), immunosuppressants such as cyclosporines, some pain killers, and some corticosteroids and anti-anxiety (Buspirone and Zoloft) or psychotic medications (benzodiazepines and Carbamazepine). As a citrus, it is a rich source for furocoumarins (psoralens) which intercalate into DNA and forms cross-links upon ultraviolet radiation. This characteristic in grapefruit is used as a form of therapy for the treatment of skin problems resembling psoriasis. Therefore, it is advisable to limit sun exposure upon grapefruit consumption. Grapefruit overconsumption was shown to be associated with the highest risk of developing malignant melanoma [645]. Grapefruit is very citric and consuming it with poor oral hygiene might cause teeth erosion [681]. Considering its richness in fiber content, overconsumption may lead to diarrhea and abdominal pain.

7 Cucumbers

Family: Cucurbits or Cucurbitaceae
Genus: *Cucumis*
Common name: Cucumber

Table 13 Cucumber nutrition facts of 100 g, from the United States Department of Agriculture (USDA SR-21)

Serving size of 100 g of raw cucumber with peel	
Per serving	% Daily value[a]
Calories 15	
Total fat 0.1 g	0
Saturated fat 0.0 g	0
Polyunsaturated fat 0.0 g	
Monounsaturated fat 0.0 g	
Total omega-3 fatty acids 5.0 mg	
Total omega-6 fatty acids 28.0 mg	
Cholesterol 0 mg	
Phytosterols 14.0 mg	
Carbohydrates 3.6 g	1
Dietary fiber 0.5 g	2
Starch 0.8 g	
Sugars 1.7 g	
Protein 0.7 g	1
Vitamins	
Vitamin A	2
Vitamin C	5
Vitamin E	0
Vitamin B-6	2
Vitamin K	21
Folic acid	2
Minerals	
Sodium	0
Potassium	4
Calcium	2
Iron	2
Magnesium	3
Manganese	4
Copper	2
Zinc	1

National Nutrient Database
[a]Based on a 2000 calorie diet

Cucumber is actually a fruit that is commonly thought to be a vegetable. There are two main kinds of cucumbers available all year around: fresh type (also known as slicing) and pickled type. Among both varieties there are different cultivars bred for specific characteristics. Cucumber is low in saturated fat, cholesterol, and sodium. It is a good source of vitamin A, magnesium, and phosphorus. It is a very rich source of fiber, vitamins C and K, and the minerals potassium and manganese. It also contains multiple B vitamins such as B-6, B-5, and B-1 (Table 13).

Cucumbers have one of the highest water contents (approx. 95%) when compared to other vegetables imparting important qualities. Water is essential for bodily function [682] since hydration affects physical performance and metabolism [683, 684]. About 40% of people obtain their total water intake from food [685]. Cucumbers keep the body hydrated. One study found that fruit and vegetable intake was associated with improvements in hydration status in children [686] and helps control body temperature as well [685]. An additional benefit of water content, i.e., improved hydration status along with magnesium presence, is that cucumber potentially alleviates headaches or migraines [687]. The high water content, low calories, and high fiber composition, specifically pectin, of cucumber make it ideal for weight management [662] and a healthy digestive system [688] overall, by preventing constipation [689–691], helping remove excess water and toxins from the body, and maintaining healthy kidney, liver, and pancreatic activity [688]. The natural diuretic effects of cucumber juice help remove the uric acid preventing gouty arthritis and kidney stones [688]. Along with vitamin K and other essential nutrients like calcium and magnesium, cucumbers' richness in silica, a compound that is instrumental in connective tissue formation and calcium assimilation, collectively contributes to maintaining healthy ligaments, tendons, cartilage, and bones [692]. The high water content and vitamin C coupled with silica make eating cucumber and applying its extracts topically a good hydrating and soothing treatment for various skin problems including burns [688].

Cucumbers' high magnesium, potassium, and fiber contents help reduce blood pressure [693]. Cucumbers can be used to control blood sugar. They contain a hormone required by the beta-cells during insulin production and in fact have a very low glycemic index [694]. In a study, cucumber peel was shown to reverse diabetes complications [695], while in another study, aqueous extracts showed protective effects and prevented diabetes-related complications as well [696].

The role of antioxidants in disease and health is well studied [542]. Cucumbers are rich in antioxidants such as flavonoids and tannins that block accumulation of free radicles and improve antioxidant function, thus reducing the risk of diseases. They impart an analgesic effect as well [697, 698]. Other antioxidant compounds include cucurbitacins, glucosides, lignans, apigenin, and firestin [688]. Cucurbitacins were found to possess anticancer activity via apoptosis pathways induction mechanism [699]. Firestin was found to possess multiple biological activities relevant to the maintenance of brain function. The molecule is neuroprotective and preserves cognitive abilities; therefore, it delays the onset of age-related decline in brain function [700].

Cucumbers are generally well tolerated, but overconsumption is linked to some unwelcome health effects. Excessive cucumber consumption can increase the diuretic effect triggering excessive loss of water and electrolytes leading to dehydration. High potassium content of cucumbers may initially lead to abdominal pain and bloating, then causing further complications that affect kidney function. High doses of vitamin C may be harmful and trigger the growth and spread of free radicals increasing the risk of cancer and premature aging. Cucumbers contain cucurbitacins or toxic triterpenoids that cause its bitter taste. Cucumbers can trigger allergic reactions in certain individuals.

8 Figs

Family: Moraceae
Genus: *Ficus*
Common name: Fig

Figs are sweet and juicy fruits when ripe. They come in multiple varieties: each with a unique taste. Types include red figs, the yellow figs, or the purple-skinned figs or green-striped figs. The paste of sweet and soft figs is used as a sugar replacement. Figs can be eaten fresh, dried, or cooked. Figs' health benefits are in the fruit, skin, pulp, and leaves and come from its high content of minerals such as calcium, iron, phosphorus, manganese, magnesium, potassium, and copper and vitamins like A, K, B-1, and B-6 and fiber (Table 14). Dried figs are thought to have more nutrients such as calcium and phenolic antioxidant compounds than fresh ones [701].

Dried fig's low fat content and high content of fiber (9.8 g/100 g) guarantee numerous health benefits. First, it ensures a healthy digestive system as it alleviates constipation and helps regulate bowel movements [702]. As a low-calorie snack replacement, fiber keeps one feeling full longer. Thus, it helps in weight control [703]. Pectin, a soluble fiber in figs, may assist in lowering cholesterol [704] and control blood sugar levels [705, 706].

The phytochemistry and pharmacology of figs have been reviewed recently [703, 707, 708]. Figs are rich in antioxidants such as flavonoids and polyphenols which prevent the damage caused by free radicals [709]. Dried figs have higher amounts of antioxidants [710, 711]. High fiber and antioxidant content in figs decreases the risk of colon [712] and breast cancers. Figs contain tirucallane-type triterpenoids that are toxic to different human cancer cell lines [713]. The bioactive compounds of fig leaves were also shown to have phototoxic capabilities and are being used to develop new photodynamic therapy for the treatment of skin cancer [714]. Fig extracts could also be used to help treat other skin conditions such as abnormal skin pigmentation, eczema, acne, psoriasis, and freckles [707, 715].

Studies have shown fig effectiveness in stimulating immune system response [716]. Furthermore, fig extract showed strong antibacterial activity against oral bacteria and fungi [708]. Moreover, fig leaves have also shown significant protective effects against certain fungal infections like *Candida albicans* and can be a strong

Table 14 Fig nutrition facts of 100 g, from the United States Department of Agriculture (USDA SR-21)

Serving size of 100 g of raw figs	
Per serving	% Daily value[a]
Calories 74	
Total fat 0.3 g	0
Saturated fat 0.1 g	0
Polyunsaturated fat 0.1 g	
Monounsaturated fat 0.1 g	
Total omega-3 fatty acids	
Total omega-6 fatty acids 144 mg	
Cholesterol 0.0 mg	
Phytosterols 31.0 mg	
Carbohydrates 19.2 g	6
Dietary fiber 3 g	12
Sugars 16.3 g	
Protein 0.7 g	1
Vitamins	
Vitamin A	3
Vitamin C	3
Vitamin E	1
Vitamin B-6	6
Vitamin K	6
Folic acid	1
Minerals	
Sodium	0
Potassium	7
Calcium	4
Iron	2
Magnesium	4
Manganese	6
Copper	4
Zinc	1

National Nutrient Database
[a]Based on a 2000 calorie diet

natural antibacterial. The methanol extract in figs showed synergistic effect when used with antibiotics [717, 718]. Meanwhile, ethanol extract has recently shown powerful body temperature reduction capability when compared to antipyretic agents [703, 719].

Figs are rich in calcium and phosphorus which are vital in bone health and reduce the risk of osteoporosis, described earlier. Low sodium and high potassium content (7–19% of daily intake) of figs helps control the blood pressure and eliminate excess water as well as uric acid from the body preventing arthritis and kidney stones per aforementioned discussions. High fiber and potassium content in fig leaves helps regulate blood glucose and prevent sugar level instability [720].

Allergic reactions to figs are uncommon. However, skin contact with the fruit or leaves can cause rashes among those with skin sensitivity. Individuals allergic to rubber, latex, or birch pollen may also be allergic to figs [721]. Dried figs are rich in vitamin K and should be consumed with caution by people taking blood-thinning medications such as warfarin. They are also high in oxalates running the risk of kidney stones. Applying fig leaves on the skin may cause skin to be more sensitive to sun. Due to its laxative effect, overconsumption of high fiber figs may lead to diarrhea.

9 Grapes

Family: Vitaceae
Genus: *Vitis*
Common name: Grape (Red Wine)

Technically, grapes are berries that grow generally in clusters. They are of two main types: table grapes, which are usually large, seedless, and with fairly thin skin, and wine grapes, which are smaller in size, usually seeded, and with comparatively thicker skin. Grapes have multiple different colors, including crimson, black, purple or blue (Concord), yellow, green, orange, red, and pink. "White" grapes are in fact green and are derived from the purple grape. Grapes can be consumed fresh as table grapes or dried as raisins. They are also used for making different wines (green grapes are used to make white wine, and purple grapes are used for red wines), vinegar, and jam. Concord grapes are used to make juice and jelly and for grape flavoring. Grape seeds are used for their oil and their extract. They are a good source of dietary fiber, potassium, and a number of vitamins and other minerals (Table 15). Grapes contain a number of well-known strong antioxidants such as vitamins A and C and manganese as well as polyphenols such as flavonoids and the non-flavonoid resveratrol, a type of natural polyphenol also known for its anti-inflammatory activity and decreasing oxidative stress.

There are countless health benefits for grapes that make them an excellent choice for a healthier diet [722]. The following discussion will focus on the antioxidants/antioxidative stress capacity of grapes [723–725] with a particular reference to res-

Table 15 Grape nutrition facts of 100 g, from the United States Department of Agriculture (USDA SR-21)

Serving size of 100 g of raw, red, or green grapes	
Per serving	% Daily value[a]
Calories 69	
Total fat 0.2 g	0
Saturated fat 0.1 g	0
Polyunsaturated fat 0.0 g	
Monounsaturated fat 0.0 g	
Total omega-3 fatty acids 11.0 mg	
Total omega-6 fatty acids 37.0 mg	
Cholesterol 0.0 mg	
Phytosterols 4.0 mg	
Carbohydrates 18.1 g	6
Dietary fiber 0.9 g	4
Sugars 15.5 g	
Protein 0.7 g	1
Vitamins	
Vitamin A	1
Vitamin C	18
Vitamin E	1
Vitamin B-6	4
Vitamin K	18
Folic acid	0
Minerals	
Sodium	0
Potassium	5
Calcium	1
Iron	2
Magnesium	2
Manganese	4
Copper	6
Zinc	0

National Nutrient Database
[a]Based on a 2000 calorie diet

veratrol [726]. Resveratrol is a natural phytoalexin found in the skin and leaves of red grapes. It is also thought to possess strong anti-inflammatory properties [727–729]. Several reviews have summarized resveratrol's therapeutic potential based on human clinical trials [730, 731]. We will pay a particular attention to chronic diseases such as cancer [732], diabetes [733], and cardiovascular disease [734]. One aspect of a future research direction for resveratrol is its activation of genes associated with aging and longevity [735, 736].

Resveratrol has been shown to slow the growth of a wide variety of cancer cells including multiple gastrointestinal cancers [737–739], breast cancer [740–743], prostate cancer [744], and melanoma and leukemia [745–747]. Quercetin, a flavonol-type flavonoid found in grapes, is known for its ability to induce apoptosis and slow cancer cell growth without affecting normal cells [748]. This is in addition to other polyphenols like anthocyanins and catechins that contribute to its anticancer properties [749]. Resveratrol is also presumed to be useful in the treatment of diabetes and its complications [750]. For example, it improves fasting sugar levels [751], increases insulin sensitivity [752], and decreases the risk of diabetic complications such as diabetic neuropathy and nephropathy [750]. Studies have also showed that consuming whole grapes, but not its juice, reduces the risk of developing type 2 diabetes [289, 753, 754]. Overall, grapes or grape products contribute to type 2 diabetes management [671, 755]. Grapes play a major role in enhancing heart and cardiovascular system health [734]. Resveratrol is believed to possess anti-atherogenic effects [756]. Along with other polyphenols in red grapes, they are thought to decrease LDL cholesterol levels, reduce lipid peroxidation and platelet aggregation, and combat oxidative stress and inflammation [756–758]. Additional benefits are derived from their high dietary fiber content that helps reduce both total cholesterol and LDL levels, further boosting heart and vascular health [75]. The value of grape flavonoids of the red or white varieties as nutraceuticals has been reviewed [723, 759] including their antioxidant and anti-inflammatory biological properties and cardioprotective activities. For instance, grape flavonols [760], in particular rutin, were found to reduce the risk of heart attacks and strokes through ERK1/2 and Akt signaling pathways [761] and by inhibiting of protein disulfide isomerase (PDI) that is associated with blood clot formation [762]. In addition, grapes' high potassium content [763] and grape seed polyphenols have been shown to have antihypertensive properties [751, 764]. Resveratrol also has major ophthalmic health effects via its numerous biological properties that have been demonstrated in in vitro and in vivo experimental studies [765]. It has been shown to decrease the risk of glaucoma, cataract, age-related ocular degeneration, and even the retinopathy secondary to diabetes [765–767]. Grapes contain the carotenoids antioxidants, lutein and zeaxanthin, which also were found to reduce age-related macular degeneration and prevent cataract [42, 445].

Resveratrol was shown to reduce cognitive impairment and play a neuroprotective role, decreasing the amyloid burden and reducing tau hyper-phosphorylation in Alzheimer's disease [768], and improve age-related memory and mood dysfunction [769, 770]. Polyphenols in grape juice were also found to improve memory in healthy older adults [771] or adults with mild cognitive impairment but not dementia, reducing the risk of Alzheimer's disease [772].

The flesh, skin, and seeds of grape extract were found to improve immune functions and have potent antimicrobial properties [773–775]. Resveratrol exhibits strong antiviral and cytoprotective activities [776] and antibacterial activity against foodborne pathogens [777]. Extracts from grape seeds have been shown to have antiviral activity against the influenza virus [778] and polio and herpes simplex viruses [779]. The topical application of pro-anthocyanidins, a class of flavonoid polyphenols [99] present in grape seeds extract, promotes wound healing. The extract cream concoction possesses anti-inflammatory and antimicrobial properties and triggers the release of vascular endothelial growth factor leading to wound contraction and closure [780].

In general, grapes are safe when consumed in moderate quantities. Some people may develop allergy against grapes and its products. Allergic reactions can range from hives and rashes to sneezing, wheezing, and difficulty in breathing. Serious anaphylactic reaction develops rarely when eating grapes or any of its products. Grapes have high vitamin K content that may increase the risk of bleeding, so it must be consumed with caution before surgical procedures and when taking anticoagulant medications. Grapes might interfere with medications like phenacetin, an analgesic fever-reducing drug, that metabolize in the liver. Consuming a lot of grapes on a regular basis may cause carbohydrate overload and promote weight gain. The high fiber content in grapes and raisins might lead to diarrhea and vomiting due to overconsumption. Those with fructose intolerance must avoid eating grapes and its products to avoid harming the liver and kidney or indigestion and abdominal pain complications.

10 Kiwi

Family: Actinidiaceae
Genus: *Actinidia*
Common name: Kiwi or Chinese Gooseberry

Kiwi is an exotic fruit, mostly oval in shape and about the size of a big chicken's egg. It has a fibrous, dull greenish-brown skin and bright green or golden flesh—depending on the type—speckled with rows of small, black, edible seeds. Kiwi has a soft texture, with a sweet and distinctive taste. Kiwifruit is full of different nutrients and phytochemicals that have enormous health benefits. It is a rich source of vitamins such as vitamins C, K, E and A and minerals such as potassium and copper in addition to sugar, fiber, and the antioxidant lutein (Table 16).

Interestingly, the kiwifruit, its skin, its seeds, and even its roots all contribute to multiple health benefits it has to offer [781]. Both green and gold kiwis contain two proteins called thaumatin-like protein and actinchinin which act against two different types of fungi and bacteria [782]. The golden kiwi seeds show antibacterial activities as well [783]. Kiwi's skin—mainly in the green variety—contains an

Table 16 Kiwi nutrition facts of 100 g, from the United States Department of Agriculture (USDA SR-21)

Serving size of 100 g of raw, fresh kiwi	
Per serving	% Daily value[a]
Calories 61	
Total fat 0.5 g	1
Saturated fat 0.0 g	0
Polyunsaturated fat 0.3 g	
Monounsaturated fat 0.0 g	
Total omega-3 fatty acids 42.0 mg	
Total omega-6 fatty acids 246 mg	
Cholesterol 0.0 mg	0
Phytosterols	
Carbohydrates 11.7 g	5
Dietary fiber 3.0 g	12
Starch 0.0 g	
Sugars 9.0 g	
Protein 1.1 g	2
Vitamins	
Vitamin A	2
Vitamin C	155
Vitamin E	7
Vitamin B-6	3
Vitamin K	50
Folic acid	6
Minerals	
Sodium	0
Potassium	9
Calcium	3
Iron	2
Magnesium	4
Manganese	5
Copper	6
Zinc	1

National Nutrient Database
[a]Based on a 2000 calorie diet

active protease enzyme called actinidain, which helps break down proteins to amino acids to ease its absorption [784, 785]. Kiwifruit supplies 275% of the daily vitamin C intake. As such, they have been shown to reduce upper respiratory tract infection symptoms [786, 787] and asthma in few studies along with decreasing wheezing [788]. Kiwi juice is a rich source of antioxidants like polyphenols including vitamins C and E, which are immunostimulatory [789] and prevent oxidative stress/DNA damage [786, 790, 791]. The high concentration of vitamin C in kiwi also enhances the absorption of iron, preventing iron deficiency, i.e., anemia [792]. The availability of vitamin C plays a critical role in the maintenance of a normal mature collagen network [793, 794]. On the other hand, the presence of lutein protects the skin from UV light [795]. Moreover, kiwi's high levels of zeaxanthin and lutein along with vitamin A contribute to its ability to prevent vision loss and age-related macular degeneration [580, 796–798]. The root extracts of the golden kiwi (abundance of antioxidants and polysaccharides) were found to have anticancerous properties against some tumors like lung, liver, colon, and oral cancer cell lines [799–801]. Furthermore, it displayed antitumor activity and tumor remission in animal studies [802].

The appreciable potassium concentration in kiwi helps reduce the risk of developing kidney stones [94]. This high potassium level coupled with low sodium content also helps regulate blood pressure [803, 804]. In addition, the high vitamin K content in kiwi contributes to a healthy cardiovascular system due to its ability to decrease triglyceride levels, increase HDL cholesterol [786, 805], and reduce platelet aggregation [806]. Kiwi's fiber also helps enhance lipid profile and is vital in weight control [805]. The high fiber composition and low glycemic index of the green kiwi make it suitable to regulate blood glucose levels [807].

In mice, green kiwi has been shown to reduce the resorption of bone minerals [808], whereas the substantial supply of vitamin K along with its content of calcium, magnesium, potassium, and phosphorus aids in bone strength, reduces the risk of bone-related injuries, and prevents diseases like osteoporosis [809].

It appears that kiwifruit consumption may improve sleep onset, duration, and efficiency in adults. Kiwi has high amounts of serotonin [810], a hormone linked to REM (rapid eye movement) sleep. Lack of serotonin in addition to folic acid and some antioxidants found in kiwi correlates with insomnia [811]. There is also suggestive evidence that serotonin may help improve memory and mood/depression disorders [812].

Kiwis are allergenic. The allergic reaction that sensitive people might develop when eating kiwi is not the only negative health effect associated with this fruit. Overconsumption of kiwi is reported to induce asthma, rash, and hives. Eating a lot of kiwis can lead to local mouth irritation and swelling of mouth, lips, and tongue. It also may lead to tingling and itching sensations in the mouth which is known as oral allergy syndrome (OAS). It may even cause skin irritation and dermatitis. Kiwi has high concentrations of potassium, serotonin, vitamin E, and vitamin C. Excess concentration of these has its own side effects. Overeating kiwi potentially affects triglyceride level in the blood leading to acute pancreatitis. Kiwi is also known to interact with some anti-fungal medications. It can increase the risk of bleeding in case of use of anti-coagulant, anti-platelet, or non-steroidal anti-inflammatory agents such as aspirin. Overeating of kiwi possibly evokes nausea, vomiting, and abdominal pain or even fainting.

11 Pomegranate

Family: Lythraceae
Genus: *Punica*
Common name: Pomegranate

Pomegranate shrub is a small tree with several spiny branches and high longevity: some types living up to 200 years. Pomegranate is a jewel-like fruit that is classified as a berry. The edible part of the pomegranate is its seeds which are known as arils. Hundreds of slightly ruby red-colored, sweet seeds lie inside the thick, brownish yellow to deep red and inedible peel. Pomegranates are a very rich source of dietary fiber, vitamins C and K, and potassium (Table 17). The pomegranate fruit, and the juice made from its seeds, is loaded with antioxidants, bioactive compounds, and sugar.

There are innumerable health benefits for pomegranates that make them an exceptional choice for a healthier diet [813]. The following discussion will focus on antioxidant and anti-inflammatory capacity of pomegranates. It is an antioxidant powerhouse showing activity three times higher than those of red wine and green tea [814]. A particular reference will be made to polyphenols such as ellagitannins (including punicalagins) and fatty acids like punicic acid considered the most abundant bioactive chemical constituents in pomegranates [815–817]. Several reviews have summarized the therapeutic potential of pomegranate [813, 816, 818–821] and proposed few mechanisms of action for these bioactives. We will pay particular attention to chronic diseases such as diabetes, cancer, and cardiovascular diseases [816, 822, 823].

There are two unique compounds in pomegranates that are mostly responsible for their health benefits. The first one is punicalagins that are ellagitannin-type phenolic compounds, a group of powerful antioxidants available in pomegranate juice and peels [824]. The second is punicic acid that is a conjugated linoleic acid, a type of polyunsaturated omega-5 fatty acid, which is the main fatty acid found in pomegranate seeds [825, 826].

Table 17 Pomegranate nutrition facts of 100 g, from the United States Department of Agriculture (USDA SR-21)

Serving size of 100 g of raw pomegranate	
Per serving	% Daily value[a]
Calories 83	
Total fat 1.2 g	2
Saturated fat 0.1 g	1
Polyunsaturated fat 0.1 g	
Monounsaturated fat 0.1 g	
Total omega-3 fatty acids	
Total omega-6 fatty acids 79.0 mg	
Cholesterol 0.0 mg	0
Phytosterols	
Carbohydrates 18.7 g	6
Dietary fiber 4 g	16
Starch	
Sugars 13.7 g	
Protein 1.7 g	3
Vitamins	
Vitamin A	0
Vitamin C	17
Vitamin E	3
Vitamin B-6	4
Vitamin K	21
Folic acid	10
Minerals	
Sodium	0
Potassium	7
Calcium	1
Iron	2
Magnesium	3
Manganese	6
Copper	8
Zinc	2

National Nutrient Database
[a]Based on a 2000 calorie diet

Punicalagins in pomegranate juice have been found to reduce inflammation and lower the risk of chronic diseases caused by inflammatory reactions [827] including type 2 diabetes mellitus [828] and digestive tract inflammation [829]. Punicic acid was also shown to possess anticancerous and anti-diabetic properties in in vitro, in vivo, and patient intervention studies [825, 830]. Numerous reviews have summarized the pre-clinical and clinical cancer studies on the use of pomegranate extracts suggesting its use as a promising chemopreventive and/or chemotherapeutic agent [831–835] with few describing molecular targets and mechanisms of the extract's major constituent polyphenols. By way of example, pomegranate extracts have been shown to affect several types of cancer, including breast [836–838], prostate [839–843], and colon cancer [844–846], in different ways.

Cell culture studies reported the cartilage protective and arthritis inhibitory effects of pomegranate extract [847, 848]. In animal studies, pomegranate polyphenol-rich extract preferentially inhibited inflammatory markers and was found to lower the risk of collagen-induced arthritis [849]. An interventional study using pomegranate juice for knee osteoarthritis patients improved physical function and stiffness, decreased breakdown cartilage enzymes, and increased antioxidant status in patients [850]. A review highlighted evidence-based studies for treatment of osteoporosis, osteoarthritis, and rheumatoid arthritis [851].

Pomegranate juice contains antioxidants and polyphenols that have been found to support a healthy cardiovascular system [852]. Pomegranate juice consumption inhibits serum ACE activity and reduces systolic blood pressure [853]. A meta-analysis of randomized controlled trials and clinical investigation of hypertensive subjects indicated control of blood pressure by employing different mechanisms [854–856]. Pomegranate seed oil contains punicic acid which has been found to have anti-atherogenic effects. In hyperlipidemic subjects, it had favorable effects as it lowered triglyceride level and improved the triglyceride:HDL ratio [857]. Both mice [858, 859] and human [860, 861] studies concluded that pomegranate juice consumption reduces oxidative stress, lipid peroxidation, and platelet aggregation even attenuated the development of atherosclerosis. Pomegranate juice was also found to lower total cholesterol and blood glucose in diabetic patients which further boosted heart and blood vessel health [862]. Pomegranate is a rich source of dietary nitrates. Besides its cardioprotective role [863, 864], it has been hypothesized to be an ergogenic, i.e., improving or enhancing blood flow and assuaging exhaustion during physical exercise [865–867].

There is a great body of in vivo animal studies and patient intervention studies evidence that support the role of pomegranate products in neuroprotection [868–871] and improvement of memory [872–874], cognition [148, 875, 876], age-related cognitive function impairment [877], and Alzheimer's disease [878–880].

Pomegranate seeds and pulp extracts display a broad range of antimicrobial properties in vitro and in vivo that enable stopping or preventing pathogenic infections [819, 881–884]. By way of example, possessing broad antibacterial activities [885–890] against antibiotic-resistant bacteria such as clinical strains of multidrug-resistant *S. epidermidis* [891], methicillin-resistant *Staphylococcus aureus* (MRSA) [892], and other β-lactamase producer species [893–895], cariogenic bacteria [896–899], antifungal activities [900–905] in particular oral *Candida* sp. [897, 906, 907] and dermatophyte fungi [908], and antiviral activities [884, 909–914]. Most of the clinical studies conducted so far have been in the oral health area to prevent dental plaques, gingivitis, or periodontitis [884, 915].

In addition, its extracts have shown good wound healing potential [916]. In rat studies, pomegranate extract accelerated the wound healing process, characterized by collagen deposition improvement, neutrophil infiltration in the wound area, angiogenesis, and fibrosis degree [917]. In another animal study, an ointment formulation of three herbal extracts that included pomegranate demonstrated its wound and topical infection healing potential [918]. It also demonstrated anti-acne and skin repair capacity [919, 920].

Additional benefits of pomegranate are in its seeds with rich fiber content that enhances digestive health and resolves constipation issues. The presence of vitamins C and E in them is vital for eye and skin health. Even the inedible peel is loaded with phytonutrients which affect health and make it a novel medicament. Pomegranate peels are used as sun block to reduce the effect of ultraviolet light and hence the potential for skin cancer. Pomegranate peel extract added to toothpaste was found to reduce plaque and gingivitis. The peel powder traditionally has been used as a medication for bronchitis, sore throat, and pulmonary tract infections.

Pomegranate could be unsafe to consume for the following reasons. Pomegranate juice or seeds may cause serious side effects due to drug interaction. Pomegranate lowers blood pressure and hence should be used cautiously with antihypertensive agents' angiotensin-converting enzyme inhibitors (ACEIs), including Capoten, Vasotec, and Prinivil, and also statins like Lipitor that lower cholesterol and blood anti-coagulants like warfarin. Pomegranate and its products should not be consumed when there are signs of food allergies. Pomegranate is rich in fiber, and its juice should be avoided in the event of diarrhea.

12 Tomatoes

Family: Solanaceae
Genus: *Solanum*
Common name: Tomato

Scientifically speaking tomato is a berry fruit that is categorized as a vegetable. With a history dating back to 500BC and over 7,500 varieties, tomato ranks as 1 of the top 3 most popular fresh market vegetables. When first cultivated, tomatoes were yellow or orange. However, through breeding, the standard color of tomatoes is now red. Furthermore, there are over ten different lower classifications of tomatoes that differ in shape, flavor, and size. Tomatoes are a good source of vitamins A, K, and C, folic acid, and other antioxidants such as lycopene and chlorogenic acid (a caffeic acid ester). It is rich in potassium and has low sodium content (Table 18). Tomato phytochemicals can vary greatly between different tomato varieties and

Table 18 Tomato nutrition facts of 100 g, from the United States Department of Agriculture (USDA SR-21)

Serving size of 100 g of raw, ripe, red tomatoes	
Per serving	% Daily valuea
Calories 18.0	
Total fat 0.2 g	0
Saturated fat 0.0 g	0
Polyunsaturated fat 0.1 g	
Monounsaturated fat 0.0 g	
Total omega-3 fatty acids 3.0 mg	
Total omega-6 fatty acids 80.0 mg	
Cholesterol 0.0 mg	0
Phytosterols 7.0 mg	
Carbohydrates 3.9 g	1
Dietary fiber 1.2 g	5
Starch 0.0 g	
Sugars 2.6 g	
Protein 0.9 g	2
Vitamins	
Vitamin A	17
Vitamin C	21
Vitamin E	3
Vitamin B-6	4
Vitamin K	10
Folic acid	4
Minerals	
Sodium	0
Potassium	7
Calcium	1
Iron	1
Magnesium	3
Manganese	6
Copper	3
Zinc	1

National Nutrient Database
[a]Based on a 2000 calorie diet

sampling periods [921, 922]. The levels of these compounds are strongly influenced by the maturity of the tomatoes.

Tomato consumption has been linked to many health benefits [923]. The water content in tomatoes is around 95%. They are also a good source of fiber (1.5 g/average size), which is mostly the insoluble type in the form of hemicellulose, cellulose, and lignin [924]. The significance of this combination has been discussed in the cucumber section. By way of example, the high content of water in tomatoes stimulates urination which increases the elimination of toxins from the body and excess of water, uric acid, and salts [925–927]. Tomatoes' fiber content prevents constipation [928], and they are widely recognized as being one of the high antioxidant foods [929]. Their health value has greatly reduced the risk of heart disease and cancer. Some of these key antioxidants are β-carotene [536, 930, 931], lycopene [932–934], naringenin [571, 935], and chlorogenic acid [936, 937]. The ensuing discussion will largely focus on cardiovascular diseases and cancer prevention by these antioxidants with a particular focus on the role of lycopene in coronary heart diseases [938] and cancer [939, 940]. Lycopene is a non-provitamin A carotenoid that has up to twice the antioxidant capacity of β-carotene in vitro and is responsible for the red color seen in tomatoes. It has been shown that low levels of lycopene in the blood are linked to increased risk of heart attacks and strokes [934, 941, 942]. Clinical studies have shown that lycopene and tomato products decreased plasma total cholesterol and LDL cholesterol and increased high-density lipoprotein cholesterol [943]. In addition, they positively affect platelet anti-aggregation activity and promote endothelial protection [944]. Parallelly, vitamin K availability in tomatoes controls bleeding and blood clot formation. Several studies have also demonstrated that lycopene possesses anti-hypertensive qualities [945–948]. Low sodium and high potassium content of tomato along with the constituent chlorogenic acid together supports vascular health and may lower blood pressure or prevent hypertension [949–952].

Epidemiologic studies consistently support the notion of lower risk of cancer being associated with higher consumption of tomatoes and tomato-based products [953, 954]. Observational studies suggest that tomatoes or tomato product consumption correlates with fewer incidents of prostate cancer [955–958]. This potential role for prostate cancer prevention is supported by clinical evidence [959]. A diet rich in carotenoids including lycopene may protect against the development of breast cancer as well [960–962]. The high quantity of lycopene in tomatoes has been shown to reduce the effects of carcinogens in cigarettes and can protect against lung cancer [963]. Beta-carotene in tomatoes is associated with lower rate of colorectal cancer [964].

Beta-carotene, lycopene, and lutein are antioxidants that have been shown to protect the eyes from harmful light effects and age-related macular degenerative changes [596].

Tomatoes are well tolerated in general. The immune system may react to the proteins in tomatoes releasing histamine and leading to joint swelling and pains. Allergic reaction may occur in some individuals due to protein cross-reactivity [965]. Tomato leaves are unsafe to eat; eating large quantities may lead to poisoning in form of nausea, vomiting, headache, muscle spasm, and even death in severe cases. Tomatoes grown in contaminated soil may contain high levels of fluoride [966]. Overconsumption of tomatoes is associated with a few undesirable side effects. Tomatoes are high in acid which can cause "heartburn" for those affected with gastro-esophageal reflux diseases. They induce gastric acid, which when taken in high amounts can flow up the esophagus. Tomato acids may irritate the bladder and exhibit symptoms of incontinence and cystitis. Tomatoes' diuretic effect is associated with increase of uric acid levels and gouty arthritis. Lycopene may have its nutritional benefits. However, it can also cause low blood pressure and increase tendency of bleeding, and it should be used with caution in individuals with gastric ulcers. Lycopene can interfere with some chemotherapy agents, and it may cause reversible lycopenodermia, the deep-orange coloration of the skin.

13 Watermelons

Family: Cucurbitaceae
Genus: *Citrullus*
Common name: Watermelon

Watermelon is a low caloric, edible summer fruit. There are more than 1200 different existing cultivars. Watermelon is a smooth, deep green to yellow color with a thick exterior rind with light green to gray stripes. The interior flesh has different colors like pink, red, or yellow depending on the variety, with numerous black seeds. Watermelon species are rich in antioxidants such as lycopene, β-carotene, lutein, zeaxanthin, and cryptoxanthin and vitamins A, C, and B-6. It is a good source of the minerals potassium, manganese, magnesium, and copper (Table 19).

The health benefits associated with watermelon consumption are numerous. Watermelon water content is nearly 92%. Its high water and electrolyte content keeps the body hydrated and protects it against heat stroke in the summer. They also have an appreciable amount of fiber [482]. The significance of this combination has been discussed in the aforementioned cucumber and tomato sections. Both the high water and fiber content in watermelon improves digestion and digestive system conditions such as constipation. These also aid in weight management. As a natural diuretic, it may help to cleanse the kidneys and bladder of impurities as well [967].

Antioxidants have been associated with a wide range of health benefits. Watermelon has been demonstrated to have high antioxidant capacity [968]. Watermelon's lycopene health-enhancing potential has been recently reviewed [969]. Another phytochemical is cucurbitacin E which is both an antioxidant and an anti-inflammatory. Watermelon antioxidants lower inflammation and oxidative

Table 19 Watermelon nutrition facts of 100 g, from the United States Department of Agriculture (USDA SR-21)

Serving size of 100 g of raw watermelon	
Per serving	% Daily value[a]
Calories 30	
Total fat 0.2 g	0
Saturated fat 0.0 g	0
Polyunsaturated fat 0.1 g	
Monounsaturated fat 0.0 g	
Total omega-3 fatty acids	
Total omega-6 fatty acids 50.0 mg	
Cholesterol 0.0 mg	0
Phytosterols 2.0 mg	
Carbohydrates 7.5 g	3
Dietary fiber 0.4 g	2
Starch 0.0 g	
Sugars 6.2 g	
Protein 0.6 g	1
Vitamins	
Vitamin A	11
Vitamin C	13
Vitamin E	0
Vitamin B-6	2
Vitamin K	0
Folic acid	1
Minerals	
Sodium	0
Potassium	3
Calcium	1
Iron	1
Magnesium	2
Manganese	2
Copper	2
Zinc	1

National Nutrient Database
[a]Based on a 2000 calorie diet

damage [969–972]. These compounds, lycopene and cucurbitacin E, in watermelon, including vitamin C, are found to have anticancerous effects. Lycopene intake is associated with a lower risk of some types of cancer such as prostate cancer [973]. Refer to the tomato section for further details. The anticancer bioactivities of cucurbitacin E were reviewed as well [974, 975]. In obesity-associated cancer, it is hypothesized that cucurbitacin E exerts its effect by lowering the insulin-like growth factor (IGF) levels [976, 977]. Vitamin C has been used to support cancer patient therapy [978]. Lycopene influence on cardiovascular health has been discussed in the tomato section [938, 979]. Watermelon supplementation was found to reduce arterial stiffness and lower blood pressure [980]. Along with other carotenoids, it reduces the risk of atherosclerosis [981, 982] and decreases cholesterol levels [969, 981]. Other health claims have been made for watermelon lycopene [969]. Citrulline is an amino acid which is found in watermelon. It is a condensation product of ornithine and carbamoyl phosphate in the urea cycle. It is also an arginine by-product of the reaction catalyzed by NOS family resulting in the concomitant release of nitric oxide. Nitric oxide plays an essential role in cardiovascular and immune functions. Watermelon juice consumption increases plasma concentration of arginine, ornithine, and citrulline. Increased nitric oxide levels help reduce the blood pressure [983] and regulate the immune system [984], which gets a boost from availability of vitamin C in watermelon [612]. The presence of potassium and magnesium in watermelon positively affects hypertension [54, 985]. Their role in reducing blood pressure has been extensively covered in the vegetable and fruit chapters. Interestingly, watermelon may control blood glucose levels as well. In animal studies, watermelon was shown to have anti-diabetic potential [986–988]. It seems that watermelon juice's antioxidative capacity resulted in the restoration of induced diabetic condition that mimics type 1 [986]. Consumption of watermelon juice is known to increase plasma concentrations of lycopene and beta-carotene in humans [989]. Another study used a type 2 animal model [988]. Here, a potential explanation was L-citrulline conversion into L-arginine. In the end, arginine availability reduced serum concentrations of cardiovascular risk factors, ameliorates vascular dysfunction, and improves glycemic control in obese animals.

In a manner similar to cherry, watermelon has the ability to promote muscle recovery and alleviate aches and pains in athletes [990]. Citrulline has been proposed as an ergogenic aid. In athletes, either natural or enriched L-citrulline watermelon juice helped reduce recovery heart rate and muscle soreness after 24 h. High doses of citrulline or watermelon juice as a pre-exercise supplement perhaps are effective in improving exercise performance [991, 992]. The presence of potassium also contributes to regulation during exercise and recovery [993].

Watermelon antioxidants beta-carotene, lutein, zeaxanthin, and vitamin C protect the eyes and prevent macular degeneration and glaucoma [994]. Watermelon seeds are rich in phytonutrients, proteins, and fats, which are beneficial overall and have protective effect on eyes and on excretory system functions [995].

A sensible amount of watermelon is well tolerated, but overconsumption can cause complication related to high vitamin concentrations. High vitamin C, sorbitol, and lycopene in watermelon may lead to diarrhea, nausea, bloating, and vomiting. Lycopene interacts with alcohol leading to increased liver oxidative stress capacity [996] and inflammation sensitivity [997]. High potassium leads to irregular heart rhythm and heart attacks. Watermelon is high in natural sugars and should be consumed cautiously under diabetic conditions.

References

1. Dreher, M. L., & Davenport, A. J. (2013). Hass avocado composition and potential health effects. *Critical Reviews in Food Science and Nutrition, 53*(7), 738–750.
2. Yoneyama, S., et al. (2007). Dietary intake of fatty acids and serum C-reactive protein in Japanese. *Journal of Epidemiology, 17*(3), 86–92.
3. Basu, A., Devaraj, S., & Jialal, I. (2006). Dietary factors that promote or retard inflammation. *Arteriosclerosis, Thrombosis, and Vascular Biology, 26*(5), 995–1001.
4. Menendez, J. A., & Lupu, R. (2006). Mediterranean dietary traditions for the molecular treatment of human cancer: Anti-oncogenic actions of the main olive oil's monounsaturated fatty acid oleic acid (18:1n-9). *Current Pharmaceutical Biotechnology, 7*(6), 495–502.
5. Blotman, F., et al. (1997). Efficacy and safety of avocado/soybean unsaponifiables in the treatment of symptomatic osteoarthritis of the knee and hip. A prospective, multicenter, three-month, randomized, double-blind, placebo-controlled trial. *Revue du Rhumatisme (English Ed.), 64*(12), 825–834.
6. DiNubile, N. A. (2010). A potential role for avocado- and soybean-based nutritional supplements in the management of osteoarthritis: A review. *The Physician and Sportsmedicine, 38*(2), 71–81.
7. Simonsen, N. R., et al. (1998). Tissue stores of individual monounsaturated fatty acids and breast cancer: The EURAMIC study. European Community Multicenter Study on Antioxidants, Myocardial Infarction, and Breast Cancer. *The American Journal of Clinical Nutrition, 68*(1), 134–141.
8. Menendez, J. A., et al. (2005). Oleic acid, the main monounsaturated fatty acid of olive oil, suppresses Her-2/neu (erbB-2) expression and synergistically enhances the growth inhibitory effects of trastuzumab (Herceptin) in breast cancer cells with Her-2/neu oncogene amplification. *Annals of Oncology, 16*(3), 359–371.
9. Ding, H., et al. (2009). Selective induction of apoptosis of human oral cancer cell lines by avocado extracts via a ROS-mediated mechanism. *Nutrition and Cancer, 61*(3), 348–356.
10. D'Ambrosio, S. M., et al. (2011). Aliphatic acetogenin constituents of avocado fruits inhibit human oral cancer cell proliferation by targeting the EGFR/RAS/RAF/MEK/ERK1/2 pathway. *Biochemical and Biophysical Research Communications, 409*(3), 465–469.
11. Lu, Q. Y., et al. (2005). Inhibition of prostate cancer cell growth by an avocado extract: Role of lipid-soluble bioactive substances. *The Journal of Nutritional Biochemistry, 16*(1), 23–30.
12. Paul, R., Kulkarni, P., & Ganesh, N. (2011). Avocado fruit (Persea americana Mill) exhibits chemo-protective potentiality against cyclophosphamide induced genotoxicity in human lymphocyte culture. *Journal of Experimental Therapeutics and Oncology, 9*(3), 221–230.
13. Duester, K. C. (2001). Avocado fruit is a rich source of beta-sitosterol. *Journal of the Academy of Nutrition and Dietetics, 101*(4), 404–405.
14. Kim, T. H., et al. (2012). Dietary supplements for benign prostatic hyperplasia: An overview of systematic reviews. *Maturitas, 73*(3), 180–185.
15. Naveh, E., et al. (2002). Defatted avocado pulp reduces body weight and total hepatic fat but increases plasma cholesterol in male rats fed diets with cholesterol. *The Journal of Nutrition, 132*(7), 2015–2018.
16. Burton-Freeman, B. (2000). Dietary fiber and energy regulation. *The Journal of Nutrition, 130*(2S Suppl), 272s–275s.
17. Wien, M., et al. (2013). A randomized 3x3 crossover study to evaluate the effect of Hass avocado intake on post-ingestive satiety, glucose and insulin levels, and subsequent energy intake in overweight adults. *Nutrition Journal, 12*, 155.
18. Pieterse, Z., et al. (2005). Substitution of high monounsaturated fatty acid avocado for mixed dietary fats during an energy-restricted diet: Effects on weight loss, serum lipids, fibrinogen, and vascular function. *Nutrition, 21*(1), 67–75.

19. Colquhoun, D. M., et al. (1992). Comparison of the effects on lipoproteins and apolipoproteins of a diet high in monounsaturated fatty acids, enriched with avocado, and a high-carbohydrate diet. *American Journal of Clinical Nutrition, 56*(4), 671–677.
20. Carranza-Madrigal, J., et al. (1997). Effects of a vegetarian diet vs. a vegetarian diet enriched with avocado in hypercholesterolemic patients. *Archives of Medical Research, 28*(4), 537–541.
21. Lopez Ledesma, R., et al. (1996). Monounsaturated fatty acid (avocado) rich diet for mild hypercholesterolemia. *Archives of Medical Research, 27*(4), 519–523.
22. Alvizouri-Munoz, M., et al. (1992). Effects of avocado as a source of monounsaturated fatty acids on plasma lipid levels. *Archives of Medical Research, 23*(4), 163–167.
23. Lerman-Garber, I., et al. (1994). Effect of a high-monounsaturated fat diet enriched with avocado in NIDDM patients. *Diabetes Care, 17*(4), 311–315.
24. Fulgoni 3rd, V. L., Dreher, M., & Davenport, A. J. (2013). Avocado consumption is associated with better diet quality and nutrient intake, and lower metabolic syndrome risk in US adults: Results from the National Health and Nutrition Examination Survey (NHANES) 2001–2008. *Nutrition Journal, 12*, 1.
25. Bentsen, H. (2017). Dietary polyunsaturated fatty acids, brain function and mental health. *Microbial Ecology in Health and Disease, 28*(sup1), 1281916.
26. Innis, S. M. (2008). Dietary omega 3 fatty acids and the developing brain. *Brain Research, 1237*, 35–43.
27. Bourre, J. M. (2004). Roles of unsaturated fatty acids (especially omega-3 fatty acids) in the brain at various ages and during ageing. *The Journal of Nutrition, Health & Aging, 8*(3), 163–174.
28. Unlu, N. Z., et al. (2005). Carotenoid absorption from salad and salsa by humans is enhanced by the addition of avocado or avocado oil. *The Journal of Nutrition, 135*(3), 431–436.
29. Cogswell, M. E., et al. (2012). Sodium and potassium intakes among US adults: NHANES 2003-2008. *The American Journal of Clinical Nutrition, 96*(3), 647–657.
30. Hoy, M. K., & Goldman, J. D. (2013). Potassium Intake of the U.S. Population, What We Eat In America, NHANES 2009–2010. *The FASEB Journal, 27*(1_supplement), 621–627.
31. Aburto, N. J., et al. (2013). Effect of increased potassium intake on cardiovascular risk factors and disease: Systematic review and meta-analyses. *BMJ, 346*, f1378.
32. van Ballegooijen, A. J., & Beulens, J. W. (2017). The role of vitamin k status in cardiovascular health: Evidence from observational and clinical studies. *Current Nutrition Reports, 6*(3), 197–205.
33. Pearson, D. A. (2007). Bone health and osteoporosis: The role of vitamin K and potential antagonism by anticoagulants. *Nutrition in Clinical Practice, 22*(5), 517–544.
34. Weber, P. (2001). Vitamin K and bone health. *Nutrition, 17*(10), 880–887.
35. Greenberg, J. A., et al. (2011). Folic ACID supplementation and pregnancy: More than just neural tube defect prevention. *Reviews in Obstetrics and Gynecology, 4*(2), 52–59.
36. Pitkin, R. M. (2007). Folate and neural tube defects. *The American Journal of Clinical Nutrition, 85*(1), 285s–288s.
37. Wilson, R. D., et al. (2003). The use of folic acid for the prevention of neural tube defects and other congenital anomalies. *Journal of Obstetrics and Gynaecology Canada, 25*(11), 959–973.
38. Iuliano, M., De Tommaso, G., & Ragone, R. (2009). Homocysteine disulphides and vascular disease. *Disease Markers, 27*(2), 55–61.
39. Aleman, G., Tovar, A. R., & Torres, N. (2001). Homocysteine metabolism and risk of cardiovascular diseases: Importance of the nutritional status on folic acid, vitamins B6 and B12. *Revista de Investigación Clínica, 53*(2), 141–151.
40. Blom, H. J., & Smulders, Y. (2011). Overview of homocysteine and folate metabolism. With special references to cardiovascular disease and neural tube defects. *Journal of Inherited Metabolic Disease, 34*(1), 75–81.
41. Ganguly, P., & Alam, S. F. (2015). Role of homocysteine in the development of cardiovascular disease. *Nutrition Journal, 14*, 6.

42. Delcourt, C., et al. (2006). Plasma lutein and zeaxanthin and other carotenoids as modifiable risk factors for age-related maculopathy and cataract: The POLA Study. *Investigative Ophthalmology and Visual Science, 47*(6), 2329–2335.
43. Palombo, P., et al. (2007). Beneficial long-term effects of combined oral/topical antioxidant treatment with the carotenoids lutein and zeaxanthin on human skin: A double-blind, placebo-controlled study. *Skin Pharmacology and Physiology, 20*(4), 199–210.
44. Pullar, J. M., Carr, A. C., & Vissers, M. C. M. (2017). The roles of vitamin C in skin health. *Nutrients, 9*(8), 866.
45. Satterfield, D., Taube, D., & Kenney, M. C. (1988). Effect of vitamin E on the production of collagen, DNA and fibronectin in keratocytes in vitro. *Ophthalmic Research, 20*(4), 227–231.
46. Sharaev, P. N., Bogdanov, N. G., & Iamaldinov, R. N. (1976). Collagen metabolism in the skin with different vitamin K regimens. *Biulleten'eksperimental'noi biologii i meditsiny, 81*(6), 665–666.
47. Korać, R. R., & Khambholja, K. M. (2011). Potential of herbs in skin protection from ultraviolet radiation. *Pharmacognosy Reviews, 5*(10), 164–173.
48. Scott, T. M., et al. (2017). Avocado consumption increases macular pigment density in older adults: A randomized, controlled trial. *Nutrients, 9*(9), 919.
49. Forster, M., et al. (2003). Distribution of nutrients in edible banana pulp. *Food Technology and Biotechnology, 41*(2), 167–171.
50. Salih, Z., et al. (2017). Physicochemical and functional properties of pulp and peel flour of dried green and ripe banana (Cavendish). *International Journal of Research in Agricultural Sciences, 4*, 348–353.
51. Alkarkhi, A., et al. (2011). Comparing physicochemical properties of banana pulp and peel flours prepared from green and ripe fruits. *Food Chemistry, 129*(2), 312–318.
52. Haddy, F. J., Vanhoutte, P. M., & Feletou, M. (2006). Role of potassium in regulating blood flow and blood pressure. *American Journal of Physiology. Regulatory, Integrative and Comparative Physiology, 290*(3), R546–R552.
53. Whelton, P. K., et al. (1997). Effects of oral potassium on blood pressure. Meta-analysis of randomized controlled clinical trials. *JAMA, 277*(20), 1624–1632.
54. Houston, M. C. (2011). The importance of potassium in managing hypertension. *Current Hypertension Reports, 13*(4), 309–317.
55. Seth, A., et al. (2014). Potassium intake and risk of stroke in women with hypertension and nonhypertension in the Women's Health Initiative. *Stroke, 45*(10), 2874–2880.
56. D'Elia, L., et al. (2011). Potassium intake, stroke, and cardiovascular disease a meta-analysis of prospective studies. *Journal of the American College of Cardiology, 57*(10), 1210–1219.
57. Weaver, C. M. (2013). Potassium and health. *Advances in Nutrition, 4*(3), 368s–377s.
58. He, F. J., & MacGregor, G. A. (2008). Beneficial effects of potassium on human health. *Physiologia Plantarum, 133*(4), 725–735.
59. Kanazawa, K., & Sakakibara, H. (2000). High content of dopamine, a strong antioxidant, in Cavendish banana. *Journal of Agricultural and Food Chemistry, 48*(3), 844–848.
60. Someya, S., Yoshiki, Y., & Okubo, K. (2002). Antioxidant compounds from banana (Musa Cavendish). *Food Chemistry, 79*(3), 351–354.
61. Fox, C., Ramsoomair, D., & Carter, C. (2001). Magnesium: Its proven and potential clinical significance. *Southern Medical Journal, 94*(12), 1195–1201.
62. Geiger, H., & Wanner, C. (2012). Magnesium in disease. *Clinical Kidney Journal, 5*(Suppl 1), i25–i38.
63. Volpe, S. L. (2013). Magnesium in disease prevention and overall health. *Advances in Nutrition, 4*(3), 378s–383s.
64. Avila, D. S., Puntel, R. L., & Aschner, M. (2013). Manganese in health and disease. *Metal Ions in Life Sciences, 13*, 199–227.
65. Aschner, J. L., & Aschner, M. (2005). Nutritional aspects of manganese homeostasis. *Molecular Aspects of Medicine, 26*(4-5), 353–362.

66. Takeda, A. (2003). Manganese action in brain function. *Brain Research. Brain Research Reviews, 41*(1), 79–87.
67. Guilarte, T. R. (2010). Manganese and Parkinson's disease: A critical review and new findings. *Environmental Health Perspectives, 118*(8), 1071–1080.
68. Buttriss, J. L., & Stokes, C. S. (2008). Dietary fibre and health: An overview. *Nutrition Bulletin, 33*(3), 186–200.
69. Anderson, J. W., et al. (2009). Health benefits of dietary fiber. *Nutrition Reviews, 67*(4), 188–205.
70. Zhang, P., et al. (2005). Banana starch: Production, physicochemical properties, and digestibility - A review. *Carbohydrate Polymers, 59*, 443–458.
71. Bird, A. R., Brown, I. L., & Topping, D. L. (2000). Starches, resistant starches, the gut microflora and human health. *Current Issues in Intestinal Microbiology, 1*(1), 25–37.
72. Topping, D. L., & Clifton, P. M. (2001). Short-chain fatty acids and human colonic function: Roles of resistant starch and nonstarch polysaccharides. *Physiological Reviews, 81*(3), 1031–1064.
73. Duan, X., et al. (2008). Modification of pectin polysaccharides during ripening of postharvest banana fruit. *Food Chemistry, 111*, 144–149.
74. Moongngarm, A. (2013). Chemical compositions and resistant starch content in starchy foods. *American Journal of Agricultural and Biological Sciences, 8*, 107–113.
75. Threapleton, D. E., et al. (2013). Dietary fibre intake and risk of cardiovascular disease: Systematic review and meta-analysis. *BMJ, 347*, f6879.
76. Satija, A., & Hu, F. B. (2012). Cardiovascular benefits of dietary fiber. *Current Atherosclerosis Reports, 14*(6), 505–514.
77. Viuda-Martos, M., et al. (2010). Role of fiber in cardiovascular diseases: A review. *Comprehensive Reviews in Food Science and Food Safety, 9*(2), 240–258.
78. Bodinham, C. L., Frost, G. S., & Robertson, M. D. (2010). Acute ingestion of resistant starch reduces food intake in healthy adults. *British Journal of Nutrition, 103*(6), 917–922.
79. Salas-Salvado, J., et al. (2008). Effect of two doses of a mixture of soluble fibres on body weight and metabolic variables in overweight or obese patients: A randomised trial. *British Journal of Nutrition, 99*(6), 1380–1387.
80. Anderson, G. H., et al. (2010). Relation between estimates of cornstarch digestibility by the Englyst in vitro method and glycemic response, subjective appetite, and short-term food intake in young men. *The American Journal of Clinical Nutrition, 91*(4), 932–939.
81. Willis, H. J., et al. (2009). Greater satiety response with resistant starch and corn bran in human subjects. *Nutrition Research, 29*(2), 100–105.
82. Higgins, J. A. (2014). Resistant starch and energy balance: Impact on weight loss and maintenance. *Critical Reviews in Food Science and Nutrition, 54*(9), 1158–1166.
83. Clark, M. J., & Slavin, J. L. (2013). The effect of fiber on satiety and food intake: A systematic review. *Journal of the American College of Nutrition, 32*(3), 200–211.
84. Robertson, M. D., et al. (2005). Insulin-sensitizing effects of dietary resistant starch and effects on skeletal muscle and adipose tissue metabolism. *The American Journal of Clinical Nutrition, 82*(3), 559–567.
85. Maki, K. C., et al. (2012). Resistant starch from high-amylose maize increases insulin sensitivity in overweight and obese men. *The Journal of Nutrition, 142*(4), 717–723.
86. Schwartz, S. E., et al. (1982). Sustained pectin ingestion delays gastric emptying. *Gastroenterology, 83*(4), 812–817.
87. Schwartz, S. E., et al. (1988). Sustained pectin ingestion: Effect on gastric emptying and glucose tolerance in non-insulin-dependent diabetic patients. *The American Journal of Clinical Nutrition, 48*(6), 1413–1417.
88. Dahl, W. J., & Stewart, M. L. (2015). Position of the Academy of Nutrition and Dietetics: Health implications of dietary fiber. *Journal of the Academy of Nutrition and Dietetics, 115*(11), 1861–1870.

89. Leonel, A. J., & Alvarez-Leite, J. I. (2012). Butyrate: Implications for intestinal function. *Current Opinion in Clinical Nutrition and Metabolic Care, 15*(5), 474–479.
90. Maclure, M., & Willett, W. (1990). A case-control study of diet and risk of renal adenocarcinoma. *Epidemiology, 1*(6), 430–440.
91. Rashidkhani, B., Lindblad, P., & Wolk, A. (2005). Fruits, vegetables and risk of renal cell carcinoma: A prospective study of Swedish women. *International Journal of Cancer, 113*(3), 451–455.
92. Olano-Martin, E., et al. (2003). Pectin and pectic-oligosaccharides induce apoptosis in in vitro human colonic adenocarcinoma cells. *Anticancer Research, 23*(1a), 341–346.
93. Leclere, L., Cutsem, P. V., & Michiels, C. (2013). Anti-cancer activities of pH- or heat-modified pectin. *Frontiers in Pharmacology, 4*, 128.
94. Ferraro, P. M., et al. (2016). Dietary protein and potassium, diet-dependent net acid load, and risk of incident kidney stones. *Clinical Journal of the American Society of Nephrology, 11*(10), 1834–1844.
95. Zerwekh, J. E., et al. (2007). Reduction of renal stone risk by potassium-magnesium citrate during 5 weeks of bed rest. *The Journal of Urology, 177*(6), 2179–2184.
96. Panigrahi, P. N., et al. (2017). Antiurolithiatic and antioxidant efficacy of Musa paradisiaca pseudostem on ethylene glycol-induced nephrolithiasis in rat. *Indian Journal of Pharmacology, 49*(1), 77–83.
97. Obrenovich, M. E., et al. (2011). Antioxidants in health, disease and aging. *CNS & Neurological Disorders Drug Targets, 10*(2), 192–207.
98. Wang, X., et al. (2014). Flavonoid intake and risk of CVD: A systematic review and meta-analysis of prospective cohort studies. *British Journal of Nutrition, 111*(1), 1–11.
99. Panche, A. N., Diwan, A. D., & Chandra, S. R. (2016). Flavonoids: An overview. *Journal of Nutritional Science, 5*, e47.
100. Kozlowska, A., & Szostak-Wegierek, D. (2014). Flavonoids--food sources and health benefits. *Roczniki Państwowego Zakładu Higieny, 65*(2), 79–85.
101. Singh, H., Bhaskar, D. J., & Rehman, R. (2014). Do ripe bananas with brown spots fight cancer? *International Journal of Dental and Medical Research, 1*(2), 4–5.
102. Iwasawa, H., & Yamazaki, M. (2009). Differences in biological response modifier-like activities according to the strain and maturity of bananas. *Food Science and Technology Research, 15*, 275–282.
103. Pfeffer, K. (2003). Biological functions of tumor necrosis factor cytokines and their receptors. *Cytokine & Growth Factor Reviews, 14*(3-4), 185–191.
104. Shaw, K., Turner, J., & Del Mar, C. (2002). Tryptophan and 5-hydroxytryptophan for depression. *Cochrane Database of Systematic Reviews, 1*, Cd003198.
105. Nieman, D. C., et al. (2012). Bananas as an energy source during exercise: A metabolomics approach. *PLoS One, 7*(5), e37479.
106. Norris Jr., F. H., Gasteiger, E. L., & Chatfield, P. O. (1957). An electromyographic study of induced and spontaneous muscle cramps. *Electroencephalography and Clinical Neurophysiology, 9*(1), 139–147.
107. Miller, K. C. (2012). Plasma potassium concentration and content changes after banana ingestion in exercised men. *Journal of Athletic Training, 47*(6), 648–654.
108. Bergeron, M. F. (2003). Heat cramps: Fluid and electrolyte challenges during tennis in the heat. *Journal of Science and Medicine in Sport, 6*(1), 19–27.
109. Stofan, J. R., et al. (2005). Sweat and sodium losses in NCAA football players: A precursor to heat cramps? *International Journal of Sport Nutrition and Exercise Metabolism, 15*(6), 641–652.
110. Schwellnus, M. P., Drew, N., & Collins, M. (2008). Muscle cramping in athletes--risk factors, clinical assessment, and management. *Clinics in Sports Medicine, 27*(1), 183–94, ix-x.
111. Hermansen, K., et al. (1992). Influence of ripeness of banana on the blood glucose and insulin response in type 2 diabetic subjects. *Diabetic Medicine, 9*(8), 739–743.

112. O'Keefe, A. W., & Ben-Shoshan, M. (2014). A 4-month-old baby boy presenting with anaphylaxis to a banana: A case report. *Journal of Medical Case Reports, 8*, 62.
113. Antal, E. J., et al. (2001). Linezolid, a novel oxazolidinone antibiotic: Assessment of monoamine oxidase inhibition using pressor response to oral tyramine. *The Journal of Clinical Pharmacology, 41*(5), 552–562.
114. Rumore, M. M., Roth, M., & Orfanos, A. (2010). Dietary tyramine restriction for hospitalized patients on linezolid: An update. *Nutrition in Clinical Practice, 25*(3), 265–269.
115. Silber, B. Y., & Schmitt, J. A. (2010). Effects of tryptophan loading on human cognition, mood, and sleep. *Neuroscience and Biobehavioral Reviews, 34*(3), 387–407.
116. Peuhkuri, K., Sihvola, N., & Korpela, R. (2012). Diet promotes sleep duration and quality. *Nutrition Research, 32*(5), 309–319.
117. Simon, L. V., & Farrell, M. W. (2018). *Hyperkalemia*. Treasure Island, FL: StatPearls Publishing.
118. Medicine IO, (2005). Potassium. In *Dietary reference intakes for water, potassium, sodium, chloride, and sulfate* (pp. 186–268). Washington, DC: The National Academies Press.
119. Wolfe, K. L., et al. (2008). Cellular antioxidant activity of common fruits. *Journal of Agricultural and Food Chemistry, 56*(18), 8418–8426.
120. Pandey, K. B., & Rizvi, S. I. (2009). Plant polyphenols as dietary antioxidants in human health and disease. *Oxidative Medicine and Cellular Longevity, 2*(5), 270–278.
121. Olas, B. (2018). Berry phenolic antioxidants - Implications for human health? *Frontiers in Pharmacology, 9*, 78.
122. Williamson, G. (2017). The role of polyphenols in modern nutrition. *Nutrition Bulletin, 42*(3), 226–235.
123. Skrovankova, S., et al. (2015). Bioactive compounds and antioxidant activity in different types of berries. *International Journal of Molecular Sciences, 16*(10), 24673–24706.
124. Manganaris, G. A., et al. (2014). Berry antioxidants: Small fruits providing large benefits. *Journal of the Science of Food and Agriculture, 94*(5), 825–833.
125. Vauzour, D., et al. (2010). Polyphenols and human health: Prevention of disease and mechanisms of action. *Nutrients, 2*(11), 1106–1131.
126. Huang, W. Y., et al. (2012). Survey of antioxidant capacity and phenolic composition of blueberry, blackberry, and strawberry in Nanjing. *Journal of Zhejiang University. Science. B, 13*(2), 94–102.
127. Kähkönen, M. P., Hopia, A. I., & Heinonen, M. (2001). Berry phenolics and their antioxidant activity. *Journal of Agricultural and Food Chemistry, 49*(8), 4076–4082.
128. Luis, A., Domingues, F., & Pereira, L. (2018). Association between berries intake and cardiovascular diseases risk factors: A systematic review with meta-analysis and trial sequential analysis of randomized controlled trials. *Food & Function, 9*(2), 740–757.
129. Zhao, C. N., et al. (2017). Fruits for prevention and treatment of cardiovascular diseases. *Nutrients, 9*(6), 598.
130. Huang, H., et al. (2016). Effects of berries consumption on cardiovascular risk factors: A meta-analysis with trial sequential analysis of randomized controlled trials. *Scientific Reports, 6*, 23625.
131. Basu, A., Rhone, M., & Lyons, T. J. (2010). Berries: Emerging impact on cardiovascular health. *Nutrition Reviews, 68*(3), 168–177.
132. Erlund, I., et al. (2008). Favorable effects of berry consumption on platelet function, blood pressure, and HDL cholesterol. *The American Journal of Clinical Nutrition, 87*(2), 323–331.
133. Vendrame, S., et al. (2016). Berry fruit consumption and metabolic syndrome. *Antioxidants (Basel), 5*(4), 34.
134. Tsuda, T. (2016). Recent progress in anti-obesity and anti-diabetes effect of berries. *Antioxidants (Basel), 5*(2), 13.
135. Kowalska, K., & Olejnik, A. (2016). Current evidence on the health-beneficial effects of berry fruits in the prevention and treatment of metabolic syndrome. *Current Opinion in Clinical Nutrition and Metabolic Care, 19*(6), 446–452.

136. Lehtonen, H. M., et al. (2010). Berry meals and risk factors associated with metabolic syndrome. *European Journal of Clinical Nutrition, 64*(6), 614–621.
137. Del Bo, C., et al. (2015). Berries and oxidative stress markers: An overview of human intervention studies. *Food & Function, 6*(9), 2890–2917.
138. Joseph, S. V., Edirisinghe, I., & Burton-Freeman, B. M. (2014). Berries: Anti-inflammatory effects in humans. *Journal of Agricultural and Food Chemistry, 62*(18), 3886–3903.
139. Joseph, S. V., Edirisinghe, I., & Burton-Freeman, B. M. (2016). Fruit polyphenols: A review of anti-inflammatory effects in humans. *Critical Reviews in Food Science and Nutrition, 56*(3), 419–444.
140. Nardi, G. M., et al. (2016). Anti-inflammatory activity of berry fruits in mice model of inflammation is based on oxidative stress modulation. *Pharmacognosy Research, 8*(Suppl 1), S42–S49.
141. McDougall, G. J., Kulkarni, N. N., & Stewart, D. (2008). Current developments on the inhibitory effects of berry polyphenols on digestive enzymes. *Biofactors, 34*(1), 73–80.
142. Kristo, A. S., Klimis-Zacas, D., & Sikalidis, A. K. (2016). Protective role of dietary berries in cancer. *Antioxidants (Basel), 5*(4), 37.
143. Zhou, Y., et al. (2016). Natural polyphenols for prevention and treatment of cancer. *Nutrients, 8*(8), 515.
144. Abdal Dayem, A., et al. (2016). The anti-cancer effect of polyphenols against breast cancer and cancer stem cells: Molecular mechanisms. *Nutrients, 8*(9), 581.
145. Lall, R. K., et al. (2015). Dietary polyphenols in prevention and treatment of prostate cancer. *International Journal of Molecular Sciences, 16*(2), 3350–3376.
146. Kammeyer, A., & Luiten, R. M. (2015). Oxidation events and skin aging. *Ageing Research Reviews, 21*, 16–29.
147. Kelly, E., Vyas, P., & Weber, J. T. (2017). Biochemical properties and neuroprotective effects of compounds in various species of berries. *Molecules, 23*(1), 26.
148. Keservani, R. K., Sharma, A. K., & Kesharwani, R. K. (2016). Medicinal effect of nutraceutical fruits for the cognition and brain health. *Scientifica (Cairo), 2016*, 3109254.
149. Shukitt-Hale, B., et al. (2015). The beneficial effects of berries on cognition, motor behaviour and neuronal function in ageing. *British Journal of Nutrition, 114*(10), 1542–1549.
150. Subash, S., et al. (2014). Neuroprotective effects of berry fruits on neurodegenerative diseases. *Neural Regeneration Research, 9*(16), 1557–1566.
151. Devore, E. E., et al. (2012). Dietary intakes of berries and flavonoids in relation to cognitive decline. *Annals of Neurology, 72*(1), 135–143.
152. Kaume, L., Howard, L. R., & Devareddy, L. (2012). The blackberry fruit: A review on its composition and chemistry, metabolism and bioavailability, and health benefits. *Journal of Agricultural and Food Chemistry, 60*(23), 5716–5727.
153. Verma, R., et al. (2014). Rubus fruticosus (blackberry) use as an herbal medicine. *Pharmacognosy Reviews, 8*(16), 101–104.
154. Oszmianski, J., et al. (2015). Analysis of phenolic compounds and antioxidant activity in wild blackberry fruits. *International Journal of Molecular Sciences, 16*(7), 14540–14553.
155. Khoo, H. E., et al. (2017). Anthocyanidins and anthocyanins: Colored pigments as food, pharmaceutical ingredients, and the potential health benefits. *Food & Nutrition Research, 61*(1), 1361779.
156. Azofeifa, G., et al. (2013). Antioxidant and anti-inflammatory in vitro activities of phenolic compounds from tropical highland blackberry (Rubus adenotrichos). *Journal of Agricultural and Food Chemistry, 61*(24), 5798–5804.
157. Wang, S. Y., & Lin, H. S. (2000). Antioxidant activity in fruits and leaves of blackberry, raspberry, and strawberry varies with cultivar and developmental stage. *Journal of Agricultural and Food Chemistry, 48*(2), 140–146.
158. Zielonka-Brzezicka, J., et al. (2016). Comparison of the antioxidant properties of selected parts of raspberry (Rubus idaeus) and blackberry (Rubus fruticosus). *Pomeranian Journal of Life Sciences, 62*(4), 52–59.

159. Wu, T., et al. (2018). Blackberry and blueberry anthocyanin supplementation counteract high-fat-diet-induced obesity by alleviating oxidative stress and inflammation and accelerating energy expenditure. *Oxidative Medicine and Cellular Longevity, 2018*, 4051232.
160. Lee, Y. M., et al. (2017). Dietary anthocyanins against obesity and inflammation. *Nutrients, 9*(10), 1089.
161. Blando, F., et al. (2018). Radical scavenging and anti-inflammatory activities of representative anthocyanin groupings from pigment-rich fruits and vegetables. *International Journal of Molecular Sciences, 19*(1), 169.
162. Esselen, M., et al. (2011). Anthocyanin-rich blackberry extract suppresses the DNA-damaging properties of topoisomerase I and II poisons in colon carcinoma cells. *Journal of Agricultural and Food Chemistry, 59*(13), 6966–6973.
163. Wang, L.-S., & Stoner, G. D. (2008). Anthocyanins and their role in cancer prevention. *Cancer Letters, 269*(2), 281–290.
164. Lin, B. W., et al. (2017). Effects of anthocyanins on the prevention and treatment of cancer. *British Journal of Pharmacology, 174*(11), 1226–1243.
165. Stoner, G. D., Wang, L.-S., & Casto, B. C. (2008). Laboratory and clinical studies of cancer chemoprevention by antioxidants in berries. *Carcinogenesis, 29*(9), 1665–1674.
166. de Sousa Moraes, L. F., et al. (2017). Anthocyanins/anthocyanidins and colorectal cancer: What is behind the scenes? *Critical Reviews in Food Science and Nutrition, 59*(1), 1–13.
167. Lippert, E., et al. (2017). Anthocyanins prevent colorectal cancer development in a mouse model. *Digestion, 95*(4), 275–280.
168. Ma, X., & Ning, S. (2019). Cyanidin-3-glucoside attenuates the angiogenesis of breast cancer via inhibiting STAT3/VEGF pathway. *Phytotherapy Research, 33*(1), 81–89.
169. Ding, M., et al. (2006). Cyanidin-3-glucoside, a natural product derived from blackberry, exhibits chemopreventive and chemotherapeutic activity. *The Journal of Biological Chemistry, 281*(25), 17359–17368.
170. Cassidy, A. (2018). Berry anthocyanin intake and cardiovascular health. *Molecular Aspects of Medicine, 61*, 76–82.
171. Wallace, T. C. (2011). Anthocyanins in cardiovascular disease. *Advances in Nutrition, 2*(1), 1–7.
172. de Pascual-Teresa, S., Moreno, D. A., & García-Viguera, C. (2010). Flavanols and anthocyanins in cardiovascular health: A review of current evidence. *International Journal of Molecular Sciences, 11*(4), 1679–1703.
173. Reis, J. F., et al. (2016). Action mechanism and cardiovascular effect of anthocyanins: A systematic review of animal and human studies. *Journal of Translational Medicine, 14*(1), 315.
174. Aghababaee, S. K., et al. (2015). Effects of blackberry (Morus nigra L.) consumption on serum concentration of lipoproteins, apo A-I, apo B, and high-sensitivity-C-reactive protein and blood pressure in dyslipidemic patients. *Journal of Research in Medical Sciences : The Official Journal of Isfahan University of Medical Sciences, 20*(7), 684–691.
175. Kimble, R., et al. (2018). Dietary intake of anthocyanins and risk of cardiovascular disease: A systematic review and meta-analysis of prospective cohort studies. *Critical Reviews in Food Science and Nutrition*, 1–12. https://doi.org/10.1080/10408398.2018.1509835
176. Cassidy, A., et al. (2016). Habitual intake of anthocyanins and flavanones and risk of cardiovascular disease in men. *The American Journal of Clinical Nutrition, 104*(3), 587–594.
177. Wallace, T. C., Slavin, M., & Frankenfeld, C. L. (2016). Systematic review of anthocyanins and markers of cardiovascular disease. *Nutrients, 8*(1), 32.
178. Serraino, I., et al. (2003). Protective effects of cyanidin-3-O-glucoside from blackberry extract against peroxynitrite-induced endothelial dysfunction and vascular failure. *Life Sciences, 73*(9), 1097–1114.
179. Murapa, P., et al. (2012). Anthocyanin-rich fractions of blackberry extracts reduce UV-induced free radicals and oxidative damage in keratinocytes. *Phytotherapy Research, 26*(1), 106–112.
180. Wang, Y., et al. (2015). The protective effects of berry-derived anthocyanins against visible light-induced damage in human retinal pigment epithelial cells. *Journal of the Science of Food and Agriculture, 95*(5), 936–944.

181. Miyake, S., et al. (2012). Vision preservation during retinal inflammation by anthocyanin-rich bilberry extract: Cellular and molecular mechanism. *Laboratory Investigation, 92*(1), 102–109.
182. Ghosh, D., & Konishi, T. (2007). Anthocyanins and anthocyanin-rich extracts: Role in diabetes and eye function. *Asia Pacific Journal of Clinical Nutrition, 16*(2), 200–208.
183. Shukitt-Hale, B., Cheng, V., & Joseph, J. A. (2009). Effects of blackberries on motor and cognitive function in aged rats. *Nutritional Neuroscience, 12*(3), 135–140.
184. Ma, H., et al. (2018). Evaluation of polyphenol anthocyanin-enriched extracts of blackberry, black raspberry, blueberry, cranberry, red raspberry, and strawberry for free radical scavenging, reactive carbonyl species trapping, anti-glycation, anti-beta-amyloid aggregation, and microglial neuroprotective effects. *International Journal of Molecular Sciences, 19*(2), 461.
185. Strathearn, K. E., et al. (2014). Neuroprotective effects of anthocyanin- and proanthocyanidin-rich extracts in cellular models of Parkinson's disease. *Brain Research, 1555*, 60–77.
186. Ames, B. N., Shigenaga, M. K., & Hagen, T. M. (1993). Oxidants, antioxidants, and the degenerative diseases of aging. *Proceedings of the National Academy of Sciences, 90*(17), 7915–7922.
187. Dasari, S., et al. (2017). Vitamin K and its analogs: Potential avenues for prostate cancer management. *Oncotarget, 8*(34), 57782–57799.
188. Kong, P., et al. (2014). Vitamin intake reduce the risk of gastric cancer: Meta-analysis and systematic review of randomized and observational studies. *PLoS One, 9*(12), e116060.
189. Jinghe, X., Mizuta, T., & Ozaki, I. (2015). Vitamin K and hepatocellular carcinoma: The basic and clinic. *World Journal of Clinical Cases: WJCC, 3*(9), 757–764.
190. Liu, B.-C., et al. (2016). Vitamin K2-induced inhibition of colorectal cancer cell proliferation and its underlying mechanisms. *International Journal of Clinical and Experimental Pathology, 9*(5), 4992–5003.
191. Ivanova, D., et al. (2018). Vitamins C and K3: A powerful redox system for sensitizing leukemia lymphocytes to everolimus and barasertib. *Anticancer Research, 38*(3), 1407–1414.
192. McGuire, K., et al. (2013). Vitamin C and K3 combination causes enhanced anticancer activity against rt-4 bladder cancer cells. *Journal of Cancer Science and Therapy, 5*(10), 325–333.
193. Hemmati, A. A., et al. (2014). Topical vitamin K1 promotes repair of full thickness wound in rat. *Indian Journal of Pharmacology, 46*(4), 409–412.
194. Gonzalez, O. A., et al. (2013). Antibacterial effects of blackberry extract target periodontopathogens. *Journal of Periodontal Research, 48*(1), 80–86.
195. Danaher, R. J., et al. (2011). Antiviral effects of blackberry extract against herpes simplex virus type 1. *Oral Surgery, Oral Medicine, Oral Pathology, Oral Radiology, and Endodontology, 112*(3), e31–e35.
196. Burton-Freeman, B. M., Sandhu, A. K., & Edirisinghe, I. (2016). Red raspberries and their bioactive polyphenols: Cardiometabolic and neuronal health links. *Advances in Nutrition, 7*(1), 44–65.
197. Rao, A. V., & Snyder, D. M. (2010). Raspberries and human health: A review. *Journal of Agricultural and Food Chemistry, 58*(7), 3871–3883.
198. Ludwig, I. A., et al. (2015). New insights into the bioavailability of red raspberry anthocyanins and ellagitannins. *Free Radical Biology and Medicine, 89*, 758–769.
199. Derosa, G., Maffioli, P., & Sahebkar, A. (2016). Ellagic acid and its role in chronic diseases. *Advances in Experimental Medicine and Biology, 928*, 473–479.
200. Jeong, H. S., et al. (2014). Effects of black raspberry on lipid profiles and vascular endothelial function in patients with metabolic syndrome. *Phytotherapy Research, 28*(10), 1492–1498.
201. Bae, J. Y., et al. (2010). Dietary compound ellagic acid alleviates skin wrinkle and inflammation induced by UV-B irradiation. *Experimental Dermatology, 19*(8), e182–e190.
202. Boukharta, M., Jalbert, G., & Castonguay, A. (1992). Biodistribution of ellagic acid and dose-related inhibition of lung tumorigenesis in A/J mice. *Nutrition and Cancer, 18*(2), 181–189.
203. Ceci, C., et al. (2016). Ellagic acid inhibits bladder cancer invasiveness and in vivo tumor growth. *Nutrients, 8*(11), E744.

204. Wang, N., et al. (2017). Direct inhibition of ACTN4 by ellagic acid limits breast cancer metastasis via regulation of beta-catenin stabilization in cancer stem cells. *Journal of Experimental & Clinical Cancer Research, 36*(1), 172.
205. Mukhtar, H., et al. (1984). Protection against 3-methylcholanthrene-induced skin tumorigenesis in Balb/C mice by ellagic acid. *Biochemical and Biophysical Research Communications, 119*(2), 751–757.
206. Kresty, L. A., Mallery, S. R., & Stoner, G. D. (2016). Black raspberries in cancer clinical trials: Past, present and future. *Journal of Berry Research, 6*(2), 251–261.
207. Shi, N., et al. (2017). Suppression of oxidative stress and NFkappaB/MAPK signaling by lyophilized black raspberries for esophageal cancer prevention in rats. *Nutrients, 9*(4), 413.
208. Casto, B. C., et al. (2002). Chemoprevention of oral cancer by black raspberries. *Anticancer Research, 22*(6c), 4005–4015.
209. Wang, L. S., et al. (2011). Modulation of genetic and epigenetic biomarkers of colorectal cancer in humans by black raspberries: A phase I pilot study. *Clinical Cancer Research, 17*(3), 598–610.
210. Wang, L. S., et al. (2009). Anthocyanins in black raspberries prevent esophageal tumors in rats. *Cancer Prevention Research (Philadelphia, Pa.), 2*(1), 84–93.
211. Noratto, G., Chew, B. P., & Ivanov, I. (2016). Red raspberry decreases heart biomarkers of cardiac remodeling associated with oxidative and inflammatory stress in obese diabetic db/db mice. *Food & Function, 7*(12), 4944–4955.
212. Smeriglio, A., et al. (2017). Proanthocyanidins and hydrolysable tannins: Occurrence, dietary intake and pharmacological effects. *British Journal of Pharmacology, 174*(11), 1244–1262.
213. Fairlie-Jones, L., et al. (2017). The effect of anthocyanin-rich foods or extracts on vascular function in adults: A systematic review and meta-analysis of randomised controlled trials. *Nutrients, 9*(8), 908.
214. Ash, M. M., et al. (2011). Unrefined and refined black raspberry seed oils significantly lower triglycerides and moderately affect cholesterol metabolism in male Syrian hamsters. *Journal of Medicinal Food, 14*(9), 1032–1038.
215. Oomah, B. D., et al. (2000). Characteristics of raspberry (Rubus idaeus L.) seed oil. *Food Chemistry, 69*, 187–193.
216. Aiyer, H. S., et al. (2008). Dietary berries and ellagic acid prevent oxidative DNA damage and modulate expression of DNA repair genes. *International Journal of Molecular Sciences, 9*(3), 327–341.
217. Patisaul, H. B., & Jefferson, W. (2010). The pros and cons of phytoestrogens. *Frontiers in Neuroendocrinology, 31*(4), 400–419.
218. Al-Anazi, A. F., et al. (2011). Preventive effects of phytoestrogens against postmenopausal osteoporosis as compared to the available therapeutic choices: An overview. *Journal of Natural Science, Biology, and Medicine, 2*(2), 154–163.
219. Montbriand, M. J. (2004). Herbs or natural products that increase cancer growth or recurrence. Part two of a four-part series. *Oncology Nursing Forum, 31*(5), E99–E115.
220. Bak, M. J., et al. (2016). Role of dietary bioactive natural products in estrogen receptor-positive breast cancer. *Seminars in Cancer Biology, 40-41*, 170–191.
221. Afrin, S., et al. (2016). Promising health benefits of the strawberry: A focus on clinical studies. *Journal of Agricultural and Food Chemistry, 64*(22), 4435–4449.
222. Basu, A., et al. (2016). Effects of dietary strawberry supplementation on antioxidant biomarkers in obese adults with above optimal serum lipids. *Journal of Nutrition and Metabolism, 2016*, 3910630.
223. Giampieri, F., et al. (2015). Strawberry as a health promoter: An evidence based review. *Food & Function, 6*(5), 1386–1398.
224. Giampieri, F., Alvarez-Suarez, J. M., & Battino, M. (2014). Strawberry and human health: Effects beyond antioxidant activity. *Journal of Agricultural and Food Chemistry, 62*(18), 3867–3876.
225. Basu, A., et al. (2014). Strawberry as a functional food: An evidence-based review. *Critical Reviews in Food Science and Nutrition, 54*(6), 790–806.

226. Ariza, M. T., et al. (2016). Strawberry achenes are an important source of bioactive compounds for human health. *International Journal of Molecular Sciences, 17*(7), 1103.
227. Giampieri, F., et al. (2017). The healthy effects of strawberry bioactive compounds on molecular pathways related to chronic diseases. *Annals of the New York Academy of Sciences, 1398*(1), 62–71.
228. Rahal, A., et al. (2014). Oxidative stress, prooxidants, and antioxidants: The interplay. *BioMed Research International, 2014*, 761264.
229. Edirisinghe, I., et al. (2011). Strawberry anthocyanin and its association with postprandial inflammation and insulin. *British Journal of Nutrition, 106*(6), 913–922.
230. Meyers, K. J., et al. (2003). Antioxidant and antiproliferative activities of strawberries. *Journal of Agricultural and Food Chemistry, 51*(23), 6887–6892.
231. Cooke, D., et al. (2005). Anthocyans from fruits and vegetables--does bright colour signal cancer chemopreventive activity? *European Journal of Cancer, 41*(13), 1931–1940.
232. Wedge, D. E., et al. (2001). Anticarcinogenic activity of strawberry, blueberry, and raspberry extracts to breast and cervical cancer cells. *Journal of Medicinal Food, 4*(1), 49–51.
233. Somasagara, R. R., et al. (2012). Extracts of strawberry fruits induce intrinsic pathway of apoptosis in breast cancer cells and inhibits tumor progression in mice. *PLoS One, 7*(10), e47021.
234. Islam, M. S., et al. (2017). An anthocyanin rich strawberry extract induces apoptosis and ROS while decreases glycolysis and fibrosis in human uterine leiomyoma cells. *Oncotarget, 8*(14), 23575–23587.
235. Amatori, S., et al. (2016). Polyphenol-rich strawberry extract (PRSE) shows in vitro and in vivo biological activity against invasive breast cancer cells. *Scientific Reports, 6*, 30917.
236. Flores, G., & Ruiz Del Castillo, M. L. (2016). Cancer-related constituents of strawberry jam as compared with fresh fruit. *Cancers (Basel), 8*(1), 16.
237. Casto, B. C., et al. (2013). Chemoprevention of oral cancer by lyophilized strawberries. *Anticancer Research, 33*(11), 4757–4766.
238. Spagnuolo, C., et al. (2016). A phenolic extract obtained from methyl jasmonate-treated strawberries enhances apoptosis in a human cervical cancer cell line. *Nutrition and Cancer, 68*(7), 1140–1150.
239. Cho, H. J., & Park, J. H. (2013). Kaempferol induces cell cycle arrest in HT-29 human colon cancer cells. *Journal of Cancer Prevention, 18*(3), 257–263.
240. Kimira, M., et al. (1998). Japanese intake of flavonoids and isoflavonoids from foods. *Journal of Epidemiology, 8*(3), 168–175.
241. Adhami, V. M., et al. (2012). Dietary flavonoid fisetin: A novel dual inhibitor of PI3K/Akt and mTOR for prostate cancer management. *Biochemical Pharmacology, 84*(10), 1277–1281.
242. Li, J., et al. (2018). Fisetin inhibited growth and metastasis of triple-negative breast cancer by reversing epithelial-to-mesenchymal transition via PTEN/Akt/GSK3beta signal pathway. *Frontiers in Pharmacology, 9*, 772.
243. Youns, M., & Hegazy, W. A. H. (2017). The natural flavonoid fisetin inhibits cellular proliferation of hepatic, colorectal, and pancreatic cancer cells through modulation of multiple signaling pathways. *PLoS One, 12*(1), e0169335.
244. Khan, N., et al. (2013). Fisetin: A dietary antioxidant for health promotion. *Antioxidants & Redox Signaling, 19*(2), 151–162.
245. Maher, P. (2015). How fisetin reduces the impact of age and disease on CNS function. *Frontiers in Bioscience (Scholar Edition), 7*, 58–82.
246. Yousefzadeh, M. J., et al. (2018). Fisetin is a senotherapeutic that extends health and lifespan. *eBioMedicine, 36*, 18–28.
247. Basu, A., et al. (2014). Freeze-dried strawberries lower serum cholesterol and lipid peroxidation in adults with abdominal adiposity and elevated serum lipids. *The Journal of Nutrition, 144*(6), 830–837.
248. Basu, A., et al. (2010). Strawberries decrease atherosclerotic markers in subjects with metabolic syndrome. *Nutrition Research, 30*(7), 462–469.

249. Ellis, C. L., et al. (2011). Attenuation of meal-induced inflammatory and thrombotic responses in overweight men and women after 6-week daily strawberry (Fragaria) intake. A randomized placebo-controlled trial. *Journal of Atherosclerosis and Thrombosis, 18*(4), 318–327.
250. Burton-Freeman, B., et al. (2010). Strawberry modulates LDL oxidation and postprandial lipemia in response to high-fat meal in overweight hyperlipidemic men and women. *Journal of the American College of Nutrition, 29*(1), 46–54.
251. Basu, A., et al. (2009). Freeze-dried strawberry powder improves lipid profile and lipid peroxidation in women with metabolic syndrome: Baseline and post intervention effects. *Nutrition Journal, 8*, 43.
252. Jenkins, D. J., et al. (2008). The effect of strawberries in a cholesterol-lowering dietary portfolio. *Metabolism, 57*(12), 1636–1644.
253. Cassidy, A., et al. (2013). High anthocyanin intake is associated with a reduced risk of myocardial infarction in young and middle aged women. *Circulation, 127*(2), 188–196.
254. Hung, C. H., et al. (2015). Quercetin is a potent anti-atherosclerotic compound by activation of SIRT1 signaling under oxLDL stimulation. *Molecular Nutrition & Food Research, 59*(10), 1905–1917.
255. Kolte, D., et al. (2014). Role of magnesium in cardiovascular diseases. *Cardiology in Review, 22*(4), 182–192.
256. Baranowska, M., et al. (2007). Potassium channels in blood vessels: Their role in health and disease. *Postępy Higieny i Medycyny Doświadczalnej (Online), 61*, 596–605.
257. Harrison, F. E. (2012). A critical review of vitamin C for the prevention of age-related cognitive decline and Alzheimer's disease. *Journal of Alzheimer's Disease, 29*(4), 711–726.
258. Del, C. V. H. M., et al. (2017). Dietary iodine exposure and brain structures and cognition in older people. Exploratory analysis in the Lothian Birth Cohort 1936. *The Journal of Nutrition, Health and Aging, 21*(9), 971–979.
259. Koide, M., et al. (2018). The yin and yang of KV channels in cerebral small vessel pathologie. *Microcirculation, 25*(1), e12436.
260. Joseph, J. A., et al. (1999). Reversals of age-related declines in neuronal signal transduction, cognitive, and motor behavioral deficits with blueberry, spinach, or strawberry dietary supplementation. *Journal of Neuroscience, 19*(18), 8114–8121.
261. Currais, A., et al. (2018). Fisetin reduces the impact of aging on behavior and physiology in the rapidly aging SAMP8 mouse. *The Journals of Gerontology. Series A, Biological Sciences and Medical Sciences, 73*(3), 299–307.
262. Gasparrini, M., et al. (2017). Strawberry-based cosmetic formulations protect human dermal fibroblasts against UVA-induced damage. *Nutrients, 9*(6), 605.
263. Ibranji, A., et al. (2015). A case report on transitory histamine intolerance from strawberry intake in a 15 month old child with acute gastroenteritis. *Clinical and Translational Allergy, 5*(Suppl 3), P61.
264. Peng, C., et al. (2014). Biology of ageing and role of dietary antioxidants. *BioMed Research International, 2014*, 13.
265. Wu, X., et al. (2004). Lipophilic and hydrophilic antioxidant capacities of common foods in the United States. *Journal of Agricultural and Food Chemistry, 52*(12), 4026–4037.
266. Torri, E., et al. (2007). Anti-inflammatory and antinociceptive properties of blueberry extract (Vaccinium corymbosum). *Journal of Pharmacy and Pharmacology, 59*(4), 591–596.
267. Wilms, L. C., et al. (2007). Impact of multiple genetic polymorphisms on effects of a 4-week blueberry juice intervention on ex vivo induced lymphocytic DNA damage in human volunteers. *Carcinogenesis, 28*(8), 1800–1806.
268. Del Bo, C., et al. (2013). A single portion of blueberry (Vaccinium corymbosum L) improves protection against DNA damage but not vascular function in healthy male volunteers. *Nutrition Research, 33*(3), 220–227.
269. Rodriguez-Mateos, A., et al. (2012). Procyanidin, anthocyanin, and chlorogenic acid contents of highbush and lowbush blueberries. *Journal of Agricultural and Food Chemistry, 60*(23), 5772–5778.

270. Mazza, G., et al. (2002). Absorption of anthocyanins from blueberries and serum antioxidant status in human subjects. *Journal of Agricultural and Food Chemistry, 50*(26), 7731–7737.
271. Kay, C. D., & Holub, B. J. (2002). The effect of wild blueberry (Vaccinium angustifolium) consumption on postprandial serum antioxidant status in human subjects. *British Journal of Nutrition, 88*(4), 389–398.
272. Tsuda, H., et al. (2013). Antioxidant activities and anti-cancer cell proliferation properties of Natsuhaze (Vaccinium oldhamii Miq.), Shashanbo (V. bracteatum Thunb.) and blueberry cultivars. *Plants (Basel), 2*(1), 57–71.
273. Johnson, S. A., & Arjmandi, B. H. (2013). Evidence for anti-cancer properties of blueberries: A mini-review. *Anti-Cancer Agents in Medicinal Chemistry, 13*(8), 1142–1148.
274. De Bont, R., & van Larebeke, N. (2004). Endogenous DNA damage in humans: A review of quantitative data. *Mutagenesis, 19*(3), 169–185.
275. Aiyer, H. S., Srinivasan, C., & Gupta, R. C. (2008). Dietary berries and ellagic acid diminish estrogen-mediated mammary tumorigenesis in ACI rats. *Nutrition and Cancer, 60*(2), 227–234.
276. Faria, A., et al. (2010). Blueberry anthocyanins and pyruvic acid adducts: Anticancer properties in breast cancer cell lines. *Phytotherapy Research, 24*(12), 1862–1869.
277. Jeyabalan, J., et al. (2014). Chemopreventive and therapeutic activity of dietary blueberry against estrogen-mediated breast cancer. *Journal of Agricultural and Food Chemistry, 62*(18), 3963–3971.
278. Adams, L. S., et al. (2011). Whole blueberry powder modulates the growth and metastasis of MDA-MB-231 triple negative breast tumors in nude mice. *The Journal of Nutrition, 141*(10), 1805–1812.
279. Simmen, F. A., et al. (2009). Lack of efficacy of blueberry in nutritional prevention of azoxymethane-initiated cancers of rat small intestine and colon. *BMC Gastroenterology, 9*, 67.
280. Tolba, M. F., & Abdel-Rahman, S. Z. (2015). Pterostilbine, an active component of blueberries, sensitizes colon cancer cells to 5-fluorouracil cytotoxicity. *Scientific Reports, 5*, 15239.
281. Wang, E., et al. (2017). Antiproliferative and proapoptotic activities of anthocyanin and anthocyanidin extracts from blueberry fruits on B16-F10 melanoma cells. *Food & Nutrition Research, 61*(1), 1325308.
282. Adom, K. K., & Liu, R. H. (2002). Antioxidant activity of grains. *Journal of Agricultural and Food Chemistry, 50*(21), 6182–6187.
283. Khurana, S., et al. (2013). Polyphenols: Benefits to the cardiovascular system in health and in aging. *Nutrients, 5*(10), 3779–3827.
284. Blacker, B. C., et al. (2013). Consumption of blueberries with a high-carbohydrate, low-fat breakfast decreases postprandial serum markers of oxidation. *British Journal of Nutrition, 109*(9), 1670–1677.
285. Riso, P., et al. (2013). Effect of a wild blueberry (Vaccinium angustifolium) drink intervention on markers of oxidative stress, inflammation and endothelial function in humans with cardiovascular risk factors. *European Journal of Nutrition, 52*(3), 949–961.
286. Basu, A., et al. (2010). Blueberries decrease cardiovascular risk factors in obese men and women with metabolic syndrome. *The Journal of Nutrition, 140*(9), 1582–1587.
287. Johnson, S. A., et al. (2015). Daily blueberry consumption improves blood pressure and arterial stiffness in postmenopausal women with pre- and stage 1-hypertension: A randomized, double-blind, placebo-controlled clinical trial. *Journal of the Academy of Nutrition and Dietetics, 115*(3), 369–377.
288. McAnulty, L. S., et al. (2014). Six weeks daily ingestion of whole blueberry powder increases natural killer cell counts and reduces arterial stiffness in sedentary males and females. *Nutrition Research, 34*(7), 577–584.
289. Muraki, I., et al. (2013). Fruit consumption and risk of type 2 diabetes: Results from three prospective longitudinal cohort studies. *BMJ, 347*, f5001.
290. Stull, A. J., et al. (2010). Bioactives in blueberries improve insulin sensitivity in obese, insulin-resistant men and women. *The Journal of Nutrition, 140*(10), 1764–1768.

291. Abidov, M., et al. (2006). Effect of Blueberin on fasting glucose, C-reactive protein and plasma aminotransferases, in female volunteers with diabetes type 2: Double-blind, placebo controlled clinical study. *Georgian Medical News, 141*, 66–72.
292. Martineau, L. C., et al. (2006). Anti-diabetic properties of the Canadian lowbush blueberry Vaccinium angustifolium Ait. *Phytomedicine, 13*(9-10), 612–623.
293. Torronen, R., et al. (2013). Berries reduce postprandial insulin responses to wheat and rye breads in healthy women. *The Journal of Nutrition, 143*(4), 430–436.
294. Wu, T., et al. (2013). Blueberry and mulberry juice prevent obesity development in C57BL/6 mice. *PLoS One, 8*(10), e77585.
295. Seymour, E. M., et al. (2011). Blueberry intake alters skeletal muscle and adipose tissue peroxisome proliferator-activated receptor activity and reduces insulin resistance in obese rats. *Journal of Medicinal Food, 14*(12), 1511–1518.
296. Giacalone, M., et al. (2011). Antioxidant and neuroprotective properties of blueberry polyphenols: A critical review. *Nutritional Neuroscience, 14*(3), 119–125.
297. Krikorian, R., et al. (2010). Blueberry supplementation improves memory in older adults. *Journal of Agricultural and Food Chemistry, 58*(7), 3996–4000.
298. Traupe, I., et al. (2018). Postoperative cognitive dysfunction and short-term neuroprotection from blueberries: A pilot study. *Minerva Anestesiologica, 84*(12), 1352–1360.
299. Boespflug, E. L., et al. (2018). Enhanced neural activation with blueberry supplementation in mild cognitive impairment. *Nutritional Neuroscience, 21*(4), 297–305.
300. Whyte, A. R., Schafer, G., & Williams, C. M. (2016). Cognitive effects following acute wild blueberry supplementation in 7- to 10-year-old children. *European Journal of Nutrition, 55*(6), 2151–2162.
301. Haskell, C. F., et al. (2010). Effects of a multi-vitamin/mineral supplement on cognitive function and fatigue during extended multi-tasking. *Human Psychopharmacology: Clinical and Experimental, 25*(6), 448–461.
302. Dutot, M., et al. (2008). Oxidative stress modulation using polyphenol-rich blueberries: Application on a human retinal cell model. *Journal Francais D'ophtalmologie, 31*(10), 975–980.
303. Jepson, R. G., & Craig, J. C. (2007). A systematic review of the evidence for cranberries and blueberries in UTI prevention. *Molecular Nutrition & Food Research, 51*(6), 738–745.
304. Ofek, I., et al. (1991). Anti-Escherichia coli adhesin activity of cranberry and blueberry juices. *The New England Journal of Medicine, 324*(22), 1599.
305. Moriwaki, S., et al. (2014). Delphinidin, one of the major anthocyanidins, prevents bone loss through the inhibition of excessive osteoclastogenesis in osteoporosis model mice. *PLoS One, 9*(5), e97177.
306. Palacios, C. (2006). The role of nutrients in bone health, from A to Z. *Critical Reviews in Food Science and Nutrition, 46*(8), 621–628.
307. Charge, S. B., & Rudnicki, M. A. (2004). Cellular and molecular regulation of muscle regeneration. *Physiological Reviews, 84*(1), 209–238.
308. Park, C. H., et al. (2018). Assessing the values of blueberries intake on exercise performance, TAS, and inflammatory factors. *Iranian Journal of Public Health, 47*(Suppl 1), 27–32.
309. Ives, S. J., et al. (2017). Effects of a combined protein and antioxidant supplement on recovery of muscle function and soreness following eccentric exercise. *Journal of the International Society of Sports Nutrition, 14*, 21.
310. McLeay, Y., et al. (2012). Effect of New Zealand blueberry consumption on recovery from eccentric exercise-induced muscle damage. *Journal of the International Society of Sports Nutrition, 9*(1), 19.
311. Zhang, H., et al. (2018). Effects of mulberry fruit (Morus alba L.) consumption on health outcomes: A mini-review. *Antioxidants (Basel), 7*(5), 69.
312. Yuan, Q., & Zhao, L. (2017). The mulberry (Morus alba L.) fruit-a review of characteristic components and health benefits. *Journal of Agricultural and Food Chemistry, 65*(48), 10383–10394.

313. Liang, L., et al. (2012). Chemical composition, nutritional value, and antioxidant activities of eight mulberry cultivars from China. *Pharmacognosy Magazine, 8*(31), 215–224.
314. Yan, F., et al. (2017). Mulberry anthocyanin extract ameliorates oxidative damage in HepG2 cells and prolongs the lifespan of caenorhabditis elegans through MAPK and Nrf2 pathways. *Oxidative Medicine and Cellular Longevity, 2017*, 7956158.
315. Ge, Q., et al. (2018). Analysis of mulberry leaf components in the treatment of diabetes using network pharmacology. *European Journal of Pharmacology, 833*, 50–62.
316. Liu, L. K., et al. (2008). Mulberry anthocyanin extracts inhibit LDL oxidation and macrophage-derived foam cell formation induced by oxidative LDL. *Journal of Food Science, 73*(6), H113–H121.
317. Jiang, Y., et al. (2017). Effects of the ethanol extract of black mulberry (Morus nigra L.) fruit on experimental atherosclerosis in rats. *Journal of Ethnopharmacology, 200*, 228–235.
318. Thaipitakwong, T., Numhom, S., & Aramwit, P. (2018). Mulberry leaves and their potential effects against cardiometabolic risks: A review of chemical compositions, biological properties and clinical efficacy. *Pharmaceutical Biology, 56*(1), 109–118.
319. Kojima, Y., et al. (2010). Effects of mulberry leaf extract rich in 1-deoxynojirimycin on blood lipid profiles in humans. *Journal of Clinical Biochemistry and Nutrition, 47*(2), 155–161.
320. Huang, H. P., Ou, T. T., & Wang, C. J. (2013). Mulberry (sang shen zi) and its bioactive compounds, the chemoprevention effects and molecular mechanisms in vitro and in vivo. *Journal of Traditional and Complementary Medicine, 3*(1), 7–15.
321. Cho, E., et al. (2017). Anti-cancer effect of cyanidin-3-glucoside from mulberry via caspase-3 cleavage and DNA fragmentation in vitro and in vivo. *Anti-Cancer Agents in Medicinal Chemistry, 17*(11), 1519–1525.
322. Chen, P. N., et al. (2006). Mulberry anthocyanins, cyanidin 3-rutinoside and cyanidin 3-glucoside, exhibited an inhibitory effect on the migration and invasion of a human lung cancer cell line. *Cancer Letters, 235*(2), 248–259.
323. Long, H. L., et al. (2018). Mulberry anthocyanins improves thyroid cancer progression mainly by inducing apoptosis and autophagy cell death. *The Kaohsiung Journal of Medical Sciences, 34*(5), 255–262.
324. Azzini, E., Giacometti, J., & Russo, G. L. (2017). Antiobesity effects of anthocyanins in preclinical and clinical studies. *Oxidative Medicine and Cellular Longevity, 2017*, 2740364.
325. Lown, M., et al. (2017). Mulberry-extract improves glucose tolerance and decreases insulin concentrations in normoglycaemic adults: Results of a randomised double-blind placebo-controlled study. *PLoS One, 12*(2), e0172239.
326. Sarikaphuti, A., et al. (2013). Preventive effects of Morus alba L. anthocyanins on diabetes in Zucker diabetic fatty rats. *Experimental and Therapeutic Medicine, 6*(3), 689–695.
327. Belwal, T., et al. (2017). Dietary anthocyanins and insulin resistance: When food becomes a medicine. *Nutrients, 9*(10), 1111.
328. Li, Y. G., et al. (2011). Hybrid of 1-deoxynojirimycin and polysaccharide from mulberry leaves treat diabetes mellitus by activating PDX-1/insulin-1 signaling pathway and regulating the expression of glucokinase, phosphoenolpyruvate carboxykinase and glucose-6-phosphatase in alloxan-induced diabetic mice. *Journal of Ethnopharmacology, 134*(3), 961–970.
329. Riche, D. M., et al. (2017). Impact of mulberry leaf extract on type 2 diabetes (Mul-DM): A randomized, placebo-controlled pilot study. *Complementary Therapies in Medicine, 32*, 105–108.
330. Banu, S., et al. (2015). Reduction of post-prandial hyperglycemia by mulberry tea in type-2 diabetes patients. *Saudi Journal of Biological Sciences, 22*(1), 32–36.
331. Stefanut, M. N., et al. (2013). Anti-hyperglycemic effect of bilberry, blackberry and mulberry ultrasonic extracts on diabetic rats. *Plant Foods for Human Nutrition, 68*(4), 378–384.
332. Eo, H. J., et al. (2014). Anti-inflammatory and anti-cancer activity of mulberry (Morus alba L.) root bark. *BMC Complementary and Alternative Medicine, 14*, 200.

333. Poschner, S., et al. (2018). Resveratrol inhibits key steps of steroid metabolism in a human estrogen-receptor positive breast cancer model: Impact on cellular proliferation. *Frontiers in Pharmacology, 9*, 742.
334. Sheth, S., et al. (2012). Resveratrol reduces prostate cancer growth and metastasis by inhibiting the Akt/MicroRNA-21 pathway. *PLoS One, 7*(12), e51655.
335. Shih, A., et al. (2002). Resveratrol induces apoptosis in thyroid cancer cell lines via a MAPK- and p53-dependent mechanism. *The Journal of Clinical Endocrinology and Metabolism, 87*(3), 1223–1232.
336. Yang, S., et al. (2015). Resveratrol elicits anti-colorectal cancer effect by activating miR-34c-KITLG in vitro and in vivo. *BMC Cancer, 15*, 969.
337. Gülçin, I. (2010). Antioxidant properties of resveratrol: A structure-activity insight. *Innovative Food Science & Emerging Technologies, 11*, 210–218.
338. Mozaffarieh, M., Sacu, S., & Wedrich, A. (2003). The role of the carotenoids, lutein and zeaxanthin, in protecting against age-related macular degeneration: A review based on controversial evidence. *Nutrition Journal, 2*, 20.
339. Kaewkaen, P., et al. (2012). Mulberry fruit extract protects against memory impairment and hippocampal damage in animal model of vascular dementia. *Evidence-Based Complementary and Alternative Medicine, 2012*, 263520.
340. Swaminathan, A., & Jicha, G. A. (2014). Nutrition and prevention of Alzheimer's dementia. *Frontiers in Aging Neuroscience, 6*, 282.
341. Lee, D., et al. (2018). Beneficial effects of bioactive compounds in mulberry fruits against cisplatin-induced nephrotoxicity. *International Journal of Molecular Sciences, 19*(4), 1117.
342. Faye, O., et al. (2018). Squamous cell carcinoma associated with use of skin-lightening cream. *Annales de Dermatologie et de Vénéréologie, 145*(2), 100–103.
343. Pillaiyar, T., Manickam, M., & Namasivayam, V. (2017). Skin whitening agents: Medicinal chemistry perspective of tyrosinase inhibitors. *Journal of Enzyme Inhibition and Medicinal Chemistry, 32*(1), 403–425.
344. Kirakosyan, A., et al. (2009). Chemical profile and antioxidant capacities of tart cherry products. *Food Chemistry, 115*, 20–25.
345. Chaovanalikit, A., & Wrolstad, R. E. (2004). Anthocyanin and polyphenolic composition of fresh and processed cherries. *Journal of Food Science, 69*(1), FCT73.
346. Carlsen, M. H., et al. (2010). The total antioxidant content of more than 3100 foods, beverages, spices, herbs and supplements used worldwide. *Nutrition Journal, 9*(1), 3.
347. Dimitriou, L., et al. (2015). Influence of a montmorency cherry juice blend on indices of exercise-induced stress and upper respiratory tract symptoms following marathon running—a pilot investigation. *Journal of the International Society of Sports Nutrition, 12*(1), 22.
348. Kang, S. Y., et al. (2003). Tart cherry anthocyanins inhibit tumor development in Apc(Min) mice and reduce proliferation of human colon cancer cells. *Cancer Letters, 194*(1), 13–19.
349. Kirakosyan, A., et al. (2018). The inhibitory potential of Montmorency tart cherry on key enzymes relevant to type 2 diabetes and cardiovascular disease. *Food Chemistry, 252*, 142–146.
350. Tjelle, T. E., et al. (2015). Polyphenol-rich juices reduce blood pressure measures in a randomised controlled trial in high normal and hypertensive volunteers. *British Journal of Nutrition, 114*(7), 1054–1063.
351. Kent, K., et al. (2016). Acute reduction in blood pressure following consumption of anthocyanin-rich cherry juice may be dose-interval dependant: A pilot cross-over study. *International Journal of Food Sciences and Nutrition, 67*(1), 47–52.
352. Wang, H., et al. (1999). Antioxidant and antiinflammatory activities of anthocyanins and their aglycon, cyanidin, from tart cherries. *Journal of Natural Products, 62*(2), 294–296.
353. Kelley, D. S., Adkins, Y., & Laugero, K. D. (2018). A review of the health benefits of cherries. *Nutrients, 10*(3), 368.
354. Min, J., et al. (2011). Neuroprotective effect of cyanidin-3-O-glucoside anthocyanin in mice with focal cerebral ischemia. *Neuroscience Letters, 500*(3), 157–161.

355. Kent, K., et al. (2017). Consumption of anthocyanin-rich cherry juice for 12 weeks improves memory and cognition in older adults with mild-to-moderate dementia. *European Journal of Nutrition, 56*(1), 333–341.
356. Traustadottir, T., et al. (2009). Tart cherry juice decreases oxidative stress in healthy older men and women. *The Journal of Nutrition, 139*(10), 1896–1900.
357. Kim, D. O., et al. (2005). Sweet and sour cherry phenolics and their protective effects on neuronal cells. *Journal of Agricultural and Food Chemistry, 53*(26), 9921–9927.
358. Jacob, R. A., et al. (2003). Consumption of cherries lowers plasma urate in healthy women. *The Journal of Nutrition, 133*(6), 1826–1829.
359. Schumacher, H. R., et al. (2013). Randomized double-blind crossover study of the efficacy of a tart cherry juice blend in treatment of osteoarthritis (OA) of the knee. *Osteoarthritis and Cartilage, 21*(8), 1035–1041.
360. Kuehl, K., et al. (2012). Efficacy of Tart Cherry Juice to Reduce Inflammation Biomarkers among Women with Inflammatory Osteoarthritis (OA). *Journal of Food Studies, 1*, 14–25.
361. Zhang, Y., et al. (2012). Cherry consumption and decreased risk of recurrent gout attacks. *Arthritis and Rheumatism, 64*(12), 4004–4011.
362. Bell, P., et al. (2014). Montmorency tart cherry (Prunus cerasus L.) concentrate lowers uric acid, independent of plasma cyanidin-3-O-glucosiderutinoside. *Journal of Functional Foods, 11*, 82–90.
363. Carson, C. A. (2015). Tart cherry juice as a treatment for peripheral neuropathy. *Integrative Medicine: A Clinician's Journal, 14*(1), 48–49.
364. Kuehl, K. S., et al. (2010). Efficacy of tart cherry juice in reducing muscle pain during running: A randomized controlled trial. *Journal of the International Society of Sports Nutrition, 7*, 17.
365. Howatson, G., et al. (2010). Influence of tart cherry juice on indices of recovery following marathon running. *Scandinavian Journal of Medicine & Science in Sports, 20*(6), 843–852.
366. Seymour, E. M., et al. (2009). Regular tart cherry intake alters abdominal adiposity, adipose gene transcription, and inflammation in obesity-prone rats fed a high fat diet. *Journal of Medicinal Food, 12*(5), 935–942.
367. Liu, A., et al. (2014). Tart cherry juice increases sleep time in older adults with insomnia (830.9). *The FASEB Journal, 28*(1_supplement), 830–839.
368. Pigeon, W. R., et al. (2010). Effects of a tart cherry juice beverage on the sleep of older adults with insomnia: A pilot study. *Journal of Medicinal Food, 13*(3), 579–583.
369. Deli, J., et al. (2001). Carotenoid composition in the fruits of red paprika (Capsicum annuum var. lycopersiciforme rubrum) during ripening; biosynthesis of carotenoids in red paprika. *Journal of Agricultural and Food Chemistry, 49*(3), 1517–1523.
370. Gomez-Garcia Mdel, R., & Ochoa-Alejo, N. (2013). Biochemistry and molecular biology of carotenoid biosynthesis in chili peppers (Capsicum spp.). *International Journal of Molecular Sciences, 14*(9), 19025–19053.
371. Kim, S., Ha, T. Y., & Kyeong Hwang, I. (2009). Analysis, bioavailability, and potential healthy effects of capsanthin, natural red pigment from Capsicum spp. *Food Reviews International, 25*, 198–213.
372. Arimboor, R., et al. (2015). Red pepper (Capsicum annuum) carotenoids as a source of natural food colors: Analysis and stability-a review. *Journal of Food Science and Technology, 52*(3), 1258–1271.
373. Rodriguez-Burruezo, A., Gonzalez-Mas Mdel, C., & Nuez, F. (2010). Carotenoid composition and vitamin A value in aji (Capsicum baccatum L.) and rocoto (C. pubescens R. & P.), 2 pepper species from the Andean region. *Journal of Food Science, 75*(8), S446–S453.
374. Guzman, I., et al. (2010). Variability of carotenoid biosynthesis in orange colored Capsicum spp. *Plant Science, 179*(1-2), 49–59.
375. Sommerburg, O., et al. (1998). Fruits and vegetables that are sources for lutein and zeaxanthin: The macular pigment in human eyes. *British Journal of Ophthalmology, 82*(8), 907–910.
376. Hornero-Mendez, D., Gomez-Ladron De Guevara, R., & Minguez-Mosquera, M. I. (2000). Carotenoid biosynthesis changes in five red pepper (Capsicum annuum L.) cultivars during

ripening. Cultivar selection for breeding. *Journal of Agricultural and Food Chemistry, 48*(9), 3857–3864.
377. Materska, M., et al. (2003). Isolation and structure elucidation of flavonoid and phenolic acid glycosides from pericarp of hot pepper fruit Capsicum annuum L. *Phytochemistry, 63*(8), 893–898.
378. Materska, M., & Perucka, I. (2005). Antioxidant activity of the main phenolic compounds isolated from hot pepper fruit (Capsicum annuum L). *Journal of Agricultural and Food Chemistry, 53*(5), 1750–1756.
379. Srinivasan, K. (2016). Biological activities of red pepper (Capsicum annuum) and its pungent principle capsaicin: A review. *Critical Reviews in Food Science and Nutrition, 56*(9), 1488–1500.
380. Fernandes, E. S., et al. (2016). Capsaicin and its role in chronic diseases. *Advances in Experimental Medicine and Biology, 929*, 91–125.
381. Gottardi, D., et al. (2016). Beneficial effects of spices in food preservation and safety. *Frontiers in Microbiology, 7*, 1394.
382. O'Neill, J., et al. (2012). Unravelling the mystery of capsaicin: A tool to understand and treat pain. *Pharmacological Reviews, 64*(4), 939–971.
383. Anand, P., & Bley, K. (2011). Topical capsaicin for pain management: Therapeutic potential and mechanisms of action of the new high-concentration capsaicin 8% patch. *British Journal of Anaesthesia, 107*(4), 490–502.
384. Peppin, J. F., & Pappagallo, M. (2014). Capsaicinoids in the treatment of neuropathic pain: A review. *Therapeutic Advances in Neurological Disorders, 7*(1), 22–32.
385. Evangelista, S. (2015). Novel therapeutics in the field of capsaicin and pain. *Expert Review of Clinical Pharmacology, 8*(4), 373–375.
386. Fattori, V., et al. (2016). Capsaicin: Current understanding of its mechanisms and therapy of pain and other pre-clinical and clinical uses. *Molecules, 21*(7), 844.
387. Chung, M. K., & Campbell, J. N. (2016). Use of capsaicin to treat pain: Mechanistic and therapeutic considerations. *Pharmaceuticals (Basel), 9*(4), 66.
388. Sharma, S. K., Vij, A. S., & Sharma, M. (2013). Mechanisms and clinical uses of capsaicin. *European Journal of Pharmacology, 720*(1-3), 55–62.
389. Cianchetti, C. (2010). Capsaicin jelly against migraine pain. *International Journal of Clinical Practice, 64*(4), 457–459.
390. Benemei, S., et al. (2014). The TRPA1 channel in migraine mechanism and treatment. *British Journal of Pharmacology, 171*(10), 2552–2567.
391. Buntinx, L., Vermeersch, S., & de Hoon, J. (2015). Development of anti-migraine therapeutics using the capsaicin-induced dermal blood flow model. *British Journal of Clinical Pharmacology, 80*(5), 992–1000.
392. Malhotra, R. (2016). Understanding migraine: Potential role of neurogenic inflammation. *Annals of Indian Academy of Neurology, 19*(2), 175–182.
393. Karrer, T., & Bartoshuk, L. (1995). Effects of capsaicin desensitization on taste in humans. *Physiology and Behavior, 57*(3), 421–429.
394. Kim, C. S., et al. (2003). Capsaicin exhibits anti-inflammatory property by inhibiting IkB-a degradation in LPS-stimulated peritoneal macrophages. *Cellular Signalling, 15*(3), 299–306.
395. Jolayemi, A. T., & Ojewole, J. A. (2013). Comparative anti-inflammatory properties of Capsaicin and ethyl-aAcetate extract of Capsicum frutescens linn [Solanaceae] in rats. *African Health Sciences, 13*(2), 357–361.
396. Deal, C. L., et al. (1991). Treatment of arthritis with topical capsaicin: A double-blind trial. *Clinical Therapeutics, 13*(3), 383–395.
397. Persson, M. S. M., et al. (2018). The relative efficacy of topical non-steroidal anti-inflammatory drugs and capsaicin in osteoarthritis: A network meta-analysis of randomised controlled trials. *Osteoarthritis and Cartilage, 26*(12), 1575–1582.
398. Guedes, V., Castro, J. P., & Brito, I. (2018). Topical capsaicin for pain in osteoarthritis: A literature review. *Reumatología Clínica, 14*(1), 40–45.

399. Laslett, L. L., & Jones, G. (2014). Capsaicin for osteoarthritis pain. *Progress in Drug Research, 68*, 277–291.
400. Mason, L., et al. (2004). Systematic review of topical capsaicin for the treatment of chronic pain. *BMJ, 328*(7446), 991.
401. Simpson, D. M., et al. (2017). Capsaicin 8% patch in painful diabetic peripheral neuropathy: A randomized, double-blind, placebo-controlled study. *The Journal of Pain, 18*(1), 42–53.
402. Musharraf, M. U., Ahmad, Z., & Yaqub, Z. (2017). Comparison of topical capsaicin and topical turpentine Oil for treatment of painful diabetic neuropathy. *Journal of Ayub Medical College, Abbottabad, 29*(3), 384–387.
403. Tandan, R., et al. (1992). Topical capsaicin in painful diabetic neuropathy. Effect on sensory function. *Diabetes Care, 15*(1), 15–18.
404. The Capsaicin Study Group. (1991). Treatment of painful diabetic neuropathy with topical capsaicin. A multicenter, double-blind, vehicle-controlled study. *Archives of Internal Medicine, 151*(11), 2225–2229.
405. Meltzer, E. O., et al. (2010). Treatment of congestion in upper respiratory diseases. *International Journal of General Medicine, 3*, 69–91.
406. Baraniuk, J. N. (2011). Subjective nasal fullness and objective congestion. *Proceedings of the American Thoracic Society, 8*(1), 62–69.
407. Fokkens, W., Hellings, P., & Segboer, C. (2016). Capsaicin for rhinitis. *Current Allergy and Asthma Reports, 16*(8), 60–60.
408. Gevorgyan, A., et al. (2015). Capsaicin for non-allergic rhinitis. *Cochrane Database of Systematic Reviews, 7*, 20150714.
409. Jones, N. L., Shabib, S., & Sherman, P. M. (1997). Capsaicin as an inhibitor of the growth of the gastric pathogen Helicobacter pylori. *FEMS Microbiology Letters, 146*(2), 223–227.
410. Lee, I. O., et al. (2007). Anti-inflammatory effect of capsaicin in Helicobacter pylori-infected gastric epithelial cells. *Helicobacter, 12*(5), 510–517.
411. Satyanarayana, M. N. (2006). Capsaicin and gastric ulcers. *Critical Reviews in Food Science and Nutrition, 46*(4), 275–328.
412. Toyoda, T., et al. (2016). Anti-inflammatory effects of capsaicin and piperine on helicobacter pylori-induced chronic gastritis in mongolian gerbils. *Helicobacter, 21*(2), 131–142.
413. Herrera-Lopez, J. A., et al. (2010). Capsaicin induction of esophageal symptoms in different phenotypes of gastroesophageal reflux disease. *Revista de Gastroenterología de México, 75*(4), 396–404.
414. Yi, C. H., et al. (2016). Influence of capsaicin infusion on secondary peristalsis in patients with gastroesophageal reflux disease. *World Journal of Gastroenterology, 22*(45), 10045–10052.
415. Zheng, J., et al. (2017). Dietary capsaicin and its anti-obesity potency: From mechanism to clinical implications. *Bioscience Reports, 37*(3), BSR20170286.
416. Varghese, S., et al. (2017). Chili pepper as a body weight-loss food. *International Journal of Food Sciences and Nutrition, 68*(4), 392–401.
417. Narang, N., Jiraungkoorskul, W., & Jamrus, P. (2017). Current understanding of antiobesity property of capsaicin. *Pharmacognosy Reviews, 11*(21), 23–26.
418. Leung, F. W. (2014). Capsaicin as an anti-obesity drug. *Progress in Drug Research, 68*, 171–179.
419. Whiting, S., Derbyshire, E. J., & Tiwari, B. (2014). Could capsaicinoids help to support weight management? A systematic review and meta-analysis of energy intake data. *Appetite, 73*, 183–188.
420. McCarty, M. F., DiNicolantonio, J. J., & O'Keefe, J. H. (2015). Capsaicin may have important potential for promoting vascular and metabolic health. *Open Heart, 2*(1), e000262.
421. Westerterp-Plantenga, M. S., Smeets, A., & Lejeune, M. P. (2005). Sensory and gastrointestinal satiety effects of capsaicin on food intake. *International Journal of Obesity, 29*(6), 682–688.
422. Janssens, P. L., Hursel, R., & Westerterp-Plantenga, M. S. (2014). Capsaicin increases sensation of fullness in energy balance, and decreases desire to eat after dinner in negative energy balance. *Appetite, 77*, 44–49.

423. van Avesaat, M., et al. (2016). Capsaicin-induced satiety is associated with gastrointestinal distress but not with the release of satiety hormones. *The American Journal of Clinical Nutrition, 103*(2), 305–313.
424. Rogers, J., et al. (2018). Capsaicinoids supplementation decreases percent body fat and fat mass: Adjustment using covariates in a post hoc analysis. *BMC Obesity, 5*, 22.
425. Josse, A. R., et al. (2010). Effects of capsinoid ingestion on energy expenditure and lipid oxidation at rest and during exercise. *Nutrition & Metabolism (London), 7*, 65.
426. Lejeune, M. P., Kovacs, E. M., & Westerterp-Plantenga, M. S. (2003). Effect of capsaicin on substrate oxidation and weight maintenance after modest body-weight loss in human subjects. *British Journal of Nutrition, 90*(3), 651–659.
427. Janssens, P. L., et al. (2013). Acute effects of capsaicin on energy expenditure and fat oxidation in negative energy balance. *PLoS One, 8*(7), e67786.
428. Ludy, M. J., Moore, G. E., & Mattes, R. D. (2012). The effects of capsaicin and capsiate on energy balance: Critical review and meta-analyses of studies in humans. *Chemical Senses 37*(2), 103–121.
429. Yoneshiro, T., et al. (2012). Nonpungent capsaicin analogs (capsinoids) increase energy expenditure through the activation of brown adipose tissue in humans. *The American Journal of Clinical Nutrition, 95*(4), 845–850.
430. Galgani, J. E., Ryan, D. H., & Ravussin, E. (2010). Effect of capsinoids on energy metabolism in human subjects. *British Journal of Nutrition, 103*(1), 38–42.
431. Song, J. X., et al. (2017). Dietary capsaicin improves glucose homeostasis and alters the gut microbiota in obese diabetic ob/ob mice. *Frontiers in Physiology, 8*, 602.
432. Kang, C., et al. (2017). Gut microbiota mediates the protective effects of dietary capsaicin against chronic low-grade inflammation and associated obesity induced by high-fat diet. *MBio, 8*(3), e00470–e00417.
433. Lu, Q. Y., et al. (2017). Prebiotic potential and chemical composition of seven culinary spice extracts. *Journal of Food Science, 82*(8), 1807–1813.
434. Qin, Y., et al. (2017). Capsaicin supplementation improved risk factors of coronary heart disease in individuals with low HDL-C levels. *Nutrients, 9*(9), 1037.
435. Yuan, L. J., et al. (2016). Capsaicin-containing chili improved postprandial hyperglycemia, hyperinsulinemia, and fasting lipid disorders in women with gestational diabetes mellitus and lowered the incidence of large-for-gestational-age newborns. *Clinical Nutrition, 35*(2), 388–393.
436. Bode, A. M., & Dong, Z. (2011). The two faces of capsaicin. *Cancer Research, 71*(8), 2809.
437. Clark, R., & Lee, S. H. (2016). Anticancer properties of capsaicin against human cancer. *Anticancer Research, 36*(3), 837–843.
438. Georgescu, S. R., et al. (2017). Capsaicin: Friend or foe in skin cancer and other related malignancies? *Nutrients, 9*(12), 1365.
439. Wang, F., et al. (2016). Capsaicin reactivates hMOF in gastric cancer cells and induces cell growth inhibition. *Cancer Biology & Therapy, 17*(11), 1117–1125.
440. Anandakumar, P., et al. (2015). The anticancer role of capsaicin in experimentally induced lung carcinogenesis. *Journal of Pharmacopuncture, 18*(2), 19–25.
441. Ramos-Torres, A., et al. (2016). The pepper's natural ingredient capsaicin induces autophagy blockage in prostate cancer cells. *Oncotarget, 7*(2), 1569–1583.
442. Demmig-Adams, B., & Adams, R. B. (2013). Eye nutrition in context: Mechanisms, implementation, and future directions. *Nutrients, 5*(7), 2483–2501.
443. Fernandez-Bedmar, Z., & Alonso-Moraga, A. (2016). In vivo and in vitro evaluation for nutraceutical purposes of capsaicin, capsanthin, lutein and four pepper varieties. *Food and Chemical Toxicology, 98*(Pt B), 89–99.
444. Koushan, K., et al. (2013). The role of lutein in eye-related disease. *Nutrients, 5*(5), 1823–1839.
445. Abdel-Aal el, S. M., et al. (2013). Dietary sources of lutein and zeaxanthin carotenoids and their role in eye health. *Nutrients, 5*(4), 1169–1185.

446. Buscemi, S., et al. (2018). The effect of lutein on eye and extra-eye health. *Nutrients, 10*(9), 1321.
447. Eisenhauer, B., et al. (2017). Lutein and zeaxanthin-food sources, bioavailability and dietary variety in age-related macular degeneration protection. *Nutrients, 9*(2), 120.
448. Tsao, R. (2010). Chemistry and biochemistry of dietary polyphenols. *Nutrients, 2*(12), 1231–1246.
449. Sora, G. T., et al. (2015). A comparative study of the capsaicinoid and phenolic contents and in vitro antioxidant activities of the peppers of the genus Capsicum: An application of chemometrics. *Journal of Food Science and Technology, 52*(12), 8086–8094.
450. Moreno-Ramirez, Y. D. R., et al. (2018). Free radical-scavenging capacities, phenolics and capsaicinoids in wild piquin chili (Capsicum annuum var. Glabriusculum). *Molecules, 23*(10), 2655.
451. Castro-Concha, L. A., et al. (2014). Antioxidant capacity and total phenolic content in fruit tissues from accessions of Capsicum chinense Jacq. (Habanero pepper) at different stages of ripening. *Scientific World Journal, 2014*, 809073.
452. Sandoval-Castro, C. J., et al. (2017). Bioactive compounds and antioxidant activity in scalded Jalapeno pepper industrial byproduct (Capsicum annuum). *Journal of Food Science and Technology, 54*(7), 1999–2010.
453. Saibabu, V., et al. (2015). Therapeutic potential of dietary phenolic acids. *Advances in Pharmacological Sciences, 2015*, 823539.
454. Sricharoen, P., et al. (2017). Phytochemicals in capsicum oleoresin from different varieties of hot chilli peppers with their antidiabetic and antioxidant activities due to some phenolic compounds. *Ultrasonics Sonochemistry, 38*, 629–639.
455. Dzialo, M., et al. (2016). The potential of plant phenolics in prevention and therapy of skin disorders. *International Journal of Molecular Sciences, 17*(2), 160.
456. Pizzino, G., et al. (2017). Oxidative stress: Harms and benefits for human health. *Oxidative Medicine and Cellular Longevity, 2017*, 8416763.
457. Perez-Ternero, C., et al. (2017). Ferulic acid, a bioactive component of rice bran, improves oxidative stress and mitochondrial biogenesis and dynamics in mice and in human mononuclear cells. *The Journal of Nutritional Biochemistry, 48*, 51–61.
458. Gerin, F., et al. (2016). The effects of ferulic acid against oxidative stress and inflammation in formaldehyde-induced hepatotoxicity. *Inflammation, 39*(4), 1377–1386.
459. Bumrungpert, A., et al. (2018). Ferulic acid supplementation improves lipid profiles, oxidative stress, and inflammatory status in hyperlipidemic subjects: A randomized, double-blind, placebo-controlled clinical trial. *Nutrients, 10*(6), 713.
460. Zych, M., et al. (2018). The effects of sinapic acid on the development of metabolic disorders induced by estrogen deficiency in rats. *Oxidative Medicine and Cellular Longevity, 2018*, 9274246.
461. Raish, M., et al. (2018). Sinapic acid ameliorates bleomycin-induced lung fibrosis in rats. *Biomedicine & Pharmacotherapy, 108*, 224–231.
462. Chen, C. (2016). Sinapic acid and its derivatives as medicine in oxidative stress-induced diseases and aging. *Oxidative Medicine and Cellular Longevity, 2016*, 3571614.
463. Srinivasan, M., Sudheer, A. R., & Menon, V. P. (2007). Ferulic acid: Therapeutic potential through its antioxidant property. *Journal of Clinical Biochemistry and Nutrition, 40*(2), 92–100.
464. Kumar, N., & Pruthi, V. (2014). Potential applications of ferulic acid from natural sources. *Biotechnology Reports (Amsterdam, Netherlands), 4*, 86–93.
465. Nićiforović, N., & Abramovič, H. (2014). Sinapic acid and its derivatives: Natural sources and bioactivity. *Comprehensive Reviews in Food Science and Food Safety, 13*(1), 34–51.
466. Szwajgier, D., Borowiec, K., & Pustelniak, K. (2017). The neuroprotective effects of phenolic acids: Molecular mechanism of action. *Nutrients, 9*(5), 477.
467. Sgarbossa, A., Giacomazza, D., & di Carlo, M. (2015). Ferulic Acid: A hope for Alzheimer's disease therapy from plants. *Nutrients, 7*(7), 5764–5782.

468. Ren, Z., et al. (2017). Ferulic acid exerts neuroprotective effects against cerebral ischemia/reperfusion-induced injury via antioxidant and anti-apoptotic mechanisms in vitro and in vivo. *International Journal of Molecular Medicine, 40*(5), 1444–1456.
469. Zare, K., et al. (2015). The neuroprotective potential of sinapic acid in the 6-hydroxydopamine-induced hemi-parkinsonian rat. *Metabolic Brain Disease, 30*(1), 205–213.
470. Lee, H. E., et al. (2012). Neuroprotective effect of sinapic acid in a mouse model of amyloid beta(1-42) protein-induced Alzheimer's disease. *Pharmacology, Biochemistry, and Behavior, 103*(2), 260–266.
471. Karakida, F., et al. (2007). Cerebral protective and cognition-improving effects of sinapic acid in rodents. *Biological & Pharmaceutical Bulletin, 30*(3), 514–519.
472. Das, S., et al. (2005). Capsicum spray injury of the eye. *International Ophthalmology, 26*(4-5), 171–173.
473. Lee, R. J., et al. (1996). Personal defense sprays: Effects and management of exposure. *Journal of the American Optometric Association, 67*(9), 548–560.
474. Gonlachanvit, S., Mahayosnond, A., & Kullavanijaya, P. (2009). Effects of chili on postprandial gastrointestinal symptoms in diarrhoea predominant irritable bowel syndrome: Evidence for capsaicin-sensitive visceral nociception hypersensitivity. *Neurogastroenterology and Motility, 21*(1), 23–32.
475. Notani, P. N., & Jayant, K. (1987). Role of diet in upper aerodigestive tract cancers. *Nutrition and Cancer, 10*(1-2), 103–113.
476. Lopez-Carrillo, L., Hernandez Avila, M., & Dubrow, R. (1994). Chili pepper consumption and gastric cancer in Mexico: A case-control study. *American Journal of Epidemiology, 139*(3), 263–271.
477. Serra, I., et al. (2002). Association of chili pepper consumption, low socioeconomic status and longstanding gallstones with gallbladder cancer in a Chilean population. *International Journal of Cancer, 102*(4), 407–411.
478. Atkinson, F. S., Foster-Powell, K., & Brand-Miller, J. C. (2008). International tables of glycemic index and glycemic load values: 2008. *Diabetes Care, 31*(12), 2281–2283.
479. Hanhineva, K., et al. (2010). Impact of dietary polyphenols on carbohydrate metabolism. *International Journal of Molecular Sciences, 11*(4), 1365–1402.
480. Lv, X., et al. (2015). Citrus fruits as a treasure trove of active natural metabolites that potentially provide benefits for human health. *Chemistry Central Journal, 9*, 68.
481. Dhingra, D., et al. (2012). Dietary fibre in foods: A review. *Journal of Food Science and Technology, 49*(3), 255–266.
482. Slavin, J. L., & Lloyd, B. (2012). Health benefits of fruits and vegetables. *Advances in Nutrition (Bethesda, MD), 3*(4), 506–516.
483. Timm, D. A., & Slavin, J. L. (2008). Dietary fiber and the relationship to chronic diseases. *American Journal of Lifestyle Medicine, 2*(3), 233–240.
484. Mackowiak, K., Torlinska-Walkowiak, N., & Torlinska, B. (2016). Dietary fibre as an important constituent of the diet. *Postępy Higieny i Medycyny Doświadczalnej (Online), 70*, 104–109.
485. Lattimer, J. M., & Haub, M. D. (2010). Effects of dietary fiber and its components on metabolic health. *Nutrients, 2*(12), 1266–1289.
486. Wisker, E., Martina Daniel, M., & Feldheim, W. (1994). Effects of fiber concentrate from citrus fruits in humans. *Nutrition Research, 14*, 361–372.
487. Kaczmarczyk, M. M., Miller, M. J., & Freund, G. G. (2012). The health benefits of dietary fiber: Beyond the usual suspects of type 2 diabetes mellitus, cardiovascular disease and colon cancer. *Metabolism, 61*(8), 1058–1066.
488. Veronese, N., et al. (2018). Dietary fiber and health outcomes: An umbrella review of systematic reviews and meta-analyses. *The American Journal of Clinical Nutrition, 107*(3), 436–444.

489. Chen, J. P., et al. (2017). Dietary fiber and metabolic syndrome: A meta-analysis and review of related mechanisms. *Nutrients, 10*(1), 24.
490. Post, R. E., et al. (2012). Dietary fiber for the treatment of type 2 diabetes mellitus: A meta-analysis. *Journal of American Board of Family Medicine, 25*(1), 16–23.
491. McRae, M. P. (2018). Dietary fiber intake and type 2 diabetes mellitus: An umbrella review of meta-analyses. *Journal of Chiropractic Medicine, 17*(1), 44–53.
492. Chen, C., et al. (2016). Therapeutic effects of soluble dietary fiber consumption on type 2 diabetes mellitus. *Experimental and Therapeutic Medicine, 12*(2), 1232–1242.
493. Wang, P. Y., et al. (2016). Higher intake of fruits, vegetables or their fiber reduces the risk of type 2 diabetes: A meta-analysis. *Journal of Diabetes Investigation, 7*(1), 56–69.
494. McRae, M. P. (2017). dietary fiber is beneficial for the prevention of cardiovascular disease: An umbrella review of meta-analyses. *Journal of Chiropractic Medicine, 16*(4), 289–299.
495. Sanchez-Muniz, F. J. (2012). Dietary fibre and cardiovascular health. *Nutrición Hospitalaria, 27*(1), 31–45.
496. Asgary, S., & Keshvari, M. (2013). Effects of Citrus sinensis juice on blood pressure. *ARYA atherosclerosis, 9*(1), 98–101.
497. Surampudi, P., et al. (2016). Lipid Lowering with Soluble Dietary Fiber. *Current Atherosclerosis Reports, 18*(12), 75.
498. Brouns, F., et al. (2012). Cholesterol-lowering properties of different pectin types in mildly hyper-cholesterolemic men and women. *European Journal of Clinical Nutrition, 66*(5), 591–599.
499. Aller, R., et al. (2004). Effect of soluble fiber intake in lipid and glucose levels in healthy subjects: A randomized clinical trial. *Diabetes Research and Clinical Practice, 65*(1), 7–11.
500. Slavin, J. L. (2005). Dietary fiber and body weight. *Nutrition, 21*(3), 411–418.
501. Buil-Cosiales, P., et al. (2017). Consumption of fruit or fiber-fruit decreases the risk of cardiovascular disease in a Mediterranean young cohort. *Nutrients, 9*(3), 295.
502. Lairon, D., et al. (2005). Dietary fiber intake and risk factors for cardiovascular disease in French adults. *The American Journal of Clinical Nutrition, 82*(6), 1185–1194.
503. Xu, H., et al. (2016). Excess protein intake relative to fiber and cardiovascular events in elderly men with chronic kidney disease. *Nutrition, Metabolism and Cardiovascular Diseases, 26*(7), 597–602.
504. Threapleton, D. E., et al. (2013). Dietary fibre and cardiovascular disease mortality in the UK Women's Cohort Study. *European Journal of Epidemiology, 28*(4), 335–346.
505. Mirmiran, P., et al. (2016). A prospective study of different types of dietary fiber and risk of cardiovascular disease: Tehran lipid and glucose study. *Nutrients, 8*(11), 686.
506. Diaz, V. A., et al. (2005). Race and diet in the overweight: Association with cardiovascular risk in a nationally representative sample. *Nutrition, 21*(6), 718–725.
507. Teramoto, T. (2017). Is "The Japan Diet" cardioprotective? *Journal of Atherosclerosis and Thrombosis, 24*(4), 388–389.
508. Yamada, T., et al. (2011). Frequency of citrus fruit intake is associated with the incidence of cardiovascular disease: The Jichi Medical School cohort study. *Journal of Epidemiology, 21*(3), 169–175.
509. Gianfredi, V., et al. (2018). Is dietary fibre truly protective against colon cancer? A systematic review and meta-analysis. *International Journal of Food Sciences and Nutrition, 69*(8), 904–915.
510. Levi, F., et al. (2001). Dietary fibre and the risk of colorectal cancer. *European Journal of Cancer, 37*(16), 2091–2096.
511. Kunzmann, A. T., et al. (2015). Dietary fiber intake and risk of colorectal cancer and incident and recurrent adenoma in the Prostate, Lung, Colorectal, and Ovarian Cancer Screening Trial. *The American Journal of Clinical Nutrition, 102*(4), 881–890.
512. Murphy, N., et al. (2012). Dietary fibre intake and risks of cancers of the colon and rectum in the European prospective investigation into cancer and nutrition (EPIC). *PLoS One, 7*(6), e39361.

513. Abdullah, M. M., et al. (2015). Dietary fibre intakes and reduction in functional constipation rates among Canadian adults: A cost-of-illness analysis. *Food & Nutrition Research, 59*, 28646.
514. Yang, J., et al. (2012). Effect of dietary fiber on constipation: A meta analysis. *World Journal of Gastroenterology, 18*(48), 7378–7383.
515. Erdogan, A., et al. (2016). Randomised clinical trial: Mixed soluble/insoluble fibre vs. psyllium for chronic constipation. *Alimentary Pharmacology and Therapeutics, 44*(1), 35–44.
516. Tiwary, C. M., Ward, J. A., & Jackson, B. A. (1997). Effect of pectin on satiety in healthy US Army adults. *Journal of the American College of Nutrition, 16*(5), 423–428.
517. Bertoia, M. L., et al. (2015). Changes in intake of fruits and vegetables and weight change in united states men and women followed for up to 24 years: Analysis from three prospective cohort studies. *PLoS Medicine, 12*(9), e1001878.
518. Burton-Freeman, B., et al. (2017). Ratios of soluble and insoluble dietary fibers on satiety and energy intake in overweight pre- and postmenopausal women. *Nutrition and Healthy Aging, 4*(2), 157–168.
519. Shreiner, A. B., Kao, J. Y., & Young, V. B. (2015). The gut microbiome in health and in disease. *Current Opinion in Gastroenterology, 31*(1), 69–75.
520. Zeng, H., Lazarova, D. L., & Bordonaro, M. (2014). Mechanisms linking dietary fiber, gut microbiota and colon cancer prevention. *World Journal of Gastrointestinal Oncology, 6*(2), 41–51.
521. Carlson, J. L., et al. (2018). Health effects and sources of prebiotic dietary fiber. *Current Developments in Nutrition, 2*(3), nzy005–nzy005.
522. Holscher, H. D. (2017). Dietary fiber and prebiotics and the gastrointestinal microbiota. *Gut Microbes, 8*(2), 172–184.
523. Kellow, N. J., Coughlan, M. T., & Reid, C. M. (2014). Metabolic benefits of dietary prebiotics in human subjects: A systematic review of randomised controlled trials. *British Journal of Nutrition, 111*(7), 1147–1161.
524. Slavin, J. (2013). Fiber and prebiotics: Mechanisms and health benefits. *Nutrients, 5*(4), 1417–1435.
525. Khan, M. S. A., et al. (2018). Fruit-derived polysaccharides and terpenoids: Recent update on the gastroprotective effects and mechanisms. *Frontiers in Pharmacology, 9*, 569.
526. Cho, K. S., et al. (2017). Terpenes from forests and human health. *Toxicological Research, 33*(2), 97–106.
527. Paduch, R., et al. (2007). Terpenes: Substances useful in human healthcare. *Archivum Immunologiae et Therapiae Experimentalis, 55*(5), 315.
528. Wagner, K. H., & Elmadfa, I. (2003). Biological relevance of terpenoids. Overview focusing on mono-, di- and tetraterpenes. *Annals of Nutrition and Metabolism, 47*(3-4), 95–106.
529. Thoppil, R. J., & Bishayee, A. (2011). Terpenoids as potential chemopreventive and therapeutic agents in liver cancer. *World Journal of Hepatology, 3*(9), 228–249.
530. Manayi, A., et al. (2016). Natural terpenoids as a promising source for modulation of GABAergic system and treatment of neurological diseases. *Pharmacological Reports, 68*(4), 671–679.
531. Kumar, S., & Pandey, A. K. (2013). Chemistry and biological activities of flavonoids: An Overview. *The Scientific World Journal, 2013*, 162750.
532. Xiao, Z. P., et al. (2011). Flavonoids health benefits and their molecular mechanism. *Mini Reviews in Medicinal Chemistry, 11*(2), 169–177.
533. Yao, L. H., et al. (2004). Flavonoids in food and their health benefits. *Plant Foods for Human Nutrition, 59*(3), 113–122.
534. Pietta, P. G. (2000). Flavonoids as antioxidants. *Journal of Natural Products, 63*(7), 1035–1042.
535. Young, A. J., & Lowe, G. L. (2018). Carotenoids-antioxidant properties. *Antioxidants (Basel, Switzerland), 7*(2), 28.

536. Fiedor, J., & Burda, K. (2014). Potential role of carotenoids as antioxidants in human health and disease. *Nutrients, 6*(2), 466–488.
537. Hammond Jr., B. R., & Renzi, L. M. (2013). Carotenoids. *Advances in Nutrition, 4*(4), 474–476.
538. Paiva, S. A., & Russell, R. M. (1999). Beta-carotene and other carotenoids as antioxidants. *Journal of the American College of Nutrition, 18*(5), 426–433.
539. Zou, Z., et al. (2016). Antioxidant activity of citrus fruits. *Food Chemistry, 196*, 885–896.
540. Oikeh, E. I., et al. (2016). Phytochemical, antimicrobial, and antioxidant activities of different citrus juice concentrates. *Food Science & Nutrition, 4*(1), 103–109.
541. Hijaz, F., et al. (2016). Nucleotides, micro- and macro-nutrients, limonoids, flavonoids, and hydroxycinnamates composition in the phloem sap of sweet orange. *Plant Signaling & Behavior, 11*(6), e1183084.
542. Pham-Huy, L. A., He, H., & Pham-Huy, C. (2008). Free radicals, antioxidants in disease and health. *International Journal of Biomedical Sciences, 4*(2), 89–96.
543. Lobo, V., et al. (2010). Free radicals, antioxidants and functional foods: Impact on human health. *Pharmacognosy Reviews, 4*(8), 118–126.
544. Roy, A., & Saraf, S. (2006). Limonoids: Overview of significant bioactive triterpenes distributed in plants kingdom. *Biological & Pharmaceutical Bulletin, 29*(2), 191–201.
545. Gualdani, R., et al. (2016). The chemistry and pharmacology of citrus limonoids. *Molecules, 21*(11), 1530.
546. Tundis, R., Loizzo, M. R., & Menichini, F. (2014). An overview on chemical aspects and potential health benefits of limonoids and their derivatives. *Critical Reviews in Food Science and Nutrition, 54*(2), 225–250.
547. Manners, G. D. (2007). Citrus limonoids: Analysis, bioactivity, and biomedical prospects. *Journal of Agricultural and Food Chemistry, 55*(21), 8285–8294.
548. Nazaruk, J., & Borzym-Kluczyk, M. (2015). The role of triterpenes in the management of diabetes mellitus and its complications. *Phytochemistry reviews: proceedings of the Phytochemical Society of Europe, 14*(4), 675–690.
549. Boshtam, M., et al. (2013). Impacts of fresh lime juice and peel on atherosclerosis progression in an animal model. *ARYA atherosclerosis, 9*(6), 357–362.
550. Kim, J., et al. (2012). Cancer chemopreventive properties of citrus limonoids. In *Emerging trends in dietary components for preventing and combating disease* (Vol. 37-50). Washington, DC: American Chemical Society.
551. Velmurugan, B. K., et al. (2018). Neuroprotective role of phytochemicals. *Molecules, 23*(10), 81.
552. Barreca, D., et al. (2017). Flavanones: Citrus phytochemical with health-promoting properties. *Biofactors, 43*(4), 495–506.
553. Testai, L., & Calderone, V. (2017). Nutraceutical value of citrus flavanones and their implications in cardiovascular disease. *Nutrients, 9*(5), 502.
554. Turner, T., & Burri, B. J. (2013). Potential nutritional benefits of current citrus consumption. *Agriculture, 3*, 170–187.
555. Assini, J. M., Mulvihill, E. E., & Huff, M. W. (2013). Citrus flavonoids and lipid metabolism. *Current Opinion in Lipidology, 24*(1), 34–40.
556. Peterson, J., et al. (2006). Flavanones in orange, tangerines (mandarins), tangors, and tangelos: A compilation and review of the data from the analytical literature. *Journal of Food Composition and Analysis, 19*, S66–S73.
557. Chanet, A., et al. (2012). Citrus flavanones: What is their role in cardiovascular protection? *Journal of Agricultural and Food Chemistry, 60*(36), 8809–8822.
558. Graf, B., Milbury, P., & Blumberg, J. (2005). Flavonols, flavones, flavanones, and human health: Epidemiological evidence. *Journal of Medicinal Food, 8*, 281–290.
559. Brahmachari, G. (2008). Naturally occurring flavanones: An overview. *Natural Product Communications, 3*, 1337–1354.

560. Peterson, J., et al. (2006). Flavanones in grapefruit, lemons, and limes: A compilation and review of the data from the analytical literature. *Journal of Food Composition and Analysis, 19*, S74–S80.
561. Khan, M. K., Zill-E-Huma, & Dangles, O. (2014). A comprehensive review on flavanones, the major citrus polyphenols. *Journal of Food Composition and Analysis, 33*(1), 85–104.
562. Tejada, S., et al. (2018). Potential anti-inflammatory effects of hesperidin from the genus Citrus. *Current Medicinal Chemistry, 25*(37), 4929–4945.
563. Benavente-Garcia, O., & Castillo, J. (2008). Update on uses and properties of citrus flavonoids: New findings in anticancer, cardiovascular, and anti-inflammatory activity. *Journal of Agricultural and Food Chemistry, 56*(15), 6185–6205.
564. Roohbakhsh, A., et al. (2014). Neuropharmacological properties and pharmacokinetics of the citrus flavonoids hesperidin and hesperetin--a mini-review. *Life Sciences, 113*(1-2), 1–6.
565. Parhiz, H., et al. (2015). Antioxidant and anti-inflammatory properties of the citrus flavonoids hesperidin and hesperetin: An updated review of their molecular mechanisms and experimental models. *Phytotherapy Research, 29*(3), 323–331.
566. Roohbakhsh, A., et al. (2015). Molecular mechanisms behind the biological effects of hesperidin and hesperetin for the prevention of cancer and cardiovascular diseases. *Life Sciences, 124*, 64–74.
567. Ahmadi, A., & Shadboorestan, A. (2016). Oxidative stress and cancer; the role of hesperidin, a citrus natural bioflavonoid, as a cancer chemoprotective agent. *Nutrition and Cancer, 68*(1), 29–39.
568. Cho, J. (2006). Antioxidant and neuroprotective effects of hesperidin and its aglycone hesperetin. *Archives of Pharmacal Research, 29*(8), 699–706.
569. Chen, R., et al. (2016). Therapeutic potential of naringin: An overview. *Pharmaceutical Biology, 54*(12), 3203–3210.
570. Alam, M. A., Kauter, K., & Brown, L. (2013). Naringin improves diet-induced cardiovascular dysfunction and obesity in high carbohydrate, high fat diet-fed rats. *Nutrients, 5*(3), 637–650.
571. Alam, M. A., et al. (2014). Effect of citrus flavonoids, naringin and naringenin, on metabolic syndrome and their mechanisms of action. *Advances in Nutrition, 5*(4), 404–417.
572. Viswanatha, G. L., Shylaja, H., & Moolemath, Y. (2017). The beneficial role of Naringin- a citrus bioflavonoid, against oxidative stress-induced neurobehavioral disorders and cognitive dysfunction in rodents: A systematic review and meta-analysis. *Biomedicine & Pharmacotherapy, 94*, 909–929.
573. Alquezar, B., Rodrigo, M., & Zacarías, L. (2008). Carotenoid biosynthesis and its regulation in citrus fruits. *Current Medicinal Chemistry, 2*, 23–35.
574. Eldahshan, O., & Singab, A. N. (2013). Carotenoids. *Journal of Pharmacognosy and Phytochemistry, 2*(1), 225–234.
575. Matsumoto, H., et al. (2007). Quantification of carotenoids in citrus fruit by LC-MS and comparison of patterns of seasonal changes for carotenoids among citrus varieties. *Journal of Agricultural and Food Chemistry, 55*(6), 2356–2368.
576. Ikoma, Y., Matsumoto, H., & Kato, M. (2016). Diversity in the carotenoid profiles and the expression of genes related to carotenoid accumulation among citrus genotypes. *Breeding Science, 66*(1), 139–147.
577. Eggersdorfer, M., & Wyss, A. (2018). Carotenoids in human nutrition and health. *Archives of Biochemistry and Biophysics, 652*, 18–26.
578. Milani, A., et al. (2017). Carotenoids: Biochemistry, pharmacology and treatment. *British Journal of Pharmacology, 174*(11), 1290–1324.
579. Woodside, J. V., et al. (2015). Carotenoids and health in older people. *Maturitas, 80*(1), 63–68.
580. Johnson, E. J. (2002). The role of carotenoids in human health. *Nutrition in Clinical Care, 5*(2), 56–65.
581. Krinsky, N. I. (2001). Carotenoids as antioxidants. *Nutrition, 17*(10), 815–817.

582. Mohammadzadeh Honarvar, N., et al. (2017). Molecular anti-inflammatory mechanisms of retinoids and carotenoids in Alzheimer's disease: A review of current evidence. *Journal of Molecular Neuroscience, 61*(3), 289–304.
583. Rubin, L. P., et al. (2017). Metabolic effects of inflammation on vitamin A and carotenoids in humans and animal models. *Advances in Nutrition (Bethesda, MD), 8*(2), 197–212.
584. Kaulmann, A., & Bohn, T. (2014). Carotenoids, inflammation, and oxidative stress--implications of cellular signaling pathways and relation to chronic disease prevention. *Nutrition Research, 34*(11), 907–929.
585. Riccioni, G. (2009). Carotenoids and cardiovascular disease. *Current Atherosclerosis Reports, 11*(6), 434–439.
586. Giordano, P., et al. (2012). Carotenoids and cardiovascular risk. *Current Pharmaceutical Design, 18*(34), 5577–5589.
587. Ciccone, M. M., et al. (2013). Dietary intake of carotenoids and their antioxidant and anti-inflammatory effects in cardiovascular care. *Mediators of Inflammation, 2013*, 782137–782137.
588. Gammone, M. A., Riccioni, G., & D'Orazio, N. (2015). Carotenoids: Potential allies of cardiovascular health? *Food & Nutrition Research, 59*, 26762–26762.
589. Soares Nda, C., et al. (2015). Anticancer properties of carotenoids in prostate cancer. A review. *Histology and Histopathology, 30*(10), 1143–1154.
590. Tanaka, T., Shnimizu, M., & Moriwaki, H. (2012). Cancer chemoprevention by carotenoids. *Molecules, 17*(3), 3202–3242.
591. Pal, D., Banerjee, S., & Ghosh, A. K. (2012). Dietary-induced cancer prevention: An expanding research arena of emerging diet related to healthcare system. *Journal of Advanced Pharmaceutical Technology & Research, 3*(1), 16–24.
592. Nishino, H., et al. (2009). Cancer prevention by carotenoids. *Archives of Biochemistry and Biophysics, 483*(2), 165–168.
593. Griffiths, K., et al. (2016). Food antioxidants and their anti-inflammatory properties: A potential role in cardiovascular diseases and cancer prevention. *Diseases (Basel, Switzerland), 4*(3), 28.
594. Sluijs, I., et al. (2015). Dietary intake of carotenoids and risk of type 2 diabetes. *Nutrition, Metabolism and Cardiovascular Diseases, 25*(4), 376–381.
595. Roohbakhsh, A., Karimi, G., & Iranshahi, M. (2017). Carotenoids in the treatment of diabetes mellitus and its complications: A mechanistic review. *Biomedicine & Pharmacotherapy, 91*, 31–42.
596. Wu, J., et al. (2015). Intakes of lutein, zeaxanthin, and other carotenoids and age-related macular degeneration during 2 decades of prospective follow-up. *JAMA Ophthalmology, 133*(12), 1415–1424.
597. Rasmussen, H. M., & Johnson, E. J. (2013). Nutrients for the aging eye. *Clinical Interventions in Aging, 8*, 741–748.
598. Cirmi, S., et al. (2016). Chemopreventive agents and inhibitors of cancer hallmarks: May citrus offer new perspectives? *Nutrients, 8*(11), 698.
599. Cirmi, S., et al. (2017). Anticancer potential of citrus juices and their extracts: A systematic review of both preclinical and clinical studies. *Frontiers in Pharmacology, 8*, 420–420.
600. Wang, A., et al. (2015). Citrus fruit intake substantially reduces the risk of esophageal cancer: A meta-analysis of epidemiologic studies. *Medicine, 94*(39), e1390–e1390.
601. Song, J.-K., & Bae, J.-M. (2013). Citrus fruit intake and breast cancer risk: A quantitative systematic review. *Journal of Breast Cancer, 16*(1), 72–76.
602. Bae, J. M., Lee, E. J., & Guyatt, G. (2009). Citrus fruit intake and pancreatic cancer risk: A quantitative systematic review. *Pancreas, 38*(2), 168–174.
603. Bae, J. M., Lee, E. J., & Guyatt, G. (2008). Citrus fruit intake and stomach cancer risk: A quantitative systematic review. *Gastric Cancer, 11*(1), 23–32.

604. Bae, J. M., Lee, E. J., & Guyatt, G. (2008). Citrus fruits intake and prostate cancer risk: A quantitative systematic review. *Journal of Preventive Medicine and Public Health, 41*(3), 159–164.
605. Wabner, C. L., & Pak, C. Y. (1993). Effect of orange juice consumption on urinary stone risk factors. *The Journal of Urology, 149*(6), 1405–1408.
606. Gul, Z., & Monga, M. (2014). Medical and dietary therapy for kidney stone prevention. *Korean Journal of Urology, 55*(12), 775–779.
607. Prezioso, D., et al. (2015). Dietary treatment of urinary risk factors for renal stone formation. A review of CLU Working Group. *Archivio Italiano di Urologia, Andrologia, 87*(2), 105–120.
608. De, S. K., Liu, X., & Monga, M. (2014). Changing trends in the American diet and the rising prevalence of kidney stones. *Urology, 84*(5), 1030–1033.
609. Odvina, C. V. (2006). Comparative value of orange juice versus lemonade in reducing stone-forming risk. *Clinical Journal of the American Society of Nephrology, 1*(6), 1269–1274
610. Granger, M., & Eck, P. (2018). Dietary vitamin C in human health. *Advances in Food and Nutrition Research, 83*, 281–310.
611. Chambial, S., et al. (2013). Vitamin C in disease prevention and cure: An overview. *Indian journal of clinical biochemistry: IJCB, 28*(4), 314–328.
612. Carr, A. C., & Maggini, S. (2017). Vitamin C and Immune Function. *Nutrients, 9*(11), 1211.
613. Wintergerst, E. S., Maggini, S., & Hornig, D. H. (2006). Immune-enhancing role of vitamin C and zinc and effect on clinical conditions. *Annals of Nutrition & Metabolism, 50*(2), 85–94.
614. Sorice, A., et al. (2014). Ascorbic acid: Its role in immune system and chronic inflammation diseases. *Mini Reviews in Medicinal Chemistry, 14*(5), 444–452.
615. Ellulu, M. S., et al. (2015). Effect of vitamin C on inflammation and metabolic markers in hypertensive and/or diabetic obese adults: A randomized controlled trial. *Drug Design, Development and Therapy, 9*, 3405–3412.
616. Moser, M. A., & Chun, O. K. (2016). Vitamin C and heart health: A review based on findings from epidemiologic studies. *International Journal of Molecular Sciences, 17*(8), 1328.
617. Sharma, P. (2013). Vitamin C rich fruits can prevent heart disease. *Indian Journal of Clinical Biochemistry, 28*(3), 213–214.
618. Carr, A. C., & Cook, J. (2018). Intravenous vitamin C for cancer therapy - identifying the current gaps in our knowledge. *Frontiers in Physiology, 9*, 1182.
619. Vissers, M. C. M., & Das, A. B. (2018). Potential mechanisms of action for vitamin C in cancer: Reviewing the evidence. *Frontiers in Physiology, 9*, 809.
620. van der Reest, J., & Gottlieb, E. (2016). Anti-cancer effects of vitamin C revisited. *Cell Research, 26*(3), 269–270.
621. Aghajanian, P., et al. (2015). The roles and mechanisms of actions of vitamin C in bone: New developments. *Journal of Bone and Mineral Research, 30*(11), 1945–1955.
622. DePhillipo, N. N., et al. (2018). Efficacy of vitamin C supplementation on collagen synthesis and oxidative stress after musculoskeletal injuries: A systematic review. *Orthopaedic Journal of Sports Medicine, 6*(10), 2325967118804544.
623. Saokar Telang, P. (2013). Vitamin C in dermatology. *Indian Dermatology Online Journal, 4*, 143–146.
624. Schagen, S. K., et al. (2012). Discovering the link between nutrition and skin aging. *Dermato-Endocrinology, 4*(3), 298–307.
625. Shibuya, S., et al. (2014). Collagen peptide and vitamin C additively attenuate age-related skin atrophy in Sod1-deficient mice. *Bioscience, Biotechnology, and Biochemistry, 78*(7), 1212–1220.
626. Crisan, D., et al. (2015). The role of vitamin C in pushing back the boundaries of skin aging: An ultrasonographic approach. *Clinical, Cosmetic and Investigational Dermatology, 8*, 463–470.
627. Ertugrul, D. T., et al. (2013). Serum holotranscobalamine, vitamin B12, folic acid and homocysteine levels in alopecia areata patients. *Cutaneous and Ocular Toxicology, 32*(1), 1–3.

628. Ballot, D., et al. (1987). The effects of fruit juices and fruits on the absorption of iron from a rice meal. *British Journal of Nutrition, 57*(3), 331–343.
629. Peneau, S., et al. (2008). Relationship between iron status and dietary fruit and vegetables based on their vitamin C and fiber content. *The American Journal of Clinical Nutrition, 87*(5), 1298–1305.
630. Kean, R. J., et al. (2015). Chronic consumption of flavanone-rich orange juice is associated with cognitive benefits: An 8-wk, randomized, double-blind, placebo-controlled trial in healthy older adults. *The American Journal of Clinical Nutrition, 101*(3), 506–514.
631. Alharbi, M. H., et al. (2016). Flavonoid-rich orange juice is associated with acute improvements in cognitive function in healthy middle-aged males. *European Journal of Nutrition, 55*(6), 2021–2029.
632. Cirmi, S., et al. (2016). Neurodegenerative diseases: Might citrus flavonoids play a protective role? *Molecules, 21*(10), 1312.
633. Elumalai, P., & Lakshmi, S. (2016). Role of quercetin benefits in neurodegeneration. *Advances in Neurobiology, 12*, 229–245.
634. Ani, P. N., & Abel, H. C. (2018). Nutrient, phytochemical, and antinutrient composition of Citrus maxima fruit juice and peel extract. *Food Science & Nutrition, 6*(3), 653–658.
635. Park, J. H., Lee, M., & Park, E. (2014). Antioxidant activity of orange flesh and peel extracted with various solvents. *Preventive Nutrition and Food Science, 19*(4), 291–298.
636. Yoshizaki, N., et al. (2014). Orange peel extract, containing high levels of polymethoxyflavonoid, suppressed UVB-induced COX-2 expression and PGE2 production in HaCaT cells through PPAR-gamma activation. *Experimental Dermatology, 23*(Suppl 1), 18–22.
637. Hakim, I. A., Harris, R. B., & Ritenbaugh, C. (2000). Citrus peel use is associated with reduced risk of squamous cell carcinoma of the skin. *Nutrition and Cancer, 37*(2), 161–168.
638. Puglia, C., et al. (2014). Protective effect of red orange extract supplementation against UV-induced skin damages: Photoaging and solar lentigines. *Journal of Cosmetic Dermatology, 13*(2), 151–157.
639. Hussain, K. A., et al. (2015). Antimicrobial effects of citrus sinensis peel extracts against periodontopathic bacteria: An in vitro study. *Roczniki Państwowego Zakładu Higieny, 66*(2), 173–178.
640. Dosoky, N. S., & Setzer, W. N. (2018). Biological activities and safety of Citrus spp. essential oils. *International Journal of Molecular Sciences, 19*(7), 1966.
641. Grobler, S. R. (1991). The effect of a high consumption of citrus fruit and a mixture of other fruits on dental caries in man. *Clinical Preventive Dentistry, 13*(4), 13–17.
642. Sovik, J. B., et al. (2015). Sour sweets and acidic beverage consumption are risk indicators for dental erosion. *Caries Research, 49*(3), 243–250.
643. Dugrand-Judek, A., et al. (2015). The distribution of coumarins and furanocoumarins in Citrus species closely Matches Citrus phylogeny and reflects the organization of biosynthetic pathways. *PLoS One, 10*(11), e0142757.
644. Sayre, R. M., & Dowdy, J. C. (2008). The increase in melanoma: Are dietary furocoumarins responsible? *Medical Hypotheses, 70*(4), 855–859.
645. Wu, S., et al. (2015). citrus consumption and risk of cutaneous malignant melanoma. *Journal of Clinical Oncology, 33*(23), 2500–2508.
646. Qin, K., et al. (2013). Characterization of chemical composition of Pericarpium citri reticulatae volatile oil by comprehensive two-dimensional gas chromatography with high-resolution time-of-flight mass spectrometry. *Evidence-based Complementary and Alternative Medicine, 2013*, 237541.
647. Apraj, V. D., & Pandita, N. S. (2016). Evaluation of skin anti-aging potential of citrus reticulata blanco peel. *Pharmacognosy Research, 8*(3), 160–168.
648. Ho, S. C., & Kuo, C. T. (2014). Hesperidin, nobiletin, and tangeretin are collectively responsible for the anti-neuroinflammatory capacity of tangerine peel (Citri reticulatae pericarpium). *Food and Chemical Toxicology, 71*, 176–182.

649. Unno, K., et al. (2011). Beta-cryptoxanthin, plentiful in Japanese mandarin orange, prevents age-related cognitive dysfunction and oxidative damage in senescence-accelerated mouse brain. *Biological & Pharmaceutical Bulletin, 34*(3), 311–317.
650. Mora, J. R., Iwata, M., & von Andrian, U. H. (2008). Vitamin effects on the immune system: Vitamins A and D take centre stage. *Nature Reviews. Immunology, 8*(9), 685–698.
651. Chouhan, S., Sharma, K., & Guleria, S. (2017). Antimicrobial activity of some essential oils-present status and future perspectives. *Medicines (Basel), 4*(3), 58.
652. Orchard, A., & van Vuuren, S. (2017). Commercial essential oils as potential antimicrobials to treat skin diseases. *Evidence-Based Complementary and Alternative Medicine, 2017*, 4517971.
653. Swamy, M. K., Akhtar, M. S., & Sinniah, U. R. (2016). Antimicrobial properties of plant essential oils against human pathogens and their mode of action: An updated review. *Evidence Based Complementary and Alternative Medicine, 2016*, 21.
654. Yap, P. S., et al. (2014). Essential oils, a new horizon in combating bacterial antibiotic resistance. *The Open Microbiology Journal, 8*, 6–11.
655. Adukwu, E. C., et al. (2016). Antimicrobial activity, cytotoxicity and chemical analysis of lemongrass essential oil (Cymbopogon flexuosus) and pure citral. *Applied Microbiology and Biotechnology, 100*(22), 9619–9627.
656. Chang, Y.-Y., Lin, C.-L., & Chang, L.-Y. (2017). The effects of aromatherapy massage on sleep quality of nurses on monthly rotating night shifts. *Evidence-Based Complementary and Alternative Medicine, 2017*, 8.
657. Choi, S. Y., et al. (2014). Effects of inhalation of essential oil of Citrus aurantium L. var. amara on menopausal symptoms, stress, and estrogen in postmenopausal women: A randomized controlled trial. *Evidence-Based Complementary and Alternative Medicine, 2014*, 7.
658. Sanchez-Vidana, D. I., et al. (2017). The effectiveness of aromatherapy for depressive symptoms: A systematic review. *Evidence-based Complementary and Alternative Medicine, 2017*, 5869315.
659. Brescoll, J., & Daveluy, S. (2015). A review of vitamin B12 in dermatology. *American Journal of Clinical Dermatology, 16*(1), 27–33.
660. Wang, E. T. (2008). Anaphylaxis caused by tangerine seeds but not tangerine fruit. *Annals of Allergy, Asthma & Immunology, 101*(5), 553–554.
661. Fujioka, K., et al. (2006). The effects of grapefruit on weight and insulin resistance: Relationship to the metabolic syndrome. *Journal of Medicinal Food, 9*(1), 49–54.
662. Stelmach-Mardas, M., et al. (2016). Link between food energy density and body weight changes in obese adults. *Nutrients, 8*(4), 229.
663. Kristensen, M., & Jensen, M. G. (2011). Dietary fibres in the regulation of appetite and food intake. Importance of viscosity. *Appetite, 56*(1), 65–70.
664. Champagne, C. M., et al. (2011). Dietary intakes associated with successful weight loss and maintenance during the Weight Loss Maintenance trial. *Journal of the American Dietetic Association, 111*(12), 1826–1835.
665. Onakpoya, I., et al. (2017). The effect of grapefruits (Citrus paradisi) on body weight and cardiovascular risk factors: A systematic review and meta-analysis of randomized clinical trials. *Critical Reviews in Food Science and Nutrition, 57*(3), 602–612.
666. Dow, C. A., et al. (2012). The effects of daily consumption of grapefruit on body weight, lipids, and blood pressure in healthy, overweight adults. *Metabolism, 61*(7), 1026–1035.
667. Silver, H. J., Dietrich, M. S., & Niswender, K. D. (2011). Effects of grapefruit, grapefruit juice and water preloads on energy balance, weight loss, body composition, and cardiometabolic risk in free-living obese adults. *Nutrition & Metabolism (London), 8*(1), 8.
668. Boeing, H., et al. (2012). Critical review: Vegetables and fruit in the prevention of chronic diseases. *European Journal of Nutrition, 51*(6), 637–663.
669. Wilcox, G. (2005). Insulin and insulin resistance. *Clinical Biochemist Reviews, 26*(2), 19–39.
670. Laville, M., & Nazare, J. A. (2009). Diabetes, insulin resistance and sugars. *Obesity Reviews, 10*(Suppl 1), 24–33.

671. Li, S., et al. (2015). Fruit intake decreases risk of incident type 2 diabetes: An updated meta-analysis. *Endocrine, 48*(2), 454–460.
672. Li, M., et al. (2014). Fruit and vegetable intake and risk of type 2 diabetes mellitus: Meta-analysis of prospective cohort studies. *BMJ Open, 4*(11), e005497.
673. Gorinstein, S., et al. (2006). Red grapefruit positively influences serum triglyceride level in patients suffering from coronary atherosclerosis: Studies in vitro and in humans. *Journal of Agricultural and Food Chemistry, 54*(5), 1887–1892.
674. Cassidy, A., et al. (2012). Dietary flavonoids and risk of stroke in women. *Stroke, 43*(4), 946–951.
675. Guo, S., & Dipietro, L. A. (2010). Factors affecting wound healing. *Journal of Dental Research, 89*(3), 219–229.
676. Han, Y. Y., et al. (2015). Diet and asthma: An update. *Current Opinion in Allergy and Clinical Immunology, 15*(4), 369–374.
677. Greenblatt, D. J., et al. (2012). Mechanism-based inhibition of human cytochrome P450-3A activity by grapefruit hybrids having low furanocoumarin content. *Xenobiotica, 42*(12), 1163–1169.
678. Seden, K., et al. (2010). Grapefruit-drug interactions. *Drugs, 70*(18), 2373–2407.
679. Bailey, D. G., Dresser, G., & Arnold, J. M. O. (2013). Grapefruit–medication interactions: Forbidden fruit or avoidable consequences? *Cmaj, 185*(4), 309–316.
680. SM, H. (2017). Grapefruit juice and some drugs don't mix. 2017 July 18, cited 2018, Retrieved from https://www.fda.gov/ForConsumers/ConsumerUpdates/ucm292276.htm
681. Lissera, R. G., Luna Maldonado, E. R., & Battellino, L. J. (1998). In vitro erosive capacity of some fruit juices and soft or low alcoholic strength beverages on human teeth. *Acta Odontológica Latinoamericana, 11*(1), 55–71.
682. Jequier, E., & Constant, F. (2010). Water as an essential nutrient: The physiological basis of hydration. *European Journal of Clinical Nutrition, 64*(2), 115–123.
683. Murray, B. (2007). Hydration and physical performance. *Journal of the American College of Nutrition, 26*(5 Suppl), 542s–548s.
684. Keller, U., et al. (2003). Effects of changes in hydration on protein, glucose and lipid metabolism in man: Impact on health. *European Journal of Clinical Nutrition, 57*(Suppl 2), S69–S74.
685. Guelinckx, I., et al. (2016). Contribution of water from food and fluids to total water intake: Analysis of a French and UK population surveys. *Nutrients, 8*(10), 630.
686. Montenegro-Bethancourt, G., Johner, S. A., & Remer, T. (2013). Contribution of fruit and vegetable intake to hydration status in schoolchildren. *The American Journal of Clinical Nutrition, 98*(4), 1103–1112.
687. Mauskop, A., & Varughese, J. (2012). Why all migraine patients should be treated with magnesium. *Journal of Neural Transmission (Vienna), 119*(5), 575–579.
688. Mukherjee, P. K., et al. (2013). Phytochemical and therapeutic potential of cucumber. *Fitoterapia, 84*, 227–236.
689. Arnaud, M. J. (2003). Mild dehydration: A risk factor of constipation? *European Journal of Clinical Nutrition, 57*(Suppl 2), S88–S95.
690. Popkin, B. M., D'Anci, K. E., & Rosenberg, I. H. (2010). Water, hydration, and health. *Nutrition Reviews, 68*(8), 439–458.
691. Xu, L., et al. (2014). Clinical benefits after soluble dietary fiber supplementation: A randomized clinical trial in adults with slow-transit constipation. *Zhonghua Yi Xue Za Zhi, 94*(48), 3813–3816.
692. Ravichandran, R., et al. (2013). Hierarchical mesoporous silica nanofibers as multifunctional scaffolds for bone tissue regeneration. *Journal of Biomaterials Science. Polymer Edition, 24*(17), 1988–2005.
693. Wright, C. I., et al. (2007). Herbal medicines as diuretics: A review of the scientific evidence. *Journal of Ethnopharmacology, 114*(1), 1–31.
694. Roman-Ramos, R., Flores-Saenz, J. L., & Alarcon-Aguilar, F. J. (1995). Anti-hyperglycemic effect of some edible plants. *Journal of Ethnopharmacology, 48*(1), 25–32.

695. Dixit, Y., & Kar, A. (2010). Protective role of three vegetable peels in alloxan induced diabetes mellitus in male mice. *Plant Foods for Human Nutrition, 65*(3), 284–289.
696. Heidari, H., et al. (2016). Protective mechanisms of Cucumis sativus in diabetes-related models of oxidative stress and carbonyl stress. *BioImpacts, 6*(1), 33–39.
697. Kumar, D., et al. (2010). Free radical scavenging and analgesic activities of Cucumis sativus L. fruit extract. *Journal of Young Pharmacists, 2*(4), 365–368.
698. Ji, L., et al. (2015). In Vivo antioxidant properties of lotus root and cucumber: A pilot comparative study in aged subjects. *The Journal of Nutrition, Health & Aging, 19*(7), 765–770.
699. Rios, J. L., et al. (2012). Cucurbitacins as inducers of cell death and a rich source of potential anticancer compounds. *Current Pharmaceutical Design, 18*(12), 1663–1676.
700. Maher, P. (2009). Modulation of multiple pathways involved in the maintenance of neuronal function during aging by fisetin. *Genes & Nutrition, 4*(4), 297–307.
701. Ammar, S., et al. (2015). Assessment of the distribution of phenolic compounds and contribution to the antioxidant activity in Tunisian fig leaves, fruits, skins and pulps using mass spectrometry based analysis. *Food & Function, 6*(12), 3663–3677.
702. Khare, C. P. (2004). *Indian herbal remedies* (pp. 213–227). Berlin: Springer.
703. Joseph, B., & Raj, J. (2010). Pharmacognostic and phytochemical properties of Ficus carica Linn –An overview. *International Journal of Pharmtech Research, 3*, 8–12.
704. Lim, T. K. (2012). *Edible medicinal and non-medicinal plants* (Vol. 1). Dordrecht: Springer.
705. Perez, C., Canal, J. R., & Torres, M. D. (2003). Experimental diabetes treated with ficus carica extract: Effect on oxidative stress parameters. *Acta Diabetologica, 40*(1), 3–8.
706. Irudayaraj, S. S., et al. (2016). Antioxidant, antilipidemic and antidiabetic effects of ficusin with their effects on GLUT4 translocation and PPARgamma expression in type 2 diabetic rats. *Chemico-Biological Interactions, 256*, 85–93.
707. Badgujar, S. B., et al. (2014). Traditional uses, phytochemistry and pharmacology of Ficus carica: A review. *Pharmaceutical Biology, 52*(11), 1487–1503.
708. Mawa, S., Husain, K., & Jantan, I. (2013). Ficus carica L. (Moraceae): Phytochemistry, traditional uses and biological activities. *Evidence-based Complementary and Alternative Medicine, 2013*, 8.
709. Vinson, J. A., et al. (2005). Dried fruits: Excellent in vitro and in vivo antioxidants. *Journal of the American College of Nutrition, 24*(1), 44–50.
710. Slatnar, A., et al. (2011). Effect of drying of figs (Ficus carica L.) on the contents of sugars, organic acids, and phenolic compounds. *Journal of Agricultural and Food Chemistry, 59*(21), 11696–11702.
711. Russo, F., et al. (2014). Phenolic compounds in fresh and dried figs from Cilento (Italy), by considering breba crop and full crop, in comparison to Turkish and Greek dried figs. *Journal of Food Science, 79*(7), C1278–C1284.
712. Vijaya Kumar Reddy, C., Dande, S., & Manchala, R. (2010). Antioxidant activity of fresh and dry fruits commonly consumed in India. *Food Research International, 43*, 285–288.
713. Jing, L., et al. (2015). Tirucallane-type triterpenoids from the fruit of Ficus carica and their cytotoxic activity. *Chemical and Pharmaceutical Bulletin, 63*(3), 237–243.
714. Conforti, F., et al. (2012). Evaluation of phototoxic potential of aerial components of the fig tree against human melanoma. *Cell Proliferation, 45*(3), 279–285.
715. Khan, H., Akhtar, N., & Ali, A. (2014). Effects of cream containing Ficus carica L. fruit extract on skin parameters: In vivo evaluation. *Indian Journal of Pharmaceutical Sciences, 76*(6), 560–564.
716. Yang, X., et al. (2015). The effects of Ficus carica polysaccharide on immune response and expression of some immune-related genes in grass carp, Ctenopharyngodon idella. *Fish & Shellfish Immunology, 42*(1), 132–137.
717. Jeong, M.-R., Kim, H.-Y., & Cha, J.-D. (2009). Antimicrobial activity of methanol extract from ficus carica leaves against oral bacteria. *Journal of Bacteriology and Virology, 39*, 97–102.

718. Keskin, D., et al. (2012). Phytochemical analysis and antimicrobial activity of different extracts of fig leaves (Ficus carica L.) from West Anatolia against some pathogenic microorganisms. *Journal of Pure and Applied Microbiology, 6*, 1105–1110.
719. Al Askari, G., et al. (2013). In vitro antimicrobial activity of aqueous and ethanolic extracts of leaves of Ficus carica collected from five different regions of Morocco. *Journal of Materials and Environmental Science, 4*, 33–38.
720. Serraclara, A., et al. (1998). Hypoglycemic action of an oral fig-leaf decoction in type-I diabetic patients. *Diabetes Research and Clinical Practice, 39*(1), 19–22.
721. Hemmer, W., et al. (2010). Identification of Bet v 1-related allergens in fig and other Moraceae fruits. *Clinical and Experimental Allergy, 40*(4), 679–687.
722. Pezzuto, J. M. (2008). Grapes and human health: A perspective. *Journal of Agricultural and Food Chemistry, 56*(16), 6777–6784.
723. Georgiev, V., Ananga, A., & Tsolova, V. (2014). Recent advances and uses of grape flavonoids as nutraceuticals. *Nutrients, 6*(1), 391–415.
724. Pisoschi, A. M., & Pop, A. (2015). The role of antioxidants in the chemistry of oxidative stress: A review. *European Journal of Medicinal Chemistry, 97*, 55–74.
725. Cantos, E., Espin, J. C., & Tomas-Barberan, F. A. (2002). Varietal differences among the polyphenol profiles of seven table grape cultivars studied by LC-DAD-MS-MS. *Journal of Agricultural and Food Chemistry, 50*(20), 5691–5696.
726. Singh, C. K., Liu, X., & Ahmad, N. (2015). Resveratrol, in its natural combination in whole grape, for health promotion and disease management. *Annals of the New York Academy of Sciences, 1348*(1), 150–160.
727. Di Lorenzo, C., et al. (2016). Evaluation of the anti-inflammatory activity of raisins (Vitis vinifera L.) in human gastric epithelial cells: A comparative study. *International Journal of Molecular Sciences, 17*(7), 1156.
728. Barona, J., et al. (2012). Grape consumption increases anti-inflammatory markers and upregulates peripheral nitric oxide synthase in the absence of dyslipidemias in men with metabolic syndrome. *Nutrients, 4*(12), 1945–1957.
729. Greenspan, P., et al. (2005). Antiinflammatory properties of the muscadine grape (Vitis rotundifolia). *Journal of Agricultural and Food Chemistry, 53*(22), 8481–8484.
730. Berman, A. Y., et al. (2017). The therapeutic potential of resveratrol: A review of clinical trials. *NPJ Precision Oncology, 1*, 35.
731. Smoliga, J. M., Baur, J. A., & Hausenblas, H. A. (2011). Resveratrol and health--a comprehensive review of human clinical trials. *Molecular Nutrition & Food Research, 55*(8), 1129–1141.
732. Zhou, K., & Raffoul, J. J. (2012). Potential anticancer properties of grape antioxidants. *Journal of Oncology, 2012*, 803294.
733. Pandey, K. B., & Rizvi, S. I. (2014). Role of red grape polyphenols as antidiabetic agents. *Integrative Medicine Research, 3*(3), 119–125.
734. Dohadwala, M. M., & Vita, J. A. (2009). Grapes and cardiovascular disease. *The Journal of Nutrition, 139*(9), 1788s–1793s.
735. McCubrey, J. A., et al. (2017). Effects of resveratrol, curcumin, berberine and other nutraceuticals on aging, cancer development, cancer stem cells and microRNAs. *Aging (Albany NY), 9*(6), 1477–1536.
736. Wahl, D., et al. (2018). Future directions of resveratrol research. *Nutrition and Healthy Aging, 4*(4), 287–290.
737. Holcombe, R. F., et al. (2015). Effects of a grape-supplemented diet on proliferation and Wnt signaling in the colonic mucosa are greatest for those over age 50 and with high arginine consumption. *Nutrition Journal, 14*, 62.
738. Zhang, C., et al. (2017). Grape seed proanthocyanidins induce mitochondrial pathway-mediated apoptosis in human colorectal carcinoma cells. *Oncology Letters, 14*(5), 5853–5860.

739. Valenzuela, M., et al. (2018). Autumn royal and ribier grape juice extracts reduced viability and metastatic potential of colon cancer cells. *Evidence-based Complementary and Alternative Medicine, 2018*, 2517080.
740. Luan, Y. Y., et al. (2015). Effect of grape seed proanthocyanidins on tumor vasculogenic mimicry in human triple-negative breast cancer cells. *Asian Pacific Journal of Cancer Prevention, 16*(2), 531–535.
741. Dinicola, S., et al. (2014). Grape seed extract suppresses MDA-MB231 breast cancer cell migration and invasion. *European Journal of Nutrition, 53*(2), 421–431.
742. Sun, T., et al. (2012). Antitumor and antimetastatic activities of grape skin polyphenols in a murine model of breast cancer. *Food and Chemical Toxicology, 50*(10), 3462–3467.
743. Signorelli, P., et al. (2015). Natural grape extracts regulate colon cancer cells malignancy. *Nutrition and Cancer, 67*(3), 494–503.
744. Burton, L. J., et al. (2015). Muscadine grape skin extract can antagonize Snail-cathepsin L-mediated invasion, migration and osteoclastogenesis in prostate and breast cancer cells. *Carcinogenesis, 36*(9), 1019–1027.
745. Gescher, A., Steward, W. P., & Brown, K. (2013). Resveratrol in the management of human cancer: How strong is the clinical evidence? *Annals of the New York Academy of Sciences, 1290*, 12–20.
746. Dybkowska, E., et al. (2018). The occurrence of resveratrol in foodstuffs and its potential for supporting cancer prevention and treatment. A review. *Roczniki Państwowego Zakładu Higieny, 69*(1), 5–14.
747. Wang, L., et al. (2013). Resveratrols in grape berry skins and leaves in vitis germplasm. *PLoS One, 8*(4), e61642.
748. Srivastava, S., et al. (2016). Quercetin, a natural flavonoid interacts with DNA, arrests cell cycle and causes tumor regression by activating mitochondrial pathway of apoptosis. *Scientific Reports, 6*, 24049.
749. Singh, C. K., et al. (2016). Combination chemoprevention with grape antioxidants. *Molecular Nutrition & Food Research, 60*(6), 1406–1415.
750. Vallianou, N. G., Evangelopoulos, A., & Kazazis, C. (2013). Resveratrol and diabetes. *The Review of Diabetic Studies, 10*(4), 236–242.
751. Urquiaga, I., et al. (2015). Wine grape pomace flour improves blood pressure, fasting glucose and protein damage in humans: A randomized controlled trial. *Biological Research, 48*, 49.
752. Sin, T. K., Yung, B. Y., & Siu, P. M. (2015). Modulation of SIRT1-Foxo1 signaling axis by resveratrol: Implications in skeletal muscle aging and insulin resistance. *Cellular Physiology and Biochemistry, 35*(2), 541–552.
753. Akaberi, M., & Hosseinzadeh, H. (2016). Grapes (Vitis vinifera) as a potential candidate for the therapy of the metabolic syndrome. *Phytotherapy Research, 30*(4), 540–556.
754. Yin, W., et al. (2015). Anti-inflammatory effects of grape seed procyanidin B2 on a diabetic pancreas. *Food & Function, 6*(9), 3065–3071.
755. Zunino, S. (2009). Type 2 diabetes and glycemic response to grapes or grape products. *The Journal of Nutrition, 139*(9), 1794s–1800s.
756. Riccioni, G., et al. (2015). Resveratrol and anti-atherogenic effects. *International Journal of Food Sciences and Nutrition, 66*(6), 603–610.
757. Murillo, A. G., & Fernandez, M. L. (2017). The relevance of dietary polyphenols in cardiovascular protection. *Current Pharmaceutical Design, 23*(17), 2444–2452.
758. Rahbar, A. R., Mahmoudabadi, M. M., & Islam, M. S. (2015). Comparative effects of red and white grapes on oxidative markers and lipidemic parameters in adult hypercholesterolemic humans. *Food & Function, 6*(6), 1992–1998.
759. Xia, E. Q., et al. (2010). Biological activities of polyphenols from grapes. *International Journal of Molecular Sciences, 11*(2), 622–646.
760. Iacopini, P., et al. (2008). Catechin, epicatechin, quercetin, rutin and resveratrol in red grape: Content, in vitro antioxidant activity and interactions. *Journal of Food Composition and Analysis, 21*(8), 589–598.

761. Lv, L., et al. (2018). Rutin inhibits coronary heart disease through ERK1/2 and Akt signaling in a porcine model. *Experimental and Therapeutic Medicine, 15*(1), 506–512.
762. Flaumenhaft, R. (2013). Protein disulfide isomerase as an antithrombotic target. *Trends in Cardiovascular Medicine, 23*(7), 264–268.
763. Yang, Q., et al. (2011). Sodium and potassium intake and mortality among US adults: Prospective data from the Third National Health and Nutrition Examination Survey. *Archives of Internal Medicine, 171*(13), 1183–1191.
764. Pons, Z., et al. (2016). Acute administration of single oral dose of grape seed polyphenols restores blood pressure in a rat model of metabolic syndrome: Role of nitric oxide and prostacyclin. *European Journal of Nutrition, 55*(2), 749–758.
765. Abu-Amero, K. K., Kondkar, A. A., & Chalam, K. V. (2016). Resveratrol and ophthalmic diseases. *Nutrients, 8*(4), 200.
766. Patel, A. K., et al. (2016). Protective effects of a grape-supplemented diet in a mouse model of retinal degeneration. *Nutrition, 32*(3), 384–390.
767. Chan, C. M., et al. (2015). Protective effects of resveratrol against UVA-induced damage in ARPE19 cells. *International Journal of Molecular Sciences, 16*(3), 5789–5802.
768. Porquet, D., et al. (2013). Dietary resveratrol prevents Alzheimer's markers and increases life span in SAMP8. *Age (Dordrecht, Netherlands), 35*(5), 1851–1865.
769. Haskell-Ramsay, C. F., et al. (2017). Cognitive and mood improvements following acute supplementation with purple grape juice in healthy young adults. *European Journal of Nutrition, 56*(8), 2621–2631.
770. Kodali, M., et al. (2015). Resveratrol prevents age-related memory and mood dysfunction with increased hippocampal neurogenesis and microvasculature, and reduced glial activation. *Scientific Reports, 5*, 8075.
771. Calapai, G., et al. (2017). A randomized, double-blinded, clinical trial on effects of a vitis vinifera extract on cognitive function in healthy older adults. *Frontiers in Pharmacology, 8*, 776.
772. Krikorian, R., et al. (2010). Concord grape juice supplementation improves memory function in older adults with mild cognitive impairment. *British Journal of Nutrition, 103*(5), 730–734.
773. Yadav, D., et al. (2015). Antimicrobial properties of black grape (Vitis vinifera L.) peel extracts against antibiotic-resistant pathogenic bacteria and toxin producing molds. *Indian Journal of Pharmacology, 47*(6), 663–667.
774. Correia, A. C., & Jordao, A. M. (2015). Antioxidant capacity, radical scavenger activity, lipid oxidation protection analysis and antimicrobial activity of red grape extracts from different varieties cultivated in Portugal. *Natural Product Research, 29*(5), 438–440.
775. Munoz-Gonzalez, I., et al. (2014). Red wine and oenological extracts display antimicrobial effects in an oral bacteria biofilm model. *Journal of Agricultural and Food Chemistry, 62*(20), 4731–4737.
776. Berardi, V., et al. (2009). Resveratrol exhibits a strong cytotoxic activity in cultured cells and has an antiviral action against polyomavirus: Potential clinical use. *Journal of Experimental & Clinical Cancer Research, 28*, 96.
777. Ma, D. S. L., et al. (2018). Resveratrol-potential antibacterial agent against foodborne pathogens. *Frontiers in Pharmacology, 9*, 102.
778. Bekhit Ael, D., et al. (2011). Antioxidant activities, sensory and anti-influenza activity of grape skin tea infusion. *Food Chemistry, 129*(3), 837–845.
779. Jayaprakasha, G. K., Selvi, T., & Sakariah, K. K. (2003). Antibacterial and antioxidant activities of grape (Vitis vinifera) seed extracts. *Food Research International, 36*(2), 117–122.
780. Hemmati, A. A., et al. (2015). The topical effect of grape seed extract 2% cream on surgery wound healing. *Global Journal of Health Science, 7*, 52–59.
781. Richardson, D. P., Ansell, J., & Drummond, L. N. (2018). The nutritional and health attributes of kiwifruit: A review. *European Journal of Nutrition, 57*(8), 2659–2676.

782. Xia, L., & Ng, T. B. (2004). Actinchinin, a novel antifungal protein from the gold kiwi fruit. *Peptides, 25*(7), 1093–1098.
783. Basile, A., et al. (1997). Antibacterial activity in Actinidia chinensis, Feijoa sellowiana and Aberia caffra. *International Journal of Antimicrobial Agents, 8*(3), 199–203.
784. Kaur, L., et al. (2010). Actinidin enhances gastric protein digestion as assessed using an in vitro gastric digestion model. *Journal of Agricultural and Food Chemistry, 58*(8), 5068–5073.
785. Morimoto, K., et al. (2006). Effects of high concentration of salts on the esterase activity and structure of a kiwifruit peptidase, actinidain. *Journal of Biochemistry, 139*(6), 1065–1071.
786. Stonehouse, W., et al. (2012). Kiwifruit: Our daily prescription for health. *Canadian Journal of Physiology and Pharmacology, 91*(6), 442–447.
787. Hunter, D. C., et al. (2012). Consumption of gold kiwifruit reduces severity and duration of selected upper respiratory tract infection symptoms and increases plasma vitamin C concentration in healthy older adults. *British Journal of Nutrition, 108*(7), 1235–1245.
788. Forastiere, F., et al. (2000). Consumption of fresh fruit rich in vitamin C and wheezing symptoms in children. SIDRIA Collaborative Group, Italy (Italian Studies on Respiratory Disorders in Children and the Environment). *Thorax, 55*(4), 283–288.
789. Iwasawa, H., et al. (2011). Anti-oxidant effects of kiwi fruit in vitro and in vivo. *Biological & Pharmaceutical Bulletin, 34*(1), 128–134.
790. Brevik, A., et al. (2011). Supplementation of a western diet with golden kiwifruits (Actinidia chinensis var.'Hort 16A':) effects on biomarkers of oxidation damage and antioxidant protection. *Nutrition Journal, 10*, 54.
791. Collins, B. H., et al. (2001). Kiwifruit protects against oxidative DNA damage in human cells and in vitro. *Nutrition and Cancer, 39*(1), 148–153.
792. Beck, K., et al. (2011). Gold kiwifruit consumed with an iron-fortified breakfast cereal meal improves iron status in women with low iron stores: A 16-week randomised controlled trial. *British Journal of Nutrition, 105*(1), 101–109.
793. Deters, A. M., Schroder, K. R., & Hensel, A. (2005). Kiwi fruit (Actinidia chinensis L.) polysaccharides exert stimulating effects on cell proliferation via enhanced growth factor receptors, energy production, and collagen synthesis of human keratinocytes, fibroblasts, and skin equivalents. *Journal of Cellular Physiology, 202*(3), 717–722.
794. Boyera, N., Galey, I., & Bernard, B. A. (1998). Effect of vitamin C and its derivatives on collagen synthesis and cross-linking by normal human fibroblasts. *International Journal of Cosmetic Science, 20*(3), 151–158.
795. Juturu, V., Bowman, J. P., & Deshpande, J. (2016). Overall skin tone and skin-lightening-improving effects with oral supplementation of lutein and zeaxanthin isomers: A double-blind, placebo-controlled clinical trial. *Clinical, Cosmetic and Investigational Dermatology, 9*, 325–332.
796. Calvo, M. M. (2005). Lutein: A valuable ingredient of fruit and vegetables. *Critical Reviews in Food Science and Nutrition, 45*(7-8), 671–696.
797. Nishiyama, I., Fukuda, T., & Oota, T. (2005). Genotypic differences in chlorophyll, lutein, and beta-carotene contents in the fruits of actinidia species. *Journal of Agricultural and Food Chemistry, 53*(16), 6403–6407.
798. Cho, E., et al. (2004). Prospective study of intake of fruits, vegetables, vitamins, and carotenoids and risk of age-related maculopathy. *Archives of Ophthalmology, 122*(6), 883–892.
799. Cheng, Q. L., et al. (2015). 2beta, 3beta, 23-trihydroxy-urs-12-ene-28-olic acid (TUA) isolated from Actinidia chinensis Radix inhibits NCI-H460 cell proliferation by decreasing NF-kappaB expression. *Chemico-Biological Interactions, 240*, 1–11.
800. Zuo, L. L., et al. (2012). Evaluation of antioxidant and antiproliferative properties of three Actinidia (Actinidia kolomikta, Actinidia arguta, Actinidia chinensis) extracts in vitro. *International Journal of Molecular Sciences, 13*(5), 5506–5518.
801. Motohashi, N., et al. (2002). Cancer prevention and therapy with kiwifruit in Chinese folklore medicine: A study of kiwifruit extracts. *Journal of Ethnopharmacology, 81*(3), 357–364.

802. Lin, P. F. (1988). Antitumor effect of actinidia chinensis polysaccharide on murine tumor. *Zhonghua Zhong Liu Za Zhi, 10*(6), 441–444.
803. Svendsen, M., et al. (2015). The effect of kiwifruit consumption on blood pressure in subjects with moderately elevated blood pressure: A randomized, controlled study. *Blood Pressure, 24*(1), 48–54.
804. McDonough, A. A., & Nguyen, M. T. (2012). How does potassium supplementation lower blood pressure? *American Journal of Physiology - Renal Physiology, 302*(9), F1224–F1225.
805. Recio-Rodriguez, J. I., et al. (2015). Effects of kiwi consumption on plasma lipids, fibrinogen and insulin resistance in the context of a normal diet. *Nutrition Journal, 14*, 97.
806. Duttaroy, A. K., & Jorgensen, A. (2004). Effects of kiwi fruit consumption on platelet aggregation and plasma lipids in healthy human volunteers. *Platelets, 15*(5), 287–292.
807. A Monro, J. (2013). Chapter Fourteen. In *Kiwifruit, carbohydrate availability, and the glycemic response* (Vol. 68). Amsterdam: Academic Press.
808. Katsumata, S., et al. (2015). Effect of kiwifruit on bone resorption in ovariectomized mice. *Journal of Nutritional Science and Vitaminology (Tokyo), 61*(4), 332–337.
809. Adams, J., & Pepping, J. (2005). Vitamin K in the treatment and prevention of osteoporosis and arterial calcification. *American Journal of Health-System Pharmacy, 62*(15), 1574–1581.
810. Feldman, J. M., & Lee, E. M. (1985). Serotonin content of foods: Effect on urinary excretion of 5-hydroxyindoleacetic acid. *The American Journal of Clinical Nutrition, 42*(4), 639–643.
811. Lin, H. H., et al. (2011). Effect of kiwifruit consumption on sleep quality in adults with sleep problems. *Asia Pacific Journal of Clinical Nutrition, 20*(2), 169–174.
812. Young, S. N., & Leyton, M. (2002). The role of serotonin in human mood and social interaction. Insight from altered tryptophan levels. *Pharmacology Biochemistry and Behavior, 71*(4), 857–865.
813. Zarfeshany, A., Asgary, S., & Javanmard, S. H. (2014). Potent health effects of pomegranate. *Advanced Biomedical Research, 3*, 100.
814. Gil, M. I., et al. (2000). Antioxidant activity of pomegranate juice and its relationship with phenolic composition and processing. *Journal of Agricultural and Food Chemistry, 48*(10), 4581–4589.
815. Akhavan, H., et al. (2015). Phenolic compounds and antioxidant activity of juices from ten Iranian pomegranate cultivars depend on extraction. *Journal of Chemistry, 2015*, 7.
816. Viladomiu, M., et al. (2013). Preventive and prophylactic mechanisms of action of pomegranate bioactive constituents. *Evidence-based Complementary and Alternative Medicine, 2013*, 18.
817. Johanningsmeier, S. D., & Harris, G. K. (2011). Pomegranate as a functional food and nutraceutical source. *Annual Review of Food Science and Technology, 2*, 181–201.
818. Elnawasany, S. (2018). Clinical Applications of Pomegranate. In *Breeding and Health Benefits of Fruit and Nut Crops* (p. 127). London: IntechOpen.
819. Jurenka, J. S. (2008). Therapeutic applications of pomegranate (Punica granatum L.): A review. *Alternative Medicine Review, 13*(2), 128–144.
820. Boroushaki, M. T., Mollazadeh, H., & Afshari, A. R. (2016). Pomegranate seed oil: A comprehensive review on its therapeutic effects. *International Journal of Pharmaceutical Sciences and Research, 7*, 430–442.
821. Rahmani, A., Ali Alsahli, M., & Abdulrahman Almatroodi, S. (2017). Active constituents of pomegranates (Punica granatum) as potential candidates in the management of health through modulation of biological activities. *Pharmacognosy Journal, 9*, 689–695.
822. Syed, Q. A., et al. (2018). Nutritional and therapeutic properties of pomegranate. *Scholarly Journal of Food and Nutrition, 1*(4), 115–120.
823. Rahimi, H. R., Arastoo, M., & Ostad, S. N. (2012). A comprehensive review of Punica granatum (Pomegranate) properties in toxicological, pharmacological, cellular and molecular biology researches. *Iranian Journal of Pharmaceutical Research, 11*(2), 385–400.
824. Heber, D. (2011). Pomegranate Ellagitannins. In I. F. F. Benzie & S. Wachtel-Galor (Eds.), *Herbal medicine: Biomolecular and clinical aspects*. Boca Raton, FL: CRC Press/Taylor & Francis Llc..

825. Shabbir, M. A., et al. (2017). Punicic acid: A striking health substance to combat metabolic syndromes in humans. *Lipids in Health and Disease, 16*(1), 99.
826. Aruna, P., et al. (2016). Health benefits of punicic acid: A review. *Comprehensive Reviews in Food Science and Food Safety, 15*(1), 16–27.
827. Danesi, F., & Ferguson, L. R. (2017). Could pomegranate juice help in the control of inflammatory diseases? *Nutrients, 9*(9), 958.
828. Sohrab, G., et al. (2014). Effects of pomegranate juice consumption on inflammatory markers in patients with type 2 diabetes: A randomized, placebo-controlled trial. *Journal of Research in Medical Sciences, 19*(3), 215–220.
829. Colombo, E., Sangiovanni, E., & Dell'agli, M. (2013). A review on the anti-inflammatory activity of pomegranate in the gastrointestinal tract. *Evidence-based Complementary and Alternative Medicine, 2013*, 247145.
830. Grossmann, M. E., et al. (2010). Punicic acid is an omega-5 fatty acid capable of inhibiting breast cancer proliferation. *International Journal of Oncology, 36*(2), 421–426.
831. Syed, D. N., et al. (2013). Pomegranate extracts and cancer prevention: Molecular and cellular activities. *Anti Cancer Agents in Medicinal Chemistry, 13*(8), 1149–1161.
832. Turrini, E., Ferruzzi, L., & Fimognari, C. (2015). Potential effects of pomegranate polyphenols in cancer prevention and therapy. *Oxidative Medicine and Cellular Longevity, 2015*, 938475–938475.
833. Vlachojannis, C., Zimmermann, B. F., & Chrubasik-Hausmann, S. (2015). Efficacy and safety of pomegranate medicinal products for cancer. *Evidence-based Complementary and Alternative Medicine: eCAM, 2015*, 258598–258598.
834. Sharma, P., McClees, S. F., & Afaq, F. (2017). Pomegranate for prevention and treatment of cancer: An update. *Molecules (Basel, Switzerland), 22*(1), 177.
835. Bassiri-Jahromi, S. (2018). Punica granatum (Pomegranate) activity in health promotion and cancer prevention. *Oncology Reviews, 12*(1), 345–345.
836. Costantini, S., et al. (2014). Potential anti-inflammatory effects of the hydrophilic fraction of pomegranate (Punica granatum L.) seed oil on breast cancer cell lines. *Molecules, 19*(6), 8644–8660.
837. Nallanthighal, S., Elmaliki, K. M., & Reliene, R. (2017). Pomegranate extract alters breast cancer stem cell properties in association with inhibition of epithelial-to-mesenchymal transition. *Nutrition and Cancer, 69*(7), 1088–1098.
838. Shirode, A. B., et al. (2014). Antiproliferative effects of pomegranate extract in MCF-7 breast cancer cells are associated with reduced DNA repair gene expression and induction of double strand breaks. *Molecular Carcinogenesis, 53*(6), 458–470.
839. Klempner, S. J., & Bubley, G. (2012). Complementary and alternative medicines in prostate cancer: From bench to bedside? *The Oncologist, 17*(6), 830–837.
840. Pantuck, A. J., et al. (2006). Phase II study of pomegranate juice for men with rising prostate-specific antigen following surgery or radiation for prostate cancer. *Clinical Cancer Research, 12*(13), 4018–4026.
841. Wang, L., & Martins-Green, M. (2014). Pomegranate and its components as alternative treatment for prostate cancer. *International Journal of Molecular Sciences, 15*(9), 14949–14966.
842. Paller, C. J., Pantuck, A., & Carducci, M. A. (2017). A review of pomegranate in prostate cancer. *Prostate Cancer and Prostatic Diseases, 20*(3), 265–270.
843. Paller, C. J., et al. (2013). A randomized phase II study of pomegranate extract for men with rising PSA following initial therapy for localized prostate cancer. *Prostate Cancer and Prostatic Diseases, 16*(1), 50–55.
844. Tortora, K., et al. (2018). Pomegranate by-products in colorectal cancer chemoprevention: Effects in Apc-mutated pirc rats and mechanistic studies in vitro and ex vivo. *Molecular Nutrition & Food Research, 62*(2), 1700401.
845. Jaganathan, S. K., et al. (2014). Role of pomegranate and citrus fruit juices in colon cancer prevention. *World Journal of Gastroenterology, 20*(16), 4618–4625.

846. Adams, L. S., et al. (2006). Pomegranate juice, total pomegranate ellagitannins, and punicalagin suppress inflammatory cell signaling in colon cancer cells. *Journal of Agricultural and Food Chemistry, 54*(3), 980–985.
847. Rasheed, Z., Akhtar, N., & Haqqi, T. M. (2010). Pomegranate extract inhibits the interleukin-1beta-induced activation of MKK-3, p38alpha-MAPK and transcription factor RUNX-2 in human osteoarthritis chondrocytes. *Arthritis Research and Therapy, 12*(5), R195.
848. Ahmed, S., et al. (2005). Punica granatum L. extract inhibits IL-1beta-induced expression of matrix metalloproteinases by inhibiting the activation of MAP kinases and NF-kappaB in human chondrocytes in vitro. *The Journal of Nutrition, 135*(9), 2096–2102.
849. Shukla, M., et al. (2008). Bioavailable constituents/metabolites of pomegranate (Punica granatum L) preferentially inhibit COX2 activity ex vivo and IL-1beta-induced PGE2 production in human chondrocytes in vitro. *Journal of Inflammation (London, England), 5*, 9–9.
850. Ghoochani, N., et al. (2016). The effect of pomegranate juice on clinical signs, matrix metalloproteinases and antioxidant status in patients with knee osteoarthritis. *Journal of the Science of Food and Agriculture, 96*(13), 4377–4381.
851. Shuid, A. N., & Mohamed, I. N. (2013). Pomegranate use to attenuate bone loss in major musculoskeletal diseases: An evidence-based review. *Current Drug Targets, 14*(13), 1565–1578.
852. Wang, D., et al. (2018). Vasculoprotective effects of pomegranate (Punica granatum L.). *Frontiers in Pharmacology, 9*, 544.
853. Stowe, C. B. (2011). The effects of pomegranate juice consumption on blood pressure and cardiovascular health. *Complementary Therapies in Clinical Practice, 17*(2), 113–115.
854. Sahebkar, A., et al. (2017). Effects of pomegranate juice on blood pressure: A systematic review and meta-analysis of randomized controlled trials. *Pharmacological Research, 115*, 149–161.
855. Asgary, S., et al. (2013). Clinical investigation of the acute effects of pomegranate juice on blood pressure and endothelial function in hypertensive individuals. *ARYA Atherosclerosis, 9*(6), 326–331.
856. Asgary, S., et al. (2014). Clinical evaluation of blood pressure lowering, endothelial function improving, hypolipidemic and anti-inflammatory effects of pomegranate juice in hypertensive subjects. *Phytotherapy Research, 28*(2), 193–199.
857. Mirmiran, P., et al. (2010). Effect of pomegranate seed oil on hyperlipidaemic subjects: A double-blind placebo-controlled clinical trial. *British Journal of Nutrition, 104*(3), 402–406.
858. Al-Jarallah, A., et al. (2013). The effect of pomegranate extract on coronary artery atherosclerosis in SR-BI/APOE double knockout mice. *Atherosclerosis, 228*(1), 80–89.
859. Kaplan, M., et al. (2001). Pomegranate juice supplementation to atherosclerotic mice reduces macrophage lipid peroxidation, cellular cholesterol accumulation and development of atherosclerosis. *The Journal of Nutrition, 131*(8), 2082–2089.
860. Aviram, M., et al. (2004). Pomegranate juice consumption for 3 years by patients with carotid artery stenosis reduces common carotid intima-media thickness, blood pressure and LDL oxidation. *Clinical Nutrition, 23*(3), 423–433.
861. Aviram, M., et al. (2000). Pomegranate juice consumption reduces oxidative stress, atherogenic modifications to LDL, and platelet aggregation: Studies in humans and in atherosclerotic apolipoprotein E-deficient mice. *The American Journal of Clinical Nutrition, 71*(5), 1062–1076.
862. Esmaillzadeh, A., et al. (2006). Cholesterol-lowering effect of concentrated pomegranate juice consumption in type II diabetic patients with hyperlipidemia. *International Journal for Vitamin and Nutrition Research, 76*(3), 147–151.
863. Alzahri, M. S., Rohra, A., & Peacock, W. F. (2016). Nitrates as a treatment of acute heart failure. *Cardiac Failure Review, 2*(1), 51–55.
864. Hord, N. G. (2011). Dietary nitrates, nitrites, and cardiovascular disease. *Current Atherosclerosis Reports, 13*(6), 484–492.

865. Crum, E. M., et al. (2017). The effect of acute pomegranate extract supplementation on oxygen uptake in highly-trained cyclists during high-intensity exercise in a high altitude environment. *Journal of the International Society of Sports Nutrition, 14*, 14–14.
866. Roelofs, E. J., et al. (2017). Effects of pomegranate extract on blood flow and vessel diameter after high-intensity exercise in young, healthy adults. *European Journal of Sport Science, 17*(3), 317–325.
867. Trexler, E. T., et al. (2014). Effects of pomegranate extract on blood flow and running time to exhaustion. *Applied Physiology, Nutrition, and Metabolism, 39*(9), 1038–1042.
868. Amri, Z., et al. (2017). Effect of pomegranate extracts on brain antioxidant markers and cholinesterase activity in high fat-high fructose diet induced obesity in rat model. *BMC Complementary and Alternative Medicine, 17*(1), 339–339.
869. Morzelle, M. C., et al. (2016). Neuroprotective effects of pomegranate peel extract after chronic infusion with amyloid-β peptide in mice. *PLoS One, 11*(11), e0166123.
870. Loren, D. J., et al. (2005). Maternal dietary supplementation with pomegranate juice is neuroprotective in an animal model of neonatal hypoxic-ischemic brain injury. *Pediatric Research, 57*(6), 858–864.
871. Braidy, N., et al. (2013). Neuroprotective effects of a variety of pomegranate juice extracts against MPTP-induced cytotoxicity and oxidative stress in human primary neurons. *Oxidative Medicine and Cellular Longevity, 2013*, 685909.
872. Sarkaki, A., et al. (2015). Pomegranate seed hydroalcoholic extract improves memory deficits in ovariectomized rats with permanent cerebral hypoperfusion/ischemia. *Avicenna Journal of Phytomedicine, 5*(1), 43–55.
873. Riaz, A., Khan, R. A., & Algahtani, H. A. (2014). Memory boosting effect of Citrus limon, Pomegranate and their combinations. *Pakistan Journal of Pharmaceutical Sciences, 27*(6), 1837–1840.
874. Bookheimer, S. Y., et al. (2013). Pomegranate juice augments memory and FMRI activity in middle-aged and older adults with mild memory complaints. *Evidence-based Complementary and Alternative Medicine, 2013*, 946298.
875. Bellone, J. A., et al. (2018). Pomegranate supplementation improves cognitive and functional recovery following ischemic stroke: A randomized trial. *Nutritional Neuroscience, 22*, 738–743.
876. Hajipour, S., et al. (2014). Motor and cognitive deficits due to permanent cerebral hypoperfusion/ischemia improve by pomegranate seed extract in rats. *Pakistan Journal of Biological Sciences, 17*(8), 991–998.
877. Ropacki, S. A., Patel, S. M., & Hartman, R. E. (2013). Pomegranate supplementation protects against memory dysfunction after heart surgery: A pilot study. *Evidence-based Complementary and Alternative Medicine, 2013*, 8.
878. Hartman, R. E., et al. (2006). Pomegranate juice decreases amyloid load and improves behavior in a mouse model of Alzheimer's disease. *Neurobiology of Disease, 24*(3), 506–515.
879. Braidy, N., et al. (2016). Consumption of pomegranates improves synaptic function in a transgenic mice model of Alzheimer's disease. *Oncotarget, 7*(40), 64589–64604.
880. Essa, M. M., et al. (2015). Long-term dietary supplementation of pomegranates, figs and dates alleviate neuroinflammation in a transgenic mouse model of Alzheimer's disease. *PLoS One, 10*(3), e0120964–e0120964.
881. Malviya, S., et al. (2014). Antioxidant and antibacterial potential of pomegranate peel extracts. *Journal of Food Science and Technology, 51*(12), 4132–4137.
882. Saeed, M., et al. (2018). The promising pharmacological effects and therapeutic/medicinal applications of Punica Granatum L. (Pomegranate) as a functional food in humans and animals. *Recent Patents on Inflammation & Allergy Drug Discovery, 12*(1), 24–38.
883. Ismail, T., Sestili, P., & Akhtar, S. (2012). Pomegranate peel and fruit extracts: A review of potential anti-inflammatory and anti-infective effects. *Journal of Ethnopharmacology, 143*(2), 397–405.

884. Howell, A. B., & D'Souza, D. H. (2013). The pomegranate: Effects on bacteria and viruses that influence human health. *Evidence-based Complementary and Alternative Medicine: eCAM, 2013*, 606212.
885. Li, Z., et al. (2015). Antimicrobial activity of pomegranate and green tea extract on propionibacterium acnes, Propionibacterium Granulosum, Staphylococcus Aureus and Staphylococcus Epidermidis. *Journal of Drugs in Dermatology, 14*(6), 574–578.
886. Duman, A. D., et al. (2009). Antimicrobial activity of six pomegranate (Punica granatum L.) varieties and their relation to some of their pomological and phytonutrient characteristics. *Molecules, 14*(5), 1808–1817.
887. Devatkal, S. K., et al. (2013). Antibacterial activity of aqueous extract of pomegranate peel against Pseudomonas stutzeri isolated from poultry meat. *Journal of Food Science and Technology, 50*(3), 555–560.
888. Choi, J.-G., et al. (2011). In vitro and in vivo antibacterial activity of Punica granatum peel ethanol extract against Salmonella. *Evidence-Based Complementary and Alternative Medicine: eCAM, 2011*, 690518.
889. Al-Zoreky, N. S. (2009). Antimicrobial activity of pomegranate (Punica granatum L.) fruit peels. *International Journal of Food Microbiology, 134*(3), 244–248.
890. Rosas-Burgos, E. C., et al. (2017). Antimicrobial activity of pomegranate peel extracts as affected by cultivar. *Journal of the Science of Food and Agriculture, 97*(3), 802–810.
891. Betanzos-Cabrera, G., et al. (2015). Antibacterial activity of fresh pomegranate juice against clinical strains of Staphylococcus epidermidis. *Food & Nutrition Research, 59*, 27620.
892. Gould, S. W., et al. (2009). Anti-microbial activities of pomegranate rind extracts: Enhancement by cupric sulphate against clinical isolates of S. aureus, MRSA and PVL positive CA-MSSA. *BMC Complementary and Alternative Medicine, 9*, 23.
893. Dey, D., et al. (2012). Pomegranate pericarp extract enhances the antibacterial activity of ciprofloxacin against extended-spectrum beta-lactamase (ESBL) and metallo-beta-lactamase (MBL) producing Gram-negative bacilli. *Food and Chemical Toxicology, 50*(12), 4302–4309.
894. Dey, D., Ray, R., & Hazra, B. (2015). Antimicrobial activity of pomegranate fruit constituents against drug-resistant Mycobacterium tuberculosis and beta-lactamase producing Klebsiella pneumoniae. *Pharmaceutical Biology, 53*(10), 1474–1480.
895. Braga, L. C., et al. (2005). Synergic interaction between pomegranate extract and antibiotics against Staphylococcus aureus. *Canadian Journal of Microbiology, 51*(7), 541–547.
896. Millo, G., et al. (2017). Antibacterial inhibitory effects of Punica granatum gel on cariogenic bacteria: An in vitro study. *International Journal of Clinical Pediatric Dentistry, 10*(2), 152–157.
897. Abdollahzadeh, S., et al. (2011). Antibacterial and antifungal activities of punica granatum peel extracts against oral pathogens. *Journal of Dentistry (Tehran, Iran), 8*(1), 1–6.
898. Aparecida Procópio Gomes, L., et al. (2016). Punica granatum L. (Pomegranate) extract: In vivo study of antimicrobial activity against Porphyromonas gingivalis in Galleria mellonella model. *The Scientific World Journal, 2016*, 8626987.
899. Ferrazzano, G. F., et al. (2017). In Vitro Antibacterial Activity of Pomegranate Juice and Peel Extracts on Cariogenic Bacteria. *BioMed Research International, 2017*, 2152749.
900. Glazer, I., et al. (2012). Partial identification of antifungal compounds from Punica granatum peel extracts. *Journal of Agricultural and Food Chemistry, 60*(19), 4841–4848.
901. Pai, M. B., et al. (2010). Antifungal efficacy of Punica granatum, Acacia nilotica, Cuminum cyminum and Foeniculum vulgare on Candida albicans: An in vitro study. *Indian Journal of Dental Research, 21*(3), 334–336.
902. Višnjevec, A. M., et al. (2017). Genetic, biochemical, nutritional and antimicrobial characteristics of pomegranate (Punica granatum L.) grown in Istria. *Food Technology and Biotechnology, 55*(2), 151–163.
903. Anibal, P. C., et al. (2013). Antifungal activity of the ethanolic extracts of Punica granatum L. and evaluation of the morphological and structural modifications of its compounds upon the cells of Candida spp. *Brazilian Journal of Microbiology, 44*(3), 839–848.

904. Endo, E. H., et al. (2010). Potent antifungal activity of extracts and pure compound isolated from pomegranate peels and synergism with fluconazole against Candida albicans. *Research in Microbiology, 161*(7), 534–540.
905. Li, Z. J., et al. (2017). Antifungal activity of gallic acid in vitro and in vivo. *Phytotherapy Research, 31*(7), 1039–1045.
906. Vasconcelos, L. C., et al. (2003). Use of Punica granatum as an antifungal agent against candidosis associated with denture stomatitis. *Mycoses, 46*(5-6), 192–196.
907. Bassiri-Jahromi, S. P., et al. (2018). In vivo comparative evaluation of the pomegranate (Punica granatum) peel extract as an alternative agent to nystatin against oral candidiasis. *Iranian Journal of Medical Sciences, 43*(3), 296–304.
908. Foss, S. R., et al. (2014). Antifungal activity of pomegranate peel extract and isolated compound punicalagin against dermatophytes. *Annals of Clinical Microbiology and Antimicrobials, 13*, 32–32.
909. Su, X., Sangster, M. Y., & D'Souza, D. H. (2011). Time-dependent effects of pomegranate juice and pomegranate polyphenols on foodborne viral reduction. *Foodborne Pathogens and Disease, 8*(11), 1177–1183.
910. Reddy, B. U., et al. (2014). Small molecule inhibitors of HCV replication from pomegranate. *Scientific Reports, 4*, 5411–5411.
911. Neurath, A. R., et al. (2004). Punica granatum(Pomegranate) juice provides an HIV-1 entry inhibitor and candidate topical microbicide. *BMC Infectious Diseases, 4*(1), 41.
912. Houston, D. M. J., et al. (2017). Potentiated virucidal activity of pomegranate rind extract (PRE) and punicalagin against Herpes simplex virus (HSV) when co-administered with zinc (II) ions, and antiviral activity of PRE against HSV and aciclovir-resistant HSV. *PLoS One, 12*(6), e0179291.
913. Arunkumar, J., & Rajarajan, S. (2018). Study on antiviral activities, drug-likeness and molecular docking of bioactive compounds of Punica granatum L. to Herpes simplex virus - 2 (HSV-2). *Microbial Pathogenesis, 118*, 301–309.
914. Lin, L.-T., et al. (2013). Broad-spectrum antiviral activity of chebulagic acid and punicalagin against viruses that use glycosaminoglycans for entry. *BMC Microbiology, 13*, 187–187.
915. Thangavelu, A., et al. (2017). Ancient seed for modern cure - pomegranate review of therapeutic applications in periodontics. *Journal of Pharmacy & Bioallied Sciences, 9*(Suppl 1), S11–S14.
916. AlMatar, M., et al. (2018). Pomegranate as a possible treatment in reducing risk of developing wound healing, obesity, neurodegenerative disorders, and diabetes mellitus. *Mini Reviews in Medicinal Chemistry, 18*(6), 507–526.
917. Yuniarti, W. M., Primarizky, H., & Lukiswanto, B. S. (2018). The activity of pomegranate extract standardized 40% ellagic acid during the healing process of incision wounds in albino rats (Rattus norvegicus). *Veterinary World, 11*(3), 321–326.
918. Elzayat, E., et al. (2018). Evaluation of wound healing activity of henna, pomegranate and myrrh herbal ointment blend. *Saudi Pharmaceutical Journal, 26*(5), 733–738.
919. Lin, T.-K., Zhong, L., & Santiago, J. L. (2017). Anti-inflammatory and skin barrier repair effects of topical application of some plant oils. *International Journal of Molecular Sciences, 19*(1), 70.
920. Lee, C.-J., et al. (2017). Multiple activities of punica granatum linne against acne vulgaris. *International Journal of Molecular Sciences, 18*(1), 141.
921. Lenucci, M. S., et al. (2006). Antioxidant composition in cherry and high-pigment tomato cultivars. *Journal of Agricultural and Food Chemistry, 54*(7), 2606–2613.
922. Kozukue, N., & Friedman, M. (2003). Tomatine, chlorophyll, β-carotene and lycopene content in tomatoes during growth and maturation. *Journal of the Science of Food and Agriculture, 83*(3), 195–200.
923. Bhowmik, D., et al. (2012). Tomato-a natural medicine and its health benefits introduction: Tomatoes are a member of. *Journal of Pharmacognosy and Phytochemistry, 1*(1), 33–43.
924. Claye, S. S., Idouraine, A., & Weber, C. W. (1996). Extraction and fractionation of insoluble fiber from five fiber source. *Food Chemistry, 57*, 305–310.

925. Guan, Y.-S., & He, Q. (2015). Plants consumption and liver health. *Evidence-based Complementary and Alternative Medicine, 2015*, 10.
926. Yamashoji, S., & Onoda, E. (2016). Detoxification and function of immature tomato. *Food Chemistry, 209*, 171–176.
927. Elvira-Torales, L. I., et al. (2018). Tomato juice supplementation influences the gene expression related to steatosis in rats. *Nutrients, 10*(9), 1215.
928. Ho, K.-S., et al. (2012). Stopping or reducing dietary fiber intake reduces constipation and its associated symptoms. *World Journal of Gastroenterology, 18*(33), 4593–4596.
929. Frusciante, L., et al. (2007). Antioxidant nutritional quality of tomato. *Molecular Nutrition & Food Research, 51*(5), 609–617.
930. Grune, T., et al. (2010). Beta-carotene is an important vitamin A source for humans. *The Journal of Nutrition, 140*(12), 2268S–2285S.
931. Kim, J. K. (2016). An update on the potential health benefits of carotenes. *EXCLI Journal, 15*, 1–4.
932. Clinton, S. K. (1998). Lycopene: Chemistry, biology, and implications for human health and disease. *Nutrition Reviews, 56*(2 Pt 1), 35–51.
933. Rao, A. V., Ray, M. R., & Rao, L. G. (2006). Lycopene. *Advances in Food and Nutrition Research, 51*, 99–164.
934. Story, E. N., et al. (2010). An update on the health effects of tomato lycopene. *Annual Review of Food Science and Technology, 1*, 189–210.
935. Bharti, S., et al. (2014). Preclinical evidence for the pharmacological actions of naringin: A review. *Planta Medica, 80*(6), 437–451.
936. Liang, N., & Kitts, D. D. (2015). Role of chlorogenic acids in controlling oxidative and inflammatory stress conditions. *Nutrients, 8*(1), 16.
937. Tajik, N., et al. (2017). The potential effects of chlorogenic acid, the main phenolic components in coffee, on health: A comprehensive review of the literature. *European Journal of Nutrition, 56*(7), 2215–2244.
938. Rao, A. V. (2002). Lycopene, tomatoes, and the prevention of coronary heart disease. *Experimental Biology and Medicine (Maywood, N.J.), 227*(10), 908–913.
939. Basu, A., & Imrhan, V. (2007). Tomatoes versus lycopene in oxidative stress and carcinogenesis: Conclusions from clinical trials. *European Journal of Clinical Nutrition, 61*(3), 295–303.
940. Riso, P., et al. (2006). Effect of a tomato-based drink on markers of inflammation, immunomodulation, and oxidative stress. *Journal of Agricultural and Food Chemistry, 54*(7), 2563–2566.
941. Karppi, J., et al. (2012). Low serum lycopene and beta-carotene increase risk of acute myocardial infarction in men. *The European Journal of Public Health, 22*(6), 835–840.
942. Karppi, J., et al. (2012). Serum lycopene decreases the risk of stroke in men: A population-based follow-up study. *Neurology, 79*(15), 1540–1547.
943. Palozza, P., et al. (2012). Effect of lycopene and tomato products on cholesterol metabolism. *Annals of Nutrition and Metabolism, 61*(2), 126–134.
944. Palomo, I., et al. (2012). Platelets and atherogenesis: Platelet anti-aggregation activity and endothelial protection from tomatoes (Solanum lycopersicum L.). *Experimental and Therapeutic Medicine, 3*(4), 577–584.
945. Mozos, I., et al. (2018). Lycopene and Vascular Health. *Frontiers in Pharmacology, 9*, 521–521.
946. Jacques, P. F., et al. (2013). Relationship of lycopene intake and consumption of tomato products to incident CVD. *The British Journal of Nutrition, 110*(3), 545–551.
947. Li, X., & Xu, J. (2013). Lycopene supplement and blood pressure: An updated meta-analysis of intervention trials. *Nutrients, 5*(9), 3696–3712.
948. Kizhakekuttu, T. J., & Widlansky, M. E. (2010). Natural antioxidants and hypertension: Promise and challenges. *Cardiovascular Therapeutics, 28*(4), e20–e32.

949. Onakpoya, I. J., et al. (2014). The effect of chlorogenic acid on blood pressure: A systematic review and meta-analysis of randomized clinical trials. *Journal of Human Hypertension, 29*, 77.
950. Zhao, Y., et al. (2012). Antihypertensive effects and mechanisms of chlorogenic acids. *Hypertension Research, 35*(4), 370–374.
951. Watanabe, T., et al. (2006). The blood pressure-lowering effect and safety of chlorogenic acid from green coffee bean extract in essential hypertension. *Clinical and Experimental Hypertension, 28*(5), 439–449.
952. Kozuma, K., et al. (2005). Antihypertensive effect of green coffee bean extract on mildly hypertensive subjects. *Hypertension Research, 28*(9), 711–718.
953. Rao, A. V., & Agarwal, S. (2000). Role of antioxidant lycopene in cancer and heart disease. *Journal of the American College of Nutrition, 19*(5), 563–569.
954. Giovannucci, E. (1999). Tomatoes, tomato-based products, lycopene, and cancer: Review of the epidemiologic literature. *Journal of the National Cancer Institute, 91*(4), 317–331.
955. Johary, A., Jain, V., & Misra, S. (2012). Role of lycopene in the prevention of cancer. *International Journal of Nutrition, Pharmacology, Neurological Diseases, 2*(3), 167–170.
956. Etminan, M., Takkouche, B., & Caamano-Isorna, F. (2004). The role of tomato products and lycopene in the prevention of prostate cancer: A meta-analysis of observational studies. *Cancer Epidemiology and Prevention Biomarkers, 13*(3), 340–345.
957. Giovannucci, E. (2002). A review of epidemiologic studies of tomatoes, lycopene, and prostate cancer. *Experimental Biology and Medicine (Maywood, N.J.), 227*(10), 852–859.
958. Gong, X., et al. (2016). Mitochondrial beta-carotene 9',10' oxygenase modulates prostate cancer growth via NF-kappaB inhibition: A lycopene-independent function. *Molecular Cancer Research, 14*(10), 966–975.
959. Holzapfel, N. P., et al. (2013). The potential role of lycopene for the prevention and therapy of prostate cancer: From molecular mechanisms to clinical evidence. *International Journal of Molecular Sciences, 14*(7), 14620–14646.
960. Sesso, H. D., et al. (2005). Dietary and plasma lycopene and the risk of breast cancer. *Cancer Epidemiology and Prevention Biomarkers, 14*(5), 1074–1081.
961. Aune, D., et al. (2012). Dietary compared with blood concentrations of carotenoids and breast cancer risk: A systematic review and meta-analysis of prospective studies. *The American Journal of Clinical Nutrition, 96*(2), 356–373.
962. Sato, R., et al. (2002). Prospective study of carotenoids, tocopherols, and retinoid concentrations and the risk of breast cancer. *Cancer Epidemiology and Prevention Biomarkers, 11*(5), 451–457.
963. Palozza, P., et al. (2011). Tomato lycopene and lung cancer prevention: From experimental to human studies. *Cancers, 3*(2), 2333–2357.
964. Okuyama, Y., et al. (2014). Inverse associations between serum concentrations of zeaxanthin and other carotenoids and colorectal neoplasm in Japanese. *International Journal of Clinical Oncology, 19*(1), 87–97.
965. Reche, M., et al. (2001). Tomato allergy in children and young adults: Cross-reactivity with latex and potato. *Allergy, 56*(12), 1197–1201.
966. Leone, I. A., Brennan, E., & Daines, R. H. (1956). Atmospheric Fluoride: Its uptake and distribution in tomato and corn plants. *Plant Physiology, 31*(5), 329–333.
967. Ogunbanwo, S. T., et al. (2013). Microbiological and nutritional evaluation of water melon juice (Citrullus lanatus). *Academic Arena, 5*(3), 36–41.
968. Ko, S. H., et al. (2005). Comparison of the antioxidant activities of nine different fruits in human plasma. *Journal of Medicinal Food, 8*(1), 41–46.
969. Naz, A., et al. (2014). Watermelon lycopene and allied health claims. *EXCLI Journal, 13*, 650–660.
970. Mohammad, M. K. A., et al. (2014). Watermelon (Citrullus lanatus (Thunb.) Matsum. and Nakai) juice modulates oxidative damage induced by low dose X-ray in mice. *BioMed Research International, 2014*, 512834.

971. Hong, M. Y., et al. (2015). Watermelon consumption improves inflammation and antioxidant capacity in rats fed an atherogenic diet. *Nutrition Research, 35*(3), 251–258.
972. Jacob, K., et al. (2008). Influence of lycopene and vitamin C from tomato juice on biomarkers of oxidative stress and inflammation. *British Journal of Nutrition, 99*(1), 137–146.
973. Chen, P., et al. (2015). Lycopene and risk of prostate cancer: A systematic review and meta-analysis. *Medicine (Baltimore), 94*(33), e1260.
974. Attard, E., & Martinoli, M. G. (2015). Cucurbitacin E, An experimental lead triterpenoid with anticancer, immunomodulatory and novel effects against degenerative diseases. A mini-review. *Current Topics in Medicinal Chemistry, 15*(17), 1708–1713.
975. Chen, X., et al. (2012). Biological activities and potential molecular targets of cucurbitacins: A focus on cancer. *Anti-Cancer Drugs, 23*(8), 777–787.
976. Bowers, L. W., et al. (2015). The role of the insulin/IGF system in cancer: Lessons learned from clinical trials and the energy balance-cancer link. *Frontiers in Endocrinology, 6*, 77.
977. Yu, H., & Rohan, T. (2000). Role of the insulin-like growth factor family in cancer development and progression. *Journal of the National Cancer Institute, 92*(18), 1472–1489.
978. Klimant, E., et al. (2018). Intravenous vitamin C in the supportive care of cancer patients: A review and rational approach. *Current Oncology (Toronto, Ont.), 25*(2), 139–148.
979. Bohm, V. (2012). Lycopene and heart health. *Molecular Nutrition & Food Research, 56*(2), 296–303.
980. Figueroa, A., et al. (2013). Effects of watermelon supplementation on arterial stiffness and wave reflection amplitude in postmenopausal women. *Menopause, 20*(5), 573–577.
981. Poduri, A., et al. (2013). Citrullus lanatus 'sentinel' (watermelon) extract reduces atherosclerosis in LDL receptor-deficient mice. *The Journal of Nutritional Biochemistry, 24*(5), 882–886.
982. Karppi, J., et al. (2013). Serum carotenoids reduce progression of early atherosclerosis in the carotid artery wall among Eastern Finnish men. *PLoS One, 8*(5), e64107.
983. Collins, J. K., et al. (2007). Watermelon consumption increases plasma arginine concentrations in adults. *Nutrition, 23*(3), 261–266.
984. Tripathi, P. (2007). Nitric oxide and immune response. *Indian Journal of Biochemistry and Biophysics, 44*(5), 310–319.
985. Houston, M. (2011). The role of magnesium in hypertension and cardiovascular disease. *The Journal of Clinical Hypertension (Greenwich), 13*(11), 843–847.
986. Oseni, O. A., Odesanmi, O. E., & Oladele, F. C. (2015). Antioxidative and antidiabetic activities of watermelon (Citrullus lanatus) juice on oxidative stress in alloxan-induced diabetic male Wistar albino rats. *Nigerian Medical Journal: Journal of the Nigeria Medical Association, 56*(4), 272–277.
987. Muraki, I., et al. (2013). Fruit consumption and risk of type 2 diabetes: Results from three prospective longitudinal cohort studies. *BMJ (Clinical Research Ed.), 347*, f5001.
988. Wu, G., et al. (2007). Dietary supplementation with watermelon pomace juice enhances arginine availability and ameliorates the metabolic syndrome in Zucker diabetic fatty rats. *The Journal of Nutrition, 137*(12), 2680–2685.
989. Edwards, A. J., et al. (2003). Consumption of watermelon juice increases plasma concentrations of lycopene and beta-carotene in humans. *The Journal of Nutrition, 133*(4), 1043–1050.
990. Tarazona-Diaz, M. P., et al. (2013). Watermelon juice: Potential functional drink for sore muscle relief in athletes. *Journal of Agricultural and Food Chemistry, 61*(31), 7522–7528.
991. Shanely, R. A., et al. (2016). Comparison of watermelon and carbohydrate beverage on exercise-induced alterations in systemic inflammation, immune dysfunction, and plasma antioxidant capacity. *Nutrients, 8*(8), 518.
992. Cutrufello, P. T., Gadomski, S. J., & Zavorsky, G. S. (2015). The effect of l-citrulline and watermelon juice supplementation on anaerobic and aerobic exercise performance. *Journal of Sports Sciences, 33*(14), 1459–1466.
993. Lindinger, M. I., & Sjogaard, G. (1991). Potassium regulation during exercise and recovery. *Sports Medicine, 11*(6), 382–401.

994. Fraser, P. D., & Bramley, P. M. (2004). The biosynthesis and nutritional uses of carotenoids. *Progress in Lipid Research, 43*(3), 228–265.
995. Sharma, P. B., et al. (1986). Studies on the nutritional quality of some cucurbit kernel proteins. *Journal of the Science of Food and Agriculture, 37*(4), 418–420.
996. Xu, Y., Leo, M. A., & Lieber, C. S. (2003). Lycopene attenuates alcoholic apoptosis in HepG2 cells expressing CYP2E1. *Biochemical and Biophysical Research Communications, 308*(3), 614–618.
997. Veeramachaneni, S., et al. (2008). High dose lycopene supplementation increases hepatic cytochrome P4502E1 protein and inflammation in alcohol-fed rats. *The Journal of Nutrition, 138*(7), 1329–1335.

Grains

Sawsan G. Mohammed, Saoud Bossa, and M. Walid Qoronfleh

Abstract The grain group is small, hard, dry seeds, known to be more durable than other staple foods. They have been a part of the human diet for tens of thousands of years. The two foremost types of commercial grain crops are cereals and legumes or pulses, discussed in Chapter 13 "Seeds." A low intake of whole grains is actually the leading dietary risk factor for death and disease in the USA. Few healthy grains are discussed in this chapter that can help prevent health problems like heart diseases, diabetes, and cancers.

Keywords Grains · Fiber · Beta-glucan · Antioxidants · Flavonoids · Avenanthramides · Phenolic alkaloids · Phenolic acids · Phytic acid · Phytoestrogens · Lignans · Glutenin · Choline · Prolamins

---

S. G. Mohammed (✉)
Qatar Research Leadership Program (QRLP), Qatar Foundation, Doha, Qatar
e-mail: sgmohammed@qf.org.qa

S. Bossa
Medical student, College of Medicine, Qatar University (QU), Doha, Qatar
e-mail: sb1908029@qu.edu.qa

M. W. Qoronfleh (✉)
Research & Policy Department, World Innovation Summit for Health (WISH), Qatar Foundation, Doha, Qatar
e-mail: wqoronfleh@qf.org.qa

M. M. Essa, M. W. Qoronfleh (eds.), *Personalized Food Intervention and Therapy for Autism Spectrum Disorder Management*, Advances in Neurobiology 24,
https://doi.org/10.1007/978-3-030-30402-7_11

1 Wheat

Family: Poaceae
Genus: *Triticum*
Common name: Wheat

Wheat is a grasslike cereal grain commonly grown for its seeds. Whole-grain kernels consist of three parts:

- First, the bran, which is the hard, outer shell. Dietary fiber comes mostly from the bran, along with minerals and antioxidants including phytic acid, lignin, and sulfur compounds
- Second, the endosperm, which is the middle layer and is mostly made up of carbohydrates
- Third, the germ which is the inner layer that contains vitamins, minerals, proteins, and other plant compounds such as lignans, stanols, and sterols.

As long as these three parts are retained in the grain preparation, they are considered a whole grain. It is classified as a refined grain when the germ and bran have been removed, thus keeping only the endosperm. Wheat grain flour can be rolled, while the grains are either crushed or cracked. Whole grains can be eaten as bread and pasta or made into breakfast cereals. Whole-grain wheat is rich in B vitamins, including thiamin (B-1) niacin (B-3), pyridoxine (B-6), and folic acid (B-9). It contains also minerals like manganese, magnesium, zinc, copper, and iron (Table 1).

Wheat, especially the whole grain, has a number of described health benefits. Whole-grain wheat contains B vitamins and complex carbohydrates, which provide energy and keep the person feeling satisfied for longer time. It has been reported that consuming whole wheat is associated with better weight control than consuming refined wheat [1]. Refined wheat tends to increase weight and was found to increase the risk of diabetes and insulin resistance [2]. Whole wheat has been potentially linked to decrease in the risk of other diseases including osteoporosis and heart diseases [3]. Wheat also contains the amino acid betaine, a trimethylglycine. The physiologic function of betaine is either as an osmolyte to protect cells under stress or as a catabolic source of methyl groups via transmethylation for use in many biochemical pathways and detoxification of homocysteine diseases [4]. Betaine has been postulated to prevent chronic diseases such as inflammation in rheumatic

Table 1 Wheat nutrition facts of 100 g, from the United States Department of Agriculture (USDA SR-21)

Serving size of 100 g of ground whole wheat flour	
Per serving	% Daily value[a]
Calories 339	
Total fat 1.9 g	3
Saturated fat 0.3 g	2
Polyunsaturated fat 0.8 g	
Monounsaturated fat 0.2 g	
Total omega-3 fatty acids 38.0 mg	
Total omega-6 fatty acids 738 mg	
Cholesterol 0.0 mg	0
Phytosterols	
Carbohydrates 72.6 g	24
Dietary fiber 12.2 g	49
Starch	
Sugars 0.4 g	
Protein 13.7 g	27
Vitamins	
Vitamin A	0
Vitamin E	4
Vitamin K	2
Vitamin C	0
Vitamin B-6	17
Folic acid	11
Minerals	
Sodium	0
Calcium	3
Magnesium	34
Copper	19
Potassium	12
Iron	22
Manganese	190
Zinc	20

National Nutrient Database
[a]Based on a 2000-calorie diet

heart, joints, and/or connective tissues and improve cardiac as well as liver functions [5]. Wheat bran however has been shown to reduce the risk of breast cancer development as it has a consistently protective effect in mammary carcinogenesis [6]. Lignans (phytochemicals), more specifically phytoestrogens, along with fiber, decrease the risk of colorectal cancer [7] and cardiovascular diseases [8]. The insoluble fiber in whole wheat has been correlated with a significant decrease of gallbladder/bile duct cancer in a Japanese population-based cohort study [9] and a reduction in blood pressure [10]. The high fiber content of whole wheat reduces constipation and gas accumulation as well. Iron, folic acid, other B vitamins, and vitamin E enhance serotonin production and reduce the risk of cognitive function deterioration of Alzheimer's disease [11]. Vitamin E, niacin, and zinc in whole wheat lower the risk of macular degeneration and cataract, whereas carotenoids, specifically lutein and zeaxanthin, improve the overall eye health status [12].

High intake of refined grains—including wheat—has been linked to some health problems. Wheat is rich in the protein gluten that is also found in barley and rye. Gluten is a storage protein composed of proteins, termed prolamins (water-soluble fraction), and glutelins (water-insoluble fraction). The wheat glutelins are called glutenin. Gluten can trigger gastrointestinal inflammatory disorders and allergies. In people with food sensitivity, gluten can damage the intestinal lining and may lead to pain, anemia, bloating, bowel irregularity, fatigue, and possibly a serious immune condition known as celiac disease [13]. Gluten sensitivity is also associated with cerebellar ataxia [14]. It has been shown that a gluten-free diet improved the condition of ataxia in gluten-sensitive patients [15]. An Autism Speaks article indicated that parents of autistic children reported behavior improvement when children were put on gluten-casein-free diet (GFCF) [16]. This improvement may be entirely attributed to healthier diet and not necessarily to being on a GFCF regimen. There is insufficient clinical evidence showing the clear benefit of such diet [17]. This controversy about effective diet intervention and favorable outcome in patients with autism continues to exist as others report the opposite with GFCF nutrition [18, 19]. Clearly, further research is needed to provide sound, compelling scientific evidence. The GFCF diet is discussed in separate chapters (Chapters 17 and 19).

Consumption of products prepared from refined grains, such as white bread, leads to spikes of blood sugar, followed by rapid glucose drop and hunger leading to weight gain. Phytic acid, a saturated cyclic acid, in both refined and whole wheat has the capacity to bind minerals such as calcium, zinc, iron, and magnesium, hence preventing their absorption [20]. Rich oxalate content in wheat may cause health problems such as gallbladder stones, kidney stones, and gout among certain individuals [21]. Other studies found that whole wheat significantly raises the unhealthy low-density lipoprotein (LDL) cholesterol [22].

2 Oats

Family: Poaceae
Genus: *Avena*
Common name: Oats

Oats are a whole-grain cereal, scientifically known as *Avena sativa*, and are more commonly eaten as oatmeal (crushed and toasted oat groats) or porridge. However, they may be used in a variety of other goods. Whole oats are the only source of a unique group of antioxidants, avenanthramides (phenolic alkaloids), which not only are believed to have protective effects against coronary heart disease but also exhibit anti-inflammatory, antiproliferative, and anti-itching activity which may provide additional protection against colon cancer and skin irritation [23]. Oats contain high amounts of many vitamins and minerals such as B vitamins, manganese, magnesium, phosphorus, copper, iron, selenium, and zinc (Table 2).

Table 2 Oats nutrition facts of 100 g, from the United States Department of Agriculture (USDA SR-21)

Serving size of 100 g of oats	
Per serving	% Daily value[a]
Calories 389	
Total fat 6.9 g	11
Saturated fat 1.2 g	6
Polyunsaturated fat 2.5 g	
Monounsaturated fat 2.2 g	
Total omega-3 fatty acids 111 mg	
Total omega-6 fatty acids 2424 mg	
Cholesterol 0.0 mg	0
Phytosterols	
Carbohydrates 66.3 g	22
Dietary fiber 10.6 g	42
Starch	
Sugars	
Protein 16.9 g	34
Vitamins	
Vitamin A	0
Vitamin E	
Vitamin K	
Vitamin C	0
Vitamin B-6	6
Folic acid	14
Minerals	
Sodium	0
Calcium	5
Magnesium	44
Copper	31
Potassium	12
Iron	26
Manganese	246
Zinc	26

National Nutrient Database
[a]Based on a 2000-calorie diet

Oats are also high in carbohydrates, starch and fiber, especially the polysaccharide beta-glucan. Oats contain more fat and protein than the other grains. Pure oats are free of gluten.

Oats are definitely among the healthiest of grains. The soluble fibers present in oats, mainly beta-glucan, affect insulin sensitivity, stabilize blood glucose levels, and lower the risk of type 2 diabetes [24]. Beta-glucan also lowers cholesterol levels and reduces blood pressure. Therefore, it protects against coronary heart disease, and furthermore, the antioxidants present in oats are very important as they contribute to enhancing the heart health [25]. Similarly, the insoluble fibers decrease constipation, improve intestinal health, lower blood pressure and along with soluble fiber, control weight as well [10]. One of the greatest advantages of consuming oats is that it keeps you feeling "full" for a longer period of time when compared with other foods [26]. For example, beta-glucan releases putative satiety or appetite-regulating-type peptides which in turn reduce calorie intake and help control the body weight [27, 28]. Beta-glucan can be used both as a treatment and as a prophylactic measure. It is thought to be a good radioprotective drug for chemotherapy and radiotherapy or in case of nuclear emergencies [29]. Oats have very high manganese content ultimately ensuring proper growth, development, and metabolism [30]. Copper helps reduce blood pressure and lower blood cholesterol [31]. Selenium has pleiotropic effects ranging from antioxidant and anti-inflammatory effects to hormone production. Selenium in whole grains has been shown to have relevance in decreasing the risk of premature death, improving cognitive condition, and boosting the immune system. It also has some beneficial effects on the risk of prostate, lung, colorectal, and bladder cancers [32]. The polyphenol avenanthramides' potent antioxidant properties were also reported to decrease inflammation, reduce the risk of coronary heart diseases and colon cancer, and control blood pressure [23]. In addition, finely ground oats are dermatologically beneficial. As a topical treatment, it can help people with dry skin and relieve symptoms of many skin conditions including itching, erythema, and eczema [33].

Despite these many health benefits, oats have their share of secondary effects. Eating too much oats can cause intestinal gas and bloating. However, eating a small portion daily may allow the body to get used to it, and these minor effects may disappear. Poorly chewed oats may also cause blockage of the intestine. So it is better for people with chewing disorders to either eliminate consumption or consume oats with plenty of water. Overeating oat bran may lead to diarrhea due to the effect of the fibers present. Oats do not contain gluten, but they do contain avenin, a protein similar to gluten. Avenin sensitivity seems to be extremely rare. Individuals allergic to avenin might experience symptoms similar to gluten intolerance. Though, oat proteins can act as both respiratory (dust allergy) and skin allergens (dermatitis). However, oat sensitivity can be explained due to frequent contamination with other grains containing gluten, and so it should not be consumed in case of gluten allergy or celiac disease [34].

3 Barley

Family: Poaceae
Genus: *Hordeum*
Common name: Barley

Being a member of the grass family, barley was one of the first grains ever cultivated. There are few types of commercial barley. These include "hulled," i.e., the whole barley grain; "hulless," the most common type of barley which is minimally processed requiring the removal of the hull covering, most of the bran layer, and the germ too; and finally, "pearled" barley type which has been processed to remove its hull and bran. During medieval times, barley was actually very common among peasants. Today, it is a main cereal grain that is commonly used to make breads, salads, and soups and brew alcoholic drinks in several cuisines worldwide. Barely is also utilized as a natural sweetener. The reason barley gained popularity over the years is largely due to its various health benefits. Whole barley provides a range of important nutrients like fiber, B vitamins like niacin (B-3) and pyridoxine (B-6), and minerals such as selenium, copper, chromium, phosphorus, and magnesium (Table 3). Moreover, barley contains several important antioxidants. Drinking barley tea or water is just as healthy as eating the grain itself.

When compared to many other grains, whole barley is higher in dietary fiber and is essentially an excellent source of both soluble and insoluble fibers. It is lower in fats and calories yet contains greater amounts of certain trace minerals. The high fiber content regulates bowel movement; prevents constipation and intrinsically, colorectal cancer; permits weight control; and reduces blood pressure [10]. Barley contributes to lowering the total amount of cholesterol especially the unhealthy LDL cholesterol and total triglycerides [35], thereby decreasing the risk of heart disease. Beta-glucan soluble fiber in barley has numerous benefits. Beta-glucan appears to control blood glucose and insulin levels effectively [36]. Beta-glucan also helps reduce the hungry feeling and aids in controlling body weight [37]. Furthermore, barley beta-glucan was reported to strengthen the immune system and reduce the chance of cold and flu [38]. Interestingly, barley tea acts as a natural diuretic and helps treat urinary tract infections and similarly shows a great ability to reduce the intensity of symptoms related to inflammatory

Table 3 Barley nutrition facts of 100 g, from the United States Department of Agriculture (USDA SR-21)

Serving size of 100 g of hulled barley	
Per serving	% Daily value[a]
Calories 354	
Total fat 2.3 g	4
Saturated fat 0.5 g	2
Polyunsaturated fat 1.1 g	
Monounsaturated fat 0.3 g	
Total omega-3 fatty acids 110 mg	
Total omega-6 fatty acids 999 mg	
Cholesterol 0.0 mg	0
Phytosterols	
Carbohydrates 73.5 g	21
Dietary fiber 17.3 g	69
Starch	
Sugars 0.8 g	
Protein 12.5 g	25
Vitamins	
Vitamin A	0
Vitamin E	3
Vitamin K	3
Vitamin C	0
Vitamin B-6	16
Folic acid	5
Minerals	
Sodium	1
Calcium	3
Magnesium	33
Copper	25
Potassium	13
Iron	20
Manganese	97
Zinc	18

National Nutrient Database
[a]Based on a 2000-calorie diet

arthritis [39]. Barley's high mineral content contributes to building and maintaining bone structure/strength. Additionally, it also plays a role in the production and maturation of collagen and in that way increases the flexibility of rigid body joints even enhancing blood vessel regeneration. For instance, the high copper concentration contributes to hemoglobin and red blood cell production [40]. Low selenium intake has been associated with increased risk of mortality, poor immune function, and cognitive decline and may adversely affect risk of cancers [32]. Actually, selenium deficiency has been associated with different types of epithelial cancers where selenium-binding protein 1 is downregulated in malignant melanoma, prostate cancer, and lung carcinoma [41]. Low selenium status may also lead to some skin disorders including dermatitis [42]. Barley can likely protect against cutaneous inflammation and cancer conditions since selenium is present in high amounts and due to its protective inflammatory and neoplastic properties. One probable mechanism is the downregulation of selenoproteins, namely glutathione peroxidase 4 (GPX4) that results in significantly improved skin lesions [43]. Barley cereal contains lignans, i.e., phytoestrogens. These are vital phytonutrients that have shown antioxidant activities even they are more potent than vitamin E [44]. This is mainly because they scavenge free radicals and protect the cells from its harmful effects. In addition, lignans exhibit strong anti-inflammatory properties in endothelial cells, at least in part, through attenuation of NF-κB and extracellular signal-regulated kinase phosphorylation [45] and accompanying decreased risk of cardiovascular disease [8]. Moreover, lignans lower the risk of ER^+ breast cancer, because of their potential for weak estrogenic or antiestrogenic effects in a woman's body [46, 47], and other forms of hormone-related cancers such as prostate cancer and endometrial cancer [48]. Choline, a water-soluble-like vitamin, is a constituent of lecithin which is found in barley. Choline derivatives like phosphatidylcholine and other classes play a role in cell membrane integrity and signaling along with acting as a source for methyl groups via its metabolite betaine [49]. It is also a precursor for the neurotransmitter acetylcholine, thus involving in vital functions including muscle movement, learning, long-term memory, and cognitive abilities [50]. It has also been associated with sleep regulation, positively influencing long sleep duration [51]. Finally, studies indicate that choline diminishes homocysteine levels associated with greater risk of cardiovascular disease, lowers the levels of several inflammatory markers (CRP, homocysteine, IL-6, and TNF), limits DNA damage and apoptosis, and accordingly decreases the risk of breast cancer [49]. Likewise, barley's antioxidants lower the risk of some cancers, like breast cancer [52], and heart disease by inhibiting inflammation processes plus the toll of aging on the body. In addition, the low sodium level in barley [53] coupled with appreciable amounts of potassium, calcium, and magnesium [54] has been found to lower hypertension and boost the heart health.

Barley is likely safe for most people. However, it is not immune to a few secondary effects. Overconsumption of barley causes unpleasant abdominal cramps, bloating, and gas formation. Since barely is not gluten-free, it triggers an immune response in the small intestine of individuals with celiac disease. Because barley lowers blood sugar levels, diabetic patients have to be cautious and consider adjusting their medication dose. Phytic acid in whole grains including barley is known to bind nutrients such as vitamins, minerals, and proteins, thereby preventing their absorption [20]. Barley flour can sometimes cause asthma [55].

4 Rye

Family: Poaceae
Genus: *Secale*
Common name: Rye

Rye is a grass widely grown as a cereal grain and a cover crop to manage soil erosion. Rye looks much like wheat, but it is longer and slenderer and ranges in color from yellowish brown to grayish green. Historically, rye was known as the poor man's grain. Rye grain is used in flour, as animal fodder, and as a fermented ingredient in some alcoholic drinks. It can be eaten as a whole grain in different types of breads. In contrast to refined wheat flour, rye flour usually retains more nutrients because it is not easy to detach the germ and bran from the endosperm layers. Rye is very rich in fiber and contains a number of minerals like copper, zinc, magnesium, and manganese. Rye is also a decent source of phenolic antioxidant compounds and some vitamins like cobalamin (B-12) and folic acid (B-9) (Table 4).

This tasty, rich grain provides numerous health benefits. Rye contains non-cellulose polysaccharides and fibers displaying high binding capacity to water that increase the satiety feeling and, as a result, help in controlling body weight. Whole grains, including rye, contain high concentration of dietary fiber and different types of antioxidants that were found to reduce the risk of multiple diseases like coronary heart diseases and certain types of cancer, especially colorectal cancer [56]. The dietary fiber in rye regulates bowel movement, prevents constipation [57], protects from colon cancer, and even decreases the risk of cholesterol gallstones [58]. It has been concluded that consuming rye products lowers the risk of type 2 diabetes [59]. Phenolic acid antioxidants in rye have significant health benefits in prevention of chronic diseases such as cardiovascular disease, diabetes, and cancer [60]. One type, the hydroxyl-cinnamate family, has been linked to a number of beneficial health effects like protecting low-density lipoprotein (LDL) from oxidation which may help prevent atherosclerosis and coronary heart disease [61]. They have also been implicated in prevention of obesity, diabetes [62], and breast cancer [63]. In other investigations, these antioxidants were observed to reduce the risk of colon (other digestive cancers), breast, and prostate cancers [64]. Whole-grain consumption, including rye, has been thought to lower the risk of childhood asthma and allergies [65].

Table 4 Rye nutrition facts of 100 g, from the United States Department of Agriculture (USDA SR-21)

Serving size of 100 g of rye	
Per serving	% Daily value[a]
Calories 335	
Total fat 2.5 g	4
Saturated fat 0.3 g	1
Polyunsaturated fat 1.1 g	
Monounsaturated fat 0.3 g	
Total omega-3 fatty acids 157 mg	
Total omega-6 fatty acids 958 mg	
Cholesterol 0.0 mg	0
Phytosterols	
Carbohydrates 69.8 g	23
Dietary fiber 14.6 g	58
Starch	
Sugars 1.0 g	
Protein 14.8 g	30
Vitamins	
Vitamin A	0
Vitamin E	6
Vitamin K	7
Vitamin C	0
Vitamin B-6	15
Folic acid	15
Minerals	
Sodium	0
Calcium	3
Magnesium	30
Copper	23
Potassium	8
Iron	15
Manganese	134
Zinc	25

National Nutrient Database
[a]Based on a 2000-calorie diet

Rye, like other grains, has a number of undesirable health consequences. As a grain containing gluten, people with gluten sensitivity or celiac disease should avoid eating rye [66]. Overconsumption of rye may cause nausea, abdominal pain, and bloating. Rye is exceptionally sensitive to a toxin-producing parasitic fungus, *Claviceps purpurea*. When ingested, some people can exhibit a range of hallucinogenic effects. Some grains, including rye, contain potentially toxic proteins termed gliadins. Gliadin proteins are a type of prolamin (i.e., proteins rich in prolines and glutamines typically found in the water-soluble component of gluten while glutenin is the insoluble fraction). These prolamins are called gliadins in wheat, hordeins in barley, avenins in oats, or secalins in rye. The rye form of gliadin that is named secalin sometimes cannot be tolerated and induces inflammatory response, e.g., TNF-α, IL-1β, and other chemokines [67]. Secalin, besides lectin, affects mucosal immunity and can cause intestinal toxicity. Secalin, for instance, elicited toxic reactions in intestinal Caco-2 epithelial cells. It induced epithelial cell layer permeability, tight junctional protein distortion, and actin reorganization similar to gliadin [68]. On the other hand, mice deficient in a C-type lectin receptor experienced aggravated fungal infection [69]. Another possible effect for secalin is red blood cell aggregation; for example, lectins are believed to contribute to the pathogenesis of acute coronary syndromes through cell aggregation/adhesion mechanisms [70]. Finally, it has the potential to provoke abnormal immune system reaction and activate zonulin (haptoglobin) signaling in the gut which is an insulin mimic that modulates permeability and cellular communication in the digestive tract. It has been implicated in the pathogenesis of celiac disease and type 1 diabetes [71].

5 Quinoa

Family: Amaranthaceae
Genus: *Chenopodium*
Common name: Quinoa

Quinoa, a flowering plant in the amaranth family, is grown as a grain crop. Quinoa is a pseudo-cereal which means that it is in fact a seed but is consumed as a grain. Quinoa is often referred to as a "super food" because of its high fiber and quality protein content (gluten-free). It is packed with micronutrients such as magnesium (helps to relax blood vessels, thus alleviating migraines and promoting blood sugar control), manganese (protects against free radicals and prevents damage of mitochondria), B vitamins (improves energy metabolism), and other minerals like iron, potassium, calcium, and phosphorus, which, generally, are higher than in most grains [72, 73]. It is high in vitamin E, an antioxidant, and plant antioxidant compounds like flavonoids, in particular, quercetin and kaempferol. In animal studies, these classes of flavonols have been shown to possess anti-inflammatory [74], anti-cancer [75], and antidepressant activities [76]. Its extracts possess strong antimicrobial activities against foodborne pathogens as well [77] (Table 5).

Table 5 Quinoa nutrition facts of 100 g, from the United States Department of Agriculture (USDA SR-21)

Serving size of 100 g of uncooked quinoa	
Per serving	% Daily value[a]
Calories 368	
Total fat 6.1 g	9
Saturated fat 0.7 g	4
Polyunsaturated fat 3.3 g	
Monounsaturated fat 1.6 g	
Total omega-3 fatty acids 307 mg	
Total omega-6 fatty acids 2977 mg	
Cholesterol 0.0 mg	0
Phytosterols	
Carbohydrates 64.2 g	21
Dietary fiber 7.0 g	28
Starch 52.2 g	
Sugars	
Protein 14.1 g	28
Vitamins	
Vitamin A	0
Vitamin E	12
Vitamin K	0
Vitamin C	
Vitamin B-6	24
Folic acid	46
Minerals	
Sodium	0
Calcium	5
Magnesium	49
Copper	30
Potassium	16
Iron	25
Manganese	102
Zinc	21

National Nutrient Database
[a]Based on a 2000-calorie diet

Quinoa is known for having all nine essential amino acids that the body cannot produce (therefore needs to be consumed), one of which is lysine that is important for normal tissue growth and repair [78]. Quinoa is even considered a complete protein by some [79]. Furthermore, using quinoa in a gluten-free diet may increase the value of antioxidants and nutrients in said diet [80]. Hence, this is an excellent alternative for individuals with autism on the gluten-casein-free (GFCF) diet. Due to its high fiber content, it helps control the body weight [81, 82], regulate blood pressure, lower blood sugar (since it is very low on the glycemic index) [83, 84], and reduce cholesterol levels [85], eventually reducing the risk of heart disease. Quinoa plays great role in handling mineral levels preventing uric acid renal stone formation [86, 87]. High fiber content in quinoa may also protect against gallbladder stones [88]. Quinoa is also high in iron, needed for hemoglobin formation, and is rich in copper, manganese, and vitamin B-2 that are vital for metabolism and development [89, 90].

Not much is known about potential overconsumption of quinoa. Like other cereal grains, quinoa contains phytic acid which interferes with absorption of some micronutrients such as iron and zinc [91]. In addition, the seeds have a coating of saponins, natural detergents, especially concentrated in the seed hull (that can be removed by rinsing well with water) giving it a bitter taste and, when digested, lead to stomach irritation [92]. Oxalate content in quinoa may lead to kidney stone formation when consumed excessively [93]. Individuals with chronic kidney disease need to limit their potassium intake in their diet [94].

References

1. Kristensen, M., Toubro, S., Jensen, M. G., Ross, A. B., Riboldi, G., Petronio, M., et al. (2012). Whole grain compared with refined wheat decreases the percentage of body fat following a 12-week, energy-restricted dietary intervention in postmenopausal women. *The Journal of Nutrition, 142*(4), 710–716.
2. Aller, E. E., Abete, I., Astrup, A., Martinez, J. A., & van Baak, M. A. (2011). Starches, sugars and obesity. *Nutrients, 3*(3), 341–369.
3. Slavin, J. (2003). Why whole grains are protective: Biological mechanisms. *The Proceedings of the Nutrition Society, 62*(1), 129–134.
4. Likes, R., Madl, R. L., Zeisel, S. H., & Craig, S. A. (2007). The betaine and choline content of a whole wheat flour compared to other mill streams. *Journal of Cereal Science, 46*(1), 93–95.
5. Craig, S. A. (2004). Betaine in human nutrition. *The American Journal of Clinical Nutrition, 80*(3), 539–549.
6. Pena-Rosas, J. P., Rickard, S., & Cho, S. (1999). Wheat bran and breast cancer: Revisiting the estrogen hypothesis. *Archivos Latinoamericanos de Nutrición, 49*(4), 309–317.
7. Qu, H., Madl, R. L., Takemoto, D. J., Baybutt, R. C., & Wang, W. (2005). Lignans are involved in the antitumor activity of wheat bran in colon cancer SW480 cells. *The Journal of Nutrition, 135*(3), 598–602.
8. Peterson, J., Dwyer, J., Adlercreutz, H., Scalbert, A., Jacques, P., & McCullough, M. L. (2010). Dietary lignans: Physiology and potential for cardiovascular disease risk reduction. *Nutrition Reviews, 68*(10), 571–603.
9. Makiuchi, T., Sobue, T., Kitamura, T., Ishihara, J., Sawada, N., Iwasaki, M., et al. (2017). The relationship between vegetable/fruit consumption and gallbladder/bile duct cancer: A population-based cohort study in Japan. *International Journal of Cancer, 140*(5), 1009–1019.
10. Behall, K. M., Scholfield, D. J., & Hallfrisch, J. (2006). Whole-grain diets reduce blood pressure in mildly hypercholesterolemic men and women. *Journal of the American Dietetic Association, 106*(9), 1445–1449.
11. Athanasopoulos, D., Karagiannis, G., & Tsolaki, M. (2016). Recent findings in Alzheimer disease and nutrition focusing on epigenetics. *Advances in Nutrition, 7*(5), 917–927.
12. Abdel-Aal el-S. M., Akhtar, H., Zaheer, K., & Ali, R. (2013). Dietary sources of lutein and zeaxanthin carotenoids and their role in eye health. *Nutrients, 5*(4), 1169–1185.
13. Di Sabatino, A., & Corazza, G. R. (2009). Coeliac disease. *Lancet, 373*(9673), 1480–1493.
14. Hadjivassiliou, M., Grünewald, R., Sharrack, B., Sanders, D., Lobo, A., Williamson, C., et al. (2003). Gluten ataxia in perspective: Epidemiology, genetic susceptibility and clinical characteristics. *Brain, 126*(Pt 3), 685–691.

15. Hadjivassiliou, M., Davies-Jones, G. A., Sanders, D. S., & Grünewald, R. A. (2003). Dietary treatment of gluten ataxia. *Journal of Neurology, Neurosurgery, and Psychiatry, 74*(9), 1221–1224.
16. Coury, D. (2014). Retrieved from https://www.autismspeaks.org/node/112986
17. Buie, T. (2013). The relationship of autism and gluten. *Clinical Therapeutics, 35*(5), 578–583.
18. Adams, J. B., Audhya, T., Geis, E., Gehn, E., Fimbres, V., Pollard, E. L., et al. (2018). Comprehensive nutritional and dietary intervention for autism spectrum disorder—A randomized, controlled 12-month trial. *Nutrients, 10*(3), E369.
19. Gogou, M., & Kolios, G. (2018). Are therapeutic diets an emerging additional choice in autism spectrum disorder management? *World Journal of Pediatrics, 14*(3), 215–223.
20. Lopez, H. W., Leenhardt, F., Coudray, C., & Remesy, C. (2002). Minerals and phytic acid interactions: Is it a real problem for human nutrition? *International Journal of Food Science & Technology, 37*(7), 727–739.
21. Fagagnini, S., Heinrich, H., Rossel, J. B., Biedermann, L., Frei, P., Zeitz, J., et al. (2017). Risk factors for gallstones and kidney stones in a cohort of patients with inflammatory bowel diseases. *PLoS One, 12*(10), e0185193.
22. Davy, B. M., Davy, K. P., Ho, R. C., Beske, S. D., Davrath, L. R., & Melby, C. L. (2002). High-fiber oat cereal compared with wheat cereal consumption favorably alters LDL-cholesterol subclass and particle numbers in middle-aged and older men. *The American Journal of Clinical Nutrition, 76*(2), 351–358.
23. Meydani, M. (2009). Potential health benefits of avenanthramides of oats. *Nutrition Reviews, 67*(12), 731–735.
24. Pereira, M. A., Jacobs, D. R. Jr., Pins, J. J., Raatz, S. K., Gross, M. D., Slavin, J. L., et al. (2002). Effect of whole grains on insulin sensitivity in overweight hyperinsulinemic adults. *The American Journal of Clinical Nutrition, 75*(5), 848–855.
25. EFSA Panel on Dietetic Products, Nutrition and Allergies (NDA). (2010). Scientific opinion on the substantiation of a health claim related to oat beta glucan and lowering blood cholesterol and reduced risk of (coronary) heart disease pursuant to Article 14 of Regulation (EC) No 1924/2006. *EFSA Journal, 8*(12), 1885.
26. Rebello, C. J., O'Neil, C. E., & Greenway, F. L. (2016). Dietary fiber and satiety: The effects of oats on satiety. *Nutrition Reviews, 74*(2), 131–147.
27. Maki, K. C., Galant, R., Samuel, P., Tesser, J., Witchger, M. S., Ribaya-Mercado, J. D., et al. (2007). Effects of consuming foods containing oat beta-glucan on blood pressure, carbohydrate metabolism and biomarkers of oxidative stress in men and women with elevated blood pressure. *European Journal of Clinical Nutrition, 61*(6), 786–795.
28. El Khoury, D., Cuda, C., Luhovyy, B. L., & Anderson, G. H. (2012). Beta glucan: Health benefits in obesity and metabolic syndrome. *Journal of Nutrition and Metabolism, 2012*, 851362.
29. Rondanelli, M., Opizzi, A., & Monteferrario, F. (2009). The biological activity of beta-glucans. *Minerva Medica, 100*(3), 237–245.
30. Aschner, M., & Dorman, D. C. (2006). Manganese: Pharmacokinetics and molecular mechanisms of brain uptake. *Toxicological Reviews, 25*(3), 147–154.
31. Allen, K. G., & Klevay, L. M. (1994). Copper: An antioxidant nutrient for cardiovascular health. *Current Opinion in Lipidology, 5*(1), 22–28.
32. Rayman, M. P. (2012). Selenium and human health. *Lancet, 379*(9822), 1256–1268.
33. Reynertson, K. A., Garay, M., Nebus, J., Chon, S., Kaur, S., Mahmood, K., et al. (2015). Anti-inflammatory activities of colloidal oatmeal (Avena sativa) contribute to the effectiveness of oats in treatment of itch associated with dry, irritated skin. *Journal of Drugs in Dermatology, 14*(1), 43–48.
34. Smulders, M. J. M., van de Wiel, C. C. M., van den Broeck, H. C., van der Meer, I. M., Israel-Hoevelaken, T. P. M., Timmer, R. D., et al. (2018). Oats in healthy gluten-free and regular diets: A perspective. *Food Research International, 110*, 3–10.
35. Talati, R., Baker, W. L., Pabilonia, M. S., White, C. M., & Coleman, C. I. (2009). The effects of barley-derived soluble fiber on serum lipids. *Annals of Family Medicine, 7*(2), 157–163.

36. Li, J., Kaneko, T., Qin, L. Q., Wang, J., & Wang, Y. (2003). Effects of barley intake on glucose tolerance, lipid metabolism, and bowel function in women. *Nutrition, 19*(11–12), 926–929.
37. Smith, K. N., Queenan, K. M., Thomas, W., Fulcher, R. G., & Slavin, J. L. (2008). Physiological effects of concentrated barley beta-glucan in mildly hypercholesterolemic adults. *Journal of the American College of Nutrition, 27*(3), 434–440.
38. Lee, K. P., Kim, C., Lee, D. W., & Apel, K. (2003). TIGRINA d, required for regulating the biosynthesis of tetrapyrroles in barley, is an ortholog of the FLU gene of Arabidopsis thaliana. *FEBS Letters, 553*(1–2), 119–124.
39. Ullah, Z., Ullah, M., Hussain, S., Kaul, H., & Lone, K. P. (2017). Determination of serum trace elements (Zn, Cu, and Fe) in Pakistani patients with rheumatoid arthritis. *Biological Trace Element Research, 175*(1), 10–16.
40. Hunter III., J. P. (2014). *Health benefits: From foods and spices* (p. 557). Washington, DC: John P. Hunter III.
41. Schott, M., de Jel, M. M., Engelmann, J. C., Renner, P., Geissler, E. K., Bosserhoff, A. K., & Kuphal, S. (2018). Selenium binding protein 1 is down-regulated in malignant melanoma. *Oncotarget, 9*(12), 10445–10456.
42. Voss, G. T., Oliveira, R. L., de Souza, J. F., Duarte, L. F. B., Fajardo, A. R., Alves, D., et al. (2018). Therapeutic and technological potential of 7-chloro-4-phenylselanyl quinoline for the treatment of atopic dermatitis-like skin lesions in mice. *Materials Science & Engineering. C, Materials for Biological Applications, 84*, 90–98.
43. Arbiser, J. L., Bonner, M. Y., Ward, N., Elsey, J., & Rao, S. (2018). Selenium unmasks protective iron armor: A possible defense against cutaneous inflammation and cancer. *Biochimica et Biophysica Acta - General Subjects*. pii: S0304-4165(18)30150-8.
44. Lu, H., & Liu, G. T. (1992). Anti-oxidant activity of dibenzocyclooctene lignans isolated from Schisandraceae. *Planta Medica, 58*(4), 311–313.
45. Spilioti, E., Holmbom, B., Papavassiliou, A. G., & Moutsatsou, P. (2014). Lignans 7-hydroxymatairesinol and 7-hydroxymatairesinol 2 exhibit anti-inflammatory activity in human aortic endothelial cells. *Molecular Nutrition & Food Research, 58*(4), 749–759.
46. Calado, A., Neves, P. M., Santos, T., & Ravasco, P. (2018). The effect of flaxseed in breast cancer: A literature review. *Frontiers in Nutrition, 5*, 4.
47. Solovyev, N. D., Fedoros, E. I., Drobyshev, E. J., Ivanenko, N. B., Pigarev, S. E., Tyndyk, M. L., et al. (2017). Anticancer activity and tissue distribution of platinum (II) complex with lignin-derived polymer of benzene-poly-carboxylic acids. *Journal of Trace Elements in Medicine and Biology, 43*, 72–79.
48. Arts, C. J., Govers, C. A., van den Berg, H., Wolters, M. G., van Leeuwen, P., & Thijssen, J. H. (1991). In vitro binding of estrogens by dietary fiber and the in vivo apparent digestibility tested in pigs. *The Journal of Steroid Biochemistry and Molecular Biology, 38*(5), 621–628.
49. Zeisel, S. H., & da Costa, K. A. (2009). Choline: An essential nutrient for public health. *Nutrition Reviews, 67*(11), 615–623.
50. Moreno, H. C., de Brugada, I., Carias, D., & Gallo, M. (2013). Long-lasting effects of prenatal dietary choline availability on object recognition memory ability in adult rats. *Nutritional Neuroscience, 16*(6), 269–274.
51. Grandner, M. A., Jackson, N., Gerstner, J. R., & Knutson, K. L. (2013). Dietary nutrients associated with short and long sleep duration. Data from a nationally representative sample. *Appetite, 64*, 71–80.
52. Kyriakopoulou, K., Kefali, E., Piperigkou, Z., Bassiony, H., & Karamanos, N. K. (2018). Advances in targeting epidermal growth factor receptor signaling pathway in mammary cancer. *Cellular Signalling, 51*, 99–109.
53. Nishimuta, M., Kodama, N., Yoshitake, Y., Shimada, M., & Serizawa, N. (2018). Dietary salt (sodium chloride) requirement and adverse effects of salt restriction in humans. *Journal of Nutritional Science and Vitaminology (Tokyo), 64*(2), 83–89.
54. Schutten, J. C., Joosten, M. M., de Borst, M. H., & Bakker, S. J. (2018). Magnesium and blood pressure: A physiology-based approach. *Advances in Chronic Kidney Disease, 25*(3), 244–250.

55. Vidal, C., & Gonzalez-Quintela, A. (1995). Food-induced and occupational asthma due to barley flour. *Annals of Allergy, Asthma & Immunology, 75*(2), 121–124.
56. Slavin, J. (2004). Whole grains and human health. *Nutrition Research Reviews, 17*(1), 99–110.
57. Hongisto, S. M., Paajanen, L., Saxelin, M., & Korpela, R. (2006). A combination of fibre-rich rye bread and yoghurt containing lactobacillus GG improves bowel function in women with self-reported constipation. *European Journal of Clinical Nutrition, 60*(3), 319–324.
58. Schwesinger, W. H., Kurtin, W. E., Page, C. P., Stewart, R. M., & Johnson, R. (1999). Soluble dietary fiber protects against cholesterol gallstone formation. *American Journal of Surgery, 177*(4), 307–310.
59. Nygren, C., Hallmans, G., & Lithner, F. (1984). Effects of high-bran bread on blood glucose control in insulin-dependent diabetic patients. *Diabète & Métabolisme, 10*(1), 39–43.
60. Van Hung, P. (2016). Phenolic compounds of cereals and their antioxidant capacity. *Critical Reviews in Food Science and Nutrition, 56*(1), 25–35.
61. Andreasen, M. F., Landbo, A. K., Christensen, L. P., Hansen, A., & Meyer, A. S. (2001). Antioxidant effects of phenolic rye (Secale cereale L.) extracts, monomeric hydroxycinnamates, and ferulic acid dehydrodimers on human low-density lipoproteins. *Journal of Agricultural and Food Chemistry, 49*(8), 4090–4096.
62. Sandberg, J. C., Björck, I. M. E., & Nilsson, A. C. J. N. J. (2017). Effects of whole grain rye, with and without resistant starch type 2 supplementation, on glucose tolerance, gut hormones, inflammation and appetite regulation in an 11–14.5 hour perspective; a randomized controlled study in healthy subjects. *Nutrition Journal, 16*(1), 25.
63. Adlercreutz, H. (2010). Can rye intake decrease risk of human breast cancer? Food & Nutrition Research, *54*. https://doi.org/10.3402/fnr.v54i0.5231
64. Adom, K. K., & Liu, R. H. (2002). Antioxidant activity of grains. *Journal of Agricultural and Food Chemistry, 50*(21), 6182–6187.
65. Tabak, C., Wijga, A. H., de Meer, G., Janssen, N. A., Brunekreef, B., & Smit, H. A. (2006). Diet and asthma in Dutch school children (ISAAC-2). *Thorax, 61*(12), 1048–1053.
66. Zeltner, D., Glomb, M. A., & Mäde, D. (2009). Real-time PCR systems for the detection of the gluten-containing cereals wheat, spelt, kamut, rye, barley and oat. *European Food Research and Technology, 228*, 321–330.
67. Gujral, N., Suh, J. W., & Sunwoo, H. H. (2015). Effect of anti-gliadin IgY antibody on epithelial intestinal integrity and inflammatory response induced by gliadin. *BMC Immunology, 16*, 41.
68. Stenman, S. M., Lindfors, K., Venäläinen, J. I., Hautala, A., Männistö, P. T., Garcia-Horsman, J. A., et al. (2010). Degradation of coeliac disease-inducing rye secalin by germinating cereal enzymes: Diminishing toxic effects in intestinal epithelial cells. *Clinical and Experimental Immunology, 161*(2), 242–249.
69. Wang, T., Pan, D., Zhou, Z., You, Y., Jiang, C., Zhao, X., et al. (2016). Dectin-3 deficiency promotes colitis development due to impaired antifungal innate immune responses in the gut. *PLoS Pathogens, 12*(6), e1005662.
70. Gorudko, I. V., Buko, I. V., Cherenkevich, S. N., Polonetsky, L. Z., & Timoshenko, A. V. (2008). Lectin-induced aggregates of blood cells from patients with acute coronary syndromes. *Archives of Medical Research, 39*(7), 674–681.
71. Fasano, A. (2011). Zonulin and its regulation of intestinal barrier function: The biological door to inflammation, autoimmunity, and cancer. *Physiological Reviews, 91*(1), 151–175.
72. Nowak, V., Du, J., & Charrondiere, U. (2015). Assessment of the nutritional composition of quinoa (Chenopodium quinoa Willd.). *Food Chemistry, 193*, 47–54.
73. Maradini-Filho, A. M. (2017). Quinoa: Nutritional aspects. *Journal of Nutraceuticals and Food Science, 2*(1), 3–7.
74. Stewart, L. K., Soileau, J. L., Ribnicky, D., Wang, Z. Q., Raskin, I., Poulev, A., et al. (2008). Quercetin transiently increases energy expenditure but persistently decreases circulating markers of inflammation in C57BL/6J mice fed a high-fat diet. *Metabolism, 57*(7 Suppl 1), S39–S46.
75. Murakami, A., Ashida, H., & Terao, J. (2008). Multitargeted cancer prevention by quercetin. *Cancer Letters, 269*(2), 315–325.

76. Hou, Y., Aboukhatwa, M. A., Lei, D. L., Manaye, K., Khan, I., & Luo, Y. (2010). Antidepressant natural flavonols modulate BDNF and beta amyloid in neurons and hippocampus of double TgAD mice. *Neuropharmacology, 58*(6), 911–920.
77. Park, J. H., Lee, Y. J., Kim, Y. H., & Yoon, K. S. (2017). Antioxidant and antimicrobial activities of quinoa (Chenopodium quinoa Willd.) seeds cultivated in Korea. *Preventive Nutrition and Food Science, 22*(3), 195–202.
78. Trackman, P. C. (2018). Functional importance of lysyl oxidase family propeptide regions. *Journal of Cell Communication and Signaling, 12*(1), 45–53.
79. Reguera, M., Conesa, C. M., Gil-Gómez, A., Haros, C. M., Pérez-Casas, M. Á., Briones-Labarca, V., et al. (2018). The impact of different agroecological conditions on the nutritional composition of quinoa seeds. *PeerJ, 6*, e4442.
80. Alvarez-Jubete, L., Arendt, E. K., & Gallagher, E. (2009). Nutritive value and chemical composition of pseudocereals as gluten-free ingredients. *International Journal of Food Sciences and Nutrition, 60*(Suppl 4), 240–257.
81. Grube, B., Chong, P. W., Lau, K. Z., & Orzechowski, H. D. (2013). A natural fiber complex reduces body weight in the overweight and obese: A double-blind, randomized, placebo-controlled study. *Obesity (Silver Spring), 21*(1), 58–64.
82. Burton-Freeman, B. (2000). Dietary fiber and energy regulation. *The Journal of Nutrition, 130*(2S Suppl), 272s–275s.
83. Ranilla, L. G., Apostolidis, E., Genovese, M. I., Lajolo, F. M., & Shetty, K. (2009). Evaluation of indigenous grains from the Peruvian Andean region for antidiabetes and antihypertension potential using in vitro methods. *Journal of Medicinal Food, 12*(4), 704–713.
84. Weickert, M. O., & Pfeiffer, A. F. (2008). Metabolic effects of dietary fiber consumption and prevention of diabetes. *The Journal of Nutrition, 138*(3), 439–442.
85. Jenkins, D. J., Wolever, T. M., Rao, A. V., Hegele, R. A., Mitchell, S. J., Ransom, T. P., et al. (1993). Effect on blood lipids of very high intakes of fiber in diets low in saturated fat and cholesterol. *The New England Journal of Medicine, 329*(1), 21–26.
86. Elsawy, A. A., Elshal, A. M., El-Nahas, A. R., Elbaset, M. A., Farag, H., & Shokeir, A. A. (2019). Can we predict the outcome of oral dissolution therapy for radiolucent renal calculi? A prospective study. *The Journal of Urology, 201*(2), 350–357.
87. Trinchieri, A., Esposito, N., & Castelnuovo, C. (2009). Dissolution of radiolucent renal stones by oral alkalinization with potassium citrate/potassium bicarbonate. *Archivio Italiano di Urologia, Andrologia, 81*(3), 188–191.
88. Sachdeva, S., Khan, Z., Ansari, M. A., Khalique, N., & Anees, A. (2011). Lifestyle and gallstone disease: Scope for primary prevention. *Indian Journal of Community Medicine, 36*(4), 263–267.
89. Ashoori, M., & Saedisomeolia, A. (2014). Riboflavin (vitamin B2) and oxidative stress: A review. *British Journal of Nutrition, 111*(11), 1985–1991.
90. Li, L., & Yang, X. (2018). The essential element manganese, oxidative stress, and metabolic diseases: Links and interactions. *Journal of Oxidative Medicine and Cellular Longevity, 2018*, 11.
91. Jancurová, M., Minarovičová, L., & Dandár, A. (2009). Quinoa—A review. *Czech Journal of Food Sciences, 27*, 71–79.
92. Wen, L., Xia, N., Tang, P., Hong, Y., Wang, Z., Liu, Y., et al. (2015). The gastrointestinal irritation of polygala saponins and its potential mechanism in vitro and in vivo. *BioMed Research International, 2015*, 1–8.
93. Holmes, R. P., & Assimos, D. G. (2004). The impact of dietary oxalate on kidney stone formation. *Urological Research, 32*(5), 311–316.
94. Fund, A. K. *Kidney-friendly diet for CKD*. Retrieved from http://www.kidneyfund.org/kidney-disease/chronic-kidney-disease-ckd/kidney-friendly-diet-for-ckd.html

Nuts

Sawsan G. Mohammed and M. Walid Qoronfleh

Abstract Nuts are fruits composed of two parts: an inedible hard shell and an edible seed. Nuts are known as an energy-dense and nutrient-rich food source. In general, nuts are recognized as a good source of fat, fiber, and protein. Nuts are extremely beneficial parts of any diet since their consumption may lower risk for some diseases, such as cardiovascular diseases and cancer. They are acknowledged for their low glycemic index owning to high unsaturated fat and protein content and relatively low carbohydrate content. They have been shown to increase cognitive function as well.

Keywords Nuts · Fiber · Antioxidants · Essential fatty acids · Monounsaturated fatty acids · Omega fatty acids · Omega-3 fatty acid · Phenolic acids · Phytic acid · Phytosterols · β-Sitosterol

---

S. G. Mohammed (✉)
Qatar Research Leadership Program (QRLP), Qatar Foundation, Doha, Qatar
e-mail: sgmohammed@qf.org.qa

M. W. Qoronfleh (✉)
Research & Policy Department, World Innovation Summit for Health (WISH), Qatar Foundation, Doha, Qatar
e-mail: wqoronfleh@qf.org.qa

M. M. Essa, M. W. Qoronfleh (eds.), *Personalized Food Intervention and Therapy for Autism Spectrum Disorder Management*, Advances in Neurobiology 24,
https://doi.org/10.1007/978-3-030-30402-7_12

1 Almonds

Family: Rosaceae
Genus: *Prunus*
Common name: Almond

Almonds are the seeds of fruits grown from the almond tree. There are two types of almonds: edible sweet ones, and bitter almonds which are used for making oil and flavoring food. They are usually sold raw, roasted, or blanched (skin removed). They can be eaten as a snack or added to salads and other dishes. Almond milk is also a tasty drink that can be an alternative to less nutritious cow's milk. Almonds can be eaten preferably on an empty stomach to increase and speed up the absorption of their nutrients. Despite the fact that they contain high amount of fat, almond is a highly nutritious nut and a rich source of healthy essential fatty acids, namely linoleic and linolenic acids, vitamin E, calcium, iron, phosphorus, manganese, and magnesium. It also contains fiber, phytic acid, zinc, selenium, copper, riboflavin (B-2), and niacin (B-3) (Table 1). The brown layer of almond skin is a very rich source of antioxidants.

Almonds have a wide array of notable health benefits. The high content of antioxidants and polyphenols in almond skin protects the body from oxidative damage that contributes to different diseases notably diabetes and coronary heart disease [1] and reduces some of the oxidative damage biomarkers, namely plasma malondialdehyde (MDA) and urinary isoprostanes (prostaglandin-like compound) [2]. The rich amount of vitamin E in almonds is linked to their protective effects against coronary heart disease [3], colon cancer [4], and age-related cognitive damage and Alzheimer's disease [5]. The low levels of carbohydrates and high levels of healthy unsaturated fats, proteins, and dietary fiber in almonds contribute to their ability to control type 2 diabetes and cardiovascular diseases [6]. Low carbs and high fiber help control body weight, decrease hunger, and promote satiety [7]. Additionally, the high fiber content of almonds regulates bowel movement and prevents constipation [8]. The high magnesium concentration in almonds is also thought to improve insulin function and decrease blood glucose [9] and lower the blood pressure [10], consequently reducing the risk of stroke and renal problems. Almonds were also found to lower the oxidation of unhealthy LDL cholesterol further boosting the heart and blood vessel health [11].

Table 1 Almond nutrition facts of 100 g, from the United States Department of Agriculture (USDA SR-21)

Serving size of 100 g of almond	
Per serving	% Daily value[a]
Calories 575	
Total fat 49.4 g	76
Saturated fat 3.7 g	19
Polyunsaturated fat 12.1 g	
Monounsaturated fat 30.9 g	
Total omega-3 fatty acids 6.0 mg	
Total omega-6 fatty acids 12,065 mg	
Cholesterol 0.0 mg	0
Phytosterols	
Carbohydrates 21.7 g	7
Dietary fiber 12.2 g	49
Starch 0.7 g	
Sugars 3.9 g	
Protein 21.2 g	42
Vitamins	
Vitamin A	0
Vitamin E	131
Vitamin K	0
Vitamin C	0
Vitamin B-6	7
Folic acid	12
Minerals	
Sodium	0
Calcium	26
Magnesium	67
Copper	50
Potassium	20
Iron	21
Manganese	114
Zinc	21

National Nutrient Database
[a]Based on a 2000-calorie diet

Like any other food, almonds have their disadvantages. They contain oxalates, and excessive consumption could cause crystallization in different tissues leading to kidney or gallbladder stones [12]. Overconsumption of almonds may cause loss of appetite, constipation, and bloating. On the other hand, the high content of fats in almonds might increase body weight especially when no regular physical activities are being performed. Overconsumption of bitter almonds might be toxic; they have high levels of the poisonous compound hydrocyanic acid (cyanide), which potentially leads to breathing difficulty, nervous breakdown, choking, and even death. It is strictly prohibited for pregnant women to eat. Allergic reactions of almonds are common, and they might lead to serious reactions or anaphylactic shock. Manganese content of almonds might interfere with certain medication such as some laxatives, antihypertensive drugs, and antibiotics.

2 Hazelnuts

Family: Betulaceae
Genus: *Corylus*
Common names: Hazelnut or cobnut or filbert

Hazelnuts are group of roughly spherical to oval or a bit elongated nuts—depending on the species—with a smooth shell covered by an outer fibrous husk. When ripe, hazelnut falls out of its husk. The marble-size seeds have thin, dark-brown-colored skin which is customarily removed before cooking. Hazelnut can be eaten raw, roasted, or utilized in recipes to prepare pastes, to give ice cream a special flavor, and to add in chocolates. Hazelnut oil is highly flavorful and can be used for cooking. Hazelnuts are rich in protein, monounsaturated fat, vitamin E, pyridoxine (B-6), folic acid (B-9), copper, manganese, and many other nutrients essential for health (Table 2).

Like other nuts, hazelnut health benefits extend to reach most if not all body systems. Hazelnuts contain a number of vitamins and minerals that boost heart and blood vessel health. Apart from the rich dietary fiber content in hazelnuts, the high amounts of fatty acids, antioxidants, potassium, and magnesium they contain help lower blood pressure [13]. The significant amount of monounsaturated fatty acids, like oleic acid, also helps reduce the unhealthy LDL cholesterol, increase healthy HDL cholesterol, and reduce inflammation [14]. Hazelnuts are loaded with different potent antioxidants like proanthocyanidins [15] in addition to some vitamins like vitamin E [16] and minerals like manganese [17] that have been demonstrated to have anticancer properties, reduce oxidative stress, and safeguard cells from damage caused by free radicals. Hazelnut extracts have been shown to protect individuals against several cancers such as cervical [18], liver, breast [19], and colon cancers [20]. β-Sitosterol, a plant phytosterol found in hazelnuts—its chemical structure similar to cholesterol—has been observed for its protective effect against few

Table 2 Hazelnut nutrition facts of 100 g, from the United States Department of Agriculture (USDA SR-21)

Serving size of 100 g of hazelnut	
Per serving	% Daily value[a]
Calories 628	
Total fat 60.7 g	93
Saturated fat 4.5 g	22
Polyunsaturated fat 7.9 g	
Monounsaturated fat 45.7 g	
Total omega-3 fatty acids 87.0 mg	
Total omega-6 fatty acids 7832 mg	
Cholesterol 0.0 mg	0
Phytosterols 96.0 mg	
Carbohydrates 16.7 g	6
Dietary fiber 9.7 g	39
Starch 0.5 g	
Sugars 4.3 g	
Protein 15.0 g	30
Vitamins	
Vitamin A	0
Vitamin E	75
Vitamin K	18
Vitamin C	11
Vitamin B-6	28
Folic acid	28
Minerals	
Sodium	0
Calcium	11
Magnesium	41
Copper	86
Potassium	19
Iron	26
Manganese	309
Zinc	16

National Nutrient Database
[a]Based on a 2000-calorie diet

cancers particularly breast [21] and prostate [22] cancers. Hazelnut's minerals have been reported to control blood glucose level and are linked to lower risk of type 2 diabetes, chief of these minerals being manganese [23] and magnesium [24]. Similarly, monounsaturated fatty acids also influence type 2 diabetes along with other inflammatory diseases and cancer [25]. Magnesium in hazelnuts is also vital in bone health as it decreases the risk of fractures and osteoporosis [26]. Hazelnuts contain elements that boost brain and nerve health as they possess some neuroprotective abilities since they improve memory and ensure healthy aging [27]. Both manganese [28] and vitamin E [29] help lower age-related cognitive deterioration and have a fundamental role in the prevention and the treatment of dementia such as Alzheimer's and Parkinson's diseases. Vitamin E also plays a protective role for the skin by shielding it from the ultraviolet ray damage, the risk of skin cancer, and the signs of premature aging [30].

However, hazelnuts have their own set of unwanted health effects they come with. They, like other nuts, contain phytic acid, which has been shown to bind some minerals, such as iron and zinc, thus preventing their absorption [31]. Hazelnut allergens induce mild-to-severe allergic reactions in sensitized individuals [32].

3 Peanuts

Family: Fabaceae
Genus: *Arachis*
Common names: Peanut, groundnut, earth nuts, or goober

Peanut is technically not a true nut. Rather, it is a legume grain. The peanut pod develops underground, and this is why it is called a groundnut. It is also classified as an oil crop due to its high oil content. Peanut is grown mostly for its seeds which are covered by a brown paperlike coat contained in an outer shell. It is mostly consumed as roasted whole peanuts with salt or as peanut butter or to a lesser extent eaten raw. Peanut products such as oil, flour, and protein are used in a variety of foods: sweets, snacks, and sauces. Peanuts are low in carbs and very rich in fiber, proteins like arachin and conarachin, and fats that are often used to make peanut oil. Peanut fat consists mostly of mono- and polyunsaturated fatty acids. They are also a main source for vitamin E, B vitamins including thiamin (B-1), riboflavin (B-2), niacin (B-3), pantothenic acid (B-5), pyridoxine (B-6), folic acid (B-9), and choline, and several minerals including copper, magnesium, manganese, and zinc (Table 3).

Peanuts contain several nutrients that are essential for health. Their rich content of B vitamins such as niacin [33] and thiamin [34] in addition to minerals like magnesium [35] and copper [36] is particularly beneficial for the heart and cardiovascular system. Peanuts contain high amounts of monounsaturated fatty acids such as oleic acid which have been shown to reduce the unhealthy LDL cholesterol and increase healthy HDL cholesterol, thereby preventing coronary artery disease [37]. The main composition of phytosterols in peanut oil is β-sitosterol. Due to its substantial amounts and structural similarity with cholesterol, it competes with cho-

Table 3 Peanut nutrition facts of 100 g, from the United States Department of Agriculture (USDA SR-21)

Serving size of 100 g of raw peanut	
Per serving	% Daily value[a]
Calories 567	
Total fat 49.2 g	76
Saturated fat 6.8 g	34
Polyunsaturated fat 15.6 g	
Monounsaturated fat 24.4 g	
Total omega-3 fatty acids 3.0 mg	
Total omega-6 fatty acids 15,555 mg	
Cholesterol 0.0 mg	0
Phytosterols 220 mg	
Carbohydrates 16.1 g	5
Dietary fiber 8.5 g	34
Starch	
Sugars 4 g	
Protein 25.8 g	52
Vitamins	
Vitamin A	0
Vitamin E	42
Vitamin K	0
Vitamin C	0
Vitamin B-6	17
Folic acid	60
Minerals	
Sodium	1
Calcium	9
Magnesium	42
Copper	57
Potassium	20
Iron	25
Manganese	97
Zinc	22

National Nutrient Database
[a]Based on a 2000-calorie diet

lesterol absorption by the gut, in essence, lowering the blood level of the unhealthy LDL cholesterol [38]. Other B vitamins contained in peanuts like biotin [39] and folic acid [40] are vital for healthy pregnancy. It has also been suggested that biotin could be helpful in treating multiple sclerosis [41]. Biotin crosses the blood–brain barrier, and deficiency may lead to brain and central nervous system disorder symptoms like ataxia, dysphagia, and dysarthria and perhaps even lead to sensory loss [42]. Biotin in combination with chromium (but not alone) was found to lower blood glucose level in diabetics [43]. Peanuts are a rich source of different types of potent antioxidants that help in preventing many diseases. They contain resveratrol, an antioxidant that prevents heart diseases and Alzheimer's disease and lowers the risk of cancer [44]. Peanuts contain p-coumaric acid, a phenolic acid antioxidant—a hydroxyl derivative of cinnamic acid—with antibacterial properties [45] that have been found to lower the risk of stomach cancer perhaps owing to its bactericidal effect against *H. pylori* [46]. Regular consumption of peanuts has been found to reduce the risk of gallstone formation possibly attributable to its cholesterol-lowering properties [47]. Peanuts are very filling as a result of their high fiber content making them very effective for weight control as they lessen the intake of other foods [48].

Peanuts present some negative health effects too. The abundant amounts of arachin and conarachin proteins conceivably cause severe allergic reactions in some people, sometimes resulting even in life-threatening anaphylactic reactions [49]. Phytic acid in peanut possibly prevents the absorption of some nutrients like iron and zinc if consumed at the same time [31]. Therefore, overconsumption of peanuts over time may lead to nutritional deficiencies of such minerals. Depending on their storage conditions, when peanuts are stored in humid conditions, they may get contaminated by fungi *Aspergillus* species producing aflatoxins. This toxin contamination can cause serious liver damage and cancer and even lead to death [50].

4 Pine Nuts

Family: Pinaceae
Genus: *Pinus*
Common name: Pine nuts, cedar nuts, pinon nuts, pinyon nuts, or pignoli

Pine nuts are not actually nuts. They are the edible, small, and elongated seeds of pine cones. Pine nuts are usually hard and go through several processing steps before becoming palatable. Pine cones are dried in a burlap clothing sack in the sun for 20 days after which they are smashed to rapidly release the seeds. The seeds are then separated by hand from the smashed fragments of the cone. Pine nuts have another shell which has to be removed before eating it. They are generally eaten as a raw or roasted snack or put in vegetable dishes and in pesto sauces. Pine nut oil has a mild flavor and a sweet aroma. It has been used in traditional medicine, for cooking, and in salad dressings and is additionally utilized as a carrier oil in aromatherapy and in cosmetic products. Pine nuts contain appreciable amounts of fiber, fats, and protein. They are a rich source of antioxidants and minerals comprising of

Table 4 Pine nut nutrition facts of 100 g, from the United States Department of Agriculture (USDA SR-21)

Serving size of 100 g of dried pine nut	
Per serving	% Daily value[a]
Calories 673	
Total fat 68.4 g	105
Saturated fat 4.9 g	24
Polyunsaturated fat 34.1 g	
Monounsaturated fat 18.8 g	
Total omega-3 fatty acids 112 mg	
Total omega-6 fatty acids 33,606 mg	
Cholesterol 0.0 mg	0
Phytosterols 141 mg	
Carbohydrates 13.1 g	4
Dietary fiber 3.7 g	15
Starch 1.4 g	
Sugars 3.6 g	
Protein 13.7 g	27
Vitamins	
Vitamin A	1
Vitamin E	47
Vitamin K	67
Vitamin C	1
Vitamin B-6	5
Folic acid	8
Minerals	
Sodium	0
Calcium	2
Magnesium	63
Copper	66
Potassium	17
Iron	31
Manganese	440
Zinc	43

National Nutrient Database
[a]Based on a 2000-calorie diet

manganese, magnesium, copper, iron, and zinc and vitamins such as vitamins E and K and folic acid (B-9) (Table 4).

Pine nuts are a very nutritious food and contain a mixture of compounds which together have plenty of positive health effects on the heart, bones, and immune system and have the ability to fight chronic diseases such as diabetes and cancer. These include monounsaturated fatty acids, magnesium, manganese, vitamin E, and vitamin K. Monounsaturated fatty acids such as pinolenic acid found in pine nut oil appear to lower the blood cholesterol [51]. Another fatty acid, oleic acid, has been observed to lower the triglycerides level in blood, lower the risk of atherosclerosis, and prevent coronary artery diseases and strokes [52]. Pine nuts are a natural source of phytosterols, namely beta-sitosterol [53] and polyphenol antioxidants such as catechin, epicatechin, and tyrosol [54] that have also been shown to lower the risk of atherosclerosis and decrease the total and the unhealthy LDL cholesterol levels in the blood. In addition, pine nuts have been observed to lower body weight, possibly by being a satiating food as it suppresses the appetite and lowers the intake of other foods [55]. Their high magnesium content seems to decrease blood pressure and lower the risk of coronary heart disease as well [56]. The appreciable amounts of magnesium and zinc are thought to boost the immune system and have been associated with the prevention and treatment of many mental health disorders, like depression, anxiety, and attention-deficit hyperactivity disorder (ADHD) [57]. Though ADHD itself is not part of *autism* spectrum disorders, the symptoms of autism and *ADHD* overlap. High content of vitamin K was found to promote blood coagulation and prevent bleeding after injuries and has also been shown to play a role in osteoporosis prevention and reducing the risk of fractures [58]. Regular consumption of pine nuts seems to lower the blood glucose level and decrease the risk of type 2 diabetes and its complications [59]. Pine nuts contain very high concentrations of antioxidants including vitamins A, B, C, D, and E and lutein. Antioxidants are believed to scavenge free radicals, thereby reducing the risk of some cancers and controlling aging symptoms. They also act as antibacterial and antiviral agents mitigating infections [60]. Lutein is a carotenoid pigment that has been associated with lowering the risk of non-Hodgkin lymphoma development [61]. Lutein has also been suggested to protect the eye from age-related macular degenerative changes and ultraviolet light and decrease the danger of glaucoma [62].

Pine nut consumption, however, may have some adverse health effects too. Some people experience harmless taste disturbance few days after raw pine nut consumption. They develop a persistent bitter, metallic aftertaste known as pine nut syndrome, which can last over a week and disappear spontaneously [63]. People sensitized to peanut or other nuts can also have allergic reaction when eating pine nuts. The reactions include vomiting, diarrhea, abdominal pain, and skin itching. Overconsumption of pine nuts may cause nausea.

5 Walnut

Family: Juglandaceae
Genus: *Juglans*
Common name: Walnut

Walnuts are rounded, single-seeded stone fruits with a wrinkly hard shell of the walnut tree commonly used for its core heart after being fully ripe. There are two major species of walnuts grown for their seeds—the Persian or English walnut and the black walnut. It can be eaten raw as a snack or added to different recipes including salads or powdered into dips and sauces, while its oil can be used in vinaigrette dressing. Walnut has strong antioxidant characteristics that mainly come from its skin which is rich in vitamin E, melatonin, and polyphenols. Walnut is a rich source of folic acid (B-9), pyridoxine (B-6), thiamin (B-1), zinc, copper, and manganese (Table 5). It also contains very high levels of fiber and fatty acids such as omega-3 and omega-6.

Table 5 Walnut nutrition facts of 100 g, from the United States Department of Agriculture (USDA SR-21)

Serving size of 100 g of walnut	
Per serving	% Daily value[a]
Calories 654	
Total fat 65.2 g	100
Saturated fat 6.1 g	31
Polyunsaturated fat 47.2 g	
Monounsaturated fat 8.9 g	
Total omega-3 fatty acids 9079 mg	
Total omega-6 fatty acids 38,092 mg	
Cholesterol 0.0 mg	0
Phytosterols 72.0 mg	
Carbohydrates 13.7 g	5
Dietary fiber 6.7 g	27
Starch 0.1 g	
Sugars 2.6 g	
Protein 15.2 g	30
Vitamins	
Vitamin A	0
Vitamin E	4
Vitamin K	3
Vitamin C	2
Vitamin B-6	27
Folic acid	25
Minerals	
Sodium	0
Calcium	10
Magnesium	40
Copper	79
Potassium	13
Iron	16
Manganese	171
Zinc	21

National Nutrient Database
[a]Based on a 2000-calorie diet

Walnuts have incredible health benefits. Numerous elements in walnut are proven to reduce inflammation and its related diseases such as type 2 diabetes, cardiovascular diseases, Alzheimer's disease and some types of cancers [64, 65]. Walnuts' significantly high level of omega-3 essential fatty acid has been shown to decrease the risk of cardiovascular diseases [66]; they lower serum cholesterol by reducing cholesterol absorption, inhibition of HMG-CoA reductase, and increased bile acid production by stimulation of 7-α-hydroxylase, especially the levels of the unhealthy LDL cholesterol [67]. It has been suggested that omega-3 fatty acids and different antioxidants contained in walnuts may boost the immune system and exhibit anticancer effects [68] against many cancers including breast [69], colorectal [70], and prostate cancers [71]. Omega-3 fatty acid, magnesium, and the amino acid arginine have been established to improve inflammation, oxidative stress, endothelial function, and blood pressure [65]. Although walnuts are high in fat, they can help in weight management. Their high fiber and protein content suppresses hunger and enhances satiety [72]. Fiber and polyphenols may also have antidiabetic effects by altering gut microflora. The unsaturated fatty acids may favorably influence glucose homeostasis and suppress appetite to aid in weight control [65]. Walnut and its oil potentially exert control over blood sugar and lower the risk of type 2 diabetes possibly due to its effect on body weight and body mass index (BMI) [73] or more directly lower blood sugar and hemoglobin A1C (HbA1c), the 3-month average of blood glucose [74]. Their high content of antioxidant polyphenols and polyunsaturated fatty acids helps reduce oxidative stress causing brain inflammation and age-related cognitive disorders [75]. Dietary supplementation with walnuts has a protective effect against Aβ-induced oxidative stress and cellular death. Moreover, findings suggest positive effect on learning skills, memory, anxiety, locomotor activity, and motor coordination [64].

Walnuts can affect the body health adversely in many ways too. Overconsumption of walnut may lead to weight gain. Allergic reactions are also common with the overconsumption of walnuts. Individuals with sensitivity may exhibit minor allergic reactions resulting in rashes or swelling, while for others, a walnut allergy can lead to life-threatening symptoms. As an allergen, walnut consumption could cause few digestive complications including diarrhea, nausea, and abdominal pain. Juglone content of black walnut may rarely cause lip and tongue cancers when overindulged over long period of time. Topical application of walnut leaf on the skin can lead to acne, eczema, ulcers, and multiple skin infections. They can also cause excessive sweating of the hands and feet.

6 Other Nuts

6.1 Cashew Nuts

Family: Anacardiaceae
Genus: *Anacardium*
Common name: Cashew

Anacardium occidentale is a tropical tree that produces the kidney-shaped cashew seed and apple. The nut is attached to the lower portion of the apple which is conically shaped. They are low-fiber nuts, packed with antioxidants and vitamins E, K, and B-6, along with minerals like copper, manganese, phosphorus, zinc, magnesium, iron, and selenium (Table 6). Cashew nuts can be eaten raw or roasted, used in different recipes, or processed to make cashew cheese or cashew butter. The cashew apple pulp can be processed into a sweet fruit drink. The by-product of liquid obtained from processing cashew nut shell is a natural source of phenols which are raw materials used in the preparation of some drugs, pesticides, paints, plastics, resins, and wood treatments.

Not only do cashews have a lower fat content than most other nuts, but also they are cholesterol-free. They may lower the risk of cardiovascular disease by bringing down systolic blood pressure [13], total cholesterol and unhealthy LDL cholesterol levels [76]. Phosphorus is crucial for the development of healthy teeth and bones as it promotes their mineralization [77]. Likewise, magnesium preserves healthy bones and decreases the risk of fractures [26]. Magnesium may regulate blood sugar levels by altering the insulin activity [78]. Iron on the other hand is important in carrying oxygen around the body and reduces the risk of anemia. Cashews are also good for eye health as the antioxidants have shielding effect against light damage. It has been shown that antioxidants such as anacardic acids, cardanols, and cardols in cashew are effective for people undergoing cancer treatment [79]. Anacardic acids in the cashew nut shell liquid have antibacterial properties [80] and were observed to be advantageous in treating skin inflammation-related conditions [81]. Frequent consumption of nuts like cashews may also lower the risk of gallstone disease [82] and lower the need to surgically remove such stones [83].

Table 6 Cashew nut nutrition facts of 100 g, from the United States Department of Agriculture (USDA SR-21)

Serving size of 100 g of raw cashew nuts	
Per serving	% Daily value[a]
Calories 553	
Total fat 43.8 g	67
Saturated fat 7.8 g	39
Polyunsaturated fat 7.8 g	
Monounsaturated fat 23.8 g	
Total omega-3 fatty acids 62.0 mg	
Total omega-6 fatty acids 7782 mg	
Cholesterol 0.0 mg	0
Phytosterols	
Carbohydrates 32.7 g	11
Dietary fiber 3.3 g	13
Starch 23.5	
Sugars 5.9 g	
Protein 18.2 g	36
Vitamins	
Vitamin A	0
Vitamin E	4
Vitamin K	43
Vitamin C	1
Vitamin B-6	21
Folic acid	6
Minerals	
Sodium	1
Calcium	4
Magnesium	73
Copper	110
Potassium	19
Iron	37
Manganese	83
Zinc	39

National Nutrient Database
[a]Based on a 2000-calorie diet

Overeating cashews can be harmful to health. Cashews do display laxative effect. However, oxalate content though moderate, dietary excess and hyper-absorption do lead to several kidney disorders including kidney stones [84]. Oxalates can also interfere with the absorption of calcium in the body. In addition, unroasted cashew can aggravate the skin and cause blisters. The nut shell oil from raw cashews is mordant and can cause contact dermatitis and skin burns [85]. Moreover, diabetics that consume cashews should keep track of their blood sugar as eating large amounts may elevate blood sugar levels. Allergic reaction to cashew nuts is quite common. People with sensitivity may develop a life-threatening anaphylactic reaction when consuming cashews.

6.2 *Pistachio*

Family: Anacardiaceae
Genus: *Pistacia*
Common name: Pistachio

Pistachio is a small tree, with pinnate (feather-like) leaves that tend to fall off in the autumn. The tree produces elongated seeds that are widely consumed as food. The fruit is a drupe that has a hard, cream-colored outer shell covering the edible seed with mauve-colored skin, light-green flesh, and a unique flavor. When the fruit ripens, the outer shell changes its color from green to yellow/red and splits partly open. The seeds are commonly eaten whole as a snack either fresh or roasted, with or without salt. They are also used to make ice cream, butter, and paste and are an ingredient of different sweets like baklava, chocolate, halva, Turkish delight (lokum), or biscotti. Pistachio is added in cold cuts such as mortadella and to salads. Pistachios are nutritionally very rich food. They are an excellent source of protein; fatty acids; dietary fiber; minerals including potassium, copper, and iron; and vitamins like pyridoxine (B-6). Pistachio also contains calcium, folic acid (B-9), and other B vitamins (Table 7).

Pistachio nuts are not only delicious but also extremely nutrient-dense. Pistachios are rich in phytosterols which are structurally similar to cholesterol and compete with dietary cholesterol to prevent their absorption [38]. They were found to lower the total as well as the unhealthy LDL blood cholesterol levels and lower the risk of heart disease [86]. Pistachios help to reduce hypertension. The high content of the amino acid L-arginine in pistachio nuts that is converted to nitric oxide in the body has been shown to help in dilating the blood vessels [87]. Pistachios contain more antioxidants than most nuts. The antioxidants and anti-inflammatory potential of polyphenols and tocopherols were shown to lower blood pressure, improve vascular health [88], and protect against a number of cancers [89] including colon [90], lung [91], breast [92], and thyroid [93] cancers. Lutein and zeaxanthin antioxidants are important for eye health since they lower the risk of age-related macular degeneration, prevent cataract and loss of vision, and protect the eyes from harmful light [94]. Likewise, vitamin E has antioxidant properties that protect the skin from the ultraviolet light damage, keep the skin moisturized, and combat antiaging skin signs [30]. The presence of biotin, a water-soluble B vitamin, enables battling nail brittleness and hair loss adequately with regular pistachio consumption [95]. The presence

Table 7 Pistachio nutrition facts of 100 g, from the United States Department of Agriculture (USDA SR-21)

Serving size of 100 g of raw pistachio	
Per serving	% Daily value[a]
Calories 557	
Total fat 44.4 g	68
Saturated fat 5.4 g	27
Polyunsaturated fat 13.5 g	
Monounsaturated fat 23.3 g	
Total omega-3 fatty acids 254 mg	
Total omega-6 fatty acids 13,200 mg	
Cholesterol 0.0 mg	0
Phytosterols 214.0 mg	
Carbohydrates 28.0 g	0
Dietary fiber 10.3 g	41
Starch 1.7 g	
Sugars 7.6 g	
Protein 20.6 g	41
Vitamins	
Vitamin A	11
Vitamin E	11
Vitamin K	
Vitamin C	8
Vitamin B-6	85
Folic acid	13
Minerals	
Sodium	0
Calcium	11
Magnesium	30
Copper	110
Potassium	29
Iron	23
Manganese	60
Zinc	39

National Nutrient Database
[a]Based on a 2000-calorie diet

of copper helps increase iron absorption, increase the hemoglobin level, and lower the risk of anemia. The combination of copper and vitamin B-6 improves blood flow and brain functions [96] and enhances immune system status [97]. Furthermore, vitamins A and E prevent inflammation-related health problems, and pistachio extract has been shown to demonstrate marked anti-inflammatory and antinociceptive activities [98]. Pistachios' rich content of dietary fibers is valuable for weight control as they increase satisfaction and decrease food intake along with maintaining digestive tract optimal capacity. Despite their high carbohydrate content, pistachios have low glycemic index. Collectively, antioxidants, magnesium, carotenoids, and phenolic compounds play a role in glucose control in type 2 diabetes and decrease the HbA1c [59, 99].

Pistachio nuts naturally have low sodium. However, salted pistachio nuts increase sodium intake and subsequently elevate blood pressure. Overconsumption of pistachio with its high content of fiber may lead to diarrhea and abdominal pain. While the fiber makes a person feel satisfied nevertheless, as it is high caloric, eating a handful of nuts at a time may result in increased weight and larger waist girth. In people sensitive to fructan, pistachio nuts can cause gastrointestinal symptoms such as bloating, flatulence, diarrhea, constipation, and abdominal cramps. In addition, individuals that suffer from nut allergies are also expected to be allergic to pistachio. Pistachio is contraindicated for allergic individuals to circumvent the possible life-threatening anaphylactic reaction. Moreover, exceedingly high levels of manganese can give rise to headaches and neurological disorders. Pistachio has high oxalate content which may stimulate kidney stone formation as well. Overconsumption of pistachios increases potassium load and may worsen the preexisting kidney problem and therefore should be eaten with caution.

6.3 Macadamia Nuts

Family: Proteaceae
Genus: *Macadamia*
Common name: Macadamia, Queensland nut, Hawaii nut, bush nut, maroochi nut, or bauple nut

The macadamia nut is the seed of the macadamia tree and was named after the Scottish-Australian scientist John Macadam. Hardest of all nut shells, the macadamia is a tough nut to crack. Moreover, the shelled nuts should be stored carefully so that they do not spoil quickly. Macadamias are a rich source of protein and B vitamins like thiamin (B-1), riboflavin (B-2), niacin (B-3), and folic acid (B-9). They also contain decent amounts of minerals like iron, zinc, copper, calcium, phosphorus, potassium, and magnesium. The macadamia nut is a good source of carbohydrates and fiber (Table 8). In addition, the nuts contain antioxidants like polyphenols, amino acids, flavones, and selenium.

The presence of monounsaturated fatty acids in macadamia nuts reduces platelet aggregation [100], regulates cholesterol metabolism, and improves blood lipid profiles [101]. Therefore, monounsaturated fatty acids are able to prevent coronary heart disease [102]. Macadamia nuts are linked to glucose and insulin metabolism.

Table 8 Macadamia nutrition facts of 100 g, from the United States Department of Agriculture (USDA SR-21)

Serving size of 100 g of raw macadamia nuts	
Per serving	% Daily value[a]
Calories 718	
Total fat 75.8 g	117
Saturated fat 12.1 g	60
Polyunsaturated fat 1.5 g	
Monounsaturated fat 58.9 g	
Total omega-3 fatty acids 206 mg	
Total omega-6 fatty acids 1296 mg	
Cholesterol 0.0 mg	0
Phytosterols 116.0 mg	
Carbohydrates 14.2 g	5
Dietary fiber 8.6 g	34
Starch 1.1 g	
Sugars 4.6 g	
Protein 7.9 g	16
Vitamins	
Vitamin A	0
Vitamin E	3
Vitamin K	
Vitamin C	2
Vitamin B-6	14
Folic acid	3
Minerals	
Sodium	0
Calcium	9
Magnesium	33
Copper	38
Potassium	11
Iron	20
Manganese	207
Zinc	9

National Nutrient Database
[a]Based on a 2000-calorie diet

Nut consumption seems to counter the diabetic effects by improving glycemia levels [59]. Omega fatty acids boost brain and nerve health since they stimulate gene expression and neuronal activity, increase synaptogenesis and neurogenesis, and prevent neuroinflammation and apoptosis. Furthermore, macadamia content of alpha-linolenic acid, a type of anti-inflammatory omega-3 fatty acid, may prevent arthritis [103] and has been shown to improve cognitive functions [104]. Erucic acid, an omega-9 fatty acid, has been found to have memory-enhancing effects on cognitive impairment and decrease the risk of Alzheimer's disease [105]. The availability of copper, vitamin B1, manganese, and magnesium aids the production of neurotransmitters to improve cognition. Oleic acid, another omega-9 fatty acid, demonstrates widespread modulatory effects from reducing blood pressure and preventing stroke to enhancing skin repair [106]. Palmitoleic acid, an omega-7 fatty acid, is an important component of myelin which is a fatty layer that protects nerve cells in the brain. It has been suggested that palmitoleic acid has hormonelike properties and improves some metabolic parameters that are impaired in obesity and type 2 diabetes by lowering insulin resistance and inducing glucose homeostasis [107]. Palmitoleic acid supposedly has other benefits such as delaying the onset of skin aging and strengthening the hair. In addition, vitamin K and its derivatives can significantly reduce the risk of certain cancers including breast [108], cervical [109], stomach [110], and prostate cancers [111] and play a role in wound healing [112]. Vitamin E enables the production of new skin cells that replace the older layers of the skin. As a result of the nut's low carbohydrate yet high fiber content, it is able to regulate bowel movement, eliminate toxins, keep the person feeling "full" for a longer time, and reduce dietary intake, thus controlling body weight. The high amount of iron in macadamia nuts can help boost the iron level and form hemoglobin. Macadamia is high in B vitamins and minerals like zinc, calcium, and magnesium which regulate calcium release in the blood and its distribution in the bones, all contributing collectively to preserving bone health [113].

In general, there are no negative side effects observed when eating macadamia nuts. If the nuts purchased are salted, they may elevate blood pressure. Macadamia nuts contain allergens. Hence, excessive intake may cause allergic reactions like skin rash and coughing. The high fiber content of the nut can cause gastrointestinal issues like gas, diarrhea, bloating, and even constipation. When the nuts are dried in order to make them suitable for consumption, the water content is lowered. So, it is advisable to drink plenty of water before munching on macadamia nuts.

References

1. Jenkins, D. J., Kendall, C. W., Josse, A. R., Salvatore, S., Brighenti, F., Augustin, L. S., et al. (2006). Almonds decrease postprandial glycemia, insulinemia, and oxidative damage in healthy individuals. *The Journal of Nutrition, 136*(12), 2987–2992.
2. Jenkins, D. J., Kendall, C. W., Marchie, A., Josse, A. R., Nguyen, T. H., Faulkner, D. A., et al. (2008). Almonds reduce biomarkers of lipid peroxidation in older hyperlipidemic subjects. *The Journal of Nutrition, 138*(5), 908–913.
3. Stampfer, M. J., Hennekens, C. H., Manson, J. E., Colditz, G. A., Rosner, B., & Willett, W. C. (1993). Vitamin E consumption and the risk of coronary disease in women. *The New England Journal of Medicine, 328*(20), 1444–1449.
4. Bostick, R. M., Potter, J. D., McKenzie, D. R., Sellers, T. A., Kushi, L. H., Steinmetz, K. A., et al. (1993). Reduced risk of colon cancer with high intake of vitamin E: The Iowa Women's Health Study. *Cancer Research, 53*(18), 4230–4237.
5. Morris, M. C., Evans, D. A., Bienias, J. L., Tangney, C. C., & Wilson, R. S. (2002). Vitamin E and cognitive decline in older persons. *Archives of Neurology, 59*(7), 1125–1132.
6. Jenkins, D. J., Srichaikul, K., Kendall, C. W., Sievenpiper, J. L., Abdulnour, S., Mirrahimi, A., et al. (2011). The relation of low glycaemic index fruit consumption to glycaemic control and risk factors for coronary heart disease in type 2 diabetes. *Diabetologia, 54*(2), 271–279.

7. Tan, S. Y., & Mattes, R. D. (2013). Appetitive, dietary and health effects of almonds consumed with meals or as snacks: A randomized, controlled trial. *European Journal of Clinical Nutrition, 67*(11), 1205–1214.
8. Miao-di, X. U. (2010). Almond and aurantii compounds in treatment of habitual constipation. *Liaoning Journal of Traditional Chinese Medicine, 1*, 94.
9. Guerrero-Romero, F., & Rodriguez-Moran, M. (2011). Magnesium improves the beta-cell function to compensate variation of insulin sensitivity: Double-blind, randomized clinical trial. *European Journal of Clinical Investigation, 41*(4), 405–410.
10. Houston, M. (2014). The role of nutrition and nutraceutical supplements in the treatment of hypertension. *World Journal of Cardiology, 6*(2), 38–66.
11. Jenkins, D. J., Kendall, C. W., Marchie, A., Parker, T. L., Connelly, P. W., Qian, W., et al. (2002). Dose response of almonds on coronary heart disease risk factors: Blood lipids, oxidized low-density lipoproteins, lipoprotein(a), homocysteine, and pulmonary nitric oxide: A randomized, controlled, crossover trial. *Circulation, 106*(11), 1327–1332.
12. Brzica, H., Breljak, D., Burckhardt, B. C., Burckhardt, G., & Sabolić, I. (2013). Oxalate: From the environment to kidney stones. *Arhiv za Higijenu Rada i Toksikologiju, 64*(4), 609–630.
13. Mohammadifard, N., Salehi-Abargouei, A., Salas-Salvadó, J., Guasch-Ferré, M., Humphries, K., & Sarrafzadegan, N. (2015). The effect of tree nut, peanut, and soy nut consumption on blood pressure: A systematic review and meta-analysis of randomized controlled clinical trials. *The American Journal of Clinical Nutrition, 101*(5), 966–982.
14. Mercanligil, S. M., Arslan, P., Alasalvar, C., Okut, E., Akgül, E., Pinar, A., et al. (2007). Effects of hazelnut-enriched diet on plasma cholesterol and lipoprotein profiles in hypercholesterolemic adult men. *European Journal of Clinical Nutrition, 61*(2), 212–220.
15. Ouedraogo, M., Charles, C., Ouédraogo, M., Guissou, I. P., Stévigny, C., & Duez, P. (2011). An overview of cancer chemopreventive potential and safety of proanthocyanidins. *Nutrition and Cancer, 63*(8), 1163–1173.
16. Alqahtani, S., & Kaddoumi, A. (2014). Vitamin E transporters in cancer therapy. *The AAPS Journal, 17*(2), 313–322.
17. Zabłocka-Słowińska, K., & Grajeta, H. (2012). The role of manganese in etiopathogenesis and prevention of selected diseases. *Postępy Higieny i Medycyny Doświadczalnej (Online), 66*, 549–553.
18. Esposito, T., Sansone, F., Franceschelli, S., Del Gaudio, P., Picerno, P., Aquino, R. P., et al. (2017). Hazelnut (Corylus avellana L.) shells extract: Phenolic composition, antioxidant effect and cytotoxic activity on human cancer cell lines. *International Journal of Molecular Sciences, 18*(2), E392.
19. Gallego, A., Metón, I., Baanante, I. V., Ouazzani, J., Adelin, E., Palazon, J., et al. (2017). Viability-reducing activity of Corylus avellana L. extracts against human cancer cell lines. *Biomedicine & Pharmacotherapy, 89*, 565–572.
20. Glei, M., Fischer, S., Lamberty, J., Ludwig, D., Lorkowski, S., & Schlörmann, W. (2018). Chemopreventive potential of in vitro fermented raw and roasted hazelnuts in LT97 colon adenoma cells. *Anticancer Research, 38*(1), 83–93.
21. Grattan Jr., B. J. (2013). Plant sterols as anticancer nutrients: Evidence for their role in breast cancer. *Nutrients, 5*(2), 359–387.
22. Cole, C., Burgoyne, T., Lee, A., Stehno-Bittel, L., & Zaid, G. (2015). Arum Palaestinum with isovanillin, linolenic acid and β-sitosterol inhibits prostate cancer spheroids and reduces the growth rate of prostate tumors in mice. *BMC Complementary and Alternative Medicine, 15*, 264–264.
23. Lee, S. H., Jouihan, H. A., Cooksey, R. C., Jones, D., Kim, H. J., Winge, D. R., et al. (2013). Manganese supplementation protects against diet-induced diabetes in wild type mice by enhancing insulin secretion. *Endocrinology, 154*(3), 1029–1038.
24. Larsson, S. C., & Wolk, A. (2007). Magnesium intake and risk of type 2 diabetes: A meta-analysis. *Journal of Internal Medicine, 262*(2), 208–214.

25. Calder, P. C. (2015). Functional roles of fatty acids and their effects on human health. *JPEN Journal of Parenteral and Enteral Nutrition, 39*(1 Suppl), 18s–32s.
26. Nieves, J. W. (2014). Maximizing bone health—Magnesium, BMD and fractures. *Nature Reviews Endocrinology, 10*(5), 255–256.
27. Bahaeddin, Z., Yans, A., Khodagholi, F., Hajimehdipoor, H., & Sahranavard, S. (2017). Hazelnut and neuroprotection: Improved memory and hindered anxiety in response to intra-hippocampal Abeta injection. *Nutritional Neuroscience, 20*(6), 317–326.
28. Takeda, A. (2003). Manganese action in brain function. *Brain Research. Brain Research Reviews, 41*(1), 79–87.
29. La Fata, G., Weber, P., & Mohajeri, M. H. (2014). Effects of vitamin E on cognitive performance during ageing and in Alzheimer's disease. *Nutrients, 6*(12), 5453–5472.
30. Keen, M. A., & Hassan, I. (2016). Vitamin E in dermatology. *Indian Dermatology Online Journal, 7*(4), 311–315.
31. Lopez, H. W., Leenhardt, F., Coudray, C., & Remesy, C. (2002). Minerals and phytic acid interactions: Is it a real problem for human nutrition? *International Journal of Food Science and Technology, 37*(7), 727–739.
32. Costa, J., Mafra, I., Carrapatoso, I., & Oliveira, M. B. (2016). Hazelnut allergens: Molecular characterization, detection, and clinical relevance. *Critical Reviews in Food Science and Nutrition, 56*(15), 2579–2605.
33. Lavigne, P. M., & Karas, R. H. (2013). The current state of niacin in cardiovascular disease prevention: A systematic review and meta-regression. *Journal of the American College of Cardiology, 61*(4), 440–446.
34. Lefevre, C., Mallett, L. H., & Wick, L. (2018). Myocardial ischemia in an adolescent secondary to nutritional thiamine deficiency. *Proceedings (Baylor University Medical Center), 31*(1), 51–52.
35. Guasch-Ferre, M., Bulló, M., Estruch, R., Corella, D., Martínez-González, M. A., Ros, E., et al. (2014). Dietary magnesium intake is inversely associated with mortality in adults at high cardiovascular disease risk. *The Journal of Nutrition, 144*(1), 55–60.
36. Kodali, H. P., Pavilonis, B. T., & Schooling, C. M. (2018). Effects of copper and zinc on ischemic heart disease and myocardial infarction: A Mendelian randomization study. *The American Journal of Clinical Nutrition, 108*(2), 237–242.
37. Barbour, J. A., Howe, P. R., Buckley, J. D., Bryan, J., & Coates, A. M. (2015). Effect of 12 weeks high oleic peanut consumption on cardio-metabolic risk factors and body composition. *Nutrients, 7*(9), 7381–7398.
38. Brauner, R., Johannes, C., Ploessl, F., Bracher, F., & Lorenz, R. L. (2012). Phytosterols reduce cholesterol absorption by inhibition of 27-hydroxycholesterol generation, liver X receptor alpha activation, and expression of the basolateral sterol exporter ATP-binding cassette A1 in Caco-2 enterocytes. *The Journal of Nutrition, 142*(6), 981–989.
39. Mock, D. M. (2014). Adequate intake of biotin in pregnancy: Why bother? *The Journal of Nutrition, 144*(12), 1885–1886.
40. Akibu, M., Tekelab, T., Amano, A., Besho, M., Grutzmacher, S., Tadese, M., et al. (2018). Adherence to prenatal iron-folic acid supplementation in low- and middle-income countries (LMIC): A protocol for systematic review and meta-analysis. *Systematic Reviews, 7*(1), 107.
41. Tourbah, A., Lebrun-Frenay, C., Edan, G., Clanet, M., Papeix, C., Vukusic, S., et al. (2016). MD1003 (high-dose biotin) for the treatment of progressive multiple sclerosis: A randomised, double-blind, placebo-controlled study. *Multiple Sclerosis, 22*(13), 1719–1731.
42. McKay, B. E., Molineux, M. L., & Turner, R. W. (2004). Biotin is endogenously expressed in select regions of the rat central nervous system. *The Journal of Comparative Neurology, 473*(1), 86–96.
43. Albarracin, C. A., Fuqua, B. C., Evans, J. L., & Goldfine, I. D. (2008). Chromium picolinate and biotin combination improves glucose metabolism in treated, uncontrolled overweight to obese patients with type 2 diabetes. *Diabetes/Metabolism Research and Reviews, 24*(1), 41–51.

44. Sales, J. M., & Resurreccion, A. V. (2014). Resveratrol in peanuts. *Critical Reviews in Food Science and Nutrition, 54*(6), 734–770.
45. Lou, Z., Wang, H., Rao, S., Sun, J., Ma, C., & Li, J. (2012). P-Coumaric acid kills bacteria through dual damage mechanisms. *Food Control, 25*(2), 550–554.
46. Kilic, I., & Yesiloglu, Y. (2013). Spectroscopic studies on the antioxidant activity of p-coumaric acid. *Spectrochimica Acta. Part A, Molecular and Biomolecular Spectroscopy, 115*, 719–724.
47. Lokko, P., Lartey, A., Armar-Klemesu, M., & Mattes, R. D. (2007). Regular peanut consumption improves plasma lipid levels in healthy Ghanaians. *International Journal of Food Sciences and Nutrition, 58*(3), 190–200.
48. Alper, C. M., & Mattes, R. D. (2002). Effects of chronic peanut consumption on energy balance and hedonics. *International Journal of Obesity and Related Metabolic Disorders, 26*(8), 1129–1137.
49. Latif, S., Pfannstiel, J., Makkar, H. P., & Becker, K. (2013). Amino acid composition, antinutrients and allergens in the peanut protein fraction obtained by an aqueous enzymatic process. *Food Chemistry, 136*(1), 213–217.
50. Kumar, P., Mahato, D. K., Kamle, M., Mohanta, T. K., & Kang, S. G. (2016). Aflatoxins: A global concern for food safety, human health and their management. *Frontiers Microbiology, 7*, 2170.
51. Xie, K., Miles, E., & Calder, P. (2016). A review of the potential health benefits of pine nut oil and its characteristic fatty acid pinolenic acid. *Journal of Functional Foods, 23*, 464–473.
52. Ros, E., & Mataix, J. (2006). Fatty acid composition of nuts—Implications for cardiovascular health. *The British Journal of Nutrition, 96*(Suppl 2), S29–S35.
53. Kornsteiner-Krenn, M., Wagner, K. H., & Elmadfa, I. (2013). Phytosterol content and fatty acid pattern of ten different nut types. *International Journal for Vitamin and Nutrition Research, 83*(5), 263–270.
54. Huseini, H. F., Anvari, M. S., Khoob, Y. T., Rabbani, S., Sharifi, F., Arzaghi, S. M., et al. (2015). Anti-hyperlipidemic and anti-atherosclerotic effects of Pinus eldarica Medw. Nut in hypercholesterolemic rabbits. *Daru, 23*(1), 32.
55. O'Neil, C. E., Fulgoni 3rd, V. L., & Nicklas, T. A. (2015). Tree nut consumption is associated with better adiposity measures and cardiovascular and metabolic syndrome health risk factors in U.S. adults: NHANES 2005-2010. *Nutrition Journal, 14*, 64.
56. Kieboom, B. C. T., Niemeijer, M. N., Leening, M. J., van den Berg, M. E., Franco, O. H., Deckers, J. W., et al. (2016). Serum magnesium and the risk of death from coronary heart disease and sudden cardiac death. *Journal of the American Heart Association, 5*(1), e002707.
57. Black, L. J., Allen, K. L., Jacoby, P., Trapp, G. S., Gallagher, C. M., Byrne, S. M., et al. (2015). Low dietary intake of magnesium is associated with increased externalising behaviours in adolescents. *Public Health Nutrition, 18*(10), 1824–1830.
58. Pearson, D. A. (2007). Bone health and osteoporosis: The role of vitamin K and potential antagonism by anticoagulants. *Nutrition in Clinical Practice, 22*(5), 517–544.
59. Hernandez-Alonso, P., Camacho-Barcia, L., Bulló, M., & Salas-Salvadó, J. (2017). Nuts and dried fruits: An update of their beneficial effects on type 2 diabetes. *Nutrients, 9*(7), pii: E673.
60. Alasalvar, C., & Bolling, B. W. (2015). Review of nut phytochemicals, fat-soluble bioactives, antioxidant components and health effects. *The British Journal of Nutrition, 113*(Suppl 2), S68–S78.
61. Chen, F., Hu, J., Liu, P., Li, J., Wei, Z., & Liu, P. (2017). Carotenoid intake and risk of non-Hodgkin lymphoma: A systematic review and dose-response meta-analysis of observational studies. *Annals of Hematology, 96*(6), 957–965.
62. Wu, J., Cho, E., Willett, W. C., Sastry, S. M., & Schaumberg, D. A. (2015). Intakes of lutein, zeaxanthin, and other carotenoids and age-related macular degeneration during 2 decades of prospective follow-up. *JAMA Ophthalmology, 133*(12), 1415–1424.
63. Munk, M. D. (2012). Pine mouth (pine nut) syndrome: Description of the toxidrome, preliminary case definition, and best evidence regarding an apparent etiology. *Seminars in Neurology, 32*(5), 525–527.

64. Muthaiyah, B., Essa, M. M., Lee, M., Chauhan, V., Kaur, K., & Chauhan, A. (2014). Dietary supplementation of walnuts improves memory deficits and learning skills in transgenic mouse model of Alzheimer's disease. *Journal of Alzheimer's Disease, 42*(4), 1397–1405.
65. Kim, Y., Keogh, J. B., & Clifton, P. M. (2017). Benefits of nut consumption on insulin resistance and cardiovascular risk factors: Multiple potential mechanisms of actions. *Nutrients, 9*(11), E1271.
66. Pan, A., Chen, M., Chowdhury, R., Wu, J. H., Sun, Q., Campos, H., et al. (2012). Alpha-linolenic acid and risk of cardiovascular disease: A systematic review and meta-analysis. *The American Journal of Clinical Nutrition, 96*(6), 1262–1273.
67. Banel, D. K., & Hu, F. B. (2009). Effects of walnut consumption on blood lipids and other cardiovascular risk factors: A meta-analysis and systematic review. *The American Journal of Clinical Nutrition, 90*(1), 56–63.
68. Hardman, W. E. (2014). Walnuts have potential for cancer prevention and treatment in mice. *The Journal of Nutrition, 144*(4 Suppl), 555s–560s.
69. Soriano-Hernandez, A. D., Madrigal-Perez, D. G., Galvan-Salazar, H. R., Arreola-Cruz, A., Briseño-Gomez, L., Guzmán-Esquivel, J., et al. (2015). The protective effect of peanut, walnut, and almond consumption on the development of breast cancer. *Gynecologic and Obstetric Investigation, 80*(2), 89–92.
70. Tsoukas, M. A., Ko, B. J., Witte, T. R., Dincer, F., Hardman, W. E., & Mantzoros, C. S. (2015). Dietary walnut suppression of colorectal cancer in mice: Mediation by miRNA patterns and fatty acid incorporation. *The Journal of Nutritional Biochemistry, 26*(7), 776–783.
71. Reiter, R. J., Tan, D. X., Manchester, L. C., Korkmaz, A., Fuentes-Broto, L., Hardman, W. E., et al. (2013). A walnut-enriched diet reduces the growth of LNCaP human prostate cancer xenografts in nude mice. *Cancer Investigation, 31*(6), 365–373.
72. Sabaté, J. (2003). Nut consumption and body weight. *The American Journal of Clinical Nutrition, 78*(3), 647S–650S.
73. Brennan, A. M., Sweeney, L. L., Liu, X., & Mantzoros, C. S. (2010). Walnut consumption increases satiation but has no effect on insulin resistance or the metabolic profile over a 4-day period. *Obesity (Silver Spring), 18*(6), 1176–1182.
74. Jackson, C. L., & Hu, F. B. (2014). Long-term associations of nut consumption with body weight and obesity. *The American Journal of Clinical Nutrition, 100*(Suppl 1), 408s–411s.
75. Zibaeenezhad, M., Aghasadeghi, K., Hakimi, H., Yarmohammadi, H., & Nikaein, F. (2016). The effect of walnut oil consumption on blood sugar in patients with diabetes mellitus type 2. *International Journal of Endocrinology and Metabolism, 14*(3), e34889–e34889.
76. Poulose, S. M., Miller, M. G., & Shukitt-Hale, B. (2014). Role of walnuts in maintaining brain health with age. *The Journal of Nutrition, 144*(4 Suppl), 561s–566s.
77. Mah, E., Schulz, J. A., Kaden, V. N., Lawless, A. L., Rotor, J., Mantilla, L. B., et al. (2017). Cashew consumption reduces total and LDL cholesterol: A randomized, crossover, controlled-feeding trial. *The American Journal of Clinical Nutrition, 105*(5), 1070–1078.
78. Penido, M., & Alon, U. S. (2012). Phosphate homeostasis and its role in bone health. *Pediatric Nephrology, 27*(11), 2039–2048.
79. Wang, J., Persuitte, G., Olendzki, B. C., Wedick, N. M., Zhang, Z., Merriam, P. A., et al. (2013). Dietary magnesium intake improves insulin resistance among non-diabetic individuals with metabolic syndrome participating in a dietary trial. *Nutrients, 5*(10), 3910–3919.
80. Trevisan, M. T., Pfundstein, B., Haubner, R., Würtele, G., Spiegelhalder, B., et al. (2006). Characterization of alkyl phenols in cashew (Anacardium occidentale) products and assay of their antioxidant capacity. *Food and Chemical Toxicology, 44*(2), 188–197.
81. Sajeevan, S. E., Chatterjee, M., Paul, V., Baranwal, G., Kumar, V. A., Bose, C., et al. (2018). Impregnation of catheters with anacardic acid from cashew nut shell prevents Staphylococcus aureus biofilm development. *Journal of Applied Microbiology, 125*(5), 1286–1295.
82. Schwarz, A., Bruhs, A., & Schwarz, T. (2017). The short-chain fatty acid sodium butyrate functions as a regulator of the skin immune system. *The Journal of Investigative Dermatology, 137*(4), 855–864.

83. Tsai, C. J., Leitzmann, M. F., Hu, F. B., Willett, W. C., & Giovannucci, E. L. (2004). A prospective cohort study of nut consumption and the risk of gallstone disease in men. *American Journal of Epidemiology, 160*(10), 961–968.
84. Tsai, C. J., Leitzmann, M. F., Hu, F. B., Willett, W. C., & Giovannucci, E. L. (2004). Frequent nut consumption and decreased risk of cholecystectomy in women. *The American Journal of Clinical Nutrition, 80*(1), 76–81.
85. Hamilton, T. K., & Zug, K. A. (1998). Systemic contact dermatitis to raw cashew nuts in a pesto sauce. *American Journal of Contact Dermatitis, 9*(1), 51–54.
86. Andreadou, I., Mitakou, S., Paraschos, S., Efentakis, P., Magiatis, P., Kaklamanis, L., et al. (2016). "Pistacia lentiscus L." reduces the infarct size in normal fed anesthetized rabbits and possess antiatheromatic and hypolipidemic activity in cholesterol fed rabbits. *Phytomedicine, 23*(11), 1220–1226.
87. Bullo, M., Juanola-Falgarona, M., Hernández-Alonso, P., & Salas-Salvadó, J. (2015). Nutrition attributes and health effects of pistachio nuts. *The British Journal of Nutrition, 113*(Suppl 2), S79–S93.
88. Hodgson, J. M., Croft, K. D., Woodman, R. J., Puddey, I. B., Bondonno, C. P., Wu, J. H., et al. (2014). Effects of vitamin E, vitamin C and polyphenols on the rate of blood pressure variation: Results of two randomised controlled trials. *The British Journal of Nutrition, 112*(9), 1551–1561.
89. Yang, C. S., Li, G., Yang, Z., Guan, F., Chen, A., & Ju, J. (2013). Cancer prevention by tocopherols and tea polyphenols. *Cancer Letters, 334*(1), 79–85.
90. Glei, M., Ludwig, D., Lamberty, J., Fischer, S., Lorkowski, S., & Schlörmann, W. (2017). Chemopreventive potential of raw and roasted pistachios regarding colon carcinogenesis. *Nutrients, 9*(12), pii: E1368.
91. Magkouta, S., Stathopoulos, G. T., Psallidas, I., Papapetropoulos, A., Kolisis, F. N., Roussos, C., et al. (2009). Protective effects of mastic oil from Pistacia lentiscus variation chia against experimental growth of Lewis lung carcinoma. *Nutrition and Cancer, 61*(5), 640–648.
92. Seifaddinipour, M., Farghadani, R., Namvar, F., Mohamad, J., & Abdul Kadir, H. (2018). Cytotoxic effects and anti-angiogenesis potential of pistachio (Pistacia vera L.) hulls against MCF-7 human breast cancer cells. *Molecules, 23*(1), pii: E110.
93. Catalani, S., Palma, F., Battistelli, S., & Benedetti, S. (2017). Oxidative stress and apoptosis induction in human thyroid carcinoma cells exposed to the essential oil from Pistacia lentiscus aerial parts. *PLoS One, 12*(2), e0172138.
94. Delcourt, C., Carrière, I., Delage, M., Barberger-Gateau, P., Schalch, W., & POLA Study Group. (2006). Plasma lutein and zeaxanthin and other carotenoids as modifiable risk factors for age-related maculopathy and cataract: The POLA study. *Investigative Ophthalmology & Visual Science, 47*(6), 2329–2335.
95. Patel, D. P., Swink, S. M., & Castelo-Soccio, L. (2017). A review of the use of biotin for hair loss. *Skin Appendage Disorder, 3*(3), 166–169.
96. Barichello, T., Generoso, J. S., Simões, L. R., Ceretta, R. A., Dominguini, D., Ferrari, P., et al. (2014). Vitamin B6 prevents cognitive impairment in experimental pneumococcal meningitis. *Experimental Biology and Medicine (Maywood, N.J.), 239*(10), 1360–1365.
97. Qian, B., Shen, S., Zhang, J., & Jing, P. (2017). Effects of vitamin B6 deficiency on the composition and functional potential of T cell populations. *Journal of Immunology Research, 2017*, 2197975.
98. Orhan, I., Küpeli, E., Aslan, M., Kartal, M., & Yesilada, E. (2006). Bioassay-guided evaluation of anti-inflammatory and antinociceptive activities of pistachio, Pistacia vera L. *Journal of Ethnopharmacology, 105*(1–2), 235–240.
99. Parham, M., Heidari, S., Khorramirad, A., Hozoori, M., Hosseinzadeh, F., Bakhtyari, L., et al. (2014). Effects of pistachio nut supplementation on blood glucose in patients with type 2 diabetes: A randomized crossover trial. *The Review of Diabetic Studies, 11*(2), 190–196.
100. Smith, R. D., Kelly, C. N., Fielding, B. A., Hauton, D., Silva, K. D., Nydahl, M. C., et al. (2003). Long-term monounsaturated fatty acid diets reduce platelet aggregation in healthy young subjects. *The British Journal of Nutrition, 90*(3), 597–606.

101. Curb, J. D., Wergowske, G., Dobbs, J. C., Abbott, R. D., & Huang, B. (2000). Serum lipid effects of a high-monounsaturated fat diet based on macadamia nuts. *Archives of Internal Medicine, 160*(8), 1154–1158.
102. Zong, G., Li, Y., Sampson, L., Dougherty, L. W., Willett, W. C., Wanders, A. J., et al. (2018). Monounsaturated fats from plant and animal sources in relation to risk of coronary heart disease among US men and women. *The American Journal of Clinical Nutrition, 107*(3), 445–453.
103. Iverson, C., Bacong, A., Liu, S., Baumgartner, S., Lundström, T., Oscarsson, J., et al. (2018). Omega-3-carboxylic acids provide efficacious anti-inflammatory activity in models of crystal-mediated inflammation. *Scientific Reports, 8*(1), 1217.
104. Rathod, R. S., Khaire, A. A., Kale, A. A., & Joshi, S. R. (2015). Beneficial effects of omega-3 fatty acids and vitamin B12 supplementation on brain docosahexaenoic acid, brain derived neurotrophic factor, and cognitive performance in the second-generation Wistar rats. *BioFactors, 41*(4), 261–272.
105. Kim, E., Ko, H. J., Jeon, S. J., Lee, S., Lee, H. E., Kim, H. N., et al. (2016). The memory-enhancing effect of erucic acid on scopolamine-induced cognitive impairment in mice. *Pharmacology, Biochemistry, and Behavior, 142*, 85–90.
106. Sales-Campos, H., Souza, P. R., Peghini, B. C., da Silva, J. S., & Cardoso, C. R. (2013). An overview of the modulatory effects of oleic acid in health and disease. *Mini Reviews in Medicinal Chemistry, 13*(2), 201–210.
107. Nunes, E. A., & Rafacho, A. (2017). Implications of palmitoleic acid (Palmitoleate) on glucose homeostasis, insulin resistance and diabetes. *Current Drug Targets, 18*(6), 619–628.
108. Kiely, M., Hodgins, S. J., Merrigan, B. A., Tormey, S., Kiely, P. A., & O'Connor, E. M. (2015). Real-time cell analysis of the inhibitory effect of vitamin K2 on adhesion and proliferation of breast cancer cells. *Nutrition Research, 35*(8), 736–743.
109. de Carvalho Scharf Santana, N., Lima, N. A., Desoti, V. C., Bidóia, D. L., de Souza Bonfim Mendonca, P., Ratti, B. A., et al. (2016). Vitamin K3 induces antiproliferative effect in cervical epithelial cells transformed by HPV 16 (SiHa cells) through the increase in reactive oxygen species production. *Archives of Gynecology and Obstetrics, 294*(4), 797–804.
110. Kong, P., Cai, Q., Geng, Q., Wang, J., Lan, Y., Zhan, Y., et al. (2014). Vitamin intake reduce the risk of gastric cancer: Meta-analysis and systematic review of randomized and observational studies. *PLoS One, 9*(12), e116060.
111. Dasari, S., Ali, S. M., Zheng, G., Chen, A., Dontaraju, V. S., Bosland, M. C., et al. (2017). Vitamin K and its analogs: Potential avenues for prostate cancer management. *Oncotarget, 8*(34), 57782–57799.
112. Hemmati, A. A., Houshmand, G., Ghorbanzadeh, B., Nemati, M., & Behmanesh, M. A. (2014). Topical vitamin K1 promotes repair of full thickness wound in rat. *Indian Journal of Pharmacology, 46*(4), 409–412.
113. Palacios, C. (2006). The role of nutrients in bone health, from A to Z. *Critical Reviews in Food Science and Nutrition, 46*(8), 621–628.

Seeds

Sawsan G. Mohammed and M. Walid Qoronfleh

Abstract A wide variety of plant species provide edible seeds. Seeds are the dominant source of human calories and protein. The most important and popular seed food sources are cereals, followed by legumes and nuts. Their nutritional content of fiber, protein, and monounsaturated/polyunsaturated fats make them extremely nutritious. They are important additions to our daily food consumption. When consumed as part of a healthy diet, seeds can help reduce blood sugar, cholesterol, and blood pressure.

Keywords Seeds · Fiber · Antioxidants · Flavonoids · Isoflavones · Phenolic acids · Essential fatty acids · Monounsaturated fatty acids · Anethole · Phytosterols · Phytoestrogens · Lignans · Phenolic acids · Lectins · Saponins

S. G. Mohammed (✉)
Qatar Research Leadership Program (QRLP), Qatar Foundation, Doha, Qatar
e-mail: sgmohammed@qf.org.qa

M. W. Qoronfleh (✉)
Research & Policy Department, World Innovation Summit for Health (WISH), Qatar Foundation, Doha, Qatar
e-mail: wqoronfleh@qf.org.qa

M. M. Essa, M. W. Qoronfleh (eds.), *Personalized Food Intervention and Therapy for Autism Spectrum Disorder Management*, Advances in Neurobiology 24,
https://doi.org/10.1007/978-3-030-30402-7_13

1 Anise

Family: Apiaceae
Genus: *Pimpinella*
Common name: Anise, aniseed, or sweet cumin

Anise is a widely planted flowering spice plant with a sweet and aromatic flavor, native to the Eastern Mediterranean and Southwest Asia regions (not to be confused with the star anise). Special anise fragrant essential oil can be acquired from its liquorice-like flavored fruit (aniseed). The active ingredient in the oil is the aromatic compound anethole. The fresh leaves of anise are occasionally used for cooking. It is utilized in mouthwashes, mouth freshening materials, and soaps. Anise has been used to flavor cakes, candies, and pastries, in herbal teas and other drinks. It is widely utilized in herbal medicine. Anise seeds contain antioxidants, many B-vitamins, and minerals including iron, calcium, copper, potassium, manganese, magnesium, and zinc (Table 1).

Anise seeds and their oil have widely recognized health benefits. The chemical composition of anise seeds has been described [1–5] and its valuable pharmacological ingredients and traits have been reviewed as well [6, 7]. Its hallmark active component is anethole (aka anise camphor), which is responsible for its unique and distinct aroma and flavor. Additionally, its next of kin are fennel seeds as they belong to the same family and hence have similar health benefits and properties [8, 9].

Anise is widely reputed for its digestive properties and is acknowledged for its antimicrobial properties. Chewing raw seeds or drinking anise tea provides immediate relief from mild indigestion, flatulence, and bloating [6, 10, 11]. It was also discovered to have anti-ulcer effects [12–14] and manage abdominal pain, nausea, and diarrhea. Intervention clinical studies have proven its usefulness in patients with dyspepsia and in circumstances of chronic constipation for safe and effective improvement in the quality of life [15–17].

Anise seed oil extract has shown powerful anti-inflammatory activities [18, 19]. They provide pain relief in cases like joint pain and arthritis [20, 21]. Preparations from aniseed have showed to be an excellent treatment for asthma and cough [22]. Essential oils are used in aromatherapy for mild depressive symptom treatment [23]. Aniseed extracts [24] and essential oil have proven to be effective in treatment

Table 1 Anise seeds nutrition facts of 100 g, from the United States Department of Agriculture (USDA SR-21)

Serving size of 100 g of anise seed	
Per serving	% Daily value[a]
Calories 337	
Total fat 16 g	24
Saturated fat 0.6 g	3
Polyunsaturated fat 3.2 g	
Monounsaturated fat 10 g	
Total omega-3 fatty acids mg	
Total omega-6 fatty acids 3150 mg	
Cholesterol 0.0 mg *Phytosterols*	
Carbohydrates 50 g	17
Dietary fiber 14.6 g	58
Starch	
Sugars	
Protein 17.6 g	35
Vitamins	
Vitamin A	6
Vitamin E	
Vitamin K	
Vitamin C	35
Vitamin B-6	33
Folic acid	2
Minerals	
Sodium	1
Calcium	65
Magnesium	43
Copper	46
Potassium	41
Iron	205
Manganese	115
Zinc	35

National Nutrient Database
[a]Based on a *2000 calorie diet*

of mood disorders [25] including the postpartum depression [15]. It has also been determined to possess anti-seizure and neuroprotective effects in laboratory animals [26]. It is well documented that aniseed is a phytoestrogen [27, 28] along with its major compound, anethole, exhibiting estrogenic-like activity [29]. Aniseeds are often used to comfort and ease menopausal symptoms like recurrent hot flashes [30] and promote breast-milk production in nursing mothers due to its structural semblance to catecholamines as a result of the ability to stimulate prolactin secretion [31, 32].

Aniseed extracts or anethole have been established to affect the activity of a number of essential enzymes involved in carbohydrates metabolism and therefore reduced blood glucose levels [33, 34]. It had also been used in recipes of traditional medicine as an anti-obesity mixture [35, 36].

Spices and essential oils are well-known for their antimicrobial properties [37–39]. Anise seeds' essential oil, its fluid extracts, and its chief component anethole are responsible for its unique antimicrobial property. They have been validated to be potent antifungals [40, 41] that also aid in the treatment of some fungal infections including athlete's foot or ringworm [42]. Its essential oil also showed strong antibacterial and antiviral properties [43–47] including bacterial pathogens or multidrug resistant strains [48, 49].

Aniseed is likely safe. However, anise might cause allergic reactions involving the skin, respiratory and gastrointestinal tracts in people allergic to other plants similar to anise. As anise has estrogenic properties and so might worsen hormone sensitive conditions including endometriosis, uterine fibroids, and some cancers like breast, uterine, and ovarian cancers [50]. It might also interfere with hormone-based medications like oral contraceptives as well as anti-cancer medication like tamoxifen. Anise oil consumption has been associated with nausea, vomiting, seizures, pulmonary edema [51], and CNS drugs interaction [52].

2 Flax

Family: Linaceae
Genus: *Linum*
Common names: Flax, common flax, or linseed

Flax is a food and a fiber crop. It is commonly grown for its seeds. Flaxseeds are traditionally roasted, powdered, and then mainly made into breads or the seeds can be pressed to produce vegetable oil. Whole flaxseeds are usually stable to store. However, ground seeds may be oxidized if left at room temperature, exposed to air, and light and become rancid. Flaxseeds come in two main colors: brown, and yellow or golden. Both are edible and generally have similar nutritional value. Flaxseeds

Table 2 Flaxseeds nutrition facts of 100 g, from the United States Department of Agriculture (USDA SR-21)

Serving size of 100 g of flaxseeds	
Per serving	% Daily value[a]
Calories 534	
Total fat 42.2 g	65
Saturated fat 3.7 g	18
Polyunsaturated fat 28.7 g	
Monounsaturated fat 7.5 g	
Total omega-3 fatty acids 22,813 mg	
Total omega-6 fatty acids 5911 mg	
Cholesterol 0.0 mg *Phytosterols*	
Carbohydrates 28.9 g	10
Dietary fiber 27.3 g	109
Starch 0.0 g	
Sugars 1.5 g	
Protein 18.3 g	37
Vitamins	
Vitamin A	0
Vitamin E	2
Vitamin K	5
Vitamin C	1
Vitamin B-6	24
Folic acid	22
Minerals	
Sodium	1
Calcium	26
Magnesium	98
Copper	61
Potassium	23
Iron	32
Manganese	124
Zinc	29

National Nutrient Database
[a]Based on a *2000 calorie diet*

are also grown for their plant fiber. Flax textiles (aka linen) are commonly used as sheets and garments. Flax oil or linseed oil is a comestible oil and is also used as a component in several wood-finishing products like paints and varnishes. A linseed meal is a protein-rich flaxseed by-product typically used as a livestock feed. Flaxseeds are a natural source of phenolic acids, plant phytoestrogens, lignans, and fatty acids. Flaxseeds are rich in dietary fiber, protein, B-vitamins, and several minerals (Table 2).

Flaxseeds are highly nutritious and potentially help in fighting off a number of health problems. This extraordinary advantage is primarily attributed to their exceptional composition of fiber, lignans, and omega-3 fatty acids [53–55]. Additionally, their protein is rich in the amino acids arginine, aspartate, and glutamate [56, 57]. Essentially, investigations on both lignans [58–63] and omega-3 fatty acids [64–74] have focused on benefits to patients with cardiovascular disease. Besides this value, we will discuss research in the areas of cancer and inflammation [75].

Plant phenolics (simple or polyphenols) possess both antioxidant and anti-inflammatory properties. Flaxseeds as a source enriched in the polyphenol lignans have several general health benefits [76] including modulating gut microbiota [77]. A recent systematic review and several other specific studies concluded that lignans lower the risk of breast cancer development, suppress metastasis [78–85], and enhance cytotoxicity of therapeutic agents [86]. Others have shown capability to suppress the proliferation of prostate cancer cells [87, 88] and colorectal cancers [89, 90]. Lignans have been also reported to have radio-protective capability against radiation exposure. They prevent DNA damage and enhance antioxidant capacity to protect normal cells from being damaged due to cancer radiotherapy [91].

Omega-3 in flaxseeds and its oil have equally lowered cancer incidence and anti-tumor activities [92–96] against several of its types such as breast [97], prostate [98], colorectal [99], and gastrointestinal cancer [100]. They have also been effective in combination therapy [101]. For instance, they were observed to induce apoptosis in cancerous cells and improve the efficacy of some anti-cancer therapies [101]. Evidence suggests that the flaxseeds' powerful composition of lignans, omega-3 fatty acids, and dietary fiber is connected to lower risk of colorectal cancer [89]. Omega-3 fatty acids in flax oil also exhibit anti-inflammatory activities and help prevent and manage chronic inflammatory disorders [102–107] including rheumatoid arthritis [104, 108, 109] and osteoarthritis [110–112].

The dietary fiber, lignans, and omega-3 fatty acids, individually or as a mixture, are proven potent anti-hypertensive and delipidemic supplements. Numerous studies have established that these lower the blood pressure effectively [113–115] and reduce the level of total and unhealthy LDL cholesterol in the blood [116–120] across ethnicity, gender, and age, thereby furthering heart health and reducing the risk of cardiovascular diseases and stroke as stated above.

Research data suggests that flaxseeds either reduce the incidence or delay the onset of diabetes [121–124]. Several lines of evidence strongly propose that they regulate sugar level [125–129], improve insulin resistance [130–133], and manage weight [134, 135]. Flax's dietary fiber, both soluble and insoluble, was affirmed to regulate blood glucose, lipid profile, and constipation [136] even among diabetes type 2 patients. It also aids in governing their weight [137]. Flaxseeds used as a substitute snack enable appetite and weight control [138–141]. Flaxseeds are gluten free and as such are good replacement for grains for people suffering from celiac disease or gluten sensitivity [142].

Flaxseed ingestion may lower the blood sugar and blood pressure and so, people on anti-hypertensive and sugar lowering medication must be cautious. Flax may slow blood clot formation and increase tendency of bleeding. It is recommended to stop eating flaxseeds before surgery. It is also recommended that they be avoided by people suffering from bleeding disorders or taking anti-coagulants. It has an estrogenic effect and might aggravate hormone-related conditions. Ground flaxseeds are easier to digest but excessive consumption of flaxseeds with inadequate amounts of water may cause bowel obstruction.

3 Legumes

Family: Fabaceae or Leguminosae
Genus: *Medicago* (Alfalfa)
Genus: *Pisum* (Pea)
Genus: *Cicer* (Chickpea)
Genus: *Lens* (Lentil)
Genus: *Glycine* (Soybean)
Genus: *Tamarindus* (Tamarind)
Common bean family:
Genus: *Phaseolus* (Lima bean, Pinto bean, Kidney bean, and Black bean)
Genus: *Vicia* (Fava bean)
Genus: *Vigna* (Cowpea, Ricebean, and Black-eyed pea)

The Fabaceae family (aka Leguminosae) is a huge, diverse, widely distributed, economically important group of crops and a staple human food. From a taxonomic, botanical standpoint, they constitute over 750 genera and about 19,000 known species with some that are designated weedy pests and/or as non-edible items due to high amounts of alkaloids. The palatable grain legumes are exceedingly nutritious and nourishing as they are a substantial source of dietary fiber, proteins, carbohydrates, macronutrients, and micronutrients [143–145]. They are also a major source of medicinal herbs as they have evolved with unique chemistries and have been exploited as part of traditional medicine since ancient times [146]. Nowadays, they are an imperative component of a regime and are on a physician's recommendation list for a plant-based diet [147, 148]. Aside from their standard health benefits, legumes, grain legumes in particular, are highly acclaimed for the prevention and treatment of chronic disease as they can be consumed in infinite combinations [149, 150]. Most intervention studies carried out are in the areas of cardiovascular diseases [151–153], cancer [154–159] metabolic syndromes [160–164], and weight management [165, 166]. Undoubtedly, more in-depth research needs to be performed on legume species that are highly consumed as part of a healthy dietary pattern.

To cover the entire common legumes collection is daunting and so, for this section, we will focus on commonly consumed legumes. Even within this we will stick only to very few that are relevant to people's conventional daily lives. We will meaningfully underscore only a minority of phytochemicals that are deemed extra important to health and have not been discussed in the Vegetables, Fruits, and Grains chapters.

3.1 Lentils

The lentil is an eatable legume with a lens-shaped seed. It mostly consists of two halves covered in a husk that contains most of its dietary fiber. Lentils can be eaten with or without the husk. There are multiple varieties including black, red, brown, yellow, macachiados, and French green lentils. Lentils are low caloric, low fat food, and contain a high protein content which make them a good choice for vegetarians. Lentils are included in many dishes all over the world and are also added to salads. They are a rich source of amino acids such as methionine, isoleucine, lysine, and cysteine as well as phytochemicals and phenols. They are also a good source of vitamins like folic acid and vitamin B-6 and minerals such as iron and copper (Table 3).

Lentil is one of the highly admired functional foods. A couple of reviews are able to truly capture the overall nutrition and health values of lentils based on in vitro, in vivo, and clinical studies with a particular emphasis on its bioactives including its polyphenols-rich content [167, 168]. Lentil consumption is associated with a lower risk of multiple lifestyle and health related conditions such as heart diseases, cancer, and type-2 diabetes. Of particular interest as well are lentil's content of the biomolecules lectins and defensins which partake in the development of innate immunity [169] and their potential theranostics application [170].

Lentils are considered a valuable source for dietary fiber including resistant starches. Besides fiber that ensures a healthy digestive system avoiding constipation [171], studies have established its positive effects on appetite control [172] and body weight management [165, 166]. Lentils have been shown to cause a favorable lipid profile lowering unhealthy LDL cholesterol and triglyceride levels and increasing the level of healthy HDL cholesterol in obese patients with type-2 diabetes [173]. Other investigations have demonstrated equally favorable metabolic features [162]. It also helps control the blood glucose level and prevent its complications in rats [174]. This glycemic response regulation was demonstrated in few small

Table 3 Lentil nutrition facts of 100 g, from the United States Department of Agriculture (USDA SR-21)

Serving size of 100 g of boiled, unsalted lentil	
Per serving	% Daily value[a]
Calories 116	
Total fat 0.4 g	1
Saturated fat 0.1 g	0
Polyunsaturated fat 0.2 g	
Monounsaturated fat 0.1 g	
Total omega-3 fatty acids 37.0 mg	
Total omega-6 fatty acids 137.0 mg	
Cholesterol 0.0 mg *Phytosterols*	
Carbohydrates 20.1 g	7
Dietary fiber 7.9 g	32
Starch	
Sugars 1.8 g	
Protein 9 g	18
Vitamins	
Vitamin A	0
Vitamin E	1
Vitamin K	2
Vitamin C	2
Vitamin B-6	9
Folic acid	45
Minerals	
Sodium	0
Calcium	2
Magnesium	9
Copper	13
Potassium	11
Iron	19
Manganese	25
Zinc	8

National Nutrient Database
[a]Based on a *2000 calorie diet*

intervention studies [172, 175, 176] while a large prospective intervention study revealed that eating lentil in the context of Mediterranean diet is inversely associated with diabetes type 2 incidence in adults [160].

Systematic reviews of clinical trials have shown that lentils reduce cardiovascular risk vis-à-vis lipid measurements [177] and lower blood pressure per animal studies [178], and per controlled feeding clinical trials [179]. One possible explanation for this hypotensive effect is that they escalate the level of arginine amino acid and several arginine-related compounds that have the potential to increase the production of the vasodilator nitric oxide (NO) thus lowering blood pressure [180]. Lentils rich source of folate and magnesium further boost heart health. Folic acid is recognized to lower the level of the homocysteine amino acid, which is a major risk factor for several heart diseases [181]. Magnesium helps relax the cardiovascular muscles, increase blood flow and the oxygenation of the body tissues, and helps naturally reduce blood pressure [182].

Lectins, carbohydrate-binding proteins widely spread in lentils, were reviewed recently [183]. They were suggested to have great potential as anti-cancer agents [159, 184–186]. Lectins have great cytotoxic activities. They affect several signaling pathways in cancer pathophysiology such as the caspase, P53, ERK, and Ras-Raf pathways leading to apoptosis which makes them a promising anti-cancer therapeutic biomolecule [185, 187]. Raw or cooked lentils have been shown to be chemopreventative (anti-carcinogenesis) particularly in the colon [188–190], prostate [156], and breast cancer recurrence [191]. Selenium, a trace element especially found in lentils, is a known antioxidant that possesses anti-cancer properties. It has been shown that selenium interferes with multiple pathways vital in the initiation, promotion, and progression of some types of cancer. This includes its role in promotion of apoptosis and prevention of angiogenesis [192]. The anti-cancer role of selenium and its effect on the development and treatment of multiple cancers, including breast cancer [193], colorectal cancer [194], and oral squamous cell carcinoma [195] were studied. Selenium is also expected to affect the liver enzymes and help detoxify the body preventing hepatocellular carcinoma [196]. Generally, selenium has been found to decrease inflammatory reactions and improve immune system vigor [197]. Undeniably, lentils' rich polyphenols content is another contributory factor as it reduces the incidence of various cancers.

The effect of nutrients on brain function has been reviewed [198]. In animal studies, red lentil extract was confirmed to be neuroprotective [199]. Lentils' content of vitamins and minerals like folic acid and potassium were shown to be equally helpful for ideal brain functioning [200]. Lectins, mainly microglial lectins family, function on central nervous system and have been suggested as effective targets to treat some neurological disorders [201].

Although lentils are the best alternative for animal protein and provide many health benefits, they also have some undesirable health effects when consumed excessively. Consuming a lot of lentils may lead to digestive complications including upset stomach, flatulence, and constipation. Lysine, an amino acid found in lentils, may lead to gallbladder stones, increase cholesterol levels, and kidney complications. High oxalate content in lentils may also result in kidney stone formation [202]. The high protein and potassium content in lentils may cause a load on the kidneys and overtime leads to renal impairment and hyperkalemia. Some people may develop an allergic reaction when eating lentils. Selenium in lentils contribute to number a of health benefits. However, consumption in large quantities could cause selenium toxicity and lead to nausea, vomiting, irritability, fatigue, discoloration and brittleness of the nails, and loss of hair [203].

3.2 *Fava Beans*

Fava bean, also known as the broad bean or fava bean, is a type of bean propagated for human consumption or "phytoremediation" as a cover crop to prevent soil erosion. They are widely popular in the Mediterranean region. The seed, after removing its outer cover, can be eaten raw or cooked. The young pod is green in color and blackish brown once matured. In younger plants, the young green pod is appetizing. Fava beans are of low fat, a rich source of dietary fiber, carbohydrate, protein, generally moderate levels of B-vitamins but a good source of vitamins like folic acid (B-9), vitamin K, and minerals such as iron, potassium, magnesium, and manganese (Table 4).

Fava bean, as a functional food, has been ascribed several health benefits with various in vivo biological activities [204, 205]. The methanolic and phenolic compound content of the fava bean pod are potent antioxidants [206] that help fight free radicles and prevent chronic diseases caused by free radicle damage [207]. They possess strong anti-inflammatory activity and work by down regulating pro-inflammatory gene expression or inhibiting inflammation related parameters [208] and even by improving arthritic condition in adults [209]. Fava beans consumption helps boost heart health. A couple of reviews have summarized the state of the epidemiological and clinical evidence for plant seeds influence on coronary heart disease including fava beans [210, 211] and their rich polyphenol content in cardioprotection along with highlighting their role in other important inflammatory diseases [212]. The complementary effect of fava beans' high dietary fiber reduces total cholesterol, the unhealthy LDL cholesterol, and triglycerides levels, helps increase the level of healthy HDL, and promotes better blood sugar regulation even in adults with type 2 diabetes [213–217]. The high fiber and protein content of fava beans allow control of body weight by suppressing appetite and increasing satiety, and tested subjects experienced improved BMI [166, 218, 219]. Fibers also promote digestive system function, e.g., preventing constipation and, along with other components, reducing the risk of colorectal cancer [220, 221]. Positive effects on prostate and breast cancers have been suggested as well by some articles [191, 222].

Fava bean is a great source of manganese, an antioxidant. It supports the immune [223], nervous [224], and endocrine systems' functions and may assist in reducing the risk of arthritis [225], osteoporosis [226], and diabetes mellitus type 2 [227]. Magnesium, potassium, and phosphorus content of fava beans are essential for controlling blood pressure levels as well.

Natural dopamine compound, L-dopa is an aromatic amino acid isolated from fava beans seedlings and is a precursor for catecholamines type neurotransmitters. It is present in high concentrations in fava beans and it was discovered to improve mood and reduce the risk of depression [228, 229]. L-dopa (aka levodopa) has been used therapeutically to help reduce the symptoms associated with Parkinson's disease [230–232]. It remains the gold standard for treatment. Fava beans have been employed as a natural product supplement to combat Parkinson's disease [233–235]. Fava bean has a rich reserve of iron which ameliorates iron deficiency anemia

Table 4 Fava beans nutrition facts of 100 g, from the United States Department of Agriculture (USDA SR-21)

Serving size of 100 g of boiled, unsalted fava beans	
Per serving	% Daily value[a]
Calories 110	
Total fat 0.4 g	1
Saturated fat 0.1 g	0
Polyunsaturated fat 0.2 g	
Monounsaturated fat 0.1 g	
Total omega-3 fatty acids 12.0 mg	
Total omega-6 fatty acids 152.0 mg	
Cholesterol 0.0 mg *Phytosterols*	
Carbohydrates 19.7 g	7
Dietary fiber 5.4 g	22
Starch	
Sugars 1.8 g	
Protein 7.6 g	15
Vitamins	
Vitamin A	0
Vitamin E	0
Vitamin K	4
Vitamin C	0
Vitamin B-6	4
Folic acid	26
Minerals	
Sodium	0
Calcium	4
Magnesium	11
Copper	13
Potassium	8
Ion	8
Manganese	21
Zinc	7

National Nutrient Database
[a]Based on a *2000 calorie diet*

and folic acid and calcium that ensure healthy pregnancy. These benefits have been discussed previously.

Eating fava beans is associated with some negative health effects. People with hereditary condition of glucose-6-phosphate dehydrogenase deficiency (G6PDD), known as favism, develop hemolytic anemia when eating fava beans due to its content of glycosides such as vicine and convicine that break the membrane of red blood cells [236, 237]. Tyramine (discussed in the fruit-banana section) is an amino acid present in high concentrations in fava beans. It may trigger migraine headaches in some people and may also interact with the antidepressant monoamine oxidase inhibitors (MAOI). MAO is an enzyme that breaks down excess tyramine in the body, dysregulating blood pressure and causing a dangerous increase [238]. Fava beans also contain levodopa (L-dopa), a prescribed treatment for Parkinson's disease patients, which interferes with the metabolism of vitamin B-6 and might result in vitamin B-6 deficiency [239, 240].

3.3 Soybean

Soybean or soya bean is a type of legume plant grown for its edible bean which, like other beans, are enclosed in pods. They are commonly green but can also be yellow, brown, or even black. Soybean is considered a meat and dairy substitute. Soya beans can be processed into a variety of food products like soymeal feed for livestock, while for humans as soya burgers, tempeh (soy cake), soya milk/cheese/yogurt, tofu (bean curd), soy-based infant formula, and soya protein (meat replacement). Soy flour, soy sauce, miso soup, and soybean oil have become one of the main ingredients of many Asian dishes. Soya beans also called edamame when fresh from the pod are a very rich source of protein and can be chosen as a meat alternative for vegetarians and vegans. Soya beans are a rich source of dietary fiber, fatty acids, B-vitamins (like vitamin B-9) and K, and minerals such as manganese, iron, copper, magnesium, potassium, calcium, and zinc. They are low in cholesterol and contain the flavonoid antioxidants, isoflavones. They contain a significant amount of phytic acid and other important phytochemicals. Soybean oil is the primary source of biodiesel fuel (Table 5).

Soybeans are exceptionally nutritious food and help maintain good health. Several articles in existing literature have tackled the soya bean health issue from different angles. For example, primary evaluation of the clinical and epidemiologic evidence of health benefits [241–243], soybean nutritional role and diet [244, 245], their topmost phytochemical isoflavones [246], or their anti-nutrient phytoestrogenic health risk feature [247, 248], to name a few. Yet a few others implicated their intake as potential factors in cardiovascular disease and cancer prevention [241, 249]. Needless to say, some research findings seem controversial or cast doubt on their efficacy. However, in the basic sense, their nutritional value is undeniable, and recent research and meta-analyses suggest that soybeans have helped reduce risk of coronary heart disease.

The role of fiber in cardiovascular health is well established [210, 211]. A review article summarized soy foods cardiovascular data [241] and a recent meta-analysis provided evidence that consumption of soya beans was inversely associated with the risk of cardiovascular disease, stroke, and coronary heart disease risk [250]. Human intervention studies have demonstrated that soya beans fiber and soya proteins have been found to reduce the levels of total cholesterol, unhealthy LDL cholesterol, and triglycerides and raise HDL levels which prevent atherosclerosis, stroke, and heart attacks [145, 251]. As a matter of fact they are on the nutritional list of

Table 5 Soybean nutrition facts of 100 g, from the United States Department of Agriculture (USDA SR-21)

Serving size of 100 g of boiled unsalted soybeans	
Per serving	% Daily value[a]
Calories 173	
Total fat 9.0 g	14
Saturated fat 1.3 g	6
Polyunsaturated fat 5.1 g	
Monounsaturated fat 2.0 g	
Total omega-3 fatty acids 598 mg	
Total omega-6 fatty acids 4466 mg	
Cholesterol 0.0 mg *Phytosterols* 161 mg	
Carbohydrates 9.9 g Dietary fiber 6.0 g	3 24
Starch	
Sugars 3.0 g	
Protein 16.6 g	33
Vitamins	
Vitamin A	0
Vitamin E	2
Vitamin K	24
Vitamin C	3
Vitamin B-6	12
Folic acid	14
Minerals	
Sodium	0
Calcium	10
Magnesium	21
Copper	20
Potassium	15
Iron	29
Manganese	41
Zinc	8

National Nutrient Database
[a]Based on a *2000 calorie diet*

recommended foods [252]. A continuous stream of information, even large cohort studies support claims of soybeans' impact on lipid profile [213, 253–255] and blood pressure [256–258] including isoflavones [259–261]. Significant amounts of linoleic and linolenic essential fatty acids in soya beans are presumed to control the function of the aortic smooth muscles [262] and help regulate blood pressure levels [263]. Other non-protein soy constituents influence cardiovascular risk factors reduction such as isoflavones, lecithins, and saponins [264, 265]. Phytosterols compounds from soya beans have been shown to reduce cholesterol absorption from the gut which further helps to lower the level of total cholesterol and augment heart health

[266–268]. Soy phytoestrogen supplementation also significantly reduced insulin resistance, hsCRP, and blood pressure in patients [269, 270].

Dietary phytoestrogens revealed a number of potential health benefits [246, 270] with the archetypal isoflavones as a potential therapeutic agent [271]. Soya bean consumption is thought to preclude a number of cancer types. Its rich antioxidant isoflavones content induces apoptosis and inhibits proteins involved in certain cell survival pathways in breast cancer [272]. In general, soybean intake is associated with lower risk of breast cancer [273–275], specifically, the isoflavones' anti-estrogenic effects are believed to reduce the risk of estrogen-dependent breast cancer [274, 276–278] and endometriosis [279]. A large cohort study showed that post-diagnosis soy intake statistically significantly reduces breast cancer relapse and improves survival [277]. Soya beans were also shown to be involved in the prevention and treatment of lung cancer [280], gastrointestinal cancer [281, 282], bladder cancer [283], prostate cancer [284, 285], and ovarian cancer [286] through multiple mechanisms including isoflavones role in DNA repair, cell cycle arrest, induction of apoptosis, and preventing angiogenesis and metastasis of cancer cells. As discussed above, legumes' dietary fiber including soybeans have been confirmed to improve the overall functions of the digestive system. Regular dietary soya bean intake that is, in essence, high content fiber and isoflavones, has been accepted to lower the risk of colorectal cancer [287–289]. Soybean contains the polypeptides lunasins and lectins glycoproteins that display anti-carcinogenic potentials [185, 290–293]. The isoflavones in soybeans are anti-inflammatory by nature though the underlying mechanisms remain vague [294]. Anti-inflammatory activities of lunasin has allegedly prevented and treated cardiovascular disorders including atherosclerosis [295]. Isoflavones in soya bean products have been linked to lowered risk of diabetes type 2 [296, 297] with a clear potential to manage the disease after diet supplementation due to blood glucose, plasma lipids, and antioxidant enzyme activity in patients [298–300]. Like other legume fibers discussed earlier, soy proteins and fiber help in weight control [166, 301–304]. In a randomized controlled trial soybean protein supplementation reduced leptin hormone levels [254]. Circulating leptin reduces appetite and obese people show resistance to the hormone (high circulating concentration) thus failing to modulate weight. By decreasing leptin levels, one improves metabolic function.

Isoflavones were also found to improve menopausal symptoms including hot flashes and excessive sweating [305–307]. They also improve bone density and are believed to lower the risk of osteoporosis [308], especially in post-menopausal women [309–313]. Phylloquinone, the vitamin K form in soya bean, is thought to control blood coagulation and boost bone health [314]. Rich soya bean content of minerals like magnesium, copper, selenium, and zinc are essential for bone health and act as a preventive and treatment measure for osteoporosis in the long term [315, 316].

Despite the fact that soybeans are highly nutritious and provide a number of health benefits, concerns about their side effects are on the rise. Soybeans' insoluble α-galactoside fibers (oligosaccharides) such as stachyose and raffinose may cause flatulence, abdominal discomfort, and diarrhea in sensitive people [317], and possibly aggravate the systemic and gastrointestinal symptoms of irritable bowel syndrome [318]. The high content of proteins, mainly glycinin and conglycinin, may cause allergic reactions in sensitive people [319–321]. In animal models, intestinal damage was observed through the expression of p38/JNK/NF-κB signaling pathway [322]. Hypothetically, phytoestrogens, in the form of isoflavones daidzein and genistein, impose some level of risk. However, the concentration is not sufficient enough to elicit a physiological response in humans. Due to their potential estrogenic effect, it is advisable to limit soya bean consumption in the cases of women who have or have had estrogen receptor positive breast cancers. These substances act as goitrogens and have been hypothesized to disturb thyroid hormones production by interfering with the iodine uptake by the thyroid gland and its activity [323, 324] Long term consumption of soybean products containing isoflavones was highly suspected in a hypothyroidism case with a susceptible individual [325]. Soya seeds contain oxalate, which when excessively consumed might lead to kidney stone formation [202]. Like other beans and seeds, soybeans contain phytic acid which impairs the absorption of minerals like manganese, zinc, and iron. This can probably be reduced by physicochemical processing of soya beans to produce low phytic acid protein isolate [326]. Overall moderate consumption of soy products does not appear to present any serious health risks.

3.4 Chickpea

Chickpea, also known by other names like chick pea, gram, Bengal gram, garbanzo, or garbanzo bean, is a small round seed with a nutty taste and buttery texture. There are two main varieties that come in different colors. The most common ones are cream or pale-brown, black, green, and red beans. It is an everyday meal in many Middle Eastern and Mediterranean countries and a key ingredient in many traditional dishes. It can be used whole during cooking, in salads, ground into in several meals, or roasted as a snack. Chickpeas and chickpea flour are moderate caloric. However, they are a very rich source of proteins, dietary fiber, several vitamins and minerals, and antioxidants (Table 6).

Chickpeas are packed with nutrients. Their nutritional profile and many health associated outcomes have been reviewed recently [327–329]. Several of the cited studies support claims of their consumption ameliorating chronic diseases and maintaining gastrointestinal health in addition to weight control. In chickpeas, the chief bioactive phytochemicals are antioxidants (isoflavones) [330, 331], phytosterols [332], oligosaccharides [333], and unsaturated fatty acids [329].

Chickpeas, as a legume in the category of pulse food (dried seeds like beans, peas, and lentil), are high-protein, high-fiber foods which impart tremendous health benefits [334–337]. Synergistic action provokes strongest reduction in hunger via hormone modulation [338, 339], helps keep appetite under control [340, 341], and permits weight management [165, 166, 342, 343]. For instance, intervention studies

Table 6 Chickpea nutrition facts of 100 g, from the United States Department of Agriculture (USDA SR-21)

Serving size of 100 g of boiled unsalted chickpeas	
Per serving	% Daily value[a]
Calories 164	
Total fat 2.6 g	4
Saturated fat 0.3 g	1
Polyunsaturated fat 1.2 g	
Monounsaturated fat 0.6 g	
Total omega-3 fatty acids 43.0 mg	
Total omega-6 fatty acids 1113 mg	
Cholesterol 0.0 mg *Phytosterols* 35.0 mg	0
Carbohydrates 27.4 g	9
Dietary fiber 7.6 g	30
Starch	
Sugars 4.8 g	
Protein 8.9 g	18
Vitamins	
Vitamin A	1
Vitamin E	2
Vitamin K	5
Vitamin C	2
Vitamin B-6	7
Folic acid	43
Minerals	
Sodium	0
Calcium	5
Magnesium	12
Copper	18
Potassium	8
Iron	16
Manganese	52
Zinc	10

National Nutrient Database
[a]Based on a *2000 calorie diet*

have revealed that chickpeas proteins and fiber combination affect body mass index [344], food choices afterwards [345] as well as calorie intake [346].

The health outcome value of chickpeas and its broad phytonutrients extends to chronic diseases as they reduce oxidative stress, counteract the effect of free radicles, and prevent a few chronic diseases including diabetes, heart diseases, and cancer [347]. Chickpeas were found to control blood glucose levels [215, 346, 348–351] and protect individuals against type-2 diabetes and its risk factors [329, 352–355] through many of its constituents: fibers, vitamins, magnesium, and zinc [338, 356, 357].

Chickpeas also contribute to hypertension control with their potassium and magnesium content [351, 358, 359]. Furthermore, they help lower total cholesterol, unhealthy LDL cholesterol, and triglycerides and numerous studies have verified these findings [348, 349, 355, 360–362]. In a Mediterranean diet cohort, lifestyle food choice plays an important role in the control and management of hypertension, lipid profile, and metabolic syndromes [363]. Propionate, a short chain fatty acid produced by the gut bacteria secondary to fiber consumption [364], was found to inhibit cholesterol synthesis [338]. Alpha-linolenic acid (ALA), an essential fatty acid with antioxidant properties is also believed to lower the risk of coronary heart disease [365]. Folic acid is known to lower the level of homocysteine amino acid which is a major risk factor for atherosclerosis, stroke, and heart attacks [366].

Soluble chickpea fiber alleviates irritable bowel disease symptoms [367] and lowers the incidence of colorectal cancer [328, 338], while the insoluble fibers are proven to maintain digestive system health, promote bowel movement, and prevent constipation [328, 345, 361, 368]. Butyrate, a fatty acid produced when consuming chickpeas arising from bacterial fermentation of dietary fiber [364, 369], is reported to have anti-cancer and anti-inflammatory properties and has been shown to reduce cell proliferation, induce apoptosis [370], and lower the risk of colorectal cancer [371]. Furthermore, chickpeas contain saponins which are plant-based natural glycosides. In chemical terms, saponins contain sugar moiety attached to a triterpenoid or steroids. They are supposed to be protective and help prevent the development of certain types of cancer [372]. Saponins display cytotoxic properties. They inhibit tumor growth, impede cell proliferation, induce apoptosis, and decrease cellular invasiveness [373, 374]. Chickpeas' vitamin content, i.e., B-vitamins and essential amino acids like methionine, is assumed to lower the risk of breast [375, 376] and lung [377] cancers. In short, regular consumption of chickpeas is associated with lesser prevalence of cancer development.

In addition to isoflavones, chickpeas contain essential amino acids like lysine, leucine, isoleucine, methionine (invariably low in sulfur amino acids), and aromatic amino acids and minerals including copper, zinc, and phosphorus that are essential for bone health and prevention of osteoporosis [310, 378]. Plant proteins are also known for their role in maintaining skeletal muscle strength [379]. Chickpeas are rich source of the chemical element molybdenum which helps with the metabolism of iron, improving its absorption, and averting anemia [380].

Although chickpeas are gluten free, contamination with other gluten sources is common during manufacturing and storage. Overeating chickpeas can lead to digestion problems like stomachache, flatulence, and diarrhea. Oligosaccharides, the type of sugar present in chickpeas, are not easy for humans to digest and can supposedly cause and worsen the symptoms of some gastrointestinal diseases including irritable bowel syndrome (IBS), Crohn's disease, and ulcerative colitis. Though uncommon, chickpeas proteins may result in few sensitive people developing an allergic reaction upon eating. The signs range from a few digestive symptoms to skin reactions, eczema, and hives to severe and even life-threatening anaphylaxis symptoms.

4 Pumpkin

Family: Cucurbitaceae
Genus: *Cucurbita*
Common name: Pumpkin seed or pepita.

A pumpkin seed (aka pepita) is the little edible seed of pumpkin or specific types of squash. A typical pumpkin seed is fairly flat and has an asymmetrical oval shape. The kernel is light green in color with a white outer covering shell. Seeds are commonly served roasted, salted, or spiced and even as an ingredient in different dishes in some cuisines. The oil of pumpkin seeds is rich in different fatty acids and is used as a cooking oil or in salads. The seeds are very nutritious, they have high content of protein, dietary fiber and numerous minerals including manganese, magnesium, iron, copper, zinc, and potassium, in addition to several vitamins like vitamin K and B-vitamins (Table 7). Pumpkin seeds have been used in folklore medicine to treat parasitic infections.

Pumpkin seeds are an emerging healthy nutraceutical snack option. Their nutritional and biochemical composition have recently been characterized [381–384]. Their herbal medicine potential has been discussed as well [385–389]. They are a good source of phytosterols [390, 391], fatty acids (omega-3 and omega-6) [392–397], and phenolic acid antioxidants such as hydroxybenzoic, caffeic, coumaric, ferulic, sinapic, protocatechuic, vanillic, and syringic acids [398–404] along with minerals and other micronutrients [382, 405, 406] that all are associated with multiple health benefits.

Pumpkin seeds/oil demonstrated antioxidant and anti-inflammatory properties in in vitro [407] and animal models [408, 409]. For example, they were shown to lower the risk of arthritis by diminishing inflammatory reactions [410]. Redox-regulation and inflammation control are critical factors to prevent and potentially treat some cardiovascular conditions [411] and other oxidative-stress related diseases [412]. For the heart, these advantages manifest themselves as being anti-hypertensive, reducing blood pressure [413, 414], and increasing HDL cholesterol [415–417]. Indeed, these animal model findings were extrapolated into human clinical trials [210, 418, 419]. The availability of fiber [420] and presence of phytosterols is

Table 7 Pumpkin seeds nutrition facts of 100 g, from the United States Department of Agriculture (USDA SR-21)

Serving size of 100 g of roasted, unsalted, pumpkin seeds (pepitas)	
Per serving	% Daily value[a]
Calories 522	
Total fat 42.1 g	65
Saturated fat 8.0 g	40
Polyunsaturated fat 19.2 g	
Monounsaturated fat 13.1 g	
Total omega-3 fatty acids 166 mg	
Total omega-6 fatty acids 19,020 mg	
Cholesterol 0.0 mg *Phytosterols* *Carbohydrates* 13.4 g	0 4
Dietary fiber 3.9 g	16
Starch	
Sugars 1.0 g	
Protein 33.0 g	66
Vitamins	
Vitamin A	8
Vitamin E	0
Vitamin K	59
Vitamin C	3
Vitamin B-6	4
Folic acid	14
Minerals	
Sodium	1
Calcium	4
Magnesium	134
Copper	69
Potassium	23
Iron	83
Manganese	151
Zinc	50

National Nutrient Database
[a]Based on a *2000 calorie diet*

another reason for their proficiency to lowering lipid absorption and reduction in total cholesterol [408, 409] and LDL cholesterol levels [421]. Equally important is pumpkin seeds protein rich content of the semi-essential amino acid, arginine. Arginine serves as a precursor substrate for nitric oxide production by vascular endothelial and immune cells. The signaling molecule nitric oxide is essential for both blood pressure and immune regulation [422]. Further benefit is derived from magnesium where it directly lowers the blood pressure [423, 424] and reduces the risk of atherosclerosis [425].

Pumpkin seed extract [426] and a diet fortified with pumpkin seeds have been shown to lower the risk of several cancers including stomach, lung, colon, and prostate cancers [427]. Pumpkin seeds/oil also contain lignans, a phytoestrogen [270, 428–430], which as an estrogen-like compound was found to play a major role in preventing and treating breast cancer [431–433]. In addition, in vivo and in vitro studies using pumpkin seeds phytosterols have supported the claims of promoting prostate gland health and reducing the symptoms associated with benign prostate hyperplasia (PBH) [426, 434–438]. Intervention studies in men confirmed that pumpkin seed oil is clinically safe and efficacious [439, 440]. Pumpkin seeds/oil content of zinc was found to maintain prostate function [441]. Pumpkin seed oil has also been found to be effective in treating and preventing urinary tract dysfunction including overactive bladder in both men and women [442, 443].

A significant body of literature exist on pumpkin seeds anti-diabetic and hypoglycemic effect. These studies were conducted in vivo [444–447] and interventional on humans [354, 448, 449]. Pumpkin seeds and juice contain high amount of magnesium which inversely correlates with diabetes risk [450, 451]. Pumpkin seeds' dietary fibers coupled with their high protein content stabilize blood sugar levels and allow proper management of body weight through a plant-based diet [149, 338, 452, 453] proven by clinical trials [454]. It has been recommended to be built-in into athletes' regimes as well [455].

Pumpkin seeds are on the list of recommended plant-based diet for its notable iron source [147]. Puzzlingly, high fortified dietary mineral intake with iron being one of them is associated with mild dementia risk and Alzheimer neuropathology [456]. Ironically, the rich content of zinc in pumpkin seeds is not only linked to improved immune system functions [457] but also to CNS functions [458–460]. It has been suggested that trace element homeostasis including zinc may possibly reduce the risk of Alzheimer's disease [461–463]. Several plant seed extracts offer protection against amyloid β-induced neurotoxicity [464] and pumpkin seeds have been suggested as a complementary treatment of AD [465]. In addition, zinc subclinical deficiency, which can be avoided by keeping pumpkin seeds as part of a routine of daily administration, might lead to memory and reasoning impairment and learning difficulties especially in children [466]. L-tryptophan, an amino acid available in pumpkin seeds which is converted into serotonin, has been suggested to improve mood disorders, alleviate depression, and is mildly sedative [467, 468]. Therefore, eating pumpkin seeds before bedtime could improve the quality of sleep as well. The tryptophan will be converted to serotonin with the help of zinc and then convert it into melatonin, a hormone that regulates the sleep cycle and commences biological rhythms. Magnesium level is also linked to better sleep pattern and efficiency [469, 470]. Magnesium supplementation intervention appears to resolve insomnia issues in the elderly [470–472] and elite athletic population [473]. Magnesium is also known to play prodigious role in keeping healthy bones, preventing fractures, and reducing the risk of osteoporosis [474].

In traditional medicine pumpkin seeds are supposed to arouse lactation in mothers, stimulate healthy digestion, relieve stomach pain, and are natural diuretics and anti-parasitics. Though pumpkin seeds are safe in moderate amounts, overconsumption might have unwanted effects on health. As a natural diuretic, it may cause improper potassium/sodium levels, increased sugar levels, and headache or dizziness in some people. Overconsumption of pumpkin seeds can cause abdominal pain due to its rich fatty oil and fiber content. As with many other seeds and nuts, pumpkin seeds contain phytic acid which can interfere with the bioavailability of some nutrients leading to nutritional deficiency [475]. Overcooking or inadequate chewing can deprive the body of its benefits as well. It is not common to develop allergies, but some people might experience skin irritation and itching when eating pumpkin seeds. Others might exhibit worse anaphylactic reaction and develop swelling in mucus membrane of the mouth, throat irritation and cough, and difficulty in breathing.

5 Others

Other seeds that are renowned for their impressive health benefits include the following:

Chia Seeds They are a rich source of protein as such a suitable substitute for vegetarians. They are high in iron and folate, as well as magnesium, omega-3 fatty acids, and soluble fiber. They also supply a good amount of manganese, phosphorus, and calcium. They are best for weight loss and keeps heart healthy.

Hemp Seeds They are loaded with protein and are a good source of polyunsaturated fats (omega-3 and omega-6 fatty acids). The seeds are rich in vitamin E (antioxidant activity), magnesium, manganese, calcium, iron, sulfur, and zinc. They also contain phytosterols that help lower cholesterol levels. It offers incredible other health benefits like helps in weight loss and boosts immunity.

Poppy Seeds They are good source of fiber and protein, plus contain a robust dose of iron, manganese, and calcium. They support weight loss and aid in chronic diseases prevention.

Sesame Seeds They contain high protein amounts and lots of fiber. They are very nutritious as they contain essential minerals such as copper, manganese, and magnesium, along with calcium and vitamin B-1. The seeds promote blood cell formation and protect against anemia.

Sunflower Seeds They are incredibly nutrient dense food containing high amounts of B-vitamins, vitamin E, and antioxidants that combats free radicals and keeps inflammation at check. Also, they possess appreciable amounts of the vitamins thiamin (B-1) and pyridoxine (B-6), and the minerals manganese and phosphorus. They positively impact cholesterol and blood sugar levels.

References

1. Taddeo, V. A., Genovese, S., Medina, P., Palmisano, R., Epifano, F., & Fiorito, S. (2017). Quantification of biologically active O-prenylated and unprenylated phenylpropanoids in dill (Anethum graveolens), anise (Pimpinella anisum), and wild celery (Angelica archangelica). *Journal of Pharmaceutical and Biomedical Analysis, 134*, 319–324.
2. Kozlowska, M., Gruczyńska, E., Ścibisz, I., & Rudzińska, M. (2016). Fatty acids and sterols composition, and antioxidant activity of oils extracted from plant seeds. *Food Chemistry, 213*, 450–456.
3. Koeduka, T., Baiga, T. J., Noel, J. P., & Pichersky, E. (2009). Biosynthesis of t-anethole in anise: Characterization of t-anol/isoeugenol synthase and an O-methyltransferase specific for a C7-C8 propenyl side chain. *Plant Physiology, 149*(1), 384–394.
4. Denev, R. V., Kuzmanova, I. S., Momchilova, S. M., & Nikolova-Damyanova, B. M. (2011). Resolution and quantification of isomeric fatty acids by silver ion HPLC: Fatty acid composition of aniseed oil (Pimpinella anisum, *Apiaceae*). *Journal of AOAC International, 94*(1), 4–8.
5. Pickrahn, S., Sebald, K., & Hofmann, T. (2014). Application of 2D-HPLC/taste dilution analysis on taste compounds in aniseed (Pimpinella anisum L.). *Journal of Agricultural and Food Chemistry, 62*(38), 9239–9245.
6. Shojaii, A., & Abdollahi Fard, M. (2012). Review of pharmacological properties and chemical constituents of Pimpinella anisum. *ISRN Pharmaceutics, 2012*, 510795–510795.
7. Yashin, A., Yashin, Y., Xia, X., & Nemzer, B. (2017). Antioxidant activity of spices and their impact on human health: A review. *Antioxidants (Basel, Switzerland), 6*(3), 70.
8. Badgujar, S. B., Patel, V. V., & Bandivdekar, A. H. (2014). Foeniculum vulgare mill: A review of its botany, phytochemistry, pharmacology, contemporary application, and toxicology. *BioMed Research International, 2014*, 842674.
9. Mohamad, R. H., El-Bastawesy, A. M., Abdel-Monem, M. G., Noor, A. M., Al-Mehdar, H. A., Sharawy, S. M., et al. (2011). Antioxidant and anticarcinogenic effects of methanolic extract and volatile oil of fennel seeds (Foeniculum vulgare). *Journal of Medicinal Food, 14*(9), 986–1001.
10. Larijani, B., Esfahani, M. M., Moghimi, M., Shams Ardakani, M. R., Keshavarz, M., Kordafshari, G., et al. (2016). Prevention and treatment of flatulence from a traditional Persian medicine perspective. *Iranian Red Crescent Medical Journal, 18*(4), e23664.
11. Babaeian, M., Naseri, M., Kamalinejad, M., Ghaffari, F., Emadi, F., Feizi, A., et al. (2015). Herbal remedies for functional dyspepsia and traditional Iranian medicine perspective. *Iranian Red Crescent Medical Journal, 17*(11), e20741.
12. Al Mofleh, I. A., Alhaider, A. A., Mossa, J. S., Al-Soohaibani, M. O., & Rafatullah, S. (2007). Aqueous suspension of anise "Pimpinella anisum" protects rats against chemically induced gastric ulcers. *World Journal of Gastroenterology, 13*(7), 1112–1118.
13. Al Mofleh, I. A. (2010). Spices, herbal xenobiotics and the stomach: Friends or foes? *World Journal of Gastroenterology, 16*(22), 2710–2719.
14. Sumbul, S., Ahmad, M. A., Mohd, A., & Mohd, A. (2011). Role of phenolic compounds in peptic ulcer: An overview. *Journal of Pharmacy & Bioallied Sciences, 3*(3), 361–367.
15. Ghoshegir, S. A., Mazaheri, M., Ghannadi, A., Feizi, A., Babaeian, M., Tanhaee, M., et al. (2015). Pimpinella anisum in the treatment of functional dyspepsia: A double-blind, randomized clinical trial. *Journal of Research in Medical Sciences : The Official Journal of Isfahan University of Medical Sciences, 20*(1), 13–21.
16. Ghoshegir, S. A., Mazaheri, M., Ghannadi, A., Feizi, A., Babaeian, M., Tanhaee, M., et al. (2014). Pimpinella anisum in modifying the quality of life in patients with functional dyspepsia: A double-blind randomized clinical trial. *Journal of Research in Medical Sciences : The Official Journal of Isfahan University of Medical Sciences, 19*(12), 1118–1123.
17. Picon, P. D., Picon, R. V., Costa, A. F., Sander, G. B., Amaral, K. M., Aboy, A. L., et al. (2010). Randomized clinical trial of a phytotherapic compound containing Pimpinella

anisum, Foeniculum vulgare, Sambucus nigra, and Cassia augustifolia for chronic constipation. *BMC Complementary and Alternative Medicine, 10*, 17.
18. Kunnumakkara, A. B., Sailo, B. L., Banik, K., Harsha, C., Prasad, S., Gupta, S. C., et al. (2018). Chronic diseases, inflammation, and spices: How are they linked? *Journal of Translational Medicine, 16*(1), 14.
19. Agarwal, A. K. (2014). Spice up your life: Adipose tissue and inflammation. *Journal of Lipids, 2014*, 182575.
20. Tas, A., Ozbek, H., Atasoy, N., Altug, M. E., & Ceylan, E. (2006). Evaluation of analgesic and antiinflammatory activity of Pimpinella anisum fixed oil extract. *The Indian Veterinary Journal, 83*(8), 840–843.
21. Ritter, A. M. V., Ames, F. Q., Otani, F., de Oliveira, R. M., Cuman, R. K., & Bersani-Amado, C. A. (2014). Effects of anethole in nociception experimental models. *Evidence-based Complementary and Alternative Medicine, 2014*, 345829.
22. Haggag, E. G., Abou-Moustafa, M. A., Boucher, W., & Theoharides, T. C. (2003). The effect of a herbal water-extract on histamine release from mast cells and on allergic asthma. *Journal of Herbal Pharmacotherapy, 3*(4), 41–54.
23. Sánchez-Vidaña, D. I., Ngai, S. P., He, W., Chow, J. K., Lau, B. W., & Tsang, H. W. (2017). The effectiveness of aromatherapy for depressive symptoms: A systematic review. *Evidence-Based Complementary and Alternative Medicine, 2017*, 5869315.
24. Shahamat, Z., Abbasi-Maleki, S., & Mohammadi Motamed, S. (2016). Evaluation of antidepressant-like effects of aqueous and ethanolic extracts of Pimpinella anisum fruit in mice. *Avicenna Journal of Phytomedicine, 6*(3), 322–328.
25. Mosaffa-Jahromi, M., Tamaddon, A. M., Afsharypuor, S., Salehi, A., Seradj, S. H., Pasalar, M., et al. (2017). Effectiveness of Anise oil for treatment of mild to moderate depression in patients with irritable bowel syndrome: A randomized active and placebo-controlled clinical trial. *Evidence-Based Complementary and Alternative Medicine, 22*(1), 41–46.
26. Karimzadeh, F., Hosseini, M., Mangeng, D., Alavi, H., Hassanzadeh, G. R., Bayat, M., et al. (2012). Anticonvulsant and neuroprotective effects of Pimpinella anisum in rat brain. *BMC Complementary and Alternative Medicine, 12*, 76.
27. Albert-Puleo, M. (1980). Fennel and anise as estrogenic agents. *Journal of Ethnopharmacology, 2*(4), 337–344.
28. Saeed, I. A., Ali, L., Jabeen, A., Khasawneh, M., Rizvi, T. A., & Ashraf, S. S. (2012). Estrogenic activities of ten medicinal herbs from the Middle East. *Journal of Chromatographic Science, 51*(1), 33–39.
29. Tabanca, N., Khan, S. I., Bedir, E., Annavarapu, S., Willett, K., Khan, I. A., et al. (2004). Estrogenic activity of isolated compounds and essential oils of Pimpinella species from Turkey, evaluated using a recombinant yeast screen. *Planta Medica, 70*(8), 728–735.
30. Nahidi, F., Kariman, N., Simbar, M., & Mojab, F. (2012). The study on the effects of Pimpinella anisum on relief and recurrence of menopausal hot flashes. *Iranian Journal of Pharmaceutical Research, 11*(4), 1079–1085.
31. Mannion, C., & Mansell, D. (2012). Breastfeeding self-efficacy and the use of prescription medication: A pilot study. *Obstetrics and Gynecology International, 2012*, 562704.
32. Drugs and Lactation Database (LactMed). (2006). *Anise*. Bethesda, MD: National Library of Medicine (US).
33. Mhaidat, N. M., Abu-zaiton, A. S., Alzoubi, K. H., Alzoubi, W., & Alazab, R. S. (2015). Antihyperglycemic properties of Foeniculum vulgare extract in streptozocin-induced diabetes in rats. *International Journal of Pharmacology, 11*, 72–75.
34. Sheikh, B. A., Pari, L., Rathinam, A., & Chandramohan, R. (2015). Trans-anethole, a terpenoid ameliorates hyperglycemia by regulating key enzymes of carbohydrate metabolism in streptozotocin induced diabetic rats. *Biochimie, 112*, 57–65.
35. Jamous, R. M., Abu-Zaitoun, S. Y., Akkawi, R. J., & Ali-Shtayeh, M. S. (2018). Antiobesity and antioxidant potentials of selected Palestinian medicinal plants. *Evidence-Based Complementary and Alternative Medicine, 2018*, 8426752.

36. Bae, J., Kim, J., Choue, R., & Lim, H. (2015). Fennel (Foeniculum vulgare) and fenugreek (Trigonella foenum-graecum) tea drinking suppresses subjective short-term appetite in overweight women. *Clinical Nutrition Research, 4*(3), 168–174.
37. Chouhan, S., Sharma, K., & Guleria, S. (2017). Antimicrobial activity of some essential oils-present status and future perspectives. *Medicines (Basel, Switzerland), 4*(3), 58.
38. D'Souza, S. P., Chavannavar, S. V., Kanchanashri, B., & Niveditha, S. B. (2017). Pharmaceutical perspectives of spices and condiments as alternative antimicrobial remedy. *Evidence-Based Complementary and Alternative Medicine, 22*(4), 1002–1010.
39. Liu, Q., Meng, X., Li, Y., Zhao, C. N., Tang, G. Y., & Li, H. B. (2017). Antibacterial and antifungal activities of spices. *International Journal of Molecular Sciences, 18*(6), 1283.
40. Shreaz, S., Bhatia, R., Khan, N., Muralidhar, S., Basir, S. F., Manzoor, N., et al. (2011). Exposure of Candida to p-anisaldehyde inhibits its growth and ergosterol biosynthesis. *The Journal of General and Applied Microbiology, 57*(3), 129–136.
41. Hitokoto, H., Morozumi, S., Wauke, T., Sakai, S., & Kurata, H. (1980). Inhibitory effects of spices on growth and toxin production of toxigenic fungi. *Applied and Environmental Microbiology, 39*(4), 818–822.
42. Kosalec, I., Pepeljnjak, S., & Kustrak, D. (2005). Antifungal activity of fluid extract and essential oil from anise fruits (Pimpinella anisum L., Apiaceae). *Acta Pharmaceutica, 55*(4), 377–385.
43. Orchard, A., & van Vuuren, S. (2017). Commercial essential oils as potential antimicrobials to treat skin diseases. *Evidence-based Complementary and Alternative Medicine, 2017*, 4517971.
44. Musa Gomaa Kdam, R., Gabra, N., & Eltayb, A. A. (2017). Identification of Anise seed oils and their antimicrobial and antioxidant activities. *Red Sea University Journal of Basic and Applied Science, 2*, 232.
45. Mohamed, H. S. A. A., Abdelgadir, W. S., & Almagboul, A. (2015). In vitro antimicrobial activity of anise seed (Pimpinella anisum L.). *International Journal of Advanced Research, 3*, 359–367.
46. Al-Bayati, F. A. (2008). Synergistic antibacterial activity between Thymus vulgaris and Pimpinella anisum essential oils and methanol extracts. *Journal of Ethnopharmacology, 116*(3), 403–406.
47. Lee, J. B., Yamagishi, C., Hayashi, K., & Hayashi, T. (2011). Antiviral and immunostimulating effects of lignin-carbohydrate-protein complexes from Pimpinella anisum. *Bioscience, Biotechnology, and Biochemistry, 75*(3), 459–465.
48. Zahid, M. S. H., Awasthi, S. P., Hinenoya, A., & Yamasaki, S. (2015). Anethole inhibits growth of recently emerged multidrug resistant toxigenic Vibrio cholerae O1 El Tor variant strains in vitro. *The Journal of Veterinary Medical Science, 77*(5), 535–540.
49. Ibrahim, M. K., Zakaria Ahmed Mattar, Z. A., Abdel-Khalek, H. H., & Azzam, Y. M. (2016). Evaluation of antibacterial efficacy of anise wastes against some multidrug resistant bacterial isolates. *Journal of Radiation Research and Applied Sciences, 10*, 34–43.
50. Montbriand, M. J. (2004). Herbs or natural products that increase cancer growth or recurrence. Part two of a four-part series. *Oncology Nursing Forum, 31*(5), E99–E115.
51. Özgüven, M. (2012). 7 - Aniseed. In K. V. Peter (Ed.), *Handbook of herbs and spices* (2nd ed., pp. 138–150). Cambridge, UK: Woodhead Publishing.
52. Samojlik, I., Mijatović, V., Petković, S., Skrbić, B., & Božin, B. (2012). The influence of essential oil of aniseed (Pimpinella anisum, L.) on drug effects on the central nervous system. *Fitoterapia, 83*(8), 1466–1473.
53. Kajla, P., Sharma, A., & Sood, D. R. (2015). Flaxseed-a potential functional food source. *Journal of Food Science and Technology, 52*(4), 1857–1871.
54. Goyal, A., Sharma, V., Upadhyay, N., Gill, S., & Sihag, M. (2014). Flax and flaxseed oil: An ancient medicine & modern functional food. *Journal of Food Science and Technology, 51*(9), 1633–1653.
55. Touré, A., & Xueming, X. (2010). Flaxseed Lignans: Source, biosynthesis, metabolism, antioxidant activity, bio-active components, and health benefits. *Comprehensive Reviews in Food Science and Food Safety, 9*(3), 261–269.

56. Rabetafika, H. N., Van Remoortel, V., Danthine, S., Paquot, M., & Blecker, C. (2011). Flaxseed proteins: Food uses and health benefits. *International Journal of Food Science and Technology, 46*(2), 221–228.
57. Shim, Y., Gui, B., Arnison, P. G., Wang, Y., & Reaney, M. J. (2014). Flaxseed (Linum usitatissimum L.) bioactive compounds and peptide nomenclature: A review. *Trends in Food Science & Technology, 38*(1), 5–20.
58. Parikh, M., Netticadan, T., & Pierce, G. N. (2018). Flaxseed: Its bioactive components and their cardiovascular benefits. *American Journal of Physiology. Heart and Circulatory Physiology, 314*(2), H146–H159.
59. Witkowska, A. M., Waśkiewicz, A., Zujko, M. E., Szcześniewska, D., Stepaniak, U., Pająk, A., et al. (2018). Are total and individual dietary lignans related to cardiovascular disease and its risk factors in postmenopausal women? A nationwide study. *Nutrients, 10*(7), 865.
60. Peterson, J., Dwyer, J., Adlercreutz, H., Scalbert, A., Jacques, P., & McCullough, M. L. (2010). Dietary lignans: Physiology and potential for cardiovascular disease risk reduction. *Nutrition Reviews, 68*(10), 571–603.
61. Prasad, K. (2009). Flaxseed and cardiovascular health. *Journal of Cardiovascular Pharmacology, 54*(5), 369–377.
62. Bloedon, L. T., Balikai, S., Chittams, J., Cunnane, S. C., Berlin, J. A., Rader, D., et al. (2008). Flaxseed and cardiovascular risk factors: Results from a double blind, randomized, controlled clinical trial. *Journal of the American College of Nutrition, 27*(1), 65–74.
63. Dodin, S., Cunnane, S. C., Mâsse, B., Lemay, A., Jacques, H., Asselin, G., et al. (2008). Flaxseed on cardiovascular disease markers in healthy menopausal women: A randomized, double-blind, placebo-controlled trial. *Nutrition, 24*(1), 23–30.
64. Rodriguez-Leyva, D., Dupasquier, C. M., McCullough, R., & Pierce, G. N. (2010). The cardiovascular effects of flaxseed and its omega-3 fatty acid, alpha-linolenic acid. *The Canadian Journal of Cardiology, 26*(9), 489–496.
65. Kones, R., Howell, S., & Rumana, U. (2018). N-3 polyunsaturated fatty acids and cardiovascular disease: Principles, practices, pitfalls, and promises - a contemporary review. *Medical Principles and Practice, 26*(6), 497–508.
66. Balk, E. M., & Lichtenstein, A. H. (2017). Omega-3 fatty acids and cardiovascular disease: Summary of the 2016 Agency of Healthcare Research and Quality Evidence Review. *Nutrients, 9*(8), 865.
67. Bowen, K. J., Harris, W. S., & Kris-Etherton, P. M. (2016). Omega-3 fatty acids and cardiovascular disease: Are there benefits? *Current Treatment Options in Cardiovascular Medicine, 18*(11), 69–69.
68. Jain, A. P., Aggarwal, K. K., & Zhang, P. Y. (2015). Omega-3 fatty acids and cardiovascular disease. *European Review for Medical and Pharmacological Sciences, 19*(3), 441–445.
69. Mohebi-Nejad, A., & Bikdeli, B. (2014). Omega-3 supplements and cardiovascular diseases. *Tanaffos, 13*(1), 6–14.
70. Harris, W. S. (2007). Omega-3 fatty acids and cardiovascular disease: A case for omega-3 index as a new risk factor. *Pharmacological Research, 55*(3), 217–223.
71. von Schacky, C., & Harris, W. S. (2007). Cardiovascular benefits of omega-3 fatty acids. *Cardiovascular Research, 73*(2), 310–315.
72. Bloedon, L. T., & Szapary, P. O. (2004). Flaxseed and cardiovascular risk. *Nutrition Reviews, 62*(1), 18–27.
73. Mori, T. A. (2014). Omega-3 fatty acids and cardiovascular disease: Epidemiology and effects on cardiometabolic risk factors. *Food & Function, 5*(9), 2004–2019.
74. Mozaffarian, D. (2005). Does alpha-linolenic acid intake reduce the risk of coronary heart disease? A review of the evidence. *Alternative Therapies in Health and Medicine, 11*(3), 24–30; quiz 31, 79.
75. Donaldson, M. S. (2004). Nutrition and cancer: A review of the evidence for an anti-cancer diet. *Nutrition Journal, 3*, 19.
76. Adlercreutz, H. (2007). Lignans and human health. *Critical Reviews in Clinical Laboratory Sciences, 44*(5–6), 483–525.

77. Duda-Chodak, A. (2012). The inhibitory effect of polyphenols on human gut microbiota. *Journal of Physiology and Pharmacology, 63*(5), 497–503.
78. Ezzat, S. M., Shouman, S. A., Elkhoely, A., Attia, Y., Elsesy, M., El Senoussy, A., et al. (2018). Anticancer potentiality of lignan rich fraction of six flaxseed cultivars. *Scientific Reports, 8*(1), 544.
79. Calado, A., Neves, P. M., Santos, T., & Ravasco, P. (2018). The effect of flaxseed in breast cancer: A literature review. *Frontiers in Nutrition, 5*, 4.
80. Delman, D. M., Fabian, C. J., Kimler, B. F., Yeh, H., & Petroff, B. K. (2015). Effects of flaxseed lignan secoisolariciresinol diglucosideon preneoplastic biomarkers of Cancer progression in a model of simultaneous breast and ovarian cancer development. *Nutrition and Cancer, 67*(5), 857–864.
81. Mason, J. K., & Thompson, L. U. (2014). Flaxseed and its lignan and oil components: Can they play a role in reducing the risk of and improving the treatment of breast cancer? *Applied Physiology, Nutrition, and Metabolism, 39*(6), 663–678.
82. Lowcock, E. C., Cotterchio, M., & Boucher, B. A. (2013). Consumption of flaxseed, a rich source of lignans, is associated with reduced breast cancer risk. *Cancer Causes & Control, 24*(4), 813–816.
83. Wang, L., Chen, J., & Thompson, L. U. (2005). The inhibitory effect of flaxseed on the growth and metastasis of estrogen receptor negative human breast cancer xenograftsis attributed to both its lignan and oil components. *International Journal of Cancer, 116*(5), 793–798.
84. Thompson, L. U., Chen, J. M., Li, T., Strasser-Weippl, K., & Goss, P. E. (2005). Dietary flaxseed alters tumor biological markers in postmenopausal breast cancer. *Clinical Cancer Research, 11*(10), 3828–3835.
85. Chen, J., Stavro, P. M., & Thompson, L. U. (2002). Dietary flaxseed inhibits human breast cancer growth and metastasis and downregulates expression of insulin-like growth factor and epidermal growth factor receptor. *Nutrition and Cancer, 43*(2), 187–192.
86. Di, Y., De Silva, F., Krol, E. S., & Alcorn, J. (2018). Flaxseed lignans enhance the cytotoxicity of chemotherapeutic agents against breast cancer cell lines MDA-MB-231 and SKBR3. *Nutrition and Cancer, 70*(2), 306–315.
87. Azrad, M., Vollmer, R. T., Madden, J., Dewhirst, M., Polascik, T. J., Snyder, D. C., et al. (2013). Flaxseed-derived enterolactone is inversely associated with tumor cell proliferation in men with localized prostate cancer. *Journal of Medicinal Food, 16*(4), 357–360.
88. Demark-Wahnefried, W., Robertson, C. N., Walther, P. J., Polascik, T. J., Paulson, D. F., & Vollmer, R. T. (2004). Pilot study to explore effects of low-fat, flaxseed-supplemented diet on proliferation of benign prostatic epithelium and prostate-specific antigen. *Urology, 63*(5), 900–904.
89. DeLuca, J. A. A., Garcia-Villatoro, E. L., & Allred, C. D. (2018). Flaxseed bioactive compounds and colorectal cancer prevention. *Current Oncology Reports, 20*(8), 59.
90. Jenab, M., & Thompson, L. U. (1996). The influence of flaxseed and lignans on colon carcinogenesis and beta-glucuronidase activity. *Carcinogenesis, 17*(6), 1343–1348.
91. Velalopoulou, A., Tyagi, S., Pietrofesa, R. A., Arguiri, E., & Christofidou-Solomidou, M. (2015). The flaxseed-derived lignan phenolic secoisolariciresinol diglucoside (SDG) protects non-malignant lung cells from radiation damage. *International Journal of Molecular Sciences, 17*(1), 7.
92. Rose, D. P., & Connolly, J. M. (1999). Omega-3 fatty acids as cancer chemopreventive agents. *Pharmacology & Therapeutics, 83*(3), 217–244.
93. Stephenson, J. A., Al-Taan, O., Arshad, A., Morgan, B., Metcalfe, M. S., & Dennison, A. R. (2013). The multifaceted effects of omega-3 polyunsaturated fatty acids on the hallmarks of cancer. *Journal of Lipids, 2013*, 261247.
94. Zhang, Y.-F., Gao, H. F., Hou, A. J., & Zhou, Y. H. (2014). Effect of omega-3 fatty acid supplementation on cancer incidence, non-vascular death, and total mortality: A meta-analysis of randomized controlled trials. *BMC Public Health, 14*, 204.
95. Laviano, A., Rianda, S., Molfino, A., & Rossi Fanelli, F. (2013). Omega-3 fatty acids in cancer. *Current Opinion in Clinical Nutrition and Metabolic Care, 16*(2), 156–161.

96. Jing, K., Wu, T., & Lim, K. (2013). Omega-3 polyunsaturated fatty acids and cancer. *Anti-Cancer Agents in Medicinal Chemistry, 13*(8), 1162–1177.
97. Fabian, C. J., Kimler, B. F., & Hursting, S. D. (2015). Omega-3 fatty acids for breast cancer prevention and survivorship. *Breast Cancer Research, 17*(1), 62.
98. Fu, Y.-Q., Zheng, J. S., Yang, B., & Li, D. (2015). Effect of individual omega-3 fatty acids on the risk of prostate cancer: A systematic review and dose-response meta-analysis of prospective cohort studies. *Journal of Epidemiology, 25*(4), 261–274.
99. Volpato, M., & Hull, M. A. (2018). Omega-3 polyunsaturated fatty acids as adjuvant therapy of colorectal cancer. *Cancer Metastasis Reviews, 37*(2–3), 545–555.
100. Park, J.-M., Kwon, S. H., Han, Y. M., Hahm, K. B., & Kim, E. H. (2013). Omega-3 polyunsaturated fatty acids as potential chemopreventive agent for gastrointestinal cancer. *Journal of Cancer Prevention, 18*(3), 201–208.
101. D'Eliseo, D., & Velotti, F. (2016). Omega-3 fatty acids and cancer cell cytotoxicity: implications for multi-targeted Cancer therapy. *Journal of Clinical Medicine, 5*(2), 15.
102. Ren, G.-Y., Chen, C. Y., Chen, G. C., Chen, W. G., Pan, A., Pan, C. W., et al. (2016). Effect of flaxseed intervention on inflammatory marker C-reactive protein: A systematic review and meta-analysis of randomized controlled trials. *Nutrients, 8*(3), 136–136.
103. Zivkovic, A. M., Telis, N., German, J. B., & Hammock, B. D. (2011). Dietary omega-3 fatty acids aid in the modulation of inflammation and metabolic health. *California Agriculture, 65*(3), 106–111.
104. Singh, S., Nair, V., & Gupta, Y. K. (2012). Linseed oil: An investigation of its antiarthritic activity in experimental models. *Phytotherapy Research, 26*(2), 246–252.
105. Calder, P. C. (2013). Omega-3 polyunsaturated fatty acids and inflammatory processes: Nutrition or pharmacology? *British Journal of Clinical Pharmacology, 75*(3), 645–662.
106. Turowski, J. B., Pietrofesa, R. A., Lawson, J. A., Christofidou-Solomidou, M., & Hadjiliadis, D. (2015). Flaxseed modulates inflammatory and oxidative stress biomarkers in cystic fibrosis: A pilot study. *BMC Complementary and Alternative Medicine, 15*, 148.
107. Kaithwas, G., Mukherjee, A., Chaurasia, A. K., & Majumdar, D. K. (2011). Anti-inflammatory, analgesic and antipyretic activities of Linum usitatissimum L. (flaxseed/linseed) fixed oil. *Indian Journal of Experimental Biology, 49*(12), 932–938.
108. Nordstrom, D. C., Friman, C., Konttinen, Y. T., Honkanen, V. E., Nasu, Y., & Antila, E. (1995). Alpha-linolenic acid in the treatment of rheumatoid arthritis. A double-blind, placebo-controlled and randomized study: Flaxseed vs. safflower seed. *Rheumatology International, 14*(6), 231–234.
109. Soeken, K. L., Miller, S. A., & Ernst, E. (2003). Herbal medicines for the treatment of rheumatoid arthritis: A systematic review. *Rheumatology (Oxford), 42*(5), 652–659.
110. Thomas, S., Browne, H., Mobasheri, A., & Rayman, M. P. (2018). What is the evidence for a role for diet and nutrition in osteoarthritis? *Rheumatology (Oxford, England), 57*(suppl_4), iv61–iv74.
111. Ameye, L. G., & Chee, W. S. S. (2006). Osteoarthritis and nutrition. From nutraceuticals to functional foods: A systematic review of the scientific evidence. *Arthritis Research & Therapy, 8*(4), R127.
112. Mosavat, S. H., Masoudi, N., Hajimehdipoor, H., Emami Meybodi, M. K., Niktabe, Z., Tabarrai, M., et al. (2018). Efficacy of topical Linum usitatissimum L. (flaxseed) oil in knee osteoarthritis: A double-blind, randomized, placebo-controlled clinical trial. *Complementary Therapies in Clinical Practice, 31*, 302–307.
113. Godos, J., Bergante, S., Satriano, A., Pluchinotta, F. R., & Marranzano, M. (2018). Dietary phytoestrogen intake is inversely associated with hypertension in a cohort of adults living in the Mediterranean area. *Molecules, 23*(2), E368.
114. Khalesi, S., Irwin, C., & Schubert, M. (2015). Flaxseed consumption may reduce blood pressure: A systematic review and meta-analysis of controlled trials. *The Journal of Nutrition, 145*(4), 758–765.
115. Rodriguez-Leyva, D., Weighell, W., Edel, A. L., Lavallee, R., Dibrov, E., Pinneker, R., et al. (2013). Potent antihypertensive action of dietary flaxseed in hypertensive patients. *Hypertension, 62*(6), 1081–1089.

116. Torkan, M., Entezari, M. H., & Siavash, M. (2015). Effect of flaxseed on blood lipid level in hyperlipidemic patients. *Reviews on Recent Clinical Trials, 10*(1), 61–67.
117. Saxena, S., & Katare, C. (2014). Evaluation of flaxseed formulation as a potential therapeutic agent in mitigation of dyslipidemia. *Biomedical Journal, 37*(6), 386–390.
118. Leyva, D. R., Zahradka, P., Ramjiawan, B., Guzman, R., Aliani, M., & Pierce, G. N. (2011). The effect of dietary flaxseed on improving symptoms of cardiovascular disease in patients with peripheral artery disease: Rationale and design of the FLAX-PAD randomized controlled trial. *Contemporary Clinical Trials, 32*(5), 724–730.
119. Patade, A., Devareddy, L., Lucas, E. A., Korlagunta, K., Daggy, B. P., & Arjmandi, B. H. (2008). Flaxseed reduces total and LDL cholesterol concentrations in Native American postmenopausal women. *Journal of Women's Health (2002), 17*(3), 355–366.
120. Kristensen, M., Jensen, M. G., Aarestrup, J., Petersen, K. E., Søndergaard, L., Mikkelsen, M. S., et al. (2012). Flaxseed dietary fibers lower cholesterol and increase fecal fat excretion, but magnitude of effect depend on food type. *Nutrition & Metabolism (London), 9*, 8.
121. Prasad, K., & Dhar, A. (2016). Flaxseed and diabetes. *Current Pharmaceutical Design, 22*(2), 141–144.
122. Nazni, P., Amirthaveni, M., & Poongodi Vijayakumar, T. (2006). Impact of flaxseed based therapeutic food on selected type II diabetic patients. *The Indian Journal of Nutrition and Dietetics, 43*(4), 141–145.
123. Pan, A., Sun, J., Chen, Y., Ye, X., Li, H., Yu, Z., et al. (2007). Effects of a flaxseed-derived lignan supplement in type 2 diabetic patients: A randomized, double-blind, cross-over trial. *PLoS One, 2*(11), e1148.
124. Wu, J. H. Y., Micha, R., Imamura, F., Pan, A., Biggs, M. L., Ajaz, O., et al. (2012). Omega-3 fatty acids and incident type 2 diabetes: A systematic review and meta-analysis. *The British Journal of Nutrition, 107*(Suppl 2), S214–S227.
125. Kapoor, S., Sachdeva, R., & Kochhar, A. (2011). Efficacy of flaxseed supplementation on nutrient intake and other lifestyle pattern in menopausal diabetic females. *Studies on Ethno-Medicine, 5*, 153–160.
126. Thakur, G., Mitra, A., Pal, K., & Rousseau, D. (2009). Effect of flaxseed gum on reduction of blood glucose and cholesterol in type 2 diabetic patients. *International Journal of Food Sciences and Nutrition, 60*(Suppl 6), 126–136.
127. Barre, D. E., Mizier-Barre, K. A., Griscti, O., & Hafez, K. (2008). High dose flaxseed oil supplementation may affect fasting blood serum glucose management in human type 2 diabetics. *Journal of Oleo Science, 57*(5), 269–273.
128. Mohammadi-Sartang, M., Sohrabi, Z., Barati-Boldaji, R., Raeisi-Dehkordi, H., & Mazloom, Z. (2018). Flaxseed supplementation on glucose control and insulin sensitivity: A systematic review and meta-analysis of 25 randomized, placebo-controlled trials. *Nutrition Reviews, 76*(2), 125–139.
129. Chen, C., Yu, X., & Shao, S. (2015). Effects of omega-3 fatty acid supplementation on glucose control and lipid levels in type 2 diabetes: A meta-analysis. *PLoS One, 10*(10), e0139565.
130. Rhee, Y., & Brunt, A. (2011). Flaxseed supplementation improved insulin resistance in obese glucose intolerant people: A randomized crossover design. *Nutrition Journal, 10*, 44.
131. Javidi, A., Mozaffari-Khosravi, H., Nadjarzadeh, A., Dehghani, A., & Eftekhari, M. H. (2016). The effect of flaxseed powder on insulin resistance indices and blood pressure in prediabetic individuals: A randomized controlled clinical trial. *Journal of Research in Medical Sciences, 21*, 70.
132. Azadbakht, L., Rouhani, M. H., & Surkan, P. J. (2011). Omega-3 fatty acids, insulin resistance and type 2 diabetes. *Journal of Research in Medical Sciences, 16*(10), 1259–1260.
133. Lepretti, M., Martucciello, S., Burgos Aceves, M. A., Putti, R., & Lionetti, L. (2018). Omega-3 fatty acids and insulin resistance: Focus on the regulation of mitochondria and endoplasmic reticulum stress. *Nutrients, 10*(3), 350.
134. Mani, U. V., Mani, I., Biswas, M., & Kumar, S. N. (2011). An open-label study on the effect of flax seed powder (Linum usitatissimum) supplementation in the management of diabetes mellitus. *Journal of Dietary Supplements, 8*(3), 257–265.

135. Wu, H., Pan, A., Yu, Z., Qi, Q., Lu, L., Zhang, G., et al. (2010). Lifestyle counseling and supplementation with flaxseed or walnuts influence the management of metabolic syndrome. *The Journal of Nutrition, 140*(11), 1937–1942.
136. Dahl, W. J., Lockert, E. A., Cammer, A. L., & Whiting, S. J. (2005). Effects of flax fiber on laxation and glycemic response in healthy volunteers. *Journal of Medicinal Food, 8*(4), 508–511.
137. Soltanian, N., & Janghorbani, M. (2018). A randomized trial of the effects of flaxseed to manage constipation, weight, glycemia, and lipids in constipated patients with type 2 diabetes. *Nutrition & Metabolism, 15*, 36.
138. Kristensen, M., Savorani, F., Christensen, S., Engelsen, S. B., Bügel, S., Toubro, S., et al. (2013). Flaxseed dietary fibers suppress postprandial lipemia and appetite sensation in young men. *Nutrition, Metabolism, and Cardiovascular Diseases, 23*(2), 136–143.
139. Ibrugger, S., Kristensen, M., Mikkelsen, M. S., & Astrup, A. (2012). Flaxseed dietary fiber supplements for suppression of appetite and food intake. *Appetite, 58*(2), 490–495.
140. Wanders, A. J., van den Borne, J. J., de Graaf, C., Hulshof, T., Jonathan, M. C., Kristensen, M., et al. (2011). Effects of dietary fibre on subjective appetite, energy intake and body weight: A systematic review of randomized controlled trials. *Obesity Reviews, 12*(9), 724–739.
141. Mohammadi-Sartang, M., Mazloom, Z., Raeisi-Dehkordi, H., Barati-Boldaji, R., Bellissimo, N., & Totosy de Zepetnek, J. O. (2017). The effect of flaxseed supplementation on body weight and body composition: A systematic review and meta-analysis of 45 randomized placebo-controlled trials. *Obesity Reviews, 18*(9), 1096–1107.
142. Shim, Y. Y., Olivia, C. M., Liu, J., Boonen, R., Shen, J., & Reaney, M. J. (2016). Secoisolariciresinol diglucoside and cyanogenic glycosides in gluten-free bread fortified with flaxseed meal. *Journal of Agricultural and Food Chemistry, 64*(50), 9551–9558.
143. Margier, M., Georgé, S., Hafnaoui, N., Remond, D., Nowicki, M., Du Chaffaut, L., et al. (2018). Nutritional composition and bioactive content of legumes: Characterization of pulses frequently consumed in France and effect of the cooking method. *Nutrients, 10*(11), 1668.
144. Marinangeli, C. P. F., Curran, J., Barr, S. I., Slavin, J., Puri, S., Swaminathan, S., et al. (2017). Enhancing nutrition with pulses: Defining a recommended serving size for adults. *Nutrition Reviews, 75*(12), 990–1006.
145. Trinidad, T. P., Mallillin, A. C., Loyola, A. S., Sagum, R. S., & Encabo, R. R. (2010). The potential health benefits of legumes as a good source of dietary fibre. *The British Journal of Nutrition, 103*(4), 569–574.
146. Ahmed, S., & Hasan, M. (2014). Legumes: An overview. *Journal of Pharmacy and Pharmaceutical Sciences, 2*, 34–38.
147. Hever, J. (2016). Plant-based diets: A physician's guide. *The Permanente Journal, 20*(3), 93–101.
148. Polak, R., Phillips, E. M., & Campbell, A. (2015). Legumes: Health benefits and culinary approaches to increase intake. *Clinical Diabetes, 33*(4), 198–205.
149. Hever, J., & Cronise, R. J. (2017). Plant-based nutrition for healthcare professionals: Implementing diet as a primary modality in the prevention and treatment of chronic disease. *Journal of Geriatric Cardiology, 14*(5), 355–368.
150. Mudryj, A. N., Yu, N., & Aukema, H. M. (2014). Nutritional and health benefits of pulses. *Applied Physiology, Nutrition, and Metabolism, 39*(11), 1197–1204.
151. Marventano, S., Izquierdo Pulido, M., Sánchez-González, C., Godos, J., Speciani, A., Galvano, F., et al. (2017). Legume consumption and CVD risk: A systematic review and meta-analysis. *Public Health Nutrition, 20*(2), 245–254.
152. Li, H., Li, J., Shen, Y., Wang, J., & Zhou, D. (2017). Legume consumption and all-cause and cardiovascular disease mortality. *BioMed Research International, 2017*, 8450618.
153. Bouchenak, M., & Lamri-Senhadji, M. (2013). Nutritional quality of legumes, and their role in cardiometabolic risk prevention: A review. *Journal of Medicinal Food, 16*(3), 185–198.
154. Sanchez-Chino, X., Jiménez-Martínez, C., Dávila-Ortiz, G., Álvarez-González, I., & Madrigal-Bujaidar, E. (2015). Nutrient and nonnutrient components of legumes, and its chemopreventive activity: A review. *Nutrition and Cancer, 67*(3), 401–410.

155. Jochems, S. H. J., Van Osch, F. H. M., Bryan, R. T., Wesselius, A., van Schooten, F. J., Cheng, K. K., et al. (2018). Impact of dietary patterns and the main food groups on mortality and recurrence in cancer survivors: A systematic review of current epidemiological literature. *BMJ Open, 8*(2), e014530.
156. Li, J., & Mao, Q.-Q. (2017). Legume intake and risk of prostate cancer: A meta-analysis of prospective cohort studies. *Oncotarget, 8*(27), 44776–44784.
157. Zhu, B., Sun, Y., Qi, L., Zhong, R., & Miao, X. (2015). Dietary legume consumption reduces risk of colorectal cancer: Evidence from a meta-analysis of cohort studies. *Scientific Reports, 5*, 8797.
158. Taha, Z., & Eltom, S. E. (2018). The role of diet and lifestyle in women with breast cancer: An update review of related research in the Middle East. *BioResearch Open Access, 7*(1), 73–80.
159. Campos-Vega, R., Oomah, B. D., Loarca-Piña, G., & Vergara-Castañeda, H. A. (2013). Common beans and their non-digestible fraction: Cancer inhibitory activity-an overview. *Foods (Basel, Switzerland), 2*(3), 374–392.
160. Becerra-Tomas, N., Díaz-López, A., Rosique-Esteban, N., Ros, E., Buil-Cosiales, P., Corella, D., et al. (2018). Legume consumption is inversely associated with type 2 diabetes incidence in adults: A prospective assessment from the PREDIMED study. *Clinical Nutrition, 37*(3), 906–913.
161. Singhal, P., Kaushik, G., & Mathur, P. (2014). Antidiabetic potential of commonly consumed legumes: A review. *Critical Reviews in Food Science and Nutrition, 54*(5), 655–672.
162. Alizadeh, M., Gharaaghaji, R., & Gargari, B. P. (2014). The effects of legumes on metabolic features, insulin resistance and hepatic function tests in women with central obesity: A randomized controlled trial. *International Journal of Preventive Medicine, 5*(6), 710–720.
163. Hosseinpour-Niazi, S., Mirmiran, P., Amiri, Z., Hosseini-Esfahani, F., Shakeri, N., & Azizi, F. (2012). Legume intake is inversely associated with metabolic syndrome in adults. *Archives of Iranian Medicine, 15*(9), 538–544.
164. Martinez, R., López-Jurado, M., Wanden-Berghe, C., Sanz-Valero, J., Porres, J. M., & Kapravelou, G. (2016). Beneficial effects of legumes on parameters of the metabolic syndrome: A systematic review of trials in animal models. *The British Journal of Nutrition, 116*(3), 402–424.
165. Kim, S. J., de Souza, R. J., Choo, V. L., Ha, V., Cozma, A. I., Chiavaroli, L., et al. (2016). Effects of dietary pulse consumption on body weight: A systematic review and meta-analysis of randomized controlled trials. *The American Journal of Clinical Nutrition, 103*(5), 1213–1223.
166. McCrory, M. A., Hamaker, B. R., Lovejoy, J. C., & Eichelsdoerfer, P. E. (2010). Pulse consumption, satiety, and weight management. *Advances in Nutrition (Bethesda, Md.), 1*(1), 17–30.
167. Faris, M. A. I. E., Takruri, H. R., & Issa, A. Y. (2013). Role of lentils (Lens culinaris L.) in human health and nutrition: A review. *Mediterranean Journal of Nutrition and Metabolism, 6*(1), 3–16.
168. Ganesan, K., & Xu, B. (2017). Polyphenol-rich lentils and their health promoting effects. *International Journal of Molecular Sciences, 18*(11), 2390.
169. Chairatana, P., & Nolan, E. M. (2017). Defensins, lectins, mucins, and secretory immunoglobulin A: Microbe-binding biomolecules that contribute to mucosal immunity in the human gut. *Critical Reviews in Biochemistry and Molecular Biology, 52*(1), 45–56.
170. Shipunova, V. O., NikitinI, M. P., Zelepukin, I. V., Nikitin, P. I., Deyev, S. M., & Petrov, R. V. (2015). A comprehensive study of interactions between lectins and glycoproteins for the development of effective theranostic nanoagents. *Doklady. Biochemistry and Biophysics, 464*, 315–318.
171. Stephen, A. M., Dahl, W. J., Sieber, G. M., van Blaricom, J. A., & Morgan, D. R. (1995). Effect of green lentils on colonic function, nitrogen balance, and serum lipids in healthy human subjects. *The American Journal of Clinical Nutrition, 62*(6), 1261–1267.

172. Mollard, R. C., Zykus, A., Luhovyy, B. L., Nunez, M. F., Wong, C. L., & Anderson, G. H. (2012). The acute effects of a pulse-containing meal on glycaemic responses and measures of satiety and satiation within and at a later meal. *The British Journal of Nutrition, 108*(3), 509–517.
173. Aslani, Z., Mirmiran, P., Alipur, B., Bahadoran, Z., & Abbassalizade Farhangi, M. (2015). Lentil sprouts effect on serum lipids of overweight and obese patients with type 2 diabetes. *Health Promotion Perspective, 5*(3), 215–224.
174. Bolsinger, J., Landstrom, M., Pronczuk, A., Auerbach, A., & Hayes, K. C. (2017). Low glycemic load diets protect against metabolic syndrome and type 2 diabetes mellitus in the male Nile rat. *The Journal of Nutritional Biochemistry, 42*, 134–148.
175. Ramdath, D. D., Wolever, T. M. S., Siow, Y. C., Ryland, D., Hawke, A., Taylor, C., et al. (2018). Effect of processing on postprandial glycemic response and consumer acceptability of lentil-containing food items. *Foods (Basel, Switzerland), 7*(5), 76.
176. Higgins, J. A. (2012). Whole grains, legumes, and the subsequent meal effect: Implications for blood glucose control and the role of fermentation. *Journal of Nutrition and Metabolism, 2012*, 829238.
177. Ha, V., Sievenpiper, J. L., de Souza, R. J., Jayalath, V. H., Mirrahimi, A., Agarwal, A., et al. (2014). Effect of dietary pulse intake on established therapeutic lipid targets for cardiovascular risk reduction: A systematic review and meta-analysis of randomized controlled trials. *CMAJ, 186*(8), E252–E262.
178. Hanson, M. G., Zahradka, P., & Taylor, C. G. (2014). Lentil-based diets attenuate hypertension and large-artery remodelling in spontaneously hypertensive rats. *The British Journal of Nutrition, 111*(4), 690–698.
179. Jayalath, V. H., de Souza, R. J., Sievenpiper, J. L., Ha, V., Chiavaroli, L., Mirrahimi, A., et al. (2014). Effect of dietary pulses on blood pressure: A systematic review and meta-analysis of controlled feeding trials. *American Journal of Hypertension, 27*(1), 56–64.
180. Hanson, M., Zahradka, P., Taylor, C. G., & Aliani, M. (2018). Identification of urinary metabolites with potential blood pressure-lowering effects in lentil-fed spontaneously hypertensive rats. *European Journal of Nutrition, 57*(1), 297–308.
181. Baszczuk, A., Kopczyński, Z., Kopczyński, J., Cymerys, M., Thielemann, A., Bielawska, L., et al. (2015). Impact of administration of folic acid on selected indicators of inflammation in patients with primary arterial hypertension. *Postępy Higieny i Medycyny Doświadczalnej (Online), 69*, 429–435.
182. Kolte, D., Vijayaraghavan, K., Khera, S., Sica, D. A., & Frishman, W. H. (2014). Role of magnesium in cardiovascular diseases. *Cardiology in Review, 22*(4), 182–192.
183. Jarpa-Parra, M. (2018). Lentil protein: A review of functional properties and food application. *An Overview of Lentil Protein Functionality, 53*(4), 892–903.
184. Yau, T., Dan, X., Ng, C. C., & Ng, T. B. (2015). Lectins with potential for anti-cancer therapy. *Molecules (Basel, Switzerland), 20*(3), 3791–3810.
185. De Mejia, E. G., & Prisecaru, V. I. (2005). Lectins as bioactive plant proteins: A potential in cancer treatment. *Critical Reviews in Food Science and Nutrition, 45*(6), 425–445.
186. Dan, X. L., & Ng, T. B. (2013). Lectins in human cancer: Both a devil and an angel? *Current Protein & Peptide Science, 14*(6), 481–491.
187. Jiang, Q. L., Zhang, S., Tian, M., Zhang, S. Y., Xie, T., Chen, D. Y., et al. (2015). Plant lectins, from ancient sugar-binding proteins to emerging anti-cancer drugs in apoptosis and autophagy. *Cell Proliferation, 48*(1), 17–28.
188. Faris, M. A., Takruri, H. R., Shomaf, M. S., & Bustanji, Y. K. (2009). Chemopreventive effect of raw and cooked lentils (Lens culinaris L) and soybeans (Glycine max) against azoxymethane-induced aberrant crypt foci. *Nutrition Research, 29*(5), 355–362.
189. Srivastava, R., & Vasishtha, H. (2013). Dietary fiber, protein and lectin contents of lentils (Lens culinaris) on soaking and cooking. *Current Advances in Agricultural Sciences, 5*, 238–241.
190. Aune, D., De Stefani, E., Ronco, A., Boffetta, P., Deneo-Pellegrini, H., Acosta, G., et al. (2009). Legume intake and the risk of cancer: A multisite case-control study in Uruguay. *Cancer Causes & Control, 20*(9), 1605–1615.

191. Braakhuis, A. J., Campion, P., & Bishop, K. S. (2016). Reducing breast cancer recurrence: The role of dietary polyphenolics. *Nutrients, 8*(9), 547.
192. Björnstedt, M., & Fernandes, A. P. (2010). Selenium in the prevention of human cancers. *The EPMA Journal, 1*(3), 389–395.
193. Babaknejad, N., Sayehmiri, F., Sayehmiri, K., Rahimifar, P., Bahrami, S., Delpesheh, A., et al. (2014). The relationship between selenium levels and breast cancer: A systematic review and meta-analysis. *Biological Trace Element Research, 159*(1–3), 1–7.
194. Huang, G., Liu, Z., He, L., Luk, K. H., Cheung, S. T., Wong, K. H., et al. (2018). Autophagy is an important action mode for functionalized selenium nanoparticles to exhibit anti-colorectal cancer activity. *Biomaterials Science, 6*(9), 2508–2517.
195. Elango, S., Samuel, S., Khashim, Z., & Subbiah, U. (2018). Selenium influences trace elements homeostasis, cancer biomarkers in squamous cell carcinoma patients administered with cancerocidal radiotherapy. *Asian Pacific Journal of Cancer Prevention, 19*(7), 1785–1792.
196. Gao, P. T., Ding, G. Y., Yang, X., Dong, R. Z., Hu, B., Zhu, X. D., et al. (2018). Invasive potential of hepatocellular carcinoma is enhanced by loss of selenium-binding protein 1 and subsequent upregulation of CXCR4. *American Journal of Cancer Research, 8*(6), 1040.
197. Duntas, L. H. (2009). Selenium and inflammation: Underlying anti-inflammatory mechanisms. *Hormone and Metabolic Research, 41*(6), 443–447.
198. Gómez-Pinilla, F. (2008). Brain foods: The effects of nutrients on brain function. *Nature Reviews. Neuroscience, 9*(7), 568–578.
199. Houshmand, G., Tarahomi, S., Arzi, A., Goudarzi, M., Bahadoram, M., & Rashidi-Nooshabadi, M. (2016). Red lentil extract: Neuroprotective effects on perphenazine induced catatonia in rats. *Journal of Clinical and Diagnostic Research: JCDR, 10*(6), FF05–FF08.
200. Bourre, J. M. (2006). Effects of nutrients (in food) on the structure and function of the nervous system: Update on dietary requirements for brain. Part 2 : Macronutrients. *The Journal of Nutrition, Health & Aging, 10*(5), 386–399.
201. Siew, J. J., & Chern, Y. (2018). Microglial lectins in health and neurological diseases. *Frontiers in Molecular Neuroscience, 11*, 158–158.
202. Massey, L. K., Palmer, R. G., & Horner, H. T. (2001). Oxalate content of soybean seeds (Glycine max: Leguminosae), soyfoods, and other edible legumes. *Journal of Agricultural and Food Chemistry, 49*(9), 4262–4266.
203. MacFarquhar, J. K., Broussard, D. L., Melstrom, P., Hutchinson, R., Wolkin, A., Martin, C., et al. (2010). Acute selenium toxicity associated with a dietary supplement. *Archives of Internal Medicine, 170*(3), 256–261.
204. Mejri, F., Selmi, S., Martins, A., Benkhoud, H., Baati, T., Chaabane, H., et al. (2018). Broad bean (Vicia faba L.) pods: A rich source of bioactive ingredients with antimicrobial, antioxidant, enzyme inhibitory, anti-diabetic and health-promoting properties. *Food & Function, 9*(4), 2051–2069.
205. Messina, V. (2014). Nutritional and health benefits of dried beans. *The American Journal of Clinical Nutrition, 100*(Suppl 1), 437s–442s.
206. Valente, I. M., Maia, M. R. G., Malushi, N., Oliveira, H. M., Papa, L., Rodrigues, J. A., et al. (2018). Profiling of phenolic compounds and antioxidant properties of European varieties and cultivars of Vicia faba L. pods. *Phytochemistry, 152*, 223–229.
207. Yao, Y., Cheng, X., Wang, L., Wang, S., & Ren, G. (2011). Biological potential of sixteen legumes in China. *International Journal of Molecular Sciences, 12*(10), 7048–7058.
208. Oomah, B. D., Corbe, A., & Balasubramanian, P. (2010). Antioxidant and anti-inflammatory activities of bean (Phaseolus vulgaris L.) hulls. *Journal of Agricultural and Food Chemistry, 58*(14), 8225–8230.
209. Reverri, E. J., Randolph, J. M., Steinberg, F. M., Kappagoda, C. T., Edirisinghe, I., & Burton-Freeman, B. M. (2015). Black beans, fiber, and antioxidant capacity pilot study: Examination of whole foods vs. functional components on postprandial metabolic, oxidative stress, and inflammation in adults with metabolic syndrome. *Nutrients, 7*(8), 6139–6154.
210. Ros, E., & Hu, F. B. (2013). Consumption of plant seeds and cardiovascular health: Epidemiological and clinical trial evidence. *Circulation, 128*(5), 553–565.

211. Johnston, C. (2009). Functional foods as modifiers of cardiovascular disease. *American Journal of Lifestyle Medicine, 3*(1 Suppl), 39S–43S.
212. Ganesan, K., & Xu, B. (2017). Polyphenol-rich dry common beans (Phaseolus vulgaris L.) and their health benefits. *International Journal of Molecular Sciences, 18*(11), 2331.
213. Anderson, J. W., & Major, A. W. (2002). Pulses and lipaemia, short- and long-term effect: Potential in the prevention of cardiovascular disease. *The British Journal of Nutrition, 88*(Suppl 3), S263–S271.
214. Winham, D. M., & Hutchins, A. M. (2007). Baked bean consumption reduces serum cholesterol in hypercholesterolemic adults. *Nutrition Research, 27*(7), 380–386.
215. Winham, D. M., Hutchins, A. M., & Thompson, S. V. (2017). Glycemic response to black beans and chickpeas as part of a rice meal: A randomized cross-over trial. *Nutrients, 9*(10), 1095.
216. Panlasigui, L. N., Panlilio, L. M., & Madrid, J. C. (1995). Glycaemic response in normal subjects to five different legumes commonly used in the Philippines. *International Journal of Food Sciences and Nutrition, 46*(2), 155–160.
217. Thompson, S. V., Winham, D. M., & Hutchins, A. M. (2012). Bean and rice meals reduce postprandial glycemic response in adults with type 2 diabetes: A cross-over study. *Nutrition Journal, 11*, 23.
218. Morenga, L. T., Williams, S., Brown, R., & Mann, J. (2010). Effect of a relatively high-protein, high-fiber diet on body composition and metabolic risk factors in overweight women. *European Journal of Clinical Nutrition, 64*(11), 1323–1331.
219. Celleno, L., Tolaini, M. V., D'Amore, A., Perricone, N. V., & Preuss, H. G. (2007). A dietary supplement containing standardized Phaseolus vulgaris extract influences body composition of overweight men and women. *International Journal of Medical Sciences, 4*(1), 45–52.
220. Vergara-Castaneda, H. A., Guevara-González, R. G., Ramos-Gómez, M., Reynoso-Camacho, R., Guzmán-Maldonado, H., Feregrino-Pérez, A. A., et al. (2010). Non-digestible fraction of cooked bean (Phaseolus vulgaris L.) cultivar Bayo Madero suppresses colonic aberrant crypt foci in azoxymethane-induced rats. *Food & Function, 1*(3), 294–300.
221. Feregrino-Perez, A. A., Piñol-Felis, C., Gomez-Arbones, X., Guevara-González, R. G., Campos-Vega, R., Acosta-Gallegos, J. A., et al. (2014). A non-digestible fraction of the common bean (Phaseolus vulgaris L.) induces cell cycle arrest and apoptosis during early carcinogenesis. *Plant Foods for Human Nutrition, 69*(3), 248–254.
222. Thompson, M. D., Thompson, H. J., Brick, M. A., McGinley, J. N., Jiang, W., Zhu, Z., et al. (2008). Mechanisms associated with dose-dependent inhibition of rat mammary carcinogenesis by dry bean (Phaseolus vulgaris, L.). *The Journal of Nutrition, 138*(11), 2091–2097.
223. Aliko, V., Qirjo, M., Sula, E., Morina, V., & Faggio, C. (2018). Antioxidant defense system, immune response and erythron profile modulation in gold fish, Carassius auratus, after acute manganese treatment. *Fish & Shellfish Immunology, 76*, 101–109.
224. Wendolowicz, A., Stefanska, E., & Ostrowska, L. (2018). Influence of selected dietary components on the functioning of the human nervous system. *Roczniki Państwowego Zakładu Higieny, 69*(1), 15–21.
225. Hapeta, B., Koczy, B., Fitowska, A., Dobrakowski, M., Kasperczyk, A., Ostałowska, A., et al. (2012). Metabolism and protein transformations in synovial membrane of a knee joint in the course of rheumatoid arthritis and degenerative arthritis. *Polish Orthopaedics and Traumatology, 77*, 53–58.
226. Landete-Castillejos, T., Molina-Quilez, I., Estevez, J. A., Ceacero, F., Garcia, A. J., & Gallego, L. (2012). Alternative hypothesis for the origin of osteoporosis: The role of Mn. *Frontiers in Bioscience (Elite Edition), 4*, 1385–1390.
227. Du, S., Wu, X., Han, T., Duan, W., Liu, L., Qi, J., et al. (2018). Dietary manganese and type 2 diabetes mellitus: Two prospective cohort studies in China. *Diabetologia, 61*(9), 1985–1995.
228. Leggio, G. M., Salomone, S., Bucolo, C., Platania, C., Micale, V., Caraci, F., et al. (2013). Dopamine D(3) receptor as a new pharmacological target for the treatment of depression. *European Journal of Pharmacology, 719*(1–3), 25–33.

229. Tavakkoli-Kakhki, M., Eslami, S., & Motavasselian, M. (2015). Nutrient-rich versus nutrient-poor foods for depressed patients based on Iranian traditional medicine resources. *Avicenna Journal of Phytomedicine, 5*(4), 298–308.
230. LeWitt, P. A. (2015). Levodopa therapy for Parkinson's disease: Pharmacokinetics and pharmacodynamics. *Movement Disorders, 30*(1), 64–72.
231. Hornykiewicz, O. (2010). A brief history of levodopa. *Journal of Neurology, 257*(Suppl 2), S249–S252.
232. Ovallath, S., & Sulthana, B. (2017). Levodopa: History and therapeutic applications. *Annals of Indian Academy of Neurology, 20*(3), 185–189.
233. Apaydin, H., Ertan, S., & Ozekmekci, S. (2000). Broad bean (Vicia faba)--a natural source of L-dopa—Prolongs "on" periods in patients with Parkinson's disease who have "on-off" fluctuations. *Movement Disorders, 15*(1), 164–166.
234. Essa, M. M., Braidy, N., Bridge, W., Subash, S., Manivasagam, T., Vijayan, R. K., et al. (2014). Review of natural products on Parkinson's disease pathology. *Journal of Aging Research & Clinical Practice, 3*(3), 127–136.
235. Ramírez-Moreno, J., Salguero Bodes, I., Romaskevych, O., & Duran-Herrera, M. C. (2015). Broad bean (Vicia faba) consumption and Parkinson's disease: A natural source of L-dopa to consider. *Neurología, 30*, 375–376.
236. Vural, N., & Sardas, S. (1984). Biological activities of broad bean (Vicia faba L.) extracts cultivated in South Anatolia in favism sensitive subjects. *Toxicology, 31*(2), 175–179.
237. Belfield, K. D., & Tichy, E. M. (2018). Review and drug therapy implications of glucose-6-phosphate dehydrogenase deficiency. *American Journal of Health-System Pharmacy, 75*(3), 97–104.
238. Kramell, R., Schmidt, J., Herrmann, G., & Schliemann, W. (2005). N-(jasmonoyl)tyrosine-derived compounds from flowers of broad beans (Vicia faba). *Journal of Natural Products, 68*(9), 1345–1349.
239. van der Steen, W., den Heijer, T., & Groen, J. (2018). Vitamin B6 deficiency caused by the use of levodopa. *Nederlands Tijdschrift voor Geneeskunde, 162*, D2818.
240. Hinz, M., Stein, A., & Cole, T. (2014). The Parkinson's disease death rate: Carbidopa and vitamin B6. *Clinical Pharmacology: Advances and Applications, 6*, 161–169.
241. Messina, M. (2016). Soy and health update: Evaluation of the clinical and epidemiologic literature. *Nutrients, 8*(12), 754.
242. D'Adamo, C. R., & Sahin, A. (2014). Soy foods and supplementation: A review of commonly perceived health benefits and risks. *Alternative Therapies in Health and Medicine, 20*(Suppl 1), 39–51.
243. Balk, E., Chung, M., & Chew, P. (2005). Effects of soy on health outcomes: Summary. In *AHRQ evidence report summaries 1998–2005*. Rockville, MD: Agency for Healthcare Research and Quality (US). https://www.ncbi.nlm.nih.gov/books/NBK11870/
244. Rizzo, G., & Baroni, L. (2018). Soy, soy foods and their role in vegetarian diets. *Nutrients, 10*(1), 43.
245. Friedman, M., & Brandon, D. L. (2001). Nutritional and health benefits of soy proteins. *Journal of Agricultural and Food Chemistry, 49*(3), 1069–1086.
246. Xiao, C. W. (2008). Health effects of soy protein and isoflavones in humans. *The Journal of Nutrition, 138*(6), 1244s–1249s.
247. Messina, M., Rogero, M. M., Fisberg, M., & Waitzberg, D. (2017). Health impact of childhood and adolescent soy consumption. *Nutrition Reviews, 75*(7), 500–515.
248. Jargin, S. V. (2014). Soy and phytoestrogens: Possible side effects. *German Medical Science, 12*, Doc18.
249. Barrett, J. R. (2006). The science of soy: What do we really know? *Environmental Health Perspectives, 114*(6), A352–A358.
250. Yan, Z., Zhang, X., Li, C., Jiao, S., & Dong, W. (2017). Association between consumption of soy and risk of cardiovascular disease: A meta-analysis of observational studies. *European Journal of Preventive Cardiology, 24*(7), 735–747.

251. Hu, X., Gao, J., Zhang, Q., Fu, Y., Li, K., Zhu, S., et al. (2013). Soy fiber improves weight loss and lipid profile in overweight and obese adults: A randomized controlled trial. *Molecular Nutrition & Food Research, 57*(12), 2147–2154.
252. Eilat-Adar, S., Sinai, T., Yosefy, C., & Henkin, Y. (2013). Nutritional recommendations for cardiovascular disease prevention. *Nutrients, 5*(9), 3646–3683.
253. Anderson, J. W., Johnstone, B. M., & Cook-Newell, M. E. (1995). Meta-analysis of the effects of soy protein intake on serum lipids. *The New England Journal of Medicine, 333*(5), 276–282.
254. Rebholz, C. M., Reynolds, K., Wofford, M. R., Chen, J., Kelly, T. N., Mei, H., et al. (2013). Effect of soybean protein on novel cardiovascular disease risk factors: A randomized controlled trial. *European Journal of Clinical Nutrition, 67*(1), 58–63.
255. Nagata, C., Wada, K., Tamura, T., Konishi, K., Goto, Y., Koda, S., et al. (2017). Dietary soy and natto intake and cardiovascular disease mortality in Japanese adults: The Takayama study. *The American Journal of Clinical Nutrition, 105*(2), 426–431.
256. Kou, T., Wang, Q., Cai, J., Song, J., Du, B., Zhao, K., et al. (2017). Effect of soybean protein on blood pressure in postmenopausal women: A meta-analysis of randomized controlled trials. *Food & Function, 8*(8), 2663–2671.
257. Dong, J. Y., Tong, X., Wu, Z. W., Xun, P. C., He, K., & Qin, L. Q. (2011). Effect of soya protein on blood pressure: A meta-analysis of randomised controlled trials. *The British Journal of Nutrition, 106*(3), 317–326.
258. He, J., Gu, D., Wu, X., Chen, J., Duan, X., Chen, J., et al. (2005). Effect of soybean protein on blood pressure: A randomized, controlled trial. *Annals of Internal Medicine, 143*(1), 1–9.
259. Gil-Izquierdo, A., Penalvo, J. L., Gil, J. I., Medina, S., Horcajada, M. N., Lafay, S., et al. (2012). Soy isoflavones and cardiovascular disease epidemiological, clinical and -omics perspectives. *Current Pharmaceutical Biotechnology, 13*(5), 624–631.
260. Lichtenstein, A. H. (1998). Soy protein, isoflavones and cardiovascular disease risk. *The Journal of Nutrition, 128*(10), 1589–1592.
261. Liu, X. X., Li, S. H., Chen, J. Z., Sun, K., Wang, X. J., Wang, X. G., et al. (2012). Effect of soy isoflavones on blood pressure: A meta-analysis of randomized controlled trials. *Nutrition, Metabolism, and Cardiovascular Diseases, 22*(6), 463–470.
262. Fan, Y. Y., Ramos, K. S., & Chapkin, R. S. (2001). Dietary gamma-linolenic acid suppresses aortic smooth muscle cell proliferation and modifies atherosclerotic lesions in apolipoprotein E knockout mice. *The Journal of Nutrition, 131*(6), 1675–1681.
263. Tsukamoto, I., & Sugawara, S. (2018). Low levels of linoleic acid and α-linolenic acid and high levels of arachidonic acid in plasma phospholipids are associated with hypertension. *Biomedical Reports, 8*(1), 69–76.
264. Ramdath, D. D., Padhi, E. M., Sarfaraz, S., Renwick, S., & Duncan, A. M. (2017). Beyond the cholesterol-lowering effect of soy protein: A review of the effects of dietary soy and its constituents on risk factors for cardiovascular disease. *Nutrients, 9*(4), 324.
265. Rivas, M., Garay, R. P., Escanero, J. F., Cia Jr., P., Cia, P., & Alda, J. O. (2002). Soy milk lowers blood pressure in men and women with mild to moderate essential hypertension. *The Journal of Nutrition, 132*(7), 1900–1902.
266. Marangoni, F., & Poli, A. (2010). Phytosterols and cardiovascular health. *Pharmacological Research, 61*(3), 193–199.
267. Lin, X., Racette, S. B., Lefevre, M., Spearie, C. A., Most, M., Ma, L., et al. (2010). The effects of phytosterols present in natural food matrices on cholesterol metabolism and LDL-cholesterol: A controlled feeding trial. *European Journal of Clinical Nutrition, 64*(12), 1481–1487.
268. Ferguson, J. J., Stojanovski, E., MacDonald-Wicks, L., & Garg, M. L. (2016). Fat type in phytosterol products influence their cholesterol-lowering potential: A systematic review and meta-analysis of RCTs. *Progress in Lipid Research, 64*, 16–29.
269. Sathyapalan, T., Manuchehri, A. M., Thatcher, N. J., Rigby, A. S., Chapman, T., Kilpatrick, E. S., et al. (2011). The effect of soy phytoestrogen supplementation on thyroid status and

cardiovascular risk markers in patients with subclinical hypothyroidism: A randomized, double-blind, crossover study. *The Journal of Clinical Endocrinology and Metabolism, 96*(5), 1442–1449.
270. Rietjens, I. M. C. M., Louisse, J., & Beekmann, K. (2017). The potential health effects of dietary phytoestrogens. *British Journal of Pharmacology, 174*(11), 1263–1280.
271. Kalaiselvan, V., Kalaivani, M., Vijayakumar, A., Sureshkumar, K., & Venkateskumar, K. (2010). Current knowledge and future direction of research on soy isoflavones as a therapeutic agents. *Pharmacognosy Reviews, 4*(8), 111–117.
272. Kaushik, S., Shyam, H., Sharma, R., & Balapure, A. K. (2018). Dietary isoflavone daidzein synergizes centchroman action via induction of apoptosis and inhibition of PI3K/Akt pathway in MCF-7/MDA MB-231 human breast cancer cells. *Phytomedicine, 40*, 116–124.
273. Trock, B. J., Hilakivi-Clarke, L., & Clarke, R. (2006). Meta-analysis of soy intake and breast cancer risk. *Journal of the National Cancer Institute, 98*(7), 459–471.
274. Ziaei, S., & Halaby, R. (2017). Dietary isoflavones and breast Cancer risk. *Medicines (Basel, Switzerland), 4*(2), 18.
275. Sarkar, F. H., & Li, Y. (2003). Soy isoflavones and cancer prevention. *Cancer Investigation, 21*(5), 744–757.
276. Takagi, A., Kano, M., & Kaga, C. (2015). Possibility of breast cancer prevention: Use of soy isoflavones and fermented soy beverage produced using probiotics. *International Journal of Molecular Sciences, 16*(5), 10907–10920.
277. Messina, M. (2016). Impact of soy foods on the development of breast cancer and the prognosis of breast cancer patients. *Forschende Komplementärmedizin, 23*(2), 75–80.
278. Douglas, C. C., Johnson, S. A., & Arjmandi, B. H. (2013). Soy and its isoflavones: The truth behind the science in breast cancer. *Anti-Cancer Agents in Medicinal Chemistry, 13*(8), 1178–1187.
279. Tsuchiya, M., Miura, T., Hanaoka, T., Iwasaki, M., Sasaki, H., Tanaka, T., et al. (2007). Effect of soy isoflavones on endometriosis: Interaction with estrogen receptor 2 gene polymorphism. *Epidemiology, 18*(3), 402–408.
280. Wu, S. H., & Liu, Z. (2013). Soy food consumption and lung cancer risk: A meta-analysis using a common measure across studies. *Nutrition and Cancer, 65*(5), 625–632.
281. Weng, K.-G., & Yuan, Y.-L. (2017). Soy food intake and risk of gastric cancer: A dose-response meta-analysis of prospective studies. *Medicine, 96*(33), –e7802.
282. Lu, D., Pan, C., Ye, C., Duan, H., Xu, F., Yin, L., et al. (2017). Meta-analysis of soy consumption and gastrointestinal cancer risk. *Scientific Reports, 7*(1), 4048.
283. Wada, K., Tsuji, M., Tamura, T., Konishi, K., Goto, Y., Mizuta, F., et al. (2018). Soy isoflavone intake and bladder Cancer risk in Japan: From the Takayama study. *Cancer Epidemiology, Biomarkers & Prevention, 27*(11), 1371–1375.
284. Mahmoud, A. M., Yang, W., & Bosland, M. C. (2014). Soy isoflavones and prostate cancer: A review of molecular mechanisms. *The Journal of Steroid Biochemistry and Molecular Biology, 140*, 116–132.
285. Applegate, C. C., Rowles, J. L., Ranard, K. M., Jeon, S., & Erdman, J. W. (2018). Soy consumption and the risk of prostate cancer: An updated systematic review and meta-analysis. *Nutrients, 10*(1), 40.
286. Park, S., Bazer, F. W., Lim, W., & Song, G. (2018). The O-methylated isoflavone, formononetin, inhibits human ovarian cancer cell proliferation by sub G0/G1 cell phase arrest through PI3K/AKT and ERK1/2 inactivation. *Journal of Cellular Biochemistry, 119*(9), 7377–7387.
287. Song, Y., Liu, M., Yang, F. G., Cui, L. H., Lu, X. Y., & Chen, C. (2015). Dietary fibre and the risk of colorectal cancer: A case- control study. *Asian Pacific Journal of Cancer Prevention, 16*(9), 3747–3752.
288. Shin, A., Lee, J., Lee, J., Park, M. S., Park, J. W., Park, S. C., et al. (2015). Isoflavone and soyfood intake and colorectal cancer risk: A case-control study in Korea. *PLoS One, 10*(11), e0143228.

289. Yu, Y., Jing, X., Li, H., Zhao, X., & Wang, D. (2016). Soy isoflavone consumption and colorectal cancer risk: A systematic review and meta-analysis. *Scientific Reports, 6*, 25939.
290. de Mejia, E. G., Bradford, T., & Hasler, C. (2003). The anticarcinogenic potential of soybean lectin and lunasin. *Nutrition Reviews, 61*(7), 239–246.
291. Hsieh, C.-C., Hernández-Ledesma, B., & de Lumen, B. (2011). Lunasin, a new breast cancer chemopreventive seed peptide. In *Breast cancer—Current and alternative therapeutic modalities* (Vol. 11, pp. 215–242). London: IntechOpen.
292. Hernández-Ledesma, B., Hsieh, C. C., Dia, V., González de Mejia, E., & de Lumen, B. (2011). Lunasin, a cancer preventive seed peptide. In *Soybean and health* (p. 145). London: IntechOpen.
293. Shi, Z., Sun, R., Yu, T., Liu, R., Cheng, L. J., Bao, J. K., et al. (2016). Identification of novel pathways in plant lectin-induced cancer cell apoptosis. *International Journal of Molecular Sciences, 17*(2), 228.
294. Yu, J., Bi, X., Yu, B., & Chen, D. (2016). Isoflavones: Anti-inflammatory benefit and possible caveats. *Nutrients, 8*(6), 361.
295. Zhu, Y., Li, H., & Wang, X. (2017). Lunasin abrogates monocytes to endothelial cells *Molecular Immunology, 92*, 146–150.
296. Nguyen, C. T., Pham, N. M., Do, V. V., Binns, C. W., Hoang, V. M., Dang, D. A., et al. (2017). Soyfood and isoflavone intake and risk of type 2 diabetes in Vietnamese adults. *European Journal of Clinical Nutrition, 71*(10), 1186–1192.
297. Ding, M., Pan, A., Manson, J. E., Willett, W. C., Malik, V., Rosner, B., et al. (2016). Consumption of soy foods and isoflavones and risk of type 2 diabetes: A pooled analysis of three US cohorts. *European Journal of Clinical Nutrition, 70*(12), 1381–1387.
298. Kwon, S. H., Ahn, I. S., Kim, S. O., Kong, C. S., Chung, H. Y., Do, M. S., et al. (2007). Anti-obesity and hypolipidemic effects of black soybean anthocyanins. *Journal of Medicinal Food, 10*(3), 552–556.
299. Chang, J. H., Kim, M. S., Kim, T. W., & Lee, S. S. (2008). Effects of soybean supplementation on blood glucose, plasma lipid levels, and erythrocyte antioxidant enzyme activity in type 2 diabetes mellitus patients. *Nutrition Research and Practice, 2*(3), 152–157.
300. Mueller, N. T., Odegaard, A. O., Gross, M. D., Koh, W. P., Yu, M. C., Yuan, J. M., et al. (2012). Soy intake and risk of type 2 diabetes in Chinese Singaporeans [corrected]. *European Journal of Nutrition, 51*(8), 1033–1040.
301. Akhlaghi, M., Zare, M., & Nouripour, F. (2017). Effect of soy and soy Isoflavones on obesity-related anthropometric measures: A systematic review and meta-analysis of randomized controlled clinical trials. *Advances in Nutrition, 8*(5), 705–717.
302. Karamali, M., Kashanian, M., Alaeinasab, S., & Asemi, Z. (2018). The effect of dietary soy intake on weight loss, glycaemic control, lipid profiles and biomarkers of inflammation and oxidative stress in women with polycystic ovary syndrome: A randomised clinical trial. *Journal of Human Nutrition and Dietetics, 31*(4), 533–543.
303. Velasquez, M. T., & Bhathena, S. J. (2007). Role of dietary soy protein in obesity. *International Journal of Medical Sciences, 4*(2), 72–82.
304. Fontaine, K. R., Yang, D., Gadbury, G. L., Heshka, S., Schwartz, L. G., Murugesan, R., et al. (2003). Results of soy-based meal replacement formula on weight, anthropometry, serum lipids & blood pressure during a 40-week clinical weight loss trial. *Nutrition Journal, 2*, 14.
305. Messina, M. (2014). Soy foods, isoflavones, and the health of postmenopausal women. *The American Journal of Clinical Nutrition, 100*(Suppl 1), 423s–430s.
306. Ahsan, M., & Mallick, A. K. (2017). The effect of soy isoflavones on the menopause rating scale scoring in perimenopausal and postmenopausal women: A pilot study. *Journal of Clinical and Diagnostic Research, 11*(9), FC13–FC16.
307. Jou, H. J., Wu, S. C., Chang, F. W., Ling, P. Y., Chu, K. S., & Wu, W. H. (2008). Effect of intestinal production of equol on menopausal symptoms in women treated with soy isoflavones. *International Journal of Gynaecology and Obstetrics, 102*(1), 44–49.

308. Zheng, X., Lee, S.-K., & Chun, O. K. (2016). Soy isoflavones and osteoporotic bone loss: A review with an emphasis on modulation of bone remodeling. *Journal of Medicinal Food, 19*(1), 1–14.
309. Powles, T. (2004). Isoflavones and women's health. *Breast Cancer Research, 6*(3), 140–142.
310. Wong, W. W., Lewis, R. D., Steinberg, F. M., Murray, M. J., Cramer, M. A., Amato, P., et al. (2009). Soy isoflavone supplementation and bone mineral density in menopausal women: A 2-y multicenter clinical trial. *The American Journal of Clinical Nutrition, 90*(5), 1433–1439.
311. Bolca, S., Bracke, M., & Depypere, H. (2012). Soy consumption during menopause. *Facts, Views & Vision in ObGyn, 4*(1), 30–37.
312. Li, L., Lv, Y., Xu, L., & Zheng, Q. (2015). Quantitative efficacy of soy isoflavones on menopausal hot flashes. *British Journal of Clinical Pharmacology, 79*(4), 593–604.
313. Lambert, M. N. T., Thybo, C. B., Lykkeboe, S., Rasmussen, L. M., Frette, X., Christensen, L. P., et al. (2017). Combined bioavailable isoflavones and probiotics improve bone status and estrogen metabolism in postmenopausal osteopenic women: A randomized controlled trial. *The American Journal of Clinical Nutrition, 106*(3), 909–920.
314. Basset, G. J., Latimer, S., Fatihi, A., Soubeyrand, E., & Block, A. (2017). Phylloquinone (vitamin K1): Occurrence, biosynthesis and functions. *Mini Reviews in Medicinal Chemistry, 17*(12), 1028–1038.
315. Lanou, A. J. (2011). Soy foods: Are they useful for optimal bone health? *Therapeutic Advances in Musculoskeletal Disease, 3*(6), 293–300.
316. Reinwald, S., & Weaver, C. M. (2010). Soy components vs. whole soy: Are we betting our bones on a long shot? *The Journal of Nutrition, 140*(12), 2312S–2317S.
317. Martinez-Villaluenga, C., Frias, J., & Vidal-Valverde, C. (2008). Alpha-galactosides: Antinutritional factors or functional ingredients? *Critical Reviews in Food Science and Nutrition, 48*(4), 301–316.
318. Ong, D. K., Mitchell, S. B., Barrett, J. S., Shepherd, S. J., Irving, P. M., Biesiekierski, J. R., et al. (2010). Manipulation of dietary short chain carbohydrates alters the pattern of gas production and genesis of symptoms in irritable bowel syndrome. *Journal of Gastroenterology and Hepatology, 25*(8), 1366–1373.
319. Bingemann, T. A., Sood, P., & Jarvinen, K. M. (2018). Food protein-induced enterocolitis syndrome. *Immunology and Allergy Clinics of North America, 38*(1), 141–152.
320. Cordle, C. T. (2004). Soy protein allergy: Incidence and relative severity. *The Journal of Nutrition, 134*(5), 1213S–1219S.
321. Cantani, A., & Lucenti, P. (1997). Natural history of soy allergy and/or intolerance in children, and clinical use of soy-protein formulas. *Pediatric Allergy and Immunology, 8*(2), 59–74.
322. Peng, C., Cao, C., He, M., Shu, Y., Tang, X., Wang, Y., et al. (2018). Soybean glycinin- and beta-conglycinin-induced intestinal damage in piglets via the p38/JNK/NF-kappaB Signaling pathway. *Journal of Agricultural and Food Chemistry, 66*(36), 9534–9541.
323. Doerge, D. R., & Sheehan, D. M. (2002). Goitrogenic and estrogenic activity of soy isoflavones. *Environmental Health Perspectives, 110*(Suppl 3), 349–353.
324. Messina, M., & Redmond, G. (2006). Effects of soy protein and soybean isoflavones on thyroid function in healthy adults and hypothyroid patients: A review of the relevant literature. *Thyroid, 16*(3), 249–258.
325. Nakamura, Y., Ohsawa, I., Goto, Y., Tsuji, M., Oguchi, T., Sato, N., et al. (2017). Soy isoflavones inducing overt hypothyroidism in a patient with chronic lymphocytic thyroiditis: A case report. *Journal of Medical Case Reports, 11*(1), 253.
326. Wang, H., Chen, Y., Hua, Y., Kong, X., & Zhang, C. (2014). Effects of phytase-assisted processing method on physicochemical and functional properties of soy protein isolate. *Journal of Agricultural and Food Chemistry, 62*(45), 10989–10997.
327. Gupta, R. K., Gupta, K., Sharma, A., Das, M., Ansari, I. A., & Dwivedi, P. D. (2017). Health risks and benefits of chickpea (Cicer arietinum) consumption. *Journal of Agricultural and Food Chemistry, 65*(1), 6–22.

328. Wallace, T. C., Murray, R., & Zelman, K. M. (2016). The nutritional value and health benefits of chickpeas and hummus. *Nutrients, 8*(12), 766.
329. Jukanti, A. K., Gaur, P. M., Gowda, C. L., & Chibbar, R. N. (2012). Nutritional quality and health benefits of chickpea (Cicer arietinum L.): A review. *The British Journal of Nutrition, 108*, S11–S26.
330. Aguilera, Y., Duenas, M., Estrella, I., Hernández, T., Benitez, V., Esteban, R. M., et al. (2011). Phenolic profile and antioxidant capacity of chickpeas (Cicer arietinum L.) as affected by a dehydration process. *Plant Foods for Human Nutrition, 66*(2), 187–195.
331. Segev, A., Badani, H., Kapulnik, Y., Shomer, I., Oren-Shamir, M., & Galili, S. (2010). Determination of polyphenols, flavonoids, and antioxidant capacity in colored chickpea (Cicer arietinum L.). *Journal of Food Science, 75*(2), S115–S119.
332. Singh, B., Singh, J. P., Shevkani, K., Singh, N., & Kaur, A. (2017). Bioactive constituents in pulses and their health benefits. *Journal of Food Science and Technology, 54*(4), 858–870.
333. Xiaoli, X., Yang, L., Shuang, H., Li, W., Yi, S., Hao, M., et al. (2008). Determination of oligosaccharide contents in 19 cultivars of chickpea (Cicer arietinum L) seeds by high performance liquid chromatograph. *Food Chemistry, 111*(1), 215–219.
334. Veronese, N., Solmi, M., Caruso, M. G., Giannelli, G., Osella, A. R., Evangelou, E., et al. (2018). Dietary fiber and health outcomes: An umbrella review of systematic reviews and meta-analyses. *The American Journal of Clinical Nutrition, 107*(3), 436–444.
335. Otles, S., & Ozgoz, S. (2014). Health effects of dietary fiber. *Acta Scientiarum Polonorum. Technologia Alimentaria, 13*(2), 191–202.
336. Wu, G. (2016). Dietary protein intake and human health. *Food & Function, 7*(3), 1251–1265.
337. Pedersen, A. N., Kondrup, J., & Børsheim, E. (2013). Health effects of protein intake in healthy adults: A systematic literature review. *Food & Nutrition Research, 57*, 21245. https://doi.org/10.3402/fnr.v57i0.21245
338. Lattimer, J. M., & Haub, M. D. (2010). Effects of dietary fiber and its components on metabolic health. *Nutrients, 2*(12), 1266–1289.
339. Wilde, P. J. (2009). Eating for life: Designing foods for appetite control. *Journal of Diabetes Science and Technology, 3*(2), 366–370.
340. Clark, M. J., & Slavin, J. L. (2013). The effect of fiber on satiety and food intake: A systematic review. *Journal of the American College of Nutrition, 32*(3), 200–211.
341. Astrup, A., Westman, E., Mattes, R. D., Wolfe, R. R., Astrup, A., & Westerterp-Plantenga, M. (2008). Protein, weight management, and satiety. *The American Journal of Clinical Nutrition, 87*(5), 1558S–1561S.
342. Slavin, J. L. (2005). Dietary fiber and body weight. *Nutrition, 21*(3), 411–418.
343. Pesta, D. H., & Samuel, V. T. (2014). A high-protein diet for reducing body fat: Mechanisms and possible caveats. *Nutrition & Metabolism, 11*(1), 53.
344. O'Neil, C. E., Nicklas, T. A., & Fulgoni III, V. L. (2014). Chickpeas and hummus are associated with better nutrient intake, diet quality, and levels of some cardiovascular risk factors: National health and nutrition examination survey 2003-2010. *Journal of Nutrition & Food Sciences, 4*(1), 1000254.
345. Murty, C. M., Pittaway, J. K., & Ball, M. J. (2010). Chickpea supplementation in an Australian diet affects food choice, satiety and bowel health. *Appetite, 54*(2), 282–288.
346. Zafar, T. A., & Kabir, Y. (2017). Chickpeas suppress postprandial blood glucose concentration, and appetite and reduce energy intake at the next meal. *Journal of Food Science and Technology, 54*(4), 987–994.
347. Li, Y., Jiang, B., Zhang, T., Mu, W., & Liu, J. (2008). Antioxidant and free radical-scavenging activities of chickpea protein hydrolysate (CPH). *Food Chemistry, 106*(2), 444–450.
348. Pittaway, J. K., Robertson, I. K., & Ball, M. J. (2008). Chickpeas may influence fatty acid and fiber intake in an ad libitum diet, leading to small improvements in serum lipid profile and glycemic control. *Journal of the American Dietetic Association, 108*(6), 1009–1013.
349. Nestel, P., Cehun, M., & Chronopoulos, A. (2004). Effects of long-term consumption and single meals of chickpeas on plasma glucose, insulin, and triacylglycerol concentrations. *The American Journal of Clinical Nutrition, 79*(3), 390–395.

350. Augustin, L. S., Chiavaroli, L., Campbell, J., Ezatagha, A., Jenkins, A. L., Esfahani, A., et al. (2016). Post-prandial glucose and insulin responses of hummus alone or combined with a carbohydrate food: A dose-response study. *Nutrition Journal, 15*, 13.
351. Kwon, Y. I., Apostolidis, E., Kim, Y. C., & Shetty, K. (2007). Health benefits of traditional corn, beans, and pumpkin: In vitro studies for hyperglycemia and hypertension management. *Journal of Medicinal Food, 10*(2), 266–275.
352. Villegas, R., Gao, Y. T., Yang, G., Li, H. L., Elasy, T. A., Zheng, W., et al. (2008). Legume and soy food intake and the incidence of type 2 diabetes in the Shanghai Women's health study. *The American Journal of Clinical Nutrition, 87*(1), 162–167.
353. Keller, U. (2011). Dietary proteins in obesity and in diabetes. *International Journal for Vitamin and Nutrition Research, 81*(2–3), 125–133.
354. Asif, M. (2014). The prevention and control the type-2 diabetes by changing lifestyle and dietary pattern. *Journal of Education and Health Promotion, 3*, 1.
355. Hosseinpour-Niazi, S., Mirmiran, P., Hedayati, M., & Azizi, F. (2015). Substitution of red meat with legumes in the therapeutic lifestyle change diet based on dietary advice improves cardiometabolic risk factors in overweight type 2 diabetes patients: A cross-over randomized clinical trial. *European Journal of Clinical Nutrition, 69*(5), 592–597.
356. Valdes-Ramos, R., Guadarrama-López, A. L., Martínez-Carrillo, B. E., & Benítez-Arciniega, A. D. (2015). Vitamins and type 2 diabetes mellitus. *Endocrine, Metabolic & Immune Disorders Drug Targets, 15*(1), 54–63.
357. Barbagallo, M., & Dominguez, L. J. (2015). Magnesium and type 2 diabetes. *World Journal of Diabetes, 6*(10), 1152–1157.
358. Castro, H., & Raij, L. (2013). Potassium in hypertension and cardiovascular disease. *Seminars in Nephrology, 33*(3), 277–289.
359. Geiger, H., & Wanner, C. (2012). Magnesium in disease. *Clinical Kidney Journal, 5*(Suppl 1), i25–i38.
360. Pittaway, J. K., Ahuja, K. D., Cehun, M., Chronopoulos, A., Robertson, I. K., Nestel, P. J., et al. (2006). Dietary supplementation with chickpeas for at least 5 weeks results in small but significant reductions in serum total and low-density lipoprotein cholesterols in adult women and men. *Annals of Nutrition & Metabolism, 50*(6), 512–518.
361. Pittaway, J. K., Ahuja, K. D., Robertson, I. K., & Ball, M. J. (2007). Effects of a controlled diet supplemented with chickpeas on serum lipids, glucose tolerance, satiety and bowel function. *Journal of the American College of Nutrition, 26*(4), 334–340.
362. Bazzano, L. A., Thompson, A. M., Tees, M. T., Nguyen, C. H., & Winham, D. M. (2011). Non-soy legume consumption lowers cholesterol levels: A meta-analysis of randomized controlled trials. *Nutrition, Metabolism, and Cardiovascular Diseases, 21*(2), 94–103.
363. Moreno Franco, B., León Latre, M., Andrés Esteban, E. M., Ordovás, J. M., Casasnovas, J. A., & Peñalvo, J. L. (2014). Soluble and insoluble dietary fibre intake and risk factors for metabolic syndrome and cardiovascular disease in middle-aged adults: The AWHS cohort. *Nutrición Hospitalaria, 30*(6), 1279–1288.
364. Graf, D., Di Cagno, R., Fåk, F., Flint, H. J., Nyman, M., Saarela, M., et al. (2015). Contribution of diet to the composition of the human gut microbiota. *Microbial Ecology in Health and Disease, 26*, –26164.
365. Grela, E. R., Samolińska, W., Kiczorowska, B., Klebaniuk, R., & Kiczorowski, P. (2017). Content of minerals and fatty acids and their correlation with phytochemical compounds and antioxidant activity of leguminous seeds. *Biological Trace Element Research, 180*(2), 338–348.
366. Chrysant, S. G., & Chrysant, G. S. (2018). The current status of homocysteine as a risk factor for cardiovascular disease: A mini review. *Expert Review of Cardiovascular Therapy, 16*(8), 559–565.
367. El-Salhy, M., Ystad, S. O., Mazzawi, T., & Gundersen, D. (2017). Dietary fiber in irritable bowel syndrome (review). *International Journal of Molecular Medicine, 40*(3), 607–613.
368. Fernando, W. M., Hill, J. E., Zello, G. A., Tyler, R. T., Dahl, W. J., & Van Kessel, A. G. (2010). Diets supplemented with chickpea or its main oligosaccharide component raffinose modify faecal microbial composition in healthy adults. *Beneficial Microbes, 1*(2), 197–207.

369. Canani, R. B., Costanzo, M. D., Leone, L., Pedata, M., Meli, R., & Calignano, A. (2011). Potential beneficial effects of butyrate in intestinal and extraintestinal diseases. *World Journal of Gastroenterology, 17*(12), 1519–1528.
370. Bultman, S. J. (2014). Molecular pathways: Gene-environment interactions regulating dietary fiber induction of proliferation and apoptosis via butyrate for cancer prevention. *Clinical Cancer Research, 20*(4), 799–803.
371. Chen, J., & Vitetta, L. (2018). Inflammation-modulating effect of butyrate in the prevention of colon cancer by dietary Fiber. *Clinical Colorectal Cancer, 17*(3), e541–e544.
372. Shi, J., Arunasalam, K., Yeung, D., Kakuda, Y., Mittal, G., & Jiang, Y. (2004). Saponins from edible legumes: Chemistry, processing, and health benefits. *Journal of Medicinal Food, 7*(1), 67–78.
373. Koczurkiewicz, P., Czyż, J., Podolak, I., Wojcik, K., Galanty, A., Janeczko, Z., et al. (2015). Multidirectional effects of triterpene saponins on cancer cells - mini-review of in vitro studies. *Acta Biochimica Polonica, 62*(3), 383–393.
374. Podolak, I., Galanty, A., & Sobolewska, D. (2010). Saponins as cytotoxic agents: A review. *Phytochemistry Reviews, 9*(3), 425–474.
375. Chou, Y. C., Chu, C. H., Wu, M. H., Hsu, G. C., Yang, T., Chou, W. Y., et al. (2011). Dietary intake of vitamin B(6) and risk of breast cancer in Taiwanese women. *Journal of Epidemiology, 21*(5), 329–336.
376. Yang, D., Baumgartner, R. N., Slattery, M. L., Wang, C., Giuliano, A. R., Murtaugh, M. A., et al. (2013). Dietary intake of folate, B-vitamins and methionine and breast cancer risk among Hispanic and non-Hispanic white women. *PLoS One, 8*(2), e54495.
377. Takata, Y., Cai, Q., Beeghly-Fadiel, A., Li, H., Shrubsole, M. J., Ji, B. T., et al. (2012). Dietary B vitamin and methionine intakes and lung cancer risk among female never smokers in China. *Cancer Causes & Control, 23*(12), 1965–1975.
378. El-Adawy, T. A. (2002). Nutritional composition and antinutritional factors of chickpeas (Cicer arietinum L.) undergoing different cooking methods and germination. *Plant Foods for Human Nutrition, 57*(1), 83–97.
379. van Vliet, S., Burd, N. A., & van Loon, L. J. (2015). The skeletal muscle anabolic response to plant- versus animal-based protein consumption. *The Journal of Nutrition, 145*(9), 1981–1991.
380. Vigani, G., & Murgia, I. (2018). Iron-requiring enzymes in the spotlight of oxygen. *Trends in Plant Science, 23*(10), 874–882.
381. Montesano, D., Blasi, F., Simonetti, M. S., Santini, A., & Cossignani, L. (2018). Chemical and nutritional characterization of seed oil from Cucurbita maxima L. (var. Berrettina) pumpkin. *Foods (Basel, Switzerland), 7*(3), 30.
382. Lestari, B., & Meiyanto, E. (2018). A review: The emerging nutraceutical potential of pumpkin seeds. *Indonesian Journal of Cancer Chemoprevention, 9*(2), 92–101.
383. Procida, G., Stancher, B., Cateni, F., & Zacchigna, M. (2013). Chemical composition and functional characterisation of commercial pumpkin seed oil. *Journal of the Science of Food and Agriculture, 93*(5), 1035–1041.
384. Glew, R. H., Glew, R. S., Chuang, L. T., Huang, Y. S., Millson, M., Constans, D., et al. (2006). Amino acid, mineral and fatty acid content of pumpkin seeds (Cucurbita spp) and Cyperus esculentus nuts in the Republic of Niger. *Plant Foods for Human Nutrition, 61*(2), 51–56.
385. Yadav, M., Jain, S., Tomar, R., Prasad, G. B., & Yadav, H. (2010). Medicinal and biological potential of pumpkin: An updated review. *Nutrition Research Reviews, 23*(2), 184–190.
386. Martha Perez Gutierrez, R. (2016). Review of Cucurbita pepo (pumpkin) its phytochemistry and pharmacology. *Medicinal Chemistry, 6*(1), 12–21.
387. Patel, S. (2013). Pumpkin (Cucurbita sp.) seeds as nutraceutic: A review on status quo and scopes. *Mediterranean Journal of Nutrition and Metabolism, 6*(3), 183–189.
388. Dar, A., Sofi, S. A., & Rafiq, S. (2017). Pumpkin the functional and therapeutic ingredient: A review. *International Journal of Food Sciences and Nutrition, 2*(6), 165–170.
389. Caili, F., Huan, S., & Quanhong, L. (2006). A review on pharmacological activities and utilization technologies of pumpkin. *Plant Foods for Human Nutrition, 61*(2), 73–80.

390. Phillips, K. M., Ruggio, D. M., & Ashraf-Khorassani, M. (2005). Phytosterol composition of nuts and seeds commonly consumed in the United States. *Journal of Agricultural and Food Chemistry, 53*(24), 9436–9445.
391. Ryan, E., Galvin, K., O'Connor, T. P., Maguire, A. R., & O'Brien, N. M. (2007). Phytosterol, squalene, tocopherol content and fatty acid profile of selected seeds, grains, and legumes. *Plant Foods for Human Nutrition, 62*(3), 85–91.
392. Stevenson, D. G., Eller, F., Wang, L., Jane, J. L., Wang, T., & Inglett, G. E. (2007). Oil and tocopherol content and composition of pumpkin seed oil in 12 cultivars. *Journal of Agricultural and Food Chemistry, 55*(10), 4005–4013.
393. Siano, F., Straccia, M. C., Paolucci, M., Fasulo, G., Boscaino, F., & Volpe, M. G. (2016). Physico-chemical properties and fatty acid composition of pomegranate, cherry and pumpkin seed oils. *Journal of the Science of Food and Agriculture, 96*(5), 1730–1735.
394. Habib, A., Biswas, S., Siddique, A., Manirujjaman, M., Uddin, B., Hasan, S., et al. (2015). Nutritional and lipid composition analysis of pumpkin seed (Cucurbita maxima Linn.). *Journal of Nutrition & Food Sciences, 5*(4), 1000374.
395. Rezig, L., Chouaibi, M., Msaada, K., & Hamdi, S. (2012). Chemical composition and profile characterisation of pumpkin (Cucurbita maxima) seed oil. *Industrial Crops and Products, 37*(1), 82–87.
396. Ganzera, M., Croom, E. M., & Khan, I. A. (1999). Determination of the fatty acid content of pumpkin seed, pygeum, and saw palmetto. *Journal of Medicinal Food, 2*(1), 21–27.
397. Murkovic, M., Hillebrand, J. A., Winkler, J., Leitner, E., & Pfannhauser, W. (1996). Variability of fatty acid content in pumpkin seeds (Cucurbita pepo L.). *Zeitschrift für Lebensmittel-Untersuchung und -Forschung, 203*(3), 216–219.
398. Peiretti, P. G., Meineri, G., Gai, F., Longato, E., & Amarowicz, R. (2017). Antioxidative activities and phenolic compounds of pumpkin (Cucurbita pepo) seeds and amaranth (Amaranthus caudatus) grain extracts. *Natural Product Research, 31*(18), 2178–2182.
399. Sabo, H. S., Sadou, H., Alma, M. M., Sidikou, R. S., Saadou, M., & Amoukou, I. A. (2014). Antioxidant activity and phenolics content of the seeds of eighteen varieties of edible cucurbitaceae of Niger. *Journal of Food Resource Science, 3*(1), 1–11.
400. Saavedra, M. J., Aires, A., Dias, C., Almeida, J. A., De Vasconcelos, M. C., Santos, P., et al. (2015). Evaluation of the potential of squash pumpkin by-products (seeds and shell) as sources of antioxidant and bioactive compounds. *Journal of Food Science and Technology, 52*(2), 1008–1015.
401. Zdunić, G., Menković, N. R., Jadranin, M. B., Novaković, M. M., Šavikin, K. P., & Živković, J. C. (2016). Phenolic compounds and carotenoids in pumpkin fruit and related traditional products. *Hemijska Industrija, 70*(4), 429–433.
402. Krimer-Malešević, V., Mađarev-Popović, S., Vaštag, Z., Radulović, L., & Peričin, D. (2011). Phenolic acids in pumpkin (Cucurbita pepo L.) seeds. In *Nuts and seeds in health and disease prevention* (pp. 925–932). Cambridge, MA: Academic.
403. Andjelkovic, M., Van Camp, J., Trawka, A., & Verhé, R. (2010). Phenolic compounds and some quality parameters of pumpkin seed oil. *European Journal of Lipid Science and Technology, 112*(2), 208–217.
404. Nawirska-Olszanska, A., Kita, A., Biesiada, A., Sokół-Łętowska, A., & Kucharska, A. Z. (2013). Characteristics of antioxidant activity and composition of pumpkin seed oils in 12 cultivars. *Food Chemistry, 139*(1–4), 155–161.
405. Murkovic, M., Hillebrand, A., Winkler, J., & Pfannhauser, W. (1996). Variability of vitamin E content in pumpkin seeds (Cucurbita pepo L.). *Zeitschrift für Lebensmittel-Untersuchung und -Forschung, 202*(4), 275–278.
406. Kim, M. Y., Kim, E. J., Kim, Y. N., Choi, C., & Lee, B. H. (2012). Comparison of the chemical compositions and nutritive values of various pumpkin (Cucurbitaceae) species and parts. *Nutrition Research and Practice, 6*(1), 21–27.
407. Xanthopoulou, M. N., Nomikos, T., Fragopoulou, E., & Antonopoulou, S. (2009). Antioxidant and lipoxygenase inhibitory activities of pumpkin seed extracts. *Food Research International, 42*(5), 641–646.

408. Fruhwirth, G. O., Wenzl, T., El-Toukhy, R., Wagner, F. S., & Hermetter, A. (2003). Fluorescence screening of antioxidant capacity in pumpkin seed oils and other natural oils. *European Journal of Lipid Science and Technology, 105*(6), 266–274.
409. Nkosi, C. Z., Opoku, A. R., & Terblanche, S. E. (2006). Antioxidative effects of pumpkin seed (Cucurbita pepo) protein isolate in CCl4-induced liver injury in low-protein fed rats. *Phytotherapy Research, 20*(11), 935–940.
410. Fahim, A. T., Abd-el Fattah, A. A., Agha, A. M., & Gad, M. Z. (1995). Effect of pumpkin-seed oil on the level of free radical scavengers induced during adjuvant-arthritis in rats. *Pharmacological Research, 31*(1), 73–79.
411. Mangge, H., Becker, K., Fuchs, D., & Gostner, J. M. (2014). Antioxidants, inflammation and cardiovascular disease. *World Journal of Cardiology, 6*(6), 462–477.
412. Pham-Huy, L. A., He, H., & Pham-Huy, C. (2008). Free radicals, antioxidants in disease and health. *International Journal of Biomedical Science: IJBS, 4*(2), 89–96.
413. El-Mosallamy, A. E., Sleem, A. A., Abdel-Salam, O. M., Shaffie, N., & Kenawy, S. A. (2012). Antihypertensive and cardioprotective effects of pumpkin seed oil. *Journal of Medicinal Food, 15*(2), 180–189.
414. Zuhair, H. A., Abd El-Fattah, A. A., & El-Sayed, M. I. (2000). Pumpkin-seed oil modulates the effect of felodipine and captopril in spontaneously hypertensive rats. *Pharmacological Research, 41*(5), 555–563.
415. Gossell-Williams, M., Hyde, C., Hunter, T., Simms-Stewart, D., Fletcher, H., McGrowder, D., et al. (2011). Improvement in HDL cholesterol in postmenopausal women supplemented with pumpkin seed oil: Pilot study. *Climacteric, 14*(5), 558–564.
416. Al-Zuhair, H., Abd el-Fattah, A. A., & Abd el Latif, H. A. (1997). Efficacy of simvastatin and pumpkin-seed oil in the management of dietary-induced hypercholesterolemia. *Pharmacological Research, 35*(5), 403–408.
417. Morrison, M. C., Mulder, P., Stavro, P. M., Suárez, M., Arola-Arnal, A., van Duyvenvoorde, W., et al. (2015). Replacement of dietary saturated fat by PUFA-rich pumpkin seed oil attenuates non-alcoholic fatty liver disease and atherosclerosis development, with additional health effects of virgin over refined oil. *PLoS One, 10*(9), e0139196.
418. Ristic-Medic, D., Perunicic-Pekovic, G., Rasic-Milutinovic, Z., Takic, M., Popovic, T., Arsic, A., et al. (2014). Effects of dietary milled seed mixture on fatty acid status and inflammatory markers in patients on hemodialysis. *ScientificWorldJournal, 2014*, 563576.
419. Witkowska, A. M., Waśkiewicz, A., Zujko, M. E., Szcześniewska, D., Pająk, A., Stepaniak, U., et al. (2017). Dietary polyphenol intake, but not the dietary total antioxidant capacity, is inversely related to cardiovascular disease in postmenopausal polish women: Results of WOBASZ and WOBASZ II studies. *Oxidative Medicine and Cellular Longevity, 2017*, 5982809.
420. Threapleton, D. E., Greenwood, D. C., Evans, C. E., Cleghorn, C. L., Nykjaer, C., Woodhead, C., et al. (2013). Dietary fibre intake and risk of cardiovascular disease: Systematic review and meta-analysis. *BMJ, 347*, f6879.
421. Gupta, A. K., Savopoulos, C. G., Ahuja, J., & Hatzitolios, A. I. (2011). Role of phytosterols in lipid-lowering: Current perspectives. *QJM: An International Journal of Medicine, 104*(4), 301–308.
422. McRae, M. P. (2016). Therapeutic benefits of l-arginine: An umbrella review of meta-analyses. *Journal of Chiropractic Medicine, 15*(3), 184–189.
423. Kass, L., Weekes, J., & Carpenter, L. (2012). Effect of magnesium supplementation on blood pressure: A meta-analysis. *European Journal of Clinical Nutrition, 66*(4), 411–418.
424. Dibaba, D. T., Xun, P., Song, Y., Rosanoff, A., Shechter, M., & He, K. (2017). The effect of magnesium supplementation on blood pressure in individuals with insulin resistance, prediabetes, or noncommunicable chronic diseases: A meta-analysis of randomized controlled trials. *The American Journal of Clinical Nutrition, 106*(3), 921–929.
425. Peacock, J. M., Ohira, T., Post, W., Sotoodehnia, N., Rosamond, W., & Folsom, A. R. (2010). Serum magnesium and risk of sudden cardiac death in the atherosclerosis risk in communities (ARIC) study. *American Heart Journal, 160*(3), 464–470.

426. Medjakovic, S., Hobiger, S., Ardjomand-Woelkart, K., Bucar, F., & Jungbauer, A. (2016). Pumpkin seed extract: Cell growth inhibition of hyperplastic and cancer cells, independent of steroid hormone receptors. *Fitoterapia, 110*, 150–156.
427. Huang, X. E., Hirose, K., Wakai, K., Matsuo, K., Ito, H., Xiang, J., et al. (2004). Comparison of lifestyle risk factors by family history for gastric, breast, lung and colorectal cancer. *Asian Pacific Journal of Cancer Prevention, 5*(4), 419–427.
428. Patisaul, H. B., & Jefferson, W. (2010). The pros and cons of phytoestrogens. *Frontiers in Neuroendocrinology, 31*(4), 400–419.
429. Bacciottini, L., Falchetti, A., Pampaloni, B., Bartolini, E., Carossino, A. M., & Brandi, M. L. (2007). Phytoestrogens: Food or drug? *Clinical Cases in Mineral and Bone Metabolism, 4*(2), 123–130.
430. Ibarreta, D., Daxenberger, A., & Meyer, H. H. (2001). Possible health impact of phytoestrogens and xenoestrogens in food. *APMIS, 109*(3), 161–184.
431. Richter, D., Abarzua, S., Chrobak, M., Vrekoussis, T., Weissenbacher, T., Kuhn, C., et al. (2013). Effects of phytoestrogen extracts isolated from pumpkin seeds on estradiol production and ER/PR expression in breast cancer and trophoblast tumor cells. *Nutrition and Cancer, 65*(5), 739–745.
432. Zaineddin, A. K., Buck, K., Vrieling, A., Heinz, J., Flesch-Janys, D., Linseisen, J., et al. (2012). The association between dietary lignans, phytoestrogen-rich foods, and fiber intake and postmenopausal breast cancer risk: A German case-control study. *Nutrition and Cancer, 64*(5), 652–665.
433. Rossi, R. E., Pericleous, M., Mandair, D., Whyand, T., & Caplin, M. E. (2014). The role of dietary factors in prevention and progression of breast cancer. *Anticancer Research, 34*(12), 6861–6875.
434. Lee, M. Y., Shin, I. S., Kyoung, H., Seo, C. S., Son, J. K., & Shin, H. K. (2014). Alpha-spinasterol from Melandrium firmum attenuates benign prostatic hyperplasia in a rat model. *Molecular Medicine Reports, 9*(6), 2362–2366.
435. Jiang, J., Loganathan, J., Eliaz, I., Terry, C., Sandusky, G. E., & Sliva, D. (2012). ProstaCaid inhibits tumor growth in a xenograft model of human prostate cancer. *International Journal of Oncology, 40*(5), 1339–1344.
436. Jiang, J., Eliaz, I., & Sliva, D. (2011). Suppression of growth and invasive behavior of human prostate cancer cells by ProstaCaid: Mechanism of activity. *International Journal of Oncology, 38*(6), 1675–1682.
437. Fruhwirth, G. O., & Hermetter, A. (2007). Seeds and oil of the Styrian oil pumpkin: Components and biological activities. *European Journal of Lipid Science and Technology, 109*(11), 1128–1140.
438. Gossell-Williams, M., Davis, A., & O'Connor, N. (2006). Inhibition of testosterone-induced hyperplasia of the prostate of sprague-dawley rats by pumpkin seed oil. *Journal of Medicinal Food, 9*(2), 284–286.
439. Hong, H., Kim, C. S., & Maeng, S. (2009). Effects of pumpkin seed oil and saw palmetto oil in Korean men with symptomatic benign prostatic hyperplasia. *Nutrition Research and Practice, 3*(4), 323–327.
440. Tantawy, S. A., Elgohary, H. M., & Kamel, D. M. (2018). Trans-perineal pumpkin seed oil phonophoresis as an adjunctive treatment for chronic nonbacterial prostatitis. *Research and Reports in Urology, 10*, 95–101.
441. Joshi, S., Nair, N., & Bedwal, R. S. (2014). Dietary zinc deficiency effects dorso-lateral and ventral prostate of Wistar rats: Histological, biochemical and trace element study. *Biological Trace Element Research, 161*(1), 91–100.
442. Nishimura, M., et al. (2014). Pumpkin seed oil extracted from Cucurbita maxima improves urinary disorder in human overactive bladder. *Journal of Traditional and Complementary Medicine, 4*(1), 72–74.
443. Vahlensieck, W., Theurer, C., Pfitzer, E., Patz, B., Banik, N., & Engelmann, U. (2015). Effects of pumpkin seed in men with lower urinary tract symptoms due to benign prostatic hyperpla-

sia in the one-year, randomized, placebo-controlled GRANU study. *Urologia Internationalis, 94*(3), 286–295.

444. Adams, G. G., Imran, S., Wang, S., Mohammad, A., Kok, M. S., Gray, D. A., et al. (2014). The hypoglycemic effect of pumpkin seeds, Trigonelline (TRG), nicotinic acid (NA), and D-Chiro-inositol (DCI) in controlling glycemic levels in diabetes mellitus. *Critical Reviews in Food Science and Nutrition, 54*(10), 1322–1329.
445. Kushawaha, D. K., Yadav, M., Chatterji, S., Srivastava, A. K., & Watal, G. (2017). Evidence based study of antidiabetic potential of C. maxima seeds - in vivo. *Journal of Traditional and Complementary Medicine, 7*(4), 466–470.
446. Makni, M., Fetoui, H., Gargouri, N. K., Garoui, M., & Zeghal, N. (2011). Antidiabetic effect of flax and pumpkin seed mixture powder: Effect on hyperlipidemia and antioxidant status in alloxan diabetic rats. *Journal of Diabetes and its Complications, 25*(5), 339–345.
447. Song, H., & Sun, Z. (2017). Hypolipidaemic and hypoglycaemic properties of pumpkin polysaccharides. *Biotech, 7*(3), 159–159.
448. Candido, F. G., de Oliveira, F. C. E., Lima, M. F. C., Pinto, C. A., da Silva, L. L., Martino, H. S. D., et al. (2018). Addition of pooled pumpkin seed to mixed meals reduced postprandial glycemia: A randomized placebo-controlled clinical trial. *Nutrition Research, 56*, 90–97.
449. Mahmoodpoor, A., Medghalchi, M., Nazemiyeh, H., Asgharian, P., Shadvar, K., & Hamishehkar, H. (2018). Effect of Cucurbita Maxima on control of blood glucose in diabetic critically ill patients. *Advanced Pharmaceutical Bulletin, 8*(2), 347–351.
450. Lopez-Ridaura, R., Willett, W. C., Rimm, E. B., Liu, S., Stampfer, M. J., Manson, J. E., et al. (2004). Magnesium intake and risk of type 2 diabetes in men and women. *Diabetes Care, 27*(1), 134–140.
451. Rodriguez-Moran, M., Mendía, L. E. S., Galván, G. Z., & Guerrero-Romero, F. (2011). The role of magnesium in type 2 diabetes: A brief based-clinical review. *Magnesium Research, 24*(4), 156–162.
452. Gannon, M. C., Nuttall, F. Q., Saeed, A., Jordan, K., & Hoover, H. (2003). An increase in dietary protein improves the blood glucose response in persons with type 2 diabetes. *The American Journal of Clinical Nutrition, 78*(4), 734–741.
453. Skerrett, P. J., & Willett, W. C. (2010). Essentials of healthy eating: A guide. *Journal of Midwifery & Women's Health, 55*(6), 492–501.
454. Tapsell, L. C., Batterham, M. J., Thorne, R. L., O'Shea, J. E., Grafenauer, S. J., & Probst, Y. C. (2014). Weight loss effects from vegetable intake: A 12-month randomised controlled trial. *European Journal of Clinical Nutrition, 68*(7), 778–785.
455. Pramuková, B., Szabadosová, V., & Soltésová, A. (2011). Current knowledge about sports nutrition. *The Australasian Medical Journal, 4*(3), 107–110.
456. Cherbuin, N., Kumar, R., Sachdev, P. S., & Anstey, K. J. (2014). Dietary mineral intake and risk of mild cognitive impairment: The PATH through life project. *Frontiers in Aging Neuroscience, 6*, 4.
457. Rink, L., & Gabriel, P. (2000). Zinc and the immune system. *The Proceedings of the Nutrition Society, 59*(4), 541–552.
458. Tyszka-Czochara, M., Grzywacz, A., Gdula-Argasińska, J., Librowski, T., Wiliński, B., & Opoka, W. (2014). The role of zinc in the pathogenesis and treatment of central nervous system (CNS) diseases. Implications of zinc homeostasis for proper CNS function. *Acta Poloniae Pharmaceutica, 71*(3), 369–377.
459. Gower-Winter, S. D., & Levenson, C. W. (2012). Zinc in the central nervous system: From molecules to behavior. *BioFactors (Oxford, England), 38*(3), 186–193.
460. Frederickson, C. J., Suh, S. W., Silva, D., Frederickson, C. J., & Thompson, R. B. (2000). Importance of zinc in the central nervous system: The zinc-containing neuron. *The Journal of Nutrition, 130*(5S Suppl), 1471s–1483s.
461. Bush, A. I., Cuajungco, M. P., Atwood, C. S., Moir, R. D., Tanzi, R. E., & Bush, A. I. (2000). Alzheimer's disease, β-amyloid protein and zinc. *The Journal of Nutrition, 130*(5), 1488S–1492S.

462. Watt, N. T., Whitehouse, I. J., & Hooper, N. M. (2010). The role of zinc in Alzheimer's disease. *International Journal of Alzheimer's Disease, 2011*, 971021.
463. Nuttall, J. R., & Oteiza, P. I. (2014). Zinc and the aging brain. *Genes & Nutrition, 9*(1), 379.
464. Okada, Y., & Okada, M. (2013). Protective effects of plant seed extracts against amyloid β-induced neurotoxicity in cultured hippocampal neurons. *Journal of Pharmacy & Bioallied Sciences, 5*(2), 141–147.
465. Ahmadian-Attari, M. M., Mosaddegh, M., Kazemnejad, A., & Noorbala, A. A. (2013). Comparison between complementary dietary treatment of Alzheimer disease in Iranian traditional medicine and modern medicine. *Iranian Journal of Public Health, 42*(12), 1414–1421.
466. Sandstead, H. H., & Freeland-Graves, J. H. (2014). Dietary phytate, zinc and hidden zinc deficiency. *Journal of Trace Elements in Medicine and Biology, 28*(4), 414–417.
467. Silber, B. Y., & Schmitt, J. A. (2010). Effects of tryptophan loading on human cognition, mood, and sleep. *Neuroscience and Biobehavioral Reviews, 34*(3), 387–407.
468. Shaw, K., Turner, J., & Del Mar, C. (2002). Tryptophan and 5-hydroxytryptophan for depression. *Cochrane Database of Systematic Reviews, 1*, Cd003198. https://doi.org/10.1002/14651858.CD003198
469. Chollet, D., Franken, P., Raffin, Y., Henrotte, J. G., Widmer, J., Malafosse, A., et al. (2001). Magnesium involvement in sleep: Genetic and nutritional models. *Behavior Genetics, 31*(5), 413–425.
470. Abbasi, B., Kimiagar, M., Sadeghniiat, K., Shirazi, M. M., Hedayati, M., & Rashidkhani, B. (2012). The effect of magnesium supplementation on primary insomnia in elderly: A double-blind placebo-controlled clinical trial. *Journal of Research in Medical Sciences : The Official Journal of Isfahan University of Medical Sciences, 17*(12), 1161–1169.
471. Held, K., Antonijevic, I. A., Künzel, H., Uhr, M., Wetter, T. C., Golly, I. C., et al. (2002). Oral Mg(2+) supplementation reverses age-related neuroendocrine and sleep EEG changes in humans. *Pharmacopsychiatry, 35*(4), 135–143.
472. Hornyak, M., Haas, P., Veit, J., Gann, H., & Riemann, D. (2004). Magnesium treatment of primary alcohol-dependent patients during subacute withdrawal: An open pilot study with polysomnography. *Alcoholism, Clinical and Experimental Research, 28*(11), 1702–1709.
473. Halson, S. L. (2014). Sleep in elite athletes and nutritional interventions to enhance sleep. *Sports Medicine (Auckland, N.Z.), 44*(Suppl 1), S13–S23.
474. Nieves, J. W. (2014). Bone. Maximizing bone health--magnesium, BMD and fractures. *Nature Reviews. Endocrinology, 10*(5), 255–256.
475. Lopez, H. W., Leenhardt, F., Coudray, C., & Remesy, C. (2002). Minerals and phytic acid interactions: Is it a real problem for human nutrition? *International Journal of Food Science & Technology, 37*(7), 727–739.

The Role of Gluten in Autism

T. Sumathi, T. Manivasagam, and A. Justin Thenmozhi

Abstract Autism spectrum disorder (ASD) is an inherited neurodevelopmental disorder of social communication and restricted, repetitive behaviors. Much remains unknown about their mechanisms of action and physiological effects. In recent years, there has been a growing interest in nutritional diets, which can be used as a form of therapeutic intervention for ASD with a recent increase in the research being carried out in this field. Selective nutrition therapy for ASD and brain function shows improvement in behavioral changes and reduction in malnutrition seemingly associated with the allergies or food intolerances to gluten. Therefore, a gluten-free diet has yielded positive outcomes giving hope in developing therapy for ASD.

Keywords Autism spectrum disorder · Models of ASD · Gluten-free diet · Diet intervention

1 Introduction

1.1 Autism

Autism spectrum disorder is a heterogeneous group of neurobehavioral and developmental disorders characterized by severe and sustained impairment in social interaction, deviance in communication, repetitive patterns of behavior, and sensitivity to sounds or textures [1, 2]. In addition, people with autism often have comorbid neurological disorders such as mental retardation and epilepsy [3].

T. Sumathi (✉)
Department of Medical Biochemistry, Dr. ALMPGIBMS, University of Madras, Chennai, Tamil Nadu, India

T. Manivasagam · A. J. Thenmozhi
Department of Biochemistry and Biotechnology, Faculty of Science, Annamalai University, Chidambaram, Tamil Nadu, India

M. M. Essa, M. W. Qoronfleh (eds.), *Personalized Food Intervention and Therapy for Autism Spectrum Disorder Management*, Advances in Neurobiology 24,
https://doi.org/10.1007/978-3-030-30402-7_14

Anxiety and mood disorders are very common features in autism [4]. All features of ASD are described as part of mental disorders in the DSM-V Fifth Edition (DSM-V). Different types of autism with multifarious levels of severity exist. People suffering from Rett syndrome display profound language alteration and severe psychomotor defects, whereas Asperger patients show stereotypic behaviors and social interaction alteration but no delay in language learning and tend to show curiosity toward their environment (DSM-V). Many children and adults with a diagnosis of ASD have comorbid features such as intellectual disability (ID) and/or epilepsy. Altogether, ASD is relatively common with a prevalence of ~1% worldwide [5, 6]. The patients can also be stressed easily by a change in habit or environment. ASDs constitute a diverse set of symptoms with multiple studies of causation indicating genetic susceptibility and interactions between genetic and environmental factors. Many studies indicate that genetic factors play an important role in the case of ASDs.

1.1.1 Genetics and Autism Spectrum Disorders

According to a recent report, the prevalence of this developmental disorder has risen to 1 in 88 children in India, and nearly one in eight have at least developed neurodevelopmental condition [7]. Familial studies demonstrate that ASD is a heritable disorder, with estimated genetic contributions accounting for >50–60% of ASD risk [8, 9]. Genetic epidemiologic studies have also shown that ASD is not a single disease but a group of symptoms involving multiple gene networks [10].

1.1.2 Pathophysiology of Autism

The pathophysiology of autism is mainly related with the changes in various systems of the brain. However, the occurrence of autism has still not been understood clearly. The pathophysiology of autism is associated with neuropsychological linkages between brain structures and behaviors [11, 12]. The cellular and nutritional bases of pathological early overgrowth include the following:

- An overabundance of neurons that cause local overconnectivity in key brain regions
- Interrupted neuronal migration during early gestation
- Unstable excitatory–inhibitory networks
- Abnormal formation of synapses and dendritic spines
- Modulation of the neurexin–neuroligin cell-adhesion system
- Poorly regulated synthesis of synaptic proteins
- Damaged synaptic development which may also contribute to epilepsy
- Deficiencies of zinc, magnesium, vitamin B, vitamin D, vitamin A, antioxidant nutrients, omega-3 fatty acids
- Food sensitivities/intolerances

- Opiate-like by products from casein and gluten
- Altered intestinal permeability
- Intestinal dysbiosis
- Poor methylation/transsulfuration
- Poor detoxification
- Inflammation in the intestine causing excessive production of pro-inflammatory cytokines
- Oxidative stress
- Mitochondrial/metabolic dysfunction
- Immune imbalances/autoimmunity
- Insufficient oxytocin

The immune system plays an important role in autism. ASD causes inflammation in both peripheral and central immune systems as indicated by increased levels of pro-inflammatory cytokines and significant activation of microglia. The unusual immune function has been associated with increased impairments in behaviors that are characteristic of the core features of autism such as deficits in social interactions and communication. The neurochemicals associated with autism have been investigated, with the most evidence for the role of serotonin and of genetic differences in its transport [13]. The metabotropic glutamate receptors (mGluRs) have been implicated in a diverse variety of neuronal functions. The Gp1 mGluRs specifically have been implicated in the pathogenesis of fragile X syndrome (FXS), the foremost inherited cause of mental retardation and the most common identified genetic cause of autism.

1.2 Models for ASD

1.2.1 Chemical Models

Valproic Acid (VPA)

Valproic acid (VPA), or 2-propylpentanoic acid, is a short-chained fatty acid widely used as an antiepileptic drug and for the treatment of bipolar disorders, migraine, headaches, and neuropathic pain. VPA is well-known for its teratogenic effects, including neural tube defects, cardiovascular anomalies, limb anomalies, craniofacial abnormalities, and neurodevelopmental delay [14, 15]. Its main mechanisms of action are the inhibition of histone deacetylase, being an epigenetic modulator and the ability to increase oxidative stress. Prenatal exposure to VPA, especially during the first trimester of pregnancy, has also been associated with reduced cognitive function and high risk for ASD among the offsprings [16, 17]. In a research study, 350 mg/kg of VPA injected into pregnant rats at the embryonic stages resulted in neural tube closure and brain stem nuclei formation causing damage to the motor cranial nerve nuclei of the fetus.

Thalidomide

Thalidomide has been shown to cause multiple birth defects such as limb reduction defects and ocular and cardiovascular anomalies. Today, thalidomide is an immunomodulatory drug still used in the treatment of leprosy and multiple myeloma [18–20]. Few animal studies have been reported the use of thalidomide exposure as a model of autism [21], and the rats induced with 500 mg/kg of thalidomide in their embryonic stage have showed that thalidomide affects the neurochemical serotonin in the hippocampus and increased the dopamine content in the frontal cortex causing hyperserotonemia.

Misoprostol

Misoprostol is a prostaglandin analogue drug commonly prescribed for the treatment of gastric ulcers. There is evidence that prenatal exposure to high doses of misoprostol during the first or second trimesters of pregnancy can lead to the occurrence of Möbius syndrome, a disorder characterized by uni- or bilateral eye–face palsy due to damage to cranial nerve nuclei, associated with muscle or skeletal malformations and a high prevalence of autism-like behavioral symptoms [22].

Propionic Acid (PPA)

Propionic acid (PPA) is a fatty acid, a metabolic end product of enteric bacteria in the gut, and a common food preservative. Several studies have indicated that PPA can cause autism-like behaviors and a neuroinflammatory response in rats [23] depicts that rats treated with PPA displayed restricted behavioral interest and impaired social behavior in addition to reactive astrogliosis and activated microglia in the brain. These results provide evidence of occurrence of ASD-like behaviors in the PPA rodent model. However, the relevance to human beings is still unknown.

1.2.2 Genetic Animal Models of Autism

Various human syndromes owing to a single gene mutation increase the risk for ASD. The more common disorders are fragile X syndrome, a mutation in FMR1; Rett syndrome, a mutation in MECP2; tuberous sclerosis, mutations in TSC1 or TSC2; and Timothy syndrome, a mutation in CACNA1C. The variants that lead to inherited maternal 15q11–13 duplication resulting in Prader–Willi syndrome and other duplications like the NPHP1 gene are also associated with autistic traits.

The development in strategies for the identification of genetic variants also led to the description of new syndromic forms of ASD and enabled the association between phenotype and genetic traits. Mice are the predominant animal model for ASD owing to their genetic tractability and their ability to demonstrate analogues of behavioral deficits associated with ASD. The following genetic models representing known human syndromes derived from a gene mutation exemplify the difficulties in establishing a valid genetic trait.

Fragile X (FMR1)

The fragile X syndrome is the most frequently inherited cause for mental retardation in ASD. The fragile mental retardation 1 locus (FMR1) resides in the X chromosome, and the expansion of triplet repeats in the untranslated region of the FMR1 gene prevents synthesis of the FMR1 gene product FMRP. FMRP is an RNA-binding protein that modulates mRNA trafficking, dendritic maturation, and synaptic plasticity. Rodents, mostly mice knocked out for the FMR1 gene, were shown to present autistic traits.

Rett Syndrome and MECP2 Mutations

Rett syndrome, an X-linked disease only affecting girls, is denoted by neurodevelopmental delay, ASD, and seizures. It is caused by mutations in gene encoding for the methyl-CpG binding protein 2 (MECP2) that binds to methylated-CpG dinucleotides and influences gene expression. MECP2 is expressed widely but is most abundant in neurons of the mature nervous system. Knockout of the Mecp2 in male animals used to study MECP2 duplication syndrome is characterized by autism, intellectual disability, motor dysfunction, anxiety, epilepsy, recurrent respiratory tract infections, and early death.

Tuberous Sclerosis TSC1 or TSC2

Tuberous sclerosis complex is an autosomal dominant disorder caused by mutations in either the TSC1 or TSC2 gene associated with cerebral cortical tubers and may be complicated by astrocytomas. Besides intellectual disability and often seizures of the type of infantile spasm, there is an increase in the rate of ASD. The Tsc2f/−; cre mice exhibit progressive Purkinje cell degeneration. Since loss of Purkinje cells was reported in patients with ASD, this model was offered for future studies [24].

Cortical Dysplasia Focal Epilepsy (CNTNAP2)

A recessive nonsense mutation in the Contactin-associated protein-like 2 (CNTNAP2) gene was shown to cause a syndromic form of ASD, Cortical Dysplasia and Focal Epilepsy Syndrome (CDFE). Cortical Dysplasia and Focal Epilepsy Syndrome (CDFE) results in epileptic seizures, language regression, intellectual disability, hyperactivity, and ASD [25].

15q11–13 Duplication Maternal/Paternal Gene

The 15q11–13 maternal/paternal gene is associated with either duplication or deletion. The loss of genomic material within the paternal 15q11.2–13 locus and deletions, unbalanced translocations, or uniparental maternal disomy results in Prader–Willi syndrome. The misplacement of maternal genomic material at the 15q11.2–13 locus results in Angelman syndrome and MECP2 duplication syndrome, characterized by autism, intellectual disability, motor dysfunction, anxiety, epilepsy, recurrent respiratory tract infections, and early death.

Phelan–McDermid Syndrome, SHANK3

Removal of the human SHANK3 gene near the terminus of chromosome 22q is associated with Phelan–McDermid syndrome and autism. The mice knock out for IB2 gene had reduced AMPA and enhanced NMDA receptor-mediated glutamatergic transmission in the cerebellum along with disturbances in the morphology of Purkinje cell dendritic arbores and autistic traits. The IB2 gene mutation has been shown to play a role in Chr22qter-associated cognitive disorders.

2 Autism and Diet

ASD can be a difficult disorder to live with on many levels, and maintaining proper dietary habits becomes essential in this regard. For children with ASD, a nutritious, balanced diet makes a lot of difference in their ability to learn, manage their emotions, and process information. Due to their severely disrupted digestion, restoration of balance in the gut, blood sugar, checking for brain-polluting heavy metals, excluding food additives, identifying food allergies and possible nutrient deficiencies, and ensuring an optimal intake of essential fat, children with ASD often have restricted diets.

2.1 Gluten-Free Diet Therapy for Autism

Gastrointestinal (GI) problems are commonly seen in children with autism. In a related study on GI pathology in children with autism by Valicenti-McDermott et al. [26], the authors reported that 70% of children with autism had GI problems compared to 42% of children with other neurodevelopmental problems such as cerebral palsy and 28% of children with typical development.

The association of autism with gluten has been implicated by historical observations, and an understanding of gluten sensitivity has been evolving over the last several decades. Celiac disease, also known as gluten-sensitive enteropathy or celiac sprue, is a genetically linked autoimmune disorder in which eating certain types of grain-based products (wheat, rye, barley, and malt) can trigger an immune reaction that causes damage to the small intestine [27]. Due to carbohydrate intolerance, autistic individuals are unable or are sensitive to digestion of a dietary product, typically considered to be an immune response allergic reaction leading to poor digestion of the product, or alteration of the microbial or immune environment due to dietary intake among other proposed mechanisms.

A wider spectrum of neurological syndromes is an extraintestinal manifestation of gluten sensitivity with or without intestinal pathology. It includes myelopathy, mononeuritis multiplex, encephalopathy, chorea, brain stem dysfunction, migraine, Guillain–Barré-like syndrome, and neuropathy with positive antiganglioside antibodies. Many neurological reports used a variety of positive results on serology tests for celiac disease without intestinal confirmation of the diagnosis.

The GI symptoms of allergy could include pain, constipation, diarrhea, rash, sleep disturbance, inflammation, and increased permeability of food-based or like-sized peptides that enter the bloodstream that might then induce allergic sensitization. This study attempted to address a theory known as the "opioid-excess" theory. The opioid-excess theory hypothesizes that autism is a result of a metabolic disorder in which opioid peptides produced through the metabolism of gluten and casein pass through an abnormally permeable intestinal membrane and then affect neurotransmission through binding with opioid receptors. So, autistic children are unusually sensitive to gluten, which results in small bowel inflammation in these children allowing these opioid peptides to enter the brain.

2.1.1 Immune Reaction to Gluten Proteins

Wheat comprises of about 100 different proteins, the majority of which are alcohol-soluble and the remaining being water-soluble. Together, α-gliadins, γ-gliadins, ω-gliadins, and low- and high-molecular-weight glutenins are the major alcohol-soluble proteins called gluten proteins representing about 75% of the total proteins of wheat grains [28]. The remainder of wheat proteins, which are generally soluble in water or salt solutions, including serine protease inhibitors (serpins), purinins, farinins, α-amylase/protease inhibitors, and globulins, are called non-gluten proteins. Intestinal T cells from celiac disease (CD) patients respond to a heterogeneous array of peptides derived from α-, γ-, and ω-gliadins and glutenins and produce a significant amount of interferon-γ [29]. The immune system recognized many peptides from single or multiple gliadin families and reacted only to one peptide. This means that many gluten epitopes may be involved in the development of gluten sensitivity.

2.1.2 Research Analysis of Gluten-Free Diet for Autism

Analysis of the blood samples revealed that a significant number of children with autism produced IgG and IgA antibodies against α-gliadin 33-mer peptide [30]. The effectiveness of a gluten-free diet was tested on children with autism, and a significant improvement in behavioral symptoms was observed in a subgroup [31]. This study on autism was conducted with α-gliadin 33-mer, especially CXCR3-binding gliadin peptide. CXCR3 is a chemokine receptor that is expressed in monocytes, eosinophils, NK cells, B cells, and T cells, particularly in CD4+ TH1 cells [32]. During the inflammatory process, CXCR3 promotes the recruitment of immune cells into the inflamed tissues by interacting with its three ligands: CXCL9, CXCL10, and CXCL11. This CXCR3 interaction with its ligands becomes overactivated in different chronic inflammatory processes such as inflammatory bowel diseases and rheumatoid arthritis [33].

A research study conducted by McCarthy and Coleman [34] involved eight children with autism. The children were given 20 g of gluten powder (added to their food) every day for 4 weeks. Each child then had a jejunal biopsy; all eight revealed entirely normal intestinal villous structure. Jejunal biopsies comprise the standard diagnosis for celiac disease. This comprises the screening of large numbers of children with autism and screening for celiac disease by blood testing of IgA and IgG antigliadin antibodies (AGA) as well as antiendomysium antibodies (AEMAb). These were the techniques used in the present study. Among our patients with autism, no cases of celiac disease were detected. In two of these patients, an abnormal increase of AGA IgG and AEMAb was found, but ensuing antibody determinations and jejunal biopsies showed normal results. Thus, our results are in line with previous research studies in this area.

Three recently published studies examined various hypotheses related to the GFCF diet in autism. Vojdani, Pangborn, Vojdani, and Cooper [30] measured the antibodies IgG, IgM, and IgA against CD26, CD69, streptokinase (SK), gliadin, casein, and ethyl mercury in 50 children diagnosed with autism. Analysis of blood samples revealed that a significant number of the children developed antibodies against casein and gliadin. In addition, SK, gliadin, casein, and ethyl mercury were shown to bind to the lymphocyte and tissue enzyme (CD26) and have been thought to perhaps trigger inflammatory and immune reactions in children with autism.

The study conducted by Arnold et al. [35] evaluated amino acid patterns of 26 children with autism on a regular diet, 10 on a gluten–casein-free diet, and 26 children with developmental delays which served as controls. The children with autism had higher deficiencies in essential amino acids compared to the control group. These findings suggest that children with autism are at high risk for amino acid deficiencies and may benefit from a structured diet.

Knivsberg et al. [36] conducted a randomized single blind study with 20 subjects to assess the effect of a gluten-free diet on children with autistic syndrome and urinary peptide abnormalities. The children in the control and experimental groups were matched according to severity of autistic symptoms, age, and cognitive level. Changes were observed in both the control and experimental group. The experimental group showed more significant changes. There was a statistically significant difference between the experimental and control group, demonstrating that the experimental group showed improvement in autistic behavior, nonverbal cognitive level, and motor problems.

These recent studies provide interesting information regarding the effects of hypothesized gluten-free diets on physiology, behavior, and cognition, though they are limited by small sample sizes. A need still exists for rigorous controlled clinical trials evaluating both physiological and behavioral effects.

3 Conclusion

With all that has been said, we can begin to conclude that a significant subset of individuals with autism are sensitive or intolerant to certain foods. Gluten-free diet therapy for autistic children with autoimmune disorder, celiac diseases, bowel inflammation allergy, or food intolerance has showed positive improvisation and changes in their behavior, performance, social interaction, and brain functions. By providing the body and brain with what they need and by eliminating that which may be causing interferences, the possibility to significantly improve overall brain functioning and, therefore, the quality of life for individuals with autism exists.

References

1. Volkmar, F. R., Reichow, B., Westphal, A., & Mandell, D. S. (2014). Autism and the autism spectrum: Diagnostic concepts. In F. R. Volkmar (Ed.), *Handbook of autism and pervasive developmental disorders* (4th ed.). Hoboken, NJ: Wiley. https://doi.org/10.1002/9781118911389.hautc01
2. Tsatsanis, K. D., Volkmar, F. R., Paul, R., Klin, A., & Cohen, D. (2014). Neuropsychological characteristics in autism and related conditions. In F. R. Volkmar, R. Paul, A. Klin, & D. Cohen (Eds.), *Handbook of autism and pervasive developmental disorders: Diagnosis, development, neurobiology, and behavior* (Vol. 1, 3rd ed., pp. 365–381). Hoboken, NJ: Wiley. https://doi.org/10.1002/9780470939345.ch13
3. McCarthy, J., Hemmings, C., Kravariti, E., Dworzynski, K., Holt, G., Bouras, N., et al. (2010). Challenging behavior and co-morbid psychopathology in adults with intellectual disability and autism spectrum disorders. *Research in Developmental Disabilities, 31*(2), 362–366.
4. Lecavalier, L. (2006). Behavioral and emotional problems in young people with pervasive developmental disorders: Relative prevalence, effects of subject characteristics, and empirical classification. *Journal of Autism and Developmental Disorders, 36*(8), 1101–1114.
5. Bourgeron, T. (2015). From the genetic architecture to synaptic plasticity in autism spectrum disorder. *Nature Reviews. Neuroscience, 16*, 551–563.
6. Ronemus, M., Iossifov, I., Levy, D., & Wigler, M. (2014). The role of de novo mutations in the genetics of autism spectrum disorders. *Nature Reviews. Genetics, 15*, 133–141.
7. Arora, N. K., MKC, N., Gulati, S., Deshmukh, V., Mohapatra, A., Mishra, D., et al. (2018). Neurodevelopmental disorders in children aged 2-9 years: Population-based burden estimates across five regions in India. *PLoS Medicine, 15*, e1002615.
8. Almeida, M. I., Reis, R. M., & Calin, G. A. (2011). MicroRNA history: Discovery, recent applications, and next frontiers. *Mutation Research/Fundamental and Molecular Mechanisms of Mutagenesis, 717*, 1–8.
9. Pickles, A., Bolton, P., Macdonald, H., Bailey, A., Le Couteur, A., Sim, C. H., et al. (1995). Latent-class analysis of recurrence risks for complex phenotypes with selection and measure-

ment error: A twin and family history study of autism. *American Journal of Human Genetics, 57*(3), 717.
10. Gaugler, T., Klei, L., Sanders, S. J., Bodea, C. A., Goldberg, A. P., Lee, A. B., et al. (2014). Most genetic risk for autism resides with common variation. *Nature Genetics, 46*(8), 881–885. https://doi.org/10.1038/ng.3039
11. Basu, S. N., Kollu, R., & Banerjee-Basu, S. (2009). AutDB: A gene reference resource for autism research. *Nucleic Acids Research, 37*, D832–D836. https://doi.org/10.1093/nar/gkn835
12. Penn, H. E. (2006). Neurobiological correlates of autism: A review of recent research. *Child Neuropsychology, 12*(1), 5779. https://doi.org/10.1080/09297040500253546
13. Hsiao, E. Y. (2013). Immune dysregulation in autism spectrum disorder. Neurobiology of autism. *International Review of Neurobiology., 113*, 269–302. https://doi.org/10.1016/B978-0-12-418700-9.00009-5
14. Levy, S. E., Mandell, D. S., & Schultz, R. T. (2009). Autism. *Lancet, 374*(9701), 1627–1638. https://doi.org/10.1016/S0140-6736(09)61376-3
15. Löscher, W. (2002). Basic pharmacology of valproate: A review after 35 years of clinical use for the treatment of epilepsy. *CNS Drugs, 16*, 669–694
16. Meador, K., Reynolds, M. W., Crean, S., Fahrbach, K., & Probst, C. (2008). Pregnancy outcomes in women with epilepsy: A systematic review and meta-analysis of published pregnancy registries and cohorts. *Epilepsy Research, 81*, 1–13.
17. Williams, G., King, J., Cunningham, M., Stephan, M., Kerr, B., & Hersh, J. H. (2001). Fetal valproate syndrome and autism: Additional evidence of an association. *Developmental Medicine and Child Neurology, 43*, 202–206.
18. Christensen, J., Gronborg, T. K., Sorensen, M. J., Schendel, D., Parner, E. T., Pedersen, L. H., et al. (2013). Prenatal valproate exposure and risk of autism spectrum disorders and childhood autism. *JAMA, 309*, 1696–1703.
19. Miller, M. T. (1991). Thalidomide embryopathy: A model for the study of congenital incomitant horizontal strabismus. *Transactions of the American Ophthalmological Society, 81*, 623–674.
20. Ito, T., Ando, H., Suzuki, T., Ogura, T., Hotta, K., Imamura, Y., et al. (2010). Identification of a primary target of thalidomide teratogenicity. *Science, 327*, 1345–1350.
21. Narita, N., Kato, M., Tazoe, M., Miyazaki, K., Narita, M., & Okado, N. (2002). Increased monoamine concentration in the brain and blood of fetal thalidomide- and valproic acid-exposed rat: Putative animal models for autism. *Pediatric Research, 52*, 576–579.
22. Gonzalez, C. H., Marques-Dias, M. J., Kim, C. A., Sugayama, S. M., Da Paz, J. A., Huson, S. M., et al. (1998). Congenital abnormalities in Brazilian children associated with misoprostol misuse in first trimester of pregnancy. *Lancet, 351*, 1624–1627.
23. MacFabe, D. F., Cain, D. P., Rodriguez-Capote, K., Franklin, A. E., Hoffman, J. E., Boon, F., et al. (2007). Neurobiological effects of intraventricular propionic acid in rats possible role of short chain fatty acids on the pathogenesis and characteristics of autism spectrum disorders. *Behavioural Brain Research, 176*, 149–169.
24. Reith, R. M., McKenna, J., Wu, H., Hashmi, S. S., Cho, S. H., Dash, P. K., et al. (2013). Loss of Tsc2 in Purkinje cells is associated with autistic-like behavior in a mouse model of tuberous sclerosis complex. *Neurobiology of Disease, 51*, 93–103.
25. Moy, S. S., Riddick, N. V., Nikolova, V. D., Teng, B. L., Agster, K. L., Nonneman, R. J., et al. (2014). Repetitive behavior profile and supersensitivity to amphetamine in the C58/J mouse model of autism. *Behavioural Brain Research, 259*, 200–214.
26. Valicenti-McDermott, M., McVicar, K., Rapin, I., Wershil, B. K., Cohen, H., & Shinnar, S. (2006). Frequency of gastrointestinal symptoms in children with autistic spectrum disorders and association with family history of autoimmune disease. *Journal of Developmental and Behavioral Pediatrics, 27*, S128–S136.
27. Catassi, C., & Fasano, A. (2008). Celiac disease. *Current Opinion in Gastroenterology, 24*, 687–691.
28. Yang, A., Chen, Y., Scherl, E., Neugut, A. I., Bhagat, G., & Green, P. H. R. (2005). Inflammatory bowel disease in patients with celiac disease. *Inflammatory Bowel Diseases, 11*, 528–523.

29. Camarca, A., Anderson, R. P., Mamone, G., Fierro, O., Facchiano, A., Costantini, S., et al. (2009). Intestinal T cell response to gluten peptides are largely heterogeneous: Implications for a peptide-based therapy in celiac disease. *Journal of Immunology, 182*, 4158–4166.
30. Vojdani, J.B. Pangborn, E. Vojdani, E.L. (2003) Cooper Infections, toxic chemicals and dietary peptides binding to lymphocyte receptors and tissue enzymes are major instigators of autoimmunity in autism. *Int. J. Immunopathol. Pharmacol, 16*, 189–199.
31. Elder, J. H., Shankar, M., Shuster, J., Theriaque, D., Burns, S., & Sherrill, L. (2006). The gluten-free, casein-free diet in autism: Results of a preliminary double blind clinical trial. *Journal of Autism and Developmental Disorders, 36*, 413–421.
32. Groom Jr., A. D. (2011) LusterCXCR3 in T cell function. *Exp. Cell Res, 317*, 620–631.
33. Hosomi, S., Oshitani, N., Kamata, N., Sogawa, M., Okazaki, H., Tanigawa, T., et al. (2011). Increased numbers of immature plasma cells in peripheral blood specifically overexpress chemokine receptor CXCR3 and CXCR4 in patients with ulcerative colitis. *Clinical & Experimental Immunology, 163*, 215–224.
34. McCarthy, D. M., & Coleman, M. (1979). Response of intestinal mucosa to gluten challenge in autistic subjects. *The Lancet, ii*, 877–878.
35. Arnold, G. L., Hyman, S. L., Mooney, R., & Kirby, R. S. (2003). Plasma amino acids profiles in children with autism: Potential risk for nutritional deficiencies. *Journal of Autism and Developmental Disorders, 33*(4), 449–454.
36. Knivsberg, A. M., Reichelt, K. L., Hoien, T., & Nodland, M. (2002). A randomized, controlled study of dietary intervention in autistic syndromes. *Nutritional Neuroscience, 5*(4), 251–261.

Food Color and Autism: A Meta-Analysis

Prabasheela Bakthavachalu, S. Meenakshi Kannan, and M. Walid Qoronfleh

Abstract Autism has been increasing dramatically since its description by Leo Kanner in 1943. The Centers for Disease Control and Prevention (CDC) in 2018 has identified 1 in 59 children (1 in 37 boys and 1 in 151 girls) has autism spectrum disorder (ASD). Autistic spectrum disorders and ADHD are complex conditions in which nutritional and environmental factors play major roles. It is important to understand how food can have an impact on their current and future health. Appealing food colors stimulate the consumption of different food products. Since 2011, it is evident that dyes are linked to harmful effects in children. Artificial dyes have neurotoxic chemicals that aggravate mental health problems. Many families with autistic children avoid food dyes in their diet in order to avoid behavioral issues. A study reported that there is a correlation between yellow dye and sleep disturbance. Food colors Blue 1 and 2, Green 3, Red 3, Yellow 5 and 6, Citrus Red 2, and Red 40 can trigger many behaviors in most kids. Artificial food color usually contains petroleum and is manufactured in a chemical process that includes formaldehyde, aniline, hydroxides, and sulfuric acids. Most impurities in the food color are in the form of salts or acids. Sometimes lead, arsenic, and mercury may be present as impurities. The U.S. FDA is yet to study the effects of synthetic dyes on behavior in children. A study conducted at Southampton University in England found a link between food dyes and hyperactive behavior in children. The research does not prove that food coloring actually causes autism spectrum disorder, but there seems to be a link. This chapter attempts to provide a broad review of the

P. Bakthavachalu (✉)
Department of Biotechnology, Aarupadai Veedu Institute of Technology, Kancheepuram, Tamil Nadu, India
e-mail: prabasheela@avit.ac.in

S. M. Kannan
Department of Biotechnology, D.G. Vaishnav College, Chennai, Tamil Nadu, India

M. W. Qoronfleh
Research & Policy Department, World Innovation Summit for Health (WISH), Qatar Foundation, Doha, Qatar
e-mail: wqoronfleh@qf.org.qa

M. M. Essa, M. W. Qoronfleh (eds.), *Personalized Food Intervention and Therapy for Autism Spectrum Disorder Management*, Advances in Neurobiology 24,
https://doi.org/10.1007/978-3-030-30402-7_15

available literature on food color and the epidemiology, etiology, prevention, and treatment of autistic spectrum disorder.

Keywords ASD · ADHD · Artificial coloring · Natural coloring · Color additives · Autism · Food colors · Food refusal · Hyperactive · Zinc deficiency · Mercury

1 Introduction

Autism is a pervasive developmental disorders (PDDs) that appears in the first 3 years of life and affects brain development impacting social and communication skills [1]. Some investigators expand the nature of autism to that of a multisystem metabolic disease, not just a brain disorder [2]. Neuronal plasticity is crucial for the creation and storage of long-term memory. Abnormal neuronal plasticity has been implicated in mental retardation and autism [3, 4]. Recent estimates suggest that 31% of children with autism spectrum disorder (ASD) also meet diagnostic criteria for attention deficit/hyperactivity disorder (ADHD) and another 24% of children with ASD exhibit subthreshold clinical symptoms for ADHD [5]. Autistic children have a ten times the number of hyperactive mast cells in most tissues. Eating and feeding is a common concern for the parents of young children. Parents of children with ASDs report many challenges with children's daily activities, behavior, and communication. This may be due to the limited number of specialists dealing with eating and feeding disorders.

2 Association Between Food Color and Autism

Food coloring, or color additive, is a dye, pigment, or substance that imparts color when added to food or drink. They are available in different forms like powders, liquids, gels, or pastes. There are two types of approved color additives—dyes and lakes. Dyes are water-soluble and usually come in the form of powders, granules, or liquids. Lakes are not water-soluble. They are found in products containing fats and oil. Some food colors are synthetically produced. Examples of these color additives include FD&C Blue Nos. 1 and 2, FD&C Green No. 3, and FD&C Red No. 40. Other food colorings come from pigments of vegetables, minerals, or animals. Examples of these natural additives include beta-carotene, grape skin extract, caramel color, and saffron. Many color additives had never been tested for toxicity or other diverse effects. Many synthesized dyes were easier and less costly to produce and were superior in coloring properties when compared to naturally derived alternatives. Some synthetic food colorants are diazo dyes, while naturally derived colors are not required to be certified by the number of regulatory bodies (including the

U.S. FDA); they still need to be approved for use in that country. For some popular natural food coloring additives and further details, see Table 1.

One of the more current controversies in the field of artificial food colors (AFCs) is concerned with their effect on children's behavior. The UK has had a voluntary ban on six food colors since 2008 as research funded by the UK Food Standard Agency (FSA) suggested that consuming mixes of the food colors and preservative sodium benzoate could increase hyperactivity in some children. The artificial colors were tartrazine (102), quinoline yellow (104), sunset yellow FCF (110), azorubine or carmoisine (122), ponceau 4R (124), and allura red AC (129) [6]. For some popular artificial food coloring additives and further details, see Table 2. These colors are used in a wide range of foods that tend to be brightly colored, including soft drinks, sweets, cakes, and ice cream. There is some evidence [7] that hyperactive children have more signs of allergy to a wide range of food than normal children, and removing allergens leads to small but significant improvements in behavior. But this kind of an individualized approach warrants further trials. Although the idea that food allergies or hypersensitivities lead to behavior and learning problems dates back to the 1920s [8], until the1970s, a specific hypothesis regarding this relationship had not been developed.

A 2012 meta-analysis [9] reported that color additives had an effect on hyperactive behavior in children, with a small subset showing more extreme behavior than others. The study concluded that the companies typically add artificial colors to make their products look more appetizing. The chemicals Yellow Nos. 5 and 6 have been in use since the early 1900s, and the FDA approved them for use in 1969 and 1986, respectively. According to the FDA, Yellow No. 5 could cause an allergic reaction for 1 out of every 10,000 people. The amount of dye the FDA has deemed acceptable for daily intake, or ADI, is 5 mg/kg of body weight per day (mg/kg bw/day) for Yellow No. 5 and 3.75 mg/kg bw/day for Yellow No. 6. In 2015 Stevens et al. [10] worked on a recommended amount of dye in servings of processed foods and found that Kraft Macaroni & Cheese contained 17.6 mg of Yellow Nos. 5 or 6 per one-cup serving. Because the chemicals are so similar in color, and thus difficult to tell apart in measurements, the researchers chose the dye that allowed the highest concentration. For a child weighing 30 kg (about 65 pounds), this translates to 0.59 mg/kg bw per serving.

Dr. Benjamin Feingold [11] proposed that pediatric hyperactivity and learning problems were due to certain foods and food additives. A study was conducted on 153 3-year-olds and 144 8/9-year-old children. For these children, the challenge drink contained sodium benzoate and a placebo mix was given. The main outcome was determined by using aggregated *z*-scores measured by observed behaviors and ratings by teachers and parents and a computerized test of attention for 8/9-year-old children in the form of global hyperactivity aggregate (GHA). According to the results, there was increased hyperactivity in 3-year-old and 8/9-year-old typical children due to diet containing artificial colors or a sodium benzoate preservative (or both) [12].

Table 1 Selected, popular natural food coloring additives

Color name	Color number	Color description	Approval USA/EU	Chemical structure	Comment
Turmeric "curcumin"	E100	Yellow-orange	Universal	Keto form 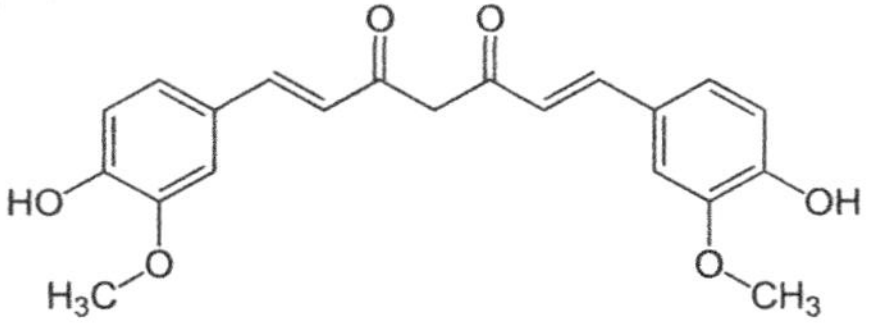 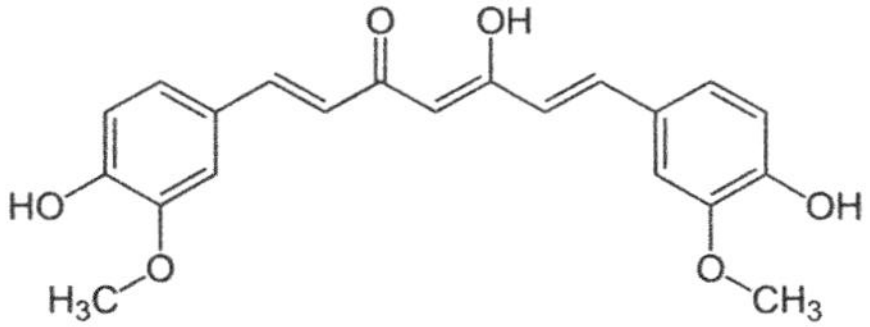 Enol form	Source: Turmeric plant root Use: Beverages, condiments, jams, jellies, marmalades, confectionery, dairy products, fish products, dietary supplements, processed meats, and vegetables
Riboflavins	E101	Yellow-orange	Universal	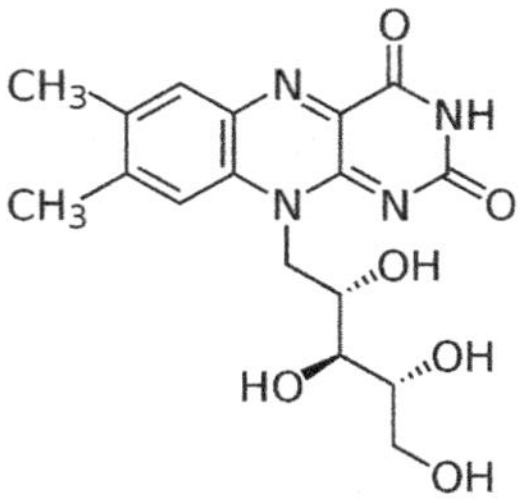	Source: Yeast extract, liver and kidney, wheat bran, eggs, meat, milk, and cheese Use: Beverages, processed meats, condiments, breakfast cereals, dairy products, fruit products, energy drinks, and dietary supplements

Carmine or carminic acid (natural red 4)	E120	Red/crimson	Prohibited by certain religions	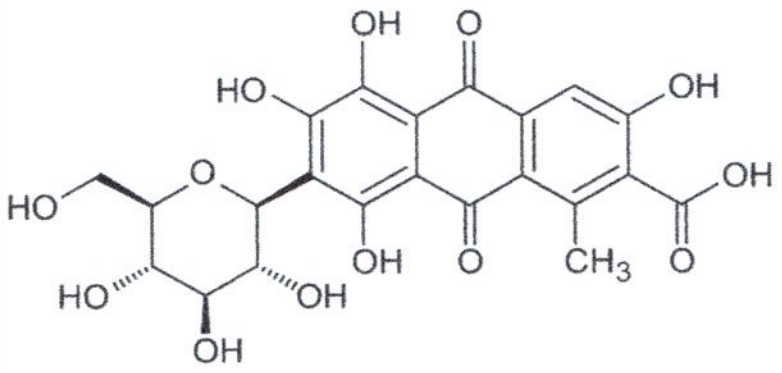	Source: The body and eggs of the cochineal insect Use: Cheeses, beverages, breakfast cereals, jams, jellies, marmalades, processed meats, fabric dye, insect repellent, cosmetics, and pharmaceuticals
Chlorophyll and chlorophyllin (Cu complex)	E140, E141	Green to olive green	E140, EU approved but is not a listed food color in the USA E141, both	Chlorophyll a universal structure 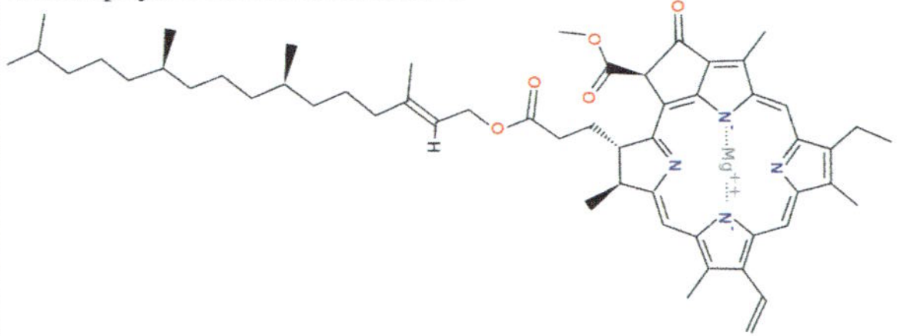 Chlorophyllin (Cu complex) 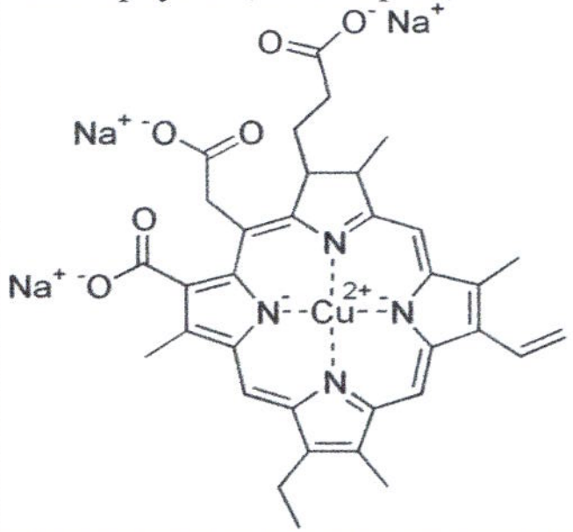	Source: E140, extracted from nettles, grass, and alfalfa E141, derived from green, leafy vegetables Use: E140, pasta, absinthe, cheeses, preserved vegetables, jams, jellies, and marmalades E141, cheeses, ice cream, soups, preserved vegetables, and fruits

(continued)

Table 1 (continued)

Color name	Color number	Color description	Approval USA/EU	Chemical structure	Comment
Caramel	E150a–d	Dark brown	150a plain and universal Others heat treated in the presence of alkali, acid, ammonia/sulfite, or combination thereof	Browning of food grade sugar by heat Caramel color divided into four classes depending on the food grade reactants used in its manufacturing • Class I is plain caramel color • Class II is caustic sulfite process caramel color • Class III is ammonia process caramel color • Class IV is sulfite ammonia process caramel color 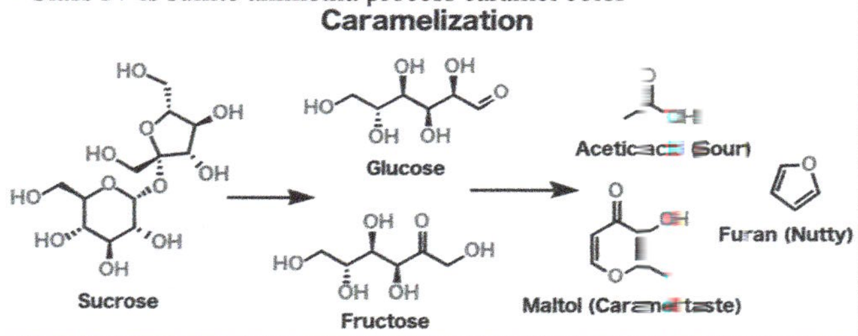	Source: Made from caramelized sugar Use: Bread, cakes, confectionery, biscuits, preserved vegetables, fish and shellfish spreads, jams, jellies, marmalades, pickles, soft drinks, vinegar, alcoholic drinks, cheeses, breakfast cereals, and processed meats
Vegetable carbon	E153	Black	USA No	Vegetable carbon (also called “vegetable black”) is a form of finely divided carbon produced by steam activation of carbonized raw material of vegetable origin	Source: Created by burning vegetable matter Use: Cheeses, concentrated fruit juices, jams, jellies, marmalades, and liquorice
Carotenoids	E160 a–f, E161 a–h, E164				

α, β, γ —carotene	E161a	Yellow-orange to brown	USA No	β-Carotene 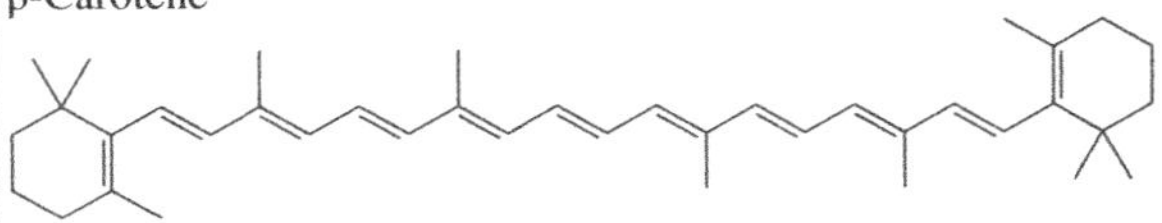	Source: Carrots, green-leafed vegetables, and tomatoes Use: Butter and soft margarines, cheeses, breakfast cereals, jams, jellies, marmalades, processed meats, preserved vegetables, coffee sponge cakes, and milk products
Annatto (bixin, norbixin)	E160b	Reddish-orange	Universal	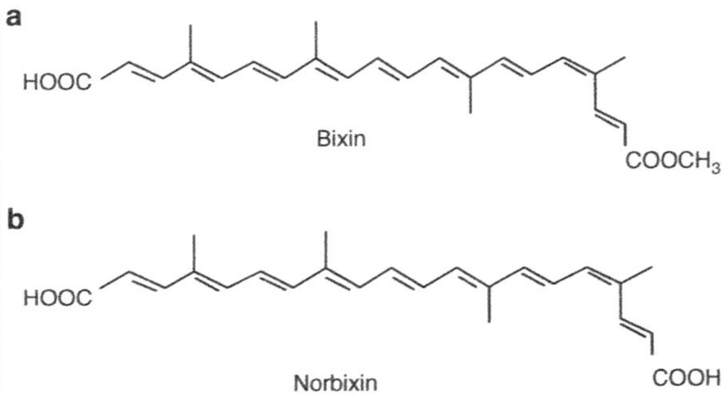	Source: The seed coat of the fruit of the Achiote shrub Use: Cheeses (e.g., cheddar, gouda and brie), margarine, butter, rice, custard powder, ice cream, cream fillings and toppings, smoked fish, breakfast cereals, liqueurs

(continued)

Table 1 (continued)

Color name	Color number	Color description	Approval USA/EU	Chemical structure	Comment
Paprika (oleoresin, capsanthin, capsorubin)	E160c	Reddish-orange	Not permitted in Canada and Australia	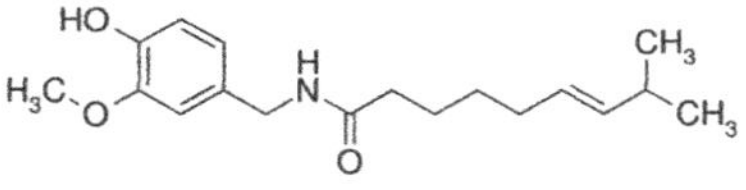 Capsaicin 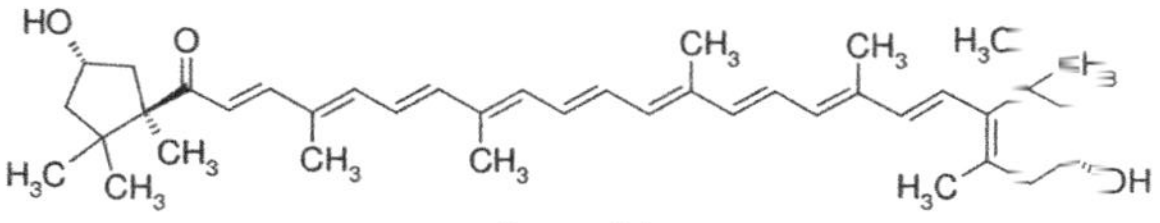 Capsanthin 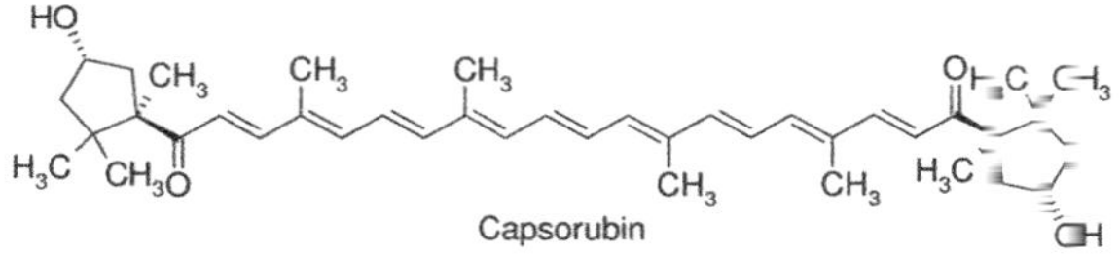 Capsorubin	Source: The fruit pod and seeds of the red pepper Use: Cheese slices, breakfast cereals, jams, jellies, marmalades, chicken pies, orange juices, spice mixtures, sauces, sweets, and processed meats
Lycopene	E160d	Bright to deep red	Not permitted in Australia	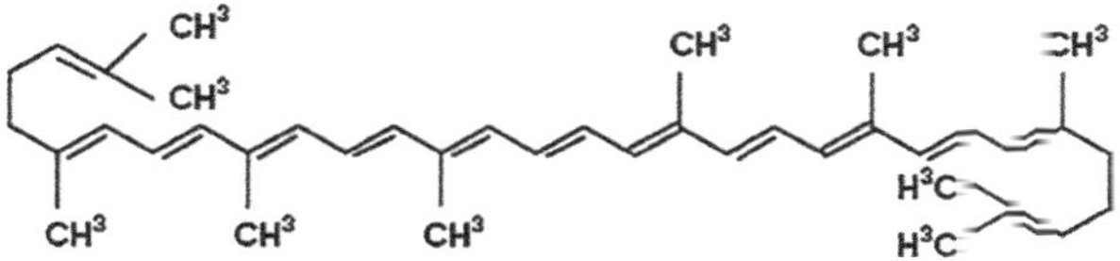	Source: Tomatoes and other red fruits and vegetables, such as red carrots, watermelons, and papayas, but not strawberries or cherries Use: Tomato-based foods such as soups, sauces, ketchup, jams, jellies, and marmalades

Beta-apo-8′-carotenal (C 30)	E160e	Orange-red to yellow	Both	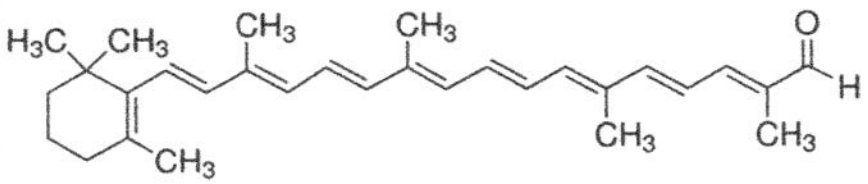	Source: Made from carotene or isolated from plants Use: Processed cheese
Lutein	E161b	Orange-red to yellow	USA No	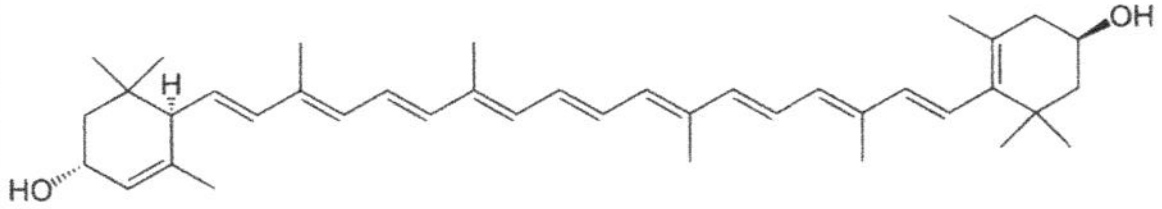	Source: Made commercially from grass, nettles or marigolds Use: Added to chicken feed to ensure the yellow color of egg yolks and chicken skin
Canthaxanthin	E161g	Violet	Not permitted in Australia and New Zealand	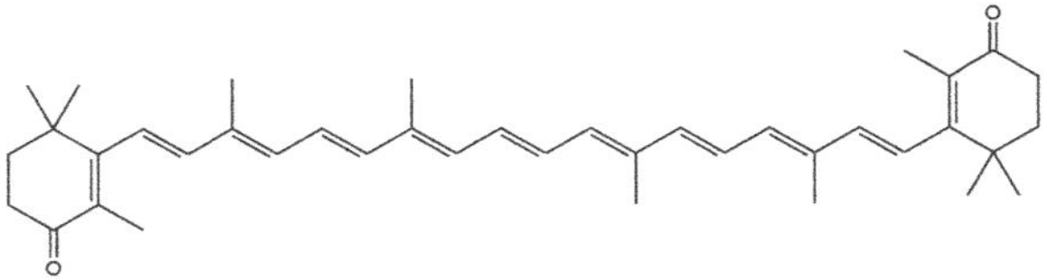	Source: Made commercially from beta-carotene Use: Restricted to Saucisse de Strasbourg, animal feeding stuff
Betanin	E162	Red	Universal	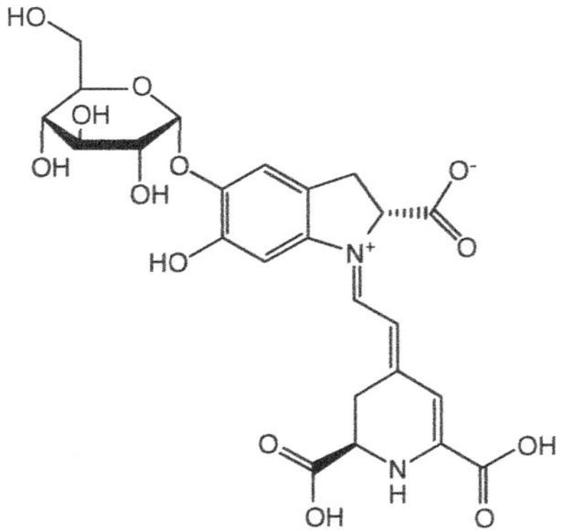	Source: Beets Use: Burgers, desserts, ice cream, jams, jellies, liquorice, oxtail soup, sauces, and sweets

(continued)

Table 1 (continued)

Color name	Color number	Color description	Approval USA/EU	Chemical structure	Comment
Anthocyanins	E163	pH-dependent May appear red, purple, blue	Universal	Basic structure and examples 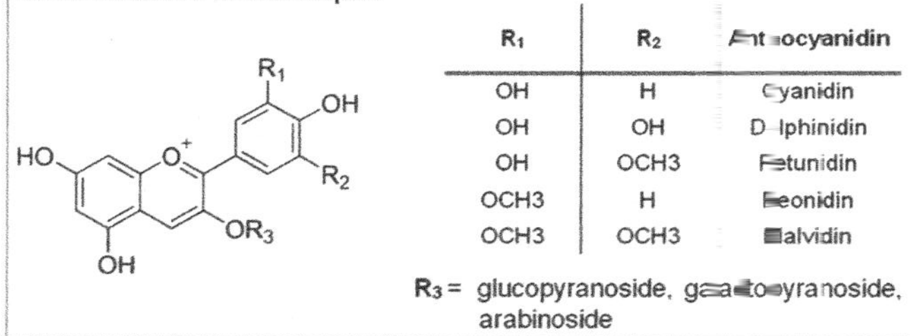	Source: Grape skins or red cabbage Use: Black cherry yoghurt, dairy products, glacé cherries, ice cream, jellies, pickles, tomato, carrot or vegetable soups, soft drinks, and sweets
Saffron (crocetin or crocin) chemically similar to carotenoid	E164	Yellow-orange-red	USA, Australia, and New Zealand approved	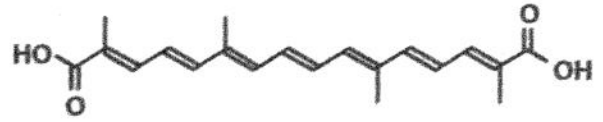 Crocetin 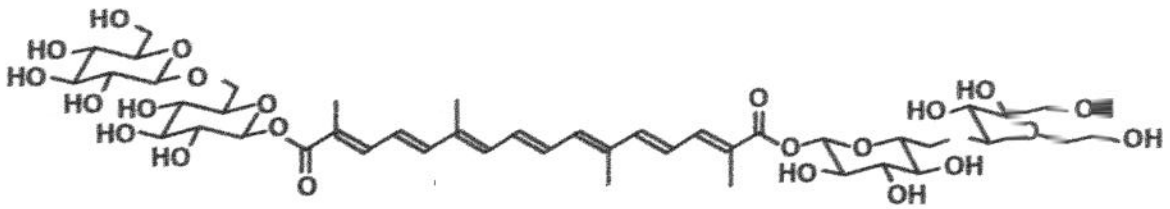 Crocin 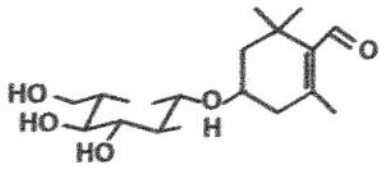 Picrocrocin 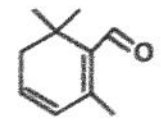 Safranal	Source: The stigmas of the flower of the crocus bulb Use: Beverages, cosmetics, and pharmaceuticals

Table 2 Selected, popular artificial food coloring additives

Color name	Color number	FD&C equivalent[a,b]	Color description	Approval USA/EU[a,c]	Chemical structure[d]	Comment
Tartrazine	E102	Yellow No. 5	Yellcw shade	Both	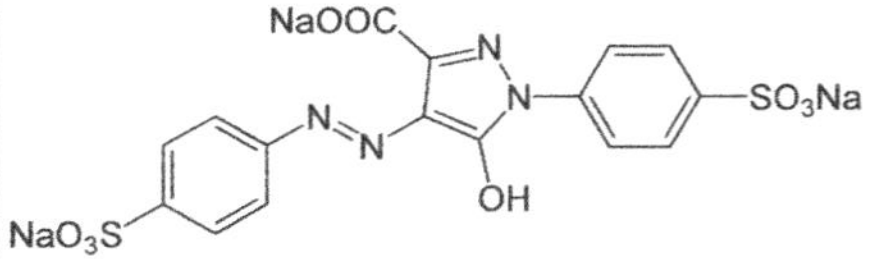	Source: An azo dye that is produced from coal tar Use: Confectionery, soft drinks, cereals, soups, sauces, preserves, processed peas, cosmetics, and pharmaceuticals People who suffer asthma may also show an allergic reaction to it Linked to ADHD based on the Southampton study[e]

(continued)

Table 2 (continued)

Color name	Color number	FD&C equivalent[a,b]	Color description	Approval USA/EU[a,c]	Chemical structure[d]	Comment
Quinoline yellow	E104		Green-yellow	EU approved. Undergoing a voluntary phaseout in the UK. Not permitted in Australia	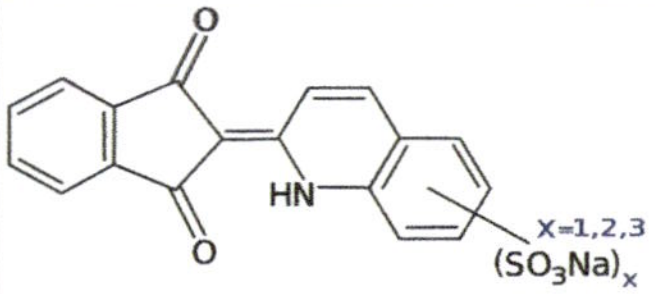	Source: A quinoline dye produced from coal tar Use: Jams, jellies, marmalades, soft drinks, confectionery, and smoked haddock Linked to ADHD based on the Southampton study[e]
Yellow 2G	E107		Yellow		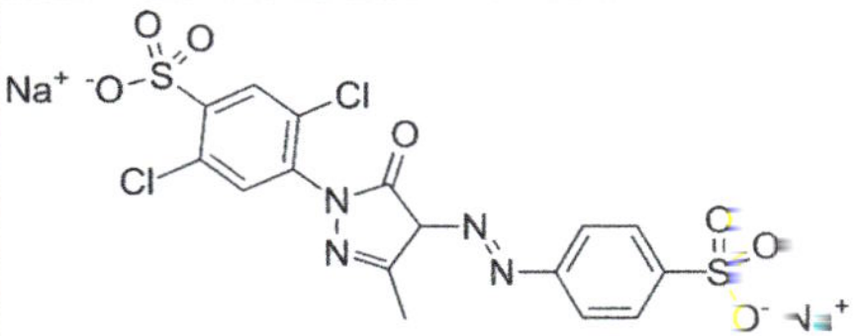	Source: An azo dye that is produced from coal tar and petroleum Use: Soft drinks, confectionary, cordials, pickles

Sunset yellow FCF	E110	Yellow No. 6	Orange shade	Approved in the EU. Banned in Norway. Products in the EU require warnings and its use is being phased out. Approved in the USA	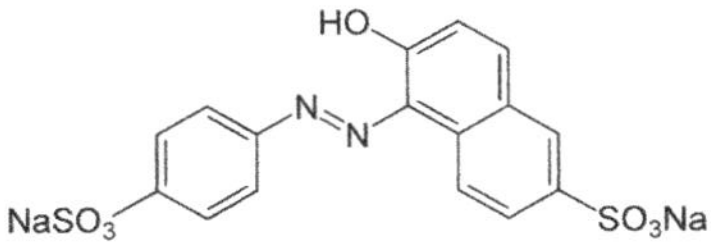	Source: An azo dye that is produced from coal tar and petroleum Use: Soft drinks, confectionery, jams, jellies, marmalades, soups, condiments, cordials, processed meats, cosmetics, and pharmaceuticals Some people show allergic reactions to it (rashes, swelling, vomiting) Linked to ADHD based on the Southampton study[e]

(continued)

Table 2 (continued)

Color name	Color number	FD&C equivalent[a,b]	Color description	Approval USA/EU[a,c]	Chemical structure[d]	Comment
Citrus red #2	E121	Red No. 2	Dark red-orange shade	Approved in the USA for limited application	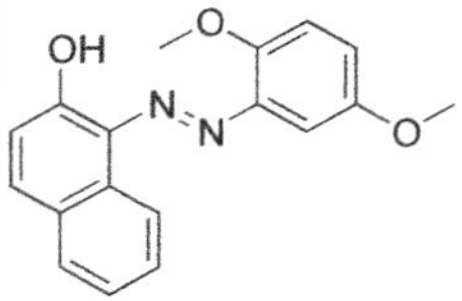	Source: An azo dye that is produced from coal tar and petroleum Use: Only for use in coloring the skin of oranges
Azorubine (carmoisine)	E122		Red		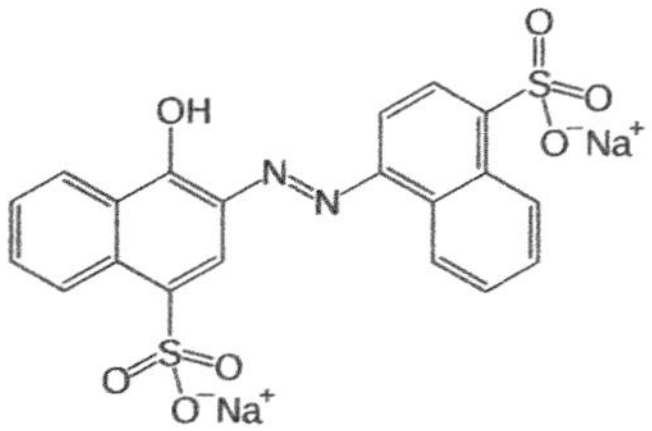	Source: An azo dye that is produced from coal tar and petroleum Use: Fabric dye, insect repellent, mouthwash, cosmetics and pharmaceuticals; confectionary, marzipan, and jelly crystals Sensitive people and asthmatics may show allergic reactions Linked to ADHD based on the Southampton study[e]

Amaranth	E123	Red No. 2	Purplish-red	Banned in the USA. Still used in Australia	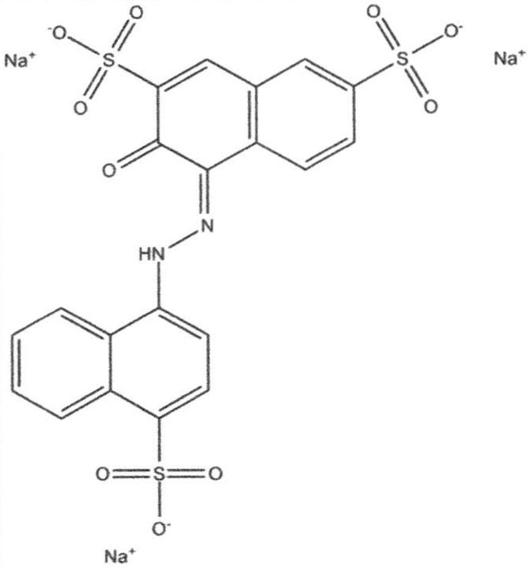	Source: An azo dye that is produced from petroleum by-products Use: Beverages, cake mixes, fruit-flavored fillings, jelly crystals alcoholic drinks, and fish roe

(continued)

Table 2 (continued)

Color name	Color number	FD&C equivalent[a,b]	Color description	Approval USA/EU[a,c]	Chemical structure[d]	Comment
Ponceau 4R (cochineal red A, brilliant scarlet 4R)	E124		Red	EU approved, must carry warning label	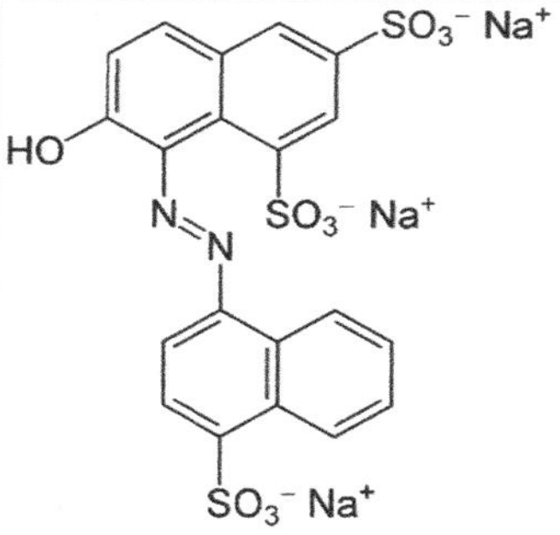	Source: An azo dye that is produced from coal tar and petroleum Use: Beverages, jams, jellies, marmalades, and processed meats Asthmatics and people with aspirin sensitivity may show allergic reactions Linked to ADHD based on the Southampton study[e]

Erythrosine	E127	Red No. 3	Pink-red shade	Both Permitted in the USA in food and ingested drugs. However, it is not permitted in cosmetics and external drugs	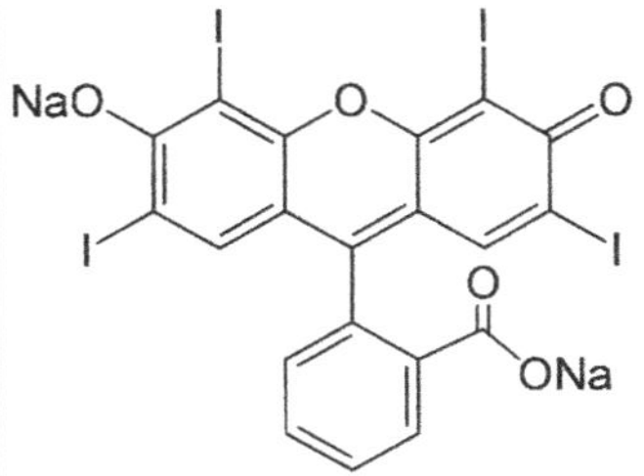	Source: A fluorone dye, which contains iodine Use: Used in the EU in cherries (e.g., candied, cocktail, and bigarreaux), canned, custard mix, biscuits only May lead to hyperthyroidism or cause sensitivity to light
Allura red AC	E129	Red No. 40	Red shade	Approved in the EU. Banned in Switzerland. Undergoing a voluntary phaseout in the UK. Approved in the USA	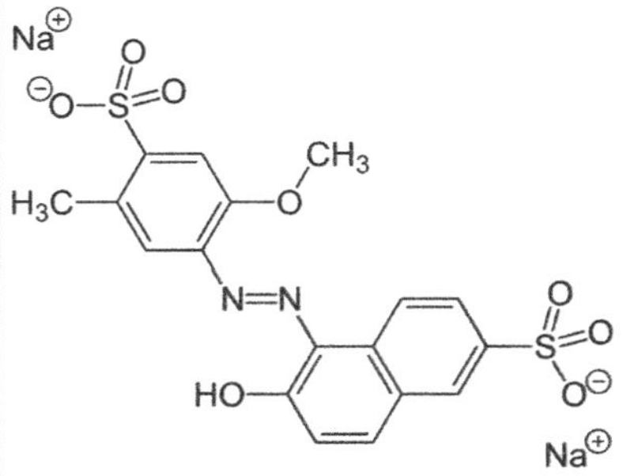	Source: An azo dye that is produced from coal tar and petroleum Use: Soft drinks, biscuits, cake mixes, fruit-flavored fillings, and processed meats Linked to ADHD based on the Southampton study[e]

(continued)

Table 2 (continued)

Color name	Color number	FD&C equivalent[a,b]	Color description	Approval USA/EU[a,c]	Chemical structure[d]	Comment
Patent blue V	E131		Dark blue	EU approved Not permitted in USA, Canada, Australia, and New Zealand	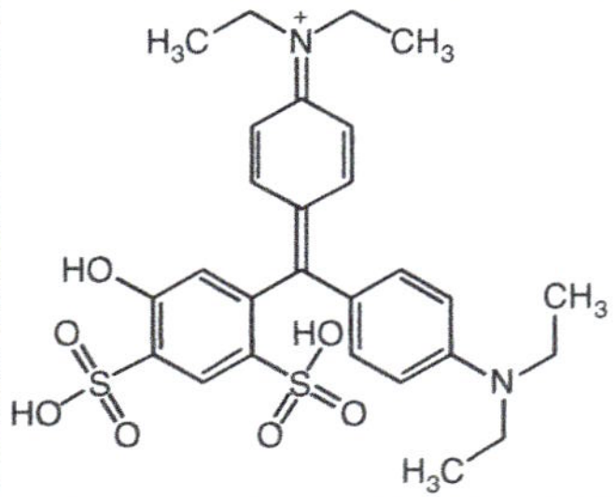	Source: Produced from coal tar Use: Outside medical use in food used in scotch eggs and jelly sweets
Indigo carmine (indigotine)	E132	Blue No. 2	Indigo	Universal	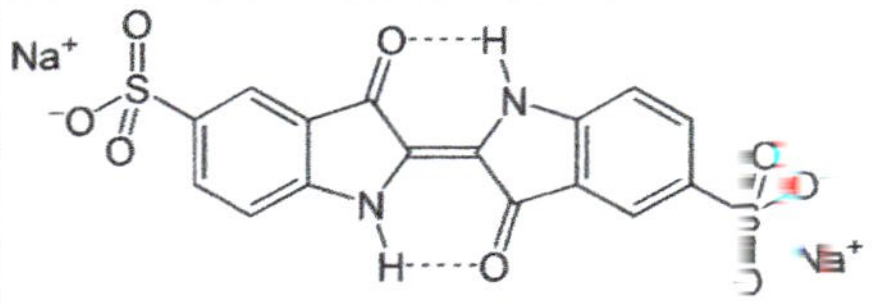	Source: Produced from coal tar Use: Ice cream, sweets, baked goods, confectionary, and biscuits People with allergies should avoid it. May cause nausea, vomiting, high blood pressure, skin rashes, breathing problems, and other allergic reactions Linked to ADHD and food allergies

Brilliant blue FCF	E133	Blue No. 1	Blue shade	Both	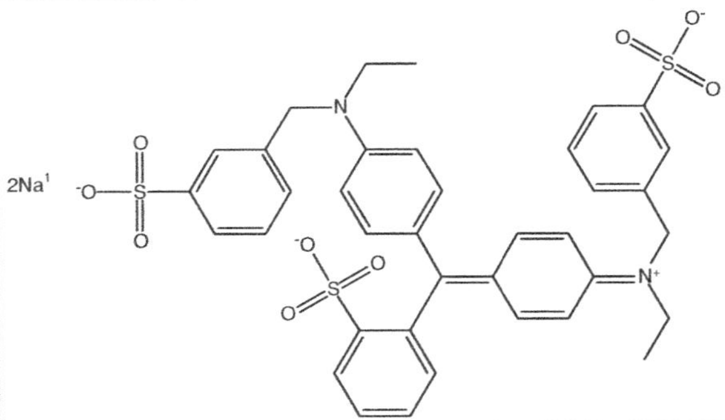	Source: Produced from petroleum Use: Ice cream, canned processed peas, packet soups, bottled food colorings, dairy products, and sweets
Green S	E142		Green	EU approved	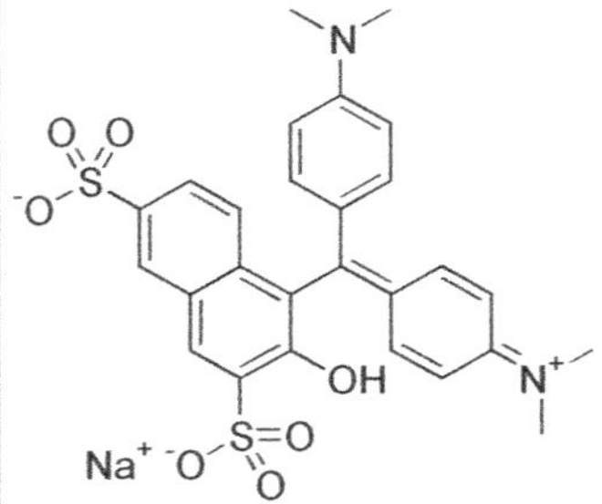	Source: Produced from coal tar Use: Desserts, processed peas, gravy granules, ice cream, mint sauce, sweets, cake mixes, jams, jellies, and marmalades

(continued)

Table 2 (continued)

Color name	Color number	FD&C equivalent[a,b]	Color description	Approval USA/EU[a,c]	Chemical structure[d]	Comment
Fast green FCF	E143	Green No. 3	Sea green-turquoise shade	USA approved Least used from the seven main FDA approved dyes Prohibited in EU	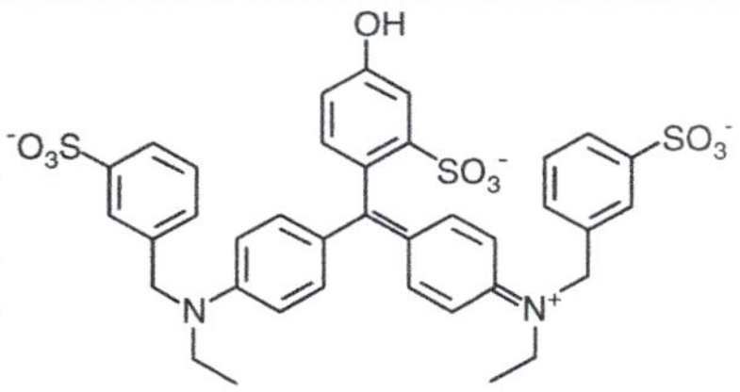	Source: Organic synthesis Use: Green peas and other vegetables, jellies, sauces, fish, desserts, and dry bakery mixes Linked to testicular and bladder cancers and tumors in lab animals and causes irritation of the gastrointestinal tract

The following are also not permitted in the USA {E180 Lithol rubine BK, E155 Chocolate Brown HT, and E151 Brilliant Black BN} though approved in the EU by European Food Safety Authority (EFSA). http://www.efsa.europa.eu/

[a]https://www.fda.gov/forindustry/coloradditives/coloradditiveinventories/ucm106626.htm

[b]The following FD&C color additives are no longer authorized or restricted by the FDA [Reds 1, 2, 3, 4, and 32; Oranges 1 and 2; Yellows 1, 2, and 3; Greens 1 and 2; and Violet 1]

[c]European Food Safety Authority (EFSA). http://www.efsa.europa.eu/

[d]PubChem. https://pubchem.ncbi.nlm.nih.gov/

[e]Southampton study—Southampton University research team investigated the effect of a mixture of six food dyes [Tartrazine E102, Quinoline Yellow E104, Sunset Yellow E110, Carmoisine E122, Ponceau 4R E124, and Allura Red E129 (dubbed the "Southampton 6")] on hyperactive behavior (ADHD). https://www.ncbi.nlm.nih.gov/pubmed/17825405

3 Effect of Food Color on Zinc Metabolism in Autism Spectrum Disorder

Zinc is an essential trace element that plays an important role in nucleic acid/protein synthesis, cell replication, and tissue growth and repair, especially in pregnant women and infants. In fact, zinc ions function as the active centers in more than 300 kinds of enzymes, and about 10% in the total gene-coded proteins [13–16]. Within the brain, especially in the hippocampus, zinc is co-stored with glutamate in presynaptic vesicles in the excitatory neuron terminal, is released from them, and controls the activity of excitatory glutamate receptors on the postsynaptic excitable membrane [17, 18]. Thus, zinc deficiency is known to be associated not only with various pathological conditions, including dyspepsia, delayed wound healing, impaired immunity, and retarded growth, but also neurodegenerative diseases and neurodevelopment disorders [19–23]. Recent investigation has reported that many infants with autistic disorder suffer from marginal to severe zinc deficiency, suggesting considerable relationship between infantile zinc deficiencies with autism [24].

Nutritional deficiencies and mercury exposure have been shown to alter neuronal functions and increase oxidative stress among children with autism. Consumption of certain artificial food color additives has also been shown to lead to zinc elimination. Dietary zinc is essential for maintaining the metabolic process required for mercury elimination. Dietary transcription factors such as zinc insufficiency or deficiency or through exposure to toxic substances found in our environment such as the heavy metals mercury and copper alter gene expression leading to adverse effects on human neurodevelopment [25]. Elimination of heavy metals requires the expression of the metallothionein (MT) gene which synthesizes the Zn-dependent metal-binding protein metallothionein [26]. With dietary zinc (Zn) loss, the metabolic processes required to eliminate heavy metals are impaired in children with autism [25]. Mercury (Hg) and other specific heavy metals, including lead (Pb), copper (Cu), cadmium (Cd), silver (Ag), and bismuth (Bi), are capable of displacing the Zn atom in the MT protein molecule [26]. If diet is deficient in Zn or the absorption of Zn is impaired, then the body may not produce enough MT protein to carry and excrete the heavy metal load [27, 28]. Children with autism may be Zn deficient and often have MT dysfunction [29–31]. Because of their diminished capacity to excrete toxic heavy metals, the severity of their condition is associated with toxic metal burden.

One study conducted on 1967 children with autistic disorders (1553 males and 414 females) analyzed hair zinc concentrations which showed considerable association between autism and zinc deficiency. A histogram of hair zinc concentration was nonsymmetric with tailing in the lower range, and 584 subjects were found to have zinc concentrations lower than two standard deviations of its reference range (86.3–193 ppm). The incidence rate of zinc deficiency in the infant group aged 0–3-year-old was estimated at 43.5% in male and 52.5% in female. The lowest zinc concentration of 10.7 ppm was detected in a 2-year-old boy, corresponding to about

1/12th of the control mean level. These findings suggest that infantile zinc deficiency may epigenetically contribute to the pathogenesis of ASD, and the nutritional approach may yield a novel hope for treatment and prevention [31].

4 Preventive Measures for Autism Spectral Disorder

It is clear that some food can trigger behavior problems, but certain foods can in fact improve these issues. The best food for people with autism is a diet including meat, nuts, and beans for protein. Complex carbohydrates, found in fresh fruits and vegetables, are also recommended. Omega-3 fatty acids can help with behavior problems, and these can be sourced from salmon, tuna, walnuts, and olive oil. Including these foods in the diet while avoiding the additives that commonly cause problem behaviors can make a significant difference for many children with autism. In addition, dietary micronutrients are also required to enhance neurological development and its function.

5 Conclusion

Popular concern regarding adverse neurobehavioral effects of food additives has recently implicated artificial colors as etiologic factors in childhood hyperactivity. AFCs are purely cosmetic, so its removal would not come at an economic or public health cost. Thus, it would be important to distinguish AFC effects from preservative effects. ASD remains poorly understood due to an unknown etiology, and there is also no particular treatment for this disorder. The prevalence of ASD is continuously increasing and has subsequently made medical management a challenging task. Food additives given in very large doses may act as a pharmacological trigger among a small percentage of children with autism. So, the FDA should insist that manufacturers include a label saying "artificial colors could cause hyperactivity in some children" to keep the public informed. Future research in this area needs to investigate these children further and also describe the mode of action of food color additives.

References

1. Volkmar, F. R., & Pauls, D. (2003). Autism. *Lancet, 362*, 1131–1141.
2. Jepson, B. (2007). *Changing the course of autism* (pp. 42–46). Boulder, CO: Sentient Publications Ed..
3. Lee, Y. S., & Silva, A. J. (2009). The molecular and cellular biology of enhancing cognition. *Nature Reviews. Neuroscience, 10*(2), 126–140.

4. Tsanov, M., & Manahan-Vaughn, D. (2008). Synaptic plasticity from visual cortex to hippocampus: System integration in spatial information processing. *The Neuroscientist, 14*(6), 584–597.
5. Yerys, B. E., Wallace, G. L., Sokoloff, J. L., Shook, D. A., James, J. D., & Kenworthy, L. (2009). Attention deficit/hyperactivity disorder symptoms moderate cognition and behavior in children with autism spectrum disorders. *Autism Research, 2*, 322–333.
6. Eur-Lex. Access to European Union law. Regulation (EC) No. 1333/2008 of the European Parliament and the Council of 16, December 2008 on Food Additives.
7. Trites, R. L., Tryphonas, H., & Ferguson, H. B. (1980). Diet treatment for hyperactive children with food allergies. In R. M. Knights & D. Bakker (Eds.), *Treatment of hyperactive and learning disordered children*. Baltimore, MD: University Park Press.
8. Nigg, J. T., Lewis, K., Edinger, T., & Falk, M. (2012). Meta-analysis of attention deficit/hyperactivity disorder or attention-deficit/hyperactivity disorder symptoms, restriction diet, and synthetic food color additives. *Journal of the American Academy of Child and Adolescent Psychiatry, 51*, 86–97.
9. Nigg, J. T., Lewis, K., Tracy Edinger, N. D., & Falk, M. (2012). Meta-analysis of attention-deficit/hyperactivity disorder or attention-deficit/hyperactivity disorder symptoms, restriction diet, and synthetic food color additives. *Journal of the American Academy of Child and Adolescent Psychiatry, 51*(1), 86–97.
10. Steven, L. J., Burgess, J. R., Stochelski, M. A., & Kuczek, T. (2015). Amounts of artificial food dyes and added sugars in foods and sweets commonly consumed by children. *Clinical Pediatrics (Phila), 54*(4), 309–321.
11. Feingold, B. F. (1973, June 24–28). *Adverse reactions to food additives*. Presented at The American Medical Association Annual Meeting, Chicago, IL.
12. McCann, D., Barrett, A., Cooper, A., Crumpler, D., Dalen, L., Grimshaw, K., et al. (2007). Food additives and hyperactive behaviour in 3-year-old and 8/9-year-old children in the community: A randomised, double-blinded, placebo controlled trial. *The Lancet, 370*, 1560–1567.
13. Dufault, R., Lukiw, W. J., Crider, R., Schnoll, R., Wallinga, D., & Deth, R. (2012). A macroepigenetic approach to identify factors responsible for the autism epidemic in the United States. *Clinical Epigenetics, 4*, 6. https://doi.org/10.1186/1868-7083-4-6.40
14. Grabrucker, A. M. (2014). A role for synaptic zinc in ProSAP/SHANK PSD scaffold malformation in autism spectrum disorders. *Developmental Neurobiology, 74*(2), 136–146. https://doi.org/10.1002/dneu.22089
15. Fukada, T., Yamasaki, S., Nishida, K., Murakami, M., & Hirano, T. (2011). Zinc homeostasis and signalling in health and diseases: Zinc signalling. *Journal of Biological Inorganic Chemistry, 16*, 1123–1134.
16. Andreini, C., Banci, L., Bertini, I., & Rosato, A. (2006). Counting the zinc-proteins encoded in the human genome. *Journal of Proteome Research, 5*, 196–201.
17. Takeda, A. (2000). Movement of zinc and its functional significance in the brain. *Brain Research Reviews, 34*, 137–148 [*International Journal of Environmental Research and Public Health, 10*, 6041 (2013)].
18. Takeda, A., Nakamura, M., Fujii, H., & Tamano, H. (2013). Synaptic Zn (2+) homeostasis and its significance. *Metallomics, 5*, 417–423.
19. Prasad, A. S. (2009). Impact of the discovery of human zinc deficiency on health. *Journal of the American College of Nutrition, 28*, 257–265.
20. Arnold, L. E., & di Silvestro, R. A. (2005). Zinc in attention-deficit/hyperactivity disorder. *Journal of Child and Adolescent Psychopharmacology, 15*, 619–627.
21. di Girolamo, A. M., & Raminez-Zea, M. (2009). Role of zinc in maternal and child mental health. *The American Journal of Clinical Nutrition, 89*, S940–S945.
22. Scheplyagina, L. A. (2005). Impact of the mother's zinc deficiency on the woman's and newborn's health status. *Journal of Trace Elements in Medicine and Biology, 19*, 29–35.
23. Plum, L. M., Rink, L., & Haase, H. (2010). The essential toxin: Impact of zinc on human health. *International Journal of Environmental Research and Public Health, 7*, 1342–1365.

24. Yasuda, H., Yoshida, K., Yasuda, Y., & Tsutsui, T. (2011). Infantile zinc deficiency: Association with autism spectrum disorders. *Scientific Reports, 1*, 1–5.
25. Dufault, R., Schnoll, R., Lukiw, W. J., LeBlanc, B., Cornett, C., Patrick, L., et al. (2009). Mercury exposure, nutritional deficiencies and metabolic disruptions may affect learning in children. *Behavioral and Brain Functions, 5*, 44.
26. Coyle, P., Philcox, J. C., Carey, L. C., & Rofe, A. M. (2002). Metallothionein: The multipurpose protein. *Cellular and Molecular Life Sciences, 59*, 627–647.
27. Shankar, A. H., & Prasad, A. S. (1998). Zinc and immune function: The biological basis of altered resistance to infection. *The American Journal of Clinical Nutrition, 68*, 447S–463S.
28. Szczurek, E. I., Bjornsson, C. S., & Taylor, C. G. (2001). Dietary zinc deficiency and repletion modulate metallothionein immunolocalization and concentration in small intestine and liver of rats. *The Journal of Nutrition, 131*, 2132–2138.
29. Yorbik, O., Akay, C., Sayal, A., Cansever, A., Sohmen, T., & Cavdar, A. O. (2004). Zinc status in autistic children. *Journal of Trace Elements in Experimental Medicine, 17*, 101–107.
30. Faber, S., Zinn, G. M., Kern, J. C., & Kingston, H. M. (2009). The plasma zinc/serum copper ratio as a biomarker in children with autism spectrum disorders. *Biomarkers, 14*, 171–180.
31. Yasuda, H., Yoshida, K., Yasuda, Y., & Tsutsui, T. (2011). Infantile zinc deficiency: Association with autism spectrum disorders. *Scientific Reports, 1*, 129.

Food Selection and Preferences of Omani Autistic Children

Najma M. Al-Kindi, Yahya M. Al-Farsi, Buthaina Al-Bulushi, Amanat Ali, Syed Gauhar Alam Rizvi, and Musthafa Mohamed Essa

Abstract *Background*: Autism spectrum disorder (ASD) is a neurodevelopmental disorder characterized by core deficits in social interactions, verbal/nonverbal communication, and restricted, repetitive, and stereotyped behaviors. Children with ASD are known to have several feeding problems that are believed to affect their nutritional and health status.

Aim: The present study was designed to assess the food preferences in Omani children diagnosed with ASD compared with controls.

Methods: A case-control study was conducted in which 375 children (males and females) aged between 4 and 13 years were recruited. The sample consisted of 163 children with ASD and a control group of 212 typically developing (TD) children. For each participant, demographic, anthropometric, and medical information and information regarding dietary intakes were gathered using the food frequency questionnaire (FFQ) to assess their food preferences.

Results: The sociodemographic characteristics of caregivers were similar in the two groups, while their perceptions based on several nutritional parameters were

N. M. Al-Kindi (✉) · Y. M. Al-Farsi · S. G. A. Rizvi
Department of Family Medicine and Public Health, College of Medicine and Health Sciences, Sultan Qaboos University, Muscat, Oman
e-mail: najma@alkindy.com

B. Al-Bulushi
Department of Food Science and Nutrition, College of Agricultural and Marine Sciences, Sultan Qaboos University, Muscat, Oman

A. Ali
Department of Food Science and Nutrition, College of Agricultural and Marine Sciences, Sultan Qaboos University, Muscat, Oman

College of Agricultural and Marine Sciences, Ajax, ON, Canada

M. M. Essa
Department of Food Science and Nutrition, College of Agricultural and Marine Sciences, Sultan Qaboos University, Muscat, Oman

Ageing and Dementia Research Group, Sultan Qaboos University, Muscat, Oman

M. M. Essa, M. W. Qoronfleh (eds.), *Personalized Food Intervention and Therapy for Autism Spectrum Disorder Management*, Advances in Neurobiology 24,
https://doi.org/10.1007/978-3-030-30402-7_16

different. Children's age and body mass index (BMI) were similar in both groups, while the number of male children was higher in ASD group ($P < 0.001$). Problematic behaviors including food refusal and selectivity were significantly higher in ASD children than in TD children. Despite that, the children with ASD were found to consume mostly traditional Omani dishes.

Conclusion: This is the first study that provides information on the eating habits and nutritional intake of Omani children diagnosed with ASD. The overall findings are promising and may contribute to further understanding of food preferences in children with ASD in Oman. Such information is highly valuable for the prevention and management of nutritional deficiencies among Omani children with autism by improving their diet quality.

Keywords ASD · Food preferences · FFQ · Oman · Feeding problems

1 Introduction

1.1 Feeding Behavior in Children with Autism Spectrum Disorder

Autism spectrum disorder (ASD) refers to a group of complex neurodevelopmental disorders characterized by repetitive and impaired social communication and interaction. ASD occurs across all socioeconomic levels, in all cultures, and across all racial and ethnic groups. However, ASD occurs more commonly in boys than in girls [1]. The severity of symptoms vary from person to person. The term ASD comprises of five different categorizations which include autistic disorder, Rett's disorder, childhood disintegrative disorder, Asperger's disorder, and pervasive developmental disorders not otherwise specified (PDD-NOS). For diagnosis of any of these, a child must exhibit a "triad of impairments" in the three areas of social communication, social interaction, and imaginative understanding [2].

Early childhood is a critical nutritional stage that involves a shift from milk-based diet to other foods [3]. It is believed that dietary exposure during this intermediate stage could contribute to future food and taste preferences [3–5]. Along with this, sensory sensitivity might also play a big role in food selection [6]. From a very early age, typically developing (TD) children can express their likes and dislikes for foods through behavior and/or speech [7]. As children grow older, individual variances appear in food-related behaviors as some children are easier to feed than others. Parental feeding practices play a major role in the development of children's eating behaviors [7–11]. Selective eating disorder (SED), more commonly known as being a "fussy" or "picky eater," can be defined as eating a narrow variety of foods and a refusal to eat or taste new foods [12].

Parents of children with ASD have reported many feeding problems including food refusal, limited dietary intake, and behavior problems at meal times, with numerous case studies over the years supporting parental concerns [2, 13–17]. Repetitive and restricted behaviors are the basic feature of autism spectrum disorder, which may play a role in food selectivity. Children with ASD often resist novel experiences, which may include tasting new foods due to their hypersensitivities toward sensory properties of foods including texture and temperature [18, 19], food presentation [20, 21], or other characteristics of foods. Understanding the content and quality of food that form part of the diet of these children could aid in remediation of food selectivity and avoid few complications related to food refusal behavior. Parents often struggle with adding new foods to their child's diet, and food selectivity along with disruptive food avoidance behavior makes mealtime stressful for the entire family. If eating difficulties are not resolved at an early age, they may lead to developing age-inappropriate eating disabilities, failure to thrive, nutrient deficiencies, problematic behavioral issues, weak parent-child interactions, and many other problems [22–24] at a later stage. The studies conducted in this area however reflect inconsistency in results which could be due to conflicting terminology, an absence of clinical instruments, and limited availability of literature on management and assessment [25, 26].

Previous studies on autism from Oman mainly focused on the association of autism with malnutrition [27], heavy metals and essential minerals in hair samples of ASD children [28] as well as on the effects of suboptimal breast-feeding [27, 29]. Information regarding mealtime behavior, food preferences, and actual intake of macro- and micronutrient levels in Omani children with ASD children is scant. In order to fill existing gaps in available literature, this case-control study was conducted with the aim of assessing the food selection criteria and preferences in children diagnosed with ASD in Oman.

2 Methodology

2.1 Participants

This cross-sectional study was conducted at various locations in the Sultanate of Oman from June 2014 to June 2015. The locations included were Sultan Qaboos University Hospital (SQUH), Developmental Medicine Clinic, Muscat Autism Center, Early Intervention Center for Children with Disability, and Al Wafa Rehabilitation Centers located in different governorates/regions of Oman. A total of 750 subjects, including both caregivers (375) and children (375), were recruited. The children aged between 4 and 13 years and comprised of both males and females. Participants were approached randomly in four governorates, Muscat, Dhofar, Musandam, and Buraymi, and from four regions, Al-Batinah, Al-Sharqiyah,

Table 1 Distribution of total and recruited ASD children from different governorates of Sultanate of Oman

Governorate	Approximate number of ASD/governorate	Proportional number of ASD/governorate	Actual number of ASD recruited cases for the study/ governorate[a]
Muscat	159	72.1	28
Dhofar	68	30.9	19
Musandam	6	2.7	3
Buraymi	8	3.6	3
Dakhiliyah	45	20.4	33
North Batinah	119	54.0	40
South Batinah	34	15.4	15
North Sharqiyah	47	21.3	10
South Sharqiyah	41	18.6	6
Dhahirah	24	10.9	6
Total	551	250	163

[a]500 subjects (children) were approached (250 ASD and 250 TD). Responses were secured from 163 ASD and 212 TD children and an equivalent number of caregivers. Therefore, the response rate among ASD children was 65% and 85% among TD children

Al-Dakhiliyah, and Al-Dhahirah. Age-range matched, typically developing children were also recruited with similar backgrounds and from the same governorates/regions. Subjects were proportionally allocated based on the number of ASD children in each governorate. Proportionate number of ASD children in a particular governorate = total number of ASD children in that governorate/total of ASD children of all governorates∗ sample size (250). The recruitment of participants was designed to produce a representative sample of children with ASD with matched ages and within the same province (Table 1).

2.1.1 Ascertainment and Selection of Cases

Selection criteria for the ASD group included all mothers with their ASD child diagnosed at SQUH according to the criteria of the Diagnostic and Statistical Manual of Mental Disorders, Fourth Edition, Text Revision (DSM-IV-TR), whose age ranged between 4 and 13 years, with the status for each child confirmed by a specialist at SQUH. All subjects diagnosed with ASD exhibited symptoms within the typical triad of autistic traits: communication impairment, social deficits, and ritualistic interests.

2.1.2 Ascertainment and Selection of Controls

Both male and female children aged between 4 and 13 years, with no signs of ASD, developmental delay, or chronic diseases (celiac disease, epilepsy, diabetes, cerebral palsy, etc.) and residing in their parents' home, were selected for the study. A written consent form in both Arabic and English was obtained from all parents. A total of 324 children and caregivers were enrolled (127 ASD and 197 TD), resulting in a total sample size of 648 participants.

2.2 Sociodemographic Factors of the Study Groups

The demographics characteristics of caregivers included age range, their relation to the child, marital status, level of education, and monthly income (in local currency that is Omani Riyals or OMR). For children, the variables related to demographics were birth weight, gender, number of siblings, and anthropometric measurements. At the time of visit, anthropometric measurements that considered weight, height, and body mass index (BMI) were carried out for every subject (both ASD and TD) and were recorded by the principle investigator herself together with six registered dietitians, according to the procedures described by the World Health Organization [30].

2.3 Evaluation of Food Intake and Preference

The semiquantitative food frequency questionnaire (FFQ) used in this study is a comprehensive list of 119 food items. It was designed to collect dietary information about the amount and frequency of consumption of various food items commonly consumed in Oman. The FFQ was developed based on the validated food FFQs according to Block et al. [31] and modified by Ali et al. [32] according to local food consumption patterns. The only modification made was the addition of a single column ("occasionally") in food consumption. This column was added to record the food item that was only consumed occasionally or during certain events. The information from the FFQ was used to evaluate food preferences in the study group. Each subject was personally interviewed to complete the questionnaire. They were asked to characterize their usual dietary intake stating how often they consumed a specific amount of each food during a period of one year, on an average. The items were rated on a five-point scale: "never" if there is no intake of the food item, "occasionally" if the food is eaten only on occasions, "daily" if the food item is taken on daily basis, "weekly" if the food is taken once a week, and "many times" if the food is eaten more than once a day. The caregivers also reported whether food items are typically served to the entire family at meal times ("yes/no"). Data indicating the frequency of consumption for 119 food items in one year was collected. The different food groups included in the questionnaire were vegetables, fruits, meat and meat substitutes, milk and dairy products, traditional Omani foods and miscellaneous dishes, breads, desserts, beverages, sandwiches, and fast food. These food items were included based on their contribution to the total intake of energy by population groups and accounted for over 90% of the total Omani population's intake [32]. Each question had several choices and the interviewing researcher marked the best appropriate choice according to the reply of the subject. Participants were asked to indicate their average consumption of each food. Dietary items included in the preference questionnaire were classified into nine groups (vegetable group, fruit group, traditional and miscellaneous dishes, bread group, meat and meat substitute group, beverage group, sandwich group, desserts group, and fast food group).

The caregivers of children with unrestricted diet checked for food items, if their child had consumed an age-appropriate portion of the served food. The food items that were not eaten remained unendorsed (not comparable). The caregivers also reported whether food items were typically served to the entire family at meal times. Scores for the nine food groups were obtained by summing food items accepted by children within each food group (e.g., the total number of vegetables typically accepted). Scores were also summed to obtain types of food items typically eaten by other family members (e.g., the total number of vegetables typically eaten). The internal consistency was found for this sample; Cronbach's α was 0.96.

2.4 Ethical Approval and Confidentiality

Written informed consent was obtained in Arabic from all parents who were willing to contribute to the study before receiving any information. A clear, detailed explanation of the study was given to all parents, and they were informed that they were free to refuse to participate or withdraw at any time during the study without any disadvantage or prejudice. Details of the tools used and the types of measurements being obtained were explained. Confidentiality and privacy of collected information was strictly ensured by giving a code number for each subject, and parents were also assured that the collected data was going to be used for scientific purposes only. This study was approved by the Medical Research Committee in the College of Medicine and Health Sciences, Sultan Qaboos University, (MREC#899).

3 Data Analysis

The collected data was reviewed for its completeness and accuracy and was statistically analyzed. The following statistical tests were used: the Student's t-test to compare sociodemographic data and the chi-squared (x^2) test to assess the statistical significance of differences among categorical data. The nonparametric Fishers exact test (two tailed) was used instead of the chi-squared test in cases of very small sample size. The odds ratios (OR) and 95% confidence intervals (CI) were calculated in comparison between ASD and TD. Analysis of variance (ANOVA) was used to evaluate the statistical significance of mean differences among continuous data, followed by post hoc multiple comparisons. All statistical analyses were performed using the Statistical Package for Social Sciences (SPSS) software (Version 20.0). A significant association is considered if the 95% CI does not include the value 1.0, and a cutoff *p*-value of less than 0.05 is used for all tests of statistical significance in this study.

4 Results

4.1 Sociodemographic Factors of Study Group

Table 2 shows the sociodemographic characteristics of the study groups. Age, monthly income, and education levels of caregivers were not significantly different. However, there was a significant difference in terms of marital status (P =0.005).

Table 2 Sociodemographic characteristics of caregivers

	ASD (N = 163)	TD (N = 212)	
Characteristic*	N (%)	N (%)	P-value
Caregivers			
Age			
≥20–30	36 (22.1)	52 (24.5)	0.261
31–40	102 (62.6)	116 (54.7)	
41–50	25 (15.3)	44 (20.8)	
Marital status			
Single	1 (0.6)	9 (4.2)	0.005
Married	159 (97.5)	197 (92.9)	
Divorced	3 (1.8)	1 (0.5)	
Widow	0 (0.0)	5 (2.4)	
Education level			
Read and write	24 (14.7%)	32 (15.1%)	0.551
School education	67 (41.1%)	96 (45.3%)	
College education	51 (31.3%)	66 (31.1%)	
Above college	21 (12.9%)	18 (8.5%)	
Monthly income level (OMR)			
<500	42 (25.8)	63 (29.7)	0.078
500–1000	78 (47.9)	107 (50.5)	
1000–1500	19 (11.7)	28 (13.2)	
>1500	24 (14.7)	14 (6.6)	

* different at α =0.05, a and b are presented by Mean (SD)

Table 3 Sociodemographic characteristics of children

Characteristic∗	ASD ($N = 163$) N (%)	TD ($N = 212$) N (%)	P-value
Children			
Gender			
Male sex	129 (79.1)	104 (49.1)	<0.001
Female sex	34 (20.9)	108 (50.9)	
Mean age (years)[a]	7.23 (2.51)	7.71 (2.63)	0.075
Mean BMI[b]	15.44 (2.48)	15.37 (2.57)	0.816
BMI percentile			
Underweight	47 (28.8)	67 (31.6)	0.301
Healthy weight	91 (55.8)	126 (59.4)	
Overweight	22 (13.3)	17 (8.0)	
Obese	3 (1.8)	2 (0.9)	

∗diagnostic groups significantly different at α =0.05, [a] and [b] are presented by Mean (SD)

Sociodemographic characteristics of ASD and TD children are shown in Table 2. The gender distribution was significantly different between the two groups (P <0.001). The ASD group included 129 males and 34 females, whereas the TD group included 104 males and 108 females, making a proportion of 79% and 49% for the ASD and TD groups, respectively. Age and BMI were nonsignificant. However, 32% of TD children were underweight (<5th percentile) as compared to 29% of ASD children. More ASD children were overweight as compared to TD children (14% versus 8%) (Table 3).

4.2 *Diversity of Food Selectivity Between the Two Study Groups*

4.2.1 A Comparison on Food Intake Preference Between ASD and TD Children

Table 4 provides a summary of the results comparing the average number of food items within each food group preferred by the two study groups. An independent t-test showed that the average number of items within every food group was significantly higher for TD children as compared to ASD children ($P < 0.001$). The results also show the percentages of the number of items consumed by the two study groups within each food group. Of the nine food groups (vegetables, fruits, miscellaneous dishes, bread, protein foods, beverages, sandwiches, fast foods, and dessert), the intake was significantly different in ASD children as compared to TD children. Less than 50% of the vegetable (47%) and fast food (36%) items were eaten by ASD children, while the TD children consumed more than 50% from every food group (63–86%).

Table 4 Percentages of food items eaten by ASD and TD children from each food group

Food group (*n*)	Food group intake (%)	Children (ASD/TD)	Mean (SE)	*P*-value*
Vegetables (19)	47	ASD	8.6 (0.5)	<0.001
	63	TD	12.1 (0.3)	
Fruits (14)	57	ASD	7.9 (0.49)	<0.001
	86	TD	12.3 (0.2)	
Miscellaneous dishes (28)	71	ASD	19.8 (0.7)	<0.001
	86	TD	23.7 (0.3)	
Bread (11)	55	ASD	6.0 (0.3)	<0.001
	73	TD	8.0 (0.1)	
Protein foods (17)	65	ASD	10.7 (0.4)	<0.001
	82	TD	14.1 (0.2)	
Beverages (6)	50	ASD	2.9 (0.2)	<0.001
	67	TD	3.6 (0.1)	
Sandwiches (7)	57	ASD	3.7 (0.2)	<0.001
	71	TD	5.1 (0.1)	
Desserts (14)	57	ASD	8.4 (0.4)	<0.001
	79	TD	10.8 (0.2)	
Fast foods (3)	36	ASD	1.1 (0.1)	<0.001
	64	TD	1.9 (0.1)	

*Significantdifferences at p-value < 0.05, ASD (127) and TD (197) children

Table 5 Intake preferences of food groups by ASD children and their families

	Children		Family		
Food group	Mean (SE)	<50% of food group	Mean (SE)	<50% of food group	*P*-value*
Vegetables	8.6 (0.5)	59.8	16.87 (0.2)	3.1	<0.001
Fruits	7.9 (0.5)	44.9	13.6 (0.2)	2.4	<0.001
Miscellaneous	19.8 (0.7)	22.0	26.8 (0.3)	0.8	<0.001
Bread	6.0 (0.3)	42.5	9.0 (0.2)	9.4	<0.001
Protein foods	10.7 (0.4)	31.5	16.3 (0.2)	1.6	<0.001
Beverage	2.9 (0.2)	41.7	5.7 (0.1)	1.6	<0.001
Sandwich	3.7 (0.2)	42.5	6.1 (0.1)	6.3	<0.001
Desserts	8.4 (0.4)	29.9	12.4 (0.2)	1.6	<0.001
Fast food	1.1 (0.1)	64.6	1.9 (0.1)	37.0	<0.001

*Paired t-test between ASD children and their families ($N = 127$). Excluded the restricted diet ($163 - 36 = 127$)

4.2.2 A Comparison on Food Intake Preference of ASD Children and Their Families

Results of paired t-tests on food preferences between children with ASD and their families showed that there were highly significant differences for all of the nine food groups ($P < 0.001$) in Table 5. From all the food groups, the ASD children not

Table 6 Intake preferences of food groups by TD children and their families

Food group	Children		Family		*P*-value
	Mean (SE)	<50% of food group	Mean (SE)	<50% of food group	
Vegetables	12.1 (0.3)	29.4	16.7 (0.2)	2.0	<0.001
Fruits	12.3 (0.2)	5.1	13.7 (0.1)	0.5	<0.001
Miscellaneous food items	23.7 (0.3)	4.6	26.6 (0.2)	0.0	<0.001
Breads	8.0 (0.1)	11.2	9.4 (0.1)	3.6	<0.001
Protein foods	14.1 (0.2)	5.1	16.1 (0.2)	1.5	<0.001
Beverages	3.6 (0.1)	22.3	5.5 (0.1)	4.1	<0.001
Sandwiches	5.1 (0.1)	15.7	6.2 (0.1)	3.0	<0.001
Desserts	10.8 (0.2)	3.0	12.2 (0.2)	2.0	<0.001
Fast foods	1.9 (0.1)	38.6	2.2 (0.1)	29.4	<0.001

only ate fewer of the listed food items but also ate lesser than their respective families. As stated in the previous section, the ASD children ate fewer than 50% of the vegetables (8.6 ± 0.5) and fast foods (1.1 ± 0.1). In contrast, the families ate more than 50% of the varieties from all food groups.

4.2.3 A Comparison of Food Intake Preference of TD Children and Their Families

The results in Table 6 show that the TD children's intake of food items from the nine food groups was significantly ($P < 0.001$) different from their families. For all food groups, the children consumed fewer listed food items than their family.

4.2.4 Food Intake Preferences of ASD Children and Their Families

Table 7 provides a list of those food items within each food group that the ASD children preferred in contrast to their families' preferences. As mentioned earlier, ASD children ate fewer varieties of vegetables compared to their families, who ate most of the 19 varieties of vegetables. More than half the ASD children ate various portions of the following vegetables: onions, mixed vegetables, cucumbers, tomatoes, carrots, and mostly potatoes. Similarly, the children with ASD had a lower intake of fruits as compared to their families. Among the 14 different types of fruits, the following were consumed by more than half the children: bananas, apples, watermelons, oranges, and mangoes. Other fruits included grapes, melons, dates, and pomegranates. On the one hand, more than 90% of the families ate all 14 listed fruits.

As expected, most of the subjects in both groups (the children and their families) ate all the food varieties listed in the traditional Omani food group. Children with

Table 7 Percentage of vegetables, fruits, and miscellaneous food items eaten by more than 50% of either ASD children or their families

Food item (%)		
Food group	Children with ASD (*N* = 163)	Family (*N* = 212)
Vegetables	Potato (80), carrot (63), tomatoes (61), cucumber (60), mixed vegetables (57), onion (51)	Tomato (97), onion (97), lettuce (96), carrot (95), cucumber) 95), garlic (95), olive (95), green pepper (95), potato (95), green onion (93), eggplant (93), M. vegetables (93), chili (91), S. potatoes (90), lady finger (88), cabbage (86), cauliflower (79), spinach (70)
Fruits	Banana (71), apple (67), watermelon (67), orange (65), grapes (56), dates (55), mango (58), melon (56), pomegranate (52)	Apple (99), orange (98), watermelon (98), banana 98)), dates (98), pomegranate (98), grapes (98), mango (97), guava (96), melon (96), pears (96), peach (95), kiwi (95), papaya (91)
Omani and miscellaneous dishes	Saloona chicken (85), saloona meat (85), biryani chicken (84), saloona fish (83), biryani fish (83), white rice (82), biryani meat (80), makboos chicken (77), qabooli meat (75), qabooli chicken (76), qabooli fish (72), makboos meat (72), makboos fish (71), macaroni (70), arsiya chicken (69), arsiya meat (69), chicken soup (66), samosa (64), meat soup (63), harees chicken (61), lentils (59), kidney beans (60), pizza (59), harees meat (58), thareed meat (52), beans (51), thareed chicken (50), chickpeas (50)	Makboos chicken (99), biryani meat (99), biryani fish (99), macaroni (99), biryani chicken (98), makboos fish (98), qabooli chicken (97), saloona fish (97), saloona meat (97), harees chicken (97), qabooli fish (97), arsiya chicken (96), lentils (96), chicken soup (96), harees meat (96), makboos meat (96), pizza (96), qabooli meat (96), saloona chicken (96), samosa (96), kidney beans (95), beans (95), chickpeas (95), white rice (94), meat soup (93), thareed chicken (87), thareed meat (85), arsiya meat (69)

ASD preferred eating fewer types of bread as compared to their families. Of the 11 types of bread listed, only 6 appeared to be the most popular among children. It included white toast bread, Lebanese bread-white, chapatti bread, burger bread, paratha, and rekhal (local thin bread). On the other hand, more than 50% of the families consumed all the listed food varieties, except Salalah bread, which is widely popular in the Dhofar region (Southern Oman). Of the 17 food items listed in the protein food group, 11 were eaten by more than 50% of ASD children. These include plain yoghurt, chicken, egg, meat, fish, cream cheese, sliced cheese, cheddar cheese, milk (whole), canned tuna, and pistachio. Most families' intake of protein foods included all the items in the food list.

Most of the families consumed all 6 of the listed beverages, whereas majority of the children consumed only two types of beverages: tea with milk and soft drinks. Similarly, most families' intake of sandwiches included all of the seven kinds listed in the food group. The majority of ASD children, however, only ate cheese sand-

Table 8 Percentages of varieties of foods eaten by more than 50% of either ASD children or their families

Food item (%)		
Food group	Children with ASD ($N = 163$)	Family ($N = 212$)
Breads	White toast bread (74), Lebanese bread-white (71), chapatti bread (59), burger bread (58), paratha (57), rekhal (55)	Lebanese bread-white (95), white bread (93), brown toast bread (78), burger bread (91), chapatti bread (91), paratha (90), rekhal (85), Lebanese bread-brown (84), unspecified bread (79), tandoor bread (71)
Proteins	Yoghurt plain (93), chicken (83%), egg (78), meat (75), fish (72), cream cheese (64), slice cheese (63), cheddar cheese (63), milk (whole) (56), tuna (canned) (55), pistachio (51)	Egg (99), pistachio (97), almonds (96), fish (96), meat (96), chicken (96), peanuts (95), cream cheese (96), cashews (95) cheddar cheese (95), milk (whole) (95), milk with chocolate ([illegible]), yoghurt with fruit ([illegible]), tuna (canned) (93), milk with fruits (91), yoghurt (plain) (78)
Beverages	Bottled fruit juice (83), tea with milk (57), soft drinks (56)	Tea with milk (98), bottled fruit juice (97), tea w/o milk (95), Omani coffee (94), soft drinks (93), instant coffee (87)
Sandwiches	Cheese sandwich (60), chicken sandwich (56), egg sandwich (51), cheeseburger (50)	Cheese sandwich (97), egg sandwich (95), chicken sandwich (94), falafel sandwich (90), cheeseburger (92), chicken fillet (81), fish fillet (55)
Desserts	Biscuit (80), candy (76), cake (70), luqaimat (66), custard (62), doughnuts (62), Omani halwa (60), croissant (54), pancake (51)	Omani halwa (97), biscuit (96), luqaimat (96), candy (96), doughnuts (95), custard (95), cake (93), pancake (93), croissant (92), pudding (81), cheese cake (80), date pie (79), apple pie (53)
Fast food	None	Pizza Hut (69), KFC (67), McDonald's (66)

∗ Significant differences at p-value < 0.05

wich, chicken sandwich, egg sandwich, and cheeseburger. The desserts appeared to be the most popular in the food groups, with 9 out of the 14 listed items being consumed by more than 50% of the children with ASD. These included biscuits, sweets/candy, sponge cake, luqaimat (Omani sweet dish), custard, doughnuts, Omani halwa, croissant, and pancake. Majority of the families ate 13 of the 14 items listed in the food group (Table 8).

5 Discussion

The current study attempted to identify the relationship between food preferences of children with ASD and their families and other classified groups of children along with providing a categorized illustration of the food preferences as per classified grouping.

5.1 Food Preferences

Although clinicians and parents widely accept that children with ASD exhibit more feeding problems than their typically developing peers, only little information is available concerning the characteristics and preferences for food items accepted by these children [20]. Generally, the children with ASD have a rigid pattern of interests and activities and show an obsessive desire to stick to their routines and order, all of which are reflected in their food behavior [33]. Severe food selectivity in ASD children can lead to limited choices of food variety in their diet that may lead to long-term nutritional challenges [34].

There are few studies on selective eating behavior in ASD children, such as selectivity based on food presentation and food types [20, 35, 36]. These studies suggest that children with ASD generally prefer a more limited variety of foods as compared to TD children, which in turn may impact the overall quality of their diet [37]. These studies have provided only preliminary information concerning the food selectivity of children with autism. However, little is known about the types of food items typically preferred by them.

The results of this study showed that as compared to their typically developing peers, the Omani children with ASD, on an average, consumed a significantly smaller variety of food items within different food groups, including vegetables (9 versus 12), fruits (8 vs. 12), traditional foods (20 vs. 24), bread types (6 vs. 8), protein foods (11 vs. 14), beverages (3 vs. 4), sandwiches (4 vs. 5), sweets and desserts (8 vs. 11), and fast foods (1 vs. 2). This indicates that Omani children with ASD consumed significantly smaller varieties of food items from each food group per year as compared to the TD children. These findings corroborated findings from previous studies [2, 20, 38].

A recent survey by Williams et al. [21] investigated 100 parents of ASD children at the age range of22 months to 10 years. They found that 67% of the children's parents recognized that their child was a "picky eater." Regardless of this, nearly 73% of the parents also reported that their child had good appetite for foods that they liked and only 6% showed poor appetite. Parents stated that the following factors influenced what their child would eat: texture (69%), appearance (58%), taste (%45) smell (36%), and temperature (22%). The most frequently reported eating problem was reluctance to try new foods (69%). Similarly, Schreck and Williams [20] stated that 72% of parents reported their children had a limited range of foods to select and 57% reported food refusal. Refusals were primarily related to food presentation (48.6%), such as use of particular utensil or different food items touching on the plate. Other factors related to food refusal and acceptance included (a) specific utensil requirements (13.8%), (b) food texture (6.5%), and (c) oral motor problems (23.2%).

In a more recent and larger-scale study [36], food selectivity in 138 children with ASD was compared to that of 298 typically developing children. Parents of the children with ASD reported that their children refused significantly more food items and had a less-varied diet than children without ASD. Children with ASD were reported to eat about half the number of food items in each food group, except for starches, where they ate about two-thirds of the number of food items as compared to typically developing children. In addition, children with ASD were significantly more likely to accept only low-texture foods such as those that had been pureed. Thus, the children with ASD had a significantly greater degree of food selectivity than typically developing children. Using the same data set in a subsequent analysis, Kimberly et al. [20] reported that the major reason for restricted food intake in children with ASD could be attributed to food presentation, i.e., different food items touching on a plate or specific utensil requirements. Across all food groups, children with ASD ate fewer types of foods than did other members of their family. However, food preference (as defined by the number of different foods eaten) was also found to be related to the family's food preferences. It is possible that the preference of many individuals with autism to eat a narrow range of food items from various food groups is due to their preference for routine and sameness. This may therefore cause learned habituation to familiar foods and rejection of any other food that is novel [20].

In line with this, our results on the BAMBI (Brief Autism Mealtime Behavior Inventory) related to selectivity, limited varieties, and food preferences showed that children with ASD displayed strong preferences for limited food varieties as compared to TD children [39].

Overall, the children consumed fewer varieties of foods than their parents in all food groups. Omani children with ASD also tended to have very low preference for vegetables and their top three most preferred vegetables were potatoes, carrots, and tomatoes. The top three Omani traditional foods were chicken saloona [stew], chicken biryani [spiced rice], and fish biryani. There was high intake of beverages (bottled fruit juice, tea with milk, and soft drinks), desserts (including biscuit, candy, and sponge cake), protein foods such as chicken, egg, and yoghurt. However, fast foods were not among the preferences, since none of the three food items was preferred by more than 50% of the children. This study also found that Omani children with ASD were indeed idiosyncratically selective in the type of food items they preferred.

Food items preferred by 50% or more of the children with autism were less, compared to that of their caregivers. The food items preferred by autistic children were foods that are high in sugar content such as bananas, mangoes, sponge cakes, biscuits, candy, doughnuts, luqaimat [sweet dumplings], custard, Omani halwa (a sugary, buttery dessert), etc. The strong preference for sweet-tasting food items of children with autism can be explained by the theory that children have a predisposition for sweet and salty foods, while rejecting the bitter or sour foods [40]. Furthermore, children tend to learn preferences for familiar foods, which is clearly illustrated in our results from the overwhelming acceptance of most varieties of traditional foods by ASD children that are presented on a daily basis at lunch and dinner, indicating food preferences.

In general, there are many dimensions of food refusal behavior in children. These include food refusal based on sensory food aversion, where children refuse to eat foods with a specific taste, such as the following: (1) meal characteristics (i.e., the selection and provision of appropriate menus and repeated exposure to different kind of foods and food textures), (2) schedule of intake (i.e., meal frequency and duration), (3) setting characteristics (i.e., altering the physical surrounding, feeding position and body support of the child, and activities before and after eating), and (4) interactions, i.e., reciprocity between child and feeder, the appliance of social contingencies by the use of behavior management procedures [41]. There is fear-based food refusal behavior, which appears or emerges after episodes of choking or severe gagging and the child's aversive reactions can range from grimacing to gagging, vomiting, or spitting out food. Some children extend their reluctance to eat one food to others that look or smell alike. For example, an aversion to green beans may extend to all green vegetables. Parents frequently report that their children are reluctant to eat new foods. Some children may even refuse to eat any food that has touched another food on the same plate, while others will only eat food prepared in a specific way [23].

ASD children, because of their food selectivity, may not reach the recommended dietary allowance (RDA) due to their refusal and limited food choices from all types of food groups. The findings of this study also indicated that ASD children ate less fruits and vegetables, which is consistent with other studies [42]. This could account for some micronutrient deficiencies and contribute to harmful health effects. Numerous studies have suggested a higher prevalence of obesity in children with ASD [43–45], and it is possible that this is related to refusal to eat healthy foods like vegetables and fruits [42, 46]. In addition, having a smaller proportion of healthy meals may cause high intake of empty calorie food. Empty calories are calories from solid fats and/or added sugars [47], and high intake is associated with obesity [48].

5.1.1 Children's Food Preferences in Relation to That of Their Family

Studies have shown that food preferences and eating patterns develop in early childhood and remain relatively stable through adolescence [49]. Therefore, early childhood may represent a sensitive window of development for establishing good eating habits and healthy food preferences that could potentially impact an individual's lifelong health [50]. Parents influence children's food preferences and intake patterns through the foods they make available to their children, the types of child feeding practices they use, and their own eating behavior [51]. Mothers have been shown to influence their daughters' fruit and vegetable intake via their own patterns of fruit and vegetable intake and by influencing their daughters' tendencies to be picky eaters [52].

Our results have shown that children with ASD in Oman preferred fewer varieties from all food groups as compared to their parents. The restrictive eating habits of children with autism spectrum disorders also tend to cause their families to restrict their own eating behavior as well (e.g., not wanting to prepare a wide variety of food items for different family members). Hence, the family's initial restricted food choices may result in the children modeling their food selectivity and restricting their exposure to new food items. Eventually, the parents adapt the family menu to better fit their child's preferences and avoid unfamiliar foods or ones they know will trigger aversive reactions. In contrast to these findings, those families that eat a wider range of foods influence their children to eat a broader range of foods as well. Parental behavior can, either positively or negatively, affect the eating patterns of children with ASD markedly [53]. Food selectivity appears to be a significant issue for many children with ASD. However, the concept of food selectivity has not been operationally defined and there are no "gold standard" measures. Food selectivity in children with ASD may occur for many reasons. Sensory sensitivity has been suggested as one of the possible mechanisms to explain, in part, food selectivity of children with ASD [15].

6 Summary and Conclusion

This case-control study was conducted in various governorates/regions of the Sultanate of Oman to compare the food preferences of 4–13-year-old children with ASD and TD children. The mean BMI and BMI percentiles were not significantly different between the two groups. A comparison of food preferences between ASD and TD children was examined using FFQ with nine food groups, such as vegetables, fruits, traditional Omani food, bread, protein, beverage, sandwiches, deserts, and fast food. In general, the mean of the total score was significantly lower in children with ASD compared to TD children in majority of food groups. The average number of items within every food group was significantly higher for TD children compared to ASD children ($P < 0.001$). Above 50% of the food consumed by ASD children were as follows: in miscellaneous dishes, saloona (chicken, meat, and fish) and biryani (chicken, meat, fish), and white rice. Likewise, yogurt and biscuits were also chosen by the same children. Whereas in the beverage section, bottled fruit juice was the top choice of ASD children. The list of most common dessert intake was biscuit, sweets, and cake. Additionally, children tended to have preferences for familiar foods which can be explained by their repetitive behaviors. This leads to the high acceptance of most varieties of traditional foods that are presented on a daily basis at lunch and dinner indicating food preferences of the ASD children. They also tend to prefer high sugar-content food, such as bananas, mangoes, sponge cakes, biscuits, candy, doughnuts, luqaimat [sweet dumplings], custard, and Omani halwa (a sugary, buttery dessert). However, less vegetable and fruit intake was observed in ASD children compared to TD children.

There are two main limitations of the FFQ used in this study. Firstly, the side effects of medication and gastrointestinal (GI) condition on the mealtime behavior of ASD children were not evaluated. Many children with autism are under medication with side effects that could adversely affect behavior and appetite. Secondly, the assessment of mealtime behavior of two groups of children (ASD and TD) was based on parental reports rather than direct observation. Despite these limitations, the food preference questionnaire is useful in that it is based on foods commonly eaten in Oman. Moreover, children on a restricted diet were excluded in order to obtain more accurate results.

Based on our results, low fruit and vegetables intake may lead to inadequate intake of vitamins and minerals which may contribute to some autistic behavior. Therefore, it is important to monitor nutritional status and dietary intake of children with ASD regularly. Furthermore, improvement in nutritional awareness status among parents and caregivers of children with ASD is of great importance. Future studies to find the link between food preferences and their role on ASD pathology are warranted.

Conflict of Interest Authors declare no conflicts of interest.

References

1. American Psychiatric Association. (2016). *What is autism spectrum* [cited 25 April 2016]. Available from: https://www.psychiatry.org/patients-families/autism/what-is-autism-spectrum-disorder
2. Zimmer, M. H., Hart, L. C., Manning-Courtney, P., Murray, D. S., Bing, N. M., & Summer, S. (2012). Food variety as a predictor of nutritional status among children with autism. *Journal of Autism and Developmental Disorders, 42*(4), 549–556.
3. Smithers, L. G., Golley, R. K., Brazionis, L., & Lynch, J. W. (2011). Characterizing whole diets of young children from developed countries and the association between diet and health: A systematic review. *Nutrition Reviews, 69*(8), 449–467.
4. Birch, L. L. (1998). Development of food acceptance patterns in the first years of life. *The Proceedings of the Nutrition Society, 57*(4), 617–624.
5. Harris, G. (2008). Development of taste and food preferences in children. *Current Opinion in Clinical Nutrition and Metabolic Care, 11*(3), 315–319.
6. Cermak, S. A., Curtin, C., & Bandini, L. (2014). *Sensory sensitivity and food selectivity in children with autism spectrum disorders* (pp. 2061–2076).
7. Cooke, L. J., & Wardle, J. (2005). Age and gender differences in children's food preferences. *The British Journal of Nutrition, 93*(5), 741–746.
8. Rigal, N., Chabanet, C., Issanchou, S., & Monnery-Patris, S. (2012). Links between maternal feeding practices and children's eating difficulties. Validation of French tools. *Appetite, 58*(2), 629–637.
9. Carruth, B. R., Skinner, J., Houck, K., Moran, J., 3rd, Coletta, F., & Ott, D. (1998). The phenomenon of "picky eater": A behavioral marker in eating patterns of toddlers. *Journal of the American College of Nutrition, 17*(2), 180–186.
10. Birch, L. L., & Davison, K. K. (2001). Family environmental factors influencing the developing behavioral controls of food intake and childhood overweight. *Pediatric Clinics of North America, 48*(4), 893–907.
11. Scaglioni, S., Salvioni, M., & Galimberti, C. (2008). Influence of parental attitudes in the development of children eating behavior. *The British Journal of Nutrition, 99*, 1.
12. Jacobi, C., Agras, W. S., Bryson, S., & Hammer, L. D. (2003). Behavioral validation, precursors, and concomitants of picky eating in childhood. *Journal of the American Academy of Child & Adolescent Psychiatry, 42*(1), 76–84.
13. Marí-Bauset, S., Llopis-González, A., Zazpe-García, I., Marí-Sanchis, A., & Morales-Suárez-Varela, M. (2015). Nutritional status of children with Autism Spectrum Disorders (ASDs): A case–control study. *Journal of Autism and Developmental Disorders, 45*(1), 203–212.
14. Hyman, S. L., Stewart, P. A., Schmidt, B., Cain, U., Lemcke, N., Foley, J. T., et al. (2012). Nutrient intake from food in children with autism. *Pediatrics, 130*(Suppl 2), S145–S153.

15. Cermak, S. A., Curtin, C., & Bandini, L. G. (2010). Food selectivity and sensory sensitivity in children with autism spectrum disorders. *Journal of the American Dietetic Association, 110*(2), 238–246.
16. Cornish, E. (1998). A balanced approach towards healthy eating in autism. *Journal of Human Nutrition and Dietetics, 11*(6), 501–509.
17. Graf-Myles, J., Farmer, C., Thurm, A., Royster, C., Kahn, P., Soskey, L., et al. (2013). Dietary adequacy of children with autism compared with controls and the impact of restricted diet. *Journal of Developmental and Behavioral Pediatrics, 34*(7), 449–459.
18. Herndon, A. C., DiGuiseppi, C., Johnson, S. L., Leiferman, J., & Reynolds, A. (2009). Does nutritional intake differ between children with autism spectrum disorders and children with typical development? *Journal of Autism and Developmental Disorders, 39*(2), 212–222.
19. Hubbard, K. L., Anderson, S. E., Curtin, C., Must, A., & Bandini, L. G. (2014). A comparison of food refusal related to characteristics of food in children with autism spectrum disorder and typically developing children. *Journal of the Academy of Nutrition and Dietetics, 114*(12), 1981–1987.
20. Schreck, K. A., & Williams, K. (2006). Food preferences and factors influencing food selectivity for children with autism spectrum disorders. *Research in Developmental Disabilities, 27*(4), 353–363.
21. Williams, P. G., Dalrymple, N., & Neal, J. (2000). Eating habits of children with autism. *Pediatric Nursing, 26*(3), 259–264.
22. Satter, E. (1995). Feeding dynamics: Helping children to eat well. *Journal of Pediatric Health Care, 9*(4), 178–184.
23. Chatoor, I., & Ganiban, J. (2003). Food refusal by infants and young children: Diagnosis and treatment. *Cognitive and Behavioral Practice, 10*(2), 138–146.
24. Bryant-Waugh, R., Markham, L., Kreipe, R. E., & Walsh, B. T. (2010). Feeding and eating disorders in childhood. *The International Journal of Eating Disorders, 43*(2), 98–111.
25. Linscheid, T. R. (2006). Behavioral treatments for pediatric feeding disorders. *Behavior Modification, 30*(1), 6–23.
26. Greer, A. J., Gulotta, C. S., Masler, E. A., & Laud, R. B. (2008). Caregiver stress and outcomes of children with pediatric feeding disorders treated in an intensive interdisciplinary program. *Journal of Pediatric Psychology, 33*(6), 612–620.
27. Al-Farsi, Y. M., Al-Sharbati, M. M., Waly, M. I., Al-Farsi, O. A., Al Shafaee, M. A., & Deth, R. C. (2011). Malnutrition among preschool-aged autistic children in Oman. *Research in Autism Spectrum Disorders, 5*(4), 1549–1552.
28. Al-Farsi, Y. M., Waly, M. I., Al-Sharbati, M. M., Al-Shafaee, M. A., Al-Farsi, O. A., Al-Khaduri, M. M., et al. (2013). Levels of heavy metals and essential minerals in hair samples of children with autism in Oman: A case-control study. *Biological Trace Element Research, 151*(2), 181–186.
29. Al-Farsi, Y. M., Al-Sharbati, M. M., Waly, M. I., Al-Farsi, O. A., Al-Shafaee, M. A., Al-Khaduri, M. M., et al. (2012). Effect of suboptimal breast-feeding on occurrence of autism: A case-control study. *Nutrition, 28*(7-8), e27–e32.
30. World Health Organization. *Measuring a child's growth* [cited 21 April 2016]. Available from: http://www.who.int/childgrowth/training/module_b_measuring_growth.pdf
31. Block, G., Hartman, A. M., & Naughton, D. (1990). A reduced dietary questionnaire: Development and validation. *Epidemiology, 1*(1), 58–64.
32. Ali, A., Al-Belushi, B. S., Waly, M. I., Al-Moundhri, M., & Burney, I. A. (2013). Dietary and lifestyle factors and risk of non-Hodgkin's lymphoma in Oman. *Asian Pacific Journal of Cancer Prevention, 14*(2), 841–848.
33. Cornish, E. (2002). Gluten and casein free diets in autism: A study of the effects on food choice and nutrition. *Journal of Human Nutrition and Dietetics, 15*(4), 261–269.
34. Kral, T. V. E., Eriksen, W. T., Souders, M. C., & Pinto-Martin, J. A. (2013). Eating behaviors, diet quality, and gastrointestinal symptoms in children with autism spectrum disorders: A brief review. *Journal of Pediatric Nursing, 28*(6), 548–556.

35. Ahearn, W. H., Castine, T., Nault, K., & Green, G. (2001). An assessment of food acceptance in children with autism or pervasive developmental disorder-not otherwise specified. *Journal of Autism and Developmental Disorders, 31*(5), 505–511.
36. Schreck, K. A., Williams, K., & Smith, A. F. (2004). A comparison of eating behaviors between children with and without autism. *Journal of Autism and Developmental Disorders, 34*(4), 433–438.
37. Schmitt, L., Heiss, C. J., & Campbell, E. E. (2008). A comparison of nutrient intake and eating behaviors of boys with and without autism. *Topics in Clinical Nutrition, 23*(1), 23–31.
38. Suarez, M. A., & Crinion, K. M. (2015). Food choices of children with autism spectrum disorders. *International Journal of School Health, 2*(3), e27502.
39. Al-Kindi, N., Al-Farsi, Y., Waly, M., Al-Shafaee, M., Bakheit, C., Al-Sharbati, M., et al. (2016). Comparative assessment of eating behavior among children with autism to typically developing children in Oman. *Canadian Journal of Clinical Nutrition, 4*, 51–64.
40. Birch, L. L. (1999). Development of food preferences. *Annual Review of Nutrition, 19*(1), 41–62.
41. Deckers, S. R. J. M., De Moor, J. M. H., & Van der Burg, J. J. W. (2011). Food preferences in young Dutch children and recommendations for feeding intervention in developmental disabilities. *Research in Developmental Disabilities, 32*(2), 630–635.
42. Bandini, L. G., Anderson, S. E., Curtin, C., Cermak, S., Evans, E. W., Scampini, R., et al. (2010). Food selectivity in children with autism spectrum disorders and typically developing children. *The Journal of Pediatrics, 157*(2), 259–264.
43. Curtin, C., Anderson, S., Must, A., & Bandini, L. (2010). The prevalence of obesity in children with autism: A secondary data analysis using nationally representative data from the National Survey of Children's Health. *BMC Pediatrics, 10*(1), 11.
44. Broder-Fingert, S., Brazauskas, K., Lindgren, K., Iannuzzi, D., & Van Cleave, J. (2014). Prevalence of overweight and obesity in a large clinical sample of children with autism. *Academic Pediatrics, 14*(4), 408–414.
45. Emond, A., Emmett, P., Steer, C., & Golding, J. (2010). Feeding symptoms, dietary patterns, and growth in young children with autism spectrum disorders. *Pediatrics, 126*(2), 2009–2391.
46. Bicer, A. H., & Alsaffar, A. A. (2013). Body mass index, dietary intake and feeding problems of Turkish children with autism spectrum disorder (ASD). *Research in Developmental Disabilities, 34*(11), 3978–3987.
47. Nutrition and Your Health. (2010). *Dietary guidelines for Americans*. Washington DC: US Government Printing Office [cited 18 April 2016].
48. Slining, M. M., & Popkin, B. M. (2013). Trends in intakes and sources of solid fats and added sugars among U.S. children and adolescents: 1994–2010. *Pediatric Obesity, 8*(4), 307–324.
49. Northstone, K., & Emmett, P. M. (2008). Are dietary patterns stable throughout early and mid-childhood? A birth cohort study. *The British Journal of Nutrition, 100*(5), 1069–1076.
50. Kaar, J. L., Shapiro, A. L. B., Fell, D. M., & Johnson, S. L. (2016). Parental feeding practices, food neophobia, and child food preferences: What combination of factors results in children eating a variety of foods? *Food Quality and Preference, 50*, 57–64.
51. Fisher, J. O., Mitchell, D. C., Smiciklas-Wright, H., & Birch, L. L. (2002). Parental influences on young girls' fruit and vegetable, micronutrient, and fat intakes. *Journal of the American Dietetic Association, 102*(1), 58–64.
52. Galloway, A. T., Fiorito, L., Lee, Y., & Birch, L. L. (2005). Parental pressure, dietary patterns, and weight status among girls who are "picky eaters". *Journal of the American Dietetic Association, 105*(4), 541–548.
53. Hill, A. J. (2002). Developmental issues in attitudes to food and diet. *The Proceedings of the Nutrition Society, 61*(2), 259–266.

Part III
Food, Dietary Intervention, and Therapy in Autism

"Tell me what you ***eat****, and I will tell you what you are."*

– Anthelme Brillat-Savarin, 1826

"From the bitterness of disease man learns the sweetness of health."

– Catalan Proverb

1.1 Overview and Reflection

Once ASD diagnosis has been established, deficits are treated with a multifaceted team approach. Approach to treatment regularly includes occupational, behavioral, speech, and play therapies. There are no pharmacological treatments for ASD, though many individuals receive medications to address comorbidities, such as seizures and attention deficit hyperactivity disorder (ADHD). In some areas, children may receive support services in schools, including special education programs that target common comorbidities including learning disorders and intellectual disabilities. For those individuals with a severe diagnosis, prognosis is far less positive. Severe cases may never learn to communicate, and they remain in a world that is withdrawn from peers and family. The focus of treatment for ASD is personalized to the needs of the individual, with a goal to improve quality life of the patient and those around them. Family and caregivers often struggle to relate due to barriers in communicating with those suffering with ASD, and are encouraged frequently as part of therapy to find shared enjoyment in activities that promote bonding.

The saying of Dr. John Christopher "Every home should have an herbalist" undoubtedly resonates with families having an ASD individual. Modern medicine and traditional medicine make unique contributions to health management. In some systems of traditional medicine, such as traditional Chinese medicine and the Ayurveda system historically rooted in India, traditional practices are supported by wisdom and experience acquired over centuries. Evidence is mounting that diet, exercise, and stress reduction can do a better job in disease/condition health management. In several North American and European countries, the production and sale of herbal medicines, dietary supplements, and the other so-called natural products have become a huge and profitable industry. In the USA alone, this industry is a $32 billion a year business.

The critical question is Do Alternative Therapies Have a Role in Autism? The efficacy of herbal medicines for the management of ASD appears to be encouraging though was inconclusive owing to low methodological quality, herbal medicine diversity, and small sample size of the examined studies. It is imperative that data supporting new personalized intervention and therapy for autism management should be scrutinized for scientific study design, clinical safety, and scientific validity, before embarking on them as modes of therapy. The most commonly used CAM treatments for ASD fall into categories of biologically based practice and manipulative and body based practices. Many parents of autistic children are turning to alternative therapies in an effort to stimulate developmental progression in language skills and social interactions.

Selected examples of intervention and therapy based practices

Body and behavioral therapy	Biological therapy
Behavioral therapy	Nutrition and diet therapies
Occupational therapy	Ayurveda/Chinese herbal medicines
Speech therapy	Homeopathy medicines
Auditory integration training (AIT)	Anti-fungal/probiotic therapy
Music therapy	Chelation therapy
Vision therapy (VT)	Hyperbaric oxygen therapy
Massage therapy	
Physical activity	
Yoga therapy	

According to the Autism Society of America, "as there is no one symptom or behavior that identifies individuals with ASD, there is no single treatment that will be effective for all people on the spectrum." There is increasing evidence that specific CAM therapies for treating ASD symptoms like vitamins, amino acids, essential fatty acids, oligoantigenic diet, and herbal medications are helpful to many ASD patients. Equally important, the capacity to manage comorbidities.

ASD is no longer the domain of psychiatrists and special educators alone. ASD therapy/management should be viewed holistically with nutritional dysfunctions being addressed. Autism remains a challenging condition for individuals and their families. Certainly, the outlook nowadays is much better than it was a generation ago. Today, with appropriate intervention and therapy, clinicians can help reduce many of the autism symptoms. Still, we believe a significant level of engagement by primary care physicians is required and should be mandated. Just as important, it is necessary that physicians increase their comprehension and knowledge base around CAM. In this section, we discuss not only phytochemicals/nutraceuticals natural products approaches but also novel technology intervention.

Lastly, taking into account the diversity of herbal medicines and varieties of integrative therapy combined with herbal medicines, we believe future research should standardize the optimal composition of herbal medicines and types of integrative therapy combination. This standardization will improve the applicability and generalization of phytochemicals/nutraceuticals utilization for healthcare management of children with ASD.

Overview of Nutritional Therapy for Autism Spectrum Disorder

Carla Vartanian

Abstract The objective of this chapter is to evaluate the latest research pertinent to nutritional management in the treatment of autism spectrum disorder (ASD) and discuss the effectiveness of dietary interventions, nutritional approaches, and supplementation in ASD. To date, the best conventional treatments for autism have been based on a combination of pharmacotherapy, behavioral treatments, and nutritional/dietary therapy, leading many parents and caregivers to opt for specific dietary interventions in the hope of alleviating the symptoms of their children and helping them cope with this disorder. Thus, the role of a registered dietitian and a nutrition specialist is crucial in planning specific nutritional and dietary interventions tailored to individual needs, to make sure the child's nutritional needs for growth and development are being met. In addition, a careful monitoring of the nutritional status and the positive or negative outcomes pertinent to the planned intervention is a must. Furthermore, numerous studies have also discussed how the maternal diet and specific dietary supplements might affect the behavioral development of children in the first few years of life. A review of the abovementioned nutrition-related key points is discussed in this chapter.

Keywords ASD · Autism · Diet · Nutritional deficiency · Nutritional therapy · Nutrition management · Dietary intervention

1 Introduction

The number of children diagnosed with autism has been significantly increasing worldwide over the last decades. Autism is considered a complex neurobehavioral condition that includes impairments in social interaction, developmental language, and communication skills combined with rigid, repetitive behaviors (Fig. 1). With both communication and behavior being greatly affected, children with autism often

C. Vartanian (✉)
Royal Society of Medicine, General Practice with Primary Healthcare Section, London, UK

M. M. Essa, M. W. Qoronfleh (eds.), *Personalized Food Intervention and Therapy for Autism Spectrum Disorder Management*, Advances in Neurobiology 24,
https://doi.org/10.1007/978-3-030-30402-7_17

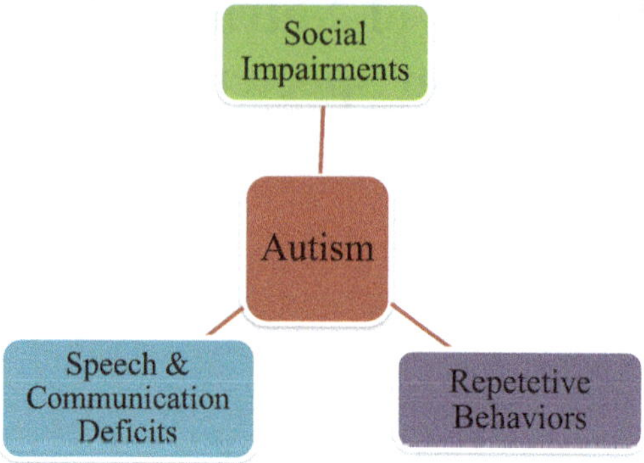

Fig. 1 The characteristics of autism

require special care since self-expression and communication in everyday life matters becomes very difficult. Furthermore, the *Diagnostic and Statistical Manual of Mental Disorders*, created by the American Psychiatric Association to diagnose mental disorders, classifies people with autism spectrum disorder to have limited interests and repetitive behaviors in addition to other symptoms such as the inability to function properly in all areas of life [1]. Autism can be diagnosed at any age; however, symptoms usually appear in the first few years of life. ASD is reported to be more common in boys than girls, with an estimated ratio of 4:1, respectively [2]. The reason for this dominance in males is still unknown; however, many therapies have been proposed, including the fact that it might be related to the direct effect of sex chromosomes [3].

2 Causes of Autism Spectrum Disorder

To date, specific underlying causes of autism spectrum disorder cannot be identified [4] and research has shown that only less than 12% of autism cases have specific identified causes [5]. Many risk factors have been investigated throughout the years, among which are genetic, infectious, nutritional, environmental/chemical, and maternal metabolic conditions such as diabetes, obesity, and the use of anti-seizure medications during pregnancy.

Most researchers think that certain combinations of genes may predispose a child to autism and genetics and environmental factors play a big role in developing autism in addition to the neurologic, metabolic, and immunologic factors [6, 7]. As a matter of fact, hereditary factors have shown to play a role in this developmental disorder with an estimated 2–8% chance [8]; raising to 12–20% in specific cases in which one of the siblings also shows other impairments associated with autism [9].

Environmental factors and the positive effect of different protective factors related to autism spectrum disorder have also been studied. Among these is the role of unsaturated fatty acids such as linoleic acid, omega-3, and omega-6, on the retinal and brain development *in* the first 2 months of pregnancy, which is considered the most critical period of embryonic physical development [10]. For instance, it was reported that that high maternal intake of omega-6 and linoleic acid is inversely associated with ASD risk in offspring, corresponding to a 34% reduction in autism risk, thus confirming that fatty acid consumption of different diets has an inverse effect on risk of autism [11]. In addition to the unsaturated fatty acids, maternal folic acid supplementation during early pregnancy was also shown to be associated with less behavioral and language development problems in offsprings during their first years of life, in addition to a lower incidence of autism [12].

3 Nutritional Management of Autism Spectrum Disorder

3.1 Special Diets

Children with autism spectrum disorder are often at risk of significant nutritional deficiencies, metabolic imbalances, and digestive problems due to feeding problems and unusual eating patterns. Furthermore, there is a lot of speculation related to the potential role of nutrition and metabolism in affecting the behavior of these children. In addition, it has been suggested that autistic children might benefit from special diets in hopes of reducing their symptoms. The most popular of these diets is the gluten-free and casein-free diet (GFCF).

3.1.1 The Gluten-Free and Casein-Free Diet

This diet basically consists of removing both gluten and casein types of proteins (found in wheat, rye, barley, and milk products, respectively) from the diet of children suffering from autism. This is due to the theory that changes in the metabolism of these specific proteins may result in high opioid peptide levels which in turn may affect the central nervous system and the brain and have a negative impact on behavior. Studies related to the effects of GFCF diets are limited because of the difficulty in monitoring the adherence of autistic children to these GFCF diets. However, available research data, to support the use of a casein-free diet, a gluten-free diet, or a combined gluten-free, casein-free diet as a primary treatment for individuals with ASD, suggests that evidence is lacking [13]. Additionally, the literature currently available suggests the need for further studies before implementing this specific diet and removing both gluten and casein.

3.1.2 Other Nutritional Interventions

Many other nutritional approaches and dietary therapies have also been proposed such as the ketogenic diet and yeast-free diet, in addition to the restriction of food allergens (Fig. 2). Other experimental therapies have included the use of dairy-free diets and the use of camel's milk [14].

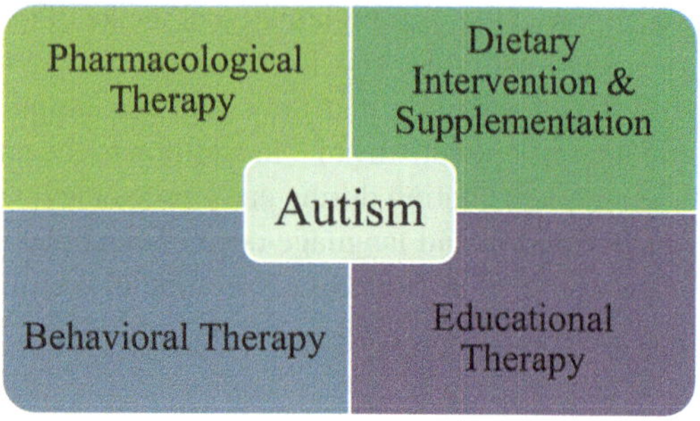

Fig. 2 Components of the treatment of autism

3.1.3 The Role of Dietary Supplements

Probiotics, digestive enzymes, and dietary supplementation with micronutrients such as such as vitamins A, C, B6, folic acid, B12, and D and minerals like magnesium, zinc, and selenium have also been studied in autism spectrum disorder; one of the most researched supplements being, the unsaturated fatty acids. For instance, two studies have reported low levels of omega-3 fatty acid observed in the blood of autistic children [15, 16]. Many studies suggest that customized vitamin/mineral supplementation is beneficial for children with autism spectrum disorder, and three studies have demonstrated that children with this disorder have impaired methylation, decreased glutathione, and increased oxidative stress [17–19]. Other studies have also reported the effectiveness of digestive enzyme supplementation in autism. One randomized, double-blind study did find that digestive enzymes were helpful for autism [20], but another similar study did not find significant benefits [21]. A recent randomized, controlled, single-blind 12-month treatment study of a comprehensive nutritional and dietary intervention had been conducted. It involved 67 children and adults with autism spectrum disorder between the ages of 3 and 58 from Arizona and 50 non-sibling neurotypical controls of similar age and gender. The test aimed at studying carnitine and homocysteine levels in the treatment group in addition to other micronutrients. Compared to the nontreatment group, the treatment group had a significant decrease in homocysteine, and a modest increase (plasma carnitine was only approximately 25%), which was less compared to previous studies. The authors hypothesized that intracellular levels may be better predictors [22].

To date, there is no cure for autism. The Food and Drug Administration (FDA) plays an important role in warning companies and taking action against those making improper claims about their products' intended use as a treatment or cure for autism or autism-related symptoms. They are also involved in informing consumers about the health consequences of potential products or treatments claiming to "cure" autism [23]. In addition, a recent review of the nutritional and dietary interventions for autism spectrum disorder showed that there is little evidence to support the use of nutritional supplements or dietary therapies for children with ASD [24].

3.2 Nutritional Concerns in Autism Spectrum Disorder

Children with ASD may have limited food intake leading to decreased consumption of nutritious meals. Many children with autism do not always get the adequate nutrition necessary for their growth and development due to many reasons [25–27]. For instance, some children may limit their food intake or only eat certain foods because of how the foods feel in their mouths. Others might try to avoid specific foods because they associate them with stomach pain or discomfort. The most common GI symptoms include chronic diarrhea, constipation, abdominal discomfort and bloating, gastroesophageal reflux disease, and a leaky gut syndrome. According to the American Academy of Nutrition and Dietetics, children with autism may have limited food selection or strong food dislikes because of their sensitivity to taste, smell, color, and texture of foods; additionally, the habit of eating small quantities since it may be hard for them to focus on eating the meal for an extended period. They may also be more prone to suffer from constipation because of their limited food choices. However, this can be resolved through a high-fiber diet with plenty of fluids, fruits and vegetables (if possible), and regular physical activity. Medication interaction is another health concern that children with autism may suffer from as some stimulant medications used with autism may lower the appetite while others may increase appetite or affect the absorption of certain vitamins and minerals [28]. In addition, it was reported that people with autism suffer more from digestive problems (such as abdominal pain or vomiting) than people without autism. However, further research is needed to confirm these findings [29, 30].

3.3 Nutrition Strategies for Children with Autism

Caring for a child with autism can be challenging on many levels, as every child is unique. The treatment of autism spectrum disorder usually consists of a comprehensive program of educational intervention, behavioral treatment, and developmental therapies in addition to dietary intervention, when recommended (Fig. 2). The characteristics of the disorder can greatly impact feeding and nutrition. Feeding problems such as unusual eating patterns and food selectivity are very common in children with ASD [31]. As a consequence, this may lead to the consumption of unbalanced meals and result in nutritional deficiencies [32].

Parents and caregivers should always work with a registered dietitian nutritionist when planning nutritional and dietary interventions to make sure the child's nutritional needs for growth and development are being met, even while on a special diet or a nutrient restriction intervention. Children who are autistic may sometimes need strategies such as positive reinforcement to increase acceptance of new foods. In addition, parents need to establish appropriate structure and rules for mealtimes to make it more fun and appealing to the child and, most importantly, decrease all potential stressors. Specialists also emphasize that continual monitoring of the diet and nutritional status of children with ASD is required [33].

4 Conclusions

While research has been growing in all aspects related to autism spectrum disorder, to date, available data has only supported the use of pharmacologic treatments and behavioral or educational interventions. In addition, most randomized control trials are limited. This is due to small sample sizes or being conducted with various populations and study groups owing to many limitations (including the long-term effects of the therapies and the potential for nutritional deficiencies as a result of long-term dietary exclusion) and the lack of long-term data and life-span data on health risk associated with specific and nutrient limiting diets. Thus, appropriate clinical and dietetic support should be considered during any attempt to make such dietary changes. More prospective controlled trials are also needed before recommendations about specific nutritional plans and supplementation can be made regarding ASD. Future research should be designed to identify medical nutrition therapies targeting this population, aim to better understand the link between ASD and nutrition, and determine the efficacy of dietary therapy approaches. In addition to this, if any supplementation is to be recommended in the future, many discrepancies and conflicting information in patients must be resolved as a safe and effective alternative approach for the treatment of ASD.

References

1. American Psychiatric Association. (2013). *Diagnostic and statistical manual of mental disorders* (5th. Ed.). Washington, DC: APA.
2. Fombonne, E. (2005). Epidemiology of autistic disorder and other pervasive developmental disorders. *The Journal of Clinical Psychiatry, 66*(suppl 10), 3–8.
3. Baron-Cohen, S., Lombardo, M. V., Auyeung, B., Ashwin, E., Chakrabarti, B., & Knickmeyer, R. (2011). Why are autism spectrum conditions more prevalent in males? *PLoS Biology, 9*, e1001081.
4. Cubala-Kucharska, M. (2010). The review of most frequently occurring medical disorders related to etiology of autism and the methods of treatment. *Acta Neurobiologiae Experimentalis, 70*, 141–146.
5. Bailey, A., Le Couteur, A., Gottesman, I., Bolton, P., Simonoff, E., Yuzda, E., et al. (1995). Autism as a strongly genetic disorder: Evidence from a British twin study. *Psychological Medicine, 25*, 63–77.
6. Packer, A. (2016). Neocortical neurogenesis and the etiology of autism spectrum disorder. *Neuroscience and Biobehavioral Reviews, 64*, 185–195.
7. Nardone, S., & Elliott, E. (2016). The interaction between the immune system and epigenetics in the etiology of autism spectrum disorder. *Frontiers in Neuroscience, 10*, 329–338.
8. Muhle, R., Trentacoste, S. V., & Rapin, I. (2004). The genetics of autism. *Pediatrics, 113*, e472–e486.
9. Sinivasan, P. (2009). A review of dietary interventions in autism. *Annals of Clinical Psychiatry, 21*(4), 237–247.
10. Karimi, P., Kamali, E., Mousavi, S. M., & Karahmadi, M. (2017). Environmental factors influencing the risk of autism. *Journal of Research in Medical Sciences : The Official Journal of Isfahan University of Medical Sciences, 22*, 27.

11. Lyall, K., Munger, K. L., O'Reilly, É. J., Santangelo, S. L., & Ascherio, A. (2013). Maternal dietary fat intake in association with autism spectrum disorders. *American Journal of Epidemiology, 178*, 209–220.
12. Schmidt, R. J., Tancredi, D. J., Ozonoff, S., Hansen, R. L., Hartiala, J., Allayee, H., et al. (2012). Maternal periconceptional folic acid intake and risk of autism spectrum disorders and developmental delay in the CHARGE (CHildhood Autism Risks from Genetics and Environment) case-control study. *The American Journal of Clinical Nutrition, 96*, 80–89.
13. Chaidez, V., Hansen, R. L., & Hertz-Picciotto, I. (2014). Gastrointestinal problems in children with autism, developmental delays or typical development. *Journal of Autism and Developmental Disorders, 44*(5), 1117–1127.
14. American Academy of Nutrition and Dietetics. (2018). Update on the use of nutritional and dietary interventions for children diagnosed with autism. https://www.eatrightpro.org/news-center/nutrition-trends/diseases-and-conditions/update-on-the-use-of-nutritional-and-dietary-interventions-for-children-diagnosed-with-autism. Accessed 22 Sept 2018.
15. Seung, H. K., Rogalski, Y., Shankar, M., & Elder, J. (2007). The gluten-and casein-free diet and autism, communication outcomes from a preliminary double blind clinical trial. *Journal of Medical Speech-Language Pathology, 15*(4), 337–345.
16. Elder, J. H. (2008). The gluten-free, casein-free diet in autism: An overview with clinical implications. *Nutrition in Clinical Practice, 23*(6), 583–588.
17. Buie, T., Campbell, D. B., Fuchs 3rd, G. J., Furuta, G. T., Levy, J., Vandewater, J., et al. (2010). Evaluation, diagnosis, and treatment of gastrointestinal disorders in individuals with ASDs: A consensus report. *Pediatrics, 125*, S1–S18.
18. James, S. J., Cutler, P., Melnyk, S., Jernigan, S., Janak, L., Gaylor, D. W., et al. (2004). Metabolic biomarkers of increased oxidative stress and impaired methylation capacity in children with autism. *The American Journal of Clinical Nutrition, 80*, 1611–1617.
19. James, S. J., Melnyk, S., Jernigan, S., Cleves, M. A., Halsted, C. H., Wong, D. H., et al. (2006). Metabolic endophenotype and related genotypes are associated with oxidative stress in children with autism. *American Journal of Medical Genetics. Part B, Neuropsychiatric Genetics, 141*, 947–956.
20. James, S. J., Melnyk, S., Fuchs, G., Reid, T., Jernigan, S., Pavliv, O., et al. (2009). Efficacy of methylcobalamin and folinic acid treatment on glutathione redox status in children with autism. *The American Journal of Clinical Nutrition, 2009*(89), 425–430.
21. Dyerberg, J., Madsen, P., Møller, J. M., Aardestrup, I., & Schmidt, E. B. (2010). Bioavailability of marine n-3 fatty acid formulations. *Prostaglandins, Leukotrienes, Essential Fatty Acids, 83*, 137–141.
22. Munasinghe, S. A., Oliff, C., Finn, J., & Wray, J. A. (2010). Digestive enzyme supplementation for autism spectrum disorders: A double-blind randomized controlled trial. *Journal of Autism and Developmental Disorders, 40*, 1131–1138.
23. Adams, J. B., Audhya, T., Geis, E., Gehn, E., Fimbres, V., Pollard, E. L., et al. (2018). Comprehensive nutritional and dietary intervention for autism spectrum disorder—A randomized, controlled 12-month trial (2018). *Nutrients, 10*, 369.
24. US Food and Drug Administration. (2017). Autism: Beware of potentially dangerous therapies and products. https://www.fda.gov/ForConsumers/ConsumerUpdates/ucm394757.htm. Accessed 22 Sept 2018.
25. Adams, J. B., Audhya, T., McDonough-Means, S., Rubin, R. A., Quig, D., Geis, E., et al. (2011). Nutritional and metabolic status of children with autism vs. neurotypical children, and the association with autism severity. *Nutrition & Metabolism, 8*(1), 34.
26. Arnold, G. L., Hyman, S. L., Mooney, R. A., & Kirby, R. S. (2003). Plasma amino acids profiles in children with autism: Potential risk of nutritional deficiencies. *Journal of Autism and Developmental Disorders, 33*(4), 449–454.
27. Zimmer, M. H., Hart, L. C., Manning-Courtney, P., Murray, D. S., Bing, N. M., & Summer, S. (2012). Food variety as a predictor of nutritional status among children with autism. *Journal of Autism and Developmental Disorders, 42*(4), 549–556.

28. Herndon, A. C., DiGuiseppi, C., Johnson, S. L., Leiferman, J., & Reynolds, A. (2009). Does nutritional intake differ between children with autism spectrum disorders and children with typical development? *Journal of Autism and Developmental Disorders, 39*(2), 212–222.
29. American Academy of Nutrition and Dietetics. (2018). Autism Spectrum Disorders and diet. https://www.eatright.org/health/diseases-andconditions/autism/nutrition-for-your-child-with-autism-spectrum-disorder-asd. Accessed 23 Sept 2018.
30. Sathe, N., Andrews, J. C., McPheeters, M. L., & Warren, Z. E. (2017). Nutritional and dietary interventions for autism spectrum disorder: A systematic review. *Pediatrics, 139*(6), pii: e20170346.
31. Sharp, W. G., Berry, R. C., McCracken, C., Nuhu, N. N., Marvel, E., Saulnier, C. A., et al. (2013). Feeding problems and nutrient intake in children with autism spectrum disorders: A meta-analysis and comprehensive review of the literature. *Journal of Autism and Developmental Disorders, 43*(9), 2159–2173.
32. Ma, N. S., Thompson, C., & Weston, S. (2016). Brief report: Scurvy as a manifestation of food selectivity in children with autism (2016). *Journal of Autism and Developmental Disorders, 46*(4), 1464–1470.
33. Kawicka, A., & Regulska-Ilow, B. (2013). How nutritional status, diet and dietary supplements can affect autism. A review. *Roczniki Państwowego Zakładu Higieny, 64*(1), 1–12.

Importance of Nutrition Intervention in Autistic Patients

Tahra ElObeid, Joyce Moawad, and Zumin Shi

Abstract Along with the issues of inflated social and financial burden associated with autism spectrum disorder (ASD), specific treatment for this disorder has also not been developed. Having a thorough look at previous trials done to treat autism, we find that nutrition intervention had been used frequently as a complementary form of therapy. Indeed, an early diagnosis of nutrition deficiency and metabolic disorders done concomitantly with accurate therapeutic interventions can be a cornerstone for improving cognitive and behavioral aptitudes of people with autism. Several studies have showed that increasing the intake of specific nutrients can reduce the symptoms and comorbidities associated with autism. Consequently, nutrition intervention and appropriate supplementation can be crucial in managing and treating autism. This paper will discuss recent literature on the significance of metabolic aspects in autistic disorder and highlight the influence of nutrition intervention on the symptoms of autism.

Keywords ASD · Autism · Diet · Nutrition deficiency · Nutrition intervention · Nutrition therapy · Metabolic disorders

1 Introduction

According to the World Health Organization (WHO), "there is no health without mental health" [1]. Indeed, psychological well-being is a fundamental aspect of total well-being, regardless of age, sex, religion, society, and race. Data collected on diets, energy consumption and their influence on emotional wellness essentially focus on three themes: psychological wellness, proposed mediations (nutritional screening and appraisal for psychological well-being), and the link between diet and psychological wellness. Miscellaneous studies have emphasized the important role

T. ElObeid (✉) · J. Moawad · Z. Shi
Human Nutrition Department, College of Health Sciences, Qatar University, Doha, Qatar
e-mail: Tahra.e@qu.edu.qa; zumin@qu.edu.qa

M. M. Essa, M. W. Qoronfleh (eds.), *Personalized Food Intervention and Therapy for Autism Spectrum Disorder Management*, Advances in Neurobiology 24,
https://doi.org/10.1007/978-3-030-30402-7_18

of nutrition intervention and the influence of individualized diets on mental disorders. Many data analyses have reported a significant interface between nutrition and many disorders like those on the schizophrenia spectrum, depressive disorders, anxiety disorders, eating disorders, neurocognitive disorders, and neurodevelopmental disorders including autism [2].

The Diagnostic and Statistical Manual of Mental Disorders-fifth edition (DSM 5) reports that autism spectrum disorder manifests as abnormalities in social communication and interaction and the occurrence of repetitive, restricted patterns of behavior or activities. According to the American Psychiatric Association, ASD represents a single continuum of impairments with a varying degree of severity [3]. A closer look at existing literature helps us find that many studies have tried to describe the root cause of ASD. However, there is no significant information on the etiology and pathogenesis of autism. Some hypotheses state that ASD is caused by genetic origins [4, 5]. Many studies have reported that environmental agents, mitochondrial disorders, parental age, infections during pregnancy, and testosterone levels can be risk factors for developing ASD [6]. On the other hand, some research has stated that ASD can result from an interaction between genetic and environmental factors with oxidative stress [7]. Other studies have showed that ASD may develop due to exposure to environmental toxins during the fetal and immediate neonatal period [8]. Furthermore, nutritional deficiencies may be a risk factor for development of ASD as well. Studies have revealed that many children with ASD are picky eaters, due either to sensitivities to certain types of food or selective eating behaviors. This can easily lead to inadequate nutrient intake [9]. In this regard, many findings reveal that nutrition intervention can significantly help some ASD patients [10–12]. Recent evidence has also suggested that probiotics, digestive enzymes, vitamins, minerals, amino acids, and specialty supplements are key components of each: the biomedical approach, primary intervention, and adjunctive measures of ASD. Of special interest to current authors is the exploration of metabolic abnormalities associated with ASD and a review of the literature that highlights the influence and relationship between nutrition, dietary intervention, and ASD.

2 Diet, Metabolic Disorders, and Autism

Metabolic disorders are the result of a defective flow of metabolic reactions in the body. Such errors may be a result of miscellaneous factors such as genetic mutation, poor diet, and unhealthy lifestyle. Metabolic disorders may affect many systems in the body and can also be fatal. It can reduce the psychomotor performance of an individual and hence requires certain dietary restrictions [13].

Although most ASD cases are not related to identifiable metabolic disorders, numerous neurometabolic disorders associated with ASD such as phenylketonuria (PKU), profound biotinidase deficiency (PBD), disorders of purine metabolism, and Smith-Lemli-Opitz syndrome (SLOS) have been recognized [14, 15].

Phenylketonuria affects approximately 1 out of 10,000 newborns in European, Chinese, and Korean populations [16]. PKU is an inborn error of metabolism that results in decreased function of phenylalanine hydroxylase, which is responsible for the transformation of phenylalanine into tyrosine [15]. If left untreated, excess phenylalanine amino acid builds up in the blood leading to brain damage and other neurological problems, such as ASD [17]. However, an early detection of PKU through newborn metabolic screening can contribute to successful treatment. Indeed, PKU cannot be treated. However, associated complications can be eradicated by following a controlled diet that limits the intake of protein products rich in phenylalanine. Nutrition intervention is crucial for prevention of brain damage and subsequent developmental and intellectual problems, thereby facilitating normal development of a child with PKU [16, 18].

The two disorders of purine metabolism related to ASD may be distinguished: adenylosuccinase deficiency and adenosine deaminase deficiency. The first is characterized by the accumulation of succinyl aminoimidazole carboxamide riboside and succinyl adenosine in body fluids, while the latter is associated with improper conversion of deoxyadenosine to the nontoxic deoxyinosine resulting in accumulation of deoxyadenosine that impairs normal immune function. Adenylosuccinase deficiency manifests as developmental delay, agitation, seizures, and autistic features (e.g., poor eye contact). Unfortunately, no effective treatment is available yet [15, 16].

Another associated disorder is profound biotinidase deficiency (PBD) which is an autosomal recessively inherited disorder related to biotin metabolism. It manifests as neurological symptoms like hypotonia, breathing problems, ataxia, sucking disorders, intractable seizures, and global developmental delay [19]. Research findings have reported that early treatment with the co-factor biotin may protect from neurologic complications.

Yet another metabolic disorder related to ASD is Smith-Lemli-Opitz syndrome (SLOS), which is an autosomal recessive disorder caused by the deficiency of 7-dehydrocholesterol reductase, a final enzyme in the cholesterol synthetic pathway. These abnormalities result in impaired embryonic and fetal somatic development, causing postnatal abnormalities of learning, growth, behavior, and language [20]. Although an elevated level of cholesterol in the blood of ASD patients is not associated with the severity of ASD symptoms, findings have showed that cholesterol supplementation may have a positive influence on autistic behaviors among children with SLOS [21].

Additionally, some studies have revealed that gastrointestinal disturbances, which may lead to behavioral impairments, are common among children with ASD. One of these metabolic abnormalities, first reported in ASD patients 30 years ago, is associated with elevated levels of peptides [22], especially gluten, casein, and gliadin.

There are also other disorders resulting from abnormalities in organic acid metabolism that are reported in ASD [23, 24]. For instance, the accumulation of organic acids in urine may indicate a disorder in metabolism, nutritional deficiencies, and bacterial overgrowth in the body [25]. Metabolic fingerprinting has indicated a correlation between the levels of succinic acid and the presence of bacterial infection [23, 26]. Urine analyses have demonstrated elevated levels of butyric acid in the urine samples of ASD patients [27].

The aforementioned issues will be described further in this paper.

3 Nutritional Disorders in Autistic Children

When the gastrointestinal tract (GIT) functions properly, enzymes break down proteins into peptides and then into amino acids. The latter is absorbed into the bloodstream and transported to the body. Some disruptions in this process are considered to be associated with ASD and are called opioid-excess theory. Opioids are a group of chemical compounds that affect the function of the brain and nervous system. These compounds influence the perception of emotion and behavior. The main assumption of this theory is that some children with ASD suffer from increased gut permeability and improper production of digestive enzymes related to gluten and casein. Inadequate levels of these enzymes result in failures to transform gluten and casein into amino acids. Consequently, increased gut permeability enables leaking into the bloodstream, where metabolites can pass through the brain–blood barrier [28] causing disruption to the normal functioning of the nervous system by regulating signal transduction in the brain. Findings have reported that elimination of gluten- and casein-containing food from the diet of such children resulted in the disappearance of the symptoms associated with ASD [29]. Gluten is found in wheat, oats, barley, and rye, and casein is a protein of animal origin found in milk and other dairy products. Inconsistently, some studies have stated that a gluten- and casein-free diet (GFGF) can lead to other nutritional deficiencies and low plasma levels of essential amino acids.

4 *Candida* and Nutrient Disorders

Children with ASD commonly have gastrointestinal (GI) problems [30]. Researchers have found that such GI problems can contribute to the severity of the ASD and the associated symptoms. Although the mechanism of this relation is not clear, abnormal gut flora and the abuse of oral antibiotics could be the culprit [31]. Findings have revealed that some ASD individuals have decreased levels of beneficial bacteria and increased levels of harmful bacteria and yeast. Indeed, the harmful bacteria and yeast, found in insufficient quantities, can contribute to disorders in mental functioning and behavior due to their ability to produce toxins such as alcohol.

Candida albicans is a yeast-like fungus present in almost all humans. It is found in the dark moist mucous membranes that line the vagina, intestinal tract, and mouth. Indeed, *Candida albicans* can cause infections, especially in immunocompromised individuals. In an ordinary situation, the fungus exists only in small colonies. However, if the natural balance is disrupted, it grows rapidly leading to undesired symptoms (e.g., white yeast infection of the mouth and tongue) [32]. Some research has suggested that toxins produced by candida have severe impacts. They can affect the brain and result in severe long-term disruptions of the immune system [33]. The overgrowth of *Candida albicans* can be associated with behavioral disorders in children with ASD, such as hyperactivity, aggression, and problems with concentration. It manifests as headaches, stomach problems, painful gases, fatigue, or depression. Studies have showed that some safe methods may be used to treat fungus overgrowth like following a low-sugar diet [34].

5 Nutritional Strategies in Autism

Studies have demonstrated significantly low levels of nutrients in blood, urine, hair, and other tissues in children with ASD. Laboratory analyses exhibit low levels of vitamins, minerals, essential fatty acids (EFA), and amino acids in children with autism. Consequently, these deficiencies result in neurological problems such as weakness of vision, speech, attention, and socialization. Consistently, such findings emphasize the significant role of vitamins and nutrition supplementation in treating autism [35–37].

Indeed, different studies have demonstrated the benefits of targeted vitamin/mineral nutritional supplementation in the improvement of neurological disorders and behavioral and cognitive gains in autistic children [10, 38]. Urine analyses of succinic, adipic, and suberic acids enable the detection of nutritional deficits in children with ASD and the introduction of appropriate supplementation [25]. Existing literature has showed that the diet of children with ASD children is unbalanced which may lead to nutritional deficits resulting in metabolic disorders. Diet analyses of children with autism reported that their diet is considerably low in vitamin C; vitamins B1, B2, B6, B9, and B12; and vitamin A [39]. Children with ASD often suffer from impaired methylation, decreased glutathione, and oxidative stress [40]. In such cases, nutritional supplementation (with vitamin methyl-B12, folinic acid, and trimethylglycine) is beneficial.

Studies have stated that magnesium and vitamin B6 can reduce symptoms of hyperexcitability (physical aggression, instability, scholar attention, hypertony, spasm, myoclonus) [41]. Magnesium has diverse essential functions. It is involved in bone formation, regulates enzyme activities included in at least 300 enzyme processes of intermediate metabolism, and is necessary in all enzyme reactions involving adenosine triphosphate. Moreover, it is included in many enzymatic reactions present in nucleic acid metabolism [42]. A number of studies have described the involvement of magnesium in the pathogenesis of autism [12, 43]. Analysis of hair and nails of children with autism determines the deficiency of micronutrients, including magnesium [44]. Vitamin B6 participates in transamination of amino acids, decarboxylation reactions, modulation of the activity of steroid hormones, and regulation of gene expression. Many studies have showed that treatments with vitamin B6 and magnesium supplementation have resulted in improvement in speech/communication, social interaction, and stereotype behavior [12]. Eating habits and nutritional deficiencies in autistic children further emphasize the necessary role of nutritional supplementation.

The levels of dicarboxylic acids after supplementation in the diet of autistic children were first described by Kałużna-Czaplińska and co-authors [45, 46]. A hypothesis on the significance of vitamins B2 and B6 and magnesium in treating autism was set due to observed dietary deficiency of the aforementioned nutrients [39, 44] and the high levels of dicarboxylic acids in the urine of autistic children. Researchers have reported that therapy with vitamin B2 can be considered a significant potential intervention in increased urinary excretion of adipic and suberic acid. Additionally, magnesium was found to be another element of therapeutic intervention in cases when high levels of succinic acid are observed [47]. Consistently, statistical analyses have showed that parents could notice an improvement of some autistic symptoms like the ability to concentrate and make eye contact.

6 Personalized Diet for Autism

For decades, medicine and diagnosis have used information obtained from analyses of metabolic profiles as they provide accurate information about the dynamics of the biological system, reflecting genetic and physiological changes [48]. The term "metabolome" refers to the complete set of metabolites, including all small-molecule metabolites and excluding proteins and nucleic acids [49]. The importance of metabolic profiling in the diagnosis of cancers and neurological and metabolic disorders has been increasing. It has also become a crucial part of the diagnostic and therapeutic process in ASD. Homovanillic (HVA) and vanillylmandelic (VMA) acids are considered to play a crucial role in the diagnosis of health problem, including neurological diseases and disorders.

The determination of 14 organic acids revealed differences in metabolic profiles of ASD and healthy individuals [24]. The levels of these compounds in children with ASD were found to be elevated. The correlation between severity of ASD symptoms and the levels of particular metabolites was implicated.

Findings have showed that some metabolic abnormalities, such as GI tract dysfunction, are associated with ASD and can lead to aggravation of symptoms [50]. Data analyses have reported that some children with ASD are diagnosed with digestive problems due to limited ability to digest proteins. Hence, children with autism follow specialized diets that are low in proteins. Proteins are built with long chains of amino acids which can be reassembled to form critical substances like neurotransmitters, enzymes, hormones, antibodies, immunoglobulins, and many others.

Consistently, many analytic studies have reported abnormally high levels of the amino acid, homocysteine (Hcy), in children with ASD. Findings have revealed that children with ASD exhibit improper metabolism of Hcy by the cerebral tissue resulting in the accumulation of this compound in the nervous system [51]. Homocysteine is considered to contribute to the neuronal damage and cell loss associated with ASD [52]. The relationship between levels of Hcy in serum samples of ASD individuals and vitamin B12 deficiency had been described first by Paşca et al. [53]. They indicated that high levels of Hcy and oxidative stress markers are associated with ASD. Additionally, other studies have also showed significantly higher levels of Hcy in ASD patients compared to healthy children. The results became a cornerstone in the preparation of individual and personalized diets for ASD patients [45, 46]. This study was independently verified by Ali et al. [54]. Studies have also revealed that the introduction of supplements such as vitamin B6, B12, and folate could reduce the Hcy levels in blood.

Some disturbances in the levels of glutamate and glutamine have been indicated. Findings showed that children with ASD have high levels of glutamate and a low level of glutamine [55]. The authors have suggested that the levels of these amino acids may serve as a distinguishing factor of high IQ between children with ASD and healthy children.

Hypotheses stating that the levels of glutamate and homocysteine are associated with aggression and irritability could be the key to treating such ASD symptoms.

Furthermore, the levels of other amino acids such as taurine [56], lysine [57, 58], and aspartic acid [59] were also found to be increased in children with ASD.

Indeed, a well-balanced diet containing all the essential and nonessential nutrients is indispensable for proper functioning of the body. Consequently, a diet rich in some nutrients and substances may contribute to the improvement of different health conditions. However, individuals with ASD exhibit many nutritional deficits due to their restrictive diets. Thus, the assumption that nutrition intervention can significantly contribute to treat ASD-associated symptoms and comorbidities is reasonably implicated.

7 Conclusion

Vitamins, minerals, amino acids, and essential fatty acids found in food are a necessity for the proper development of a child's brain. Any deficiency in such nutrients can affect the production of neurotransmitters and disturb normal visual and cognitive processing. Nutrition deficiencies, allergies, sensitivities, and gastrointestinal disorders are often reported in children with autism. Autistic children have restrictive eating behaviors and problem feeding behaviors that put them at risk for poor nutrition intake. Many researchers believe that improving nutritional intake of the autistic child can help improve overall health, behavior, and brain function. In existing literature, there are reports of the beneficial effects of vitamins and different nutritional supplements in the treatment of autism. This has also been ascertained by parents of autistic children who have reported improvement of some autistic symptoms like concentration and eye contact after introducing nutrition supplementation. Consequently, there is an increasing interest in an individualized diet and supplementation for autistic children.

Nutritional strategies and personalized diet have been necessary and helpful for many autistic children with a disorder in metabolism, nutritional deficiencies, and bacterial overgrowth. Indeed, nutritional interventions are a matter of parent/caregiver choice.

Many studies have been conducted by scientists across different countries and scientific specialties to explore nutritional strategies and personalized diets for those with ASD. Continued observations and research on the association between nutrition intervention and ASD and its comorbidities are required.

References

1. World Health Organization. (2010). *Mental health: Strengthening our response*. Fact Sheet Noe, 220.
2. Eissa, N., Al-Houqani, M., Sadeq, A., Ojha, S. K., Sasse, A., & Sadek, B. (2018, May 16). Current enlightenment about etiology and pharmacological treatment of autism spectrum disorder. *Frontiers in Neuroscience, 12*, 304. https://doi.org/10.3389/fnins.2018.00304
3. American Psychiatric Association. (2013). *Diagnostic and statistical manual of mental disorders* (5th ed.). Arlington, VA: American Psychiatric Association.
4. Folstein, S. E., & Rosen-Sheidley, B. (2001). Genetic of autism: Complex aetiology for a heterogeneous disorder. *Nature Reviews Genetics, 2*(12), 943–955.
5. Freitag, C. M. (2007). The genetics of autistic disorders and its clinical relevance: A review of the literature. *Molecular Psychiatry, 12*(1), 2–22.
6. Ratajczak, H. V. (2011). Theoretical aspects of autism: Causes–a review. *Journal of Immunotoxicology, 8*, 68–79.
7. Ming, X., Stein, T. P., Brimacombe, M., Johnson, W. G., Lambert, G. H., & Wagner, G. C. (2005). Increased excretion of a lipid peroxidation biomarker in autism. *Prostaglandins, Leukotrienes & Essential Fatty Acids, 73*, 379–384.
8. Landrigan, P. J. (2010). What causes autism? Exploring the environmental contribution. *Current Opinion in Pediatrics, 22*, 219–225.
9. Feucht, S., Ogata, B., & Lucas, B. (2010). Nutrition concerns of children with autism spectrum disorders. *Nutrition Focus for Children with Special Health Needs, 25*(4), 1–13.
10. Adams, J. B., & Holloway, C. (2004). Pilot study of a moderate dose multivitamin/mineral supplement for children with autistic spectrum disorder. *The Journal of Alternative and Complementary Medicine, 10*, 1033–1039.
11. Mousain-Bosc, M., Roche, M., Polge, A., Pradal-Prat, D., Rapin, J., & Bali, J. P. (2006). Improvement of neurobehavioral disorders in children supplemented with magnesium – Vitamin B6. *Magnesium Research, 19*, 46–52.
12. Rossignol, D. A. (2009). Novel and emerging treatments for autism spectrum disorders: A novel systematic review. *Annals of Clinical Psychiatry, 21*, 213–236.
13. National Center for Biotechnology Information (US). (1998). The NCBI Handbook [Internet]. 2nd edition. Bethesda, MD: National Center for Biotechnology Information (US).
14. Sikora, D. M., Pettit-Kekel, K., Penfield, J., Merkens, L. S., & Steiner, R. D. (2006). The near universal presence of autism spectrum disorders in children with Smith-Lemli-Opitz syndrome. *American Journal of Medical Genetics Part A, 140A*, 1511–1518.
15. Zecavati, N., & Spence, S. J. (2009). Neurometabolic disorders and dysfunction in autism spectrum disorders. *Current Neurology and Neuroscience Reports, 9*(2), 129–136.
16. Manzi, B., Loizzo, A. L., Giana, G., & Curatolo, P. (2008). Autism and metabolic diseases. *Journal of Child Neurology, 23*(3), 307–314.

17. Donlon, J., Levy, H., & Scriver, C. R. (2008). *Hyperphenylalaninemia: Phenylalanine hydroxylase deficiency. Online metabolic & molecular bases of inherited disease*. New York, NY: McGraw-Hill. (Chapter 77).
18. Baieli, S., Pavone, L., Meli, C., Fiumara, A., & Coleman, M. (2003). Autism and phenylketonuria. *Journal of Autism and Developmental Disorders, 33*(2), 201–204.
19. Weber, P., Scholl, S., & Baumgartner, E. R. (2004). Outcome in patients with profound biotinidase deficiency: Relevance of newborn screening. *Developmental Medicine & Child Neurology, 46*(7), 481–484.
20. Marcos, J., Guo, L. W., Wilson, W. K., Porter, F. D., & Shackleton, C. (2004). The implications of 7-dehydrosterol-7-reductase deficiency (Smith-Lemli-Opitz syndrome) to neurosteroid production. *Steroids, 69*(1), 51–60.
21. Tierney, E., Bukelis, I., Thompson, R. E., Ahmed, K., Aneja, A., Kratz, L., et al. (2006). Abnormalities of cholesterol metabolism in autism spectrum disorders. *American Journal of Medical Genetics Part B: Neuropsychiatric Genetics, 141B*(6), 666–668.
22. Reichelt, K. L., Hole, K., Hamberger, A., Saelid, G., Edminson, P. D., Braestrup, C. B., et al. (1981). Biologically active peptide-containing fractions in schizophrenia and childhood autism. *Advances in Biochemical Psychopharmacology, 28*, 627–643.
23. Kałużna-Czaplińska, J., Zurawicz, E., & Jóźwik, J. (2014). Chromatographic techniques coupled with mass spectrometry for the determination of organic acids in the study of autism. *Journal of Chromatography B Analytical Technologies in Biomedical Life Sciences, 964*, 128–135.
24. Kałużna-Czaplińska, J., Zurawicz, E., Struck, W., & Markuszewski, M. (2014). Identification of organic acids as potential biomarkers in the urine of autistic children using gas chromatography/mass spectrometry. *Journal of Chromatography B Analytical Technologies in Biomedical Life Sciences, 966*, 70–76.
25. Kałużna-Czaplińska, J. (2011). Noninvasive urinary organic acids test to assess biochemical and nutritional individuality in autistic children. *Clinical Biochemistry, 44*, 686–691.
26. Reig, M., Molina, D., Loza, E., Ledesma, M. A., & Meseguer, M. A. (1983). Gas-liquid chromatography in routine processing of blood cultures for detecting anaerobic bacteraemia. *Journal of Clinical Pathology, 34*(2), 189–193.
27. Clark-Taylor, T., & Clark-Taylor, B. E. (2004). Is autism a disorder of fatty acid metabolism? Possible dysfunction of mitochondrial beta-oxidation by long chain acyl-CoA dehydrogenase. *Medical Hypotheses, 62*(6), 970–975.
28. Mulloy, A., Lang, R., O'Reilly, M., Sigafoos, J., Lancioni, G., & Rispoli, M. (2010). Gluten-free and casein-free diets in the treatment of autism spectrum disorders: A systematic review. *Research in Autism Spectrum Disorders, 4*(3), 328–339.
29. Kałużna-Czaplińska, J., Michalska, M., & Rynkowski, J. (2010). Determination of tryptophan in urine of autistic and healthy children by gas chromatography/mass spectrometry. *Medical Science Monitor, 16*, 488–492.
30. Buie, T., Fuchs 3rd, G. J., Furuta, G. T., Kooros, K., Levy, J., Lewis, J. D., et al. (2010). Recommendations for evaluation and treatment of common gastrointestinal problems in children with ASDs. *Pediatrics, 125*(Suppl. 1), S19eS29. https://doi.org/10.1542/peds.2009-1878
31. Adams, J. B., Johansen, L. J., Powell, L. D., Quig, D., & Rubin, R. A. (2011). Gastrointestinal flora and gastrointestinal status in children with autism-comparisons to typical children and correlation with autism severity. *BMC Gastroenterology, 16*, 11–22. https://doi.org/10.1186/1471-230X-11-22
32. Denfert, C., & Hube, B. (Eds.). (2007). *Candida: Comparative and functional genomics*. Wymondham: Caister Academic Press.
33. Emam, A. M., Mamdouh, M. E., & Abdelrahim, A. S. (2012). Candida albicans infection in autism. *Journal of American Science, 8*(12), 739.
34. Shaw, W., Kassen, E., & Chaves, E. (2000). Assessment of antifungal drug therapy in autism by measurement of suspected microbial metabolites in urine with gas chromatography-mass spectrometry. *Clinical Practice of Alternative Medicine, 1*, 15–26.

35. Adams, J. B. (2007). *Summary of biomedical treatments for autism*. San Diego, CA: ARI Publication.
36. Adams, J. B., Audhya, T., McDonough-Means, S., Rubin, R. A., Quig, D., Geis, E., et al. (2011). Effect of a vitamin/mineral supplement on children and adults with autism. *BMC Pediatrics, 12*(11), 111. https://doi.org/10.1186/1471-2431-11-111
37. Santhanam, B., & Kendler, B. (2012). Nutritional factors in autism: An overview of nutritional factors in the etiology and management of autism. *Integrative Medicine, 11*(1), 46–49.
38. Chez, M. G., Buchanan, C. P., Aimonovitch, M. C., Becker, M., Schaefer, K., Black, C., et al. (2002). Double-blind, placebo-controlled study of L-carnosine supplementation in children with autistic spectrum disorders. *Journal of Child Neurology, 17*, 833–837.
39. Xia, W., Zhou, Y., Sun, C., Wang, J., & Wu, L. (2010). A preliminary study on nutritional status and intake in Chinese children with autism. *European Journal of Pediatrics, 169*, 1201–1206.
40. James, S. J., Cutler, P., Melnyk, S., Jernigan, S., Janak, L., Gaylor, D. W., et al. (2004). Metabolic biomarkers of increased oxidative stress and impaired methylation capacity in children with autism. *The American Journal of Clinical Nutrition, 80*, 1611–1617.
41. Mousain-Bosc, M., Roche, M., Rapin, J., & Bali, J. P. (2004). Magnesium VitB6 intake reduces central nervous system hyperexcitability in children. *The Journal of the American College Nutrition, 23*, 545–548.
42. Strambi, M., Longini, M., Hayek, J., Berni, S., Macucci, F., Scalacci, E., et al. (2006). Magnesium profile in autism. *Biological Trace Element Research, 109*, 97–104.
43. Rimland, B. (2005). Puberty, aggression, and seizures. *Autism Research Review International, 19*, 3.
44. Lakshmi Priya, M. D., & Geetha, A. (2011). Level of trace elements (copper, zinc, magnesium and selenium) and toxic elements (lead and mercury) in the hair and nail of children with autism. *Biological Trace Element Research, 142*, 148–158.
45. Kałużna-Czaplińska, J., Michalska, M., & Rynkowski, J. (2011). Homocysteine level in urine of autistic and healthy children. *Acta Biochimica Polonica, 58*(1), 31e34.
46. Kałużna-Czaplińska, J., Michalska, M., & Rynkowski, J. (2011). Vitamin supplementation reduces the levels of homocysteine in the urine of autistic children. *Nutrition Research, 31*, 318–321.
47. Bralley, J., & Lord, R. (2001). *Laboratory evaluations in molecular medicine: Nutrients, toxicants, and cell regulators* (pp. 175–208). Norcross, GA: The Institute for Advances in Molecular Medicine. 173e221.
48. Clarke, C. J., & Haselden, J. N. (2008). Metabolic profiling as a tool for understanding mechanisms of toxicity. *Toxicologic Pathology, 36*, 140–147.
49. Lewis, G. D., Asnani, A., & Gerszten, R. E. (2008). Application of metabolomics to cardiovascular biomarker and pathway discovery. *Journal of the American College of Cardiology, 52*, 117–123.
50. Williams, T. (2011). *Autism spectrum disorders e from genes to environment*. Rijeka, Croatia: InTech.
51. Kałużna-Czaplińska, J., Zurawicz, E., Michalska, M., & Rynkowski, J. (2013). A focus on homocysteine in autism. *Acta Biochimica Polonica, 60*(2), 137–142.
52. Kruman, I. I., Culmsee, C., Chan, S. L., Kruman, Y., Guo, Z., Penix, L., et al. (2000). Homocysteine elicits a DNA damage response in neurons that promotes apoptosis and hypersensitivity to excitotoxicity. *The Journal of Neuroscience, 20*(18), 6920–6926.
53. Paşca, S. P., Nemeş, B., Vlase, L., Gagyi, C. E., Dronca, E., Miu, A. C., et al. (2006). High levels of homocysteine and low serum paraoxonase 1 arylesterase activity in children with autism. *Life Science, 78*(19), 2244–2248.
54. Ali, A., Waly, M. I., Al-Farsi, Y. M., Essa, M. M., Al-Sharbati, M. M., & Deth, R. C. (2011). Hyperhomocysteinemia among Omani autistic children: A case-control study. *Acta Biochimica Polonica, 58*(4), 547–551.
55. Shimmura, C., Suda, S., Tsuchiya, K. J., Hashimoto, K., Ohno, K., Matsuzaki, H., et al. (2011). Alteration of plasma glutamate and glutamine levels in children with high-functioning autism. *PLoS One, 6*(10), 25340. https://doi.org/10.1371/journal.pone.0025340

56. Moreno-Fuenmayor, H., Borjas, L., Arrieta, A., Valera, V., & Socorro-Candanoza, L. (1996). Plasma excitatory amino acids in autism. *Journal of Clinical Investigation, 37*(2), 113–128.
57. Aldred, S., Moore, K. M., Fitzgerald, M., & Waring, R. H. (2003). Plasma amino acid levels in children with autism and their families. *Journal of Autism and Developmental Disorders, 33*(1), 93–97.
58. Arnold, G. L., Hyman, S. L., Mooney, R. A., & Kirby, R. S. (2003). Plasma amino acids profiles in children with autism: Potential risk of nutritional deficiencies. *Journal of Autism and Developmental Disorders, 33*(4), 449–454.
59. Tu, W. J., Chen, H., & He, J. (2012). Application of LC-MS/MS analysis of plasma amino acids profiles in children with autism. *Journal of Clinical Biochemistry and Nutrition, 51*(3), 248–249. https://doi.org/10.3164/jcbn.12-45

Dietary Approaches to the Management of Autism Spectrum Disorders

Richard E. Hartman and Dhira Patel

Abstract This chapter reviews the literature surrounding autism spectrum disorders (ASD) and their relation to gastrointestinal (GI), behavioral, neurological, and immunological functioning. Individuals with ASD often have poor GI health, including bowel motility issues, autoimmune and/or other adverse responses to certain foods, and lack of necessary nutrient absorption. These issues may be caused or exacerbated by restrictive behavioral patterns (e.g., preference for sweet and salty foods and/or refusal of healthy foods). Those individuals with GI issues tend to demonstrate more behavioral deficits (e.g., irritability, agitation, hyperactivity) and also tend to have an imbalance in overall gut microbiome composition, thus corroborating several studies that have implicated brain–gut pathways as potential mediators of behavioral dysfunction.

We examine the literature regarding dietary approaches to managing ASDs, including elimination diets for gluten, casein, or complex carbohydrates, a ketogenic diet, and a low oxalate diet. We also explore the research examining dietary supplements such as fatty acids, pro- and prebiotics, vitamins, minerals, glutathione, phytochemicals, and hormones. The research on dietary approaches to managing ASDs is limited and the results are mixed. However, a few approaches, such as the gluten-free/casein-free diet, fatty acid supplementation, and pre/probiotics have generally demonstrated improved GI and associated behavioral symptoms. Given that GI issues seem to be overrepresented in ASD populations, and that GI issues have been associated with a number behavioral and neurological deficits, dietary manipulation may offer a cheap and easily implemented approach to improve the lives of those with ASD.

Keywords ASD · Autism · Diet · Eating behaviors · Supplements · Nutrition intervention · Neurodevelopment · Neurological disorders · Gastrointestinal dysfunction · Gut–brain axis · Nutraceuticals

---

R. E. Hartman (✉) · D. Patel
Department of Psychology, Loma Linda University, Loma Linda, CA, USA

M. M. Essa, M. W. Qoronfleh (eds.), *Personalized Food Intervention and Therapy for Autism Spectrum Disorder Management*, Advances in Neurobiology 24,
https://doi.org/10.1007/978-3-030-30402-7_19

1 Overview of Autism Spectrum Disorders

1.1 Prevalence and Common Features

Autism spectrum disorders (ASDs) are characterized by sustained deficits with social communication/interactions and repetitive/restricted behavioral patterns that may interfere with activities of daily living. By some accounts, approximately 60 million people are affected worldwide by ASDs. The Autism and Developmental Disabilities Monitoring (ADDM) Network estimates that the prevalence among 8-year-old children has increased from approximately one in 150 children during the years 2000 to 2002 to one in 68 children during the years 2010 to 2012. According to the Centers for Disease Control and Prevention, about one in 59 children currently have a diagnosis of ASD [1]. This increase in ASD diagnoses may be at least partially due to the relatively recent inclusion of milder disorders, such as Asperger syndrome and pervasive developmental disorder (PDD), along with autism in the *Diagnostic and Statistical Manual of Mental Disorders-5*'s definition of an ASD. Other contributors to the increase in ASD diagnoses may include changes in referral practices and public awareness but may also include increased exposure to environmental risk factors. A number of ASD-related behavioral, neurological, immunological, and gastrointestinal (GI) features have been described.

ASD symptoms often gradually manifest within the first 2 or 3 years of life. About 50% of parents first notice the ASD-related symptoms by 1.5 years of age, whereas about 80% notice something unusual by 2 years [2, 3]. From an early age, individuals with ASDs tend to lack social–emotional reciprocity. Rather than reflecting, commenting, sharing feelings, and generally participating in a conversation, they may be more prone to simply requesting or labeling. Although adults can develop compensatory strategies to overcome these challenges, they may still suffer from the anxiety and effort of continuously evaluating appropriate social interactions.

Individuals with ASDs may also experience delayed language, speech comprehension deficits, echoed speech, superfluous language, or even a complete lack of language development. Others may have an impaired ability to communicate with others, despite possessing vocabulary or grammar skills. Nonverbal behaviors essential to communication (e.g., eye contact, gestures, facial expressions, body orientation, and speech intonation) are often diminished, absent, or atypical relative to cultural standards.

In addition to social interaction and communication deficits, individuals with ASD often display restricted and/or repetitive behaviors such as finger flicking, continuously using the same objects, repetitive speech, and rigidity to routine. This may manifest differently depending on the age of the individual. Deviance from a structured routine may cause distress, as they maintain these patterns of behavior with abnormal intensity. Symptoms of distress may include shutting down communication, aggression, tantrums, and/or self-injurious behavior [4]. Other common features include intellectual disability, temporal processing deficits, affective disorders (e.g., anxiety, depression), attention deficit disorders (with or without hyperactivity), oppositional defiant disorders, sleep disorders, epilepsy, and Tourette syndrome and/or related disorders, and increased incidence of metabolic disorders such as phenylketonuria [5–9].

Several neuroanatomical features have been described, including an excess of neurons (perhaps resulting from lack of normal apoptosis/pruning during the brain's development; [2, 10]), ectopic neuronal arrangement (perhaps resulting from abnormal neuronal migration during the brain's development), and abnormal synaptic development [11, 12]. Interestingly, a pattern of brain overgrowth early in life followed by slower-than-normal growth has been described in some ASD individuals [13] that would be consistent with diminished normal early pruning. These characteristics have been associated specifically with communication dysfunction of the brain's neural networks [14], particularly in the so-called mirror neuron network, which consists of widespread cortical neurons that spike when an individual performs an action and when others perform a similar action. This network has been hypothesized to modulate imitation, empathy, social awareness, and communication. Other studies have suggested neurochemical imbalances in excitatory/inhibitory neurotransmitters and/or their receptors [15–19] and impaired mitochondrial functioning [20].

Finally, ASDs have also been associated with a number of inflammatory/immunological factors, including increased expression of pro-inflammatory cytokines and increased activation of microglia in the brain [21–23]. In some cases, increased severity of these symptoms has been associated with more severe behavioral deficits. Presumably, maternal exposure to infectious agents or environmental toxins during pregnancy can lead to early overactivation of the fetal immune system, leading to problems with nervous system development [24].

1.2 Potential Causes and Risk Factors

Autism and other ASDs have been associated with a number of environmental and genetic risk factors. For example, some evidence suggests that environmental insults during gestation (e.g., infections and/or exposure to teratogens such as drugs and pollution) may play a role in the development of ASDs [25–27]. Some studies have reported significantly higher levels of heavy metals in ASD children compared to matched controls [28, 29]. However, other studies have reported lower levels of heavy metals or no significant differences between ASD individuals and controls [30, 31].

The high degree of heritability [32–36] suggests a strong genetic influence, and genetic disorders such as Fragile X syndrome have a strong association with ASDs and related behavioral symptoms. Links between ASDs and schizophrenia have also been suggested, especially in individuals with abnormalities on chromosome 1 (i.e., 1q21.1 deletion syndrome; [37]). Postmortem analyses of brains from ASD individuals have recently shown a significant decrease in RNA editing, especially with synaptic genes across several brain regions, and similar patterns of dysregulated RNA editing were observed in the brains of Fragile X individuals [38].

Simply having a Y chromosome seems to be one of the greatest risk factors for a diagnosis of ASD, in that they occur about twice as often in males when intellectual disabilities are also present, and more than 5× as often without intellectual disabilities [39]. One main hypothesis for this phenomenon suggests that females require a greater etiologic load to display impairments consistent with an ASD diagnosis [40]. For example, in a subject pool that presented with ASDs and/or developmental/intellectual disabilities, Jacquemont et al. [41] reported that females had more deleterious autosomal variants (copy number and single nucleotide) than males. A higher number of these deleterious variants was associated with lower performance IQ. The authors hypothesized that increased mutational burden (more deleterious variants) and worse presentation of symptoms (lower IQ scores) are required for females to meet the ASD threshold. Another study demonstrated that females with ASD have more problems with social interaction, communication, externalizing behavioral problems, irritability, feelings of lethargy, and lower IQ/language processing abilities compared to ASD males of the same age range, and that the differences grew larger with age [42].

However, some studies have not corroborated ASD-related sex differences in social communication, cognitive functioning, or adaptive behaviors [43, 44]. Furthermore, one study reported that young ASD females had significantly better social skills than young ASD males [45], but that this may be due to general female behavioral traits/tendencies in maintaining social relationships (e.g., empathy and care taking).

1.3 Eating Behaviors

Problems with eating can decrease the quality of nutrient intake and GI health in those with ASDs [46, 47]. Excessive rigidity regarding routines in individuals with ASDs can lead to extreme behavioral reactions to foods (particularly regarding texture) and/or rituals around food packaging, presentation, preparation, and/or eating patterns. Therefore, it is not surprising that children with ASD were reported to consume a significantly lower volume of food compared to their non-ASD siblings [48]. In one study, at least 78% of ASD children omitted one or more food groups and displayed problematic mealtime behaviors, such as pushing away food, turning away their head, crying, leaving the table, making negative statements, and/or displaying aggression toward caregivers [49]. Schreck et al. [50] reported that children with ASDs generally accepted a relatively narrow range of presented food options and refused food significantly more often than children without ASDs, but that they were more likely to accept food that was paired with specific preferred utensils. Similarly, Bandini et al. [51] reported that children with ASDs refused more foods and had a smaller food repertoire than typically developing children of the same age range. These findings can be partially explained by behavioral rigidity and intolerance to new foods. However, it is not clear how much of an effect their medications, such as stimulants prescribed for attention deficits and/or hyperactivity, may play a role in this reduced food intake. Furthermore, motor behaviors in individuals with ASDs, such as weak sucking, tongue thrusting, and poor lip closure, can further affect eating patterns and reactions to food [52]. For example, caregivers reported that children with ASDs required more supervision during mealtimes because they had tendencies to gag, vomit, cough, or choke [48].

Interestingly, an fMRI study reported a significantly positive correlation between taste reactivity and response to sweeter tastes versus neutral tastes in the primary gustatory cortex of children with ASDs compared to typically developing children, and that children with ASDs who reported more taste-related symptoms had a greater cortical response [53]. Nevertheless, caregiver education can positively impact the eating behaviors of children with ASDs. For example, parents of 3–6-year-old children in Japan were provided education on factors contributing to food selectivity and approaches for coping with problems of selective eating [54]. By the end of the study, the range of acceptable foods significantly increased.

1.4 Characteristics of Gastrointestinal Dysfunction

Individuals with ASDs seem to be more susceptible to GI issues, such as chronic abdominal pain, impaired peristaltic reflexes, bowel motility disorders (e.g., constipation and/or chronic loose stools), and/or bloating. Compared to children with ASD but without GI symptoms, those with GI symptoms are more likely to be irritable, agitated, socially withdrawn, lethargic, hyperactive, and/or noncompliant [4, 55–60].

Constipation seems to be most strongly correlated with dairy intake, indicating that specific foods may be incongruent with the GI makeup of individuals with ASD. Those that experience diarrhea, loose stools, and/or gaseousness tend to exhibit lower than normal activity of digestive enzymes such as disaccharidase, lactase, maltase, sucrase, palatinase, and glucoamylase, as well as higher pancreatobiliary fluid output following secretin stimulation. Furthermore, protein intake that is significantly higher than the recommended dietary allowance is associated with increased bowel motility issues.

Endoscopic examination has revealed increased blood flow (hyperemia) consistent with gastroesophageal reflux (e.g., esophageal swelling, gastritis duodenitis, and colitis). Other tests have indicated incomplete digestion of dietary gluten and casein, low levels of gastric acid, excessive levels of abnormal gut bacteria, increased intestinal permeability (“leaky gut”), increased absorption of incompletely hydrolyzed peptides, and elevated serotonin concentrations in the GI associated with GI inflammation [4, 49, 51, 61–66]. These GI-related symptoms, combined with restricted and rigid eating patterns/food preferences, can lead to inefficient and/or ineffective nutrient absorption [4, 51, 66]. Indeed, some studies have reported significant differences in nutrient intake between children with and without ASDs [49, 61–63, 66].

For example, Schreck et al. [50] reported that children with ASDs generally consume fewer vegetables, fruits, and starches than children without ASDs. Others have demonstrated inadequate consumption of dietary fiber, minerals such as potassium, iron, zinc, magnesium, and calcium [61, 62, 67, 68] and vitamins such as the retinoids (vitamin A), riboflavin (vitamin B_2), folate/folic acid (vitamin B_9), ascorbic acid (vitamin C), cyanocobalamin (vitamin B_{12}), and vitamin D [4, 51, 56, 61, 62, 66–68]. Other studies have reported higher than average consumption of calories from monosaturated fats (perhaps due to a preference for crunchy/crispy/fried snacks; [49, 56, 61]) and niacin (vitamin B_3; [66]).

Physical ramifications of imbalanced nutrient intake include reports of scurvy, presumably caused by a paucity of fruits and vegetables in the diet [69] as well as lower bone mass density scores in males with ASD [68]. Children with ASDs are also more likely to be obese, presumably due to increased preference for snack foods and/or decreased ability to exercise from poor motor skills, low-muscle tone, and/or unstable posture [59, 70].

1.5 Mechanisms of Gastrointestinal Dysfunction

The brain and the gut can interact via multiple pathways, including those mediated by the vagus nerve, immune responses, and metabolites [71–75]. Many of the GI issues that children with ASD endorse seem to be associated with "leaky gut." Normally, the small intestinal mucosa acts a luminal barrier to prohibit substances from entering the bloodstream. However, in individuals with ASDs, this luminal barrier is impaired, allowing larger molecules that normally cannot cross the membrane, to pass via various ways through the compromised membrane [76]. These enterocolitis specific issues seem to mediate the neurobehavioral features observed in children with ASD [58].

The gut contains a microbial collection composed of various bacteria, viruses, and fungi that develops and grows during infancy. Changes in gut microbiota composition can impact cognitive behaviors (e.g., depression, anxiety, increased stress levels), but these symptoms may also be reversed by replacing beneficial microbes in the gut through probiotic supplementation [56, 74]. Hoban et al. [77] recently showed that gut microbes can regulate the expression of microRNA in the amygdala and prefrontal cortex, providing at least one mechanism by which the microbiota could influence cognition, affect, and behavior.

Interestingly, there is often an overall imbalance in the gut microbe composition of individuals with ASDs. A healthy gut normally contains species of *Bifidobacteria*, *Lactobacillus*, *Prevotella*, *Coprococcus*, and *Veillonellaceae*, which help break down carbohydrates and control the expression of inflammatory cytokines (e.g., TNF-α). But Mezzelani et al. [78] reported that individuals with ASDs tend to have decreased levels of these bacteria. Deficiencies in beneficial gut microbes, which may be attributed to the poor dietary patterns often exhibited by individuals with ASDs, can foster the growth of potentially harmful bacterial species such as *Clostridia*, *Desulfovibrio*, and *Bacteroides*. It is therefore not surprising that these species are found more often in stool samples of individuals with ASDs than those without ASDs [56, 79–82]. Lipopolysaccharides (LPS) are endotoxins found in Gram-negative bacteria (e.g., *Desulfovibrio* and *Bacteroides*) that can disrupt the blood–brain barrier and interfere with neuroimmunological communication, and such pathways have been found to be disrupted in individuals with ASD [78, 83]. Prenatal exposure to bacterial LPS via maternal infections that occur during pregnancy may also play a role in the development of ASDs [84].

Other potential explanations for the GI issues observed in individuals with ASD include autoimmune responses against the gut epithelium and/or allergic reactions/sensitivities to certain foods. Specifically, chronic gastritis is associated with an increased number of lymphoid aggregates in the mucosa and an increased number of local immune defense cells. Cow's milk (for example) may cause an allergic reaction, which results in an antigen induced distal constipation [55]. Whether caused by poor diet, gut microbiome imbalances, or autoimmune/allergic responses, chronic inflammation in the GI is also associated with inflammation in other organs, including the brain [57].

1.6 Current Treatment Options

The overarching treatment goal for individuals with ASDs is to increase quality of life, including functional independence, increased social interaction, and improved language skills. Improvements in these areas of life will often reduce stress for both the individual and the family, but specific treatment goals depend on the range and severity of impairment. Generally, higher IQ and earlier intervention have both been associated with better overall outcomes. As reviewed by Poleg et al. [85], most available treatments are tailored toward behavioral impairments and psychoeducation [85, 86]. Perhaps due to the widely variable nature of ASD symptoms, no specific treatment strategy has been proven as reliably effective [87]. Behavioral intervention and positive support will not "cure" ASDs but can mask or reduce the presentation of ASD symptoms.

Similarly, no pharmaceutical treatments have been shown to reliably improve the social and language problems central to a diagnosis of ASD. However, overall brain function, repetitive behaviors, and comorbid/secondary symptoms such as attention deficits, hyperactivity, irritability, depression, and/or anxiety have been targets of pharmaceutical intervention. Over 50% of children with an ASD diagnosis are prescribed psychoactive drugs such as antipsychotics (e.g., risperidone and apriprazole), antidepressants (e.g., serotonin/norepinephrine reuptake inhibitors), stimulants (e.g., methylphenidate, norepinephrine reuptake inhibitors), antihypertensives (e.g., beta-blockers, guanfacine), depressants (e.g., GABAergic drugs), and hormones (e.g., oxytocin, vasopressin) [88–92]. However, one recent study that assessed a cohort of Danish ASD children born between 1992 and 2011 reported that only about 30% of the sample used ADHD medications (e.g., methylphenidate), antipsychotics (e.g., risperidone), antidepressants (e.g., sertraline), and/or hormones (e.g., melatonin) [93]. These data suggest regional differences in either ASD diagnoses, symptom presentation, and/or treatment protocols.

One recent study found that postnatal administration of an antidiabetic drug (pioglitazone) improved ASD-like social impairments in a rat model of autism (prenatal LPS exposure; [84]). Unfortunately, although certain psychoactive drugs may provide some relief from symptoms such as repetitive behaviors (e.g., antidepressants), irritability, aggression, self-injurious behaviors (e.g., antipsychotics), or attention deficits/hyperactivity (e.g., stimulants, antihypertensives) [90, 94, 95], individuals with ASDs (and children in general) can often respond atypically. Additionally, these drugs have a number of unpleasant and/or harmful side effects, including weight gain, lethargy, and dyskinesias [96].

2 Dietary Approaches to Autism Spectrum Disorder Management

The relative lack of efficacy for either behavioral or pharmaceutical treatment strategies for ASDs has led to an increased interest in the use of complementary or alternative medicine in the treatment of autism [97–99]. The beneficial effects of dietary interventions for neurological disorders and injuries have been reported many times. For example, papers from our laboratory have demonstrated improvements in neuropathology and/or behaviors in mouse studies of irradiation [100] and Alzheimer's disease [101] and in human studies of recovery from coronary artery bypass surgery [102] and stroke [103]. Due in part to the lack of available empirically validated therapies, there has been an increasing trend toward using similar strategies in children with ASD [104]. In a study conducted by Hall and Riccio [105], parents commonly resorted to trying elimination diets (e.g., gluten-free/casein-free) and/or dietary supplements including probiotics, omega-3 fatty acids, and melatonin.

2.1 Elimination Diets

Based on the hypothesis that some ASD symptoms are at least partly caused by dietary hypersensitivities that may be exacerbated by the GI issues mentioned above, "elimination diets" aim to improve behavioral symptoms by restricting intake of the problem-causing component(s) [4, 104].

2.1.1 Casein and Gluten

Dietary proteins such as casein, found in milk, and the gluten, found in grains such as wheat, rye, and barley, have been linked to heightened inflammatory and immune responses [106, 107]. In a study by Jyonouchi et al. [83], children with ASD expressed more proinflammatory cytokines and LPS (endotoxins produced by pathogenic microbial intestinal flora) after the consumption of cow's milk or gliadin (a component of wheat gluten). Similarly, casein and gluten can induce expression of immunoglobulin A and G antibodies in subsets of individuals with ASDs, which could exacerbate symptoms [108, 109].

Another explanation behind some of the atypical behavior observed in children with ASD is the "excess opioid" hypothesis, which proposes that gluten and casein are metabolized in the gut into short-chain peptides called gluteomorphins and caseomorphins (respectively) that are structurally similar to endorphins and have opiate agonist properties. Normally, the small intestinal mucosa acts as a luminal barrier to prohibit such metabolites from entering the bloodstream. However, in individuals with ASD, with increased intestinal permeability ("leaky gut") due to inflammation, this luminal barrier is impaired, potentially allowing these "exorphin" (exogenous opioid) metabolites into the bloodstream and ultimately into the brain to activate opiate receptors [52, 57, 76, 78, 104, 110].

Regardless of the specific mechanisms by which these dietary proteins may exacerbate ASD symptoms, some evidence suggests that reducing their consumption may help. Ghalichi et al. [111] conducted a randomized controlled trial in which children with ASDs were assigned to a gluten-free diet or their regular diet. Those in the "gluten-free" group exhibited a significant decrease in GI symptoms and stereotyped behaviors with slightly improved communication and social interaction, whereas those who maintained a regular diet actually showed a significant increase in their GI symptoms after 6 weeks. Another study [112] reported that, compared to controls, children with ASD had higher levels of casein-specific antibodies, and their ASD symptoms seemed to improve after 8 weeks on the cow's milk elimination diet. However, some studies have shown no significant differences in intestinal permeability and behavioral symptoms between individuals who were on a gluten-/casein-free diet and those who were not [113, 114]. A systemic review by the Cochrane group suggested that there is little hard evidence for the effectiveness of casein- and gluten-free diets in ASDs, but that only large-scale, randomized trials would yield more conclusive data [115].

2.1.2 Specific Carbohydrate Diet

The specific carbohydrate diet eliminates ingestion of complex carbohydrates (e.g., sugars, grains, starches, and dairy), allowing only those requiring minimal digestion. Nutrients in the diet come from monosaccharides (e.g., fruit, some vegetables, honey, meat, eggs, natural cheeses, homemade yogurt, nuts, soaked lentils, and beans). The idea behind the diet is that complex carbohydrates take longer to break down and digest in the GI system, thereby becoming a foundation for pathogenic intestinal microflora to breed [4, 52, 116].

Individuals with colonic and ileocolonic Crohn's disease who followed the diet for nearly 3 years found that their symptoms generally improved [117]. Although these individuals did not have ASDs, their GI symptoms were similar to those found in individuals with ASDs, so the results may be generalizable across populations. In a 2018 case study, Barnhill et al. [118] observed a 4-year-old male who was diagnosed with ASD and followed the specific carbohydrate diet for 4 months. He showed significant improvement in stool consistency and level of irritability when passing stool. His symptoms, including sensory, repetitive, and ritualistic behaviors, receptive and expressive language problems, and learning and memory also significantly improved.

2.1.3 Ketogenic Diet

The ketogenic diet (and the similar modified Atkins diet), which generally prescribes low-carbohydrate, moderate protein, and high-fat intake, forces the metabolism of ketones rather than glucose, and its overall effects include increased blood ketones, reduced blood glucose, and improved mitochondrial function [119]. The diet has shown some effectiveness in treating individuals with refractory epilepsy (which is more common in individuals with ASD than those without) and other neurological disorders. Indeed, in one study that administered a ketogenic diet to ASD children who presented with seizures, it was found that the sample showed an overall decrease in seizures, along with improved learning ability and social skills [120].

Animal models of ASDs have yielded some promising results using the ketogenic diet. For example, Ruskin et al. [119] assessed the BTBR mouse model of ASD, which has severely reduced interhemispheric communication due an absent corpus callosum and a diminished hippocampal commissure, and displayed behaviors similar to those seen in humans with ASDs (abnormal social interactions, play behaviors, and vocalizations). They reported that the ketogenic diet increased sociability, decreased repetitive behaviors, and improved social communication. Another mouse study [121] that examined the offspring of C57B1/6 mice given an infection during pregnancy reported that male, but not female, offspring exhibited behavioral patterns similar to those seen in humans with ASDs, and that a ketogenic diet attenuated these behaviors. Other studies using rodent models of autism have yielded mixed results. For example, the diet significantly improved sociability in glut3$^{+/-}$ (but not wild-type) mice, and significantly improved spatial cognition in wild-type (but not glut3$^{+/-}$) mice [122]. Kasprowska-Liśkiewicz et al. [123] also demonstrated increased social interaction, but no differences in locomotor activity, anxiety, or working memory, in male Long-Evans rats, and that the administration of exogenous ketones did not affect social behavior.

Human studies have also demonstrated some modest improvements from the ketogenic diet. For example, the aforementioned study [120] not only reported fewer seizures, but also improved learning ability and social skills. Additionally, Evangeliou et al. [20] implemented the ketogenic diet in 18 children with autistic behavior for 6 months. They reported significant improvement in two subjects, average improvement in eight subjects, and minor improvement in eight subjects, with the individuals on the lower end of the ASD spectrum showing the most improvement. In a study comparing a modified Atkins/ketogenic diet to a gluten-/casein-free diet, El-Rashidy et al. [124] found that a ketogenic diet improved cognition and sociability significantly more than the gluten-free/casein-free diet. Finally, a case study of a 6-year-old ASD patient with glucose hypometabolism showed that 1 month of a ketogenic diet improved hyperactivity, attention span, abnormal reactions to stimuli, communication skills, fear, anxiety, and emotional reactions [125].

2.1.4 Low Oxalate Diet

One study reported a 3× higher concentration of plasma oxalate and more than a 2.5× higher concentration of urinary oxalate than the recommended value in urine among a sample of children with ASD compared to healthy peers [126]. High concentrations of oxalates are found in spinach, beets, cocoa, black tea, and certain fruits, grains and nuts. Related compounds such as oxalic acid, in conjunction with GI system dysfunction, have been linked to impaired neurological development and abnormalities in the nervous system [4, 127]. However, there have been no empirical studies demonstrating the effectiveness of a low oxalate diet on individuals with ASD.

2.2 *Supplements*

In addition to eliminating problem-causing compounds (e.g., certain proteins, carbohydrates, oxalates) from the diet, some studies suggest that supplementing the diet with beneficial compounds may improve ASD symptoms.

2.2.1 Fatty Acids

Insufficient omega-3 fatty acid intake has been implicated in the abnormal development of the nervous system [4], and children with ASDs often have decreased plasma levels of phospholipid fatty acids [52]. Thus, a reduced omega-3 index (the proportion of omega-3 fatty acids to the total amount of fatty acids in the brain) may be a biomarker for ASDs, because neurons should be rich in polyunsaturated fatty acids to ensure normal development, membrane fluidity, and functional properties.

Nevertheless, there have been mixed reviews regarding the results of omega-3 fatty acid supplementation [4]. Parellada et al. [128] found that children aged 5–17 years who supplemented their diets with omega-3 fatty acids for approximately 2 months showed a significant improvement in social motivation as reported by parents. Another study that provided supplemental omega-3 fatty acids for 3 months showed a significant improvement of atypical sensory processing from baseline (again, as reported by parents; [129]). They also found increased levels of long-chain omega-3 fatty acids eicosapentaenoic acid (EPA) and docosahexaenoic acid (DHA), which promote healthy brain development.

The administration of omega-3 fatty acids along with vitamin B_{12} has also been found to increase the growth of a Gram-positive bacterium (*Staphylococcus*) and reduce survival of a Gram-negative bacterium (*Clostridia*; [130]). Another combination study administered medium chain triglycerides (which also improve fatty acid levels) along with a ketogenic diet and gluten-free diet to children with ASD for 3 months [131]. The combination of multiple dietary therapies showed significant improvement in core features of ASD and symptom severity.

A meta-analysis conducted by Cheng et al. [132] suggested that omega-3 fatty acid supplementation in ASD children induced only borderline improvements in hyperactivity, but significant improvements in stereotypic behavior. When Agostoni et al. [133] reviewed studies on omega-3 fatty acid supplementation across developmental psychopathologies, they found mixed effects on ASD, with nonsignificant trends for beneficial effects on impaired behavior.

A fatty acid-like compound derived from the cannabis plant may also offer some relief from ASD symptoms. Since cannabidiol (CBD) has anticonvulsive, sedative, hypnotic, antipsychotic, anti-inflammatory, and neuroprotective properties, it may benefit individuals with ASDs [85, 134, 135]. For example, one study demonstrated that CBD reduced seizures by 70% in a mouse model of an epileptic disorder (Dravet syndrome), and when administered in low doses, significantly increased social interaction [136].

2.2.2 Pro- and Pre-biotics

Probiotics are live microorganisms (e.g., *Lactobacillus*, *Bifidobacterium* spp.) that naturally occur in certain (often fermented) foods such as yogurt and sauerkraut or can be added to the diet via supplemental capsules [137]. *Pre*biotics are compounds found in (often high fiber) foods that selectively promote the growth and colonization of healthy gut probiotics. Improving gut health via ingestion of dietary probiotics or prebiotics may ameliorate some of the gut-related issues associated with the ASDs [138, 139].

Probiotics have been shown to alleviate GI dysfunction commonly associated with ASD by a number of mechanisms [52]. For example, probiotics may reduce gut permeability and reconstruct or stabilize the intestinal barrier via increased mucin production [56, 140]. They also produce digestive enzymes that metabolize potentially toxic/irritating compounds (such as casein and the glutens; [141]), synthesize antioxidants that protect the gut from pathogens [137], and modulate immune responses. Supplemental consumption of beneficial probiotic bacteria such as *Lactobacillus* can normalize the gut microbiome and influence social/sensory/

cognitive behaviors [142]. A probiotic mixture of *Lactobacillus*, *Streptococcus*, and *Bifidobacterium* species attenuated elevated levels of *Clostridia* and reversed persistent ASD-like behaviors induced by propionic acid (a neurotoxin) in young, male golden Syrian hamsters [143]. In another study using a rodent model of ASDs, Shank3b−/− mice were administered *Lactobacillus*, resulting in fewer unsocial/aggressive behaviors in males, and fewer stereotypical repetitive behaviors in both males and females [144].

In a double-blind, placebo-controlled, crossover-designed feeding sample, a sample of 17 ASD subjects took a probiotic (*Lactobacillus plantarum WCSF1)* for 12 weeks. The supplement significantly increased *Lactobacilli* and *Enterococci* bacteria and reduced *Clostridia* bacteria in the gut and significantly decreased behavioral and emotional disturbances [145]. The first study to evaluate prebiotic supplementation in ASD demonstrated that a galacto-oligosaccharide, in combination with a gluten-free/casein-free diet, improved beneficial bacteria growth and increased gut microbiota diversity [146].

2.2.3 Vitamins

These organic compounds are essential nutrients that play a wide role in general life functions. Insufficient consumption of vitamins can lead to a number of psychiatric issues [147] and can potentially exacerbate the issues already present in the ASDs. A recent literature review found mixed results on the overall effectiveness of vitamin supplementation among population with ASD, most likely due to the heterogeneity of methodological aspects (e.g., type of vitamin, dosage, sample size, treatment duration; [148]). Nevertheless, several studies have been published that suggest therapeutic potential for supplementing with certain vitamins.

For example, vitamin B_6 (pyridoxine) is involved in the synthesis of serotonin, dopamine, and norepinephrine [148], and supplementation has been shown to improve behavioral symptoms, sleep, and GI symptoms [66, 149, 150]. Individuals with ASDs often have less vitamin B_9 (folic acid) in their cerebrospinal fluid, because autoantibodies block folic acid synthesis by binding to folate receptors and inhibiting folate transport. Therefore, dietary supplementation with folic acid has been suggested for ASD individuals with cerebral folate deficiency syndrome. In a recent study, Alfawaz et al. [130] reported that dietary supplementation with vitamin B_{12} (cobalamin) or omega-3 fatty acid equally alleviated ASD-like symptoms in a rat model of ASD (neurotoxic propionic acid administration). Methyl B_{12} (methylcobalamin) administration has also induced improvements in methylation, antioxidant capacity, and clinician-rated global symptoms in an ASD sample [151].

Individuals with ASDs were also reported to have lower levels of vitamin C (ascorbic acid), suggesting that supplementation may provide some benefit [150]. Some ASD individuals have even presented with scurvy (a symptom of vitamin C deficiency), presumably caused by a paucity of fruits and vegetables in their diet [69]. Maternal vitamin deficiencies (e.g., vitamin D) during pregnancy may increase the infant's risk of developing an ASD, suggesting that prenatal supplementation may also provide some benefit [4, 150]. Furthermore, vitamin D deficiencies in individuals with ASDs may exacerbate symptoms. Patrick and Ames [64] proposed that vitamin D supplementation could lower the elevated levels of serotonin and the associated GI inflammation in ASD subjects.

2.2.4 Minerals

These inorganic compounds are essential nutrients that play a wide role in general life functions. Like vitamin B6, magnesium has been implicated in improving behavioral symptoms, sleep, and GI symptoms [66, 149, 150]. It is also involved in serotonin, dopamine, and norepinephrine synthesis [148]. Zinc has been implicated in neuronal genesis, plasticity, fetal growth, cellular differentiation and reproduction, tissue repair, and immunity. Adams and Holloway [150] reported a significantly lower zinc to copper ratio in children with ASD, suggesting that increasing zinc levels (and/or reducing levels of copper) may aid this population. In corroboration of this idea, the administration of zinc reversed the effects of impaired vocalization and improved social behavior in a rodent model of autism (prenatal valproic acid exposure; [152]).

2.2.5 Glutathione

Oxidative stress is a common biomarker in ASD populations and is congruent with GI dysfunction. Lower antioxidant capacity has been implicated as a potential contribution to ASD pathophysiology and social impairment. Glutathione, an antioxidant molecule synthesized in the liver, has been reported as deficient in populations with ASD [153]. Although glutathione has poor oral bioavailability, ingestion of *N*-acetylcysteine (NAC), which is metabolized to one of its precursors (L-cysteine), can replenish glutathione levels. NAC supplementation has demonstrated mixed results. Although it seems to increase levels of glutathione in individuals with ASD [154–156], a recent study in which NAC was administered to subjects with ASDs for 6 months reported no significant improvements in sociability or repetitive behaviors compared to controls [157]. However, NAC seemed to improve the effects of Risperidone treatment on irritability and hyperactivity among subjects with ASDs [154, 155].

2.2.6 Phytochemicals

Sulforaphane is an organosulfur phytochemical (an organic plant-derived compound that contains sulfur) found in cruciferous vegetables such as broccoli seeds and sprouts. It has a number of reported physiological effects, including antioxidant/anti-inflammatory properties [158]. It has also been found to regulate the expression of cytoprotective responses through long-lasting mediation of a transcription factor, making it a potentially efficient dietary therapeutic [159].

One study [160] examined whether dietary treatment with sulforaphane might reduce the severity of socially impaired behavior among a sample of young males with ASDs. After an 18-week period, they found that the sample's social interactions, aberrant behavior, and verbal communication significantly improved, with symptoms starting to change around 1 month after initiating treatment. In a 2018 study, Bent and colleagues investigated whether treating a sample of ASD adolescents with sulforaphane would improve behavioral impairments and metabolic output. They found that social communication and symptom severity significantly improved, and that metabolites involved in oxidative stress, amino acid/gut microbiome, neurotransmitters, hormones/stress response, and sphingomyelin metabolism were significantly different in the ASD sample. In a study of subjects who took sulforaphane for a few years, there was considerable improvement, and subjects reported that it worked better than pharmacological interventions such as aripiprazole and levetiracetam [161]. However, one recent study [162] reported that a significant number of subjects in a study of individuals without ASDs experienced upset stomach while taking sulforaphane supplements. Finally, prenatal administration of another phytochemical, resveratrol (a stilbenoid found in grapes, berries, nuts, etc.) was reported to prevent social impairments in a valproic acid exposure rodent model of ASDs [163].

2.2.7 Hormones

Secretin is a hormone that stimulates pancreatic secretion and inhibits gastric acid secretion, thereby maintaining the pH of the intestinal luminal fluid in the GI system. Individuals with ASDs tend to produce lower levels of this hormone, so their gastric acid secretion is higher and pancreatic secretion is lower, which increases luminal acidity and permeability [76]. Therefore, secretin supplementation may be a viable therapeutic agent to improve GI dysfunction in individuals with ASDs. One case study found a significant improvement in diet and behavioral symptoms after a 6-month intravenous administration of secretin [164]. However, other studies have concluded that secretin showed no improvement in the core features of ASDs after single or multiple doses [165, 166].

Melatonin is a hormone secreted by the pineal gland, GI system, lungs, renal cortex, and retina and is responsible for regulating circadian rhythm, GI motility, and influencing immune and reproductive systems [167]. The secretion pattern of melatonin is different in individuals with ASDs, which could explain the common symptom of sleep problems. The administration of melatonin can ameliorate sleep disturbances [168–170]. Mothers of children with ASDs have been reported to have lower levels of melatonin in their urine compared to mothers of children in control groups, suggesting that parental melatonin levels could be a potential contributor to the development of ASDs in their offspring [171]. Finally, the GI system has a high concentration of melatonin, which exerts both excitatory and inhibitory effects on the gut muscles and modulates inflammatory responses. Although there is no empirical evidence of melatonin alleviating GI issues in ASD populations, its role in the GI system may lead to future research developing this theory.

3 Conclusion

Although there has been extensive research on the symptoms and potential causes of ASDs, the role of GI dysfunction is an emerging topic of interest. Not only do individuals with ASDs display rigid eating patterns, but they are more likely to suffer from GI issues such as diarrhea, constipation, and irritable bowel syndrome. Ultimately, the dietary approaches discussed throughout this chapter may ameliorate, at least partially, both GI and behavioral impairments. Dietary approaches may be cheaper, easier to implement, and better tolerated with fewer side effects than pharmaceutical interventions. Future research should also determine whether these diets can be generalizable to different populations and if they are feasible in different settings (areas with fewer resources, lower socioeconomic areas, countries with different dietary restrictions, etc.).

References

1. Baio, J. (2018). Prevalence of autism spectrum disorder among children aged 8 years — Autism and developmental disabilities monitoring network, 11 sites, USA, 2014.
2. Courchesne, E., Pierce, K., Schumann, C. M., Redcay, E., Buckwalter, J. A., Kennedy, D. P., et al. (2007). Mapping early brain development in autism. *Neuron, 56*, 399–413.
3. Landa, R. J. (2008). Diagnosis of autism spectrum disorders in the first 3 years of life. *Nature Clinical Practice. Neurology, 4*, 138–147.
4. Kawicka, A., & Regulska-Ilow, B. (2013). How nutritional status, diet and dietary supplements can affect autism. A review. *Roczniki Państwowego Zakładu Higieny, 64*, 1–12.
5. Allman, M. J. (2011). Deficits in temporal processing associated with autistic disorder. *Frontiers in Integrative Neuroscience, 5*, 1–2.
6. Bejerot, S., Eriksson, J. M., & Mörtberg, E. (2014). Social anxiety in adult autism spectrum disorder. *Psychiatry Research, 220*, 705–707.
7. Manzi, B., Loizzo, A. L., Giana Grazia, G., & Curatolo, P. (2008). Autism and metabolic diseases. *Journal of Child Neurology, 23*, 307–314.
8. Simonoff, E., Pickles, A., Charman, T., Chandler, S., Loucas, T., & Baird, G. (2008). Psychiatric disorders in children with autism spectrum disorders: Prevalence, comorbidity, and associated factors in a population-derived sample. *Journal of the American Academy of Child and Adolescent Psychiatry, 47*, 921–929.
9. Zafeiriou, D. I., Ververi, A., & Vargiami, E. (2007). Childhood autism and associated comorbidities. *Brain and Development, 29*, 257–272.
10. Lewis, J. D., Theilmann, R. J., Townsend, J., & Evans, A. C. (2013). Network efficiency in autism spectrum disorder and its relation to brain overgrowth. *Frontiers in Human Neuroscience, 7*, 845.
11. Bear, M. F., Dölen, G., Osterweil, E., & Nagarajan, N. (2008). Fragile X: Translation in action. *Neuropsychopharmacology, 33*, 84–87.

12. Kelleher, R. J., & Bear, M. F. (2008). The autistic neuron: Troubled translation? *Cell, 135*, 401–406.
13. Geschwind, D. H. (2009). Advances in autism. *Annual Review of Medicine, 60*, 367–380.
14. Wickelgren, I. (2005). Neurology. Autistic brains out of synch? *Science, 308*, 1856–1858.
15. Kumar, H., & Sharma, B. (2016). Memantine ameliorates autistic behavior, biochemistry & blood brain barrier impairments in rats. *Brain Research Bulletin, 124*, 27–39.
16. McTighe, S. M., Neal, S. J., Lin, Q., Hughes, Z. A., & Smith, D. G. (2013). The BTBR mouse model of autism spectrum disorders has learning and attentional impairments and alterations in acetylcholine and kynurenic acid in prefrontal cortex. *PLoS One, 8*, e62189.
17. Nikvarz, N., Alaghband-Rad, J., Tehrani-Doost, M., Alimadadi, A., & Ghaeli, P. (2017). Comparing efficacy and side effects of memantine vs. risperidone in the treatment of autistic disorder. *Pharmacopsychiatry, 50*, 19–25.
18. Persico, A. M., & Bourgeron, T. (2006). Searching for ways out of the autism maze: Genetic, epigenetic and environmental clues. *Trends in Neurosciences, 29*(7), 349–358.
19. Wei, H., Dobkin, C., Sheikh, A. M., Malik, M., Brown, W. T., & Li, X. (2012). The therapeutic effect of memantine through the stimulation of synapse formation and dendritic spine maturation in autism and fragile X syndrome. *PLoS One, 7*, e36981.
20. Evangeliou, A., Vlachonikolis, I., Mihailidou, H., Spilioti, M., Skarpalezou, A., Makaronas, N., et al. (2003). Application of a ketogenic diet in children with autistic behavior: Pilot study. *Journal of Child Neurology, 18*, 113–118.
21. Marchezan, J., Winkler Dos Santos, E. G. A., Deckmann, I., & Riesgo, R. D. S. (2018). Immunological dysfunction in autism spectrum disorder: A potential target for therapy. *Neuroimmunomodulation, 903*, 1–20.
22. Onore, C., Careaga, M., & Ashwood, P. (2012). The role of immune dysfunction in the pathophysiology of autism. *Brain, Behavior, and Immunity, 26*, 383–392.
23. Rossignol, D. A., & Frye, R. E. (2014). Evidence linking oxidative stress, mitochondrial dysfunction, and inflammation in the brain of individuals with autism. *Frontiers in Physiology, 5*, 150.
24. Patterson, P. H. (2011). Maternal infection and immune involvement in autism. *Trends in Molecular Medicine, 17*(7), 389–394.
25. Atladóttir, H. O., Thorsen, P., Østergaard, L., Schendel, D. E., Lemcke, S., Abdallah, M., et al. (2010). Maternal infection requiring hospitalization during pregnancy and autism spectrum disorders. *Journal of Autism and Developmental Disorders, 40*, 1423–1430.
26. Lee, B. K., Magnusson, C., Gardner, R. M., Blomström, Å., Newschaffer, C. J., Burstyn, I., et al. (2015). Maternal hospitalization with infection during pregnancy and risk of autism spectrum disorders. *Brain, Behavior, and Immunity, 44*, 100–105.
27. Ornoy, A., Weinstein-Fudim, L., & Ergaz, Z. (2015). Prenatal factors associated with autism spectrum disorder (ASD). *Reproductive Toxicology, 56*, 155–169.
28. Al-Farsi, Y. M., Waly, M. I., Al-Sharbati, M. M., Al-Shafaee, M. A., Al-Farsi, O. A., Al-Khaduri, M. M., et al. (2013). Levels of heavy metals and essential minerals in hair samples of children with autism in Oman: A case-control study. *Biological Trace Element Research, 151*, 181–186.
29. El-Ansary, A., Bjørklund, G., Tinkov, A. A., Skalny, A. V., & Al Dera, H. (2017). Relationship between selenium, lead, and mercury in red blood cells of Saudi autistic children. *Metabolic Brain Disease, 32*, 1073–1080.
30. Skalny, A. V., Simashkova, N. V., Klyushnik, T. P., Grabeklis, A. R., Radysh, I. V., Skalnaya, M. G., et al. (2017). Assessment of serum trace elements and electrolytes in children with childhood and atypical autism. *Journal of Trace Elements in Medicine and Biology, 43*, 9–14.
31. Tschinkel, S., Fabiana, P., Bjørklund, G., Conón, L. Z. Z., Chirumbolo, S., & Nascimento, V. A. (2018). Plasma concentrations of the trace elements copper, zinc and selenium in Brazilian children with autism spectrum disorder. *Biomedicine & Pharmacotherapy, 106*, 605–609.

32. Freitag, C. M., Staal, W., Klauck, S. M., Duketis, E., & Waltes, R. (2010). Genetics of autistic disorders: Review and clinical implications. *European Child & Adolescent Psychiatry, 19*, 169–178.
33. Miles, J. H. (2011). Autism spectrum disorders – A genetics review. *Genetics in Medicine, 13*, 278–294.
34. Morrow, E. M., Yoo, S. Y., Flavell, S. W., Kim, T. K., Lin, Y., Hill, R. S., et al. (2008). Identifying autism loci and genes by tracing recent shared ancestry. *Science, 321*, 218–223.
35. Steyaert, J. G., & De La Marche, W. (2008). What's new in autism? *European Journal of Pediatrics, 167*, 1091–1101.
36. Sutcliffe, J. S. (2008). Insights into the pathogenesis of autism. *Science, 321*, 208–209.
37. Kirov, G. (2015). CNVs in neuropsychiatric disorders. *Human Molecular Genetics, 24*, R45–R49.
38. Tran, S. S., Jun, H.-I., Bahn, J. H., Azghadi, A., Ramaswami, G., Van Nostrand, E. L., et al. (2019). Widespread RNA editing dysregulation in brains from autistic individuals. *Nature Neuroscience, 22*, 25–36.
39. Newschaffer, C. J., Croen, L. A., Daniels, J., Giarelli, E., Grether, J. K., Levy, S. E., et al. (2007). The epidemiology of autism spectrum disorders. *Annual Review of Public Health, 28*, 235–258.
40. Robinson, E. B., Lichtenstein, P., Anckarsäter, H., Happé, F., & Ronald, A. (2013). Examining and interpreting the female protective effect against autistic behavior. *Proceedings of the National Academy of Sciences, 110*, 5258–5262.
41. Jacquemont, S., Coe, B. P., Hersch, M., Duyzend, M. H., Krumm, N., Bergmann, S., et al. (2014). A higher mutational burden in females supports a "female protective model" in neurodevelopmental disorders. *American Journal of Human Genetics, 94*, 415–425.
42. Frazier, T. W., Georgiades, S., Bishop, S. L., & Hardan, A. Y. (2014). Behavioral and cognitive characteristics of females and males with autism in the Simons Simplex Collection. *Journal of the American Academy of Child and Adolescent Psychiatry, 53*, 329-40.e1-3.
43. Dean, M., Kasari, C., Shih, W., Frankel, F., Whitney, R., Landa, R., et al. (2014). The peer relationships of girls with ASD at school: Comparison to boys and girls with and without ASD. *Journal of Child Psychology and Psychiatry, 55*, 1218–1225.
44. Reinhardt, V. P., Wetherby, A. M., Schatschneider, C., & Lord, C. (2015). Examination of sex differences in a large sample of young children with autism spectrum disorder and typical development. *Journal of Autism and Developmental Disorders, 45*(3), 697–706.
45. Head, A. M., McGillivray, J. A., & Stokes, M. A. (2014). Gender differences in emotionality and sociability in children with autism spectrum disorders. *Molecular Autism, 5*, 1–9.
46. Kral, T. V. E., Eriksen, W. T., Souders, M. C., & Pinto-Martin, J. A. (2013). Eating behaviors, diet quality, and gastrointestinal symptoms in children with autism spectrum disorders: A brief review. *Journal of Pediatric Nursing, 28*, 548–556.
47. Ranjan, S., & Nasser, J. A. (2015). Nutritional status of individuals with autism spectrum disorders: Do we know enough? *Advances in Nutrition, 6*, 397–407.
48. Nadon, G., Feldman, D. E., Dunn, W., & Gisel, E. (2011). Mealtime problems in children with autism spectrum disorder and their typically developing siblings: A comparison study. *Autism, 15*, 98–113.
49. Sharp, W. G., Berry, R. C., McCracken, C., Nuhu, N. N., Marvel, E., Saulnier, C. A., et al. (2013). Feeding problems and nutrient intake in children with autism spectrum disorders: A meta-analysis and comprehensive review of the literature. *Journal of Autism and Developmental Disorders, 43*, 2159–2173.
50. Schreck, K. A., Williams, K., & Smith, A. F. (2004). A comparison of eating behaviors between children with and without autism. *Journal of Autism and Developmental Disorders, 34*, 433–438.
51. Bandini, L. G., Anderson, S. E., Curtin, C., Cermak, S., Evans, E. W., Scampini, R., et al. (2010). Food selectivity in children with autism spectrum disorders and typically developing children. *The Journal of Pediatrics, 157*, 259–264.

52. Geraghty, M. E., Bates-Wall, J., Ratliff-Schaub, K., & Lane, A. E. (2010). Nutritional interventions and therapies in autism: A spectrum of what we know: Part 2. *ICAN Infant Child & Adolescent Nutrition, 2*, 120–133.
53. Avery, J. A., Ingeholm, J. E., Wohltjen, S., Collins, M., Riddell, C. D., Gotts, S. J., et al. (2018). Neural correlates of taste reactivity in autism spectrum disorder. *NeuroImage Clinical, 19*, 38–46.
54. Miyajima, A., Tateyama, K., Fuji, S., Nakaoka, K., Hirao, K., & Higaki, K. (2017). Development of an intervention programme for selective eating in children with autism spectrum disorder. *Hong Kong Journal of Occupational Therapy, 30*, 22–32.
55. Afzal, N., Murch, S., Thirrupathy, K., Berger, L., Fagbemi, A., & Heuschkel, R. (2003). Constipation with acquired megarectum in children with autism. *Pediatrics, 112*, 939–942.
56. Berding, K., & Donovan, S. M. (2018). Diet can impact microbiota composition in children with autism spectrum disorder. *Frontiers in Neuroscience, 12*, 1–16.
57. Horvath, K., & Perman, J. A. (2002). Autism and gastrointestinal symptoms. *Current Gastroenterology Reports, 4*, 251–258.
58. Ibrahim, S. H., Voigt, R. G., Katusic, S. K., Weaver, A. L., & Barbaresi, W. J. (2009). Incidence of gastrointestinal symptoms in children with autism: A population-based study. *Pediatrics, 124*, 680–686.
59. Kazek, B., Jamroz, E., Grzybowska-chlebowczyk, U., & Kajor, M. (2010). Gastrointestinal disturbances in patients with autistic spectrum disorders. *Neurol Dziecięca, 19*, 27–31.
60. McElhanon, B. O., McCracken, C., Karpen, S., & Sharp, W. G. (2014). Gastrointestinal symptoms in autism spectrum disorder: A meta-analysis. *Pediatrics, 133*, 872–883.
61. Graf-Myles, J., Farmer, C., Thurm, A., Royster, C., Kahn, P., Soskey, L., et al. (2013). Dietary adequacy of children with autism compared with controls and the impact of restricted diet. *Journal of Developmental and Behavioral Pediatrics, 34*, 449–459.
62. Herndon, A. C., DiGuiseppi, C., Johnson, S. L., Leiferman, J., & Reynolds, A. (2009). Does nutritional intake differ between children with autism spectrum disorders and children with typical development? *Journal of Autism and Developmental Disorders, 39*, 212–222.
63. Levy, S. E., Souders, M. C., Ittenbach, R. F., Giarelli, E., Mulberg, A. E., & Pinto-Martin, J. A. (2007). Relationship of dietary intake to gastrointestinal symptoms in children with autistic spectrum disorders. *Biological Psychiatry, 61*, 492–497.
64. Patrick, R. P., & Ames, B. N. (2014). Vitamin D hormone regulates serotonin synthesis. Part 1: Relevance for autism. *FASEB Journal, 28*(6), 2398–2413.
65. Veenstra-VanderWeele, J., Muller, C. L., Iwamoto, H., Sauer, J. E., Owens, W. A., Shah, C. R., et al. (2012). Autism gene variant causes hyperserotonemia, serotonin receptor hypersensitivity, social impairment and repetitive behavior. *Proceedings of the National Academy of Sciences, 109*, 5469–5474.
66. Xia, W., Zhou, Y., Sun, C., Wang, J., & Wu, L. (2010). A preliminary study on nutritional status and intake in Chinese children with autism. *European Journal of Pediatrics, 169*, 1201–1206.
67. Meguid, N. A., Anwar, M., Bjørklund, G., Hashish, A., Chirumbolo, S., Hemimi, M., et al. (2017). Dietary adequacy of Egyptian children with autism spectrum disorder compared to healthy developing children. *Metabolic Brain Disease, 32*, 607–615.
68. Neumeyer, A. M., Cano Sokoloff, N., McDonnell, E. I., Macklin, E. A., McDougle, C. J., Holmes, T. M., et al. (2018). Nutrition and bone density in boys with autism spectrum disorder. *Journal of the Academy of Nutrition and Dietetics, 118*, 865–877.
69. Saavedra, M. J., Aziz, J., & San Román, N. C. (2018). Scurvy due to restrictive diet in a child with autism spectrum disorder. Case report. *Archivos Argentinos de Pediatría, 116*, 684–687.
70. Curtin, C., Anderson, S. E., Must, A., & Bandini, L. (2010). The prevalence of obesity in children with autism: A secondary data analysis using nationally representative data from the National Survey of Children's Health. *BMC Pediatrics, 10*, 0–4.
71. Desbonnet, L., Clarke, G., Traplin, A., O'Sullivan, O., Crispie, F., Moloney, R. D., et al. (2015). Gut microbiota depletion from early adolescence in mice: Implications for brain and behaviour. *Brain, Behavior, and Immunity, 48*, 165–173.

72. Dinan, T. G., Stilling, R. M., Stanton, C., & Cryan, J. F. (2015). Collective unconscious: How gut microbes shape human behavior. *Journal of Psychiatric Research, 63*, 1–9.
73. Foster, J. A., & McVey Neufeld, K. A. (2013). Gut-brain axis: How the microbiome influences anxiety and depression. *Trends in Neurosciences, 36*, 305–312.
74. Liu, X., Cao, S., & Zhang, X. (2015). Modulation of gut microbiota-brain axis by probiotics, prebiotics and diet. *Journal of Agricultural and Food Chemistry, 63*, 7885–7895.
75. Rogers, G. B., Keating, D. J., Young, R. L., Wong, M.-L., Licinio, J., & Wesselingh, S. (2016). From gut dysbiosis to altered brain function and mental illness: Mechanisms and pathways. *Molecular Psychiatry, 21*, 1–11.
76. White, J. F. (2003). Intestinal pathophysiology in autism. *Experimental Biology and Medicine, 228*, 639–649.
77. Hoban, A. E., Stilling, R. M., Moloney, G. M., Moloney, R. D., Shanahan, F., Dinan, T. G., et al. (2017). Microbial regulation of microRNA expression in the amygdala and prefrontal cortex. *Microbiome, 5*, 102.
78. Mezzelani, A., Landini, M., Facchiano, F., Raggi, M. E., Villa, L., Molteni, M., et al. (2015). Environment, dysbiosis, immunity and sex-specific susceptibility: A translational hypothesis for regressive autism pathogenesis. *Nutritional Neuroscience, 18*, 145–161.
79. Kang, D.-W., Park, J. G., Ilhan, Z. E., Wallstrom, G., Labaer, J., Adams, J. B., et al. (2013). Reduced incidence of Prevotella and other fermenters in intestinal microflora of autistic children. *PLoS One, 8*, e68322.
80. Kantarcioglu, A. S., Kiraz, N., & Aydin, A. (2016). Microbiota-gut-brain axis: Yeast species isolated from stool samples of children with suspected or diagnosed autism spectrum disorders and in vitro susceptibility against Nystatin and Fluconazole. *Mycopathologia, 181*, 1–7.
81. Kushak, R. I., Winter, H. S., Buie, T. M., Cox, S. B., Phillips, C. D., & Ward, N. L. (2017). Analysis of the duodenal microbiome in autistic individuals: Association with carbohydrate digestion. *Journal of Pediatric Gastroenterology and Nutrition, 64*, e110–e116.
82. Rose, D. R., Yang, H., Serena, G., Sturgeon, C., Ma, B., Careaga, M., et al. (2018). Differential immune responses and microbiota profiles in children with autism spectrum disorders and co-morbid gastrointestinal symptoms. *Brain, Behavior, and Immunity, 70*, 354–368.
83. Jyonouchi, H., Geng, L., Ruby, A., & Zimmerman-Bier, B. (2005). Dysregulated innate immune responses in young children with autism spectrum disorders: Their relationship to gastrointestinal symptoms and dietary intervention. *Neuropsychobiology, 51*, 77–85.
84. Kirsten, T. B., Casarin, R. C., Bernardi, M. M., & Felicio, L. F. (2018). Pioglitazone abolishes autistic-like behaviors via the IL-6 pathway. *PLoS One, 13*, 1–14.
85. Poleg, S., Golubchik, P., Offen, D., Weizman, A., Daniel, O., & Weizman, A. (2019). Cannabidiol as a suggested candidate for treatment of autism spectrum disorder. *Progress in Neuro-psychopharmacology & Biological Psychiatry, 89*, 90–96.
86. Seida, J. K., Ospina, M. B., Karkhaneh, M., Hartling, L., Smith, V., & Clark, B. (2009). Systematic reviews of psychosocial interventions for autism: An umbrella review. *Developmental Medicine and Child Neurology, 51*, 95–104.
87. Smith, T., & Iadarola, S. (2015). Evidence base update for autism spectrum disorder. *Journal of Clinical Child and Adolescent Psychology, 44*, 897–922.
88. Doyle, C. A., & McDougle, C. J. (2012). Pharmacologic treatments for the behavioral symptoms associated with autism spectrum disorders across the lifespan. *Dialogues in Clinical Neuroscience, 14*, 263–279.
89. Frye, R. E. (2018). Social skills deficits in autism spectrum disorder: Potential biological origins and progress in developing therapeutic agents. *CNS Drugs, 32*, 713–734.
90. Ji, N. Y., & Findling, R. L. (2016). Pharmacotherapy for mental health problems in people with intellectual disability. *Current Opinion in Psychiatry, 29*, 103–125.
91. Oswald, D. P., & Sonenklar, N. A. (2007). Medication use among children with autism spectrum disorders. *Journal of Child and Adolescent Psychopharmacology, 17*(3), 348–355.
92. Sanchack, K. E., & Thomas, C. A. (2016). Autism spectrum disorder: Primary care principles. *American Family Physician, 94*, 972–979.

93. Rasmussen, L., Bilenberg, N., Thomsen Ernst, M., Abitz Boysen, S., & Pottegård, A. (2018). Use of psychotropic drugs among children and adolescents with autism spectrum disorders in Denmark: A nationwide drug utilization study. *Journal of Clinical Medicine, 7*, 10.
94. Johnson, C. P., & Myers, S. M. (2007). Identification and evaluation of children with autism spectrum disorders. *Pediatrics, 120*, 1183–1215.
95. Leskovec, T. J., Rowles, B. M., & Findling, R. L. (2008). Pharmacological treatment options for autism spectrum disorders in children and adolescents. *Harvard Review of Psychiatry, 16*, 97–112.
96. Williamson, E. (2017). Medical therapies for children with autism spectrum disorder—An update. AHRQ Comparative Effectiveness Reviews 189.
97. Alanazi, A. S. (2013). The role of nutraceuticals in the management of autism. *Saudi Pharmaceutical Journal, 21*, 233–243.
98. Nath, D. (2017). Complementary and alternative medicine in the school-age child with autism. *Journal of Pediatric Health Care, 31*, 393–397.
99. Sathe, N., Andrews, J. C., McPheeters, M. L., & Warren, Z. E. (2017). Nutritional and dietary interventions for autism spectrum disorder: A systematic review. *Pediatrics, 139*, e20170346.
100. Dulcich, M. S., & Hartman, R. E. (2013). Pomegranate supplementation improves affective and motor behavior in mice after radiation exposure. *Evidence-Based Complementary and Alternative Medicine, 2013*, 940830.
101. Hartman, R. E., Shah, A., Fagan, A. M., Schwetye, K. E., Parsadanian, M., Schulman, R. N., et al. (2006). Pomegranate juice decreases amyloid load and improves behavior in a mouse model of Alzheimer's disease. *Neurobiology of Disease, 24*, 506–515.
102. Ropacki, S., Patel, S. M., & Hartman, R. E. (2013). Pomegranate supplementation protects against memory dysfunction after heart surgery: A pilot study. *Evidence-Based Complementary and Alternative Medicine, 2013*(10), 932401.
103. Bellone, J. A., Murray, J. R., Jorge, P., Fogel, T. G., Kim, M., Wallace, D. R., et al. (2019). Pomegranate supplementation improves cognitive and functional recovery following ischemic stroke: A randomized trial. *Nutritional Neuroscience, 22*, 738–743.
104. Christison, G. W., & Ivany, K. (2006). Elimination diets in autism spectrum disorders: Any wheat amidst the chaff? *Journal of Developmental and Behavioral Pediatrics, 27*, S162–S171.
105. Hall, S. E., & Riccio, C. A. (2012). Complementary and alternative treatment use for autism spectrum disorders. *Complementary Therapies in Clinical Practice, 18*, 159–163.
106. Buie, T. (2013). The relationship of autism and gluten. *Clinical Therapeutics, 35*, 578–583.
107. Jyonouchi, H., Sun, S., & Itokazu, N. (2002). Innate immunity associated with inflammatory responses and cytokine production against common dietary proteins in patients with autism spectrum disorder. *Neuropsychobiology, 46*, 76–84.
108. Lau, N. M., Green, P. H. R., Taylor, A. K., Hellberg, D., Ajamian, M., Tan, C. Z., et al. (2013). Markers of Celiac disease and gluten sensitivity in children with autism. *PLoS One, 8*, 6–11.
109. Marí-Bauset, S., Zazpe, I., Mari-Sanchis, A., Llopis-González, A., & Morales-Suárez-Varela, M. (2014). Evidence of the gluten-free and casein-free diet in autism spectrum disorders: A systematic review. *Journal of Child Neurology, 29*, 1718–1727.
110. Saad, K., Eltayeb, A. A., Mohamad, I. L., Al-Atram, A. A., Elserogy, Y., Bjørklund, G., et al. (2015). A randomized, placebo-controlled trial of digestive enzymes in children with autism spectrum disorders. *Clinical Psychopharmacology and Neuroscience, 13*, 188–193.
111. Ghalichi, F., Ghaemmaghami, J., Malek, A., & Ostadrahimi, A. (2016). Effect of gluten free diet on gastrointestinal and behavioral indices for children with autism spectrum disorders: A randomized clinical trial. *World Journal of Pediatrics, 12*, 436–442.
112. Lucarelli, S., Zingoni, A. M., Ferruzzi, F., Giardini, O., Frediani, T., Zingoni, A. M., et al. (1995). Food allergy and infantile autism. *Panminerva Medica, 37*, 137–141.
113. Harris, C., & Card, B. (2012). A pilot study to evaluate nutritional influences on gastrointestinal symptoms and behavior patterns in children with autism spectrum disorder. *Complementary Therapies in Medicine, 20*, 437–440.

114. Navarro, F., Pearson, D. A., Fatheree, N., Mansour, R., Hashmi, S. S., & Rhoads, J. M. (2015). Are "leaky gut" and behavior associated with gluten and dairy containing diet in children with autism spectrum disorders? *Nutritional Neuroscience, 18*, 177–185.
115. Millward, C., Ferriter, M., Calver, S. J., & Connell-Jones, G. G. (2008). Gluten- and casein-free diets for autistic spectrum disorder. *Cochrane Database of Systematic Reviews*, 1–29.
116. Gottschall, E. (2004). Digestion-gut-autism connection: The specific carbohydrate diet. *Medical Veritas The Journal of Medical Truth, 1*, 261–271.
117. Kakodkar, S., Farooqui, A. J., Mikolaitis, S. L., & Mutlu, E. A. (2015). The specific carbohydrate diet for inflammatory bowel disease: A case series. *Journal of the Academy of Nutrition and Dietetics, 115*, 1226–1232.
118. Barnhill, K., Devlin, M., Moreno, H. T., Potts, A., Richardson, W., Schutte, C., et al. (2018). Brief report: Implementation of a specific carbohydrate diet for a child with autism spectrum disorder and Fragile X Syndrome. *Journal of Autism and Developmental Disorders*. https://doi.org/10.1007/s10803-018-3704-9
119. Ruskin, D. N., Svedova, J., Cote, J. L., Sandau, U., Rho, J. M., Kawamura, M., et al. (2013). Ketogenic diet improves core symptoms of autism in BTBR mice. *PLoS One, 8*, 4–9.
120. Napoli, E., Dueñas, N., & Giulivi, C. (2014). Potential therapeutic use of the ketogenic diet in autism spectrum disorders. *Frontiers in Pediatrics, 2*, 1–9.
121. Ruskin, D. N., Murphy, M. I., Slade, S. L., & Masino, S. A. (2017). Ketogenic diet improves behaviors in a maternal immune activation model of autism spectrum disorder. *PLoS One, 12*, e0171643.
122. Dai, Y., Zhao, Y., Tomi, M., Shin, B.-C., Thamotharan, S., Mazarati, A., et al. (2017). Sex-specific life course changes in the neuro-metabolic phenotype of Glut3 null heterozygous mice: Ketogenic diet ameliorates electroencephalographic seizures and improves sociability. *Endocrinology, 158*, 936–949.
123. Kasprowska-Liśkiewicz, D., Liśkiewicz, A. D., Nowacka-Chmielewska, M. M., Nowicka, J., Małecki, A., & Barski, J. J. (2017). The ketogenic diet affects the social behavior of young male rats. *Physiology & Behavior, 179*, 168–177.
124. El-Rashidy, O., El-Baz, F., El-Gendy, Y., Khalaf, R., Reda, D., & Saad, K. (2017). Ketogenic diet versus gluten free casein free diet in autistic children: A case-control study. *Metabolic Brain Disease, 32*, 1935–1941.
125. Żarnowska, I., Chrapko, B., Gwizda, G., Nocuń, A., Mitosek-Szewczyk, K., & Gasior, M. (2018). Therapeutic use of carbohydrate-restricted diets in an autistic child; A case report of clinical and 18FDG PET findings. *Metabolic Brain Disease, 33*, 1187–1192.
126. Konstantynowicz, J., Porowski, T., Zoch-Zwierz, W., Wasilewska, J., Kadziela-Olech, H., Kulak, W., et al. (2012). A potential pathogenic role of oxalate in autism. *European Journal of Paediatric Neurology, 16*, 485–491.
127. Cristiano, C., Lama, A., Lembo, F., Mollica, M. P., Calignano, A., & Mattace Raso, G. (2018). Interplay between peripheral and central inflammation in autism spectrum disorders: Possible nutritional and therapeutic strategies. *Frontiers in Physiology, 9*, 184.
128. Parellada, M., Llorente, C., Calvo, R., Gutierrez, S., Lázaro, L., Graell, M., et al. (2017). Randomized trial of omega-3 for autism spectrum disorders: Effect on cell membrane composition and behavior. *European Neuropsychopharmacology, 27*, 1319–1330.
129. Boone, K. M., Gracious, B., Klebanoff, M. A., Rogers, L. K., Rausch, J., Coury, D. L., et al. (2017). Omega-3 and -6 fatty acid supplementation and sensory processing in toddlers with ASD symptomology born preterm: A randomized controlled trial. *Early Human Development, 115*, 64–70.
130. Alfawaz, H., Bhat, R. S., Al-Mutairi, M., Alnakhli, O. M., Al-Dbass, A., Alonazi, M., et al. (2018). Comparative study on the independent and combined effects of omega-3 and vitamin B12 on phospholipids and phospholipase A2 as phospholipid hydrolyzing enzymes in PPA-treated rats as a model for autistic traits. *Lipids in Health and Disease, 17*, 1–7.
131. Lee, R. W. Y., Corley, M. J., Pang, A., Arakaki, G., Abbott, L., Nishimoto, M., et al. (2018). A modified ketogenic gluten-free diet with MCT improves behavior in children with autism spectrum disorder. *Physiology & Behavior, 188*, 205–211.

132. Cheng, Y. S., Tseng, P. T., Chen, Y. W., Stubbs, B., Yang, W. C., Chen, T. Y., et al. (2017). Supplementation of omega 3 fatty acids may improve hyperactivity, lethargy, and stereotypy in children with autism spectrum disorders: A meta-analysis of randomized controlled trials. *Neuropsychiatric Disease and Treatment, 13*, 2531–2543.
133. Agostoni, C., Nobile, M., Ciappolino, V., Delvecchio, G., Tesei, A., Turolo, S., et al. (2017). The role of omega-3 fatty acids in developmental psychopathology: A systematic review on early psychosis, autism, and ADHD. *International Journal of Molecular Sciences, 18*, 1–25.
134. Anderson, C., Evans, V., DeMarse, T., Febo, M., Johnson, C., & Carney, P. (2017). Cannabidiol for the treatment of drug-resistant epilepsy in children: Current state of research. *Journal of Pediatric Neurology, 15*, 143–150.
135. Gu, B. (2017). Cannabidiol provides viable treatment opportunity for multiple neurological pathologies of autism spectrum disorder. *Global Drugs and Therapeutics, 2*, 1–4.
136. Kaplan, J. S., Stella, N., Catterall, W. A., & Westenbroek, R. E. (2017). Cannabidiol attenuates seizures and social deficits in a mouse model of Dravet syndrome. *Proceedings of the National Academy of Sciences of the United States of America, 114*, 11229–11234.
137. Kałużna-Czaplińska, J., & Błaszczyk, S. (2012). The level of arabinitol in autistic children after probiotic therapy. *Nutrition, 28*, 124–126.
138. Doenyas, C. (2018). Dietary interventions for autism spectrum disorder: New perspectives from the gut-brain axis. *Physiology & Behavior, 194*, 577–582.
139. Hsiao, E. Y., McBride, S. W., Hsien, S., Sharon, G., Hyde, E. R., McCue, T., et al. (2013). Microbiota modulate behavioral and physiological abnormalities associated with neurodevelopmental disorders. *Cell, 155*, 1451–1463.
140. Anderson, R. C., Cookson, A. L., McNabb, W. C., Kelly, W. J., & Roy, N. C. (2010). Lactobacillus plantarum DSM 2648 is a potential probiotic that enhances intestinal barrier function. *FEMS Microbiology Letters, 309*, 184–192.
141. Lindfors, K., Blomqvist, T., Juuti-Uusitalo, K., Stenman, S., Venäläinen, J., Mäki, M., et al. (2008). Live probiotic Bifidobacterium lactis bacteria inhibit the toxic effects induced by wheat gliadin in epithelial cell culture. *Clinical and Experimental Immunology, 152*, 552–558.
142. Patusco, R., & Ziegler, J. (2018). Role of probiotics in managing gastrointestinal dysfunction in children with autism spectrum disorder: An update for practitioners. *Advances in Nutrition, 9*, 637–650.
143. El-Ansary, A., Bacha, A. B., Bjørklund, G., Al-Orf, N., Bhat, R. S., Moubayed, N., et al. (2018). Probiotic treatment reduces the autistic-like excitation/inhibition imbalance in juvenile hamsters induced by orally administered propionic acid and clindamycin. *Metabolic Brain Disease, 33*, 1155–1164.
144. Tabouy, L., Getselter, D., Ziv, O., Karpuj, M., Tabouy, T., Lukic, I., et al. (2018). Dysbiosis of microbiome and probiotic treatment in a genetic model of autism spectrum disorders. *Brain, Behavior, and Immunity, 73*, 310–319.
145. Parracho, H. M. R. T., Gibson, G. R., Knott, F., Bosscher, D., & Kleerebezem, M. (2010). A double-blind, placebo-controlled, crossover-designed probiotic feeding study in children diagnosed with autistic spectrum disorders. *International Journal of Probiotics and Prebiotics, 5*, 69–74.
146. Grimaldi, R., Gibson, G. R., Vulevic, J., Giallourou, N., Castro-Mejía, J. L., Hansen, L. H., et al. (2018). A prebiotic intervention study in children with autism spectrum disorders (ASDs). *Microbiome, 6*, 1–13.
147. Mehl-Madrona, L., & Cornish, S. (2008). The role of vitamins and minerals in psychiatry. *Integrative Medicine Insights, 3*, 33–423.
148. Gogou, M., & Kolios, G. (2017). The effect of dietary supplements on clinical aspects of autism spectrum disorder: A systematic review of the literature. *Brain Dev, 39*, 656–664.
149. Adams, J. B., Audhya, T., McDonough-Means, S., Rubin, R. A., Quig, D., Geis, E., et al. (2011). Nutritional and metabolic status of children with autism vs. neurotypical children, and the association with autism severity. *Nutrition and Metabolism, 8*, 34.

150. Adams, J. B., & Holloway, C. (2004). Pilot study of a moderate dose multivitamin/mineral supplement for children with autistic spectrum disorder. *Journal of Alternative and Complementary Medicine, 10*, 1033–1039.
151. Hendren, R. L., James, S. J., Widjaja, F., Lawton, B., Rosenblatt, A., & Bent, S. (2016). Randomized, placebo-controlled trial of methyl B12 for children with autism. *Journal of Child and Adolescent Psychopharmacology, 26*, 774–783.
152. Cezar, L. C., Kirsten, T. B., da Fonseca, C. C. N., de Lima, A. P. N., Bernardi, M. M., & Felicio, L. F. (2018). Zinc as a therapy in a rat model of autism prenatally induced by valproic acid. *Progress in Neuro-Psychopharmacology & Biological Psychiatry, 84*, 173–180.
153. Yui, K., Tanuma, N., Yamada, H., & Kawasaki, Y. (2017). Decreased total antioxidant capacity has a larger effect size than increased oxidant levels in urine in individuals with autism spectrum disorder. *Environmental Science and Pollution Research, 24*, 9635–9644.
154. Ghanizadeh, A., & Moghimi-Sarani, E. (2013). A randomized double blind placebo controlled clinical trial of N-Acetylcysteine added to risperidone for treating autistic disorders. *BMC Psychiatry, 13*, 196.
155. Nikoo, M., Radnia, H., Farokhnia, M., Mohammadi, M.-R., & Akhondzadeh, S. (2015). N-Acetylcysteine as an adjunctive therapy to risperidone for treatment of irritability in autism. *Clinical Neuropharmacology, 38*, 11–17.
156. Wink, L. K., Adams, R., Wang, Z., Klaunig, J. E., Plawecki, M. H., Posey, D. J., et al. (2016). A randomized placebo-controlled pilot study of N-acetylcysteine in youth with autism spectrum disorder. *Molecular Autism, 7*, 26.
157. Dean, O. M., Gray, K. M., Villagonzalo, K.-A., Dodd, S., Mohebbi, M., Vick, T., et al. (2017). A randomised, double blind, placebo-controlled trial of a fixed dose of N-acetyl cysteine in children with autistic disorder. *The Australian and New Zealand Journal of Psychiatry, 51*, 241–249.
158. Bent, S., Lawton, B., Warren, T., Widjaja, F., Dang, K., Fahey, J. W., et al. (2018). Identification of urinary metabolites that correlate with clinical improvements in children with autism treated with sulforaphane from broccoli. *Molecular Autism, 9*, 35.
159. Panjwani, A. A., Liu, H., & Fahey, J. W. (2018). Crucifers and related vegetables and supplements for neurologic disorders: What is the evidence? *Current Opinion in Clinical Nutrition and Metabolic Care, 21*(6), 451–457.
160. Singh, K., Connors, S. L., Macklin, E. A., Smith, K. D., Fahey, J. W., Talalay, P., et al. (2014). Sulforaphane treatment of autism spectrum disorder (ASD). *Proceedings of the National Academy of Sciences, 111*, 15550–15555.
161. Lynch, R., Diggins, E. L., Connors, S. L., Zimmerman, A. W., Singh, K., Liu, H., et al. (2017). Sulforaphane from Broccoli reduces symptoms of autism: A follow-up case series from a randomized double-blind study. *Global Advances in Health and Medicine, 6*, 2164957X1773582.
162. Fahey, J. W., Wade, K. L., Wehage, S. L., Holtzclaw, W. D., Liu, H., Talalay, P., et al. (2017). Stabilized sulforaphane for clinical use: Phytochemical delivery efficiency. *Molecular Nutrition & Food Research, 61*, 1–10.
163. Bambini, V., Zanatta, G., Della Flora Nunes, G., Mueller de Melo, G., Michels, M., Fontes-Dutra, M., et al. (2014). Resveratrol prevents social deficits in animal model of autism induced by valproic acid. *Neuroscience Letters, 583*, 176–181.
164. Pallanti, S., Lassi, S., La Malfa, G., Campigli, M., Di Rubbo, R., Paolini, G., et al. (2005). Short report: Autistic gastrointestinal and eating symptoms treated with secretin: A subtype of autism. *Clinical Practice and Epidemiology in Mental Health, 1*, 24.
165. Krishnaswami, S., McPheeters, M. L., & Veenstra-VanderWeele, J. (2011). A systematic review of secretin for children with autism spectrum disorders. *Pediatrics, 127*, e1322–e1325.
166. Williams, K., Wray, J. A., & Wheeler, D. M. (2012). Intravenous secretin for autism spectrum disorders (ASD). *The Cochrane Database of Systematic Reviews*, CD003495.
167. Gagnon, K., & Godbout, R. (2018). Melatonin and comorbidities in children with autism spectrum disorder. *Current Developmental Disorders Reports, 5*, 197–206.

168. Cuomo, B. M., Vaz, S., Lee, E. A. L., Thompson, C., Rogerson, J. M., & Falkmer, T. (2017). Effectiveness of sleep-based interventions for children with autism spectrum disorder: A meta-synthesis. *Pharmacotherapy, 37*, 555–578.
169. Damiani, J. M., Sweet, B. V., & Sohoni, P. (2014). Melatonin: An option for managing sleep disorders in children with autism spectrum disorder. *American Journal of Health-System Pharmacy, 71*, 95–101.
170. Rossignol, D. A., & Frye, R. E. (2011). Melatonin in autism spectrum disorders: A systematic review and meta-analysis. *Developmental Medicine and Child Neurology, 53*, 783–792.
171. Braam, W., Ehrhart, F., Maas, A. P. H. M., Smits, M. G., & Curfs, L. (2018). Low maternal melatonin level increases autism spectrum disorder risk in children. *Research in Developmental Disabilities, 82*, 79–89.

Protein Nutrition in Autism

Saravana Babu Chidambaram, Abid Bhat, Arehally Marappa Mahalakshmi, Bipul Ray, Sunanda Tuladhar, B. S. Sushmitha, B. Saravanan, Manivasagam Thamilarasan, Arokiasamy Justin Thenmozhi, Musthafa Mohamed Essa, Gilles J. Guillemin, and M. Walid Qoronfleh

Abstract Autism is a developmental disorder that affects communication and behavior. Although autism can be diagnosed at any age, it is said to be a "developmental disorder" because symptoms generally appear in the first 2 years of life. The primary cause of autism is still not clear and therapy is currently restricted to controlling behavioral abnormalities. However, emerging studies have shown a link between mitochondrial dysfunction and autism. Dietary supplements that promote mitochondrial biogenesis and inhibit the production of oxidative stress have been used to treat autism patients. Dietary adjustments in treating autism is a novel approach to suppress autistic symptoms. Supplementation with antioxidants has been found to not only inhibit cognitive decline but also improve behavioral symptoms in autism. Dietary supplements fortified with vitamins should only be given

S. B. Chidambaram (✉) · A. Bhat · B. Ray · S. Tuladhar
Department of Pharmacology, JSS College of Pharmacy, JSS Academy of Higher Education & Research, Mysuru, India

Central Animal Facility, JSS Academy of Higher Education & Research, Mysuru, India

A. M. Mahalakshmi · B. S. Sushmitha · B. Saravanan
Department of Pharmacology, JSS College of Pharmacy, JSS Academy of Higher Education & Research, Mysuru, India

M. Thamilarasan · A. Justin Thenmozhi
Department of Biochemistry and Biotechnology, Annamalai University, Chidambaram, Tamil Nadu, India

M. M. Essa
Department of Food Science and Nutrition, CAMS, Sultan Qaboos University, Muscat, Oman

Ageing and Dementia Research Group, Sultan Qaboos University, Muscat, Oman

G. J. Guillemin
Department of Biomedical Sciences, Faculty of Medicine and Health Sciences, Macquarie University, Sydney, NSW, Australia

M. W. Qoronfleh
Research & Policy Department, World Innovation Summit for Health (WISH), Qatar Foundation, Doha, Qatar

M. M. Essa, M. W. Qoronfleh (eds.), *Personalized Food Intervention and Therapy for Autism Spectrum Disorder Management*, Advances in Neurobiology 24,
https://doi.org/10.1007/978-3-030-30402-7_20

under the supervision of a physician. A wide range of nutraceuticals are under clinical trials to understand whether they physiologically target mitochondrial pathways and improve the quality of life in autism.

Keywords Autism · ASD · Nutritional imbalance · Dietary therapy · Mitochondrial dysfunction · Protein maldigestion · Malabsorption · Amino acids · Peptides · Gluten-free, casein-free (GFCF) diet

1 Introduction

Autism a neurodevelopmental disorder characterized by disrupted social and communicational interactions with stereotyped and repetitive behavior characterized by differing levels of severity. Though genetics play a vital role in the cause of the disease, research in the recent years has strongly suggested nutritional deficiencies and imbalances as contributing and aggravating factors of autism. ASD has a clear biological basis with features of known medical disorders. Available evidence links oxidative stress, mitochondrial dysfunction, and immune dysregulation/inflammation in the brain of ASD individuals to the disorder abnormalities. The brain regions found to contain these physiological abnormalities in individuals with ASD are involved in speech and auditory processing, social behavior, memory, and sensory and motor coordination.

Many children with ASD have selective eating that goes beyond the usual "picky eating" behavior seen in most children at specific developmental stages. These types of self-limiting diets are usually a direct result of the disorder. The diet may be limited to as few as two or three foods or have food preferences. The most common gastrointestinal (GI) symptoms include chronic diarrhea, abdominal distention, discomfort and bloating, gastroesophageal reflux disease (GERD), excessive gas, constipation, fecal impaction, food regurgitation, and a leaky gut syndrome. Children with autism are also at risk for many other nutritional problems such as nutrient deficiencies, food allergies, food intolerances, and feeding problems.

Briefly, the most common dietary "treatments" or approaches involves the list below. Some of these nutrients have been studied to see if providing/restricting children with these supplements may help with autism symptoms. However, we believe more research is needed and the situation should be judged on individual basis.

- Gluten-free, casein-free (GFCF) diet
- Exclusion of phenolic compounds and foods high in salicylates
- Exclusion of food additives
- Yeast-free diet
- Supplements (in particular antioxidants like vitamins and minerals)
- Fish oils and other supplements rich in omega-3 fats
- Probiotics and enzymes

2 Protein Maldigestion in Autistic Children

Many studies have revealed that digestive impairment in autistic children contribute to immune and GI impairment, evident in clinical cases. But there is lack of data for the direct assessment of proteolytic enzyme levels and activity in these children. Results from one study also indicated decreased levels of saccharolytic digestive enzymes. Characterization of upper gastrointestinal tract in these children revealed decreased activity of glucoamylase in 58.3% of children with autism compared to healthy controls [1]. Another study found decreased levels of ileal sucrose isomaltose, maltase glucose amylase, and lactase mRNA in children with ASD and gastrointestinal symptoms when compared to non-autistic children presenting similar gastrointestinal symptoms. Hence, it can be concluded that these symptoms are autism-specific. As these enzymes are located on the brush border of the enterocyte membrane, reduction in their levels is attributed to chronic inflammation of the GI tract which may result in the impairment of other brush border enzymes such as peptidases [2].

3 Role of Amino Acids/Peptides in Autism

Increased levels of urinary peptides of dietary origin provides evidence that autistic children experience impairment in protein digestion with increased intestinal permeability. But these findings are viewed carefully as the opioid excess theory hypothesizes that breakdown products of certain dietary proteins such as casein and gluten are potent agonists of opioid receptors that can have systemic effects and are able to cross the blood–brain barrier [1]. The presence of urinary peptides in children with ASD is classified as exorphins: exogenous opioids including casomorphins, gliadinomorphins, gluteomorphins, deltorphin, and dermorphin. The same population undergoing treatment with a gluten-free, casein-free diet (GFCF, this regimen is the removal of all wheat protein (gluten) and milk protein (casein) from the diet), i.e., GFCF diet for 2–4 years showed significant reduction in urinary peptide levels with improved behavioral metrics, despite a high intake of meat and fish protein. Many animal studies have revealed that gut peptidase inhibition specifically dipeptidyl peptidase IV results in increased levels of urinary peptides of dietary origin. This indicates that a set of autistic children will have reduced digestive enzyme activity, and undigested dietary peptides can cross the intestinal mucosal barrier exerting biological effects systemically [1].

Reports from various studies have shown altered amino acid profiles. Some studies have shown elevated plasma amino acid levels such as glutamic acid, aspartic acid, and taurine, while other studies have shown decreased levels of amino acids, specifically glutamine. A study by Arnold et al. showed frequent amino acid deficiencies in children with ASD given both restricted (GFCF) and unrestricted diets. Children with autism on unrestricted diets showed deficiency in valine, leucine, phenylalanine, and lysine, while children on restricted diets were deficient in isoleucine. In another study, Adams et al. found decreased levels of isoleucine, phenylalanine, tryptophan, and taurine which might be due to decreased protein intake or impaired protein digestion. Low levels of plasma amino acids in autistic children suggest an impaired capacity for protein digestion and increased passage of dietary peptides into systemic circulation by compromised intestinal integrity [1].

On a concluding remark, many plasma amino acids serve either as neurotransmitters (glutamate, aspartate) or precursors (tryptophan and tyrosine) for important neurotransmitters such as serotonin and dopamine. Disturbance or deficiencies in these systems are found to be common among children with autism and contribute significantly to autistic symptoms [1].

4 Dietary Therapies in Autism

Dietary adjustments in treating autism is a newer approach to suppress autistic symptoms. Many physicians have proposed different diets to improve behavior and other symptoms among autistic children.

Some of the proportionate diets are discussed below.

4.1 Elimination Diet

Allergenic foods or substances in food have shown increase in the levels of IgG, IgE, and IgA antibody classes, leading to immune dysregulation in autistic patients. Hence, as the name implies, this diet entails an elimination of those foods that are found to be allergenic. This diet needs careful monitoring as elimination of allergenic foods in the diet may also lead to malnutrition which in turn leads to an increase of the symptoms of the disease. Studies have shown greater improvement in autistic patients' clinical symptoms after adopting an elimination diet.

One such important form of elimination diet is the GFCF diet. This diet necessitates the complete removal of major sources of gluten and casein. However, one source of casein is cow's milk and other dairy products, the removal of which may also simultaneously lead to deficiency in calcium: a major nutrient aiding the maintenance of bone and teeth health. Goat or sheep milk is an often-suggested alternative but that may require the body to tackle new allergens. Hence, specialists suggest soy or rice milk and yeast flakes with added molasses that can serve as substitute for cheese [3].

4.2 *Feingold Diet (A Food Restriction Diet)*

The principles of this diet fall in line with the elimination diet. The primary recommendation is to avoid and eliminate any potential sources of preservatives, food additives, food enhancers and dye additives from the diet which can cause allergies or intolerance and may have carcinogenic and mutagenic properties. These include the presence of salicylates in toothpastes, mouth wash, and cough syrups as well as in natural foods like apples, grapes, cucumbers and any ready-to-eat packet food and fast foods [3].

4.3 *The Ketogenic Diet*

This is a high-fat, low-protein, and low-carbohydrate diet initially developed to treat children with frequent epileptic attacks. The main source of energy production in this form of diet is from fat accounting to 90%, the remaining is from protein and carbohydrates. After the initiation of the diet, it has to be continued for 2–3 years with an initial period of fasting. During the diet, the body will be in the state of ketosis where the metabolic shift occurs from glucose to the main source of energy being ketone compounds formed from fatty acids in the blood.

Many groups of researchers have found that ketogenic diet is one of the best forms of dietary therapy for autism. It is also hypothesized that the diet has a positive impact on the mechanisms of neurological diseases. But adopting this diet requires greater expertise from both the physician and dietician as the uncontrolled concentration of ketone bodies in the blood serum leads to high risk of metabolic disorders [3].

4.4 *The Specific Carbohydrate Diet (SCD)*

The diet was introduced by *Gottschall* as a method to treat autism. The basic premise of this diet is the alleviation of malabsorption thereby preventing the growth of pathogenic intestinal microflora. The diet recommends the intake of only monosaccharides sourced from fruits, vegetables, and honey avoiding complex polysaccharides as the digestion of polysaccharides takes longer time. This eventually disturbs the gastrointestinal tract leading to difficulty in absorption where the residual food becomes a breeding ground for pathogenic intestinal flora. The aim of this diet is to restore normal functions of intestine and to prevent the development of intestinal pathogenic microorganisms. Recommended foods are meat, eggs, natural cheese, homemade yogurt, vegetables (cabbage, cauliflower, onions, spinach, pepper), fresh fruits, nuts (almonds, walnuts), soaked lentils, and beans [3].

4.5 Low Oxalate Diet

The clinical symptoms of autism, a genetically predisposed disease, are aggravated by gastrointestinal disturbances including high levels of oxalates in the blood serum. The acceptable daily intake of dietary oxalates in an adult is 250 mg/day, and it goes up to about 1000 mg/day in a Western diet. However, patients with autism are required to restrict their intake of dietary oxalates to just 40–50 mg/day. Foods rich in oxalates are spinach, beetroot, cocoa, black tea, figs, lemon zest, green apples, black grapes, kiwis, oats, wheat, peanuts, cashew nuts, and blueberries [3].

5 Different Nutritional Imbalance Affecting Growth and Development in Autism

Rapid brain and nerve growth occur during childhood and continue through adolescence and adulthood. Evidence has shown the importance of nutritional supplements for the cause of autism. Many studies have suggested that low scores on behavioral assessment tests have consistently been correlated with low nutritional levels, and when supplemented with specific nutritional additives, the hyperactivity, impulsiveness, and inability to pay attention improve dramatically.

5.1 Omega-3 Fatty Acids

Omega-3 fatty acids, also known as ω-3 fatty acids or n-3 fatty acids, are polyunsaturated fatty acids (PUFA). Omega-3 fatty acids occur naturally in two forms, triglycerides and phospholipids. The major composition of brain nerve cells is fat. Three omega-3 fatty acids are important in human physiology. These are α-linolenic acid (ALA), eicosapentaenoic acid (EPA), and docosahexaenoic acid (DHA). The chemical structure of these key PUFA is presented in Table 1. They are found primarily in fish, eggs, and flax seeds. The most widely available dietary source of EPA and DHA is fish. PUFA are an absolute necessity for human health. Their concentrations in the brain also play an important role in neurological disorders like autism. The ability of the brain to create neural signals in response to new experiences and learning environment is known as neuronal plasticity, a crucial step in long term memory and learning. Proper levels of omega-3 fatty acids and DHA are important for membrane fluidity that maintains learning ability and enhances neuronal plasticity [5].

Table 1 Chemical structures of vital polyunsaturated fatty acids (PUFA): omega-3 fatty acids

PUFA[a]	Chemical structure(s)	References
α-Linolenic acid (ALA) It is an essential fatty acid found in seeds, nuts, and many common vegetable oils	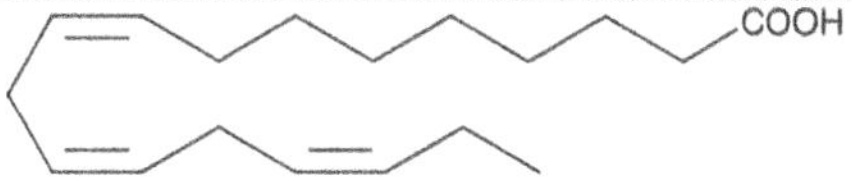 	PubChem ALA was first isolated by Rollett [4]
Eicosapentaenoic acid (EPA) Found in oily fish. A portion of absorbed ALA is converted into EPA. Acts as a precursor for prostaglandin-3 and thromboxane-3 families and inhibits arachidonic acid conversion into the thromboxane-2 and prostaglandin-2 families	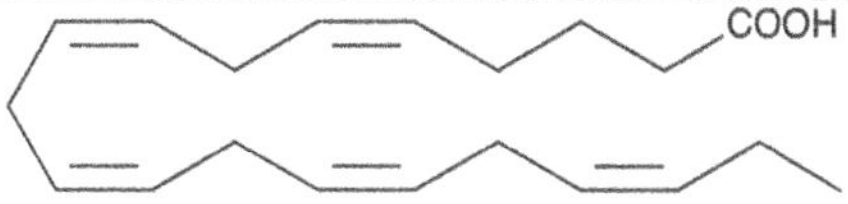 	PubChem Omega-3 fatty acids, particularly EPA, have been studied for their effect on ASD patients
Docosahexaenoic acid (DHA) It is abundant in seafood. It can be synthesized from ALA. It is a primary structural component of the human brain and major component in the retina. It is widely used as a food supplement	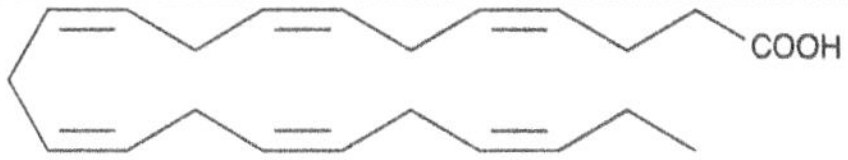 	PubChem

[a]In 1929 George and Mildred Burr reported that dietary fatty acid was required to prevent a deficiency disease that occurred in rats fed a fat-free diet. The discovery of essential fatty acids was a paradigm-changing finding [The Journal of Lipid Research, 56, 11–21 (2015)]. Original work [Journal of Biological Chemistry, 82, 345–367 (1929) and Journal of Biological Chemistry, 86, 587–621 (1930)]

5.2 Zinc

Zinc, the mood mineral, plays an important role as it serves as a cofactor to several neurotransmitters affecting mood and learning. Specifically, the production of dopamine, an important neurotransmitter in learning and emotions like motivation and pleasure, is disturbed with low zinc levels. Zinc also plays an important role in clearing toxic chemicals from the brain tissue called mercury, the improper elimination of which leads to the accumulation of toxicants leading to rise in conditions such as autism [6].

5.3 Vitamins

All the vitamins are necessary at optimal levels for the healthy development of the brain. Specifically, vitamin D supplementations have been shown to reverse some autistic behaviors. Vitamins possess strong antioxidant activity and help in countering the deleterious effects of free radicals on cellular and mitochondrial function [7]. They also act as cofactors and play an important role in many biological processes. They play a vital role in DNA synthesis and control lipid and protein metabolism. Researchers have also found that low folate levels in pregnancy were associated with hyperactivity in children. Vitamin B1 has also shown clinical benefit in autistic children. Vitamin C acts as double-edged sword in autism. It is essential for the synthesis of certain neurotransmitters and it also exhibits antioxidant properties. Absorption of vitamin C takes place in the small intestine. The higher levels of consumption of vitamin C (around 1 g/day) show reduced absorption efficiency. Researchers have also shown absence of a specific gene encoding for a protein essential in vitamin A synthesis. Supplementation of vitamin A in clinical studies of autism patients has significantly improved language skills and eye contact. But it is very important that vitamin A supplementation has to be taken under the supervision of physicians [7].

5.4 *Iron*

Iron deficiency in patients with autism might be caused by malabsorption of the nutrient in the gastrointestinal tract. Thus, the deficiency of iron is said to produce negative impact on sleep and neuroprotection. Some of clinical studies have suggested that cognitive impairment, reduced growth, disturbance in concentration, and mood changes in autistic children are associated with anemia [8].

5.5 *Magnesium*

Magnesium and vitamin B6 work together in improving clinical symptoms of autism. When a group of autistic children were supplemented with magnesium and vitamin B6, 70% of the children showed improvement in social interaction and communication [9].

5.6 *Probiotics*

Probiotics help rejuvenate certain healthy strains of microorganisms which in turn leads to better utilization of food ingredients along with alleviating the development of pathogenic organisms thereby increasing immunomodulatory effects in the body. This also helps overcome gastrointestinal complications associated with autism such as constipation, acute diarrhea, inflammatory intestinal disease, and irritable bowel syndrome [3]. A chapter in this book has been dedicated to probiotics discussion.

6 Must Avoid Foods in Autism

- Gluten. In part II the role of gluten in autism is explained.
- Dairy products. The GFCF diet is discussed above.
- Sugars.
- Corn.
- Artificial ingredients include dyes, flavoring agents, taste enhancers, and preservatives. In part II a chapter details popular natural food coloring additives.

7 Mitochondrial Dysfunction in Autism

Autism refers to a group of neurodegenerative diseases. The primary cause of autism is still not clear, and therapy is currently restricted to controlling behavioral abnormalities. However, emerging studies have shown a link between mitochondrial dysfunction and autism [10]. Coleman and Blass [11] and Lombard have postulated that autism is possibly because of mitochondrial dysfunction which results in neuronal oxidative phosphorylation within the central nervous system. These hypotheses are based on outcome that lactic acidosis, increase in Krebs cycle metabolites levels, plasma carnitine deficiency, and diminished brain glucose utilization and reduced ATP levels have been found in autistic children. Furthermore, Lombard [12] postulated that autism may be a disorder because of mitochondrial dysfunction [11, 12]. Several clinical studies have confirmed the involvement of mitochondrial dysfunction in autism, where they found an increase in the lactate content in the blood of autistic patients [13–15]. An MRI scan of autistic patients showed changes in brain energy and phospholipid metabolism in autism that lead to impairments in learning and memory [16]. Richard et al. [17] found that children with autism have higher levels of in tumor necrosis factor-α which inhibits mitochondrial functions in lymphocytes and the brain [18, 19]. An increase in the levels of pyruvate, ubiquinone, and acylcarnitines, the markers of mitochondrial function, has also been reported among children with autism [20, 21]. Analysis of postmortem brain samples of autistic patients has shown the presence of oxidative stress, decrease in major cellular antioxidants, altered proteins, and lipid metabolism, and it also affects the functions of important enzymes. GSH levels were found to be reduced in autism brains which promotes oxidative stress, immune dysfunction, and apoptosis promoting the development of autism [22, 23]. Mitochondria can be a critical target for the therapeutic management of autism.

8 Anti-Oxidants in Targeting Autism

Dietary supplements that promote mitochondrial biogenesis and inhibit the production of oxidative stress have been used to treat autism patients [24]. Supplementation with antioxidants has been found to not only inhibit cognitive decline but also

Table 2 Chemical structures of supplements that targets mitochondrial pathways

Antioxidant supplement	Chemical structure	References
L-Carnitine (vitamin B_T) It is a quaternary ammonium compound involved in metabolism in most mammals, plants, and some bacteria	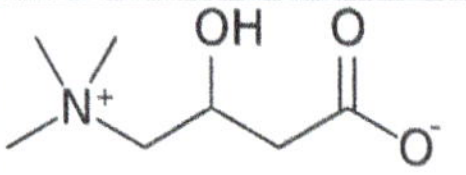	PubChem Discovered 1905 in meat extract. Structure was established 1927 Bremer [26]
L-Creatine It is biosynthesized from the amino acids glycine and arginine	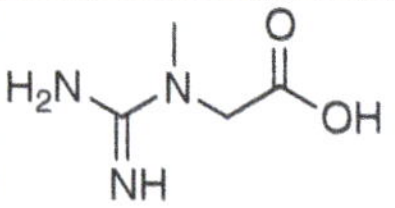	PubChem Identified in 1832 by Michel Eugène Chevreul Cannan and Shore [27]
CoQ_{10} (ubiquinone 10) It is a 1,4-benzoquinone also known as ubidecarenone or coenzyme Q	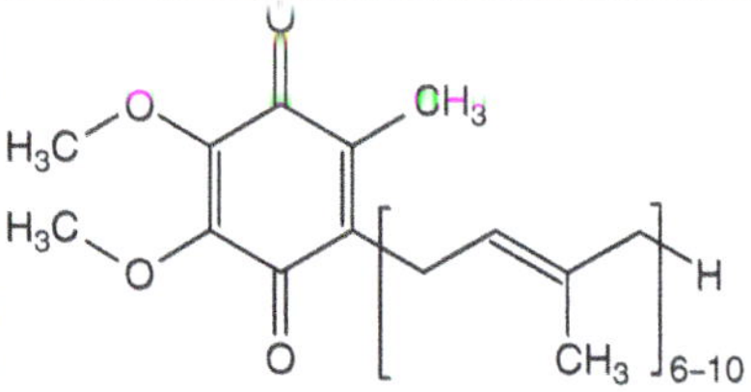	PubChem Identified in 1958 by Donald E. Wolf and Karl Folkers Crane [28]
L-Arginine It is an α-amino acid that is used in the biosynthesis of proteins	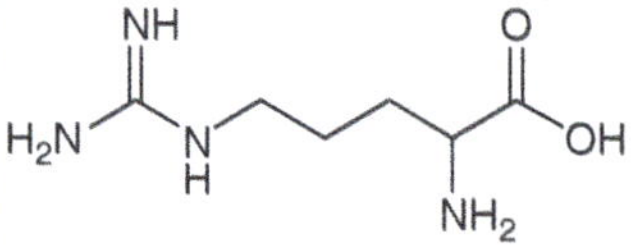	PubChem First isolated in 1886 from lupin and pumpkin seedlings by the German chemist Ernst Schulze Apel [29]

improve behavioral symptoms in autism [25]. Dietary supplements fortified with vitamins should only be given under the supervision of a physician [7]. A wide range of nutraceuticals are under clinical trials to understand whether they physiologically target mitochondrial pathways and improve the quality of life in autism. Chemical structures of selected antioxidant supplement that targets mitochondrial pathways are presented below (see Table 2).

8.1 L-Carnitine

L-Carnitine is a cellular compound that has an important role in the metabolism of lipids in mitochondria. Carnitine relocates long-chain fatty acids along the mitochondrial inner membrane as acylcarnitine esters. These esters are oxidized to acetyl-CoA which takes part in the Krebs cycle causing oxidative phosphorylation which in turn results in the production of ATP. Carnitine inhibits CoA exhaustion and eliminates acyl compounds that are toxic in nature. Till date, no pharmacological intervention which can increase the levels of CoA is available. Diet is the major source of carnitine, although some quantity is also synthesized by the muscle, liver, and kidneys. Skeletal muscles are rich in carnitine [30]. Levels of carnitine in blood are controlled by its active reabsorption in the proximal renal tubules. L-Carnitine fortified food supplements for mitochondrial dysfunction are expected to improve the free carnitine content and remove toxic compounds from the body. Carnitine is available in supplements fortified with vitamins and other cofactors and is given to autistic patients either orally or parentally [7].

8.2 Coenzyme Q10

Coenzyme Q10 (also known as ubiquinone) endogenously produced in mammalian mitochondria is a critical component of the mitochondrial electron transport chain. CoQ10 exists in all cellular and organelle membranes. It plays a significant role in intracellular signaling and also acts as a strong antioxidant. CoQ10 controls the mitochondrial permeability transition pore involved in apoptosis and leads to the activation of uncoupling proteins. Any alterations in the CoQ10 biosynthetic results in human mitochondrial disease like neonatal encephalopathy with nephropathy (COQ2), Leigh syndrome, lactic acidosis, and nephropathy (PDSS2) infantile nephropathy, hepatopathy, retardation (PDSS1) and recessive ataxia, cerebellar atrophy ± retardation, lactic acidosis, and exercise intolerance (ADCK3). Exogenous administration of CoQ10 revived in 2007 has been found to improve the cognitive and behavioral conditions in mitochondrial disease like autism [7].

8.3 L-Creatine

L-Creatine is a compound existing in all cells, which, in the presence of phosphate, forms phosphocreatine in mitochondria. It is released through anaerobic metabolism and is considered a high-energy source. It plays an important role during the relocation of high-energy phosphates from mitochondria to cytoplasm for the production of ATP. High levels of creatine have been found in tissues like the skeletal muscle and brain because of their higher energy demand. Creatine levels are maintained by endogenous production and through diet. Decrease in the levels of phosphocreatine has been found in the skeletal muscles and brain of patients with mitochondrial dysfunction. Clinical trials have shown that exogenous administration of creatine in mitochondrial cytopathies improves the quality of life in patients with mitochondrial cytopathies. Similarly, it has been found that creatine given to pediatric patients with mitochondrial encephalomyopathies improves the behavior and cognitive decline [7].

L-Arginine L-Arginine is a semi-essential amino acid playing an important role in growth, urea elimination, and synthesis of creatine. Nitric oxide acting as a neurotransmitter and having vasodilatory properties is produced by L-arginine.

Administration of L-arginine (500 mg/kg/dose) reduces the severity of stroke-like symptoms and improves microcirculation in patients with mitochondrial encephalomyopathy, lactic acidosis, and stroke-like episodes (MELAS). Arginine has been successfully used in stroke and other types of mitochondrial diseases like autism to reduce the progression and severity of the disease [7].

9 Conclusion

Over the past two decades, research on the effect of diet and nutrition on ASD has been increasing. Particular attention has focused on the role of food additives, refined sugar, food allergies, and fatty acid metabolism. Many studies have also proposed that the increase in oxidative stress is linked to the pathogenesis of this neurocognitive disorder. The increased production of reactive oxygen species (ROS) both centrally (in the brain) and peripherally (in the plasma) may result in the reduction of brain cell numbers leading to apoptosis and contributing to autism development. On one hand systemic review analysis results on diet/ nutritional intervention which however seems to be conflicting and not conclusive while on the other hand controlled, long-term (12 months) nutritional and dietary intervention clinical trials revealed that there was a significant improvement in nonverbal intellectual ability in the treatment group compared to the non-treatment group. Treatments used antioxidants like vitamin/mineral supplements, essential fatty acids, carnitine, GFCF diet, etc. It appears that a comprehensive nutritional and dietary intervention is effective at ameliorating nutritional status, nonverbal IQ, autism symptoms, and other symptoms in most individuals with ASD reflecting potential therapeutic benefit of antioxidants (by enhancing antioxidants capacity) in improving social communication, unusual behaviors, and self-regulation behaviors of children with ASD. It is also possible that certain antioxidants balance neurotransmitter levels in the brain, which decreases the presentation of some features of autism. Generally, the treatment group had significantly increased amounts of EPA; DHA; carnitine; vitamins A, B2, B5, B6, B12, and folic acid; and coenzyme Q10. The word treatment should be used with caution and should be interpreted as interventions that are intended to help autistic individuals to adjust more effectively to their surroundings.

References

1. Sanctuary, M. R., Kain, J. N., Angkustsiri, K., & German, J. B. (2018). Dietary considerations in autism spectrum disorders: The potential role of protein digestion and microbial putrefaction in the gut-brain axis. *Frontiers in Nutrition, 5*, 40. https://doi.org/10.3389/fnut.2018.00040
2. McElhanon, B. O., McCracken, C., Karpen, S., & Sharp, W. G. (2014). Gastrointestinal symptoms in autism spectrum disorder: A meta-analysis. *Pediatrics, 133*, 872–883. https://doi.org/10.1542/peds.2013-3995

3. Kawicka, A., & Regulska-Ilow, B. (2013). How nutritional status, diet and dietary supplements can affect autism. A review. *Roczniki Państwowego Zakładu Higieny, 64*, 1–12.
4. Rollett, A. (1909). Zur Kenntnis der Linolensäure und des Leinöls. *Zeitschrift für physiologische Chemie, 62*(5–6), 422–431.
5. Cheng, Y. S., Tseng, P. T., Chen, Y. W., Stubbs, B., Yang, W. C., Chen, T. Y., et al. (2017). Supplementation of omega 3 fatty acids may improve hyperactivity, lethargy, and stereotypy in children with autism spectrum disorders: A meta-analysis of randomized controlled trials. *Neuropsychiatric Disease and Treatment, 13*, 2531–2543. https://doi.org/10.2147/NDT.S147305
6. Hagmeyer, S., Sauer, A. K., & Grabrucker, A. M. (2018). Prospects of zinc supplementation in autism spectrum disorders and shankopathies such as Phelan McDermid syndrome. *Frontiers in Synaptic Neuroscience, 10*, 11. https://doi.org/10.3389/fnsyn.2018.00011
7. Parikh, S., Saneto, R., Falk, M. J., Anselm, I., Cohen, B. H., Haas, R., et al. (2009). A modern approach to the treatment of mitochondrial disease. *Current Treatment Options in Neurology, 11*, 414–430.
8. Gunes, S., Ekinci, O., & Celik, T. (2017). Iron deficiency parameters in autism spectrum disorder: Clinical correlates and associated factors. *Italian Journal of Pediatrics, 43*, 86. https://doi.org/10.1186/s13052-017-0407-3
9. Mousain-Bosc, M., Roche, M., Polge, A., Pradal-Prat, D., Rapin, J., & Bali, J. P. (2006). Improvement of neurobehavioral disorders in children supplemented with magnesium-vitamin B6. II. Pervasive developmental disorder-autism. *Magnesium Research, 19*, 53–62.
10. Griffiths, K. K., & Levy, R. J. (2017). Evidence of mitochondrial dysfunction in autism: Biochemical links, genetic-based associations, and non-energy-related mechanisms. *Oxidative Medicine and Cellular Longevity, 2017*, 4314025. https://doi.org/10.1155/2017/4314025
11. Coleman, M., & Blass, J. P. (1985). Autism and lactic acidosis. *Journal of Autism and Developmental Disorders, 15*, 1–8.
12. Lombard, J. (1998). Autism: A mitochondrial disorder? *Medical Hypotheses, 50*, 497–500.
13. Bernier, F. P., Boneh, A., Dennett, X., Chow, C. W., Cleary, M. A., & Thorburn, D. R. (2002). Diagnostic criteria for respiratory chain disorders in adults and children. *Neurology, 59*, 1406–1411. https://doi.org/10.1212/01.wnl.0000033795.17156.00
14. Correia, C., Coutinho, A. M., Diogo, L., Grazina, M., Marques, C., Miguel, T., et al. (2006). Brief report: High frequency of biochemical markers for mitochondrial dysfunction in autism: No association with the mitochondrial aspartate/glutamate carrier SLC25A12 gene. *Journal of Autism and Developmental Disorders, 36*, 1137–1140. https://doi.org/10.1007/s10803-006-0138-6
15. Oliveira, G., Diogo, L., Grazina, M., Garcia, P., Ataíde, A., Marques, C., et al. (2005). Mitochondrial dysfunction in autism spectrum disorders: A population-based study. *Developmental Medicine and Child Neurology, 47*, 185–189.
16. Minshew, N. J., Goldstein, G., Dombrowski, S. M., Panchalingam, K., & Pettegrew, J. W. (1993). A preliminary 31P MRS study of autism: Evidence for undersynthesis and increased degradation of brain membranes. *Biological Psychiatry, 33*, 762–773. https://doi.org/10.1016/0006-3223(93)90017-8
17. Richard E. F, & Daniel A. R. (2011). Mitochondrial dysfunction can connect the diverse medical symptoms associated with autism spectrum disorders. *Pediatric Research, 69*, 5. https://doi.org/10.1203/PDR.0b013e318212f16b
18. Chez, M. G., Dowling, T., Patel, P. B., Khanna, P., & Kominsky, M. (2007). Elevation of tumor necrosis factor-alpha in cerebrospinal fluid of autistic children. *Pediatric Neurology, 36*, 361–365. https://doi.org/10.1016/j.pediatrneurol.2007.01.012
19. Malik, M., Sheikh, A. M., Wen, G., Spivack, W., Brown, W. T., & Li, X. (2011). Expression of inflammatory cytokines, Bcl 2 and cathepsin D are altered in lymphoblasts of autistic subjects. *Immunobiology, 216*, 80–85. https://doi.org/10.1016/j.imbio.2010.03.001
20. Kurup, R. K., & Kurup, P. A. (2003). A hypothalamic digoxin-mediated model for autism. *The International Journal of Neuroscience, 113*, 1537–1559. https://doi.org/10.1080/00207450390231482

21. Siddiqui, M. F., Elwell, C., & Johnson, M. H. (2016). Mitochondrial dysfunction in autism spectrum disorders. *Autism-Open Access, 6*, 1000190. https://doi.org/10.4172/2165-7890.1000190
22. Chauhan, A., Audhya, T., & Chauhan, V. (2012). Brain region-specific glutathione redox imbalance in autism. *Neurochemical Research, 37*, 1681–1689. https://doi.org/10.1007/s11064-012-0775-4
23. Rose, S., Melnyk, S., Pavliv, O., Bai, S., Nick, T. G., Frye, R. E., et al. (2012). Evidence of oxidative damage and inflammation associated with low glutathione redox status in the autism brain. *Translational Psychiatry, 2*, e134. https://doi.org/10.1038/tp.2012.61
24. Frye, R. E., & Rossignol, D. A. (2014). Treatments for biomedical abnormalities associated with autism spectrum disorder. *Frontiers in Pediatrics, 2*, 66. https://doi.org/10.3389/fped.2014.00066
25. James, S. J., Melnyk, S., Fuchs, G., Reid, T., Jernigan, S., Pavliv, O., et al. (2009). Efficacy of methylcobalamin and folinic acid treatment on glutathione redox status in children with autism. *The American Journal of Clinical Nutrition, 89*, 425–430. https://doi.org/10.3945/ajcn.2008.26615
26. Bremer, J. (1983). Carnitine–metabolism and functions. *Physiological Review, 63*, 1420–1479.
27. Cannan, R. K., & Shore, A. (1928). The creatine-creatinine equilibrium. The apparent dissociation constants of creatine and creatinine. *Biochemical Journal, 22*(4), 920–929.
28. Crane, F. L. (2007). Discovery of ubiquinone (coenzyme Q) and an overview of function. *Mitochondrion, 7*(Suppl), S2–S7.
29. Apel, F. (2015, July). *Biographie von Ernst Schulze*. http://www.arginium.de/wp-content/uploads/2015/09/Biographie-Ernst-Schulze-Juli-2015.pdf.
30. Tein, I. (2003). Carnitine transport: Pathophysiology and metabolism of known molecular defects. *Journal of Inherited Metabolic Disease, 26*, 147–169.

Autism and Gut–Brain Axis: Role of Probiotics

Saravana Babu Chidambaram, Sunanda Tuladhar, Abid Bhat, Arehally Marappa Mahalakshmi, Bipul Ray, Musthafa Mohamed Essa, Muhammed Bishir, Srinivasa Rao Bolla, Nandakumar Dalavaikodihalli Nanjaiah, Gilles J. Guillemin, and M. Walid Qoronfleh

Abstract Characterized by a wide range of behavioural, social and language problems, autism is a complex developmental disability that affects an individual's capacity to communicate and interact with others. Although the real causes that lead to the development of autism are still unclear, the gastrointestinal tract has been found to play a major role in the development of autism. Alterations in macrobiotic compositions have been reported in autistic children. Irregularities in carbohydrate

S. B. Chidambaram (✉) · S. Tuladhar · A. Bhat · B. Ray
Department of Pharmacology, JSS College of Pharmacy, JSS Academy of Higher Education & Research, Mysuru, India

Central Animal Facility, JSS Academy of Higher Education & Research, Mysuru, India

A. M. Mahalakshmi · M. Bishir
Department of Pharmacology, JSS College of Pharmacy, JSS Academy of Higher Education & Research, Mysuru, India

M. M. Essa
Department of Food Science and Nutrition, CAMS, Sultan Qaboos University, Muscat, Oman

Ageing and Dementia Research Group, Sultan Qaboos University, Muscat, Oman

S. R. Bolla
Department of Anatomy, College of Medicine, Imam Abdulrahman Bin Faisal University, Dammam, Saudi Arabia

N. D. Nanjaiah
Department of Neurochemistry, National Institute of Mental Health and Neurosciences (NIMHANS), Bengaluru, India

G. J. Guillemin
Department of Biomedical Sciences, Faculty of Medicine and Health Sciences, Macquarie University, Sydney, NSW, Australia

M. W. Qoronfleh
Research & Policy Department, World Innovation Summit for Health (WISH), Qatar Foundation, Doha, Qatar

M. M. Essa, M. W. Qoronfleh (eds.), *Personalized Food Intervention and Therapy for Autism Spectrum Disorder Management*, Advances in Neurobiology 24,
https://doi.org/10.1007/978-3-030-30402-7_21

digestion and absorption could also explain some of the gastrointestinal problems reported in autistic patients, although their role in the neurological and behavioural problems remains uncertain. A relationship between improved gut health and decrease of symptoms in autism has been reported as well. Studies done to evaluate the gluten-free diets, casein-free diets, pre- and probiotic and multivitamin supplementation have shown promising results. Probiotics have been thought to alleviate the progression of autism and reduce cognitive and behavioural deficits.

Keywords Autism · ASD · Cognitive and behavioural deficits · Gut–brain axis · GI dysfunction · Barrier pathway · Microbiome · Probiotics

1 Introduction

Autism spectrum disorder (ASD) is a neurodevelopmental disorder characterized by disrupted social and communication interactions with stereotyped and repetitive behaviour of different levels of severity. ASD has traditionally been framed as a behavioural disorder. However, evidence is accumulating that ASD is characterized by certain physiological abnormalities, including oxidative stress, mitochondrial dysfunction and immune dysregulation/inflammation. The brain regions found to contain these physiological abnormalities in individuals with ASD are involved in speech and auditory processing, social behaviour, memory and sensory and motor coordination. Though genetics play a vital role in the cause of the disease, recent investigations have strongly suggested nutritional deficiencies and imbalances as also contributing to and aggravating autism. Studies have shown that people with autism often have abnormal digestive health conditions. Research indicates there is a strong link between the functioning of the brain and the gut where postnatal development of a child depends on the microbiome. Some experts claim that several types of food and diet interventions can treat (social and behavioural management) children and adults with ASD. In this chapter, we discuss various theories as well as the effectiveness of diets and probiotics, the so-called "friendly bacteria", in helping ease the symptoms of autism. Also, we present some notable findings, demonstrating probiotic success in relieving GI symptoms in autistic kids as well as its efficacy in controlling the children's anxiety and oversensitivity to stimuli.

2 Altered Gut–Brain Axis in Autism

The occurrence of gastrointestinal (GI) problems due to alterations in the gut microbiota has been documented in autism based on existing patient observations. This complexity in the crosstalk between the gut and the brain has been discussed broadly as the "gut–brain axis" or GBA. The triad of GBA, immune system and GI

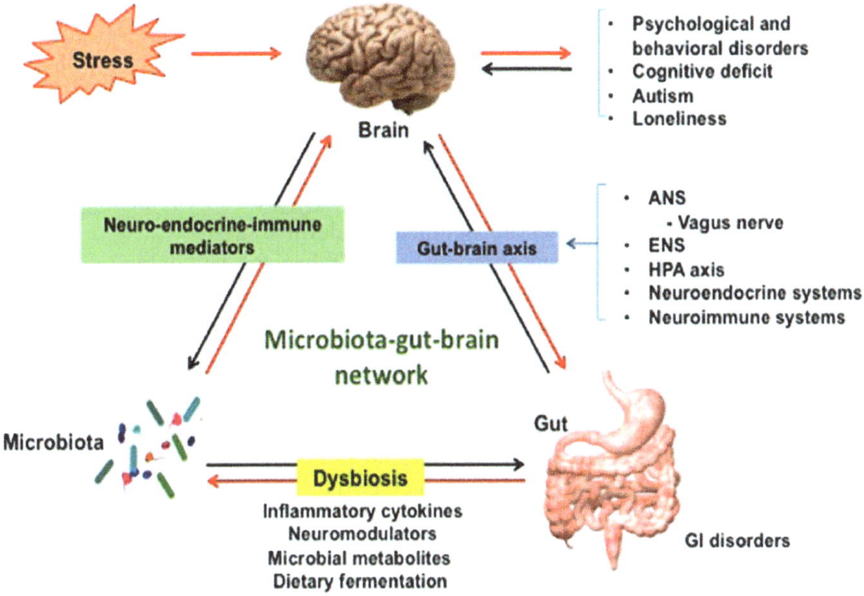

Fig. 1 The triad of gut–brain axis, immune system and GI microbiome. Adapted from Sirisinha [1] and Thakur et al. [2]

microflora cross-communication is illustrated in Fig. 1. Various studies involving animal models of different behavioural disorders such as autism, anxiety and cognitive dysfunction have shown that the constitution of the gut microbiota has an influence in the brain functions. It functions by monitoring and integrating the gut functions along with the emotional and cognitive centres of the brain and the peripheral intestinal functions and mechanisms such as immune activation, intestinal permeability, enteric reflex and entero-endocrine signalling [3]. Cognitive and behavioural alterations, induced due to various neuroactive compounds in the intestinal lumen crossing the blood–brain barrier [4]. The link between GI symptoms and neurodevelopmental disorders has been supported by the following observations:

- The onset of disease usually follows antimicrobial therapy.
- At the advent of the disease, frequent persistence of a number of gastrointestinal abnormalities has been observed.
- Autistic symptoms have sometimes been reduced by oral vancomycin treatment, while relapse occurs following cessation of treatment.

The gut–brain axis accesses the signal from the gut microbiota, influences the brain functions and vice versa. The bidirectional communication acts via the neuroendocrine and neuroimmune mechanism which involves both the autonomic nervous system (ANS) and the enteric nervous system (ENS). The fundamental morphologic components of the brain to gut microbiota signalling are the sympathetic and parasympathetic nerves of the ANS [5]. The sympathetic system inhibits the intestinal motor functions and decreases gut secretion. Under conditions of stress, the sympathetic system is over activated, the integrity of the gut epithelium is destroyed and gut motility and secretions are changed [6]. Stress-induced changes of the gut alter the habitat of resident bacteria and promote alterations to microbiota composition or activity [7]. The hypothalamus–pituitary–adrenal (HPA) axis is another critical mechanism by which the brain influences the composition of the gut microbiota. When the HPA axis is overactivated, the levels of circulating cortisol and pro-inflammatory cytokines are significantly elevated [8].

The human GI tract contains approximately 10^4 bacteria belonging to approximately 1000 species. The healthy adult GI tract is most dominated by *Bacteroidetes* and *Firmicutes* phyla (both account for up to 70–90% of total bacteria), followed by *Actinobacteria*, *Proteobacteria* and *Verrucomicrobia* [9]. The symbiotic microbiota dwelling in the gut have long been appreciated for the various beneficial effects they offer to the host, including providing essential nutrients by metabolizing indigestible dietary compounds, defending the gut against opportunistic pathogen colonization by nutrient competition and antimicrobial substance production and contributing to the intestinal epithelial barrier. Moreover, studies on the immune defects in germ-free (GF) mice have suggested that gut microbiome is essential to the host immune system [10]. A recent review also indicated that gut bacterial colonization could drive maturation and functionality of the host's adaptive immune system [11, 12].

Bifidobacterium, a beneficial bacteria appeared to be reduced in children with autism [13], while other probiotics, i.e. *Lactobacillus*, *Bacteroides* and *Desulfovibrio*, were reported to be present in higher concentrations amongst them [14]. Consistent with this, the abundances of *Blautia*, *Dialister*, *Prevotella*, *Turicibacter* and *Veillonella* were all decreased [15]. Children with autism had much lower levels of *Bifidobacterium* (−45%, $p = 0.002$), slightly lower levels of *Enterococcus* (−16%, $p = 0.05$ per Wilcox) and much higher levels of *Lactobacillus* (+100%, $p = 0.00003$) [14].

Potentially harmful *Clostridium* species were observed to be abundant in faeces of children with autism [16]. Recent studies related to faecal microbial profiles of autistic patients have also indicated tenfold higher counts of *Clostridium spp*., which produce neurotoxins and contribute to the development of autistic behaviours, compared with healthy controls [17].

In addition, De Angelis et al. indicated that *Oscillospira* decreased and *Roseburia* increased in Alzheimer's Disease (AD). Meanwhile it was also observed that opportunistic pathogens like *Enterobacter* and *Shigella* were elevated in the case of AD patients [18]. *Faecalibacterium* [19] and *Ruminococcus* [20] were also reported to increase in patients with autism. Notably, these particular species are known to be versatile carbohydrate metabolizers [21]. *Blautia* plays an important role in nutrient assimilation and gut maturation in children [22]. The reduction of these beneficial bacteria in autism patients may be implicated in the pathogenesis of the disease. Short-chain fatty acids (SCFAs), the critical mediators within the microbiota–gut–brain axis, can cross the blood–brain barrier (BBB) and modulate brain activity directly [4].

Therefore, a number of possible mechanisms have been postulated relating the gut microbiome and the brain axis in autism. Few of these pathways are discussed below.

2.1 The Barrier Pathway

An increase in intestinal permeability was found in patients with autism, and this was measured by the lactulose/mannitol test [23]. One particular study showed that the impaired intestinal and blood–brain barrier function in autism decreased the level of intestinal tight junction (TJ) components and caused an increase in the Claudin level in the autism brain when compared to a group of controls [24]. The microbiota along with its metabolites contributes to the regulation of the intestinal barrier. The dysbiosis in the case of autism is a result of increased permeability of the gastrointestinal tract which is referred to as the "leaky gut". The "leaky gut" allows bacterial metabolites, metabolites that do not naturally cross this barrier and are potentially neuroactive, to readily cross the intestinal barrier. Studies have shown evidence of increased metabolites in urine and systemic circulation in autism [25]. Zonulin has structural similarities with several growth factors known to affect intercellular TJ integrity. This enzyme regulating intestinal permeability was seen to be significantly increased in subjects with autism bearing GI symptoms also showing hampered intestinal permeability in the disease condition [26]. Hence, a disrupted intestinal barrier allows endotoxins to enter the bloodstream. For instance, lipopolysaccharide (LPS) is a potent endotoxin which alters neuronal and microglial activity in the amygdala, a region involved in control of emotions [27]. In patients with autism, the serum LPS levels were significantly high when compared to healthy individuals, and this could be correlated with impairment in social behavioural scores [28]. Targeting improvement in the epithelial barrier in autism can reduce the entrance of the microbial endotoxins, thus normalizing the gut–brain pathway. The BBB acts as a shield against the infiltration of pathogens and other endotoxins entering the brain. In order to maintain brain functions and development, it is necessary to maintain the integrity of this barrier. BBB dysfunction can be caused by multiple prenatal and postnatal risk factors that are also evident in autism [29]. Several of these risk factors are detailed in Chap. "Overview and Introduction to Autism Spectrum Disorder (ASD)".

2.2 The Serotonin Pathway

The serotonin functions in the brain both for regulation of mood and cognition and for regulating intestinal secretion, motility and pain perception. Its synthesis in the intestine and the brain depends upon the intake of dietary tryptophan [30].

Serotonin synthesis in the brain is decreased in patients with autism. A recent study demonstrated the correlation between whole blood serotonin level and the intestinal symptoms in autism [31]. Inflammation in the intestinal tract leads to production of serotonin by the enterochromaffin cells and intestinal mast cells. This leads to alteration in motility, vasodilation and an increase in vascular permeability, causing functional intestinal dysmotility. During intestinal tract inflammation, there is increased consumption of dietary tryptophan, causing low concentrations to be available for the brain. Thus, brain serotonin levels will be reduced causing mood and cognitive dysfunction in autism. On depleting dietary tryptophan, indeed an increase in autistic behaviour was observed in autism patients. Also, the availability of tryptophan was seen to be affected in the case of intestinal dysbiosis in autism [32].

In a murine model of autism induced by prenatal exposure to valproic acid (VPA), impairments in social behaviour were associated with intestinal inflammation and a disturbed serotonergic system in the brain and intestinal tract [32]. In the prefrontal cortex as well as in the amygdale, reduced levels of serotonin and increased turnover were found in VPA-exposed male offspring. The reduction in intestinal serotonin in VPA-exposed mice was attributable to reduced number of serotonin-positive cells (possibly enterochromaffin cells) in the small intestine [33].

2.3 Immune System Pathway

The gut microbiota can also be related to cerebral dysfunctions by modulating the host immune response. Pathogenic and bacterial microbiota stimulate the secretion of pro-inflammatory cytokines like IL-1, IL-6 and IL-8 by the intestinal epithelial cells, dendritic cells and macrophages [34], which account for various neuropsychiatric disorders including anxiety, schizophrenia as well as autism [35]. Parents of autism children report more often food allergies than parents of healthy children [36]. The persistent default state of mucosal immune tolerance observed in food allergy is strongly associated with a changed microbiota composition such as enhanced *Bacteroidetes* and *Enterobacter*. The majority of allergies are characterized by a T-helper 2-type immune response with the characteristic cytokines interleukin (IL) 4, IL5 and IL13. Supporting the role of allergy in autism, children produced significantly higher levels of the mentioned cytokines [37]. In addition, less IL-10-producing T cells are present in the periphery and intestinal mucosa as well as reduced plasma levels of tumour necrosis factorβ in autism patients suffering from intestinal problems [38]. Taken together, there seems to be a disturbed T-cell balance in the intestinal tract of autism patients.

2.4 *Neuronal Pathway*

Another possible mechanism by which the microbiota–gut–brain axis mediates communication may be through the use of established neuronal circuits. Vagal afferents are critical neuronal pathways allowing information flow from the viscera to the CNS. Gut microbiota can deliver their signals to the brain via the vagus nerve [5]. In a study with autism patients suffering from epilepsy, besides reducing the seizure frequency, stimulation of the vagus nerve resulted in improved verbal skills, mood and alertness [39]. Epilepsy has been observed in about 30% of autistic patients [40].

It might be that microbial neurotransmitters affect the ENS and afferent nerve function directly or via the intestinal epithelium. Based on the fact that stress-related host neurotransmitter release increases the proliferation rate and the activity of intestinal microbiota [41], it has been postulated that microbiota-derived neurotransmitters have a primary role in the sustainability of the microbes themselves in the intestinal tract in stressful situations [29]. In fact, neurochemical and behavioural effects were not present in vagotomized mice, identifying the vagus as the major modulatory constitutive communication pathway between the microbiota and the brain [42]. These data suggest that vagus stimulation, possibly through a "healthy" microbiome, might be beneficial in autism. Taken together, the role of the ENS, the vagus nerve and bacterial neuroactive metabolites and molecular pathways in relation to the microbiome–gut–brain axis remains to be established in autism.

Recent research has indicated that the effect of the gut microbiota extends much beyond the modulation of the gut itself. Metabolites derived from the microbiota can be absorbed and transported by the blood before crossing the BBB to modulate cerebral function. For example, strains of *Lactobacillus rhamnosus* YS9 are able to produce gamma-aminobutyric acid (GABA), an important inhibitory neurotransmitter in the brain [26].

A large percentage of autism patients have a history of extensive antibiotic use. Oral antibiotics (i.e. β-lactams) disrupt the protective microbiota and cause the proliferation of anaerobic bacteria in the gut. For example, *Clostridia*, *Bacteroidetes* and *Desulfovibrio* are common bacteria that may promote GI symptoms and autistic behaviours in autism [43].

3 Probiotics in Autism

The internationally accepted definition of probiotics is "live microorganisms which when administered in adequate amounts confer a health benefit on the host". Probiotics are typically administered as a food supplement promoting various health benefits to the host by maintaining the stability and composition of the intestinal and gut microbiota and increasing resistance against various pathological infections [44].

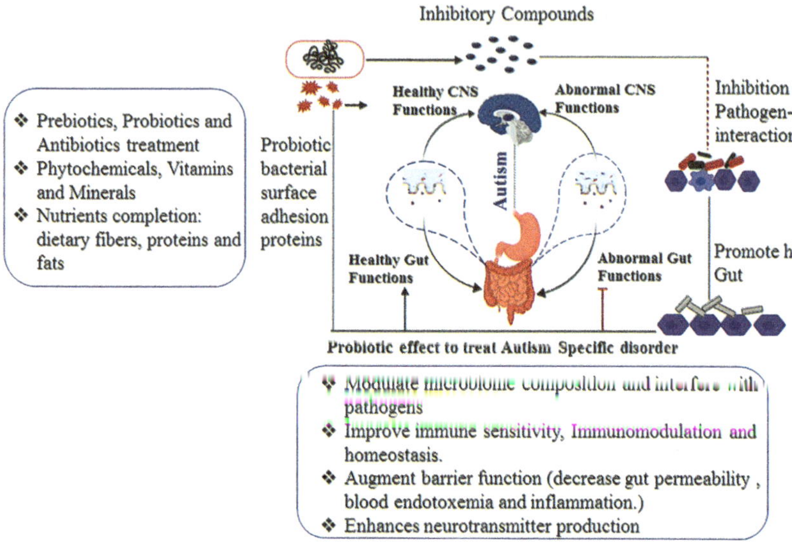

Fig. 2 Probiotic influence of gut microflora and ASD behavioural abnormalities. Adapted from Poornachandra Rao and Sreenivasa [47] and Sánchez et al. [48]

The potentially synergistic combinations of pro- and prebiotics are called symbiotics [45]. Various preclinical and clinical findings have suggested that treatment with probiotics can help improve gastrointestinal health, thereby stabilizing behavioural abnormalities in adults and children with autism [46]. The probiotic influence of gut microbiome and ASD behavioural abnormalities is depicted in Fig. 2.

Currently, the consumption of probiotic cells via food products has been categorized as functional foods, the worldwide market for which had been predicted to increase from 33 billion in 2000 to 176.7 billion in 2013. About 60–70% of the total food market comprises of probiotic foods [49]. There has been remarkable success in the past few decades in producing dairy products like ice cream, flavoured liquid milk, fermented milk, milk powder, baby food, frozen dairy products, buttermilk, cheese and many others which contain probiotics and can be safely administered in all these forms. One of the key aspects for probiotics to gain such rapid momentum is that they are safe, comparatively cheap and an accessible target for microbial infections. The World Health Organization (WHO) in 1994 considered probiotics to be used as an effective immune defence system in the cases of antibiotic resistance. This treatment was termed as microbial interference therapy [50].

The probiotic microbes are artificially introduced into the food at the time of its production. Most of the cultures are commercially available in extremely concentrated form as either freeze-dried powders or highly concentrated frozen cultures. Some of the popularly used probiotic microorganisms are *Lactobacillus rhamnosus*, *Lactobacillus reuteri*, *Bifidobacteria* and certain strains of *Lactobacillus casei* and *Lactobacillus acidophilus* group. Other microbes include *Bacillus coagulans*;

Escherichia coli strains; certain enterococci, especially *Enterococcus faecium* SF68; and the yeast *Saccharomyces boulardii*. Probiotic products may contain either a single strain or a mixture of two or more strains. The effect of probiotics is very strain specific and cannot be generalized. A single strain may exhibit different benefits when used individually and in combination. The benefits of a probiotic formulation also differ by the patient group [45]. Once probiotics are incorporated in the food, its effectiveness depends upon the total number of viable cells per ml and also the number of active cells present on being consumed by the individual [51].

It is also noted that on addition of the probiotic microbes, the aroma and the flavours in the food are modified due to the formation of metabolic components during fermentation such as the synthesis of acetic acid by *Bifidobacterium spp.* Therefore, in order to attain product quality and patient competence for its administration, necessary steps are taken in order to eradicate the smell and aftertaste [52].

The known mechanisms by which probiotic bacteria have an impact on the gut microbiota may be as follows:

1. Competition for dietary ingredients as growth substrates
2. Bioconversion of, for example, sugars into fermentation products with inhibitory properties
3. Production of growth substrates
4. Direct effect on pathogens
5. Competitive exclusion for binding sites
6. Improved barrier function
7. Reduction of inflammation, thus altering intestinal properties for colonization and persistence within and
8. Stimulation of innate immune response

In a recent study using a rodent model of autism, the alteration in the gut microbiota and the related alteration in serum metabolites were considered to play an important role in the behavioural manifestation of autism-like behaviour and subsequent GI function alteration. However, these changes were seen to be rapidly reversible by ingestion of probiotics [53].

There have been interesting findings made in human autism research where the main microbiota intervention in the clinical study is the probiotic administration. One such study used faecal transplantation. In this study, it was observed that on being treated with probiotics for over 6 months ($n = 6$), children diagnosed with autism showed a decrease in the severity of diarrhoea and constipation. Each participant received a 6-month supply of DelPro® containing 10 billion CFUs of different probiotic strains including *L. acidophilus, L. casei, Lactobacillus delbrueckii, Bifidobacterium longum* and *B. bifidum* and 8 mg of Del-Immune V® powder, containing peptidoglycan, muramyl peptides and nucleotide-containing components or DNA motifs that is derived from the *L. rhamnosus V* strain. Any other probiotics were advised to be discontinued. An Autism Treatment Evaluation Checklist (ATEC) score, which reflects the changes occurring in autism patients and accesses various domains like speech/language/communication, sociability, sensory/cognitive awareness and health/physical/behaviour, was studied. Thus, in this study following probiotic treatment, an 88% improvement in the above-mentioned domains of autism was observed overall. The mean ATEC value decreased from 72.8 to 58.3 [54].

Parracho et al. carried out a randomized, double blind, controlled study with children from age 3 to 16 suffering from autism and divided them into a placebo and probiotic group. The probiotic group was given 4.5 × 1010 CFU *Lactobacillus plantarum* WCFS1 daily over a period of 6 weeks. Group I received placebo during the first feeding period (3 weeks) and probiotic during the second feeding period (3 weeks), and vice versa for group II (i.e. probiotic first). Improvements in destructive and antisocial behaviour, anxiety and communication problems were observed in the children with autism who were treated with probiotics [55].

A previous study showed that in the case of the offspring of an immune activated mother, gut permeability was affected, an abnormal increase in the level of cytokines and gut dysbiosis was observed and hence changes in neuropathological and behavioural autism features due to changes caused by *Clostridia* and *Bacteroidia* in the gut environment were seen. On the treatment with probiotic *Bacteroides fragilis,* intestinal permeability was improved in MIA, and this specifically increased pro-inflammatory cytokine IL-6 in the colon. It was also seen that this treatment could restore 6 out of 67 bacterial species units which are compromised in the case of autism patients. An improvement in communication, repetitive sensorimotor and anxiety of the MIA offspring was achieved. However, they also found that the effect of *B. fragilis* on autism behaviour was seen when treated with *Bacteroides thetaiotaomicron* and not on treating with *Enterococcus faecalis.* Thus, from this study it was inferred that treatment with probiotics in autism relieved certain symptoms by reducing inflammation, improving the gut permeability, restoring microbial imbalances and ameliorating nonsocial autism symptoms [56].

In another clinical study, all the autism patients suffering from severe GI problems were grouped. The participants received probiotic capsules of *L. acidophilus* (strain Rosell-11, containing 5 × 109 CFU/g) orally, twice daily for 2 months. The use of antibiotics during therapy was restrained, and urine samples were collected for each of the participants. Prior to treatment, the level of D-arabinitol was significantly higher in the urine of children with autism and was seen to decrease thereafter. The autistic symptoms such as concentration and following out orders also improved after the probiotic therapy [57].

In a trial with autism patients ($n = 11$), an oral liquid dose of vancomycin 500 mg/day was given, thrice a day for 8 weeks. This was followed by probiotic therapy given orally for 4 weeks, comprising of a mixture of *L. acidophilus*, *L. bulgaricus* and *B. bifidum* (40 × 109 CFUs/ml). This suggested that multiple probiotic therapy led to short-term pre- and post-therapy improvement in communication as well as pattern behaviours [58].

From the results obtained by researchers on examining a group of children with autism ($n = 22$) of ages four to ten where the patients were administered a sugar-free diet and probiotic capsules of *L. acidophilus* (5 × 109 CFU/g) for a period of 2 months, twice daily, major changes were observed in the behavioural domains with significant improvement in concentration and the ability to follow instructions. However, there was no improvement in other distinct behaviours and the ability to make eye contact [59].

4 Conclusion

Perturbation of GI tract bacterial microflora may play an important role in the pathophysiology of some digestive tract disorders. Probiotics have been used as a treatment modality for over a century. Microbial modification with the use of antibiotics, probiotics and faecal transplantation has been effective in the treatment of GI conditions. They may restore normal bacterial microflora and effect the functioning of the GI tract by a variety of mechanisms. Gut microbiome-related changes are seen in children with autism compared to normally developed children. Virtually all of the GI functions postulated to be impaired in ASD have been shown to be improved by probiotics in animal studies. Evidence suggests that probiotics can have beneficial effects for people with autism as well. However, many questions regarding the use of probiotics in GI disorders remain to be answered in future studies, such as most optimal doses, duration of treatment, physiological and immunological effects, efficacy of specific probiotics in specific disease states and safety in debilitated patients, since there is a complex interplay in these conditions between GI function (motility, secretion, permeability), the immune system and the microbiota.

References

1. Sirisinha, S. (2016). The potential impact of gut microbiota on your health: Current status and future challenges. *Asian Pacific Journal of Allergy and Immunology, 34*(4), 249–264.
2. Thakur, A., Shakya, A., Husain, G., Emerald, M., & Kumar, V. (2014). Gut-microbiota and mental health: Current and future perspectives. *Journal of Pharmacology and Clinical Toxicology, 2*, 1–15.
3. Rhee, S. H., Pothoulakis, C., & Mayer, E. A. (2009, May). Principles and clinical implications of the brain-gut-enteric microbiota axis. *Nature Reviews. Gastroenterology & Hepatology, 6*(5), 306–314.
4. Rizk, M. Z. (2019, May 18). Role of gut-brain axis in the aetiology of neurodevelopmental disorders with reference to autism. *Journal Clinical Toxicology, S6*, 005.
5. Li, Q., & Zhou, J.-M. (2016, June). The microbiota–gut–brain axis and its potential therapeutic role in autism spectrum disorder. *Neuroscience, 324*, 131–139.
6. Snoek, S. A., Verstege, M. I., Boeckxstaens, G. E., van den Wijngaard, R. M., & de Jonge, W. J. (2010, October). The enteric nervous system as a regulator of intestinal epithelial barrier function in health and disease. *Expert Review of Gastroenterology & Hepatology, 4*(5), 637–651.
7. Collins, S. M., Surette, M., & Bercik, P. (2012, November). The interplay between the intestinal microbiota and the brain. *Nature Reviews. Microbiology, 10*(11), 735–742.

8. Dinan, T. G., Quigley, E. M. M., Ahmed, S. M. M., Scully, P., O'Brien, S., O'Mahony, L., et al. (2006, February). Hypothalamic-pituitary-gut axis dysregulation in irritable bowel syndrome: Plasma cytokines as a potential biomarker? *Gastroenterology, 130*(2), 304–311.
9. Donaldson, G. P., Lee, S. M., & Mazmanian, S. K. (2016, January). Gut biogeography of the bacterial microbiota. *Nature Reviews. Microbiology, 14*(1), 20–32.
10. Falk, P. G., Hooper, L. V., Midtvedt, T., & Gordon, J. I. (1998, December). Creating and maintaining the gastrointestinal ecosystem: What we know and need to know from gnotobiology. *Microbiology and Molecular Biology Reviews, 62*(4), 1157–1170.
11. Round, J. L., & Mazmanian, S. K. (2009, May). The gut microbiota shapes intestinal immune responses during health and disease. *Nature Reviews Immunology, 9*(5), 313–323.
12. Yang, Y., Tian, J., & Yang, B. (2018, February). Targeting gut microbiome: A novel and potential therapy for autism. *Life Sciences, 194*, 111–119.
13. Wang, L., Christophersen, C. T., Sorich, M. J., Gerber, J. P., Angley, M. T., & Conlon, M. A. (2011, September). Low relative abundances of the mucolytic bacterium Akkermansia muciniphila and Bifidobacterium spp. in feces of children with autism. *Applied and Environmental Microbiology, 77*(18), 6718–6721.
14. Adams, J. B., Johansen, L. J., Powell, L. D., Quig, D., & Rubin, R. A. (2011, March 16). Gastrointestinal flora and gastrointestinal status in children with autism—Comparisons to typical children and correlation with autism severity. *BMC Gastroenterology, 11*, 22.
15. Liu, F., Li, J., Wu, F., Zheng, H., Peng, Q., & Zhou, H. (2019, January 29). Altered composition and function of intestinal microbiota in autism spectrum disorders: A systematic review. *Translational Psychiatry, 9*(1), 43.
16. Song, Y., Liu, C., & Finegold, S. M. (2004, November). Real-time PCR quantitation of Clostridia in feces of autistic children. *Applied and Environmental Microbiology, 70*(11), 6459–6465.
17. Sekirov, I., Russell, S. L., Antunes, L. C. M., & Finlay, B. B. (2010, July). Gut microbiota in health and disease. *Physiological Reviews, 90*(3), 859–904.
18. Angelis, M. D., Francavilla, R., Piccolo, M., Giacomo, A. D., & Gobbetti, M. (2015, May 4). Autism and intestinal microbiota. *Gut Microbes, 6*(3), 207–213.
19. Kang, D.-W., Ilhan, Z. E., Isern, N. G., Hoyt, D. W., Howsmon, D. P., Shaffer, M., et al. (2018, February 1). Differences in fecal microbial metabolites and microbiota of children with autism spectrum disorders. *Anaerobe, 49*, 121–131.
20. Wang, L., Christophersen, C. T., Sorich, M. J., Gerber, J. P., Angley, M. T., & Conlon, M. A. (2013, November 4). Increased abundance of Sutterella spp. and Ruminococcus torques in feces of children with autism spectrum disorder. *Molecular Autism, 4*(1), 42.
21. Navarro, F., Liu, Y., & Rhoads, J. M. (2016, December 14). Can probiotics benefit children with autism spectrum disorders? *World Journal of Gastroenterology, 22*(46), 10093–10102.
22. Eren, A. M., Sogin, M. L., Morrison, H. G., Vineis, J. H., Fisher, J. C., Newton, R. J., et al. (2015). A single genus in the gut microbiome reflects host preference and specificity. *The ISME Journal, 9*, 90–100.
23. de Magistris, L., Familiari, V., Pascotto, A., Sapone, A., Frolli, A., Iardino, P., et al. (2010, October). Alterations of the intestinal barrier in patients with autism and in their first-degree relatives. *Journal of Pediatric Gastroenterology and Nutrition, 51*(4), 418–424.
24. Fiorentino, M., Sapone, A., Senger, S., Camhi, S. S., Kadzielski, S. M., Buie, T. M., et al. (2016). Blood-brain barrier and intestinal epithelial barrier alterations in autism spectrum disorders. *Molecular Autism, 7*, 49.
25. Ming, X., Stein, T. P., Barnes, V., Rhodes, N., & Guo, L. (2012, December 7). Metabolic perturbance in autism spectrum disorders: A metabolomics study. *Journal of Proteome Research, 11*(12), 5856–5862.
26. Esnafoglu, E., Cırrık, S., Ayyıldız, S. N., Erdil, A., Ertürk, E. Y., Daglı, A., et al. (2017). Increased serum zonulin levels as an intestinal permeability marker in autistic subjects. *The Journal of Pediatrics, 188*, 240–244.
27. Haba, R., Shintani, N., Onaka, Y., Wang, H., Takenaga, R., Hayata, A., et al. (2012, March 17). Lipopolysaccharide affects exploratory behaviors toward novel objects by impairing cognition

and/or motivation in mice: Possible role of activation of the central amygdala. *Behavioural Brain Research, 228*(2), 423–431.

28. Emanuele, E., Orsi, P., Boso, M., Broglia, D., Brondino, N., Barale, F., et al. (2010, March). Low-grade endotoxemia in patients with severe autism. *Neuroscience Letters, 471*(3), 162–165.
29. Kraneveld, A. D., Szklany, K., de Theije, C. G. M., & Garssen, J. (2016). Gut-to-brain axis in autism spectrum disorders. *International Review of Neurobiology, 131*, 263–287.
30. Costedio, M. M., Hyman, N., & Mawe, G. M. (2007, March). Serotonin and its role in colonic function and in gastrointestinal disorders. *Diseases of the Colon and Rectum, 50*(3), 376–388.
31. McDougle, C. J., Naylor, S. T., Cohen, D. J., Aghajanian, G. K., Heninger, G. R., & Price, L. H. (1996, November). Effects of tryptophan depletion in drug-free adults with autistic disorder. *Archives of General Psychiatry, 53*(11), 993–1000.
32. de Theije, C. G. M., Wu, J., da Silva, S. L., Kamphuis, P. J., Garssen, J., Korte, S. M., et al. (2011, September). Pathways underlying the gut-to-brain connection in autism as future targets for disease management. *European Journal of Pharmacology, 668*(Suppl 1), S70–S80.
33. de Theije, C. G. M., Wopereis, H., Ramadan, M., van Eijndthoven, T., Lambert, J., Knol, J., et al. (2014, March). Altered gut microbiota and activity in a murine model of autism spectrum disorders. *Brain, Behavior, and Immunity, 37*, 197–206.
34. Maynard, C. L., Elson, C. O., Hatton, R. D., & Weaver, C. T. (2012, September 13). Reciprocal interactions of the intestinal microbiota and immune system. *Nature, 489*(7415), 231–241.
35. Petra, A. I., Panagiotidou, S., Hatziagelaki, E., Stewart, J. M., Conti, P., & Theoharides, T. C. (2015, May 1). Gut-microbiota-brain axis and its effect on neuropsychiatric disorders with suspected immune dysregulation. *Clinical Therapeutics, 37*(5), 984–995.
36. Chandler, S., Carcani-Rathwell, I., Charman, T., Pickles, A., Loucas, T., Meldrum, D., et al. (2013, December). Parent-reported gastro-intestinal symptoms in children with autism spectrum disorders. *Journal of Autism and Developmental Disorders, 43*(12), 2737–2747.
37. Molloy, C. A., Morrow, A. L., Meinzen-Derr, J., Schleifer, K., Dienger, K., Manning-Courtney, P., et al. (2006, March). Elevated cytokine levels in children with autism spectrum disorder. *Journal of Neuroimmunology, 172*(1–2), 198–205.
38. Ashwood, P., & Wakefield, A. J. (2006, April). Immune activation of peripheral blood and mucosal CD3+ lymphocyte cytokine profiles in children with autism and gastrointestinal symptoms. *Journal of Neuroimmunology, 173*(1–2), 126–134.
39. Park, Y. D. (2003, June 1). The effects of vagus nerve stimulation therapy on patients with intractable seizures and either Landau–Kleffner syndrome or autism. *Epilepsy & Behavior, 4*(3), 286–290.
40. Berney, T. P. (2000, January). Autism – An evolving concept. *The British Journal of Psychiatry, 176*(1), 20–25.
41. Lyte, M. (2014, December). The effect of stress on microbial growth. *Animal Health Research Reviews, 15*(2), 172–174.
42. Carabotti, M., Scirocco, A., Maselli, M. A., & Severi, C. (2015). The gut-brain axis: Interactions between enteric microbiota, central and enteric nervous systems. *Annals of Gastroenterology, 28*, 203–209.
43. MacFabe, D. F. (2015, May 29). Enteric short-chain fatty acids: Microbial messengers of metabolism, mitochondria, and mind: Implications in autism spectrum disorders. *Microbial Ecology in Health & Disease, 26*, 28177.
44. Begum, P., Madhavi, G., Rajagopal, S., Viswanath, B., Razak, M., & Venkataratnamma, V. (2017, January 1). Probiotics as functional foods: Potential effects on human health and its impact on neurological diseases. *International Journal of Nutrition, Pharmacology, Neurological Diseases, 7*, 23.
45. Pandey Kavita, R., Naik Suresh, R., & Vakil Babu, V. (2015, December). Probiotics, prebiotics and synbiotics – A review. *Journal of Food Science and Technology, 52*(12), 7577–7587.
46. Slattery, J., MacFabe, D. F., & Frye, R. E. (2016). The significance of the enteric microbiome on the development of childhood disease: A review of prebiotic and probiotic therapies in disorders of childhood. *Clinical Medicine Insights. Pediatrics, 10*, 91–107.

47. Poornachandra Rao, K., & Sreenivasa, M. Y. (2017). Probiotic Lactobacillus strains. The future biological missiles to treat autism spectrum disorder: A short communication. *Current Nutrition & Food Science, 13*(1), 3–5. https://doi.org/10.2174/1573401313666161118162040
48. Sánchez, B. R., Delgado, S. A., Blanco-Míguez, A., Lourenço, A., Gueimonde, M., & Margolles, A. (2017). Probiotics, gut microbiota, and their influence on host health and disease. *Molecular Nutrition & Food Research, 61*(1). https://doi.org/10.1002/mnfr.201600240
49. Kołożyn-Krajewska, D., & Dolatowski, Z. J. (2012, December 1). Probiotic meat products and human nutrition. *Process Biochemistry, 47*(12), 1761–1772.
50. Zhou, J. S., Pillidge, C. J., Gopal, P. K., & Gill, H. S. (2005, February 1). Antibiotic susceptibility profiles of new probiotic Lactobacillus and Bifidobacterium strains. *International Journal of Food Microbiology, 98*(2), 211–217.
51. Korbekandi, H., Mortazavian, A., & Iravani, S. (2011). Technology and stability of probiotic in fermented milks containing probiotics and prebiotics. Probiotic and Prebiotic Foods: Technology, Stability and Benefits to Human Health. Nova Science Publishers, Inc. USA
52. Mohammadi, R., Mortazavian, A. M., Khosrokhavar, R., & da Cruz, A. G. (2011, September 1). Probiotic ice cream: Viability of probiotic bacteria and sensory properties. *Annales de Microbiologie, 61*(3), 411–424.
53. Hsiao, E. Y., McBride, S. W., Hsien, S., Sharon, G., Hyde, E. R., McCue, T., et al. (2013, December 19). Microbiota modulate behavioral and physiological abnormalities associated with neurodevelopmental disorders. *Cell, 155*(7), 1451–1463.
54. *Improvements in Gastrointestinal Symptoms among Children with Autism Spectrum Disorder Receiving the Delpro® Probiotic and Immunomodulator Formulation* [Internet]. (2019, June 12). Retrieved from https://www.omicsonline.org/improvements-in-gastrointestinal-symptoms-among-children-with-autism-spectrum-disorder-receiving-the-delpro-probiotic-and-immunomodulator-formulation-2329-8901.1000102.php?aid=13384
55. Parracho, H. M. R. T., Gibson, G. R., Knott, F., Bosscher, D., Kleerebezem, M., & McCartney, A. L. (2010, May). A double-blind, placebo-controlled, crossover-designed probiotic feeding study in children diagnosed with autistic spectrum disorders. *International Journal of Probiotics and Prebiotics, 5*, 69–74.
56. Doenyas, C. (2018). Gut microbiota, inflammation, and probiotics on neural development in autism spectrum disorder. *Neuroscience, 15*(374), 271–286.
57. Kałużna-Czaplińska, J., & Błaszczyk, S. (2012, February). The level of arabinitol in autistic children after probiotic therapy. *Nutrition, 28*(2), 124–126.
58. Sandler, R. H., Finegold, S. M., Bolte, E. R., Buchanan, C. P., Maxwell, A. P., Väisänen, M. L., et al. (2000, July). Short-term benefit from oral vancomycin treatment of regressive-onset autism. *Journal of Child Neurology, 15*(7), 429–435.
59. Romeo, M. G., Romeo, D. M., Trovato, L., Oliveri, S., Palermo, F., Cota, F., et al. (2011, January). Role of probiotics in the prevention of the enteric colonization by Candida in preterm newborns: Incidence of late-onset sepsis and neurological outcome. *Journal of Perinatology, 31*(1), 63–69.

Natural Products and Their Therapeutic Effect on Autism Spectrum Disorder

Satarupa Deb, Banashree Chetia Phukan, Ankumoni Dutta, Rajib Paul, Pallab Bhattacharya, Thamilarasan Manivasagam, Arokiasamy Justin Thenmozhi, Chidambaram Saravana Babu, Musthafa Mohamed Essa, and Anupom Borah

Abstract Autism is a complex neurodevelopmental disorder that is evident in early childhood and can persist throughout the entire life. The disease is basically characterized by hurdles in social interaction where the individuals demonstrate repetitive and stereotyped interests or patterns of behavior. A wide number of neuroanatomical studies with autistic patients revealed alterations in brain development which lead to diverse cellular and anatomical processes including atypical neurogenesis, neuronal migration, maturation, differentiation, and degeneration. Special education programs, speech and language therapy, have been employed for the amelioration

Satarupa Deb, Banashree Chetia Phukan, Ankumoni Dutta, Rajib Paul, Pallab Bhattacharya and Anupom Borah contributed equally with all other contributors.

S. Deb · B. C. Phukan · A. Dutta · A. Borah (✉)
Cellular and Molecular Neurobiology Laboratory, Department of Life Science and Bioinformatics, Assam University, Silchar, Assam, India

R. Paul
Department of Zoology, Pandit Deendayal Upadhyaya Adarsha Mahavidyalaya (PDUAM), Karimganj, Assam, India

P. Bhattacharya
Department of Pharmacology and Toxicology, National Institute of Pharmaceutical Education and Research (NIPER)-Ahmedabad, Gandhinagar, Gujarat, India

T. Manivasagam · A. Justin Thenmozhi
Department of Biochemistry and Biotechnology, Faculty of Science, Annamalai University, Chidambaram, Tamil Nadu, India

C. S. Babu
Department of Pharmacology, JSS College of Pharmacy, JSS Academy of Higher Education and Research (JSSAHER), Mysuru, Karnataka, India

M. M. Essa
Department of Food Science and Nutrition, CAMS, Sultan Qaboos University, Muscat, Oman

Ageing and Dementia Research Group, Sultan Qaboos University, Muscat, Oman

Food and Brain Research Foundation, Chennai, Tamil Nadu, India

M. M. Essa, M. W. Qoronfleh (eds.), *Personalized Food Intervention and Therapy for Autism Spectrum Disorder Management*, Advances in Neurobiology 24,
https://doi.org/10.1007/978-3-030-30402-7_22

of behavioral deficits in autism. Although commonly prescribed antidepressants, antipsychotics, anticonvulsants, and stimulants have revealed satisfactory responses in autistic individuals, adverse side effects and increased risk of several other complications including obesity, dyslipidemia, diabetes mellitus, thyroid disorders, etc. have compelled the researchers to turn their attention toward herbal remedies. Alternative approaches with natural compounds are on continuous clinical trial to confirm their efficacy and to understand their potential in autism treatment. This chapter aims to cover the major plant-based natural products which hold promising outcomes in the field of reliable therapeutic interventions for autism.

Keywords ASD, autism · Neurodevelopmental disorder · Herbal remedies · Neurotherapeutics, natural products, nutrition therapy, neuronal migration · Antidepressants · Antipsychotics

Abbreviations

CTIP2	Chicken ovalbumin upstream promoter transcription factor-interacting protein 2
GABA	Gamma-aminobutyric acid
IL-6	Interleukin-6
LTP	Long-term potentiation
NMDA	*N*-methyl-D-aspartic acid
STAT3	Signal transducer and activator of transcription 3
TBR1	T-box brain 1
TNF-α	tumor necrosis factor-α

1 Introduction

Autism spectrum disorder (ASD) is a cumulative neurological disorder distinguished by difficulties in social interactions and communication, language development, and restricted or stereotyped interest and preferences [1, 2]. Autism has multiple subtypes, e.g., Rett syndrome, Asperger's syndrome, and childhood disintegration, and each individual with this rare disorder faces unique challenges [3]. Highly variable symptoms usually show onset by the age of 6 months, become evident by 2 or 3 years, and have a tendency to continue through adulthood. The median wide-scale pervasiveness of autism is around 0.62–0.70%, but estimates of 1–2% are made in the latest surveys [4, 5].

The precise cause behind the prevalence of autism is not yet comprehensible. Both genetics and environmental exposures have their individual roles to play [6]. Some of the suspected risks include gene-level mutations, being born to older parents, having a closely related family member with autism, severe viral infections, metabolic imbalances, and vulnerability to harmful chemicals, toxins, and heavy metals [6–8]. In autistic children, processing information in the brain is affected by alterations in the association between neurons and their synapses. Case studies and animal model experiments have also suggested that altered neuroimmune responses and autoimmunity lead to phenotypic defects apparent in autism [9]. Increased levels of pro-inflammatory cytokines were detected in brain specimen samples, cerebrospinal fluid, and peripheral blood isolated from autistic subjects.

For the benefit of autistic children, many treatment approaches have been made through the development of different education techniques, rehabilitation training, sensory integration, and distinctive dietary approaches [10, 11]. Nutritional enhancement with modified diet and supplements of vitamin and minerals, along with the practice of Epsom salt bath, cutting down on sugar, gluten, and casein, intake of probiotics, exposure to greenery, and increase in reading habits may have several beneficial impacts in long duration. Herbal medicines and acupuncture displayed promising results in behavioral and developmental improvements in affected children [12]. In this chapter, we highlighted the natural compounds and plant-based drugs which have experimentally proved to be worthwhile with regard to autism therapeutics.

2 Characteristics of Autism

Autism cannot be distinguished only by a single symptom. There is a large range of abnormal manifestations associated with it. Variations in age, features, and cognitive ability are correlated with the development of altered behavioral patterns in different individuals [13].

Abnormal communication skills and social attachment become apparent in autistic infants who develop relatively less attention to social stimuli, have problems maintaining eye contact and turn-taking, and fail to employ facial and bodily gestures to express themselves to others [14]. In some cases, autistic children with intellectual disability exhibit destructive nature and excessive aggression. Repetitive patterns of behavior are observed in autism, such as placing objects according to a specific order, being intolerant to changes, hand rolling, and body rocking. About 60–80% of affected individuals display poor muscle tone and motor planning as well as an unvarying pattern of everyday routine [15].

The early signs of autism in preschool children include delayed speech development, speaking in a voice that sounds very monotonous or flat, frequent repetitions of a particular set of words or phrases, preferring to communicate using limited words, and ignoring their names being called despite possessing normal hearing ability [16]. Individuals with autism are intolerant of people entering their own personal space and are generally unable to distinctly express their thoughts and desires to others [17].

3 Causes of Autism

3.1 Genetics

The complexity of autism arises due to interactions among multiple genes, the environment as well as the epigenetic factors [18]. Although the genetic basis of autism is complex and elusive, escalating evidences from genome sequencing have revealed the link between genetic alterations and development of the disorder [19]. Siblings of an autistic individual are 25% more likely to develop the condition. Twin studies have also reported increased heritability of autism. The exact cause behind genetic alterations could not be revealed yet; however, the disease cannot be linked with single-gene mutation or any single chromosomal abnormality [20].

3.2 Environmental Causes

Fetal exposure to air pollutants containing heavy metals, toxins (such as thalidomide, valproic acid, retinoic acid), and suspended particulates elevates the risk of autism [21]. Prenatal stress, unhealthy lifestyle and diet as well as a familial history of infectious diseases are some of the factors leading to behavioral anomalies visible in autistic infants [22, 23]. Perinatal factors of autism include low birth weight, preterm delivery, and asphyxia during birth [24]. Genetic heterogeneity may also rise due to environmental influences, which in turn can be associated with enzymatic deficits in autism. Gene-environment interactions are complex and its understanding is still in the root level.

3.3 Neuropathogical Complications

Electrophysiological detections and neuroimaging of test subjects suggest that autism is associated with atypical neural connectivity, leading to altered information processing [25]. Neural networks in autism involve decreased fronto-posterior and enhanced parietal-occipital connectivity and reduced long-range and increased short-range connectivity to temporal binding deficits. Increase in total brain volume is reported as one of the neuroanatomical features in autism [26]. An enlarged amygdala and a significantly minimized volume of corpus callosum have consistently been revealed in affected individuals [27]. Moreover, an alteration in GABA and serotonin neurotransmitter levels has also been associated with autism [28]. Several hypotheses for the neurological basis of this disorder have been put forward which include impaired neuronal migration during early gestation, abnormal spacing between the neurons and disordered synapses, over-connectivity due to excessive neuronal outgrowth, and disturbed excitatory as well as inhibitory networks [29].

4 Recent Advances in Treatment Therapy and Management of Autism

Although a few psychosocial interventions have proved to be beneficial, no single treatment strategy for autism can be distinguished as the best one. A wide range of symptoms are exhibited by the affected individuals that differ to a huge degree. Treatment methods are being tailored to the needs of each patient. Special training, educational programs, and behavioral therapies may assist in improving self-care, maturity, and job skills, while medications can ameliorate anxiety and irritability [10]. Among the useful interventions, applied behavior analysis (ABA) depends on unique one-on-one teaching tasks using the behaviorist principles of stimulus, response, and reward [30]. Discrete trial training (DTT) uses a slightly different technique to teach fundamental skills, such as attention, compliance, and imitation. Pivotal response training (PRT) improves self-management and social attachment in autistic individuals. Diagnosed children are generally prescribed with antidepressants, anticonvulsants, stimulants, and antipsychotics, such as risperidone or aripiprazole [31]. However, side effects of long-term intake need to be widely investigated as each individual responds uniquely to such drugs. Modulation of gene functioning also proved to be an effective approach in the management of autism, but the process requires precision and expertise [32]. Thus, an uncompromising search for alternative strategies is evident through the recent reports accumulating on autism.

5 Natural Products in the Treatment of Autism

The need for safe and reliable medications for the successful management of autism has led to the exploration of various plant-based natural products which bear therapeutic potential. Effective herbal medicines taken along with conventional rehabilitation and training programs may improve the core symptoms with fewer side effects [33]. Here, we have compiled the most important natural compounds that have proven to be very beneficial in ameliorating the pathophysiological conditions involved in autism.

5.1 Luteolin

Microglia are one of the primary macrophages of the central nervous system (CNS) whose main function is detailed scanning and activation during insults, such as damage, disease, or infection [34]. Their activation also implicates inflammatory responses of the CNS [35]. Maternal immune activation and resulting microglial dysfunction in the developing brain is associated with the occurrence of autism [36]. Several etiological theories with different degrees of evidence that have been proposed to target microglial activation in order to regulate these inflammatory cascades can have positive effects in autism treatment [37–39].

Luteolin is a naturally occurring flavonoid found in edible plants. In a human cell-based model of maternal immune activation, luteolin treatment significantly counteracted IL-6 induced increment of glial fibrillary acidic protein (GFAP) in astrocytes [40]. GFAP are usually surplus in proliferative glial scars [41]. In addition, a marked depletion in the levels of phosphorylated transcription factor STAT3 was noted [40]. Excessive phosphorylation of STAT3 is an indication of heightened activity of the cytokine and growth factors that often lead to inflammation [42]. Regulated levels of TBR1-positive and CTIP2-positive cells are also evident on luteolin administration [40]. TBR1 and CTIP2 expression is important for normal cortical development in initial stages [43, 44]. In a study reported by Bertolino et al. [45], the flavonoid luteolin along with the fatty acid palmitoylethanolamide together proved to be neuroprotective and anti inflammatory. Co-ultramicronized luteolin and palmitoylethanolamide showed beneficial outcomes in autistic murine models. Subsequent examinations on its effects in a 10-year-old male child improved the clinical scenario to a remarkable extent. Another luteolin formulation, commercially known as NeuroProtek®, was shown to be equally beneficial to a large cohort of children with ASD [46].Thus, luteolin might be regarded as a safe and effective medication for the management of autistic behavior [47].

Since autism is represented as a condition associated with neuroinflammation, high levels of interleukin-6 (IL-6) and tumor necrosis factor (TNF) are also visualized in the serum of affected individuals [48]. But autistic children who routinely took a dietary formulation with luteolin showed improved social attachment and behavior. The serum levels of IL-6, TNF, and other cytokines also favorably diminished on luteolin intake [49]. Luteolin also inhibits mast cell-dependent stimulation of activated T cells and minimizes histamine, leukotrienes as well as other inflammatory molecules [50].

5.2 *Green Tea Extract (*Camellia sinensis*)*

There is a positive correlation between increased oxidative stress and the development of autism [51]. Increased levels of lipid peroxidation and major antioxidant serum proteins, altered status of glutathione, and major antioxidant enzymes such as superoxide dismutase, glutathione peroxidase, and catalase are observed in children with autism. This demonstrates the linkage between the mechanism of elevated oxidative stress and development of autism [52].

Camellia sinensis is an important dietary source of polyphenols, principally flavonoids, whose antioxidant potentials are well recognized. Green tea has been experimentally proved to exert a wide range of favorable health effects [53]. Flavonoids can cross blood-brain barrier and display a multitude of neuroprotective properties [54]. Daily intake of green tea extract (75 and 300 mg/kg) after postnatal day 14 for 40 consecutive days has shown a substantial improvement in motor coordination, nociceptive response, locomotion, anxiety, exploratory activity, and cognition in valproate-induced autistic mice. Formation of oxidative stress markers was also downregulated on the administration of the extract in both the doses. Histological findings at 300 mg/kg of the extract confirmed its neuroprotective potential [55]. Bioactive components in green tea can directly regulate the level of neurotransmitters in brain, particularly dopamine and serotonin in specific brain regions.

L-theanine, an amino acid of tea, displays anti-stress effect and is capable of NMDA-independent increase in long-term potentiation (LTP), thereby improving memory [56]. Consumption of epigallocatechin-3-gallate, the major type of catechin in green tea, can reverse the cardinal behavioral alterations in sodium-valproate-treated autistic rat model [57]. One of the key pathological findings in autism involves the loss of Purkinje cell integrity in the cerebellum region [58]. Histological findings in subjects that consumed 300 mg/kg of green tea extract regularly showed a gradual regeneration of distinct Purkinje layer and cells, which suggests its neuroprotective potential in the treatment of autism [59].

5.3 Piperine

Piperine, chemically an N-acylpiperidine, is the major alkaloid isolated from black pepper (*Piper nigrum*) and long pepper (*Piper longum*). The compound is capable of activating heat and acidity-sensing ion channels on the pain-sensing nerve cells, particularly known as nociceptors [60]. Its action on the nervous system is critically acclaimed since it is traditionally used to treat seizure disorders and displays considerable anti-oxidative effects along with notable enhancement of memory and cognition [61].

Pretreatment with piperine in cultured hippocampal neurons showed protection against loss of cell viability due to glutamatergic upsurge. The mechanism of its action has been hypothesized to be associated with the regulation of Ca^{2+} ion entry into the neurons and pre-synaptic release of glutamine [62]. Sodium valproate-induced autistic Balb/C mice have been experimentally treated with 20 mg/kg of piperine, following which they were subjected to behavioral evaluation, histopathological observation, and biochemical assessment after postnatal day 14. The results demonstrated that piperine is capable of inducing favorable neurorescue effects, as evident through its antioxidant activity, memory improvement, and neuroprotective attributes [63]. The compound also executes anxiolytic effect, for which it holds the potential to act as a medication for anti-stress and relaxation. So, clinical trials with piperine are progressing to unveil its prospective beneficial effects in autistic children [61].

5.4 Curcumin

Curcumin is the principal curcuminoid of turmeric, *Curcuma longa*, which is well-known for its neuroprotective properties. It is reported to target multiple pathways related to cellular signaling and extend its role in regulating nitrosative or oxidative stress, mitochondrial functioning, as well as protein aggregation [64]. Curcumin exhibits a wide range of anti-inflammatory effects and is able to cross the blood-brain barrier easily [65].

In a study conducted to assess the effect of prenatal valproic acid exposure, it has been revealed that curcumin supplements help diminish dysfunctions and significantly improve the level of antioxidant enzymes [66]. Up to 200 mg/kg of curcumin administered in rats displaying autistic phenotype can attenuate oxidative stress, mitochondrial dysfunction, release of tumor necrosis factor (TNF-α), and matrix metalloproteinases. Thus, curcumin has been reported to act as neuro-psychopharmacotherapeutic adjunct for autism spectrum disorders [67]. As a direct treatment or adjunct, curcumin is capable of reducing several inflammatory markers in various diseases and has consistently demonstrated in vitro and in vivo antioxidant radical scavenging activities [68, 69]. Increase in synaptic plasticity, leading to enhancement of cognition is possible with the regular intake of curcumin in diet [70]. Although there are no convincing reports on clinical studies on the usefulness of curcumin in human clinical trials, evidence for curcumin as a neuroprotective agent is sufficient for it to be employed in upcoming research related to autism and other related disorders.

5.5 Cannabinoids

Medical use of *Cannabis* is being explored in various neurological disorders currently, and different levels of efficacy are seen with its utilization [71]. Tetrahydrocannabinol (THC), the phytocannabinoid, which forms the main psychoactive component of *Cannabis sativa*, can exacerbate several neurological disorders, when employed in the adequate quantity. Cannabidiol (CBD) is reported to be sufficiently effective in suppressing autistic behavior [72]. The compound could promise therapeutic options like immunomodulation, antioxidant defense, and neuroprotection, with little or no side effects [73]. Cannabidivarin (CBDV) has also displayed satisfactory potential to ameliorate behavioral alterations, and clinical trials with this compound have shown immense improvement of autistic condition. Moreover, 10 mg/kg/day of CBDV for 12 weeks has been approved for further assessment to confirm its tolerability and safety level [74].

The endocannabinoid (EC) system represents a major neuromodulatory system that can regulate the emotional responses and behavioral reactivity for desired level of social interaction. In most cases, the EC system is found to be affected in patients diagnosed with autism spectrum disorders [75]. Group of endogenous molecules like signaling compounds consisting of arachidonic acid compound derived and associated enzymes are able to bind and activate the EC receptors, resulting in upregulation of RNA and protein levels [76]. However, a fault in this system disturbs the normal metabolic pathways and leads to neuroinflammation. Therefore, the activation of the EC system with natural cannabinoid phyto-products could regulate immune responses, display antioxidant activity, and help in ameliorating the plethora of autistic symptoms [75].

5.6 Ginkgo biloba *Extract*

Important compounds present in the standardized extract of *Ginkgo biloba* leaves, specifically EGb 761 [Ginkgo biloba extract EGb 761, Rökan, Tanakan, Tebonin], include approximately 24% flavone glycosides (primarily quercetin, kaempferol, and isorhamnetin) and 6% terpene lactones (2.8–3.4% ginkgolides A, B, and C and 2.6–3.2% bilobalide) [77]. Ginkgolide B and bilobalide account for about 0.8% and 3% of the total extract. Other constituents include proanthocyanadins, glucose, rhamnose, organic acids, D-glucaric acid, and ginkgolic acid [78].

The terpenoids, organic acids, and flavonoids present in the extract facilitate its neuroprotective effects against disorders such as ischemic stroke, Parkinson's disease, and Alzheimer's disease [79]. An observational study showed that 100 mg/kg twice a day of *Ginkgo biloba* extract is sufficient to improve aberrant behavior and symptoms in autistic individuals. The extract is capable of ameliorating behavioral irritability, hyperactivity, inadequate eye contact, and inappropriate speech in autism [80]. For the treatment of autism, *Ginkgo biloba* extract is used as an adjunct to risperidone at 80 mg/day for patients under 30 kg and 120 mg/day for patients above 30 kg. The treated group showed less adverse effect as compared to the placebo. The literature related to the pharmacokinetics and bioavailability, in relation to the central nervous system (CNS), is still sparse. Further research is required to be carried out to assess the probable efficacy of *Ginkgo biloba* to improve neurological conditions, including autism [81].

6 Conclusion

Plant-based drugs exhibit promising therapeutic effects against a range of complications, including neurodevelopmental disorders like autism. Ongoing research conducted over multiple years has come to the conclusion that the benefits and adverse effects of these natural products are yet to be established, proven, and/or recommended in near future. These natural products that are recognized as possible drug entities can serve as chemical models or templates for the synthesis or modification of novel substances for treating autism. Above all, plant resources can prove to be reliable pharmacological treatments for diminishing the behavioral issues in autistic individuals. More than half of the patients are prescribed psychoactive drugs or anticonvulsants, more specifically, synthetically prepared antidepressants, stimulants, and antipsychotics, which bear numerous side effects evident from their long-term consumption. Although improved learning techniques, substitute therapies, and interventions which have improved the scenario of autism are available nowadays, herbal treatments still emerge as a trustworthy alternative among them all. All we need now is for their potential to work against autism needs to be concluded as long-standing and firm through further research.

References

1. Lai, M.-C., Lombardo, M. V., & Baron-Cohen, S. (2014). Autism. *Lancet, 383*, 896–910. https://doi.org/10.1016/S0140-6736(13)61539-1
2. Rutter, M. (1978). Diagnosis and definition of childhood autism. *Journal of Autism and Childhood Schizophrenia, 8*, 139–161.
3. McPartland, J., & Volkmar, F. R. (2012). Autism and related disorders. *Handbook of Clinical Neurology, 106*, 407–418. https://doi.org/10.1016/B978-0-444-52002-9.00023-1
4. Adachi, M., Takahashi, M., Takayanagi, N., Yoshida, S., Yasuda, S., Tanaka, M., et al. (2018). Adaptation of the Autism Spectrum Screening Questionnaire (ASSQ) to preschool children. *PLoS One, 13*, e0199590. https://doi.org/10.1371/journal.pone.0199590
5. Elsabbagh, M., Divan, G., Koh, Y.-J., Kim, Y. S., Kauchali, S., Marcín, C., et al. (2012). Global prevalence of autism and other pervasive developmental disorders. *Autism Research, 5*, 160–179. https://doi.org/10.1002/aur.239
6. Ratajczak, H. V. (2011). Theoretical aspects of autism: Causes—A review. *Journal of Immunotoxicology, 8*, 68–79. https://doi.org/10.3109/1547691X.2010.545086
7. Chaste, P., & Leboyer, M. (2012). Autism risk factors: Genes, environment, and gene-environment interactions. *Dialogues in Clinical Neuroscience, 14*, 281–292.
8. Kern, J. K., Geier, D. A., Sykes, L. K., Haley, B. E., & Geier, M. R. (2016). The relationship between mercury and autism: A comprehensive review and discussion. *Journal of Trace Elements in Medicine and Biology, 37*, 8–24. https://doi.org/10.1016/j.jtemb.2016.06.002
9. Xiao, Z., Qiu, T., Ke, X., Xiao, X., Xiao, T., Liang, F., et al. (2014). Autism spectrum disorder as early neurodevelopmental disorder: Evidence from the brain imaging abnormalities in 2–3 years old toddlers. *Journal of Autism and Developmental Disorders, 44*, 1633–1640. https://doi.org/10.1007/s10803-014-2033-x
10. Bent, S., & Hendren, R. L. (2015). Complementary and alternative treatments for autism Part 1: Evidence-supported treatments. *AMA Journal of Ethics, 17*, 369–374. https://doi.org/10.1001/journalofethics.2015.17.4.sect1-1504
11. Toscano, C. V. A., Carvalho, H. M., & Ferreira, J. P. (2018). Exercise effects for children with autism spectrum disorder: Metabolic health, autistic traits, and quality of life. *Perceptual and Motor Skills, 125*, 126–146. https://doi.org/10.1177/0031512517743823
12. Nath, D. (2017). Complementary and alternative medicine in the school-age child with autism. *Journal of Pediatric Health Care, 31*, 393–397. https://doi.org/10.1016/j.pedhc.2016.12.001
13. Weber, W., & Newmark, S. (2007). Complementary and alternative medical therapies for attention-deficit/hyperactivity disorder and autism. *Pediatric Clinics of North America, 54*, 983–1006. https://doi.org/10.1016/j.pcl.2007.09.006
14. de Veld, D. M. J., Howlin, P., Hoddenbach, E., Mulder, F., Wolf, I., Koot, H. M., et al. (2017). Moderating effects of parental characteristics on the effectiveness of a theory of mind training for children with autism: A randomized controlled trial. *Journal of Autism and Developmental Disorders, 47*, 1987–1997. https://doi.org/10.1007/s10803-017-3117-1

15. Yenkoyan, K., Grigoryan, A., Fereshetyan, K., & Yepremyan, D. (2017). Advances in understanding the pathophysiology of autism spectrum disorders. *Behavioural Brain Research, 331*, 92–101. https://doi.org/10.1016/j.bbr.2017.04.038
16. Bhat, S., Acharya, U. R., Adeli, H., Bairy, G. M., & Adeli, A. (2014). Autism: Cause factors, early diagnosis and therapies. *Reviews in the Neurosciences, 25*, 841–850. https://doi.org/10.1515/revneuro-2014-0056
17. Gupta, S., Aggarwal, S., Rashanravan, B., & Lee, T. (1998). Th1- and Th2-like cytokines in CD4+ and CD8+ T cells in autism. *Journal of Neuroimmunology, 85*, 106–109. https://doi.org/10.1016/S0165-5728(98)00021-6
18. Tordjman, S., Somogyi, E., Coulon, N., Kermarrec, S., Cohen, D., Bronsard, G., et al. (2014). Gene × Environment interactions in autism spectrum disorders: Role of epigenetic mechanisms. *Frontiers in Psychiatry, 5*, 53. https://doi.org/10.3389/fpsyt.2014.00053
19. Gardener, H., Spiegelman, D., & Buka, S. L. (2011). Perinatal and neonatal risk factors for autism: A comprehensive meta-analysis. *Pediatrics, 128*, 344–355. https://doi.org/10.1542/peds.2010-1036
20. Chaidez, V., Fernandez y Garcia, E., Wang, L. W., Angkustsiri, K., Krakowiak, P., Hertz-Picciotto, I., et al. (2018). Comparison of maternal beliefs about causes of autism spectrum disorder and association with utilization of services and treatments. *Child: Care, Health and Development, 44*, 916–925. https://doi.org/10.1111/cch.12612
21. Yamaguchi, H., Hara, Y., Ago, Y., Takano, E., Hasebe, S., Nakazawa, T., et al. (2017). Environmental enrichment attenuates behavioral abnormalities in valproic acid-exposed autism model mice. *Behavioural Brain Research, 333*, 67–73. https://doi.org/10.1016/j.bbr.2017.06.035
22. Kinney, D., Munir, K., Crowley, D., & Miller, A. (2008). Prenatal stress and risk for autism. *Neuroscience and Biobehavioral Reviews, 32*, 1519–1532. https://doi.org/10.1016/j.neubiorev.2008.06.004
23. Lyall, K., Schmidt, R. J., & Hertz-Picciotto, I. (2014). Maternal lifestyle and environmental risk factors for autism spectrum disorders. *International Journal of Epidemiology, 43*, 443–464. https://doi.org/10.1093/ije/dyt282
24. Zachariah, S., Oommen, S., & Koshy, B. (2017). Clinical features and diagnosis of autism spectrum disorder in children. *Current Medical Issues, 15*, 6. https://doi.org/10.4103/0973-4651.200297
25. Sigman, M., Spence, S. J., & Wang, A. T. (2006). Autism from developmental and neuropsychological perspectives. *Annual Review of Clinical Psychology, 2*, 327–355. https://doi.org/10.1146/annurev.clinpsy.2.022305.095210
26. O'Hearn, K., Asato, M., Ordaz, S., & Luna, B. (2008). Neurodevelopment and executive function in autism. *Development and Psychopathology, 20*, 1103. https://doi.org/10.1017/S0954579408000527
27. Schumann, C. M., Hamstra, J., Goodlin-Jones, B. L., Lotspeich, L. J., Kwon, H., Buonocore, M. H., et al. (2004). The amygdala is enlarged in children but not adolescents with autism; the hippocampus is enlarged at all ages. *The Journal of Neuroscience, 24*, 6392–6401. https://doi.org/10.1523/JNEUROSCI.1297-04.2004
28. Fatemi, S. H., Aldinger, K. A., Ashwood, P., Bauman, M. L., Blaha, C. D., Blatt, G. J., et al. (2012). Consensus paper: Pathological role of the cerebellum in autism. *The Cerebellum, 11*, 777–807. https://doi.org/10.1007/s12311-012-0355-9
29. Silver, W. G., & Rapin, I. (2012). Neurobiological basis of autism. *Pediatric Clinics of North America, 59*, 45–61. https://doi.org/10.1016/j.pcl.2011.10.010
30. Myers, S. M., Johnson, C. P., & American Academy of Pediatrics Council on Children With Disabilities. (2007). Management of children with autism spectrum disorders. *Pediatrics, 120*, 1162–1182. https://doi.org/10.1542/peds.2007-2362
31. Coury, D. L., Anagnostou, E., Manning-Courtney, P., Reynolds, A., Cole, L., McCoy, R., et al. (2012). Use of psychotropic medication in children and adolescents with autism spectrum disorders. *Pediatrics, 130*(Suppl 2), S69–S76. https://doi.org/10.1542/peds.2012-0900D

32. Benger, M., Kinali, M., & Mazarakis, N. D. (2018). Autism spectrum disorder: Prospects for treatment using gene therapy. *Molecular Autism, 9*, 39. https://doi.org/10.1186/s13229-018-0222-8
33. Bang, M., Lee, S. H., Cho, S.-H., Yu, S.-A., Kim, K., Lu, H. Y., et al. (2017). Herbal medicine treatment for children with autism spectrum disorder: A systematic review. *Evidence-Based Complementary and Alternative Medicine, 2017*, 1–12. https://doi.org/10.1155/2017/8614680
34. Casano, A. M., & Peri, F. (2015). Microglia: Multitasking specialists of the brain. *Developmental Cell, 32*, 469–477. https://doi.org/10.1016/j.devcel.2015.01.018
35. Streit, W. J., Mrak, R. E., & Griffin, W. S. T. (2004). Microglia and neuroinflammation: A pathological perspective. *Journal of Neuroinflammation, 1*, 14. https://doi.org/10.1186/1742-2094-1-14
36. Kim, J. W., Hong, J. Y., & Bae, S. M. (2018). Microglia and autism spectrum disorder: Overview of current evidence and novel immunomodulatory treatment options. *Clinical Psychopharmacology and Neuroscience, 16*, 246–252. https://doi.org/10.9758/cpn.2018.16.3.246
37. Gottfried, C., Bambini-Junior, V., Francis, F., Riesgo, R., & Savino, W. (2015). The impact of neuroimmune alterations in autism spectrum disorder. *Frontiers in Psychiatry, 6*, 121. https://doi.org/10.3389/fpsyt.2015.00121
38. Marchezan, J., Winkler Dos Santos, E. G. A., Deckmann, I., & Riesgo, R. D. S. (2018). Immunological dysfunction in autism spectrum disorder: A potential target for therapy. *Neuroimmunomodulation, 25*, 300–319. https://doi.org/10.1159/000492225
39. Solek, C. M., Farooqi, N., Verly, M., Lim, T. K., & Ruthazer, E. S. (2018). Maternal immune activation in neurodevelopmental disorders. *Developmental Dynamics, 247*, 588–619. https://doi.org/10.1002/dvdy.24612
40. Zuiki, M., Chiyonobu, T., Yoshida, M., Maeda, H., Yamashita, S., Kidowaki, S., et al. (2017). Luteolin attenuates interleukin-6-mediated astrogliosis in human iPSC-derived neural aggregates: A candidate preventive substance for maternal immune activation-induced abnormalities. *Neuroscience Letters, 653*, 296–301. https://doi.org/10.1016/J.NEULET.2017.06.004
41. Gullotta, F., Schindler, F., Schmutzler, R., & Weeks-Seifert, A. (1985). GFAP in brain tumor diagnosis: Possibilities and limitations. *Pathology, Research and Practice, 180*, 54–60. https://doi.org/10.1016/S0344-0338(85)80075-3
42. Chen, E., Xu, D., Lan, X., Jia, B., Sun, L., Zheng, J. C., et al. (2013). A novel role of the STAT3 pathway in brain inflammation-induced human neural progenitor cell differentiation. *Current Molecular Medicine, 13*, 1474–1484.
43. Golonzhka, O., Leid, M., Indra, G., & Indra, A. K. (2007). Expression of COUP-TF-interacting protein 2 (CTIP2) in mouse skin during development and in adulthood. *Gene Expression Patterns, 7*, 754–760. https://doi.org/10.1016/j.modgep.2007.06.002
44. Notwell, J. H., Heavner, W. E., Darbandi, S. F., Katzman, S., McKenna, W. L., Ortiz-Londono, C. F., et al. (2016). TBR1 regulates autism risk genes in the developing neocortex. *Genome Research, 26*, 1013–1022. https://doi.org/10.1101/gr.203612.115
45. Bertolino, B., Crupi, R., Impellizzeri, D., Bruschetta, G., Cordaro, M., Siracusa, R., et al. (2017). Beneficial effects of co-ultramicronized palmitoylethanolamide/luteolin in a mouse model of autism and in a case report of autism. *CNS Neuroscience & Therapeutics, 23*, 87–98. https://doi.org/10.1111/cns.12648
46. Theoharides, T. C., Asadi, S., & Panagiotidou, S. (2012). A case series of a luteolin formulation (NeuroProtek®) in children with autism spectrum disorders. *International Journal of Immunopathology and Pharmacology, 25*, 317–323. https://doi.org/10.1177/039463201202500201
47. Chen, H.-Q., Jin, Z.-Y., Wang, X.-J., Xu, X.-M., Deng, L., & Zhao, J.-W. (2008). Luteolin protects dopaminergic neurons from inflammation-induced injury through inhibition of microglial activation. *Neuroscience Letters, 448*, 175–179. https://doi.org/10.1016/j.neulet.2008.10.046
48. Xu, N., Li, X., & Zhong, Y. (2015). Inflammatory cytokines: Potential biomarkers of immunologic dysfunction in autism spectrum disorders. *Mediators of Inflammation, 2015*, 531518. https://doi.org/10.1155/2015/531518

49. Tsilioni, I., Taliou, A., Francis, K., & Theoharides, T. C. (2015). Children with autism spectrum disorders, who improved with a luteolin-containing dietary formulation, show reduced serum levels of TNF and IL-6. *Translational Psychiatry, 5*, e647–e647. https://doi.org/10.1038/tp.2015.142
50. Kritas, S. K., Saggini, A., Varvara, G., Murmura, G., Caraffa, A., Antinolfi, P., et al. (2013). Luteolin inhibits mast cell-mediated allergic inflammation. *Journal of Biological Regulators and Homeostatic Agents, 27*, 955–959.
51. Rossignol, D. A., & Frye, R. E. (2014). Evidence linking oxidative stress, mitochondrial dysfunction, and inflammation in the brain of individuals with autism. *Frontiers in Physiology, 5*, 150. https://doi.org/10.3389/fphys.2014.00150
52. Chauhan, A., & Chauhan, V. (2006). Oxidative stress in autism. *Pathophysiology, 13*, 171–181. https://doi.org/10.1016/j.pathophys.2006.05.007
53. Schimidt, H. L., Garcia, A., Martins, A., Mello-Carpes, P. B., & Carpes, F. P. (2017). Green tea supplementation produces better neuroprotective effects than red and black tea in Alzheimer-like rat model. *Food Research International, 100*, 442–448. https://doi.org/10.1016/j.foodres.2017.07.026
54. Cabrera, C., Artacho, R., & Giménez, R. (2006). Beneficial effects of green tea—A review. *Journal of the American College of Nutrition, 25*, 79–99.
55. Banji, D., Banji, O. J. F., Abbagoni, S., Hayath, M. S., Kambam, S., & Chiluka, V. L. (2011). Amelioration of behavioral aberrations and oxidative markers by green tea extract in valproate induced autism in animals. *Brain Research, 1410*, 141–151. https://doi.org/10.1016/j.brainres.2011.06.063
56. Takeda, A., Sakamoto, K., Tamano, H., Fukura, K., Inui, N., Suh, S. W., et al. (2011). Facilitated neurogenesis in the developing hippocampus after intake of theanine, an amino acid in tea leaves, and object recognition memory. *Cellular and Molecular Neurobiology, 31*, 1079–1088. https://doi.org/10.1007/s10571-011-9707-0
57. Kumaravel, P., Melchias, G., Vasanth, N., & Manivasagam, T. (2017). Epigallocatechin Gallate Attenuates Behavioral Defects in Sodium Valproate Induced Autism Rat Model. *Research Journal of Pharmacy and Technology, 10*, 1477. https://doi.org/10.5958/0974-360X.2017.00260.8
58. Sundberg, M., & Sahin, M. (2015). Cerebellar development and autism spectrum disorder in tuberous sclerosis complex. *Journal of Child Neurology, 30*, 1954–1962. https://doi.org/10.1177/0883073815600870
59. Urdaneta, K. E., Castillo, M. A., Montiel, N., Semprún-Hernández, N., Antonucci, N., & Siniscalco, D. (2018). Autism spectrum disorders: Potential neuro-psychopharmacotherapeutic plant-based drugs. *Assay and Drug Development Technologies, 16*, 433–444. https://doi.org/10.1089/adt.2018.848
60. McNamara, F. N., Randall, A., & Gunthorpe, M. J. (2005). Effects of piperine, the pungent component of black pepper, at the human vanilloid receptor (TRPV1). *British Journal of Pharmacology, 144*, 781–790. https://doi.org/10.1038/sj.bjp.0706040
61. Wattanathorn, J., Chonpathompikunlert, P., Muchimapura, S., Priprem, A., & Tankamnerdthai, O. (2008). Piperine, the potential functional food for mood and cognitive disorders. *Food and Chemical Toxicology, 46*, 3106–3110. https://doi.org/10.1016/j.fct.2008.06.014
62. Fu, M., Sun, Z.-H., & Zuo, H.-C. (2010). Neuroprotective effect of piperine on primarily cultured hippocampal neurons. *Biological & Pharmaceutical Bulletin, 33*, 598–603.
63. Pragnya, B., Kameshwari, J. S. L., & Veeresh, B. (2014). Ameliorating effect of piperine on behavioral abnormalities and oxidative markers in sodium valproate induced autism in BALB/C mice. *Behavioural Brain Research, 270*, 86–94. https://doi.org/10.1016/j.bbr.2014.04.045
64. Ak, T., & Gülçin, İ. (2008). Antioxidant and radical scavenging properties of curcumin. *Chemico-Biological Interactions, 174*, 27–37. https://doi.org/10.1016/J.CBI.2008.05.003
65. Cole, G. M., Teter, B., & Frautschy, S. A. (2007). Neuroprotective effects of curcumin. *Advances in Experimental Medicine and Biology, 595*, 197–212. https://doi.org/10.1007/978-0-387-46401-5_8

66. Al-Askar, M., Bhat, R. S., Selim, M., Al-Ayadhi, L., & El-Ansary, A. (2017). Postnatal treatment using curcumin supplements to amend the damage in VPA-induced rodent models of autism. *BMC Complementary and Alternative Medicine, 17*, 259. https://doi.org/10.1186/s12906-017-1763-7
67. Bhandari, R., & Kuhad, A. (2015). Neuropsychopharmacotherapeutic efficacy of curcumin in experimental paradigm of autism spectrum disorders. *Life Sciences, 141*, 156–169. https://doi.org/10.1016/j.lfs.2015.09.012
68. Panahi, Y., Badeli, R., Karami, G.-R., & Sahebkar, A. (2015). Investigation of the efficacy of adjunctive therapy with bioavailability-boosted curcuminoids in major depressive disorder. *Phytheraphy Research, 29*, 17–21. https://doi.org/10.1002/ptr.5211
69. Panahi, Y., Saadat, A., Beiraghdar, F., & Sahebkar, A. (2014). Adjuvant therapy with bioavailability-boosted curcuminoids suppresses systemic inflammation and improves quality of life in patients with solid tumors: A randomized double-blind placebo-controlled trial. *Phytheraphy Research, 28*, 1461–1467. https://doi.org/10.1002/ptr.5149
70. Dong, S., Zeng, Q., Mitchell, E. S., Xiu, J., Duan, Y., Li, C., et al. (2012). Curcumin enhances neurogenesis and cognition in aged rats: Implications for transcriptional interactions related to growth and synaptic plasticity. *PLoS One, 7*, e31211. https://doi.org/10.1371/journal.pone.0031211
71. Solimini, R., Rotolo, M. C., Pichini, S., & Pacifici, R. (2017). Neurological disorders in medical use of cannabis: An update. *CNS Neurological Disorders Drug Targets, 16*, 527–533. https://doi.org/10.2174/1871527316666170413105421
72. Salgado, C. A., & Castellanos, D. (2018). Autism spectrum disorder and cannabidiol: Have we seen this movie before? *Global Pediatric Health, 5*, 2333794X18815412. https://doi.org/10.1177/2333794X18815412
73. Nagarkatti, P., Pandey, R., Rieder, S. A., Hegde, V. L., & Nagarkatti, M. (2009). Cannabinoids as novel anti-inflammatory drugs. *Future Medicinal Chemistry, 1*, 1333–1349. https://doi.org/10.4155/fmc.09.93
74. Perucca, E. (2017). Cannabinoids in the treatment of epilepsy: Hard evidence at last? *Journal of Epilepsy Research, 7*, 61–76. https://doi.org/10.14581/jer.17012
75. Zamberletti, E., Gabaglio, M., & Parolaro, D. (2017). The endocannabinoid system and autism spectrum disorders: Insights from animal models. *International Journal of Molecular Sciences, 18*, 1916. https://doi.org/10.3390/ijms18091916
76. Brigida, A., Schultz, S., Cascone, M., Antonucci, N., & Siniscalco, D. (2017). Endocannabinoid signal dysregulation in autism spectrum disorders: A correlation link between inflammatory state and neuro-immune alterations. *International Journal of Molecular Sciences, 18*, 1425. https://doi.org/10.3390/ijms18071425
77. Uebel-von Sandersleben, H., Rothenberger, A., Albrecht, B., Rothenberger, L. G., Klement, S., & Bock, N. (2014). Ginkgo biloba extract EGb 761 ® in children with ADHD. *Zeitschrift für Kinder- und Jugendpsychiatrie und Psychotherapie, 42*, 337–347. https://doi.org/10.1024/1422-4917/a000309
78. Ude, C., Schubert-Zsilavecz, M., & Wurglics, M. (2013). Ginkgo biloba extracts: A review of the pharmacokinetics of the active ingredients. *Clinical Pharmacokinetics, 52*, 727–749. https://doi.org/10.1007/s40262-013-0074-5
79. Fang, W., Deng, Y., Li, Y., Shang, E., Fang, F., Lv, P., et al. (2010). Blood brain barrier permeability and therapeutic time window of Ginkgolide B in ischemia–reperfusion injury. *European Journal of Pharmaceutical Sciences, 39*, 8–14. https://doi.org/10.1016/j.ejps.2009.10.002
80. Niederhofer, H. (2009). First preliminary results of an observation of *Ginkgo Biloba* treating patients with autistic disorder. *Phytheraphy Research, 23*, 1645–1646. https://doi.org/10.1002/ptr.2778
81. Hasanzadeh, E., Mohammadi, M.-R., Ghanizadeh, A., Rezazadeh, S.-A., Tabrizi, M., Rezaei, F., et al. (2012). A double-blind placebo controlled trial of ginkgo biloba added to risperidone in patients with autistic disorders. *Child Psychiatry and Human Development, 43*, 674–682. https://doi.org/10.1007/s10578-012-0292-3

Dietary Phytochemicals as Neurotherapeutics for Autism Spectrum Disorder: Plausible Mechanism and Evidence

Ranjana Bhandari, Jyoti K. Paliwal, and Anurag Kuhad

Abstract Autism spectrum disorder (ASD) is a complex neurodevelopmental disorder with symptoms ranging from lack of social interaction and communication deficits to rigid, repetitive, and stereotypic behavior. It has also been associated with comorbidities such as anxiety, aggression, epilepsy, deficit in sensory processing, as well as ADHD (attention deficit hyperactivity disorder). Apart from several behavioral and cognitive complications arising as a result of central nervous system dysfunction, there are various physiological comorbidities such as immune system deregulation, neuroinflammation, oxidative stress, mitochondrial dysfunction, and gastrointestinal complications which can worsen existing behavioral complications. There are no available treatments for these physiological comorbidities. The prevalence of gastrointestinal complications in ASD ranges from 9% to 70% and it correlates with behaviors consistent with the autistic endophenotype indicating that these are one of the major comorbidities associated with ASD. A strong connection of gut–brain cross talk occurs as a result of gut dysbiosis responsible for excessive production of short-chain fatty acids such as propanoic acid (PPA) by abnormal gut flora in ASD patients. This worsens behavioral, neurochemical, and mitochondrial dysfunction occurring in ASD. These physiological comorbidities are responsible for the generation of free radical species that cause immune system dysfunction leading to synthesis of various pro-inflammatory cytokines and chemokines. This in turn causes activation of microglia. Dietary phytochemicals are thought to be safer and useful as an alternative neurotherapeutic moiety. These compounds provide neuroprotection by modulating signaling pathways such as Nrf2, NF-κB, MAPK pathway or Sirtuin-FoxO pathway. There has been recent evidence in scientific literature regarding the modulation of gut–brain cross talk responsible for behavioral, biochemical, and mitochondrial dysfunction as well as cellular and behavioral sensory alterations by dietary phytochemicals such as curcumin, resveratrol, naringenin,

R. Bhandari · J. K. Paliwal · A. Kuhad (✉)
Pharmacology Research Laboratory, University Institute of Pharmaceutical Sciences, UGC-Centre of Advanced Study, Panjab University, Chandigarh, India
e-mail: anurag.kuhad@pu.ac.in

M. M. Essa, M. W. Qoronfleh (eds.), *Personalized Food Intervention and Therapy for Autism Spectrum Disorder Management*, Advances in Neurobiology 24,
https://doi.org/10.1007/978-3-030-30402-7_23

and sulforaphane. These dietary phytochemicals can be formulated in novel brain-targeted delivery systems which overcome their limitation of low oral bioavailability and short half-life leading to prolonged action. Till date, not much work has been done on the development of brain-targeted neurotherapeutics for ASD. In this chapter we discuss plausible mechanisms and evidence from our own and other scientific research for the utilization of curcumin, resveratrol, naringenin, and sulforaphane as neurotherapeutics for ASD.

Keywords Autism spectrum disorder · ASD · Nutrition therapy · Dietary phytochemicals · Neurotherapeutics · Herbal remedies · Curcumin · Naringenin · Resveratrol · Sulforaphane

1 Introduction

Autism spectrum disorder (ASD) is a complex neurodevelopmental disorder with behavioral symptoms manifesting at 3 years of age. Behavioral symptoms range from deficits in social interaction and communication skills to restrictive, rigid, pervasive, and stereotypic behavior. It has also been associated with comorbidities such as anxiety, depression, aggression, as well as ADHD (attention deficit hyperactivity disorder), epilepsy and deficit in sensory processing [1–4]. According to the National Institute of Mental Health, 2.41% of children in US have ASD. Prevalence of ASD has been reported by US Center for Disease Control and Prevention (CDC) to be 1 in 68 children in 2014 in comparison to 2010 reports of 1 in 88. World Health Organization (WHO) reports of 2017 show that worldwide 1 in 160 children are suffering from ASD [5]. ASDs show significant skewness for occurrence in boys with a sex ratio of 4:1 [6–9]. Complex amalgamation of genetic, epigenetic, and environmental factors exists in ASD. It includes complex interaction between pre-existing genetic factors and environmental factors [10–13].

Currently, there is no cure that exists for the core symptoms associated with ASD with even the basic disturbances not being modifiable. Associated symptoms such as anxiety, depression, irritability, epilepsy, and some mood disturbances can, however, be decreased with pharmacological agents like antipsychotics, antidepressants, mood stabilizers, medications for ADHD, NMDA receptor antagonists, melatonin, oxytocin, and omega-3 fatty acids [14, 15]. Two antipsychotic drugs—resperidone and aripripazole—have been shown to lead to improvement of associated symptoms in clinical trials and have been approved for symptomatic treatment of ASD by the US-FDA [16, 17].

However, these agents are not used as part of pharmacotherapy for the core symptoms of social interaction, repetitive and restricted behavior as they are untreatable. The primary research focus now is to develop novel treatments that target these core symptoms, many of which are under clinical trials; for these disorders as well for the core symptoms drugs like IGF-1, Evorolimus, Arbaclofen, Fenobam, Memantine, and Lithium. These are under clinical trials for the treatment of aggression, irritability, social withdrawal, and sensory gating. Currently under phase I, Memantine is being explored as a treatment for the core symptoms of ASD. Oxytocin is thought to be implicated in ASD when considering the two core symptoms of social development and repetitive behavior. It has been observed in animal models that oxytocin, when centrally administered, improves social deficits, enhances social novelty preference, and decreases aggression [14, 15].

Apart from several behavioral and cognitive complications arising as a result of central nervous system dysfunction, there are many physiological comorbidities associated with ASD which can worsen behavioral complications. Research and clinical studies have indicated many physiological comorbidities like immune system deregulation, environmental toxicant exposures, oxidative stress, mitochondrial dysfunction, and gastrointestinal complications [18–21]. There are no treatments currently available for these physiological comorbidities.

The prevalence of gastrointestinal complications in ASD ranges from 9% to 70%. It correlates with behaviors consistent with autistic endophenotype indicating that these are one of the major comorbidities associated with ASD [22–24]. Song et al. [25] found significant increase in *Clostridium bolteae* and *Clostridium clusters I and XI*. Finegold et al. [26] noted an increase of *Desulfovibrio* spp. and Wang et al. [27] have observed higher levels of *Sutterella* and *Ruminococcus* spp. in individuals with ASD compared to control [28, 29]. These gut bacteria produce short-chain fatty acids (SCFAs) such as propanoic acid (PPA) as a consequence of metabolism of dietary carbohydrates as well as amino acids [30]. Scientific literature has suggested that behavioral and gastrointestinal complications worsen in autistic children after intake of a diet rich in carbohydrates or foods that use PPA as a preservative [23, 31–34].

Propanoic acid (PPA) is an organic acid which can cross blood–brain barrier (BBB) [35] and cause alterations in the levels of serotonin, dopamine, and glutamate by stimulating calcium release [36]. It causes disruption of glutamate: GABAergic transmission which simulates what occurs in autism [37–39]. It releases pro-inflammatory cytokines, depletes endogenous antioxidants, and elevates lipid peroxidase leading to oxidative stress [40, 41].

Immune system dysregulation and generation of reactive oxygen species (ROS) leads to synthesis of various pro-inflammatory cytokines and chemokines causing activation of microglia. In order to tackle this, dietary phytochemicals were thought to be a safer and more useful alternative as an adjunct neurotherapeutic moiety. These compounds provide neuroprotection by modulation of various signaling pathways [42, 43]. The signaling pathways which are modulated by phytochemicals are either the Nrf2 pathway [44], the NF-κB signaling pathway [45, 46], MAPK pathway [47, 48] or Sirtuin-FoxO pathway [49, 50]. Hence, in this chapter we bring forth a variety of dietary phytochemicals such as curcumin, resveratrol, naringenin, and sulforaphane and the plausible mechanism and evidence for their use as neurotherapeutics for autism spectrum disorders.

2 Dietary Phytochemicals as Neurotherapeutic

Phytopharmaceuticals are a safer alternative and, owing to their therapeutic potential, are recently being considered important in medicine. These natural antioxidants provide a safe path for protecting the body against free radicals. Our central nervous system has a mechanism for combating oxidative stress utilizing our endogenous oxidant system. However, this endogenous defense mechanism needs a support system. This system is modulated by dietary phytochemicals such as various polyphenols, alkaloids, flavonoids, terpenoids, saponins, polyunsaturated fatty acids, other phytochemicals such as sulforaphane, curcumin, resveratrol, and allicin. There is vast scientific and empirical evidence in support of the use of antioxidants as neurotherapeutic.

These dietary phytochemicals may have a beneficial effect in reducing neuronal damage thereby having neuroprotective action. A number of studies have explored their use in slowing down neuronal loss in various neurodegenerative disorders like multiple sclerosis, amyotrophic lateral sclerosis, stroke, Parkinson's disease, Alzheimer's disease, and Huntington's disease [51]. The role of dietary phytochemicals as neurotherapeutic for autism spectrum disorders has not been explored much. In this chapter we will delve into understanding the mechanism of their beneficial effect in ASD.

3 Why Are Dietary Phytochemicals Beneficial in Autism Spectrum Disorder (ASD)?

3.1 Neuroinflammation and Oxidative Stress Associated with ASD

ASD involves neuroinflammation as indicated by immune system dysregulation and microglial activation [52, 53]. Though microglial cells have protective properties, sustained microglia activation can result in the damage of neurons. ASD shows sustained activation of microglial cells as indicated by neuroinflammation found in brains of ASD patients obtained during postmortem studies [53]. Immune system dysregulation as well as increased levels of inflammatory cytokines like TNF-α, IL-6, IL-β, IL-2, IL-4, IL-13, have been found in the brain of individuals with ASD [54, 55]. Onore et al. [56] have shown that the levels of adhesion molecules are reduced in children having ASD.

It is found that patients with ASD have increased levels of TNF-α, INF-γ as well as microglia activation in CSF, plasma, and amniotic fluid [55, 57–59]. Clinical studies have shown elevated levels of MMP-9 in the samples of amniotic fluid. MMP-9 is involved in processing of pro-inflammatory cytokines and genes associated with ASD [60]. Growth factors such as TGF-1β are decreased and BDNF is increased in the brain (postmortem) and plasma of children with autism [61]. These inflammatory cytokines are responsible for the generation of ASD-like behaviors [62, 63].

3.2 *Oxidative Stress and Mitochondrial Dysfunction in ASD*

Reactive oxygen species (ROS) is responsible for oxidative stress involved in the pathogenesis of autism [64, 65]. Children suffering from autism have low levels of coenzyme Q10 along with high levels of malondialdehyde (MDA) [66]. González-fraguela et al. [67] have indicated that total GSH was lowered in ASD patients while there was increase in the levels of catalase, MDA, and 8-hydroxy-2deoxyguanosine (8OHdG) in the blood samples of children suffering from autism. Figure 1 describes the mechanism of oxidative stress in autism spectrum disorders (ASD).

Development of oxidative stress in ASD as a result of free radical generation is responsible for mitochondrial dysfunction. It can lead to the generation of inflammatory cytokines and activation of microglia. This sustained activation of microglia will be responsible for neuroinflammation as a result of neuronal dysfunction and manifest as behavioral symptoms associated with ASD.

Disruption of the electron transport chain of mitochondria and oxidative stress are major physiological disturbances occurring as part of ASD [68, 69]. Impairment of mitochondrial energy metabolism is one of the primary pathological consequences of autism. Meta-analysis of three population-based studies by Rossignol and Frye [69] has revealed that 30% of children among the ASD population show

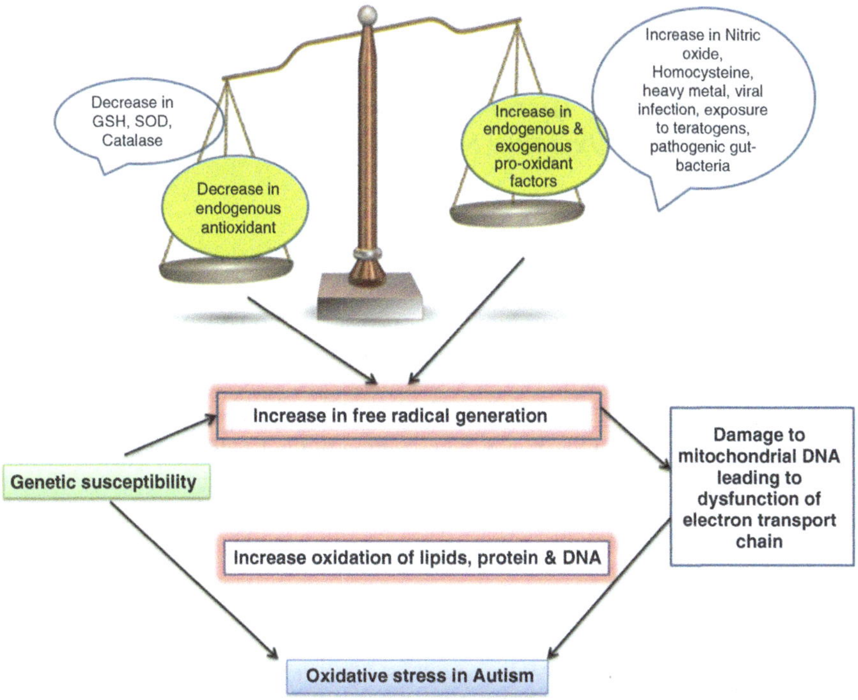

Fig. 1 Oxidative stress mechanism in autism spectrum disorder (ASD)

similar biomarkers to those occurring in mitochondrial disease. Its prevalence in ASD was at 5.0% which is significantly higher than the prevalence across the general population (~0.01%) indicating that there is a close connection between ASD and mitochondrial dysfunction (MD) [70]. Acquired mitochondrial dysfunction can occur as a result of exposure to environmental pollutants like pesticides and biphenyls which lead to epigenetic changes and the release of pro-inflammatory cytokines by dysregulation of the immune system and oxidative stress. Neuroinflammatory response stimulated by matrix metalloproteinases (MMPs) also plays an important role in the development of autistic phenotype as MMPs stimulate release of pro-inflammatory cytokines along with mitochondrial dysfunction. This leads to neuronal dysfunction and consequently, the development of autistic phenotype [68, 71]. Dysregulation of the immune system may release pro-inflammatory cytokines in individuals with autism which may lead to ROS production. While prenatal exposure may lead to functional disconnection and lack of integration of information processed by the brain, postnatal exposure leads to mitochondrial dysfunction and oxidative stress resulting in glial cell activation. Sustained glial cell activation has been observed to cause neuroinflammation in brains of individuals suffering from ASD [53, 54, 72]. The overexpression of HSP-70 can be a protective mechanism as a result of prevention of misfolding of proteins. It can also play a significant role as a potential biomarker induced as a result of neuroinflammation and oxidative stress [73].

3.3 Role of Gut-Microbiota in ASD: How Are They Responsible for Microglial Activation, Mitochondrial Dysfunction, Oxidative Stress, and Neuroinflammation in ASD?

Gastrointestinal complications occur among 70% of autistic patients. They have abnormal gut flora. It has been observed that gut bacteria such as *Clostridia*, *Desulfovibrio*, *Sutterella*, and *Ruminococcus* species produce short-chain fatty acids (SCFAs) such as PPA because of the metabolism of dietary carbohydrates and amino acids [26, 27, 29, 30].

Propanoic acid (PPA) is responsible for the generation of pro-inflammatory cytokines like TNF-α, IL-6, INF-ϒ and the reduction in the levels of endogenous antioxidants such as glutathione, superoxide dismutase as well the elevation of lipid peroxidase [41]. PPA causes disruption of GAP junction coupling as a result of increase in the levels of neurotransmitters such as serotonin, dopamine, and glutamate [74–76] (Fig. 2).

A strong gut–brain cross talk exists in ASD which is responsible for worsening of behavioral and gastrointestinal symptoms after consumption of diet rich in carbohydrates or food in which PPA is present as a preservative [23, 31]. These short-chain fatty acids result in the enhanced production of serotonin from enteric neurons of the gut and cause severe contractions of the smooth muscles of the gut [75]. In

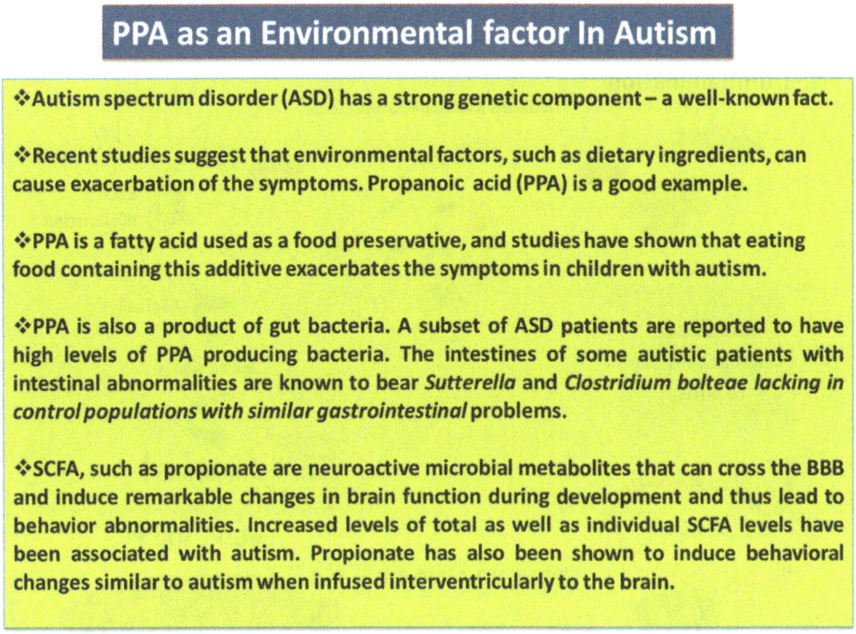

Fig. 2 An overview of PPA as an environmental factor in ASD

their study aimed at understanding the effect of CNS exposure of PPA on the behavioral, biochemical, and neurological pathology, MacFabe et al. [75, 77] administered an intracerebroventricular injection of PPA to adolescent rats. PPA inhibits mitochondrial Complex I, II, III, and IV activity as it can enter Kreb's cycle as propionyl-CoA and cause disruption of ETC by reducing production of ATP [75, 78–81].

Mitochondrial dysfunction is a result of disruption of the electron transport chain (ETC). This leads to damage of mitochondrial DNA further leading to the generation of a reactive oxygen species (ROS) as a result of oxidative stress. This in turn will cause the generation of inflammatory cytokines responsible for activation of the transcription factor NF-κB leading to neuroinflammation and further behavioral changes characteristic to ASD (Fig. 3).

Hence, gut dysbiosis and genetic susceptibility along with various environmental factors are responsible for immune system activation, oxidative stress, and mitochondrial dysfunction in patients suffering from ASD. All of these physiological comorbidities associated with autism can worsen the behavioral complications. Thus, dietary phytochemicals can prove to be a safer alternative in these patients. They can help mitigate the oxidative stress and neuroinflammation occurring as part of ASD. We will now delve further into the mechanistic details of the beneficial aspect of dietary phytochemicals such as curcumin, resveratrol, naringenin, and sulforaphane which have, till date, been explored by the scientific community as a neurotherapeutic in ASD.

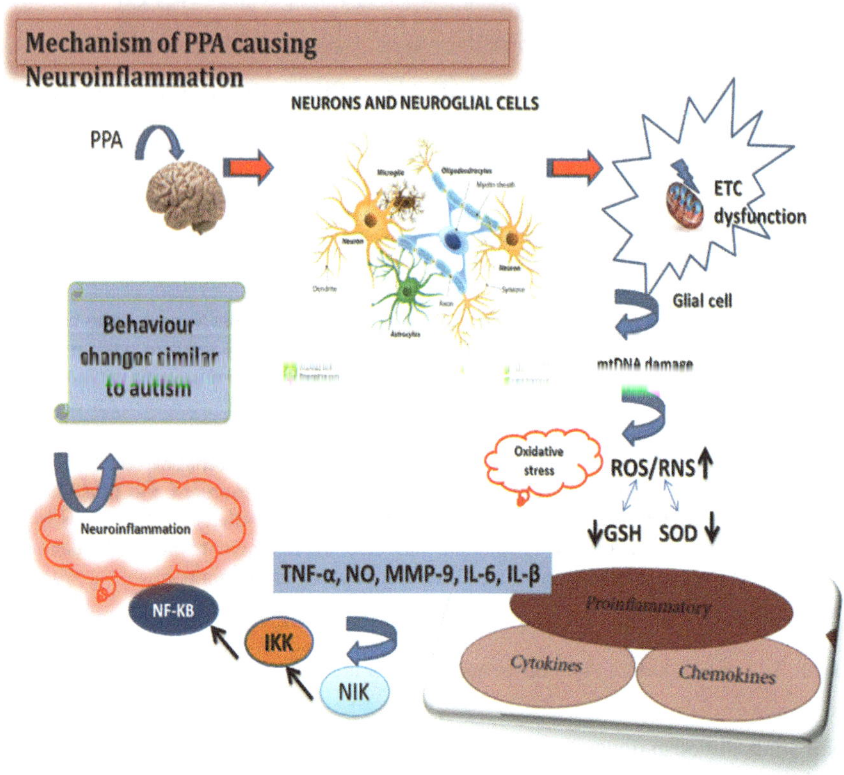

Fig. 3 Mechanism of neuroinflammation caused by PPA

4 Antioxidant Potential of Curcumin, Resveratrol, Naringenin, and Sulforaphane: Cross Talk Between Nrf2–NF-κB Pathways

Nrf2 is a transcription factor protecting cells from damage by oxidative stress. It does so through the activation of various genes expressed in the antioxidant response element (ARE) such as NADPH, GSH, SOD, catalase, heme-oxygenae-1 (HO-1), and NQO1. It binds to ARE and is responsible for maintaining cellular homeostasis by balancing of redox pathways. Nrf2 function can cause cellular dysfunction and apoptosis. Activity of Nrf2 is regulated by its inhibitor protein Keap1 (Kelch-like ECH-associated protein 1) present near the plasma membrane sequestering Nrf2 inside the cell. Keap1 acts as an important link for interaction of Nrf2 with Cul3-Rbx E3 ubiquitin ligase complex. This interaction will eventually lead to Nrf2 ubiquitination and cause proteasomal degradation of Nrf2 so reduced levels of Nrf2 are maintained. During oxidative stress, reactive oxygen species (ROS) cause modification of cysteine residues of Keap1 and hence, can no longer cause repression of Nrf2. This leads to Nrf2 translocation into the nucleus. Nrf2 gets associated with Maf proteins. After getting associated with Maf proteins it binds to the promoter region of the genes present on the ARE, protecting them from cellular stress. This leads to the initiation of transcription [82, 83]. NF-κB is a major transcription factor regulating the activation of the immune system. In response to enhanced oxidative stress, it initiates the release of pro-inflammatory cytokines such as TNF-α, IL-β, Il-6, and LPS. It consists of p65, p50, p52, and RelB. Under basal conditions, inhibitor of NF-κB (IkB-α) sequesters NF-κB in the cell. Increased oxidative stress causes activation of IkB kinase, which phosphorylates NF-κB inhibitor. This results in proteasomal degradation of IkB-α and translocation of NF-κB to the nucleus where it binds with the genome at the k region. NF-κB causes transcription of various pro-inflammatory cytokines with the help of histone acetyl transferases (HAT). There is strong molecular cross talk between Nrf2 and NF-κB pathway [82, 83]. There is inhibition of the NF-κB pathway activation by Nrf2–ARE pathway by increasing the expression of antioxidant genes as well as HO-1. This prevents cellular stress and apoptosis by neutralization of free radicals. Another process which causes inhibition of transcription of various pro-inflammatory cytokines by NF-κB is through Keap1. Once there is translocation of Nrf2 inside the nucleus in response to oxidative stress, Keap1 binds to IkB kinase and reduces the degradation of IkB-α. Hence, Nrf2 pathway inhibits NF-κB activation. NF-κB pathway activation also ameliorates Nrf2 pathway activation as a result of reduction in the transcription of genes present on ARE. NF-κB also facilitates the binding of HDAC3 (histone deacetylase3) to the antioxidant response element region as a result of its binding to Maf proteins. This results in repression of transcription caused by Nrf2 [84]. There are various phytochemicals that can activate the Nrf2 pathway and interact with Keap1 like curcumin [85, 86], resveratrol [87], naringenin [88], and sulforaphane [89–91]. Figure 4 summarizes both the molecular cross talk between Nrf2/ARE and NF-κB pathway occurring as part of ASD and the activation of Nrf2/ARE pathway by curcumin, resveratrol, naringenin, and sulforaphane by acting on Keap1.

5 CURCUMIN: Potential of Indian Solid Gold as Neurotherapeutic in ASD

5.1 CURCUMIN: Structure, Description, and Physicochemical Properties

Curcumin is the primary curcuminoid present in the Indian spice, turmeric (Curcuma longa) and is regarded as "Indian Solid Gold" (Fig. 5). It has several anti-inflammatory and antioxidant activities and affects angiogenesis and cell adhesion. It also has anticarcinogenic properties evident in its action on three primary cell-signaling pathways, i.e., Akt, NF-κB, and PI3K [92–94]. Curcumin is known to be

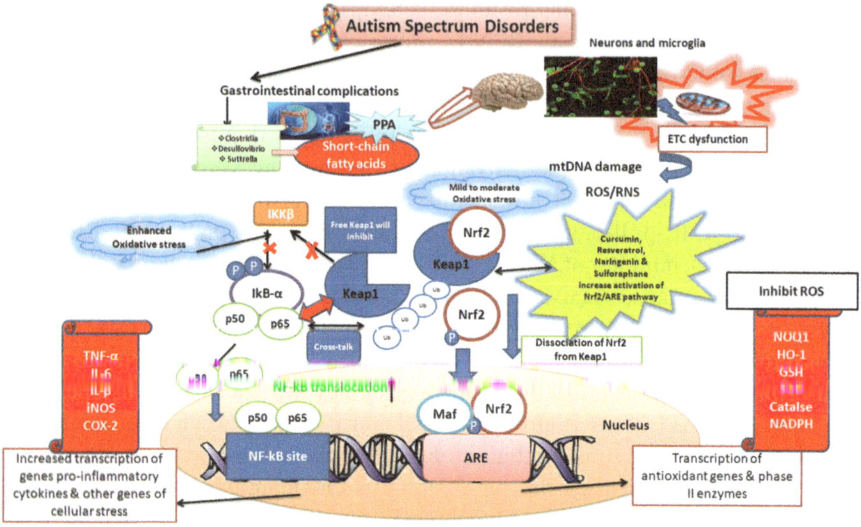

Fig. 4 Molecular cross talk between Nrf2 and NF-κB pathways in ASD. ASD is accompanied by gastrointestinal complications. As a result, there is enhanced production by short-chain fatty acids such as PPA (propanoic acid) and through abnormal gut flora such as *Clostridia*, *Desulfovibrio*, and *Suttrela* species. PPA can cross the blood–brain barrier (BBB) and can cause mitochondrial dysfunction as a result of disruption of the electron transport chain. There will be generation of reactive oxygen species (ROS) as a result of oxidative stress. In case of mild–moderate oxidative stress there is activation of the Nrf2/ARE pathway which will result in dissociation of Nrf2 from its inhibitor Keap1 and its translocation in nucleus. In the nucleus, Nrf2 will associate with Maf protein and bind itself to the antioxidant response element (ARE). This will lead to transcription of antioxidant genes and phase II enzymes which will inhibit ROS. Enhanced oxidative stress will activate IKKβ causing phosphorylation of IkB-α. IkB-α is an inhibitor of NF-κB that causes proteasomal degradation of IkB-α. Thus, NF-κB will migrate to the nucleus and bind to its region. This will cause transcription of pro-inflammatory cytokines and other genes such as TNF-α, Il-β, Il-6, iNOS, and COX-2. There is molecular cross talk between the Nrf2/ARE pathway and the NF-κB pathway as free Keap1 prevents degradation of IkB-α leading to inhibition of NF-κB pathway. The p65 subunit of NF-κB also inhibits Keap1 from interfering with facilitation of transcription by Nrf2. Dietary phytochemicals like curcumin, resveratrol, naringenin, and sulforaphane enhance the activation of Nrf2/ARE pathway by interacting with Keap1. Hence, they inhibit the activation of the NF-κB pathway leading to release of various pro-inflammatory cytokines and other cellular stress mediators

beneficial because of its neuroprotective action in several neurodegenerative diseases like Alzheimer's, Huntington's, Parkinson's, and peripheral neuropathy [95]. It primarily exerts its neuroprotective action because of its antiproliferative effect on activated microglia and reactive astrocytes which can lead to release of cytokines and other reactive substances. This leads to exacerbation of these pathologies [96]. Curcumin acts as a potent neuroinflammatory agent which protects against oxidative stress by inducing heme oxygenase-1 (HO-1) resulting in the increased activity of the enzyme, heme oxygenase [97]. Karlstetter et al. [98] have shown that curcumin can modulate transcription of microglial cells. It markedly reduces their

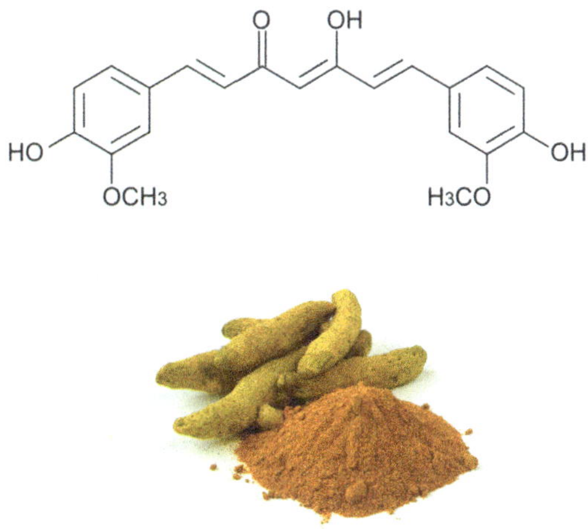

Fig. 5 Structure of curcumin

migration by inhibiting NF-κB signaling resulting in suppression of neuroinflammation, a key component of various neurodegenerative diseases. It has also shown to be protective in the case of axonal degeneration of neurons [99]. Curcumin has shown to be neuroprotective as a result of upregulation of the Nrf2 gene [86]. Curcumin also shows its protective effect in restoring cognitive deficits and adult neurogenesis in a rat model of Alzheimer's disease [100, 101].

There is a variety of mechanisms targeted by curcumin like mitochondrial dysfunction, oxidative stress, mTOR pathway, TLR-4 receptors, MAPK pathway, and molecular chaperon dysfunction. The primary pharmacological benefit of curcumin is its anti-inflammatory effects because of which it is able to show its impact on several pathologies. This effect is due to its multifactorial nature of regulating several transcription factors, cytokines and enzymes associated with the NF-κB pathway [102]. It modulates a wide variety of inflammatory targets like TNF-α, COX-2, Wnt/β-catenin, 5-LOX, IL-6, IL-1, MMP-9, iNOS, and PPAR-ϒ [103, 104]. Various studies have revealed that curcumin exerts its anti-TNF-α effect and MMP-9 inhibitory effects through inhibition of the NF-κB and MAPK pathway [105, 106]. Oral administration of curcumin has been centrally neuroprotective [107]. Table 1 describes the physicochemical profile of curcumin.

5.2 Curcumin as Neurotherapeutic in ASD: Evidence from Preclinical Studies

There have not been many preclinical studies on the use of curcumin as a neurotherapeutic in ASD. Bhandari and Kuhad [108] have explored the neurotherapeutic potential of curcumin in autism spectrum disorders. 1M Propanoic acid (PPA) (4 μl) was infused over 10 min into the anterior portion of lateral ventricle to induce

Table 1 Physicochemical properties of curcumin

S. no.	Physicochemical properties	
1.	Molecular formula	$C_{21}H_{20}O_6$
2.	Molecular weight	368.385 g/mol
3.	Chemical name	Diferuloylmethane
4.	IUPAC name	(1E,6E)-1,7-bis(4-hydroxy-3-methoxyphenyl) hepta-1,6-diene-3,5-dione
5.	Color	Yellow-orange
6.	Physical description	Crystalline powder
7.	Solubility	Insoluble in water, ether; soluble in alcohol, glacial acetic acid. Freely soluble in ethanol, acetic acid
8.	Melting point	179–182 °C
9.	Log *P* (polarity)	Log K_{ow} = 3.29
10.	Stability	Should be stored in a tightly closed container in a dry and well-ventilated place at a temperature of −20 °C

autism-like behavior in adolescent rats. PPA is an enteric product of gut bacteria in individuals with autism. In order to observe how this enteric product affects behavior and causes biochemical and mitochondrial dysfunction by immune system activation and producing inflammatory cytokines, this model was validated in our lab. Oral administration of curcumin (50, 100, and 200 mg/kg) was carried out starting from the second day of induction and was continued up to the 28th day. Rats induced with ASD-like phenotype were tested for various neurobehavioral aspects such as reciprocal social interaction, stereotypy, locomotor activity, anxiety, novelty, depression, spatial learning and memory as well as for repetitive and pervasive behavior between the 7th and the 28th day. Additionally, estimation of biochemical parameters, activity of mitochondrial complexes, TNF-α and MMP-9 were also carried out.

The findings of this study suggested that MMP-9 release mediated mitochondrial dysfunction and release of pro-inflammatory cytokines. This was responsible for the development of the characteristic behavioral and biochemical phenotypic profile associated with autism. Curcumin (50, 100, and 200 mg/kg) could significantly and dose-dependently restore this pathological alteration as a result of its strong antioxidant, anti-inflammatory, anti-TNF-α, and anti-MMP-9 potential. Hence, curcumin can be utilized as a potential neurotherapeutic for ameliorating the neurobehavioral, biochemical, and molecular alterations occurring in ASD.

Al-Askar et al. [109] have investigated the beneficial effect of curcumin in ameliorating the neurodevelopmental brain deficit resulting in autism after exposure of mothers to valproic acid (VPA), an antiepileptic drug during the first trimester of pregnancy. The researchers used valproic acid rat model of autism in which rat fetuses were exposed to VPA (600 mg/kg, intraperitoneal injection) on the 12.5th day post conception. At 7 days from their birth, the animals were administered a single dose of curcumin (1 g/kg). It was observed that rats administered with VPA showed delay in maturation and a reduction in body and brain weight along with several signs of toxicity in brain. There was also a reduction in endogenous antioxidants like reduced glutathione and changes in neurotransmitters like depletion of serotonin and glutamine. There was an increase in oxidative stress parameters such as increase in lipid peroxidation, oxidized glutathione, IL-6, and excitatory neurotransmitters such as glutamate. Curcumin supplementation resulted in moderate correction of these endogenous dysfunctions and there was significant improvement in maturation delay and reduction in body weight.

Shu-juan et al. [110] also demonstrated the neurotherapeutic potential of curcumin in ameliorating autistic behavior and increase in levels of brain-derived neurotropic factor (BDNF) in sodium valproate rat model of autism by administering 600 mg/kg VPA, i.p. on 12.5 day after gestation. Curcumin was administered at a concentration of 10 g/l for 2 weeks to 35-day old rat pups. There was significant improvement in behaviors like social interaction and reduction in repetitive behavior. Curcumin significantly increased levels of BDNF in temporal cortex as well.

6 Resveratrol: Potential of Red Grape Constituent as Neurotherapeutic in ASD

6.1 *Resveratrol: Structure, Description, and Physicochemical Properties*

Resveratrol is a polyphenolic stilbenoid produced naturally by several plants when attacked by bacteria and fungi [111]. It is found in food sources such as grapes, nuts, and berries. It is multifactorial and interacts with various targets. It acts as a cyclooxygenase (COX) inhibitor, PPAR-α activator, eNOS inducer, and SIRT1 activator [50, 112, 113]. Resveratrol is an allosteric modulator of the regulatory target SIRT1. It enhances AMPK phosphorylation and decreases the oxidative damage occurring in F2 hybrid mice [114]. Literature reports have suggested that LPS-induced activation of NF-κB is suppressed by resveratrol in C6 microglia [115]. Resveratrol ameliorates the increased levels of MMP-9 induced as a result of cerebral ischemia-reperfusion injury in mice [116]. It also causes inhibition of levels of MMP-9 by upregulation of PPAR-α expression in an oxygen glucose deprivation-exposed neuron model [117]. Social deficits in the sociability tests evaluated in an animal model of ASD, induced using valproic acid, were prevented by resveratrol [118]. Resveratrol reduces MMP-9 levels and induces immune responses which make the brain resilient to the deposition of β-amyloid. Resveratrol slows down the cognitive deficit in Alzheimer's disease which may also arrest apoptosis of neurons [119]. It also reverses neuroinflammation caused by morphine by reversing the expression of HDAC1 [120]. Resveratrol modulates many calcium signaling pathways. Hence, resveratrol acts not only by immunomodulation but also by lowering hyperexcitability of the membrane [121]. Thus, resveratrol can be used as a therapeutic for various diseases such as neurodegeneration, autoimmune disorders, heart diseases, and cancer (Fig. 6). Table 2 describes the physicochemical profile of resveratrol.

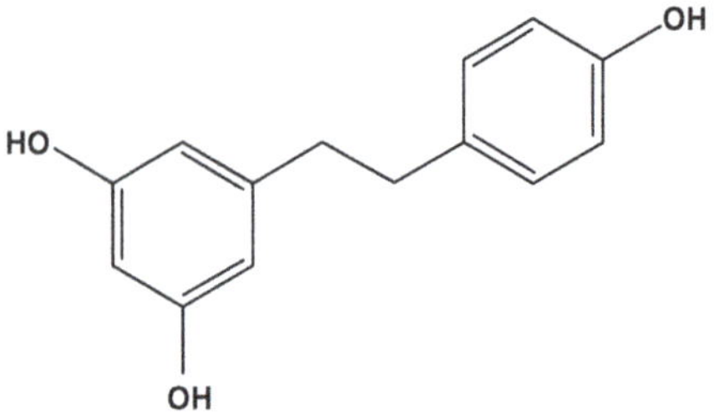

Fig. 6 Structure of resveratrol

Table 2 Physicochemical properties of resveratrol

S. no.	Physicochemical properties	
1.	Molecular formula	$C_{14}H_{12}O_3$
2.	Molecular weight	228.247 g/mol
3.	Chemical name	Trans-resveratrol; 3,4′,5-Trihydroxystilbene
4.	IUPAC name	5-[(E)-2-(hydroxyphenyl)ethenyl]benzene-1,3-diol
5.	Color	Off-white powder
6.	Physical description	Solid
7.	Melting point	254 °C
8.	Solubility	Solubility in water is 3 mg/100 ml; soluble in ethanol, DMSO and DMF, solubility in phosphate buffer saline (pH 7.4) is 100 μg/ml
9.	Log *P* (polarity)	Log $K_{o/w}$ = 3.10
10.	Stability	Should be stored in a tightly closed container in a dry and well-ventilated place at a temperature of −20 °C

6.2 *Resveratrol as Neurotherapeutic in ASD: Evidence from Preclinical Studies*

Fontes-Dutra et al. [122] have explored the neurotherapeutic potential of resveratrol in the valproic acid (VPA) animal model of autism. The primary aim of these researchers was to understand the neurodevelopmental deficit occurring as a result of exposure to valproic acid prenatally and whether resveratrol could be used as a probable intervention. Effects of resveratrol on sensory behavior were evaluated after induction of autism. Sensory brain regions were studied for their localization of GABAergic parvalbumin (PVC) neurons and expressions of excitatory and inhibitory synapses. Treatment with resveratrol (3.6 mg/kg) was done in pregnant dams from gestation day E6.5 to E18.5 and they were administered valproic acid (600 mg/kg) at E12.5. Behavioral parameters such as nest seeking (NS) behavior and behavior during the whisker nuisance task (WNT) were evaluated in male pups. Brain tissues were removed on postnatal day 30 and were analyzed for protein expression as well as localization of PVC neurons. Their results indicated that there is change in localization of PVC neurons in the sensory cortex as well as the amygdala. The treatment with RSV showed significant prevention of the alterations occurring after valproic acid exposure.

Scientific literature has documented that if pregnant women are exposed to progesterone in the form of oral contraceptive pills, in food or drink or any preterm birth drug, it can lead to development of autism in the child. Hence, in order to ameliorate autistic behavior induced as a result of prenatal/postnatal exposure to progestin, use of resveratrol was evaluated as plausible therapeutic intervention by Xie et al. [123]. Their results indicated that there was significant improvement in autistic behavior after oral administration as a result of activation of the ERβ pathway in the amygdala. The mechanism deciphered from their results indicated histone and DNA demethylation of the promotor region of ERβ resulting in the activation of the ERβ pathway. This results in reduction of oxidative stress, mitochondrial dysfunction, and lipid peroxidation in the brain leading to amelioration of autistic behavior.

The ameliorative potential of resveratrol on neuroinflammation was studied in rats induced with ASD-like phenotype using PPA by Bhandari and Kuhad [124]. Resveratrol was administered in doses (5, 10, and 15 mg/kg) starting from the 2nd day post-surgery and continued up to the 28th day. Rats were tested for various behavioral paradigms between the 7th and 28th day. Behavioral tests included tests for sociability, repetitive behavior, anxiety, depression, novel object recognition, and the Morris water maze test for perseverative behavior. Biochemical tests for oxidative stress, mitochondrial complexes, TNF-α, and MMP-9 were also assessed. The findings of this study suggested that MMP-9 activation resulted in mitochondrial dysfunction and inflammatory cytokine release leading to the development of phenotypic profile similar to that present in ASD. Resveratrol (5, 10, and 15 mg/kg) dose dependency restored characteristic neuropathological, behavioral, and mitochondrial dysfunction in PPA-administered rats induced with ASD. Therefore, resveratrol can be explored clinically as a neurotherapeutic agent for ameliorating the neurobehavioral, biochemical, and molecular alterations occurring in ASD.

7 Naringenin: Potential of Grapefruit Constituent as Neurotherapeutic in ASD

7.1 Naringenin: Structure, Description, and Physicochemical Properties

(±)-Naringenin (Fig. 7) is a flavanone abundantly present in grapefruit, oranges, and tomato skin [125, 126] (Fig. 8). Naringenin shows inhibition of CYP1A2 isoform of human cytochrome P450 metabolizing enzymes [127]. Naringenin exerts an antioxidant effect by reducing oxidative damage to DNA induced by exposure to

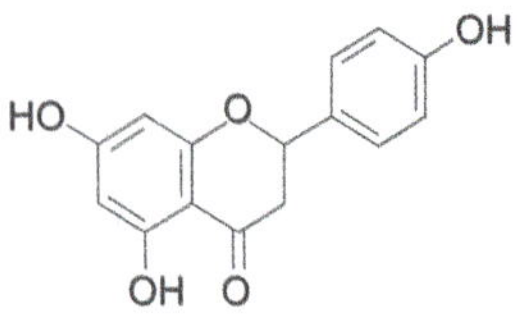

Fig. 7 Structure of naringenin

Fig. 8 Source of naringenin

radiations in mice as a result of inhibition of the NF-κB pathway [128]. It also shows an antihyperlipidemic effect by inhibiting very low-density lipoproteins (VLDL) secretion [129]. It also shows antidepressant potential in chronic unpredictable mild stress by BDNF signaling [130]. Naringenin has antiproliferative effects in breast, colon, and uterus cancer cells as a result of its ability to hamper cell proliferation by binding to estrogen receptors (ER) [131]. It has been observed to have antiestrogenic effects by the regulation of palmitoylation of estrogen receptor-α. It also showed beneficial effects in osteoporosis, cancer, and cardiovascular diseases [132]. Naringenin also plays a role in the suppression of neuroinflammation by triggering the suppression of cytokine signaling 3 expression (SOCS)-3 in glial cells [133]. It shows a neuroprotective effect in the middle cerebral artery occlusion (MCAO) model of ischemic stroke as a result of NF-κB pathway inactivation [134]. Table 3 describes the physicochemical profile of naringenin.

7.2 Naringenin as Neurotherapeutic in ASD: Evidence from Preclinical Studies

To this day, there have not been any studies done on the evaluation of neurotherapeutic efficacy of naringenin and its brain-targeted nanocarriers in ASD. Recently, Bhandari et al. [135] explored, in their study, the neurotherapeutic potential of naringenin, naringenin-loaded glutathione, and Tween-80 coated nanocarriers in ASD. The primary objective of the current study was to evaluate the neurotherapeutic potential of naringenin and its brain-targeted nanoformulation in an experimental paradigm of ASD. A 1M propanoic acid (PPA) (4 μl) was infused into the anterior portion of the lateral ventricle in Sprague Dawley rats to induce an ASD-like phenotype. Naringenin in doses of 25, 50, and 100 mg/kg, naringenin-loaded poly(lactic-co-glycolic acid)

Table 3 Physicochemical properties of naringenin

S. no.	Physicochemical properties	
1.	Molecular formula	$C_{15}H_{12}O_5$
2.	Molecular weight	272.256 g/mol
3.	Chemical name	(S)-5,7-Dihydroxy-2-(4-hydroxyphenyl)chroman-4-one, Naringenin
4.	IUPAC name	(2S)-5,7-Dihydroxy-2-(4-hydroxyphenyl)-2,3-dihydrochromen-4-one
5.	Color	Off-white powder
6.	Physical description	Crystalline solid
7.	Melting point	251 °C
8.	Solubility	Sparingly soluble in water and aqueous buffers; soluble in ethanol, DMSO, and DMF
9.	Log *P* (polarity)	Log $K_{o/w}$ = 2.52
10.	Stability	Should be stored in a tightly closed container in a dry and well-ventilated place at a temperature of −20 °C

(PLGA) nanoparticles (25 mg/kg), glutathione (GSH), Tween-80 coated naringenin nanoparticles (25 mg/kg), and minocycline (50 mg/kg) were given per-orally thrice daily (8 hourly) for 29 days. A battery of neurobehavioral tests and biochemical, blood–brain barrier permeability, TNF-α, MMP-9, HSP-70, and P-glycoprotein tests were performed at different points of time to study the autistic phenotype. The Pearson correlation was applied between various neurobehavioral tests and neuroinflammatory markers like TNF-α, MMP-9, HSP-70, and P-glycoprotein levels. The primary mechanism associated with the neuroinflammatory cascade was mitochondrial complex inhibition and generation of ROS as indicated by biochemical markers and mediated by MMP. There was also an increase in the plasma levels of circulating antibodies to heat shock protein 70, TNF-α, and an upregulation of efflux transporters such as P-glycoprotein (P-gp). Naringenin was effective in its unencapsulated form only at a higher dose of 100 mg/kg. In the unencapsulated form, it cannot cross BBB efficiently due to its low bioavailability and P-glycoprotein efflux. In the experiment, glutathione and Tween-80 coated naringenin nanocarriers served as multifactorial neurotherapeutic agents which restored neuropathology generated as a result of PPA administration, by circumventing low oral bioavailability of naringenin and enhancing its brain uptake at a low oral dose of 25 mg/kg. Therefore, these brain-targeted nanocarriers of naringenin can be utilized in clinics as a neurotherapeutic for ASD.

The main aim of the research group was to increase oral bioavailability of naringenin by developing naringenin-loaded poly(lactic-co-glycolic acid) (PLGA) nanocarriers and provide sustained drug release. The nanoprecipitation method was used to prepare NGN-PLGA nanocarriers that were coated with 1% polysorbate 80 or 1% glutathione to produce Tween-80-NGN-PLGA or GLU-NGN-PLGA nanoparticles. The morphology was examined by optical microscopy, florescent microscopy, field emission scanning emission microscopy (FE-SEM), and transmission electron microscopy (TEM). Particle size and zeta potential of the formulations were determined using photon correlation spectroscopy. Total drug content and encapsulation efficiency (EE) was determined using dialysis membrane, centrifugation method, and ultraviolet spectroscopy, respectively. In vitro release studies were performed using sample and separate method (SS). Fourier transform infrared spectroscopy (FT-IR), differential scanning calorimetry (DSC), powder X-ray diffraction studies (PXRD), and nuclear magnetic resonance (NMR) were carried out to confirm the encapsulation of drug into nanoparticles and the coating of nanocarriers with glutathione and Tween-80. Real-time and accelerated stability studies of both nanosuspension and lyophilized nanoparticles were carried out at 4 °C and 25 °C for 6 months. Other studies such as 3-(4,5-dimethylthiazol-2-yl)-2,5-diphenyl-tetrazolium bromide (MTT) assay, in vivo brain uptake using coumarin-6 loaded nanoparticles and confocal laser scanning microscope (CLSM) and single dose pharmacokinetics were performed to quantify the drug in the plasma and brain of rats.

It was found that both the coated as well as uncoated nanoparticles were spherical with particle size less than 200 nm, had high drug content and entrapment efficiency, and sustained-release profile with initial burst release and exhibited Korsmeyer–Peppas release kinetics with Fickian release mechanism. Optimized uncoated and coated NGN-PLGA nanoparticles were characterized using DSC, FT-IR, PXRD, and NMR to confirm that naringenin was encapsulated in polymeric nanoparticles in an amorphous form and the presence of coating over the nanoparticles was confirmed from the shift in characteristic peaks of glutathione and Tween-80 in FT-IR and NMR studies. These nanoparticles were stable at 4 °C in both nanosuspension and the lyophilized form for 6 months. But only the lyophilized form was stable at 25 °C as, after 6 months, nanosuspension showed increase in size and decrease in entrapment efficiency. The MTT cell viability assay using human liver cancer cell line (HepG2) showed that the PLGA nanoparticles were nontoxic. In vivo brain uptake studies were done by examining frozen brain sections after administration of coumarin-6 loaded coated and uncoated nanoparticles. They indicated enhanced uptake for glutathione and Tween-80 coated nanocarriers in comparison to uncoated naringenin nanocarriers. Further, single-dose pharmacokinetic studies showed that there was a significant improvement in the relative bioavailability of naringenin from uncoated and coated naringenin nanoparticles. This was in comparison to naringenin suspension as well as enhanced brain uptake after coating with both glutathione and Tween-80 as indicated by the enhanced brain-to-plasma ratio. Hence, this work has showed that PLGA nanoparticles are an efficient delivery system not only for improving oral bioavailability of poorly water-soluble drugs such as naringenin but also for enhancement of their brain uptake by coating with ligands such as glutathione (GSH) and Tween-80. They do so either by inhibiting P-gp efflux, receptor-mediated endocytosis, or the presence of glutathione transporters at BBB. This is part of the author's unpublished data.

8 Sulforaphane: Potential of Broccoli Constituent as Neurotherapeutic in ASD

8.1 Sulforaphane: Structure, Description, and Physicochemical Properties

Sulforaphane is chemically known as 1-isothiocyanato-4-(methylsulfinyl) butane and this phytochemical belongs to the isothiocyanate group. Sulforaphane is produced when glucoraphanin is metabolized by the catalytic action of myrosinase. Glucoraphanin is a precursor of sulforaphane and is present in vegetables of the cruciferous family like broccoli and cauliflower. Isolated by Dr. Paul Talalay and Yuesheng Zhang, sulforaphane has been known to show beneficial effect in the attenuation of oxidative stress and mitochondrial dysfunction [136, 137]. It is a neuroprotective compound and protects hippocampal neurons from apoptotic death by free radical generation and oxidative stress [90, 138]. It also has antidiabetic [139] and anticarcinogenic effects [140] and helps in the reduction of infarct volume after ischemic stroke [141]. It acts by action of both the Nrf2-dependent and the independent pathways. It is known to activate Nrf2 response element in astrocytes [142]. It also upregulates heat shock protein 27 [143]. Table 4 describes the physicochemical properties of sulforaphane (Fig. 9).

Table 4 Physicochemical properties of sulforaphane

S. no.	Physicochemical properties	
1.	Molecular formula	$C_6H_{11}NOS_2$
2.	Molecular weight	177.28 g/mol
3.	Chemical name	Sulforaphane
4.	IUPAC name	1-Isothiocyanato-4-methylsulfinylbutane
5.	Color	Off-white powder
6.	Physical description	Solid
7.	Melting point	68.93 °C
8.	Solubility	Soluble in DMSO (>5 mg/ml), 100% ethanol, methanol, chloroform, and ethyl acetate.
9.	Log *P* (polarity)	Log $K_{o/w}$ = 1.8
10.	Stability	Should be stored in a tightly closed container in a dry and well-ventilated place at a temperature of −20 °C

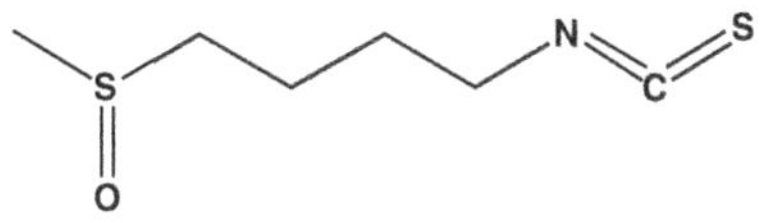

Fig. 9 Structure of sulforaphane

8.2 *Sulforaphane as Neurotherapeutic in ASD: Evidence from Clinical Studies*

Singh et al. [144] evaluated the neurotherapeutic action of sulforaphane in young men between 13 and 27 years of age, suffering from moderate to severe ASD. This was a randomized double-blind and placebo-controlled clinical trial. Patients received sulforaphane at a dose of 50–150 μmol/day for 18 weeks and was followed by a 4-week period of drug holiday. There were 29 patients of ASD and 15 subjects in a placebo control group. They were assessed for their behavior utilizing behavioral rating scales after 18 weeks of treatment. Those receiving placebo showed negligible change while there was significant reduction in autistic behavior after treatment with sulforaphane for 18 weeks. This was indicated by improvement in behavioral scores assessment using behavioral rating scales such as Aberrant Behavior Checklist (ABC), Social Responsiveness scale (SRS), and Clinical Global Impression Improvement Scale (CGI-I). It was observed that sulforaphane treatment resulted in improvement in social interaction ability and common deficit. Hence, sulforaphane present in broccoli can reduce oxidative stress, neuroinflammation, and DNA damage.

Bent et al. [145] conducted an open label study (NCT02654743) to understand the antioxidant mechanism of sulforaphane in improving social interaction and communication deficit in children suffering from autism. The urine of autistic children contains specific metabolites which indicate biochemical and mitochondrial dysfunction occurring as a result of gut dysbiosis and neuroinflammation. Bent and his team of researchers wished to unearth the potential of sulforaphane in changing the metabolites excreted through urine. They enrolled school children suffering from ASD in a 12-week study. Fasting urine samples were collected and behavior was assessed using behavior scales before starting treatment and after it ended.

9 Bioavailability Issues of Curcumin, Naringenin, Resveratrol, and Sulforaphane and the Encapsulation of These in Novel Brain-Targeted Delivery Systems: *Potential and Use in ASD*

Dietary phytochemicals have low oral bioavailability. Their bioavailability is dependent on their chemical structures and the dietary form in which they are taken. They undergo first-pass metabolism resulting in low oral bioavailability [146, 147].

All four drugs discussed in this chapter have low oral bioavailability. Though all of these are absorbed well, they undergo first-pass metabolism and hence require higher dose to show effects. There are several documented research reports on novel drug delivery systems developed for curcumin [101, 148–151], resveratrol [152], naringenin [153–158], and sulforaphane [159]. These research reports indicate the increase in oral bioavailability of these drugs after encapsulation in novel drug delivery systems such as PLGA-based nanoparticles, lipid nanoparticles or liposomes or any other method to increase bioavailability. Till date, these systems have been developed and evaluated for their use in several neurodegenerative diseases such as Alzheimer's, stroke, Parkinson's, depression as well as in cancer. But there has been no documented literature regarding their use as a targeted delivery system for ASD. In our studies, we have developed glutathione and Tween-80 coated naringenin-encapsulated PLGA-based nanocarriers and evaluated their potential in ASD [135]. These ligand-coated nanocarriers gave promising results through enhanced brain delivery and improvement in bioavailability as compared to unencapsulated naringenin. There was a reduction in dose, i.e., similar to the effect observed among 25 mg/kg utilizing nanocarriers as observed at 100 mg/kg dose of unencapsulated naringenin. The P-gp efflux, enhanced as a result of neuroinflammation, was mitigated by both glutathione and Tween-80 coated nanocarriers. Thus, these dietary phytochemicals are a safer alternative for autistic patients and their potential as a neurotherapeutic can be enhanced if we develop novel brain-targeted delivery systems. It will not only overcome the low bioavailability issues but also improve patient compliance by providing sustained action, thus requiring only a single dose.

10 Future Prospects of Use of Dietary Phytochemicals as Potential Neurotherapeutic in ASD

Immune system deregulation, neuroinflammation, environmental toxicant exposures, oxidative stress, mitochondrial dysfunction and gastrointestinal complications are physiological comorbidities occurring in individuals with autism spectrum disorders (ASD) [18, 19, 21, 69] and can worsen behavioral complications. Gastrointestinal complications can lead to immune system dysregulation and result in generation of oxidative stress due to generation of short-chain fatty acids by abnormal gut flora [29]. This leads to synthesis of various pro-inflammatory cytokines and chemokines causing activation of microglia. There is an unmet need to develop such neuro-psychopharmaco-therapeutic interventions for ASD that are safer alternatives in comparison to the existing drugs which are only meant to mitigate associated symptoms of ASD such as anxiety, depression, ADHD, and aggression. To this day, there are no therapeutic interventions which target the physiological comorbidities associated with ASD, especially gastrointestinal complications and immune system dysregulation. Phytochemicals can act as a useful neurotherapeutic for the attenuation of oxidative stress that initiates mitochondrial dysfunction in ASD. It is clear that curcumin, resveratrol, naringenin, and sulforaphane have shown positive results in preclinical studies. Currently, clinical studies on sulforaphane and Enhansa® (enhanced absorption curcumin) have shown positive results in ameliorating behavior and biochemical alterations occurring among these individuals [160]. We are now required to explore more phytochemicals for their neurotherapeutic potential, both preclinically and clinically. These phytochemicals also have low oral bioavailability [161–163].

Blood–brain barrier (BBB) permeability is another important factor which needs to be taken into consideration as it prevents the penetration of various neurotherapeutics [164, 165]. There are several ABC efflux transporters across the BBB such as the P-gp, encoded by the ABCB1 gene, is responsible for regulating the uptake and efflux of drugs across the BBB [166, 167]. Thus, we need to develop such brain-targeted neurotherapeutics utilizing dietary phytochemicals as they hold enormous potential to mitigate various physiological comorbidities associated with autism spectrum disorders. Developing such brain-targeted delivery systems will not only enhance their therapeutic potential further, but also show sustained action at a lower dose resulting in patient compliance.

11 Conclusions

Autism spectrum disorder is a disorder with multifactorial origins. There is an involvement of complex genetic mutations and epigenetic changes. It is a complex disorder involving behavioral complications and comorbidities such as anxiety, epilepsy, gastrointestinal complications causing gut dysbiosis, oxidative stress, mitochondrial dysfunction, and biochemical alterations. The physiological comorbidities have the capacity to worsen behavioral and biochemical complications. Till date, no neurotherapeutic agent that can ameliorate the physiological comorbidities like gastrointestinal complications, oxidative stress, mitochondrial dysfunction, and neuroinflammation exist. Dietary phytochemicals provide a new source of hope among all the uncertainty surrounding this disorder which currently offers very limited agents for therapeutic intervention. The experimental studies discussed in this chapter indicate that dietary phytochemicals such as curcumin, resveratrol, naringenin, and sulforaphane hold enormous potential as neurotherapeutic agents for autism spectrum disorder. With high healthcare costs and burden on the caregivers of ASD patients, these represent a safe and inexpensive approach to mitigate oxidative stress, mitochondrial dysfunction, neuroinflammation occurring as a result of gut dysbiosis in autistic patients which influence and worsen behavior. Hence, the need of the hour is to develop brain-targeted delivery systems for these dietary phytochemicals and explore their potential clinically.

Acknowledgments Research grants sanctioned by SERB, Department of Science and Technology (grant no SB/FT/LS-284/2012), All India Council of Technical Education (11-25/RIFD/CAYT/POL-II/2013-14) and University Grants Commission [20-29(12)/2012(BSR)], New Delhi to Dr. Anurag Kuhad are gratefully acknowledged. Senior Research Fellowship sanctioned by the Indian Council of Medical Research, New Delhi to Ms. Ranjana Bhandari is also gratefully acknowledged.

Conflict of Interest The authors declare no conflicts of interest.

References

1. American Psychiatric Association. (2013). *Diagnostic and statistical manual of mental disorders*. Author.
2. Estabillo, J. A., Matson, J. L., & Cervantes, P. E. (2018). Autism symptoms and problem behaviors in children with and without developmental regression. *Journal of Developmental and Physical Disabilities, 30*, 17–26. https://doi.org/10.1007/s10882-017-9573-x
3. Frye, R. E., & Rossignol, D. A. (2016). Identification and treatment of pathophysiological comorbidities of autism spectrum disorder to achieve optimal outcomes. *Clinical Medicine Insights. Pediatrics, 10*, 43–56. https://doi.org/10.4137/CMPed.S38337
4. Helverschou, S. B., Bakken, T. L., & Martinsen, H. (2011). Psychiatric disorders in people with autism spectrum disorders: Phenomenology and recognition. In *International handbook of autism and pervasive developmental disorders* (pp. 53–74). New York: Springer.
5. WHO. (2017). *Autism spectrum disorders*. Author.
6. Fombonne, E. (2005). Epidemiology of autistic disorder and other pervasive developmental disorders. *The Journal of Clinical Psychiatry, 66*(Suppl 10), 3–8.
7. Holt, R., & Monaco, A. P. (2011). Links between genetics and pathophysiology in the autism spectrum disorders. *EMBO Molecular Medicine, 3*, 438–450. https://doi.org/10.1002/emmm.201100157
8. Santangelo, S. L., & Tsatsanis, K. (2005). What is known about autism: Genes, brain, and behavior. *American Journal of Pharmacogenomics, 5*, 71–92.
9. Werling, D. M., & Geschwind, D. H. (2013). Sex differences in autism spectrum disorders. *Current Opinion in Neurology, 26*, 146–153. https://doi.org/10.1097/WCO.0b013e32835ee548
10. Ciernia, A. V., Laufer, B. I., Dunaway K. W., et al. (2018). Epigenomic convergence of genetic and immune risk factors in autism brain. bioRxiv 270827. https://doi.org/10.1101/270827
11. Jones, E. J. H., Gliga, T., Bedford, R., et al. (2014). Developmental pathways to autism: A review of prospective studies of infants at risk. *Neuroscience and Biobehavioral Reviews, 39*, 1–33. https://doi.org/10.1016/j.neubiorev.2013.12.001
12. LaSalle, J. M. (2013). Epigenomic strategies at the interface of genetic and environmental risk factors for autism. *Journal of Human Genetics, 58*, 396–401. https://doi.org/10.1038/jhg.2013.49
13. Modabbernia, A., Velthorst, E., & Reichenberg, A. (2017). Environmental risk factors for autism: An evidence-based review of systematic reviews and meta-analyses. *Molecular Autism, 8*, 13. https://doi.org/10.1186/s13229-017-0121-4
14. Benvenuto, A., Battan, B., Porfirio, M. C., & Curatolo, P. (2013). Pharmacotherapy of autism spectrum disorders. *Brain & Development, 35*, 119–127. https://doi.org/10.1016/j.braindev.2012.03.015
15. Young, N. J., & Findling, R. L. (2015). An update on pharmacotherapy for autism spectrum disorder in children and adolescents. *Current Opinion in Psychiatry, 28*, 91–101. https://doi.org/10.1097/yco.0000000000000132
16. Ichikawa, H., Mikami, K., Okada, T., et al. (2017). Aripiprazole in the treatment of irritability in children and adolescents with autism spectrum disorder in Japan: A randomized, double-blind, placebo-controlled study. *Child Psychiatry and Human Development, 48*, 796–806. https://doi.org/10.1007/s10578-016-0704-x
17. Maneeton, N., Maneeton, B., Putthisri, S., et al. (2017). Risperidone versus placebo in the treatment of children and adolescents with autism spectrum disorders: A meta-analysis and systematic review. *European Neuropsychopharmacology, 27*, S1107–S1108. https://doi.org/10.1016/S0924-977X(17)31922-3
18. Kerin, T., Volk, H., Li, W., et al. (2018). Association between air pollution exposure, cognitive and adaptive function, and ASD severity among children with autism spectrum disorder. *Journal of Autism and Developmental Disorders, 48*, 137–150. https://doi.org/10.1007/s10803-017-3304-0

19. Nadeem, A., Ahmad, S. F., Attia, S. M., et al. (2018). Activation of IL-17 receptor leads to increased oxidative inflammation in peripheral monocytes of autistic children. *Brain, Behavior, and Immunity, 67*, 335–344. https://doi.org/10.1016/J.BBI.2017.09.010
20. Rossignol, D. A., & Frye, R. E. (2012). A review of research trends in physiological abnormalities in autism spectrum disorders: Immune dysregulation, inflammation, oxidative stress, mitochondrial dysfunction and environmental toxicant exposures. *Molecular Psychiatry, 17*, 389–401. https://doi.org/10.1038/mp.2011.165
21. Rossignol, D. A., & Frye, R. E. (2014). Evidence linking oxidative stress, mitochondrial dysfunction, and inflammation in the brain of individuals with autism. *Frontiers in Physiology, 5*, 150. https://doi.org/10.3389/fphys.2014.00150
22. Chaidez, V., Hansen, R. L., & Hertz-Picciotto, I. (2014). Gastrointestinal problems in children with autism, developmental delays or typical development. *Journal of Autism and Developmental Disorders, 44*, 1117–1127. https://doi.org/10.1007/s10803-013-1973-x
23. Horvath, K., Papadimitriou, J. C., Rabsztyn, A., et al. (1999). Gastrointestinal abnormalities in children with autistic disorder. *The Journal of Pediatrics, 135*, 559–563.
24. Vuong, H. E., & Hsiao, E. Y. (2017). Emerging roles for the gut microbiome in autism spectrum disorder. *Biological Psychiatry, 81*, 411–423. https://doi.org/10.1016/j.biopsych.2016.08.024
25. Song, Y., Liu, C., & Finegold, S. M. (2004). Real-time PCR quantitation of clostridia in feces of autistic children. *Applied and Environmental Microbiology, 70*(11), 6459–6465.
26. Finegold, S. M., Molitoris, D., Song, Y., et al. (2002). Gastrointestinal microflora studies in late-onset autism. *Clinical Infectious Diseases, 35*, S6–S16. https://doi.org/10.1086/341914
27. Wang, L., Christophersen, C. T., Sorich, M. J., et al. (2013). Increased abundance of Sutterella spp. and Ruminococcus torques in feces of children with autism spectrum disorder. *Molecular Autism, 4*, 42. https://doi.org/10.1186/2040-2392-4-42
28. Luna, R. A., Oezguen, N., Balderas, M., et al. (2017). Distinct microbiome-neuroimmune signatures correlate with functional abdominal pain in children with autism Spectrum disorder. *Cellular and Molecular Gastroenterology and Hepatology, 3*, 218–230. https://doi.org/10.1016/J.JCMGH.2016.11.008
29. Toh, M. C., & Allen-Vercoe, E. (2015). The human gut microbiota with reference to autism spectrum disorder: Considering the whole as more than a sum of its parts. *Microbial Ecology in Health and Disease, 26*. https://doi.org/10.3402/mehd.v26.26309
30. Haskå, L., Andersson, R., & Nyman, M. (2011). The effect of dietary fiber from wheat processing streams on the formation of carboxylic acids and microbiota in the hindgut of rats. *Journal of Agricultural and Food Chemistry, 59*, 3406–3413. https://doi.org/10.1021/jf104380f
31. Jyonouchi, H., Sun, S., & Itokazu, N. (2002). Innate immunity associated with inflammatory responses and cytokine production against common dietary proteins in patients with autism spectrum disorder. *Neuropsychobiology, 46*, 76–84.
32. Macfabe, D. F. (2015). Enteric short-chain fatty acids: Microbial messengers of metabolism, mitochondria, and mind: Implications in autism spectrum disorders. *Microbial Ecology in Health and Disease, 26*, 28177. https://doi.org/10.3402/mehd.v26.28177
33. Morris, G., Berk, M., Carvalho, A., et al. (2017). The role of the microbial metabolites including tryptophan catabolites and short chain fatty acids in the pathophysiology of immune-inflammatory and neuroimmune disease. *Molecular Neurobiology, 54*, 4432–4451. https://doi.org/10.1007/s12035-016-0004-2
34. Strasser, L. (2016). The effects of propionic acid on locomotor, repetitive and anxiety-related behaviors in female adolescent rats. *Western Undergraduate Psychology Journal, 4*.
35. Karuri, A. R., Dobrowsky, E., & Tannock, I. F. (1993). Selective cellular acidification and toxicity of weak organic acids in an acidic microenvironment. *British Journal of Cancer, 68*, 1080–1087.
36. DeCastro, M., Nankova, B. B., Shah, P., et al. (2005). Short chain fatty acids regulate tyrosine hydroxylase gene expression through a cAMP-dependent signaling pathway. *Brain Research. Molecular Brain Research, 142*, 28–38. https://doi.org/10.1016/j.molbrainres.2005.09.002

37. Cannizzaro, C., Monastero, R., Vacca, M., & Martire, M. (2003). [3H]-DA release evoked by low pH medium and internal H+ accumulation in rat hypothalamic synaptosomes: Involvement of calcium ions. *Neurochemistry International, 43*, 9–17.
38. El-Ansary, A. K., Ben Bacha, A., & Kotb, M. (2012). Etiology of autistic features: The persisting neurotoxic effects of propionic acid. *Journal of Neuroinflammation, 9*, 661. https://doi.org/10.1186/1742-2094-9-74
39. Sziray, N., Leveleki, C., Levay, G., et al. (2007). Mechanisms underlying the long-term behavioral effects of traumatic experience in rats: The role of serotonin/noradrenaline balance and NMDA receptors. *Brain Research Bulletin, 71*, 376–385. https://doi.org/10.1016/j.brainresbull.2006.10.006
40. Choi, J., Lee, S., Won, J., et al. (2018). Pathophysiological and neurobehavioral characteristics of a propionic acid-mediated autism-like rat model. *PLoS One, 13*, e0192925. https://doi.org/10.1371/journal.pone.0192925
41. Wajner, M., Latini, A., Wyse, A. T. S., & Dutra-Filho, C. S. (2004). The role of oxidative damage in the neuropathology of organic acidurias: Insights from animal studies. *Journal of Inherited Metabolic Disease, 27*, 427–448. https://doi.org/10.1023/B:BOLI.0000037353.13085.e2
42. Davinelli, S., Maes, M., Corbi, G., et al. (2016). Dietary phytochemicals and neuro-inflammaging: From mechanistic insights to translational challenges. *Immunity & Ageing, 13*, 16. https://doi.org/10.1186/s12979-016-0070-3
43. Lee, J., Jo, D.-G., Park, D., et al. (2014). Adaptive cellular stress pathways as therapeutic targets of dietary phytochemicals: Focus on the nervous system. *Pharmacological Reviews, 66*, 815–868. https://doi.org/10.1124/pr.113.007757
44. Sandberg, M., Patil, J., D'Angelo, B., et al. (2014). NRF2-regulation in brain health and disease: Implication of cerebral inflammation. *Neuropharmacology, 79*, 298–306. https://doi.org/10.1016/j.neuropharm.2013.11.004
45. O'Neill, L. A., & Kaltschmidt, C. (1997). NF-kappa B: A crucial transcription factor for glial and neuronal cell function. *Trends in Neurosciences, 20*, 252–258.
46. Snow, W. M., Stoesz, B. M., Kelly, D. M., & Albensi, B. C. (2014). Roles for NF-κB and gene targets of NF-κB in synaptic plasticity, memory, and navigation. *Molecular Neurobiology, 49*, 757–770. https://doi.org/10.1007/s12035-013-8555-y
47. Chen, H., & Liu, R. H. (2018). Potential mechanisms of action of dietary phytochemicals for cancer prevention by targeting cellular signaling transduction pathways. *Journal of Agricultural and Food Chemistry, 66*(13), 3260–3276. https://doi.org/10.1021/acs.jafc.7b04975
48. Duvarci, S., Nader, K., & LeDoux, J. E. (2005). Activation of extracellular signal-regulated kinase- mitogen-activated protein kinase cascade in the amygdala is required for memory reconsolidation of auditory fear conditioning. *The European Journal of Neuroscience, 21*, 283–289. https://doi.org/10.1111/j.1460-9568.2004.03824.x
49. Hori, Y. S., Kuno, A., Hosoda, R., & Horio, Y. (2013). Regulation of FOXOs and p53 by SIRT1 modulators under oxidative stress. *PLoS One, 8*, e73875. https://doi.org/10.1371/journal.pone.0073875
50. Xiong, S., Salazar, G., Patrushev, N., & Alexander, R. W. (2011). FoxO1 mediates an autofeedback loop regulating SIRT1 expression. *The Journal of Biological Chemistry, 286*, 5289–5299. https://doi.org/10.1074/jbc.M110.163667
51. Kumar, G. P., & Khanum, F. (2012). Neuroprotective potential of phytochemicals. *Pharmacognosy Reviews, 6*, 81–90. https://doi.org/10.4103/0973-7847.99898
52. Frick, L. R., Williams, K., & Pittenger, C. (2013). Microglial dysregulation in psychiatric disease. *Clinical & Developmental Immunology, 2013*, 608654. https://doi.org/10.1155/2013/608654
53. Vargas, D. L., Nascimbene, C., Krishnan, C., et al. (2005). Neuroglial activation and neuroinflammation in the brain of patients with autism. *Annals of Neurology, 57*, 67–81. https://doi.org/10.1002/ana.20315

54. Bjorkland, G., Saad, K., Chirumbolo, S., et al. (2016). Immune dysfunction and neuroinflammation in autism spectrum disorder. *Acta Neurobiologiae Experimentalis, 76*, 257–268. https://doi.org/10.5772/22318
55. Croonenberghs, J., Bosmans, E., Deboutte, D., et al. (2002). Activation of the inflammatory response system in autism. *Neuropsychobiology, 45*, 1–6.
56. Onore, C. E., Nordahl, C. W., Young, G. S., et al. (2012). Levels of soluble platelet endothelial cell adhesion molecule-1 and P-selectin are decreased in children with autism spectrum disorder. *Biological Psychiatry, 72*, 1020–1025. https://doi.org/10.1016/j.biopsych.2012.05.004
57. Chez, M. G., Dowling, T., Patel, P. B., et al. (2007). Elevation of tumor necrosis factor-alpha in cerebrospinal fluid of autistic children. *Pediatric Neurology, 36*, 361–365. https://doi.org/10.1016/j.pediatrneurol.2007.01.012
58. Connolly, A. M., Chez, M., Streif, E. M., et al. (2006). Brain-derived neurotrophic factor and autoantibodies to neural antigens in sera of children with autistic spectrum disorders, Landau-Kleffner syndrome, and epilepsy. *Biological Psychiatry, 59*, 354–363. https://doi.org/10.1016/j.biopsych.2005.07.004
59. Nakagawa, Y., & Chiba, K. (2016). Minireviews involvement of neuroinflammation during brain development in social cognitive deficits in autism spectrum disorder and schizophrenia. *Journal of Pharmacology and Experimental Therapeutics, 358*(3), 504–515.
60. Abdallah, M. W., & Michel, T. M. (2013). Matrix metalloproteinases in autism spectrum disorders. *Journal of Molecular Psychiatry, 1*, 16. https://doi.org/10.1186/2049-9256-1-16
61. Gottfried, C., & Savino, W. (2015). The impact of neuroimmune alterations in autism spectrum disorder. *Frontiers in Psychiatry, 6*, 1–16. https://doi.org/10.3389/fpsyt.2015.00121
62. Goines, P. E., & Ashwood, P. (2013). Cytokine dysregulation in autism spectrum disorders (ASD): Possible role of the environment. *Neurotoxicology and Teratology, 36*, 67–81. https://doi.org/10.1016/j.ntt.2012.07.006.Cytokine
63. Masi, A., Glozier, N., Dale, R., & Guastella, A. J. (2017). The immune system, cytokines, and biomarkers in autism spectrum disorder. *Neuroscience Bulletin, 33*, 194–204. https://doi.org/10.1007/s12264-017-0103-8
64. Chauhan, A., & Chauhan, V. (2006). Oxidative stress in autism. *Pathophysiology, 13*, 171–181. https://doi.org/10.1016/j.pathophys.2006.05.007
65. Chauhan, A., Chauhan, V., Brown, W. T., & Cohen, I. (2004). Oxidative stress in autism: Increased lipid peroxidation and reduced serum levels of ceruloplasmin and transferrin--the antioxidant proteins. *Life Sciences, 75*, 2539–2549. https://doi.org/10.1016/j.lfs.2004.04.038
66. Meguid, N. A., Azab, S. N., Saber, A. S., et al. (2016). Impact of oxidative stress on autism spectrum disorder behaviors in children with autism. *International Journal of Pharmaceutical and Clinical Research, 8*, 193–197.
67. González-fraguela, M. E., Hung, M. D., Vera, H., et al. (2013). Oxidative stress markers in children with autism spectrum disorders. *British Journal of Medicine and Medical Research, 3*, 307–317.
68. Rose, S., Frye, R. E., Slattery, J., et al. (2015). Oxidative stress induces mitochondrial dysfunction in a subset of autistic lymphoblastoid cell lines. *Translational Psychiatry, 5*, e526. https://doi.org/10.1038/tp.2015.29
69. Rossignol, D. A., & Frye, R. E. (2012). Mitochondrial dysfunction in autism spectrum disorders: A systematic review and meta-analysis. *Molecular Psychiatry, 17*, 290–314. https://doi.org/10.1038/mp.2010.136
70. Skladal, D., Halliday, J., & Thorburn, D. R. (2003). Minimum birth prevalence of mitochondrial respiratory chain disorders in children. *Brain, 126*, 1905–1912.
71. Palmieri, L., & Persico, A. M. (1797). Mitochondrial dysfunction in autism spectrum disorders: Cause or effect? *Biochimica et Biophysica Acta*, 1130–1137. https://doi.org/10.1016/j.bbabio.2010.04.018
72. Blaha, C. D., Blatt, G. J., Chauhan, A., et al. (2013). Consensus paper: Pathological role of the cerebellum in autism. *Cerebellum, 11*, 777–807. https://doi.org/10.1007/s12311-012-0355-9

73. Geraci, F., Turturici, G., & Sconzo, G. (2011). Hsp70 and its molecular role in nervous system diseases. *Biochemistry Research International, 2011*, 618127. https://doi.org/10.1155/2011/618127
74. Aldbass, A. M., Bhat, R. S., & El-Ansary, A. (2013). Protective and therapeutic potency of N-acetyl-cysteine on propionic acid-induced biochemical autistic features in rats. *Journal of Neuroinflammation, 10*, 42. https://doi.org/10.1186/1742-2094-10-42
75. MacFabe, D. F., Cain, D. P., Rodriguez-Capote, K., et al. (2007). Neurobiological effects of intraventricular propionic acid in rats: Possible role of short chain fatty acids on the pathogenesis and characteristics of autism spectrum disorders. *Behavioural Brain Research, 176*, 149–169. https://doi.org/10.1016/j.bbr.2006.07.025
76. Shultz, S. R., Macfabe, D. F., Martin, S., et al. (2009). Intracerebroventricular injections of the enteric bacterial metabolic product propionic acid impair cognition and sensorimotor ability in the Long-Evans rat: Further development of a rodent model of autism. *Behavioural Brain Research, 200*, 33–41. https://doi.org/10.1016/j.bbr.2008.12.023
77. MacFabe, D. F., Rodríguez-Capote, K., Hoffman, J. E., et al. (2008). A novel rodent model of autism: Intraventricular infusions of propionic acid increase locomotor activity and induce neuroinflammation and oxidative stress in discrete regions of adult rat brain. *American Journal of Biochemistry and Biotechnology, 4*, 146–166. https://doi.org/10.3844/ajbbsp.2008.146.166
78. Frye, R. E., Melnyk, S., & Macfabe, D. F. (2013). Unique acyl-carnitine profiles are potential biomarkers for acquired mitochondrial disease in autism spectrum disorder. *Translational Psychiatry, 3*, e220. https://doi.org/10.1038/tp.2012.143
79. Frye, R. E., Rose, S., Chacko, J., et al. (2016). Modulation of mitochondrial function by the microbiome metabolite propionic acid in autism and control cell lines. *Translational Psychiatry, 6*, e927. https://doi.org/10.1038/tp.2016.189
80. Macfabe, D. F. (2012). Short-chain fatty acid fermentation products of the gut microbiome: Implications in autism spectrum disorders. *Microbial Ecology in Health and Disease, 23*, 19260. https://doi.org/10.3402/mehd.v23i0.19260
81. Ramamoorthy, H., Abraham, P., & Isaac, B. (2014). Mitochondrial dysfunction and electron transport chain complex defect in a rat model of tenofovir disoproxil fumarate nephrotoxicity. *Journal of Biochemical and Molecular Toxicology, 28*, 246–255. https://doi.org/10.1002/jbt
82. Stefanson, A. L., & Bakovic, M. (2014). Dietary regulation of Keap1/Nrf2/ARE pathway: Focus on plant-derived compounds and trace minerals. *Nutrients, 6*, 3777–3801. https://doi.org/10.3390/nu6093777
83. Wardyn, J. D., Ponsford, A. H., & Sanderson, C. M. (2015). Dissecting molecular cross-talk between Nrf2 and NF-κB response pathways. *Biochemical Society Transactions, 43*, 621–626. https://doi.org/10.1042/BST20150014
84. Yu, M., Li, H., Liu, Q., et al. (2011). Nuclear factor p65 interacts with Keap1 to repress the Nrf2-ARE pathway. *Cellular Signalling, 23*, 883–892. https://doi.org/10.1016/j.cellsig.2011.01.014
85. González-Reyes, S., Guzmán-Beltrán, S., Medina-Campos, O. N., & Pedraza-Chaverri, J. (2013). Curcumin pretreatment induces Nrf2 and an antioxidant response and prevents hemin-induced toxicity in primary cultures of cerebellar granule neurons of rats. *Oxidative Medicine and Cellular Longevity, 2013*, 801418. https://doi.org/10.1155/2013/801418
86. Yang, C., Zhang, X., Fan, H., & Liu, Y. (2009). Curcumin upregulates transcription factor Nrf2, HO-1 expression and protects rat brains against focal ischemia. *Brain Research, 1282*, 133–141. https://doi.org/10.1016/j.brainres.2009.05.009
87. Cheng, L., Jin, Z., Zhao, R., et al. (2015). Resveratrol attenuates inflammation and oxidative stress induced by myocardial ischemia-reperfusion injury: Role of Nrf2/ARE pathway. *International Journal of Clinical and Experimental Medicine, 8*, 10420–10428. https://doi.org/10.1016/j.jbtep.2015.11.003
88. Lou, H., Jing, X., Wei, X., et al. (2014). Naringenin protects against 6-OHDA-induced neurotoxicity via activation of the Nrf2/ARE signaling pathway. *Neuropharmacology, 79*, 380–388. https://doi.org/10.1016/j.neuropharm.2013.11.026

89. Dinkova-Kostova, A. T., Fahey, J. W., Kostov, R. V., & Kensler, T. W. (2017). KEAP1 and done? Targeting the NRF2 pathway with sulforaphane. *Trends in Food Science and Technology, 69*, 257–269. https://doi.org/10.1016/J.TIFS.2017.02.002
90. Holloway, P. M., Gillespie, S., Becker, F., et al. (2016). Sulforaphane induces neurovascular protection against a systemic inflammatory challenge via both Nrf2-dependent and independent pathways. *Vascular Pharmacology, 85*, 29–38. https://doi.org/10.1016/j.vph.2016.07.004
91. Lin, W., Wu, R. T., Wu, T., et al. (2008). Sulforaphane suppressed LPS-induced inflammation in mouse peritoneal macrophages through Nrf2 dependent pathway. *Biochemical Pharmacology, 76*, 967–973. https://doi.org/10.1016/j.bcp.2008.07.036
92. Kim, J. M., Araki, S., Kim, D. J., et al. (1998). Chemopreventive effects of carotenoids and curcumins on mouse colon carcinogenesis after 1,2-dimethylhydrazine initiation. *Carcinogenesis, 19*, 81–85.
93. Sharma, S., Kulkarni, S. K., & Chopra, K. (2006). Curcumin, the active principle of turmeric (Curcuma longa), ameliorates diabetic nephropathy in rats. *Clinical and Experimental Pharmacology & Physiology, 33*, 940–945. https://doi.org/10.1111/j.1440-1681.2006.04468.x
94. Srimal, R. C., & Dhawan, B. N. (1973). Pharmacology of diferuloyl methane (curcumin), a non-steroidal anti-inflammatory agent. *The Journal of Pharmacy and Pharmacology, 25*, 447–452.
95. Cole, G. M., Teter, B., & Frautschy, S. A. (2007). Neuroprotective effects of curcumin. In *The molecular targets and therapeutic uses of Curcumin in health and disease* (pp. 197–212). Boston: Springer.
96. Tizabi, Y., Hurley, L., Qualls, Z., & Akinfiresoye, L. (2014). Relevance of the anti-inflammatory properties of Curcumin in neurodegenerative diseases and depression. *Molecules, 19*, 20864–20879. https://doi.org/10.3390/molecules191220864
97. Motterlini, R., & Foresti, R. (2000). BR and GC. *Fast Track Paper., 28*, 1303–1312. https://doi.org/10.1038/eye.2013.315
98. Karlstetter, M., Lippe, E., Walczak, Y., et al. (2011). Curcumin is a potent modulator of microglial gene expression and migration. *Journal of Neuroinflammation, 8*, 125. https://doi.org/10.1186/1742-2094-8-125
99. Tegenge, M. A., Rajbhandari, L., Shrestha, S., et al. (2014). Curcumin protects axons from degeneration in the setting of local neuroinflammation. *Experimental Neurology, 253*, 102–110. https://doi.org/10.1016/j.expneurol.2013.12.016
100. Bassani, T. B., Turnes, J. M., Moura, E. L. R., et al. (2017). Effects of curcumin on short-term spatial and recognition memory, adult neurogenesis and neuroinflammation in a streptozotocin-induced rat model of dementia of Alzheimer's type. *Behavioural Brain Research, 335*, 41–54. https://doi.org/10.1016/j.bbr.2017.08.014
101. Tiwari, S. K., Agarwal, S., Seth, B., et al. (2014). Curcumin-loaded nanoparticles potently induce adult neurogenesis and reverse cognitive deficits in Alzheimer's disease model *via* canonical Wnt/β-catenin pathway. *ACS Nano, 8*, 76–103. https://doi.org/10.1021/nn405077y
102. Lee, W.-H., Loo, C.-Y., Bebawy, M., et al. (2013). Curcumin and its derivatives: Their application in neuropharmacology and neuroscience in the 21st century. *Current Neuropharmacology, 11*, 338–378. https://doi.org/10.2174/1570159X11311040002
103. Aggarwal, B. B., Gupta, S. C., & Sung, B. (2013). Curcumin: An orally bioavailable blocker of TNF and other pro-inflammatory biomarkers. *British Journal of Pharmacology, 169*, 1672–1692. https://doi.org/10.1111/bph.12131
104. Jacob, A., Wu, R., Zhou, M., & Wang, P. (2007). Mechanism of the anti-inflammatory effect of Curcumin: PPAR- γ activation. *PPAR Research, 2007*, 1–5. https://doi.org/10.1155/2007/89369
105. Saja, K., Babu, M. S., Karunagaran, D., & Sudhakaran, P. R. (2007). Anti-inflammatory effect of curcumin involves downregulation of MMP-9 in blood mononuclear cells. *International Immunopharmacology, 7*, 1659–1667. https://doi.org/10.1016/j.intimp.2007.08.018

106. Zhong, Y., Yu, W., Feng, J., et al. (2014). Curcumin suppresses tumor necrosis factor-α-induced matrix metalloproteinase-2 expression and activity in rat vascular smooth muscle cells via the NF-κB pathway. *Experimental and Therapeutic Medicine, 7*, 1653–1658. https://doi.org/10.3892/etm.2014.1647
107. Rajakrishnan, V., Viswanathan, P., Rajasekharan, K. N., & Menon, V. P. (1999). Neuroprotective role of curcumin from curcuma longa on ethanol-induced brain damage. *Phytotherapy Research, 13*, 571–574.
108. Bhandari, R., & Kuhad, A. (2015). Neuropsychopharmacotherapeutic efficacy of curcumin in experimental paradigm of autism spectrum disorders. *Life Sciences, 141*, 156–169. https://doi.org/10.1016/j.lfs.2015.09.012
109. Al-Askar, M., Bhat, R. S., Selim, M., et al. (2017). Postnatal treatment using curcumin supplements to amend the damage in VPA-induced rodent models of autism. *BMC Complementary and Alternative Medicine, 17*, 1–11. https://doi.org/10.1186/s12906-017-1763-7
110. Shu-juan, C., Zhi-mei, J., Shi-ling, Z., Qi-feng, S., & Lan-min, G. W. P. (2012). Effect of Curcumin on behavior of autism rats and the expression of brain derived neurotrophic factor. *Journal of Applied Clinical Pediatrics*.
111. Frémont, L. (2000). Biological effects of resveratrol. *Life Sciences, 66*, 663–673. https://doi.org/10.1016/S0024-3205(99)00410-5
112. Fullerton, M. D., & Steinberg, G. R. (2010). SIRT1 takes a backseat to AMPK in the regulation of insulin sensitivity by resveratrol. *Diabetes, 59*, 551–553. https://doi.org/10.2337/db09-1732
113. Wendeburg, L., de Oliveira, A. C. P., Bhatia, H. S., et al. (2009). Resveratrol inhibits prostaglandin formation in IL-1beta-stimulated SK-N-SH neuronal cells. *Journal of Neuroinflammation, 6*, 26. https://doi.org/10.1186/1742-2094-6-26
114. Wong, Y. T., Gruber, J., Jenner, A. M., et al. (2009). Elevation of oxidative-damage biomarkers during aging in F2 hybrid mice: Protection by chronic oral intake of resveratrol. *Free Radical Biology & Medicine, 46*, 799–809. https://doi.org/10.1016/j.freeradbiomed.2008.12.016
115. Kim, Y. A., Kim, G.-Y., Park, K.-Y., & Choi, Y. H. (2007). Resveratrol inhibits nitric oxide and prostaglandin E2 production by lipopolysaccharide-activated C6 microglia. *Journal of Medicinal Food, 10*, 218–224. https://doi.org/10.1089/jmf.2006.143
116. Gao, D., Zhang, X., Jiang, X., et al. (2006). Resveratrol reduces the elevated level of MMP-9 induced by cerebral ischemia-reperfusion in mice. *Life Sciences, 78*, 2564–2570. https://doi.org/10.1016/j.lfs.2005.10.030
117. Cheng, G., Zhang, X., Gao, D., et al. (2009). Resveratrol inhibits MMP-9 expression by up-regulating PPAR alpha expression in an oxygen glucose deprivation-exposed neuron model. *Neuroscience Letters, 451*, 105–108. https://doi.org/10.1016/j.neulet.2008.12.045
118. Bambini-Junior, V., Zanatta, G., Della Flora Nunes, G., et al. (2014). Resveratrol prevents social deficits in animal model of autism induced by valproic acid. *Neuroscience Letters, 583*, 176–181. https://doi.org/10.1016/j.neulet.2014.09.039
119. Moussa, C., Hebron, M., Huang, X., et al. (2017). Resveratrol regulates neuro-inflammation and induces adaptive immunity in Alzheimer's disease. *Journal of Neuroinflammation, 14*(1). https://doi.org/10.1186/s12974-016-0779-0
120. Tsai, R. Y., Wang, J. C., Chou, K. Y., et al. (2016). Resveratrol reverses morphine-induced neuroinflammation in morphine-tolerant rats by reversal HDAC1 expression. *Journal of the Formosan Medical Association, 115*, 445–454. https://doi.org/10.1016/j.jfma.2015.05.010
121. McCalley, A. E., Kaja, S., Payne, A. J., & Koulen, P. (2014). Resveratrol and calcium signaling: Molecular mechanisms and clinical relevance. *Molecules, 19*, 7327–7340. https://doi.org/10.3390/molecules19067327
122. Fontes-Dutra, M., Santos-Terra, J., Deckmann, I., et al. (2018). Resveratrol prevents cellular and behavioral sensory alterations in the animal model of autism induced by valproic acid. *Frontiers in Synaptic Neuroscience, 10*, 1–12. https://doi.org/10.3389/fnsyn.2018.00009

123. Xie, W., Ge, X., Li, L., et al. (2018). Resveratrol ameliorates prenatal progestin exposure-induced autism-like behavior through ERβ activation. *Molecular Autism, 9*, 1–13. https://doi.org/10.1186/s13229-018-0225-5
124. Bhandari, R., & Kuhad, A. (2017). Resveratrol suppresses neuroinflammation in the experimental paradigm of autism spectrum disorders. *Neurochemistry International, 103*, 8–23. https://doi.org/10.1016/j.neuint.2016.12.012
125. Felgines, C., Texier, O., Morand, C., et al. (2000). Bioavailability of the flavanone naringenin and its glycosides in rats. *American Journal of Physiology. Gastrointestinal and Liver Physiology, 279*, G1148–G1154.
126. Vallverdu, A., Odriozola-serrano, I., Oms-oliu, G., et al. (2012). Changes in the polyphenol profile of tomato juices processed by pulsed electric fields. *Journal of Agricultural and Food Chemistry, 60*, 9667–9672.
127. Fuhr, U., Klittich, K., & Staib, A. H. (1993). Inhibitory effect of grapefruit juice and its bitter principal, naringenin, on CYP1A2 dependent metabolism of caffeine in man. *British Journal of Clinical Pharmacology, 35*, 431–436. https://doi.org/10.1111/j.1365-2125.1993.tb01162.x
128. Kumar, S., & Tiku, A. B. (2016). Biochemical and molecular mechanisms of radioprotective effects of Naringenin, a phytochemical from citrus fruits. *Journal of Agricultural and Food Chemistry, 64*(8), 1676–1685. https://doi.org/10.1021/acs.jafc.5b05067
129. Nahmias, Y., Goldwasser, J., Casali, M., et al. (2008). Apolipoprotein B-dependent hepatitis C virus secretion is inhibited by the grapefruit flavonoid naringenin. *Hepatology, 47*, 1437–1445. https://doi.org/10.1002/hep.22197
130. Yi, L.-T., Liu, B.-B., Li, J., et al. (2014). BDNF signaling is necessary for the antidepressant-like effect of naringenin. *Progress in Neuro-Psychopharmacology & Biological Psychiatry, 48*, 135–141. https://doi.org/10.1016/j.pnpbp.2013.10.002
131. Birt, D. F., Hendrich, S., & Wang, W. (2001). Dietary agents in cancer prevention: Flavonoids and isoflavonoids. *Pharmacology & Therapeutics, 90*, 157–177. https://doi.org/10.1016/S0163-7258(01)00137-1
132. Galluzzo, P., Ascenzi, P., Bulzomi, P., & Marino, M. (2008). The nutritional flavanone naringenin triggers antiestrogenic effects by regulating estrogen receptor alpha-palmitoylation. *Endocrinology, 149*, 2567–2575. https://doi.org/10.1210/en.2007-1173
133. Wu, L. H., Lin, C., Lin, H. Y., et al. (2016). Naringenin suppresses neuroinflammatory responses through inducing suppressor of cytokine signaling 3 expression. *Molecular Neurobiology, 53*, 1080–1091. https://doi.org/10.1007/s12035-014-9042-9
134. Raza, S. S., Khan, M. M., Ahmad, A., et al. (2013). Neuroprotective effect of naringenin is mediated through suppression of NF-κB signaling pathway in experimental stroke. *Neuroscience, 230*, 157–171. https://doi.org/10.1016/j.neuroscience.2012.10.041
135. Bhandari, R., Paliwal, J. K., & Kuhad, A. (2018). Naringenin and its nanocarriers as potential phytotherapy for autism spectrum disorders. *Journal of Functional Foods, 47*, 361–375. https://doi.org/10.1016/j.jff.2018.05.065
136. Guerrero-Beltrán, C. E., Calderón-Oliver, M., Pedraza-Chaverri, J., & Chirino, Y. I. (2012). Protective effect of sulforaphane against oxidative stress: Recent advances. *Experimental and Toxicologic Pathology, 64*, 503–508. https://doi.org/10.1016/j.etp.2010.11.005
137. Negrette-Guzmán, M., Huerta-Yepez, S., Tapia, E., & Pedraza-Chaverri, J. (2013). Modulation of mitochondrial functions by the indirect antioxidant sulforaphane: A seemingly contradictory dual role and an integrative hypothesis. *Free Radical Biology & Medicine, 65*, 1078–1089. https://doi.org/10.1016/j.freeradbiomed.2013.08.182
138. Soane, L., Li Dai, W., Fiskum, G., & Bambrick, L. L. (2010). Sulforaphane protects immature hippocampal neurons against death caused by exposure to hemin or to oxygen and glucose deprivation. *Journal of Neuroscience Research, 88*, 1355–1363. https://doi.org/10.1002/jnr.22307
139. Wang, G., Fang, H., Zhen, Y., et al. (2016). Sulforaphane prevents neuronal apoptosis and memory impairment in diabetic rats. *Cellular Physiology and Biochemistry, 39*, 901–907. https://doi.org/10.1159/000447799

140. Zhang, Y., Kensler, T. W., Cho, C. G., et al. (1994). Anticarcinogenic activities of sulforaphane and structurally related synthetic norbornyl isothiocyanates. *Proceedings of the National Academy of Sciences, 91*, 3147–3150. https://doi.org/10.1073/pnas.91.8.3147
141. Zhao, J., Kobori, N., Aronowski, J., & Dash, P. K. (2006). Sulforaphane reduces infarct volume following focal cerebral ischemia in rodents. *Neuroscience Letters, 393*, 108–112. https://doi.org/10.1016/j.neulet.2005.09.065
142. Kraft, A. D., Johnson, D. A., & Johnson, J. A. (2004). Nuclear factor E2-related factor 2-dependent antioxidant response element activation by tert-butylhydroquinone and sulforaphane occurring preferentially in astrocytes conditions neurons against oxidative insult. *The Journal of Neuroscience, 24*, 1101–1112. https://doi.org/10.1523/JNEUROSCI.3817-03.2004
143. Gan, N., Wu, Y. C., Brunet, M., et al. (2010). Sulforaphane activates heat shock response and enhances proteasome activity through up-regulation of Hsp27. *The Journal of Biological Chemistry, 285*, 35528–35536. https://doi.org/10.1074/jbc.M110.152686
144. Singh, K., Connors, S. L., Macklin, E. A., et al. (2014). Sulforaphane treatment of autism spectrum disorder (ASD). *Proceedings of the National Academy of Sciences, 111*, 15550–15555. https://doi.org/10.1073/pnas.1416940111
145. Bent, S., Lawton, B., Warren, T., et al. (2018). Identification of urinary metabolites that correlate with clinical improvements in children with autism treated with sulforaphane from broccoli. *Molecular Autism, 9*, 35. https://doi.org/10.1186/s13229-018-0218-4
146. D'Archivio, M., Filesi, C., Varì, R., et al. (2010). Bioavailability of the polyphenols: Status and controversies. *International Journal of Molecular Sciences, 11*, 1321–1342. https://doi.org/10.3390/ijms11041321
147. Selby-Pham, S. N. B., Miller, R. B., Howell, K., et al. (2017). Physicochemical properties of dietary phytochemicals can predict their passive absorption in the human small intestine. *Scientific Reports, 7*, 1931. https://doi.org/10.1038/s41598-017-01888-w
148. Jithan, A., Madhavi, K., Madhavi, M., & Prabhakar, K. (2011). Preparation and characterization of albumin nanoparticles encapsulating curcumin intended for the treatment of breast cancer. *International Journal of Pharmaceutical Investigation, 1*, 119–125. https://doi.org/10.4103/2230-973X.82432
149. Mathew, A., Fukuda, T., Nagaoka, Y., et al. (2012). Curcumin loaded-PLGA nanoparticles conjugated with Tet-1 peptide for potential use in Alzheimer's disease. *PLoS One, 7*, e32616. https://doi.org/10.1371/journal.pone.0032616
150. Ranjan, A. P., Mukerjee, A., Helson, L., & Vishwanatha, J. K. (2012). Scale up, optimization and stability analysis of Curcumin C3 complex-loaded nanoparticles for cancer therapy. *Journal of Nanobiotechnology, 10*, 38. https://doi.org/10.1186/1477-3155-10-38
151. Xie, X., Tao, Q., Zou, Y., et al. (2011). PLGA nanoparticles improve the oral bioavailability of Curcumin in rats: Characterizations and mechanisms. *Journal of Agricultural and Food Chemistry, 59*, 9280–9289. https://doi.org/10.1021/jf202135j
152. Singh, G., & Pai, R. S. (2014). Optimized PLGA nanoparticle platform for orally dosed *trans* -resveratrol with enhanced bioavailability potential. *Expert Opinion on Drug Delivery, 11*, 647–659. https://doi.org/10.1517/17425247.2014.890588
153. Shulman, M., Cohen, M., Soto-Gutierrez, A., et al. (2011). Enhancement of naringenin bioavailability by complexation with hydroxypropoyl-β-cyclodextrin. *PLoS One, 6*, e18033. https://doi.org/10.1371/journal.pone.0018033
154. Tsai, M. J., Huang, Y. B., Fang, J. W., et al. (2015). Preparation and characterization of naringenin-loaded elastic liposomes for topical application. *PLoS One, 10*, 1–12. https://doi.org/10.1371/journal.pone.0131026
155. Wang, Y., Wang, S., Firempong, C. K., et al. (2017). Enhanced solubility and bioavailability of naringenin via liposomal nanoformulation: Preparation and in vitro and in vivo evaluations. *AAPS PharmSciTech, 18*, 586–594. https://doi.org/10.1208/s12249-016-0537-8
156. Yang, L.-J., Ma, S.-X., Zhou, S.-Y., et al. (2013). Preparation and characterization of inclusion complexes of naringenin with β-cyclodextrin or its derivative. *Carbohydrate Polymers, 98*, 861–869. https://doi.org/10.1016/j.carbpol.2013.07.010

157. Yen, F.-L., Wu, T.-H., Lin, L.-T., et al. (2008). Naringenin-loaded nanoparticles improve the physicochemical properties and the hepatoprotective effects of naringenin in orally-administered rats with CCl 4 -induced acute liver failure. *Pharmaceutical Research, 26*(4), 893–902. https://doi.org/10.1007/s11095-008-9791-0
158. Yen, F.-L., Wu, T.-H., Lin, L.-T., et al. (2009). Naringenin-loaded nanoparticles improve the physicochemical properties and the hepatoprotective effects of naringenin in orally-administered rats with CCl(4)-induced acute liver failure. *Pharmaceutical Research, 26*, 893–902. https://doi.org/10.1007/s11095-008-9791-0
159. Danafar, H., Sharafi, A., Manjili, H. K., & Andalib, S. (2017). Sulforaphane delivery using mPEG–PCL co-polymer nanoparticles to breast cancer cells. *Pharmaceutical Development and Technology, 22*(5), 642–651. https://doi.org/10.3109/10837450.2016.1146296
160. Demio, P. C., & Finley-Belgrad, E. (2011). A clinical study of effects of Enhansa® (enhanced absorption curcumin) on immunologic and cognitive/metabolic disorders – An overview of results
161. Manach, C., Scalbert, A., Morand, C., et al. (2004). Bioavailability, polyphenols: Food sources and. *The American Journal of Clinical Nutrition, 79*, 727–747. https://doi.org/10.1038/nature05488
162. Teng, H., & Chen, L. (2018). Polyphenols and bioavailability: An update. *Critical Reviews in Food Science and Nutrition*, 00–00. https://doi.org/10.1080/10408398.2018.1437023
163. Thilakarathna, S. H., & Vasantha Rupasinghe, H. P. (2013). Flavonoid bioavailability and attempts for bioavailability enhancement. *Nutrients, 5*, 3367–3387. https://doi.org/10.3390/nu5093367
164. Pardridge, W. M. (2001). *Brain drug targeting: The future of brain drug development*. Cambridge University Press.
165. Pardridge, W. M., Oldendorf, W. H., Cancilla, P., & Frank, H. J. (1986). Blood-brain barrier: Interface between internal medicine and the brain. *Annals of Internal Medicine, 105*, 82–95.
166. Lin, J. H. (2004). How significant is the role of P-glycoprotein in drug absorption and brain uptake? *Drugs Today (Barc), 40*, 5–22.
167. Löscher, W., & Potschka, H. (2005). Blood-brain barrier active efflux transporters: ATP-binding cassette gene family. *NeuroRx, 2*, 86–98. https://doi.org/10.1602/neurorx.2.1.86

Regulation of Dietary Amino Acids and Voltage-Gated Calcium Channels in Autism Spectrum Disorder

Shubham Singh, Supraj Raja Sangam, and Rajagopal Senthilkumar

Abstract Autism, or autism spectrum disorders (ASD), is one of the complex genetic diseases and its etiology is unknown for majority of the patients. It is characterized by deterioration in social interaction, communication, interests, imagination, and activities. As autism is a highly heterogeneous disorder, the symptoms can vary greatly in each affected individual. Oxidative stress implicates major pathogenesis of neurological disorders like ASD. Nutrients and dietary supplements play an important role in the health of an individual and there are several lines of evidence suggesting the role of dietary factors in the development or pathogenesis of ASD. The amino acids supplement has been found to reduce symptoms as they act as the precursors of neurotransmitters which in turn may extenuate mental disorders. The biosynthesis of amino acids in the brain is regulated by the concentration of amino acids in plasma. Amino acids are also considerable entities as they themselves, or peptides consisting of them, have profound antioxidant activities. Dietary constituents have an effect on the transport of amino acids across the blood–brain barrier (BBB) thus indirectly modulating the therapeutic value of amino acids. Among the other factors, voltage-gated calcium channels are directly linked to ASD as per results of genetic studies. Malfunctioning of these calcium channels causes ASD. The intricate biochemical and molecular machinery contributing to neurological disorders is still unknown. Here we discuss the preventive role of dietary amino acids against and regulation of voltage-gated calcium channels on ASD.

Keywords ASD · Autism · Diet · Amino acids · Neurotransmitters · Antioxidants · Blood–brain barrier · Calcium channels · Neurological disorders · Oxidative stress

---

S. Singh
Department of Biological Sciences, IISER, Pune, Maharashtra, India

S. R. Sangam
Department of Biotechnology, JNTU, Hyderabad, Telangana, India

R. Senthilkumar (✉)
Department of Biochemistry, Rayalaseema University, Kurnool, Andhra Pradesh, India

M. M. Essa, M. W. Qoronfleh (eds.), *Personalized Food Intervention and Therapy for Autism Spectrum Disorder Management*, Advances in Neurobiology 24,
https://doi.org/10.1007/978-3-030-30402-7_24

1 Introduction

According to Johns Hopkins University Bloomberg School of Public Health and the CDC 1–2% of children in the USA are affected by neurodevelopmental diseases such as ASD and it is characterized by different levels of severity and occurs across all ethnic groups. Recent studies have shown that 1 out of 88 children aged 8 years will develop an ASD (2012 report that looked at 2008 data), with males more at risk than females. In a 2016 report that looked at 2012 data the estimate is 1 in 68 children. Blumberg et al. showed that the prevalence of ASD had risen 75% from 2007 to 2012 in the USA.

Nutrition is one of the key components of one's mental health and physiological well being. A large number of essential nutrients are provided by our diet while a few can be synthesized in the body constitutively or on demand. Several mental disorders, where nutrients play a pivotal role, start to take shape at the early stages of development. Mutational or genetic disorders cannot be treated completely by diet but smart diet plans prove to be promising in the management of symptomatic effects and can improve the quality of life by rescuing the phenotype partially. There has been increasing awareness of and need for research toward functional foods which offer the hope of a better life for patients. Apart from direct contribution toward decreasing severity of several pathological conditions, dietary nutrients modulate the efficacy of drugs and medication through various means like increasing bioavailability of drugs or facilitating delivery across blood–brain barrier. Treatment of neurological disorders is still a challenging task due to poor understanding and difficulties in drug-delivery across blood–brain barrier. Amino acids have been shown to have health benefits in several neuropsychiatric disorders like schizophrenia [1, 2], mood disorders like mania [3, 4], cognitive impairments, and autistic disorders [5]. Amino acids modulate the permeability of blood–brain barrier and few neuroactive amino acids can further reduce or aggravate neurological diseases.

Autism is a neurodevelopmental disorder that had been identified in the early twentieth century. However, its exact molecular mechanism or methods for treatment and prevention are still not available [6]. It is a developmental disorder as symptoms are observable at an early age. Notably, early-age dietary intake can potentially influence the lifestyle and pathogenesis of ASD. Few biochemical studies have suggested an association of neuroactive amino acids and comorbidities of ASD but conclusive implications of neuroactive amino acids on ASD are still lacking [7–9]. Glutamate serum levels are high in ASD subjects [10] while few amino acids like methionine, phenylalanine, valine, tryptophan, leucine, and isoleucine have been reported to be reduced in ASD patients [11]. Amino acids metabolism is also found to be dysregulated in children with ASD and CAMP (Children's Autism Metabolome Project) is currently trying to push toward identification of amino acid-based metabotypes as early markers of autism. CAMP's effort toward mapping diagnostic markers has provided us with some concrete clues regarding branched-chain amino acid dysregulation and ASD. It has also been reported that ornithine and glutamine metabolism relate more to ASD in males than in females, suggesting differential metabolism of amino acids across the different sexes [12]. One of the classical studies on Branched-Chain Ketoacid Dehydrogenase Kinase (BCKDK) KO mice, showing reduced levels of branched-chain amino acids in the plasma, led to autism, gait disorder, and epilepsy in mice. The ASD-like phenotype was partially rescued by feeding the experimental animals with branched-chain amino acids [13]. Supplying dietary branched amino acids to human subjects further extended this work and it could normalize the levels of branched amino acids in serum [13, 14]. This work provides strong evidence in favor of the use of dietary amino acids as treatment for patients carrying mutation in BCKDK. Other work by Novarino group in Austria has shown a strong link between branched-chain amino acid levels in the murine brain and ASD [15]. Solute carrier transporter 7a5 (Slc7a5) protein is a neutral amino acid transporter enriched at blood–brain barrier and it maintains levels of branched-chain amino acids in brain. Homozygous mutation in this gene leads to ASD symptoms and the loss of motor coordination, which can partially be rescued by injection of amino acids like leucine and isoleucine in Slc7a5 KO adult mice [15]. There is growing evidence, based on animal model studies and cohort studies on ASD subject matters, that amino acids might help in improving the lifestyle of ASD patients and can also be potential biomarkers. Development of biochemical markers will make ASD diagnosis easy, currently done only by trained psychiatrists (Fig. 1).

1.1 Types of Autism Symptoms and Disorders Related to Autism Types

Among several neurodevelopmental disorders, autism is a lifelong disorder, within which patients have impaired communication, repetitive behaviors, hyperactivity, and increased interest in a particular subject. Symptoms can be seen at early stages of childhood, i.e., approximately 2–3 years. Autism severity and progression is contributed to by both complex genetic predisposition and environmental factors. It was identified as a psychological disorder in the early twentieth century and was often confused with schizophrenia. Swiss psychiatrist Eugen Bleuler used "autism" to refer to a set of schizophrenia-related symptoms [16]. Autism is a Greek word originating from the word "autos" meaning "self." Given its origins, this name was probably chosen as patients exhibited poor social interaction and were acting withdrawn from society [16]. Further research and identification of patients lead to the expansion of autism into autism spectrum disorder that broadly includes three different sets of developmental disabilities, namely, autistic disorder, Asperger's syndrome, and pervasive developmental disorder not otherwise specified (PDD-NOS) or atypical autism [17]. Patients with autistic disorder exhibit classical autism symptoms like delayed learning of language, communication, and social skills with significant level of intellectual disability. They develop very deep interest in some topics and exhibit unusual behaviors.

Fig. 1 Differnt signs of autism infographic

The second classification under autism spectrum disorder is Asperger's syndrome, reported independently by Hans Asperger—Austrian pediatrician—in 1944 [18]. Asperger noted behaviors similar to autism in his subjects but the severity of symptoms was less and intellectual abilities were uncompromised or rather increased in some of them [19].

The third type is pervasive developmental disorder not otherwise specified or atypical autism. Individuals belonging to this group do not show all symptoms of autistic disorder or Asperger's syndrome. Deficits are mostly limited to social and communication skills but do not meet all the criteria to qualify for specific pervasive development disorder.

1.2 Prevalence of Autism Across Age, Race, and Sex

According to the WHO, it is estimated that, on an average, 1 in every 160 children has ASD across the world, but these numbers are inaccurate due to the lack of any available data from low-economy and developing countries. Improved awareness, diagnostic tools, and criteria have caused a gradual increase in the number of ASD cases reported year by year. Individuals with ASD often do not have any physical characteristics that differentiate them from healthy individuals. It can be diagnosed on the basis of their behavior, communication, and socially withdrawn nature. Centre for Disease Control and Prevention (CDC) is actively tracking the prevalence of ASD across USA since 1998 through its active surveillance system of The Autism and Developmental Disabilities Monitoring Network (ADDM) [20, 21]. A 15% increase in the prevalence of autism was reported: from 1 out of every 68 children in 2014 to 1 in 59 children in 2018 across the USA. Getting the exact statistics of its prevalence in developing countries is difficult due to the lack of any such programs and active affordable diagnosis for such disorders [22] (Fig. 2).

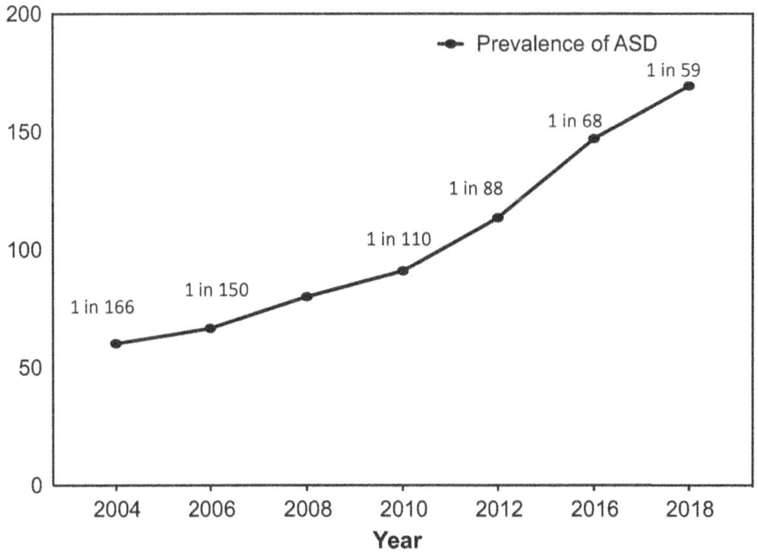

Fig. 2 ASD prevalence data obtained from CDC website. Plotted here is the year-wise prevalence rate of ASD, i.e., one patient with ASD diagnosis per fixed number of population

The ADDM survey in 2014 also found ASD to be four times more likely to occur in males than in females and in non-Hispanic than Hispanic children. Within this population, it was more prevalent among non-Hispanic white children compared to non-Hispanic black children. Intellectual impairment was highly variable among subjects across race, sex, and ethnicity [22]. These numbers provide an idea about the prevalence of ASD but these may be far less than the actual numbers as tracking ASD is quite challenging due to the variability of symptoms across sex, race, and ethnicity. There are also the issues of significant overlap of symptoms with other psychiatric disorders, changes in diagnostic criteria, and lastly due to the paucity of any robust biochemical diagnostic marker [23]. The Diagnostic and Statistical Manual of Mental Disorders, Fifth edition (DSM-V) did away with the social and communication subtypes of DSM-IV-TR and brought them into one combined domain. As per the DSM-V, individuals with ASD must meet all three criteria under the domain of social communication and interaction and at least two of the four criteria under the restrictive/repetitive behavior domain. Social communication and interaction domain includes following deficits [24]:

1. Deficits in social–emotional reciprocity
2. Deficits in nonverbal communicative behaviors
3. Deficits in developing, understanding, and maintaining relationships.

Four criteria included in restrictive and repetitive interaction includes:

1. Repetitive speech or motor movements
2. Insistence or sameness
3. Restricted interests
4. Unusual response to sensory input.

2 Dietary Food

2.1 Functional Food and Diet: The Relation to ASD

Functional food helps improve health. It is a kind of food in which a new ingredient can be added to a food and the new product thus acquires a new function aiding health promotion or disease prevention. Functional foods are one of the fastest growing segments of the food industry. In some countries, functional foods have already become part of the dietary landscape. Functional foods reduce the risk of chronic diseases and have physiological benefits beyond the traditional nutrients it contains [25, 26].

The American Dietetic Association (ADA) defines functional foods as foods "that include whole foods and fortified, enriched or enhanced foods that have a potentially beneficial effect on health when consumed as a part of varied diet on regular basis, at effective levels." The ADA breaks down functional foods into four categories: conventional foods, modified foods, medical foods, and foods for special dietary use [27].

2.1.1 Conventional Foods

Conventional foods are the most basic of the functional foods as they have not been modified by enrichment or fortification; they are still in their natural state. Most whole fruits and vegetables fall into this category because they are rich in phytochemicals such as lycopene and lutein as well as other beneficial compounds [28, 29].

2.1.2 Modified Foods

Modified foods have been enriched, fortified, or enhanced with nutrients or other beneficial ingredients. Calcium-fortified orange juice, folic acid-enriched breads, and margarine enhanced with plant sterols are functional foods that have been modified. Energy drinks that have been enhanced with herbs such as ginseng and guarana, and other potentially controversial foods also fall into this category [30, 31].

2.1.3 Medical Foods

The FDA defines medical food as "food that is formulated for consumption or administration entirely under the supervision of a physician and is intended for the specific dietary management of a disease or condition for which distinctive nutritional requirements, based on recognized scientific principles, are established by medical evaluation" [32, 33].

2.1.4 Foods for Special Dietary Use

Foods for special dietary use are similar to medical foods but they are available commercially and do not require the supervision of a healthcare provider.

2.2 Nutritional Deficits of Autism

Impairment in social interaction, communication, behavior as well as sensory challenges are prominent features of ASD. Substantial evidence supports benefits of specific behavioral, educational, and some pharmacologic interventions for children with ASD. However, given the limited availability of treatment in improving core and associated symptoms of ASD, substantial challenges in accessing evidence-based treatment approaches and perceptions regarding lessened risks of treatment, many families pursue dietary and nutritional approaches as components of treatment. With limitations in the existing empirical evidence, families and providers alike often struggle to understand the safety and potential benefit of such approaches

[34, 35]. Evidence supporting specific theories, however, is lacking. Studies have also explored differences in nutrient status in children with and without ASD and potential correlations with ASD symptoms as well as the effects of vitamin supplementation. The results of these studies have been inconclusive [36].

2.3 Amino Acids and Other Natural Compounds for Preventive Effects

Milk and dairy products have been found to be beneficial in ASD as they contain bioactive peptides, probiotic bacteria, antioxidants, vitamins, specific proteins, oligosaccharides, organic acids, highly absorbable calcium, conjugated linoleic acid, and other biologically active components [37, 38].

2.3.1 Omega 3 Fatty Acid Supplementation

Not much evidence supports the effectiveness of omega 3 supplementation in improving core or associated ASD symptoms. Three randomized controlled trials (RCTs) of omega 3s versus placebo reported no significant group differences on most measures of challenging behavior, communication, language, and adaptive behavior. One study reported significantly improved scores in the placebo group compared with the omega 3 group in externalizing behaviors after 6 months of treatment and another reported a significant improvement in parent ratings of stereotypy and lethargy in children receiving omega 3 supplements compared with those receiving placebo, teacher ratings were not significantly different. Another RCT of dietary docosahexaenoic acid (DHA) supplementation versus placebo reported improvement in parent-rated social skills in children receiving placebo versus those receiving DHA, while teachers rated communication as improving more in the treatment group compared with placebo. Scores on other measures did not differ significantly between groups [39–41].

2.4 Digestive Enzyme Supplementation

Evidence is inadequate to assess the effects of short-term digestive enzyme supplements. Two RCTs addressed digestive enzyme supplements compared with placebo: one evaluated a proteolytic enzyme supplement (Peptizyde) and the other, a digestive enzyme supplement (Neo-Digestin); both supplements contained papain and pepsin or peptidase. The Peptizyde RCT reported no significant differences in measures of behavior, sleep quality, or gastrointestinal symptoms and no significant differences in adverse effects. In a 3 month trial of Neo-Digestin versus placebo, symptom severity scores improved significantly in the treatment group compared with placebo [42, 43].

2.5 *Other Supplements*

Two RCTs addressed methyl B12 supplementation. Clinical Global Impression (CGI) scores improved significantly in the methyl B12 group in one RCT but studies reported few other significant group differences in measures of behavior or communication. Two RCTs of levocarnitine reported improvements in symptom severity in the levocarnitine group compared with placebo, but scores on other behavioral measures or adverse effects did not differ between groups. In the second RCT, symptom severity did not differ between groups after 6 months of treatment [44].

3 Amino Acids

Eating a variety of vegetarian and animal proteins throughout the day fulfills the daily need for our body's amino acids. Amino acids are the building blocks of proteins in our body. Our body can synthesize certain amino acids, but others, called essential amino acids, must come from protein-containing foods in our diet. We rely on amino acids to build and repair tissues, digest food, for the formation and function of enzymes, and to transport molecules like oxygen throughout our body [45].

3.1 *Neuroactive Amino Acids*

Neuroactive amino acids play an important role in central brain functions based on their availability, metabolism, and/or receptor activity. They are associated with the pathogenesis and/or pharmacotherapy of several psychiatric disorders that have symptoms, such as cognitive impairment and problems with social interactions, in common with ASD. Other amino acids could also be involved and hence, it will be important to conduct comprehensive studies in which a number of these amino acids are investigated simultaneously. Due to the potential role of neuroactive amino acids in the pathogenesis and treatment of ASD, monitoring changes in their concentrations in body fluids is also important in the case that they are relevant to early diagnosis and intervention among patients with ASD [10, 46, 47].

4 Voltage-Gated Calcium Channels Function/Malfunction in ASD

ASD is a syndrome that affects normal brain development and is characterized by impaired social interaction, verbal and nonverbal communication and repetitive, stereotypic behavior. ASD is a complex disorder arising from a combination of multiple genetic and environmental factors that are independent from race, ethnicity, and socioeconomical status. The high heritability of ASD suggests a strong genetic basis for the disorder. Additionally, growing evidence has revealed the role of various ion channel gene defects (channelopathies) in the pathogenesis of autism. Indeed, recent genome-wide association, and whole exome- and whole-genome resequencing studies have linked polymorphisms and rare variants in calcium, sodium, and potassium channels and their subunits with susceptibility to ASD, much as they do with bipolar disorder, schizophrenia, and other neuropsychiatric disorders. Moreover, animal models with these genetic variations recapitulate endophenotypes considered to be correlates of autistic behavior seen in patients. An ion flux across the membrane regulates a variety of cell functions, from generation of action potentials to gene expression and cell morphology. Thus, it is not surprising that channelopathies have profound effects on brain functions [48–51].

Voltage-gated calcium channels (VOCC) are among the factors linked to ASD, shown by results of genetic studies. Mutations of nearly all pore-forming and some auxiliary subunits of VOCC have been revealed in studies investigating ASD patients and populations. Though there are only few electrophysiological characterizations of VOCC mutations found in autistic patients, these studies suggest their functional relevance. Ion channel subunits comprise of the single largest gene family underlying disorders of heart, muscle, and brain and the most frequently tested for precision clinical diagnosis of a broad phenotypic spectrum of central nervous system disease. These disorders collectively constitute an enormous public health burden, with a greater number of diminishing life years than cancer. The significance of each variant, which may spell the difference between lifelong disability and sudden death, requires the most accurate functional interpretation to assign causality, stimulate drug discovery, and guide the use of gene variant-specific therapies. Chromatin modification/transcription regulation, MAP kinase/cellular signaling, and neuronal development/axon guidance pathways are involved in ASD with variation of copy number and deleterious single nucleotides variations [52–58] (Table 1).

5 Conclusion

Amino acids are the chemical building blocks of key neurotransmitters that act on the brain to influence mood and behavior. For this reason, proper balance of these nutrients is essential for healthy emotional and cognitive development in children. NeuroScience's Targeted Amino Acid Therapies provide patients with amino acid supplements specifically formulated to rebuild depleted inhibitory transmitters, while simultaneously supporting healthier levels of excitatory neurotransmitters, to help children pay attention and improve their cognitive, speech, and social skills. ASD is not a simple pathology and is associated with a large spectrum of other diseases. In addition to this, defective regulation of ion flux through the cell membrane caused by altered kinetics of ion channels and transporters appears to cause an imbalance of excitation/inhibition in neural function. This may lead to defective

Table 1 Calcium channels and calcium channel subunits implicated in ASD (adapted from [59])

Protein	Description	Normal function	Disease association
CACNA1C	Voltage-regulated L-type calcium channel, alpha 1C subunit	Regulates entry of Ca^{2+} into excitable cells: muscle contraction, hormone/ neurotransmitter release, gene expression, cell cycle	Timothy syndrome, ASD, psychiatric diseases
CACNA1D	Voltage-regulated calcium channel, alpha 1D subunit	High-voltage activated, long-lasting calcium activity	Sinoatrial node dysfunction and deafness, ASD, psychiatric diseases
CACNA1E	Voltage-regulated R-type calcium channel, alpha 1E subunit	High-voltage activated, rapidly inactivating	ASD, psychiatric diseases
CACNA1F	Voltage-regulated L-type calcium channel, alpha 1F subunit	Regulates entry of Ca^{2+} into excitable cells: muscle contraction, hormone/ neurotransmitter release, gene expression, cell cycle	ASD and X-linked congenital stationary night blindness
CACNA1G	Voltage-regulated T-type calcium channel, alpha 1G subunit	Regulates entry of Ca^{2+} into excitable cells: muscle contraction, hormone/ neurotransmitter release, gene expression, cell cycle	ASD; intellectual disability; juvenile myoclonic epilepsy
CACNA1H	Voltage-regulated T-type calcium channel, alpha 1H subunit	Regulates neuronal and cardiac pacemaker activity	Familial autism; childhood absence epilepsy
CACNA1I	Voltage-regulated T-type calcium channel, alpha 1I subunit	Characterized by a slower activation and inactivation compared to other T-channels	Possibly implicated ASD
CACNA2D3	Voltage-regulated calcium channel, alpha 2/delta 3 subunit	Accessory calcium channel subunit; regulates entry of Ca^{2+} into excitable cells	ASD
CACNA2D4	Voltage-regulated calcium channel, alpha 2/delta 4 subunit	Accessory calcium channel subunit; regulates entry of Ca^{2+} into excitable cells	Gene deletion along with CACNA1C leads to ASD
CACNB2	Accessory calcium channel beta-2 subunit	Contributes to the function of calcium channels. Modulates voltage dependence of activation and inactivation and controls trafficking of the calcium channel family	ASD, psychiatric diseases

neuronal circuit formation and physiological response. Restoring ion dynamics to their physiological equilibrium may represent a promising therapeutic strategy for this neurodevelopmental psychiatric disorder.

Acknowledgement The support provided by the Department of Biotechnology, Ministry of Science and Technology, Government of India (To Dr. Senthilkumar Rajagopal—No: BT/RLF/Reentry/42/2012) to complete this book in a successful manner is gratefully acknowledged.

References

1. Coyle, J. T. (2006). Glutamate and schizophrenia: Beyond the dopamine hypothesis. *Cellular and Molecular Neurobiology, 26*, 365–384.
2. Labrie, V., Lipina, T., & Roder, J. C. (2008). Mice with reduced NMDA receptor glycine affinity model some of the negative and cognitive symptoms of schizophrenia. *Psychopharmacology, 200*, 217–230.
3. Ongur, D., Jensen, J. E., Prescot, A. P., Stork, C., Lundy, M., Cohen, B. M., et al. (2008). Abnormal glutamatergic neurotransmission and neuronal-glial interactions in acute mania. *Biological Psychiatry, 64*, 718–726.
4. Yuksel, C., & Ongur, D. (2010). Magnetic resonance spectroscopy studies of glutamate-related abnormalities in mood disorders. *Biological Psychiatry, 68*, 785–794.
5. Lam, K. S., Aman, M. G., & Arnold, L. E. (2006). Neurochemical correlates of autistic disorder: A review of the literature. *Research in Developmental Disabilities, 27*, 254–289.
6. Blaylock, R. L. (2008). A possible central mechanism in autism spectrum disorders, part 1. *Alternative Therapies in Health and Medicine, 14*, 46–53.
7. Maynard, T. M., & Manzini, M. C. (2017). Balancing act: Maintaining amino acid levels in the autistic brain. *Neuron, 93*, 476–479.
8. Zheng, H. F., Wang, W. Q., Li, X. M., Rauw, G., & Baker, G. B. (2017). Body fluid levels of neuroactive amino acids in autism spectrum disorders: A review of the literature. *Amino Acids, 49*, 57–65.
9. Żurawicz, E., & Kałużna-Czaplińska, J. (2015). Analysis of amino acids in autism spectrum disorders. *TrAC Trends in Analytical Chemistry, 73*, 91–118.
10. Vargason, T., Kruger, U., McGuinness, D., Adams, J., Geis, E., Gehn, E., et al. (2018). Investigating plasma amino acids for differentiating individuals with autism spectrum disorder and typically developing peers. *Research in Autism Spectrum Disorders, 50*, 60–72.
11. Li, C., Shen, K., Chu, L., Liu, P., Song, Y., & Kang, X. (2018). Decreased levels of urinary free amino acids in children with autism spectrum disorder. *Journal of Clinical Neuroscience, 54*, 45–49.
12. Duman, R., Aghajanian, G., Sanacora, G., & Krystal, J. (2016). Synaptic plasticity and depression: New insights from stress and rapid-acting antidepressants. *Nature Medicine, 22*, 238–249.
13. Novarino, G., El-Fishawy, P., Kayserili, H., Meguid, N. A., Scott, E. M., Schroth, J., et al. (2012). Mutations in BCKD-kinase lead to a potentially treatable form of autism with epilepsy. *Science, 338*, 394–397.
14. Garcia, P., Kolesky, S., & Andrew, J. (2010). General anesthetic actions on $GABA_A$ receptors. *Current Neuropharmacology, 8*, 2–9.
15. Tarlungeanu, D. C., Deliu, E., Dotter, C. P., Kara, M., Janiesch, P. C., Scalise, M., et al. (2016). Impaired amino acid transport at the blood brain barrier is a cause of autism spectrum disorder. *Cell, 167*(1481–1494), e1418.
16. Evans, B. (2013). How autism became autism: The radical transformation of a central concept of child development in Britain. *History of the Human Sciences, 26*, 3–31.
17. Sharma, A., Kumar, V., Giridhar, P., & Ravishankar, G. (2008). Induction of *in vitro* flowering in *Capsicum frutescens* under the influence of silver nitrate and cobalt chloride and pollen transformation. *Plant Biotechnology Journal, 11*, 1–8.
18. Faras, H., Al Ateeqi, N., & Tidmarsh, L. (2010). Autism spectrum disorders. *Annals of Saudi Medicine, 30*, 295–300.
19. Klin, A. (2003). Asperger syndrome: An update. *Revista Brasileira de Psiquiatria, 25*, 103–109.
20. Bertrand, J., Mars, A., Boyle, C., Bove, F., Yeargin-Allsopp, M., & Decoufle, P. (2001). Prevalence of autism in a United States population: The Brick Township, New Jersey, investigation. *Pediatrics, 108*, 1155–1161.

21. Yeargin-Allsopp, M., Rice, C., Karapurkar, T., Doernberg, N., Boyle, C., & Murphy, C. (2003). Prevalence of autism in a US metropolitan area. *Journal of the American Medical Association, 289*, 49–55.
22. Baio, J., Wiggins, L., Christensen, D. L., Maenner, M. J., Daniels, J., Warren, Z., et al. (2018). Prevalence of autism spectrum disorder among children aged 8 years — Autism and developmental disabilities monitoring network, 11 sites, United States, 2014. *Morbidity and Mortality Weekly Report. Surveillance Summaries (Washington, D.C.: 2002), 67*(6), 1–23.
23. Rice, C., Baio, J., Van Naarden Braun, K., Doernberg, N., Meaney, F., & Kirby, R. (2007). A public health collaboration for the surveillance of autism spectrum disorders. *Paediatric and Perinatal Epidemiology, 21*, 179–190.
24. Swedo, S. E., Baird, G., Cook Jr., E. H., Happe, F. G., Harris, J. C., Kaufmann, W. E., et al. (2012). Commentary from the DSM-5 workgroup on neurodevelopmental disorders. *Journal of the American Academy of Child and Adolescent Psychiatry, 51*, 347–349.
25. Thomas, L., & Morrison, C. (2013). Amino acid-dependent regulation of food intake: Is protein more than the sum of its parts? *Journal of Physiology, 591*, 5417–5418.
26. Tome, D. (2004). Protein, amino acids and the control of food intake. *British Journal of Nutrition, 92*, S27–S30.
27. Sanahuja, J., & Harper, A. (1962). Effect of amino acid imbalance on food intake and preference. *The American Journal of Physiology, 202*, 165–170.
28. Baudry, J., Touvier, M., Alles, B., Peneau, S., Mejean, C., Galan, P., et al. (2016). Typology of eaters based on conventional and organic food consumption: Results from the NutriNet-Sante cohort study. *British Journal of Nutrition, 116*, 700–709.
29. Eisinger-Watzl, M., Wittig, F., Heuer, T., & Hoffmann, I. (2015). Customers purchasing organic food - do they live healthier? Results of the German National Nutrition Survey II. *European Journal of Nutrition & Food Safety, 5*, 59–71.
30. Liu, Y., & Wang, H. (2013). Medium-chain triglyceride ketogenic diet, an effective treatment for drug-resistant epilepsy and a comparison with other ketogenic diets. *Biomedical Journal, 36*, 9–15.
31. Neal, E., Chaffe, H., Schwartz, R., Lawson, M., Edwards, N., Fitzsimmons, G., et al. (2009). A randomized trial of classical and medium-chain triglyceride ketogenic diets in the treatment of childhood epilepsy. *Epilepsia, 50*, 1109–1117.
32. Bollinger, H. (2001). Functional drinks with dietary fibre. *Fruit Process, 12*, 252–254.
33. Dhingra, D., Michael, M., Rajput, H., & Patil, R. (2012). Dietary fibre in foods: A review. *Journal of Food Science and Technology, 49*, 255–266.
34. Corvey, K., Menear, K., Preskitt, J., Goldfarb, S., & Menachemi, N. (2016). Obesity, physical activity and sedentary behaviors in children with an autism spectrum disorder. *Maternal and Child Health Journal, 20*, 466–476.
35. Linda, B., Carol, C., Sarah, P., Sarah, E., Melissa, M., & Aviva, M. (2017). Changes in food selectivity in children with autism spectrum disorder. *Journal of Autism and Developmental Disorders, 47*, 439–446.
36. Nila, S., Andrews, J., McPheeters, M., & Warren, Z. (2017). Nutritional and dietary interventions for autism spectrum disorder: A systematic review. *Pediatrics, 139*, e20170346.
37. Cox, D. (2012). From interdisciplinary to integrated care of the child with autism: The essential role for a code of ethics. *Journal of Autism and Developmental Disorders, 42*, 2729–2738.
38. Goin-Kochel, R., Mackintosh, V., & Myers, B. (2009). Parental reports on the efficacy of treatments and therapies for their children with autism spectrum disorders. *Research in Autism Spectrum Disorders, 3*, 528–537.
39. Annio, P., & Paola, V. (2016). Omega-3 supplementation in autism spectrum disorders: A still open question? *Journal of Pediatric Neurosciences, 11*, 225–227.
40. James, S., Montgomery, P., & Williams, K. (2011). Omega-3 fatty acids supplementation for autism spectrum disorders (ASD). *Cochrane Database Systematic Review, 11*, CD007992.
41. Meyer, B., Mann, N., Lewis, J., Milligan, G., Sinclair, A., & Howe, P. (2003). Dietary intakes and food sources of omega-6 and omega-3 polyunsaturated fatty acids. *Lipids, 38*, 391–398.

42. Munasinghe, S., Oliff, C., Finn, J., & Wray, J. (2010). Digestive enzyme supplementation for autism spectrum disorders: A double-blind randomized controlled trial. *Journal of Autism and Developmental Disorders, 40*, 1131–1138.
43. Yong-Jiang, L., Jian-Jun, O., Ya-Min, L., & Da-Xiong, X. (2017). Dietary supplement for core symptoms of autism spectrum disorder: Where are we now and where should we go? *Frontiers in Psychiatry, 8*, 155.
44. Geier, D., Kern, J., Davis, G., King, P., Adams, J., Young, J., et al. (2011). A prospective double-blind, randomized clinical trial of levocarnitine to treat autism spectrum disorders. *Medical Science Monitor, 17*, PI15–PI23.
45. Bala, K., Dogan, M., Mutluer, T., Kaba, S., Aslan, O., Balahoroglu, S., et al. (2016). Plasma amino acid profile in autism spectrum disorder (ASD). *European Review for Medical and Pharmacological Sciences, 20*, 923–929.
46. Flugge, K. (2017). Impaired amino acid metabolism in autism spectrum disorders. *Biomarkers in Medicine, 11*, 711–712.
47. Zheng, H., Wang, W., Li, X., Rauw, G., & Baker, G. (2016). Body fluid levels of neuroactive amino acids in autism spectrum disorders: A review of the literature. *Amino Acids, 49*, 57–65.
48. Arrigoni, C., Rohaim, D., Findeisen, S., Stein, R., Nurva, S., Mishra, S., et al. (2016). Unfolding of a temperature-sensitive domain controls voltage-gated channel activation. *Cell, 164*, 922–936.
49. Bomben, V., Aiba, I., Qian, J., Mark, M., Herlitze, S., & Noebels, J. L. (2016). Isolated P/Q calcium channel deletion in layer VI corticothalamic neurons generates absence epilepsy. *Journal of Neuroscience, 36*, 405–418.
50. Burgess, D., Jones, J., Meisler, M., & Noebels, J. (1997). Mutation of the Ca^{2+} channel β subunit gene Cchb4 is associated with ataxia and seizures in the lethargic (lh) mouse. *Cell, 88*, 385–392.
51. Lipscombe, D., & Andrade, A. (2015). Calcium channel CaVα1 splice isoforms – Tissue specificity and drug action. *Current Molecular Pharmacology, 8*, 22–31.
52. Boland, L., Morrill, J., & Bean, B. (1994). ω-Conotoxin block of N-type calcium channels in frog and rat sympathetic neurons. *Journal of Neuroscience, 14*, 5011–5027.
53. Nalli, A., Senthilkumar, R., Sunila, M., Grider, J., & Murthy, K. (2015). Inhibition of RhoA-dependent pathway and contraction by endogenous hydrogen sulfide in rabbit gastric smooth muscle cells. *American Journal of Physiology. Cell Physiology, 308*, C485–C495.
54. Samuel, H., Pratta, W., Elliott, R., Shehrazade, D., Laurent, F., Owen, M., et al. (2015). Genetic disruption of voltage-gated calcium channels in psychiatric and neurological disorders. *Progress in Neurobiology, 134*, 36–54.
55. Senthilkumar, R. (2013). *Amino acids from food as medicine: Valuable function*. New York: Nova Publishers.
56. Senthilkumar, R., Supraj, S., Singh, S., & Venkateswara Rao, J. (2016). Modulatory effects of dietary amino acids on neurodegenerative diseases. In M. Essa, J. Gilles, & M. Akbar (Eds.), *The benefits of natural products for neurodegenerative diseases* (pp. 401–414). Cham, Switzerland: Springer.
57. Senthilkumar, R., Venkateshwara Rao, J., Srinivasan, T., & Veerappan, R. (2014). Beneficial effects of dietary amino acids on brain health. In M. Essa & M. Memon (Eds.), *Food and brain function* (pp. 221–242). New York: Nova Publisher.
58. Singh, S., Sangam, S., & Senthilkumar, R. (2016). Role of dietary amino acids on Parkinson's diseases. In M. Essa, T. ManiVasagam, J. Thenmozhi, & A. Mohammed (Eds.), *Food and Parkinson's disease* (pp. 1–14). New York: Nova Publisher.
59. Schmunk, G., & Gargus, J. (2013). Channelopathy pathogenesis in autism spectrum disorders. *Frontiers in Genetics, 4*, 222.

Bioactive Metabolites from Marine Ascidians: Future Treatment for Autism Spectrum Disorder

Manigandan Venkatesan, Velusamy Arumugam, Rathinam Ayyasamy, Selvakumar Murugesan, Nishakavya Saravanan, Umamaheswari Sundaresan, Saravanan Ramachandran, Thamilarasan Manivasagam, Arokiasamy Justin Thenmozhi, and M. Walid Qoronfleh

Abstract Autism spectrum disorder (ASD) is a developmental disorder that influences communication and behavior. Numerous researches propose that genes can act together with manipulations from the environment to affect development in ways that lead to ASD. The broad range of issues facing people with ASD means that there is no single proper drug and treatment for ASD. Numerous shortcomings associated with the present conventional therapeutic strategies have forced researchers to venture into alternative natural sources for effective compounds. The marine environment has emerged as an alternate search environment due to its versatile conditions where organisms employ various biodefense mechanisms for their survival. Ascidians are an excellent source for unique bioactive compounds with nutritive and therapeutic content and it still holds credit for being an underused source from marine animals. Bioactive compounds isolated from ascidians have

M. Venkatesan · N. Saravanan · S. Ramachandran
Department of Medical Biotechnology, Chettinad Academy of Research and Education, Chennai, Tamil Nadu, India

V. Arumugam (✉) · U. Sundaresan
Department of Environmental Biotechnology, School of Environmental Sciences, Bharathidasan University, Tiruchirappalli, Tamil Nadu, India

R. Ayyasamy
Department of Animal Science, Centre for Pheromone Technology, Bharathidasan University, Tiruchirappalli, Tamil Nadu, India

S. Murugesan
Department of Biotechnology, Anna University, Tiruchirappalli, Tamil Nadu, India

T. Manivasagam · A. Justin Thenmozhi
Department of Biochemistry and Biotechnology, Annamalai University, Chidambaram, Tamil Nadu, India

M. W. Qoronfleh
Research & Policy Department, World Innovation Summit for Health (WISH), Qatar Foundation, Doha, Qatar

M. M. Essa, M. W. Qoronfleh (eds.), *Personalized Food Intervention and Therapy for Autism Spectrum Disorder Management*, Advances in Neurobiology 24,
https://doi.org/10.1007/978-3-030-30402-7_25

various commendable biomedical applications due to their unique chemical structures. The present chapter will focus on the potential of bioactive compounds derived from ascidians for the treatment of the neurologic disorder—ASD.

Keywords Marine environment · Ascidians · Alkaloids · Neuroprotection · Autism spectrum disorder

1 Introduction

Autism spectrum disorder (ASD) covers a range of neurodevelopmental disorders involving deficits in social interaction with self-focus, communication, and nonsocial features such as restricted and stereotyped behaviors [1, 2]. ASD poses great challenges for both afflicted individuals and their caregivers or family, impacting their ability to participate in standardized education, have meaningful peer interactions, hold employment, and overall deterioration in the basic quality of daily life (Fig. 1) [3]. The symptoms of ASD are categorized into two types: the primary and the secondary symptoms. The primary symptoms include reduced language skills and social interaction and the presence of repetitive and stereotypic behaviors [4, 5]

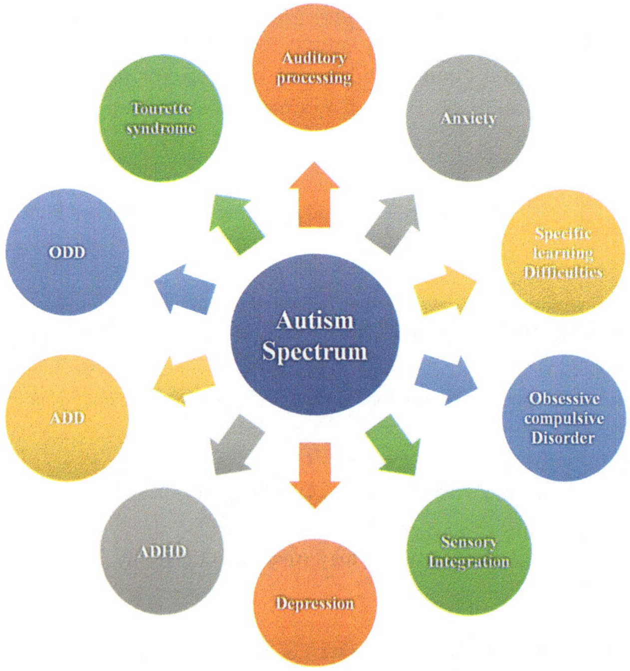

Fig. 1 Autistic spectrum disorder. ODD oppositional defiant disorder, ADD attention deficit disorder, ADHD attention deficit hyperactivity disorder

while secondary symptoms include self-injury, hyperactivity, aggression, depression, and co-occurring psychiatric disorders (anxiety) [6]. Due to a combination of higher incidence and/or improved screening, the rate of ASD diagnosis also appears to be increasing rapidly in all countries where prevalence studies have been conducted. Globally, ASD now affects at least one out of roughly every 100 children [7–9], and one out of every 68 children in the USA [10]. Thus, early ASD diagnosis and effective treatment, combined with an understanding of its association to the child or infant neurocognitive development are of critical concern for both science and public health.

2 Early Intervention

When the diagnosis is made, parents are urged to start early intervention. This consists of applied behavioral analysis (ABA), speech therapy, occupational therapy, psychomotor therapy, and special education. Toddlers should be placed in regular day care to increase interaction with neuro-typically normal children. This emphasis is due to the neural plasticity still present at this age [11] and early intervention in autistic patients between the ages 18–48 months has a major positive effect on later outcomes.

Speech therapy produces improvement in communication skills. It is most effective when the therapist adopts a collaborative approach including the family, peers, teachers, and special educators. It is often advised to limit the use of language to just one at home and at school. Occupational therapy promotes self-care skills, organization, and attention and play skills. Sensory integration remediates deficits in neurological processing of sensory information, thereby improving adaptation of the child to the environment [12].

3 Pharmacotherapy for Autism

Many drugs have been investigated for the alleviation of symptoms. While only few drugs have proven to be useful, many others are still undergoing in clinical trials. In several controlled studies, risperidone proved to be efficacious in treating irritability and aggression in ASD patients of all ages. Risperidone is FDA approved for the treatment of irritability in ASD children and adolescents [13]. One multicenter, double-blind placebo controlled trial conducted by McCracken et al. assessing risperidone use for irritability in ASD showed a 69% response rate, with more efficient reduction of irritability, stereotypic behaviors, hyperactivity, and noncompliance, particularly when combined with parent training [14].

Aripiprazole (known as Abilify), a third generation antipsychotic drug used for the treatment of schizophrenia, major depression, and psychosis, has also been approved by the FDA for use in the treatment of irritability in individuals with autism. Randomized controlled trials using aripiprazole in autistic patients resulted in less irritability, hyperactivity, and stereotypies compared to placebo. Side effects included weight gain, tremors, and sedation [15].

4 Marine Environment

There has been recent spike in the interest in drug discoveries based on natural products, leading to the exploration of the marine resources. Covering above 70% of the earth's surface, the marine environment represents the largest unexplored, wealthy resource for the exploration of natural products in the treatment of various diseases and disorders [16]. The marine environment contains more than 200,000 organisms and has survived the unusual conditions of low to zero light, high pressure, high or low temperature, and high salt content. The unique conditions require marine organisms to produce a wealth of chemical compounds for adaption [17] that have not been found in terrestrial creatures. Natural products have been part of the well-known pharmacological sources for the past 50 years and these bioactive compounds are derived from terrestrial and marine resources such as plants, animals, or microorganisms [18]. The marine environment is structurally diverse with unique pharmacologically active compounds along with a number of novel metabolites that carry beneficial pharmacological properties. Some of the most interesting marine creatures include bacteria, fungi, algae, sponges, soft corals, tunicates, molluscs, and bryozoans that have been reported to possess numerous biologically active compounds, globally [19].

5 Marine Ascidians

Ascidians belong to the phylum *Chordata* (Class: Ascidiacea). It comprises of more than 3000 species reported worldwide [20] with more than 400 species being reported just in the Indian coast (Fig. 2, [21]). Ascidians represent the most unique invertebrate group of animals commonly investigated by chemists for natural marine products [22]. Thousands of compounds were reported in marine ascidians [23]; they are made up of more than 80% nitrogen and 70% of nitrogenous compounds and are divided into two types of structure-based groups of polycyclic aromatic alkaloids and peptides [24]. Ascidians have been reported to have several pharmacologically active compounds like antibiotics, cyclin-dependent kinases, cytotoxic compounds and display immunosuppressive activities and inhibition of topoisomerases among other activities [25, 26]. A recent update has showed more than 300 alkaloids being reported from the ascidians and their occurrence, structural type and biological function have been clearly investigated [18]. The notable ascidian compounds are trabectedin (Yondelis®) alkaloids derived from *Ecteinascidia turbinata* and another alkaloid of Plitidepsin (Aplidin®) isolated from *Aplidium*

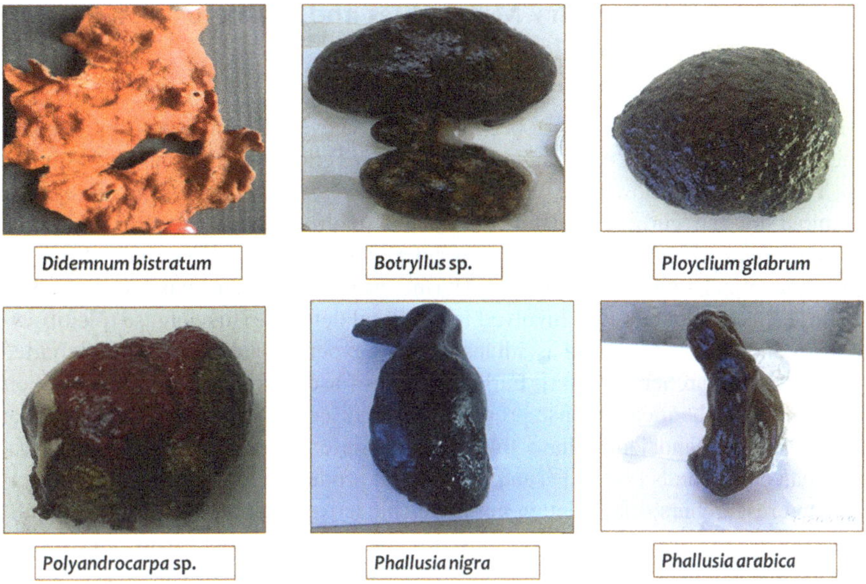

Fig. 2 Some important ascidians collected from the Gulf of Mannar

albicans [27]. The last 40 years have seen massive advancements in drug discovery and the development of natural products from marine sources which has subsequently grabbed the attention of various pharmaceutical companies and research bodies that have the advanced techniques of HPLC, the NMR, and mass techniques [28, 29].

6 Extraction and Characterization of Secondary Metabolites and Alkaloids

The isolation of secondary metabolites and alkaloids from ascidians uses different methods like the Stas-Otto method that involves the distribution of alkaloidal bases between aqueous or acid solution and immiscible carbon-based solvent. In another common method using water soluble organic solvents, different polar and nonpolar solvents were used for extraction [30]. Water or aqueous alcohol method, Soxhlet extraction process [31], and Kippenberger's process are commonly used in the extraction of alkaloids [32]. The separation techniques are thin layer chromatography (TLC) [33] and the high performance liquid chromatograph technique (HPLC) for qualitative and quantitative estimation of alkaloids [34] and high-performance thin-layer chromatography (HPTLC) [35]. Structure elucidation can be done using different methods like UV spectroscopy [36], IR spectroscopy, nuclear magnetic resonance spectroscopy [37], mass spectroscopy (GC-MS and LC-MS) [38], FAB (expand), and MALDI-TOF MS [39].

7 Sources of Secondary Metabolites and Alkaloids from Marine Ascidians

Marine animals and plants have been reported to have different types of secondary metabolites due to the environmental and oceanographic conditions they survive in. The increasing knowledge of marine natural chemistry, ecology, and biology are inspired by the current researcher attracted to drug development. The marine resources are wealth resources for new interest for development of new drugs for pharmaceutical industries [40, 41]. Marine ascidians are an abundant source of toxic secondary metabolites involved in chemical defense. This defense role of several secondary compounds of ascidians has been experimentally proven and reported by several researchers [42, 43]. Earlier reports of ascidian larvae have shown their antipredatory chemical defense properties. Few researchers have also reported that adult ascidians and larvae have this chemical defense, particularly from tropical ascidian compounds [44, 45]. Tunicates natural products are attracting many chemists and researchers; they contain numerous bioactive compounds like carotenoids, macrolides, alkaloids, and tunichromes [46]. Secondary metabolites are not required for the development, reproduction, and growth of these organisms and it is essential for an organism to adapt to its surrounding environment. This is complemented by the need for defense against predators to survive in a particular environment [47].

An intensive research effort during the last 25 years has led to a number of alkaloids being derived from marine ascidians (Fig. 3). Most of the alkaloids have been derived from genus *Eudistoma*, *Pseudodistoma*, *Ritterella*, *Synoicum*, *Lissoclinum*, and *Didemnum* [21]. Currently, 300 alkaloid structures were isolated and structurally characterized as ascidians and their occurrence, structural type, and pharmacological activity have been discussed. The different types of alkaloids reported as ascidians are shown in Fig. 4 [24]. The recent trends in tunicates have indicated that they are an important and major source of biomedical compounds for metabolic disorders. The important alkaloid compounds of aplidin, trabectedin, and other alkaloids isolated from tunicates have been identified as new and promising anticancer drugs [48, 49].

8 Biological Activities of Secondary Metabolites and Alkaloids from Marine Ascidians

8.1 Antioxidant Properties of Ascidians

Reactive oxygen species (ROS), also called oxygen free radicals, have been involved in many pathological disorders, especially in diabetes, arthritis, and cancer. Most of the ROS have unpaired electron valence or unstable bonds and they originate from mitochondrial metabolism. ROS generate oxidative stress and it is associated with the aging process and cell death, affecting all major organ systems [50].

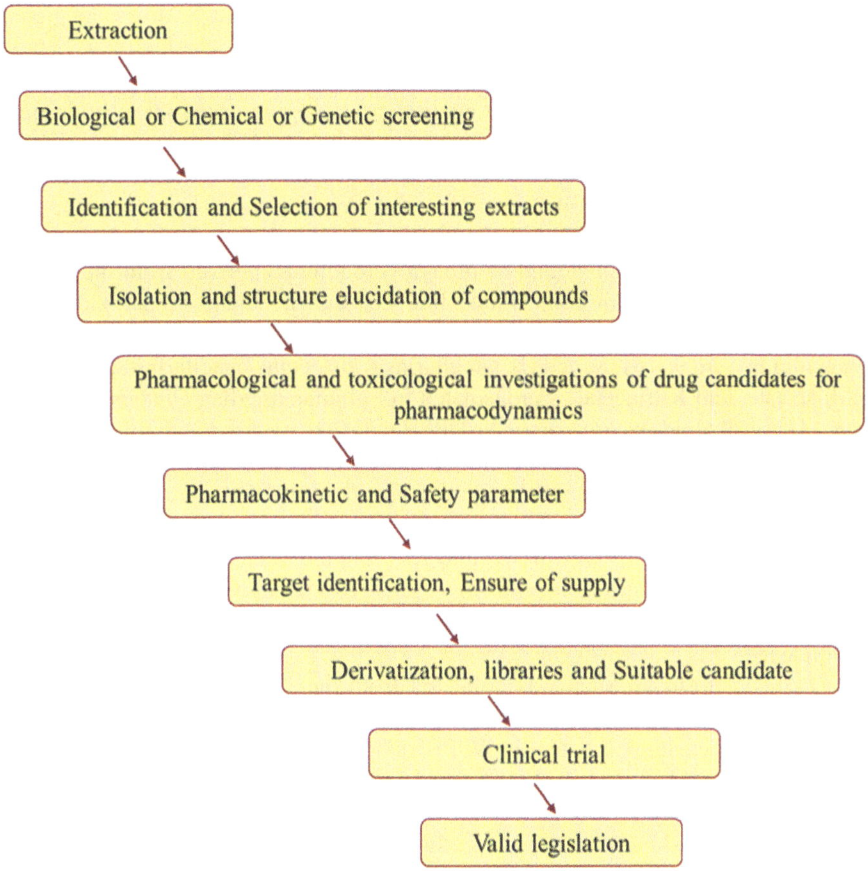

Fig. 3 The drug development process from marine ascidians

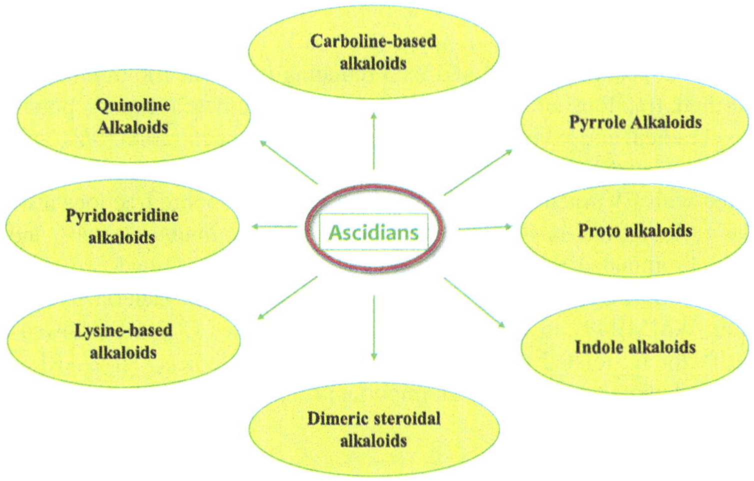

Fig. 4 Important alkaloids isolated from marine ascidians

Many research studies support the relationship between oxidative stress and global health. These ROS include the superoxide anion radical (O_2), the hydroxyl radical (OH), the peroxyl radicals (ROO), the alkoxyl radical (RO), and peroxynitrite (HOONO). Oxidative species have significant damaging potential to biological targets with different reactivity and formation pathways [51].

The higher consumption of nutraceutical diets (plants and marine derivatives) that contain valuable antioxidants has been linked with lower disease rates and preventive protection. Marine alkaloids have been used to treat oxidative stress [52]. The aromatic organic compound of prenylhydroquinone derived from the marine colonial tunicate *Aplidium californicum* has been shown to inhibit superoxide anion production in rat alveolar macrophages [53]. Eleven different ascidian alkaloids of lamellarins (γ, α, ε, I, M, K, and U), lamellarins K-diacetate, lamellarins K-triacetate, C-diacetate, and X-triacetate were isolated from Indian ascidian *Didemnum obscurum.* These alkaloids have reported the DPPH radical scavenging activity. Lamellarin γ, lamellarins (K, U, and I), lamellarin γ-monoacetate, and lamellarin C-diacetate were shown to have potential scavenging properties with the appropriate IC_{50} concentrations (3.28 mM, 5.63 mM, 5.80 mM, 6.70 mM, 2.96 mM, and 10.87 mM). *Didemnum albidum* ascidians were collected from the Mandapam region. The secondary metabolites were isolated using organic 70% ethanol. The crude extract of ascidian inhibits free radicals of nitric oxide on the RAW 264.7 human monocytic cell line [54].

Phallusia nigra ascidian was collected from the Tuticorin Port in India and these crude extract of alcoholic and aqueous secondary metabolites (Phenolics and Flavanoids) showed potential DPPH radical scavenging activities [55]. The two ascidians *Eudistoma viride* (EV) and *Didemnum psammathodes* (DP) were collected from Hare Island Tuticorin Coast, India and the bioactive metabolites were isolated by solvent extraction method. Fractions of partially purified ascidian extracts (using Sephadex LH-20 gel filtration and TLC fraction) were examined for antioxidant potential and showed high scavenging activity. In a nitrogen-centered free radical of DPPH assay, EV fraction and DP fraction showed 93% and 96% scavenging potential at higher concentrations. Two fractions were reduced from the ferric ions complex (Fe^{3+} to Fe^{2+}) by donating electrons with appropriate concentration (2.3 and 2.23 reducing power in 100 μg/mL). The partial purified fractions of ascidians showed maximum scavenging potential on Hydroxyl radical antioxidant assay with increasing concentration (96% and 98%). In the H_2O_2 scavenging assay, DP fraction (89%) showed higher radical inhibition compared with EV fraction (78%). The secondary metabolite fractions also inhibited the nitrite/nitrate ions (NO_2^-/NO_3^-) with moderate inhibition (64% and 72%) [56]. The brominated analogues of halomadurones C and D and two chlorinated pyrenes were derived from *Ecteinascidia turbinata* associated bacteria *Actinomadura* sp. strain WMMB499 in Florida Keys. Halomadurones C and D showed potent nuclear factor E2-related factor antioxidant response element (Nrf2-ARE) activation. These factors together play an important role in the treatment of neurodegenerative diseases [23, 57].

8.2 Anti-inflammatory Compounds from Ascidians

Inflammation is one of the two factors of pathological and physiological processes affected by injury, infection, and stress. The process involves the impairment of the immune system when tenacious for a long period of time as activated macrophages produce toxic factors [58]. Benslimane et al. [59] discussed the chloromethylenic extract of ascidian *Aplidium antillense* and the mixture compound of natural and synthetic of cordiachromene and reported the anti-inflammatory properties in induced rat-paw edema methods. Both compounds showed moderated activity at IC_{50} values of 4.6 mg/ml in natural isomer and 15 mg/ml in racemic mixture. Previous studies discussed New Zealand biota that have numerous anti-inflammatory and natural products from the ascidian *Aplidium* sp. revealing strong inhibition of superoxide production by human neutrophils stimulated with PMA [60, 61]. A novel dimeric alkaloid, lissoclin disulfoxide, has been isolated from the South African ascidian *Lissoclinum* sp. and these ascidian alkaloid have been reported to show activity against inhibition of IL-8 Rα and IL-8 Rβ receptors with IC_{50} values of 0.6 and 0.82 μM [62].

Appleton et al. [63] reported information regarding 2, 2, 5-trisubstituted imidazolone-containing alkaloids of kottamides A–D derived from New Zealand ascidian *Pycnoclavella kottae*. These novel alkaloids of kottamides D have shown anti-inflammatory properties at IC_{50} values of 2–200 μM and antimetabolic activity IC_{50} at values of 6–10 μM. Novel indole alkaloids of Conicamin were derived from the MeOH extract of Mediterranean ascidian *Aplidium conicum* and they have shown histamine-antagonistic activity at a concentration of 10^{-6} to 10^{-5} M in an ex vivo model with guinea pig ileum [64]. The anti-inflammatory properties of two new tricyclic thiazine-containing quinoline quinone alkaloids, ascidiathiazones A and B (Fig. 5), were reported from the New Zealand ascidian *Aplidium* sp. The two compounds were inhibited with the superoxide by PMA-stimulated human neutrophils with IC_{50} value of 1.55 ± 0.32 and 0.44 ± 0.09 μM in in vitro. The in vivo model (murine model) of gout with oral doses of 25.6 μmol/kg in both compounds [61]. The new anti-inflammatory halogenated furanone compound of rubrolide O as a mixture of E/Z isomers was derived from the New Zealand ascidians *Synoicum* sp. and rubrolide O showed inhibition against human neutrophil free radical release at an IC_{50} value of 35 μM [65] .

Pearce et al. [66] have discussed the ascidians alkaloids, dihydroxystyrylguanidine alkaloid of tubastrine, Orthidines A–C, E and Orthidine F derived from New Zealand ascidian *Aplidium orthium*. These six alkaloids show in vitro production of superoxide by PMA-stimulated human neutrophils in a dose-dependent manner with IC_{50} values of 10–36 μM and also conformed to the in vivo murine model of gouty inflammation at a concentration of 25 μmol/kg. Pearce et al. [67] reported that two tetracyclic alkaloids, Distomadines A and B, were derived from New Zealand ascidian *Pseudodistoma aureum*. The alkaloids Distomadines A demonstrated antifungal activity at 600 μg/ml and also weak activity shown in anti-inflammatory, cytotoxicity, antitumor, and antimycobacterial tests.

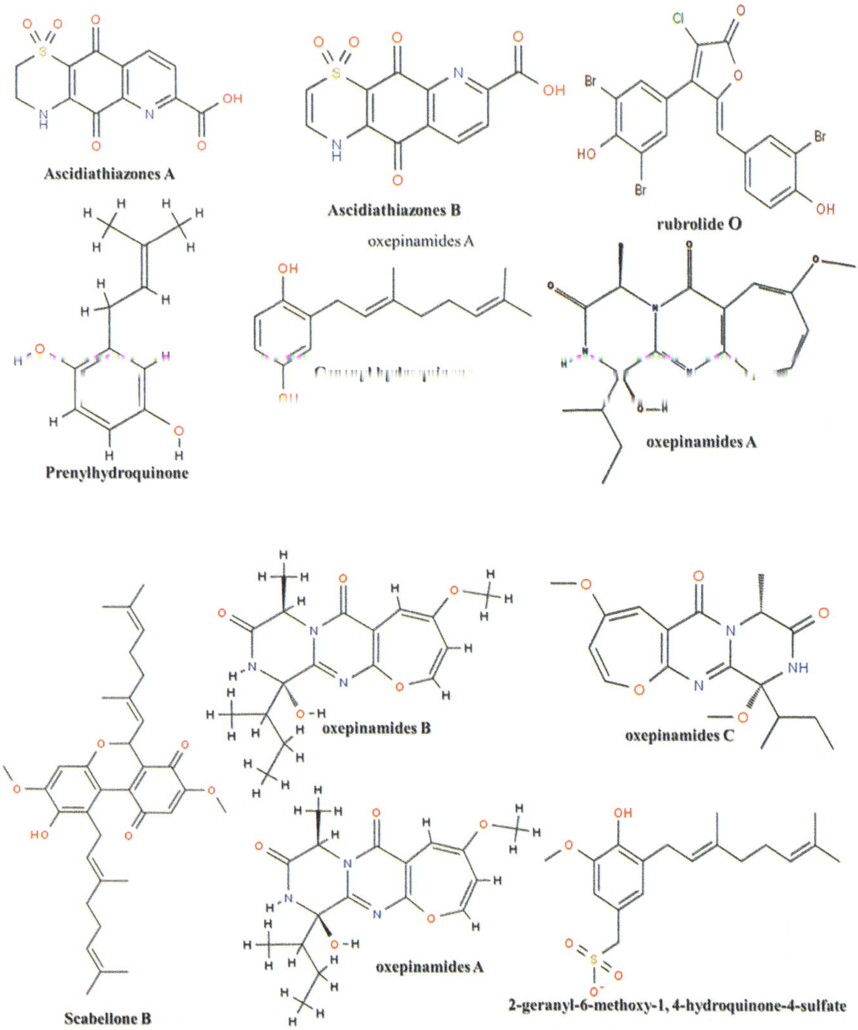

Fig. 5 Important bioactive metabolite reported from the marine ascidians

Gompel et al. [68] reported that Meridianins, brominated 3-(2-aminopyrimidine)-indoles derived from the *Aplidium meridianum* from the South Atlantic, also inhibit several protein kinases such as cyclic nucleotide-dependent kinases, casein kinase1, cyclin-dependent kinases, and glycogen synthase kinase-3. Meridianins prevent cell proliferation and induce apoptosis, a demonstration of their ability to enter cells and to interfere with the activity of kinases important for cell division and cell death. The anti-inflammatory properties of chondroitin sulfate were derived from ascidians *Styela clava* and it has shown inflammation activity in a mouse skin in vivo model. Chondroitin sulfate was inhibited by TPA-induced NF-κB activation, VCAM-1, COX-2, and inflammation cytokines suppressed by the 1KK and Akt/PKB signals at a concentration of 2 mg/ml [69]. The Caribbean ascidian *Ecteinascidia turbinate*, associated with bacteria of *Acremonium* sp., was used to isolate three new oxepin-containing oxepinamides A–C and its associated bacterial compound, oxepinamide A, showed notable anti-inflammatory activities in a topical RTX-induced mouse ear oedema assay [23, 70]. The meroterpene derivatives, Rossinones A and B, were isolated from an Antarctic ascidian *Aplidium* sp. and showed anti-inflammatory activity in the human peripheral blood neutrophils by inhibiting superoxide production [71]. Chan and his colleagues reported anti-inflammatory activities of meroterpenoids, 2-geranyl-6-methoxy-1, 4-hydroquinone-4-sulfate and scabellone B derived from the New Zealand Ascidian *Aplidium scabellum*. It inhibited superoxide production by PMA-stimulated human neutrophils in vitro at IC_{50} values of 21 and 125 μM [72]. Anti-inflammatory properties of MeOH extract compounds isolated from the ascidian *Eudistoma viride* and crude compounds are exhibited in various concentrations at 200 mg/kg compared to the Diclofenac standard drug [73]. Another study on same ascidian *Eudistoma viride* reported anti-inflammatory properties of methanol extract concentration at 100 and 200 mg/kg body weight in rat models [74]. Bertanha and his colleagues reported that geranyl hydroquinone and prenyl hydroquinone derived from the *Aplidium* sp. exhibited in vitro anti-inflammatory assay with activated human peripheral blood neutrophils by inhibiting superoxide production [75].

8.3 CNS Depressant Properties of Marine Ascidians

The crude compound of CNS depressant compounds was isolated from *Distaplia nathensis* at a concentration of 100 mg/kg reported by [76]. Meenakshi and her coworkers reported that EtOH crude extract isolated from Indian ascidians of *Microcosmus exasperatus* has shown attributed activity of CNS depressant properties with a concentration at 150 mg/kg [77]. The new tyrosine derivatives, botryllamides K, L with 6 known compounds, botryllamides A–C, botryllamide G and perspicamides A and B were derived from the Australian ascidian *Aplidium altarium*. These derivatives were reported to display cytotoxicity against tumor cell lines, SF268 (central nervous system), MCF-7 (breast), and H460 (lung) with a concentration at 10 μM [78, 79]. Rajesh and Murugan [73] discovered CNS depressant compounds from the ascidian *Eudistoma viride* and reported that they the moderated activity of MeOH extract concentration at 200 mg/kg showed the activity of 90.7 ± 1.2% was comparable to the positive control chlorpromazine (99.4 ± 1.1%).

8.4 Ascidians Use as Functional Foods for ASD Patients in the Future

The biodiversity of the marine environment and its associated chemical diversity constitute practically an unlimited resource for the development of new bioactive products. Marine natural products are a valuable source of bioactive compounds that are responsible for many biological activities such as anti-inflammatory, antioxidant, and neuroprotective properties. Marine organisms like sponges, tunicates, bryozoans, molluscs, bacteria, cyanobacteria, fungi, microalgae, and macroalgae have been utilized recently in medical biotechnology. Bioactive compounds isolated from tunicates (ascidians) appears to be more effective and highly specific for neurological diseases like autism, Parkinson's disease, and Alzheimer's. Numerous marine products isolated from ascidian are in various phases of preclinical and clinical studies (e.g., *Ecteinascidia turbinata* ecteinascidin 743 antitumor activity and Aplidine from the ascidian *Aplidium albicans*, which shows promise in shrinking tumors in pancreatic, stomach, bladder, and prostate cancers). These marine nutraceutical valuable products offer several benefits like promoting body health, lowering body burdens of toxins, reducing excitotoxicity, improving antioxidant capacity, enhancing immunomodulatory system, and minimizing the stress conditions that may help to manage, reduce, and prevent ASD symptoms. The marine ascidians possess vast bioactive assets like secondary metabolites, proteins, peptides, polysaccharides, polyunsaturated fatty acids, vitamins, minerals and many other bioactive compounds (polyphenols, flavonoids, terpenes, alkaloids, etc.). These bioactive metabolites as valuable nutraceutical food additives and dietary supplements to effectively manage ASD. They can be utilized and successfully integrated with current treatment to achieve desired interventional results. Omega-3 is a polyunsaturated fatty acid, present at high concentrations in marine ascidians. Omega-3 fats play an important vital role in brain development and neurological function. Evidence supports the effectiveness of Omega-3 supplementation to reduce and control the core or associated ASD symptoms (stereotypy and lethargy). This experimental ascidians work and utility is similar to the exploration of therapeutic benefits of probiotics and enzyme products supplementation being carried out by researchers.

Globally, numerous pharmacologically active compounds have been reported from ascidians with diverse biological properties such as antibacterial, antifungal, anti-inflammatory, antioxidant, antitumor, anticancer, antiviral, antidiabetic, antiproliferative, and antiparasitic properties [80]. Palanisamy et al. [78] discussed the ascidian metabolites of approximately 580 compounds reported from 1994 to 2014. The ascidians families Polyclinidae, Didemnidae, Polycitoridae, and Styelidae are the greatest prolific producers of bioactive compounds of potential therapeutic activity against diseases [78]. Moreover, ascidians are consumed in various cuisine preparations by many countries like France, Italy, Chile, and Korea. The family Pyuridae species of *Halocynthia roretzi* and *H. auranlum* are cultured and eaten in Japan [81]. In Chile, Sweden, and Japan *Pyura chilensis* is are used as a delicacy by humans and in European countries *Microcosmus sabatieri* and *M. vulgaris* are used for food as well [82]. Karthikeyan et al. [83] discussed the composition and nutritive value in solitary ascidian *Microcosmus* exasperates. The components were for protein (24.7 ± 3.65%), carbohydrate (4.97 ± 2.82%), and lipids (2.64 ± 1.11%)—values reported in dry weight—including 18 essential and nonessential amino acids in ascidians muscles. The saturated fatty acid (SAFA), monounsaturated fatty acid (MUFA), and polyunsaturated fatty acid (PUFA) contents ranged from 0.131% to 1.612%, 1.304% to 1.546%, and 1.021% to 1.732% of total fatty acids, respectively.

Ananthan et al. [84] discussed the ten most commonly available Southeast coast Indian ascidians species that have notable amounts of lipids (1.05–2.97%), protein (3.8–20.01%), and carbohydrate (2.2–8.29%) where the authors recommended them for human consumption especially solitary ascidians. Others reported on the nutritional composition of the two ascidians *Eudistoma viride* and *Didemnum psammathodes*. The total nutritional content was determined to be the following: lipid (0.23 and 0.32 μg/mL) carbohydrate (2.15 and 2.2 μg/mL), protein (13.78 and 3.62 μg/mL), total free amino acid content (3.2 and 3.9 μg/mL), and crude fiber (9.2 and 7.9 μg/mL) [85]. On the other hand, the tunic of *Halocynthia roretzi* and *H. roretzi* contain total carbohydrate amount of 46% and crude protein content of 40% [86]. Another study of the tunic *Styela clava* revealed rich content of protein (8.1 ± 0.1%), lipid (0.4 ± 0.1%), and carbohydrate (16.7 ± 0.2%) [87]. Typically, the nutritional content values vary from one region to another and fluctuates with the season. Several reports concluded that ascidians have rich nutritional value and used as food sources for many countries and it could be useful for the ASD patients in the future.

9 Conclusion

This chapter mainly focused on novel secondary metabolites and alkaloids from tunicates and treatment for ASD. The bioactive metabolites of the ascidians were reported to show various biological properties, i.e., antioxidant, anti-inflammatory, CNS depressant, anticancer, antitumor, and antiviral activities. Ascidians have also proved to show commendable activity against autism spectrum disorders. In the recent years, ascidians have been shown to be potential candidates for ameliorating diseases. Further, many studies across the globe reported the beneficial effects of ascidians. The current focus on pharmacopoeia warrants the exploration of bioactive molecules with unique structures from marine ascidians for future use in the treatment of autism spectrum disorder (ASD).

Acknowledgments The chapter was supported by the Science and Engineering Research Board, India under grant SB/YS/LS-374/2013, CSIR—SRF, UGC-FIST and UGC-SAP, New Delhi.

Competing Interests Authors disclose no potential conflicts of interest.

References

1. Woolfenden, S., Sarkozy, V., Ridley, G., & Williams, K. (2012). A systematic review of the diagnostic stability of autism spectrum disorder. *Research in Autism Spectrum Disorders, 6*(1), 345–354.
2. Worley, J. A., & Matson, J. L. (2012). Comparing symptoms of autism spectrum disorders using the current DSM-IV-TR diagnostic criteria and the proposed DSM-V diagnostic criteria. *Research in Autism Spectrum Disorders, 6*(2), 965–970.

3. Frith, U. (2008). *Autism: A very short introduction*. New York: Oxford University Press.
4. American Psychiatric Association. (2013). *Diagnostic and statistical manual of mental disorders* (5th ed.). Washington, DC: Author.
5. Weitlauf, A. S., Gotham, K. O., Vehorn, A. C., & Warren, Z. E. (2014). Brief report: DSM-5 "levels of support:" A comment on discrepant conceptualizations of severity in ASD. *Journal of Autism and Developmental Disorders, 44*(2), 471–476.
6. Kaat, A. J., Gadow, K. D., & Lecavalier, L. (2013). Psychiatric symptom impairment in children with autism spectrum disorders. *Journal of Abnormal Child Psychology, 41*(6), 959–969.
7. Elsabbagh, M., & Johnson, M. H. (2016). Autism and the social brain: The first-year puzzle. *Biological Psychiatry, 80*(2), 94–99.
8. Mattila, M. L., Kielinen, M., Linna, S. L., Jussila, K., Ebeling, H., Bloigu, R., et al. (2011). Autism spectrum disorders according to DSM-IV-TR and comparison with DSM-5 draft criteria: An epidemiological study. *Journal of the American Academy of Child & Adolescent Psychiatry, 50*(6), 583–592.
9. Saemundsen, E., Magnússon, P., Georgsdóttir, I., Egilsson, E., & Rafnsson, V. (2013). Prevalence of autism spectrum disorders in an Icelandic birth cohort. *BMJ Open, 3*(6), e002748.
10. Center for Disease Control. (2014). Autism and Developmental Disabilities Monitoring Network Surveillance Year 2010 Principal Investigators. Prevalence of autism spectrum disorder among children aged 8 years—Autism and developmental disabilities monitoring network, 11 sites, United States, 2010. *Morbidity and Mortality Weekly Report: Surveillance Summaries 2014, 63*(2), 1–21.
11. Reichow, B., Barton, E. E., Boyd, B. A., & Hume, K. (2012). Early intensive behavioral intervention (EIBI) for young children with autism spectrum disorders (ASD). *Cochrane Database of Systematic Reviews, 17*, 10.
12. Myers, S. M., & Johnson, C. P. (2007). The Council on Children with Disabilities. Management of children with autism spectrum disorders. *Pediatrics, 120*(5), 1162–1182.
13. Doyle, C. A., & McDougle, C. J. (2012). Pharmacologic treatments for the behavioral symptoms associated with autism spectrum disorders across the lifespan. *Dialogues in Clinical Neuroscience, 14*, 263–279.
14. McCracken, J. T., McGough, J., Shah, B., Cronin, P., Hong, D., Aman, M. G., et al. (2002). Risperidone in children with autism and serious behavioral problems. *New England Journal of Medicine, 347*(5), 314–321.
15. Marcus, R. N., Owen, R., Kamen, L., Manos, G., McQuade, R. D., Carson, W. H., et al. (2009). A placebo-controlled, fixed-dose study of aripiprazole in children and adolescents with irritability associated with autistic disorder. *Journal of the American Academy of Child & Adolescent Psychiatry, 48*(11), 1110–1119.
16. Montaser, R., & Luesch, H. (2011). Marine natural products: A new wave of drugs? *Future Medicinal Chemistry, 3*(12), 1475–1489.
17. Lei, J., & Zhou, J. (2002). A marine natural product database. *Journal of Chemical Information and Computer Sciences, 42*(3), 742–748.
18. Nastrucci, C., Cesario, A., & Russo, P. (2012). Anticancer drug discovery from the marine environment. *Recent Patents on Anti-Cancer Drug Discovery, 7*(2), 218–232.
19. Faulkner, D. J. (2000). Marine natural products. *Natural Product Reports, 19*, 1–48.
20. Shenkar, N., & Swalla, B. J. (2011). Global diversity of Ascidiacea. *PLoS One, 6*(6), e20657.
21. Ali, H. A., & Tamilselvi, M. (2016). *Ascidians in coastal water: A comprehensive inventory of Ascidian Fauna from the Indian Coast*. Cham, Switzerland: Springer.
22. Arumugam, V., Venkatesan, M., Ramachandran, S., & Sundaresan, U. (2018). Bioactive peptides from marine ascidians and future drug development – A review. *International Journal of Peptide Research and Therapeutics, 24*(1), 13–18.
23. Chen, L., Hu, J. S., Xu, J. L., Shao, C. L., & Wang, G. Y. (2018). Biological and chemical diversity of Ascidian-associated microorganisms. *Marine Drugs, 16*(10), 362.
24. Menna, M., Fattorusso, E., & Imperatore, C. (2011). Alkaloids from marine ascidians. *Molecules, 16*(10), 8694–8732.

25. Faulkner, D. J. (2002). Marine natural products: Metabolites of marine invertebrates. *Natural Product Reports, 19*, 1–48.
26. Seleghim, M. H., de Lira, S. P., Campana, P. T., Berlinck, R. G., & Custódio, M. R. (2007). Localization of granulatimide alkaloids in the tissues of the ascidian *Didemnum granulatum. Marine Biology, 150*(5), 967–975.
27. Mayer, A. M., Glaser, K. B., Cuevas, C., Jacobs, R. S., Kem, W., Little, R. D., et al. (2010). The odyssey of marine pharmaceuticals: A current pipeline perspective. *Trends in Pharmacological Sciences, 31*(6), 255–265.
28. Blunt, J. W., Copp, B. R., Hu, W. P., Munro, M. H., Northcote, P. T., & Prinsep, M. R. (2008). Marine natural products. *Natural Product Reports, 25*(1), 35–94.
29. Menna, M. (2009). Antitumor potential of natural products from Mediterranean ascidians. *Phytochemistry Reviews, 8*(2), 461–472.
30. Uddin, J., Ueda, K., Siwu, E. R., Kita, M., & Uemura, D. (2006). Cytotoxic labdane alkaloids from an ascidian *Lissoclinum* sp.: Isolation, structure elucidation, and structure–activity relationship. *Bioorganic & Medicinal Chemistry, 14*(20), 6954–6961.
31. Valentin, B. B., Vinod, V., & Beulah, M. C. (2011). Biopotential of secondary metabolites isolated from marine sponge *Dendrilla nigra. Asian Pacific Journal of Tropical Disease, 1*(4), 299–303.
32. Madhumitha, G., & Fowsiya, J. A. (2015). *Handbook on: Semi micro technique for extraction of alkaloids*. Indore, India: International Science Congress Association.
33. Marcelo, A., Geronimo, R. M., Vicente, C. J., Callanta, R. B., Bennett, R. M., Ysrael, M. C., et al. (2018). TLC screening profile of secondary metabolites and biological activities of *Salisapiliatartarea* S1YP1 isolated from Philippine Mangroves. *Journal of Oleo Science, 67*(12), 1585–1595.
34. Pooja, S., Aditi, T., Naine, S. J., & Devi, C. S. (2017). Bioactive compounds from marine *Streptomyces* sp. VITPSA as therapeutics. *Frontiers in Biology, 12*(4), 280–289.
35. Kadam, N. S., Naik, A. A., Doshi, P. J., & Nikam, T. D. (2019). High-performance thin-layer chromatography method for simultaneous determination of antipsychotic and medicinally important five β-carboline alkaloids. *Journal of Chromatographic Science, 57*(4), 312–322.
36. Bontemps, N., Bry, D., López-Legentil, S., Simon-Levert, A., Long, C., & Banaigs, B. (2010). Structures and antimicrobial activities of pyridoacridine alkaloids isolated from different chromotypes of the ascidian *Cystodytes dellechiajei. Journal of Natural Products, 73*(6), 1044–1048.
37. Van Wagoner, R. M., Jompa, J., Tahir, A., & Ireland, C. M. (1999). Trypargine alkaloids from a previously undescribed *Eudistoma* sp. ascidian. *Journal of Natural Products, 62*(5), 794–797.
38. Pimenta, A. T., Jimenez, P. C., Costa-Lotufo, L. V., Braz-Filhoc, R., & Lima, M. A. (2014). New unusual alkaloids from the ascidian *Eudistoma vannamei. Natural Product Communications, 9*(12), 1713–1715.
39. Makarieva, T. N., Dmitrenok, A. S., Dmitrenok, P. S., Grebnev, B. B., & Stonik, V. A. (2001). Pibocin B, the first N-O-methylindole marine alkaloid, a metabolite from the far-eastern ascidian Eudistoma species. *Journal of Natural Products, 64*(12), 1559–1561.
40. Gul, W., & Hamann, M. T. (2005). Indole alkaloid marine natural products: An established source of cancer drug leads with considerable promise for the control of parasitic, neurological and other diseases. *Life Sciences, 78*(5), 442–453.
41. Singh, K. S., & Majik, M. S. (2016). *Bioactive alkaloids from marine sponges. InMarine sponges: Chemicobiological and biomedical applications* (pp. 257–286). New Delhi, India: Springer.
42. Joullié, M. M., Leonard, M. S., Portonovo, P., Liang, B., Ding, X., & La Clair, J. J. (2003). Chemical defense in ascidians of the Didemnidae family. *Bioconjugate Chemistry, 14*(1), 30–37.
43. Tianero, M. D., Kwan, J. C., Wyche, T. P., Presson, A. P., Koch, M., Barrows, L. R., et al. (2015). Species specificity of symbiosis and secondary metabolism in ascidians. *The ISME Journal, 9*(3), 615.

44. Lindquist, N., Hay, M. E., & Fenical, W. (1992). Defense of ascidians and their conspicuous larvae: Adult vs. larval chemical defenses. *Ecological Monographs, 62*(4), 547–568.
45. Pisut, D. P., & Pawlik, J. R. (2002). Anti-predatory chemical defenses of ascidians: Secondary metabolites or inorganic acids? *Journal of Experimental Marine Biology and Ecology, 270*(2), 203–214.
46. Nakamura, A., Ashino, T., & Yamamoto, M. (1991). Structure determination of a very unusual peroxide from solitary ascidians, *Phallusia mammillata, ascidia ahodori, styelapricata* and *halocynthia* roretzi. *Tetrahedron Letters, 32*(34), 4355–4358.
47. Dias, D. A., Urban, S., & Roessner, U. (2012). A historical overview of natural products in drug discovery. *Metabolites, 2*(2), 303–336.
48. Tohme, R., Darwiche, N., & Gali-Muhtasib, H. (2011). A journey under the sea: The quest for marine anti-cancer alkaloids. *Molecules, 16*(11), 9665–9696.
49. Zubía, E., Ortega, M. J., & Salvá, J. (2005). Natural products chemistry in marine ascidians of the genus Aplidium. *Mini-Reviews in Organic Chemistry, 2*(4), 389–399
50. Schieber, M., & Chandel, N. S. (2014). ROS function in redox signaling and oxidative stress. *Current Biology, 24*(10), R453–R462.
51. Bebianno, M. J., Company, R., Serafi, A., Camus, L., Cosson, P., & Fiala-Medoni, A. (2005). Antioxidant systems and lipid peroxidation in *Bathymodiolus azoricus* from mid-Atlantic ridge hydrothermal vent fields. *Ecotoxicology and Environmental Safety, 75*, 354–373.
52. Farooqi, A., Fayyaz, S., Hou, M. F., Li, K. T., Tang, J. Y., & Chang, H. W. (2014). Reactive oxygen species and autophagy modulation in non-marine drugs and marine drugs. *Marine Drugs, 12*(11), 5408–5424.
53. Cotelle, N., Moreau, S., Bernier, J. L., Catteau, J. P., & Henichart, J. P. (1991). Antioxidant properties of natural hydroquinones from the marine colonial tunicate *Aplidium californicum. Free Radical Biology and Medicine, 11*(1), 63–68.
54. Ananthan, G., & Iyappan, K. (2014). Immunomodulatory activity of ethanol extract of the ascidian *Didemnum Albidum. World Journal of Pharmacy and Pharmaceutical Sciences, 3*(12), 745–755.
55. Priya, D. S. (2015). Antioxidant activity of the simple ascidian *Phallusia nigra* of Thoothukudi Coast. *International Journal of Pharmaceutical Chemistry, 05*(12), 410–412.
56. Viride, E. (2017). In vitro antioxidant studies on colonial ascidians eudistoma viride and *Didemnum psammathodes. IJPSR, 8*(7), 3170–3179.
57. Wyche, T. P., Standiford, M., Hou, Y., Braun, D., Johnson, D. A., Johnson, J. A., et al. (2013). Activation of the nuclear factor E2-related factor 2 pathway by novel natural products halomadurones A–D and a synthetic analogue. *Marine Drugs, 11*(12), 5089–5099.
58. Kim, S. K., Perera, U. M., Rajapakse, N., & Kim, S. (2016). *Seafood processing by-products*. New York: Springer.
59. Benslimane, A. F., Pouchus, Y. F., Verbist, J. F., Petit, J. Y., Khettab, E. N., Welin, L., et al. (1992). Marine bioactive compounds: Stereospecific anti-inflammatory activity of natural and synthetic Cordiachromene A. *The Journal of Clinical Pharmacology, 32*(1), 37–40.
60. McNamara, C. E., Larsen, L., Perry, N. B., Harper, J. L., Berridge, M. V., Chia, E. W., et al. (2005). Anti-inflammatory Sesquiterpene-quinones from the New Zealand Sponge Dysidea cf. c ristagalli. *Journal of Natural Products, 68*(9), 1431–1433.
61. Pearce, A. N., Chia, E. W., Berridge, M. V., Clark, G. R., Harper, J. L., Larsen, L., et al. (2007). Anti-inflammatory thiazine alkaloids isolated from the New Zealand ascidian *Aplidium* sp.: Inhibitors of the neutrophil respiratory burst in a model of gouty arthritis. *Journal of Natural Products, 70*(6), 936–940.
62. Patil, A. D., Freyer, A. J., Killmer, L., Zuber, G., Carte, B., Jurewicz, A. J., et al. (1997). Lissoclin disulfoxide, a novel dimeric alkaloid from the ascidian *Lissoclinum* sp.: Inhibitor of interleukin-8 receptors. *Natural Product Letters, 10*(3), 225–229.
63. Appleton, D. R., Page, M. J., Lambert, G., Berridge, M. V., & Copp, B. R. (2002). Kottamides A–D: Novel bioactive Imidazolone-containing alkaloids from the New Zealand Ascidian *Pycnoclavella kottae. The Journal of Organic Chemistry, 67*(15), 5402–5404.

64. Aiello, A., Borrelli, F., Capasso, R., Fattorusso, E., Luciano, P., & Menna, M. (2003). Conicamin, a novel histamine antagonist from the mediterranean tunicate *Aplidium conicum*. *Bioorganic & Medicinal Chemistry Letters, 13*(24), 4481–4483.
65. Pearce, A. N., Chia, E. W., Berridge, M. V., Maas, E. W., Page, M. J., Webb, V. L., et al. (2007). E/Z-rubrolide O, an anti-inflammatory halogenated furanone from the New Zealand ascidian *Synoicum* n. sp. *Journal of Natural Products, 70*(1), 111–113.
66. Pearce, A. N., Chia, E. W., Berridge, M. V., Maas, E. W., Page, M. J., Harper, J. L., et al. (2008). Orthidines A–E, tubastrine, 3, 4-dimethoxyphenethyl-β-guanidine, and 1, 14-sperminedihomovanillamide: Potential anti-inflammatory alkaloids isolated from the New Zealand ascidian *Aplidium orthium* that act as inhibitors of neutrophil respiratory burst. *Tetrahedron, 64*(24), 5748–5755.
67. Pearce, A. N., Appleton, D. R., Babcock, R. C., & Copp, B. R. (2003). Distomadines A and B, novel 6-hydroxyquinoline alkaloids from the New Zealand ascidian, *Pseudodistoma aureum*. *Tetrahedron Letters, 44*(20), 3897–3899.
68. Gompel, M., Leost, M., Joffe, E. B., Puricelli, L., Franco, L. H., Palermo, J., et al. (2004). Meridianins, a new family of protein kinase inhibitors isolated from the ascidian Aplidium meridianum. *Bioorganic & Medicinal Chemistry Letters, 14*(7), 1703–1707.
69. Xu, C. X., Jin, H., Chung, Y. S., Shin, J. Y., Woo, M. A., Lee, K. H., et al. (2008). Chondroitin sulfate extracted from the Styela clava tunic suppresses TNF-α-induced expression of inflammatory factors, VCAM-1 and iNOS by blocking Akt/NF-κB signal in JB6 cells. *Cancer Letters, 264*(1), 93–100.
70. Belofsky, G. N., Anguera, M., Jensen, P. R., Fenical, W., & Köck, M. (2000). Oxepinamides A-C and Fumiquinazolines H-I: Bioactive metabolites from a marine isolate of a fungus of the genus Acremonium. *Chemistry–A European Journal, 6*(8), 1355–1360.
71. Appleton, D. R., Chuen, C. S., Berridge, M. V., Webb, V. L., & Copp, B. R. (2009). Rossinones A and B, biologically active meroterpenoids from the Antarctic ascidian, Aplidium species. *The Journal of Organic Chemistry, 74*(23), 9195–9198.
72. Chan, S. T., Pearce, A. N., Januario, A. H., Page, M. J., Kaiser, M., McLaughlin, R. J., et al. (2011). Anti-inflammatory and antimalarial meroterpenoids from the New Zealand ascidian *Aplidium scabellum*. *The Journal of Organic Chemistry, 76*(21), 9151–9156.
73. Rajesh, R. P., & Murugan, A. (2013). Central nervous system depressant, anti-inflammatory analgesic and antipyretic activity of the ascidian *Eudistoma virde*. *Pharmacologia, 65*, 69.
74. Ansari, T., Ravichandran, V., & Suba, V. (2012). Anti-inflammatory property of the methanol extracts of the Ascidian *Eudistoma* viride in rat models. *Journal of Pharmacy Research, 5*(11), 5131–5133.
75. Bertanha, C., Januário, A., Alvarenga, T., Pimenta, L., Silva, M., Cunha, W., et al. (2014). Quinone and hydroquinone metabolites from the ascidians of the genus Aplidium. *Marine Drugs, 12*(6), 3608–3633.
76. Rajasekaran, A., Murugan, A., Anand, P. R., Vijayakumar, P., Kumaresan, T., & Ramasamy, M. S. (2003). CNS depressant activity of the methanolic extract of the ascidian *Distaplia nathensis*. *International Journal of Chemical Science, 1*, 13–16.
77. Meenakshi, V. K., Delighta Mano Joyce, M. I., Paripooranaselvi, M., & Gomathy, S. (2013). CNS depressant activity of the simple ascidian Microcosmus exasperatus Heller, 1878. *International Journal of Current Microbiology and Applied Sciences, 2*(10), 16–25.
78. Palanisamy, S. K., Rajendran, N. M., & Marino, A. (2017). Natural products diversity of marine ascidians (tunicates; ascidiacea) and successful drugs in clinical development. *Natural Products and Bioprospecting, 7*(1), 1–11.
79. Yin, S., Cullinane, C., Carroll, A. R., Quinn, R. J., & Davis, R. A. (2010). Botryllamides K and L, new tyrosine derivatives from the Australian ascidian Aplidium altarium. *Tetrahedron Letters, 51*(26), 3403–3405.
80. Watters, D. (2018). Ascidian toxins with potential for drug development. *Marine Drugs, 16*(5), 162.
81. Nanri, K., Ogawa, J., & Nishikawa, T. (1992). Tunic of a pyurid ascidian Microcosmus hartmeyeri Oka is eaten locally in Japan. *Nanki Seibutu, 34*(2), 135.

82. Davis, A. R. (1995). Over-exploitation of *Pyura chilensis* (Ascidiacea) in southern Chile: The urgent need to establish marine reserves. *Revista Chilena de Historia Natural, 68*(1), 7–1.
83. Karthikeyan, M. M., Ananthan, G., & Balasubramanian, T. (2011). Biochemical components of a solitary ascidian *Microcosmus exasperatus* Heller, 1878 (Ascidiacea: Pyuridae). *Journal of the Marine Biological Association of India, 53*(1), 139–141.
84. Ananthan, G., Karthikeyan, M. M., Selva, P. A., & Raghunathan, C. (2012). Studies on the seasonal variations in the proximate composition of ascidians from the Palk Bay, southeast coast of India. *Asian Pacific Journal of Tropical Biomedicine, 2*(10), 793–797.
85. Kumaran, N. S., & Bragadeeswaran, S. (2014). Nutritional composition of the colonial Ascidian *Eudistoma viride* and *Didemnum psammathodes*. *Biosciences, Biotechnology Research Asia, 1*, 331–338.
86. Lee, K. H., Hong, B. I., Choi, B. D., Kang, S. J., Ruck, J. H., & Jung, B. C. (1998). Utilization of pigments and tunic components of ascidian as an improved feed aids for aquaculture 1. Effective extraction methods of crude polysaccharides in ascidian (Halocpthia roretzi) tunic. *Korean Journal of Fisheries and Aquatic Sciences, 31*(3), 423–428.
87. Ahn, S. H., Jung, S. H., Kang, S. J., Jeong, T. S., & Choi, B. D. (2003). Extraction of glycosaminoglycans from Styela clava tunic. *Korean Journal of Biotechnology and Bioengineering, 18*(3), 180–185.

Reality-Based Technologies for Children with Autism Spectrum Disorder: A Recommendation for Food Intake Intervention

Bilikis Banire, Kamran Khowaja, Bilal Mansoor, Marwa Qaraqe, and Dena Al Thani

Abstract Food selectivity by children with autism spectrum disorder (ASD) is relatively high as compared to typical children and consequently puts them at risk of nutritional inadequacies. Thus, there is a need to educate children with ASD on food types and their benefits in a simple and interesting manner that will encourage food acceptance and enable a move toward healthy living. The use of technological intervention has proven to be an effective tool for educating children with ASD in maintaining attention and mastering new skills as compared to traditional methods. Some of the popularly used technologies are computer-based intervention and robotics which do not support ecological validity (i.e., mimicking natural scenario). Consideration of natural factors is essential for better learning outcomes and generalized skills which can easily be incorporated into reality-based technologies such as virtual reality, augmented reality, and mixed reality. These technologies provide evidence-based support for ecological validation of intervention and sustaining the attention of children with ASD. The main objective of this study is to review existing reality-based technology intervention for children with ASD and investigate the following: (1) commonly used reality-based technology, (2) types of intervention targeted with reality-based technology, and (3) what subjects' inclusion types are used in the reality-based interventions. These objective statements have guided our recommendation of reality-based technology that can support ecological validity of food intake intervention.

Keywords Virtual reality · Augmented reality · Mixed reality · Attention · Ecological validity · ASD · Autism · Food intake · Food intervention

---

B. Banire · K. Khowaja · M. Qaraqe · D. Al Thani (✉)
Department of Information, and Computing Technology, Hamad Bin Khalifa University, Doha, Qatar
e-mail: dalthani@hbku.edu.qa

B. Mansoor
Mechanical Engineering Program, Texas A&M University at Doha, Doha, Qatar

M. M. Essa, M. W. Qoronfleh (eds.), *Personalized Food Intervention and Therapy for Autism Spectrum Disorder Management*, Advances in Neurobiology 24,
https://doi.org/10.1007/978-3-030-30402-7_26

1 Introduction

The definition of autism spectrum disorder (ASD) has been modified over time. The most recent definition has been provided by the *Diagnostic and Statistical Manual of Mental Disorders* (DSM-5) wherein ASD is a neurodevelopmental disorder characterized by deficits in social communication and repetitive or restricted pattern of behavior [1]. They also exhibit atypical patterns of food intake (i.e., they limit themselves to selected food types) which is as high as 90% [2, 3]. Food selectivity is the most prominent problem among the feeding issues faced by children with ASD [4], to the extent that it was initially used as one of the diagnostic measures of ASD [5]. This has been identified as a major concern leading to several problems like nutritional deficiency (such as inadequate vitamins and minerals required by the body [2]), medical issues (like increased constipation [6]), and poor bone development [7]. The consequences of food selectivity are not limited to health issues but also behavioral problems like throwing tantrums and strong emotional reactions when given the food they do not like to take [8]. This behavioral problem with regard to food is of great concern to families of children with ASD as it is associated with unhealthy living and unwanted behaviors. Although being picky with food is not a deliberate act on their part, studies have related feeding selectivity with their sensory processing disorder such as taste, smell, texture, or sight [9, 10]. One of the consequences of unusual attitudes toward food selection may lead to further dissociation and stigmatization in terms of social acceptance. Moreover, this feeding problem could be a long-term issue if not corrected at an early age and might continue to pose the risk of health complications. Therefore, it is imperative to find a way of assisting individuals with ASD in selecting the right choice of food which is good for their health.

One of the evidenced-based solutions identified by some authors is behavioral intervention [11, 12]. This intervention is a way of preventing atypical behaviors in children with ASD and children with other behavioral issues. Food selectivity is one of the problems faced by children with ASD; refer to this study [13] for other problem behaviors faced by young children with ASD. There has been proof of a substantial improvement in feeding habits when behavioral intervention method was used by therapists training the children on food types and its benefits [8, 14]. Despite the success rate of this method, there is the persisting problem of therapist–client ratio in hospitals [15]. According to a report by the Centers for Disease Control and Prevention (CDC) of the USA, the prevalence of children with ASD increased from 1 in 110 children in the year 2000 to 1 in 68 children in year 2018 [16], indicating an increasing population that needs to be catered to. Thus, technological support that can simulate the real life of a therapist engaging in behavioral intervention at the hospital may be a good option. The technology that has the potential to do so, as of today, is reality-based technology like virtual reality, augmented reality, and mixed reality. Reality-based technology can be used as a medium in creating a desired interactive intervention for teaching skills needed to support the deficits of children with ASD such as social interaction and communication [17]. Furthermore, reality-based technology can be used to automate behavioral therapy sessions which are typically time-consuming. It also provides the opportunity of carrying out the intervention at home or any other location [18, 19].

A number of systematic literature reviews have looked at reality-based technology and other technologies developed to support the common needs of children with ASD in the aspects of communication, social learning, and imitation skills as well as other skills (like doing exercise). Examples of these reviews are referenced here [20–22]. However, these reviews have not investigated the degree of inclusion of the subjects with regard to behavioral intervention and how reality-based technology can be applied to feeding problems in children with ASD. Therefore, the objective of this study is to review recent studies on how reality-based technology has been used for children with ASD, specifically for behavioral intervention and the impact of such intervention. We will also investigate the type of reality-based technology that can support feeding habit intervention in children with ASD.

2 Reality-Based Intervention

Reality-based intervention involves technologies that consider the ecological validity of an intervention. Ecological validity is a mechanism of using technological intervention in the real world while considering possible factors associated with the treatment. The importance of the ecological validity of intervention is the generalized, pragmatic knowledge gained. Thus, participants are able to usually extend the skills learned on a computer to the real world. There are different categories of technologies used for reality-based intervention. These categories are explained below.

2.1 Virtual Reality-Based Intervention

Virtual reality-based intervention applies virtual reality (VR) tools in the treatment of common deficits or training for specific skills that can be generalized and used in the real world. The advantage of this technology is the simulation of a desired scenario for users [23]. There are three types of VR. The first is a desktop virtual environment where the VR contents are displayed on a monitor and the users interact with computer input devices like the mouse, keyboard, etc. Many studies with ASD find this safe for children as there is no risk of "cyber-sickness" like feeling nauseated or dizzy [24]. The second type is the head-mounted display (HMD); this involves the use of oculus rift and a similar device which gives the effect of immersion in the scene used for intervention while viewing. Some studies that have used HMD have reported positive results [25, 26], while others have found it inconvenient for children with ASD and typical peers [22, 27]. The third type is cave automatic virtual environment (CAVE) which involves full immersion of the users by displaying the computer application on a wall. Despite the robustness of the technique of VR application for intervention, there are scenarios that may be risky for children with autism in terms of fear and anxiety [28].

2.2 Augmented Reality-Based Intervention

This type of intervention makes use of augmented-reality technologies which are a combination of real-world and computer-generated 3D animations or objects which provide real interactivity [29]. Other studies refer to the combination of computer-generated 3D objects and real-world object as mixed reality [20, 30, 31]. There are three types of display used with AR applications [32]. The first type is the see-through head-mounted display where users can view the real world from virtual objects overlay. The second type is the AR display which is a projection-based display; this allows users to view the virtual and real world from a projection without the need to wear any head gear. The third type is handheld display. This display is a good alternative as compared to the other types due to its size and portability options. The type of display used will also be based on the type of intervention.

3 Review Objectives

The main objective for this study is to review the existing reality-based technology intervention for children with ASD. The rest are as follows:

1. *Reality-based technology intervention that is commonly used*: There are different types of technologies that can be used for reality-based intervention which could be virtual reality, augmented reality, or mixed reality. The participants usually serve as a major influence on the type of reality-based technology used. Sensory processing disorder is common in children with ASD. This disorder affects the way they react to touch and visuals. Thus, selecting the type of reality-based technology requires the involvement of the children. As such, we want to know which reality-based technology is commonly used by children with ASD.
2. *The types of intervention that have been targeted with reality-based technology*: Intervention varies from one group to another, and different types of technologies are used. In this context, the goal of intervention usually influences the type of reality-based technology to use here. Hence, we would like to know the type of intervention that influences the type of reality-based technology used by children with ASD.
3. *The types of inclusion used for the children in reality-based interventions*: Inclusion simply means the type of immersion users experience with reality-based technology. In general, inclusion is immersive; users interact directly with 3D objects without conscious awareness of the immediate environment. The second type is non-immersive, where the user interacts with 3D objects with conscious awareness of the immediate environment. The third type is semi-immersive; the user interacts with 3D objects as well as other objects in the environment as seen in the projection of 3D content on the wall.

These three objective statements have guided our search, review, and recommendation of reality-based technology that can support the ecological validity of food intake intervention for children with ASD.

4 Search Methods

We have searched three large databases, PubMed, Scopus, and Web of Science, using the search terms as virtual reality OR augmented reality OR mixed reality AND behavioral interventions AND children AND (autis∗ OR ASD) within the span of 11 years between 2007 and 2017. The search result gave us a total of 41 articles. Fourteen articles were indexed from PubMed, 17 from Scopus, and 10 from Web of Science. The search results showed us that the number of studies that used reality-based intervention for children with ASD has been rising and falling during this 11-year period as shown in Fig. 1. There were 9 articles that were duplicated and hence removed, leaving us with 37 articles. To ensure the reliability of our article selection in meeting our objective, two authors independently selected the articles that meet inclusion criteria. The inclusion criteria used for the selection of articles for this study were:

1. Studies published between 2007 and 2017
2. Studies that have used any of the reality-based technology for behavioral intervention and evaluated their behaviors in real-world scenario after the intervention
3. Studies whose participants were young children (less than 13 years) with ASD

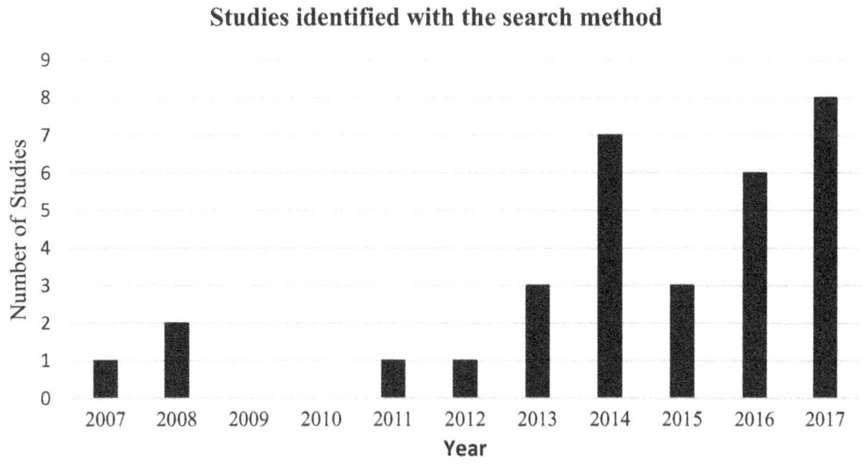

Fig. 1 Studies on behavioral intervention using reality-based technology for children with ASD retrieved from Scopus, Web of Science, and PubMed with search query (virtual reality OR augmented reality OR mixed reality) AND behavioral interventions AND children AND (autis∗ OR ASD) between 2007 and 2017

We reviewed the titles and abstracts of the 32 articles following our inclusion criteria to select related papers for this study after which we applied the inclusion criteria and added studies referenced from other articles. We excluded studies on reality-based behavioral interventions for caregivers to use and have evaluated its usability on the therapist rather than the children with ASD. Therefore, we reviewed 12 full text articles that have reported successful intervention with regard to generalizing behaviors to real-life scenarios. We then analyzed these articles based on the core areas of behavioral interventions and participant inclusion type.

The rise and fall in reality-based technology intervention for children with ASD occurred majorly between years 2007 and 2012 when desktop VR application was popular. Two years after, there was a clear increase in reality based application. This could be interpreted as a trend of technological advancement and acceptance of reality-based intervention like oculus rift, cave automatic virtual environment (CAVE), etc. as compared to the popular desktop VR. Additionally, the perceived positive impact of these technologies on children with learning and other disabilities has increased its applicability. However, there was a fall in the number of studies in 2015 as compared to 2014 with a 40% decrease. This could be related to the cost of acquiring and maintaining reality-based applications. The influx of cheap and affordable reality-based tools created opportunities for more people to choose reality-based application for training and intervention. Therefore, there has been an increase in the number studies using reality-based intervention since 2015 to 2017, and we can also predict more studies in this area as a result of its positive impact and low cost.

4.1 Data Extraction

Twelve out of 47 articles identified were reviewed to investigate the types of reality-based intervention used for children with ASD as well as their inclusion. The idea of this investigation is to recommend reality-based technology that enhances behavioral intervention to correct food habits among the children. The complete summary of the 12 papers reviewed can be found in Table 1.

4.2 Result

We have highlighted the results obtained from reviewing the 12 articles based on our 3 objective statements that guided this study.

Table 1 The types of reality-based intervention used for children with ASD across the reviewed articles

Author	Year	ASD diagnosis	Subjects (children)	Age	Skills target	Reality-based technology	Inclusion classification	Result
Self et al. [33]	2007	ASD	$n = 8$	6–12	Safety skills against fire and tornado	VR/desktop VR	Non-immersive	The participants who used the VR approach were able to transfer fire safety skills to the real-world situation in half the time it took the other group to respond
Fabri et al. [34]	2007	Severe and mild	$n = 34$	7–16	Social emotion recognition	VR/computer with virtual avatar	Non-immersive	The children showed improved emotion recognition
Tartaro et al. [35]	2008	Mild	$n = 6$	7–11	Safety skills against fire	VR/desktop VR	Non-immersive	Increased contingency with virtual peers than human peers
Josman et al. [36]	2008	ASD and typical peers	$n = 12$	Not known	Street crossing skill	VR/desktop VR	Non-immersive	3 out of 6 children with ASD made considerable improvement in real-life street crossing, while others did not
Herrera G et al. [37]	2008	ASD	$n = 2$	8–15	Purchasing skill from a supermarket	VR/desktop VR with touch screen effect	Non-immersive	Participants showed considerable progress in structured pretend play, displaying improvement and generalization of knowledge gain
Cheng et al. [38]	2010	ASD	$n = 3$	7–8	Social competence like emotion recognition	VR/desktop VR with mouse	Non-immersive	Participants exhibited increased social competence that improved after the intervention
Ke et al. [39]	2013	ASD	$n = 4$	9–10	Facial expression and body gesture recognition during conversation and greetings	VR/desktop VR with adult as participatory VR agent	Non-immersive	The children showed improvement in facial expression and body gesture recognition during and after the intervention
Wang et al. [40]	2013	Mild to moderate	$n = 4$	5–10	Contextual processing of objects	VR/screen projection system	Semi-immersive	All subjects showed statistically significant improvements in contextual processing and cognitive flexibility

(continued)

Table 1 (continued)

Author	Year	ASD diagnosis	Subjects (children)	Age	Skills target	Reality-based technology	Inclusion classification	Result
Cai et al. [28]	2013	ASD	$n = 15$	6–17	Joint attention and non-verbal communication	VR/immersive virtual reality	Immersive	Among the participants, 3 were able to actively learn and function with minimal supervision, 3 were able to learn and function but required prompting, 5 were overwhelmed by the VR experience and were supported with parental mediation, and 4 were overwhelmed by the VR experience despite parental mediation
Escobedo et al. [41]	2014	Severe	$n = 16$	3–8	Selective and sustained attention during object discrimination	AR/real-time object recognition using accelerometer to augment the object	Non-immersive with real world	The AR "Mobis" application helped increase students' engagement with people and objects
Zhen Bai et al. [42]	2015	Mild to moderate	$n = 12$	4–7	Pretense play	AR/pretense play with car, train, and airplane	Non-immersive with real world	Significant improvement in pretend play in terms of frequency, duration, and relevance
Liu et al. [43]	2017	Mild to moderate	$n = 2$	8–9	Behavioral self-regulation through emotion	AR/gamified augmented reality applications with smart glasses	Non-immersive with real world	Response improvements in irritability, lethargy, stereotypy, hyper-activity/non-compliance, and inappropriate speech

Mild requiring support, *moderate* requiring substantial support, *severe* requiring very substantial support [1]

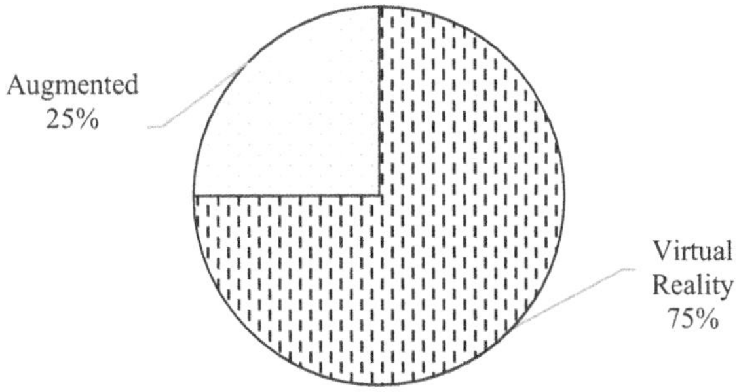

Fig. 2 Reality-based intervention for children with ASD

4.2.1 Objective Statement 1: The Reality-Based Technology that Is Commonly Used

There are different reality-based technologies that are being used for behavioral intervention for improving deficit skills of children with ASD. These technologies range from virtual reality using head-mounted displays (HMD), screen-wall projection, and CAVE to desktop VR. Another reality-based technology is augmented reality or mixed reality which is usually the combination of VR and the real world. Some of the devices used vary from smart glasses to mobile displays, etc. We can infer from our review that the most common type of reality-based technology used was virtual reality as depicted in Fig. 2.

4.2.2 Objective Statement 2: The Types of Intervention that Have Been Targeted with the Reality-Based Technology

We have identified that the reality-based technologies are commonly used for intervention in the core deficit area for children with ASD such as social interaction [25, 34, 41] and communication skills [28, 42]. None has been targeted toward the feeding problem. Hence, there is a need for implementing more reality-based technology intervention to teach the importance of taking the right meal and seeing how they can tolerate the sensory feeling of food.

4.2.3 Objective Statement 3: The Subjects' Inclusion Types Used in the Reality-Based Interventions

Four inclusion methods using reality-based intevention for children with ASD were identified. They are as follows.

4.3 Immersive Inclusion

One of the user's inclusions in reality-based technology was immersive which was used only by 1 of the 12 studies. This inclusion involves direct interaction of the user with 3D objects through some other input devices with little or no awareness of the immediate environment. The study by [28] used immersive inclusion to train children with ASD in improving their joint attention and non-verbal communication skills using dolphin-assisted therapy. Joint attention therapy was implemented with a virtual dolphinarium scene in an immersive room. This room had a 3D screen spanning at 320 degrees and was made to display images from five projectors that were mounted on the ceiling. The children interacted with the simulated dolphinarium as a trainer using hand gestures with dolphins. This simulation was made to teach non-verbal communication skills using hand gestures to the participants. The findings from their study showed that among the 15 participants who took part in the study, 3 participants were able to actively learn and use gestures with little supervision, 3 learned the gesture functions but required prompting, and 5 were overwhelmed by the VR experience and were supported with parental mediation, while 4 others were overwhelmed by the VR experience despite mediation and support from their parents. We can infer from the findings that immersive inclusion may require some sort of support for the children to overcome the overwhelming effect since 9 out of 15 were affected by the same in immersive inclusion.

4.4 Non-immersive Inclusion

Non-immersive inclusion was one of the other methods used in the reality-based intervention where participants interacted with the virtual environment through a desktop computer. Seven out of the 12 studies reviewed have used this type of inclusion for improving different behavioral skills such as safe road crossing [36], social emotion recognition [34, 38, 39], safety skills against fire [33, 35], and purchasing skills [37].

4.5 Semi-immersive Inclusion

Only one of the studies used semi-immersive inclusion for training on behavioral skills for improving contextual processing of objects [44] so that they could recognize objects irrespective of the form they were being presented with. This type of inclusion used a larger and wider area of display as compared to desktop display. The display is projected from a laptop onto the walls surrounding the child to prevent other environmental distractions. The VR application required the child to drag virtual objects to a specific location on the laptop screen. Eighteen randomized test items were given to the children to match the pairs of objects with similar functional characteristics. The software application and the authors independently marked the correct match made by each child, and the score was not shown to the children to make sure their learning was not influenced.

4.6 Non-immersive with Real-World Inclusion

The last of the inclusion types used was non-immersive with real-world interaction. Three out of the 12 reviewed articles used this method which involved the combination of virtual reality with real-world interaction to improve behavioral skills. The study conducted by [41] emphasized object discrimination using a mobile application called "Mobis" with an accelerometer. The children used Mobis which automatically detected objects that the child was viewing, and a message was sent to the therapist to identify which object the child had been focusing on. The second study taught pretense play using AR objects: three foam blocks, a cardboard box with markers attached, and computer with a 24-inch monitor screen on which they viewed the AR objects to engage in pretend play [42]. The third study imparted emotional self-regulation with AR smart glasses used by the children for a face game to identify emotions [43].

5 Discussion

The results showed that reality-based technology for behavioral intervention in children with ASD leads to a successful outcome. Additionally, several methods of inclusion were used for different trainings. This study identified four different methods of subject's inclusion: immersive, non-immersive, semi-immersive, and non-immersive with real-world interaction. However, non-immersive inclusion of the subjects is mostly used in reality-based intervention for children with ASD. The reason for its popularity may be linked to the common effect of "cyber-sickness." We have also seen from the results that the subjects' inclusion methods were based on the type of intervention that is to be achieved. For example, "non-immersive with real-world" interaction was used as an intervention for the children to help them differentiate one object from another as seen in the study of [41]. This method can be justified as the object's texture in real world will provide more information as compared to a visual description alone. Furthermore, it may have been chosen in order to prevent the children from "cyber-sickness" or other inconveniences. It is important to know that food selectivity in children with ASD is primarily attached to the smell, texture, and color of the food [9]. Therefore, a good description of food with its nutritional value needs to be thought of with real-world features such as texture and smell. Hence, augmented reality-based application using "non-immersive with real-world interaction" is recommended for enhancing food habit intervention for children with ASD. However, there are three main issues highlighted by [45], namely learning, pedagogical, and technical issues due to its multitasking content, inflexibility of content, and bulky equipment, respectively. These issues are minimal or absent when it comes to intervention tools which are going to be used with a therapist or caregivers who are going to be assisting the user during the process instead of leaving them to do it on their own.

6 Conclusion

This study reviewed the existing reality-based technologies (i.e., virtual, augmented, and mixed reality) used as behavioral interventions for children with ASD to investigate the technology types and approach that suit the behavioral intervention for food selectivity in these children. Studies have shown that feeding problem in these children is of great concern as it is risky for their well-being and mental development. Hence, behavioral intervention by a therapist has been identified as an evidenced-based solution. However, the problem of therapist–client ratio (in the context of ASD) still exists. Therefore, the procedures for technological support need to be in place. We have seen the different technological approaches of developing technological intervention for children with ASD using robotics, dedicated system, and telehealth [20]. However, the need for reality-based technology is mainly because of its potential for ecological validity of the target behavior and repetitive functionalities based on the reviewed and analyzed 12 articles that met our inclusion criteria. Despite the findings from this study, some of its limitations are the limited number of databases used and the fact that our search queries may not have covered all the articles on reality-based intervention for children with ASD. However, we used some related articles cited from the searched articles in order to increase the amount of relevant data.

Acknowledgment The authors would like to express their gratitude to Hamad Bin Khalifa University and Qatar Foundation for their financial and moral support.

References

1. A. P. Association. (2013). *Diagnostic and statistical manual of mental disorders (DSM-5®)*. Philadelphia, PA: American Psychiatric Publication.
2. Bandini, L. G., Anderson, S. E., Curtin, C., Cermak, S., Evans, E. W., Scampini, R., et al. (2010). Food selectivity in children with autism spectrum disorders and typically developing children. *The Journal of Pediatrics, 157*, 259–264. https://doi.org/10.1016/j.jpeds.2010/08/01/2010
3. Sharp, W. G., Berry, R. C., McCracken, C., Nuhu, N. N., Marvel, E., Saulnier, C. A., et al. (2013). Feeding problems and nutrient intake in children with autism spectrum disorders: A meta-analysis and comprehensive review of the literature. *Journal of Autism and Developmental Disorders, 43*, 2159–2173.

4. Dominick, K. C., Davis, N. O., Lainhart, J., Tager-Flusberg, H., & Folstein, S. (2007). Atypical behaviors in children with autism and children with a history of language impairment. *Research in Developmental Disabilities, 28*, 145–162.
5. Ritvo, E. R., & Freeman, B. J. (1978). Introduction: The National Society for Autistic Children's definition of the syndrome of autism. *Journal of the American Academy of Child Psychiatry, 17*, 565–575.
6. Ibrahim, S. H., Voigt, R. G., Katusic, S. K., Weaver, A. L., & Barbaresi, W. J. (2009). Incidence of gastrointestinal symptoms in children with autism: A population-based study. *Pediatrics, 124*, 680–686.
7. Hediger, M. L., England, L. J., Molloy, C. A., Kai, F. Y., Manning-Courtney, P., & Mills, J. L. (2008). Reduced bone cortical thickness in boys with autism or autism spectrum disorder. *Journal of Autism and Developmental Disorders, 38*, 848–856.
8. Sharp, W. G., Jaquess, D. L., Morton, J. F., & Miles, A. G. (2011). A retrospective chart review of dietary diversity and feeding behavior of children with autism spectrum disorder before and after admission to a day-treatment program. *Focus on Autism and Other Developmental Disabilities, 26*, 37–48.
9. Sharp, W. G., Burrell, T. L., & Jaquess, D. L. (2014). The autism MEAL plan: A parent-training curriculum to manage eating aversions and low intake among children with autism. *Autism, 18*, 712–722.
10. Suarez, M. A. (2017). Laboratory food acceptance in children with autism spectrum disorder compared with children with typical development. *American Journal of Occupational Therapy, 71*, 7106220020p1–7106220020p6.
11. Sharp, W. G., Jaquess, D. L., Morton, J. F., & Herzinger, C. V. (2010). Pediatric feeding disorders: A quantitative synthesis of treatment outcomes. *Clinical Child and Family Psychology Review, 13*, 348–365.
12. Matson, J. L., & Fodstad, J. C. (2009). The treatment of food selectivity and other feeding problems in children with autism spectrum disorders. *Research in Autism Spectrum Disorders, 3*, 455–461. https://doi.org/10.1016/j.rasd.2008.09.005
13. Horner, R. H., Carr, E. G., Strain, P. S., Todd, A. W., & Reed, H. K. (2002). Problem behavior interventions for young children with autism: A research synthesis. *Journal of Autism and Developmental Disorders, 32*, 423–446.
14. Laud, R. B., Girolami, P. A., Boscoe, J. H., & Gulotta, C. S. (2009). Treatment outcomes for severe feeding problems in children with autism spectrum disorder. *Behavior Modification, 33*, 520–536.
15. Zablotsky, B., Pringle, B. A., Colpe, L. J., Kogan, M. D., Rice, C., & Blumberg, S. J. (2015). Service and treatment use among children diagnosed with autism spectrum disorders. *Journal of Developmental and Behavioral Pediatrics: JDBP, 36*, 98.
16. CDC. (2014). *CDC estimates 1 in 68 children has been identified with autism spectrum disorder*. https://www.cdc.gov/media/releases/2014/p0327-autism-spectrum-disorder.html
17. Welch, K. C., Lahiri, U., Liu, C., Weller, R., Sarkar, N., & Warren, Z. (2009). An affect-sensitive social interaction paradigm utilizing virtual reality environments for autism intervention. In *International conference on human-computer interaction* (pp. 703–712). Berlin: Springer.
18. Strickland, D., Marcus, L. M., Mesibov, G. B., & Hogan, K. (1996). Brief report: Two case studies using virtual reality as a learning tool for autistic children. *Journal of Autism and Developmental Disorders, 26*, 651–659.
19. Dautenhahn, K., & Werry, I. (2004). Towards interactive robots in autism therapy: Background, motivation and challenges. *Pragmatics & Cognition, 12*, 1–35.
20. Aresti-Bartolome, N., & Garcia-Zapirain, B. (2014). Technologies as support tools for persons with autistic spectrum disorder: A systematic review. *International Journal of Environmental Research and Public Health, 11*, 7767–7802.
21. Wang, M., & Reid, D. (2011). Virtual reality in pediatric neurorehabilitation: Attention deficit hyperactivity disorder, autism and cerebral palsy. *Neuroepidemiology, 36*, 2–18.

22. Bellani, M., Fornasari, L., Chittaro, L., & Brambilla, P. (2011). Virtual reality in autism: State of the art. *Epidemiology and Psychiatric Sciences, 20*, 235–238.
23. Mitchell, P., Parsons, S., & Leonard, A. (2007). Using virtual environments for teaching social understanding to 6 adolescents with autistic spectrum disorders. *Journal of Autism and Developmental Disorders, 37*, 589–600.
24. Holden, M. K. (2005). Virtual environments for motor rehabilitation. *Cyberpsychology & Behavior, 8*, 187–211.
25. Cheng, Y., & Huang, R. (2012). Using virtual reality environment to improve joint attention associated with pervasive developmental disorder. *Research in Developmental Disabilities, 33*, 2141–2152.
26. Ip, H. H. S., Lai, C. H.-Y., Wong, S. W. L., Tsui, J. K. Y., Li, R. C., Lau, K. S.-Y., et al. (2017). Visuospatial attention in children with autism spectrum disorder: A comparison between 2-D and 3-D environments. *Cogent Education, 4*, 1307709.
27. Didehbani, N., Allen, T., Kandalaft, M., Krawczyk, D., & Chapman, S. (2016). Virtual reality social cognition training for children with high functioning autism. *Computers in Human Behavior, 62*, 703–711.
28. Cai, Y., Chia, N. K., Thalmann, D., Kee, N. K., Zheng, J., & Thalmann, N. M. (2013). Design and development of a virtual dolphinarium for children with autism. *IEEE Transactions on Neural Systems and Rehabilitation Engineering, 21*, 208–217.
29. Zheng, Z., Fu, Q., Zhao, H., Swanson, A., Weitlauf, A., Warren, Z., et al. (2015). Design of a computer-assisted system for teaching attentional skills to toddlers with ASD. In *International conference on universal access in human-computer interaction* (Vol. 9177, pp. 721–730). Cham: Springer.
30. Milgram, P., & Kishino, F. (1994). A taxonomy of mixed reality visual displays. *IEICE Transactions on Information and Systems, 77*, 1321–1329.
31. Botella, C., Juan, M. C., Baños, R. M., Alcañiz, M., Guillén, V., & Rey, B. (2005). Mixing realities? An application of augmented reality for the treatment of cockroach phobia. *Cyberpsychology & Behavior, 8*, 162–171.
32. Zhou, F., Duh, H. B.-L., & Billinghurst, M. (2008). Trends in augmented reality tracking, interaction and display: A review of ten years of ISMAR. In *Proceedings of the 7th IEEE/ACM international symposium on mixed and augmented reality* (pp. 193–202).
33. Self, T., Scudder, R. R., Weheba, G., & Crumrine, D. (2007). A virtual approach to teaching safety skills to children with autism spectrum disorder. *Topics in Language Disorders, 27*, 242–253.
34. Fabri, M., Elzouki, S. Y. A., & Moore, D. (2007). Emotionally expressive avatars for chatting, learning and therapeutic intervention. In *International conference on human-computer interaction* (pp. 275–285).
35. Merryman, J., Tartaro, A., Arie, M., & Cassell, J. (2008). Designing virtual peers for assessment and intervention for children with autism. In *Proceedings of the 7th international conference on interaction design and children, IDC 2008* (pp. 81–84).
36. Josman, N., Ben-Chaim, H. M., Friedrich, S., & Weiss, P. L. (2008). Effectiveness of virtual reality for teaching street-crossing skills to children and adolescents with autism. *International Journal on Disability and Human Development, 7*, 49–56.
37. Herrera, G., Alcantud, F., Jordan, R., Blanquer, A., Labajo, G., & De Pablo, C. (2008). Development of symbolic play through the use of virtual reality tools in children with autistic spectrum disorders: Two case studies. *Autism, 12*, 143–157.
38. Cheng, Y., & Ye, J. (2010). Exploring the social competence of students with autism spectrum conditions in a collaborative virtual learning environment: The pilot study. *Computers & Education, 54*, 1068–1077. https://doi.org/10.1016/j.compedu.2009.10.011
39. Ke, F., & Im, T. (2013). Virtual-reality-based social interaction training for children with high-functioning autism. *The Journal of Educational Research, 106*, 441–461.
40. Wang, M., & Reid, D. (2013). Using the virtual reality-cognitive rehabilitation approach to improve contextual processing in children with autism. *Scientific World Journal, 2013*, 716890.

41. Escobedo, L., Tentori, M., Quintana, E., Favela, J., & Garcia-Rosas, D. (2014). Using augmented reality to help children with autism stay focused. *IEEE Pervasive Computing, 13*, 38–46.
42. Zhen, B., Blackwell, A. F., & Coulouris, G. (2015). Using augmented reality to elicit pretend play for children with autism. *IEEE Transactions on Visualization and Computer Graphics, 21*, 598–610.
43. Liu, R., Salisbury, J. P., Vahabzadeh, A., & Sahin, N. T. (2017). Feasibility of an autism-focused augmented reality Smartglasses system for social communication and behavioral coaching. *Frontiers in Pediatrics, 5*, 145.
44. Wang, M., & Reid, D. (2013). Using the virtual reality-cognitive rehabilitation approach to improve contextual processing in children with autism. *The Scientific World Journal, 2013*, 716890.
45. Wu, H.-K., Lee, S. W.-Y., Chang, H.-Y., & Liang, J.-C. (2013). Current status, opportunities and challenges of augmented reality in education. *Computers & Education, 62*, 41–49.

The manufacturer's authorised representative in the EU is Springer Nature Customer Service Centre GmbH, Europaplatz 3, 69115 Heidelberg, Germany. If you have any concerns regarding our products, please contact ProductSafety@springernature.com

Printed and bound by CPI Group (UK) Ltd, Croydon, CR0 4YY
30/09/2026
02225495-0018